“十四五”时期国家重点出版物出版专项规划项目

现代数学基础丛书 207

凸分析基础

杨新民　孟志青　编著

科学出版社

北　京

内 容 简 介

本书系统介绍了凸分析基础的五个核心部分. ①涉及与凸集理论有关的线性子空间、仿射集、超平面、凸包、单纯形、闭包、内部、相对内部、凸集分离和支撑超平面等基本性质和一些重要定理. ②涵盖了与凸锥有关的顶点锥、锥包、凸锥包、回收锥、共轭锥(正极锥)、负极锥、法锥与切锥、障碍锥、凸锥分离、多面体、多面锥和多面体集等基本性质和重要定理. ③细述了实值(有限值)凸函数、可微凸函数、正常与非正常凸函数、复合凸函数、半连续凸函数、闭凸函数、连续凸函数和 Lipschitz 连续凸函数、共轭凸函数、支撑凸函数、规范凸函数、严格凸函数、半严格凸函数、显凸函数等性质和定理. ④阐述了拟凸函数、半严格拟凸函数、显拟凸函数、伪凸函数、二次可微广义凸函数和广义单调性等广义凸函数的基本理论与性质. ⑤讨论了凸函数的微分学基本理论, 其中主要包含了凸函数的可微性判定定理、方向导数与次微分的关系, 凸函数的中值定理与若干运算性质, Dini 方向导数与拟凸函数之间的关系等内容.

本书适用于高年级本科生和研究生的凸分析课程学习, 也可供研究最优化和其他相关领域的专业人员参考.

图书在版编目(CIP)数据

凸分析基础 / 杨新民, 孟志青编著. -- 北京 : 科学出版社, 2025. 3
(现代数学基础丛书). -- ISBN 978-7-03-081331-2
I. O174.13
中国国家版本馆 CIP 数据核字第 2025B4H489 号

责任编辑: 李静科 李 萍 / 责任校对: 彭珍珍
责任印制: 张 伟 / 封面设计: 陈 敬

科学出版社 出版
北京东黄城根北街 16 号
邮政编码: 100717
http://www.sciencep.com
北京中石油彩色印刷有限责任公司印刷
科学出版社发行 各地新华书店经销
*
2025 年 3 月第 一 版 开本: 720×1000 1/16
2025 年 3 月第一次印刷 印张: 25 1/2
字数: 509 000

定价: 148.00 元

(如有印装质量问题, 我社负责调换)

“现代数学基础丛书”序

在信息时代，数学是社会发展的一块基石.

由于互联网，现在人们获得数学知识和信息的途径之多和便捷性是以前难以想象的. 另一方面，人们通过搜索在互联网获得的数学知识和信息很难做到系统深入，也很难保证在互联网上阅读到的数学知识和信息的质量.

在这样的背景下，高品质的数学书就变得益发重要.

科学出版社组织出版的“现代数学基础丛书”旨在对重要的数学分支和研究方向或专题作系统的介绍，注重基础性和时代性. 丛书的目标读者主要是数学专业的高年级本科生、研究生以及数学教师和科研人员，丛书的部分卷次对其他与数学联系紧密的学科的研究生和学者也是有参考价值的.

本丛书自 1981 年面世以来，已出版 200 卷，介绍的主题广泛，内容精当，在业内享有很高的声誉，深受尊重，对我国的数学人才培养和数学研究发挥了非常重要的作用.

这套丛书已有四十余年的历史，一直得到数学界各方面的大力支持，科学出版社也十分重视，高专业标准编辑丛书的每一卷. 今天，我国的数学水平不论是广度还是深度都已经远远高于四十年前，同时，世界数学的发展也更为迅速，我们对跟上时代步伐的高品质数学书的需求从而更为迫切. 我们诚挚地希望，在大家的支持下，这套丛书能与时俱进，越办越好，为我国数学教育和数学研究的继续发展做出不负期望的重要贡献.

席南华

2024 年 1 月

前　言

凸分析是基础数学、应用数学、运筹学、经济与管理等学科的重要理论基础之一, 在集合论、拓扑学、函数论、广义微分、向量优化、流形优化、微分方程等现代数学领域的理论与应用研究中随处可见. 众所周知, 凸分析中凸集与凸函数具有许多独特的理论性质. 例如, 凸函数在闭凸集上具有全局最优性, 因此在算法设计中找到平稳点即可得到凸函数的全局最优解; 又如, 凹凸函数的鞍点存在性等. 另外, 随着大数据、机器学习和人工智能等新一代信息技术的快速发展, 凸分析理论被广泛应用于解决这些领域的许多问题, 为这些领域的理论研究与应用提供了重要的数学基础和工具支持, 并在解决相关问题中发挥了重要作用. 例如, 在深度学习和强化学习的算法设计中, 凸分析理论能有效提高算法性能, 有利于算法快速收敛. 凸分析理论在众多理论研究与实际问题中的应用, 充分表明了学习和掌握凸分析基础理论的重要性.

学习优化理论和设计全局收敛算法都离不开凸分析基础. 随着凸分析理论的不断发展, 新的理论与方法不断涌现, 亟须对现有的凸分析著作进行补充和完善. 相较目前已有的凸分析专著而言, 本书的优点在于更加全面系统地介绍了凸分析理论, 尽可能考虑到内容的全面性以及知识的广度和深度, 及时跟踪凸分析领域新的研究成果和发展趋势, 将新的理论和方法纳入书中, 保证了本书的基础性、时效性和前沿性. 针对以往著作习题量不足的情况, 本书还在每一章最后精心设计了一定数量的难易适中的习题, 以帮助本科生、研究生和研究者熟练掌握相关的方法与技巧.

国内外已出版了众多凸分析著作和教材, 由于出版时间和作者对内容的侧重点的差异, 这些专著和教材在理论内容深度和广度上都不尽相同. 国际上, 最著名的凸分析的经典专著是由 Rockafellar 在 1970 年出版的 *Convex Analysis*, 该书的凸分析基本内容是 1970 年以前的研究成果. 此后, 陆续有更多的凸分析专著得到了出版. 例如: 1993 年 Panik 出版了 *Fundamentals of Convex Analysis*, 该书主要介绍凸集和凸锥等基础理论, 但未涉及凸函数理论; 2001 年 Hiriart-Urruty 和 Lemarechal 出版了 *Fundamentals of Convex Analysis*, 该书包含凸集、凸函数、次微分和共轭函数等基本内容, 而关于凸锥的内容较少; 2003 年 Bertsekas 等出版了 *Convex Analysis and Optimization*, 该书主要包含 2003 年以前的凸分析和凸优化理论成果; 2003 年 Magaril-Ilyaev 和 Tikhomirov 出版了 *Convex*

Analysis: Theory and Applications, 主要包含凸函数及优化的一些基本内容, 未涉及凸锥理论; 2006 年 Borwein 和 Lewis 出版了 *Convex Analysis and Nonlinear Optimization: Theory and Examples*, 主要内容侧重于凸函数和凸优化理论; 2013 年 Cheridito 出版了 *Convex Analysis*, 主要包含基本的凸集、凸函数和凸优化等内容, 但凸锥基本理论不多; 2015 年 Soltan 出版了 *Lectures on Convex Sets*, 主要内容是凸集凸锥的理论, 未介绍凸函数理论; 2022 年 Mordukhovich 和 Nam 出版了 *Convex Analysis and Beyond*, 内容以拓扑向量空集的凸分析和凸优化等理论成果为主. 而国内学者出版的凸分析基础方面的专著则主要集中在 2000 年以前, 如史树中 (1990) 出版的《凸分析》、刘光中 (1991) 出版的《凸分析与集值问题》、冯德兴 (1995) 出版的《凸分析基础》等. 2000 年以后出版的凸分析基础专著非常少, 其中代表作是李庆娜等 (2019) 出版的《凸分析讲义》, 该书主要包含仿射集、凸集及凸集上的运算、凸集的拓扑性质、凸函数及其运算等; 刘歆和刘亚峰 (2024) 出版的《凸分析》, 该书主要内容包括凸包、相对拓扑、锥近似、投影、Moreau 分解和分离定理等, 以及凸函数的仿射下界、Moreau 包络、连续性、对偶理论、次微分等. 因此, 还没有专著对近二十年凸分析的一些基础的研究成果进行系统、全面的介绍, 其中包括广义凸函数基本理论.

事实上, 广义凸函数是凸分析的另一个重要分支. 一些广义凸函数有着与凸函数接近的性质. 例如, 伪凸函数也有全局最优性, 拟凸函数有 Dini 方向导数与次微分等理论; 还有不变凸、预不变凸、半严格不变凸等广义凸函数也具有好的性质. 因此, 国内外学者出版了有关广义凸函数方面的专著, 如 Cambini 和 Martein(2009) 出版的 *Generalized Convexity and Optimization: Theory and Applications* 的广义凸性部分主要介绍了广义非可微拟凸、广义可微拟凸函数和广义单调性内容; Avrnial 和 Zag(1974) 出版的 *Generalized Convex Functions with Applications on Nonlinear Programming* 主要内容是广义凸函数及应用; Mishra 等 (2009) 出版的 *Generalized Convexity and Vector Optimization* 主要是针对的广义凸性和向量优化理论; 杨新民和戎卫东 (2016) 出版的《广义凸函数及应用》包含了比较全面的广义凸函数理论和近些年一些新的研究成果. 已出版的广义凸函数专著几乎都涵盖了拟凸函数、半严格拟凸函数、伪凸函数和广义单调性等理论. 然而, 由于这些专著侧重于广义凸函数的理论成果, 它们并不适用于基础凸函数的学习和掌握. 针对此, 本书在上述内容的基础上, 更全面地介绍了一些常见的广义凸函数的理论与性质, 并对一些关键性的结论进行了证明. 力求通过深入浅出地讲解, 帮助研究者更好地理解相关理论.

本书作为一本关于凸分析基础的专著, 内容涵盖凸集、凸锥、凸函数、广义凸函数和凸函数微分等基本理论. 全书共六章, 尽量做到内容独立, 范围适当, 但并未包含一些比较特别的性质, 或普适性比较差的理论结果. 全书尽可能全面地介绍近年来关于凸分析理论的一些新研究成果, 尤其是广义凸函数的理论成果. 本书对 (广义) 凸函数的微分理论部分的介绍相对较少, 如有需要深入学习的读者, 可通过本书最后列出的参考文献进一步学习. 本书的具体内容介绍如下.

第 1 章为预备知识. 介绍了书中用到的各种主要运算符号和基本运算定义形式 (本书中所定义的符号可参见书末的符号列表), 给出了距离和范数的定义; 还介绍了本书中常用的拓扑知识, 如邻域、序列、开集、闭集和紧集等概念, 并列出了与它们相关的一些重要性质. 定义了实函数连续性、导数和微分等概念, 并给出了与本书相关的一些重要基本定理, 如中值定理和泰勒展开等. 最后补充了一些线性代数知识, 如行列式、矩阵、线性无关等有关概念和性质.

第 2 章凸集是本书的重要章节. 首先介绍了在凸集、凸函数的理论中常用到的仿射集、线性子空间和超平面等基本理论知识. 然后给出了凸集的定义和基本性质, 其中包含凸集判定定理; 还讨论了凸包及其各种性质、单纯形与仿射集之间的关系、凸集的几种代数运算, 重点讨论了凸集的闭包、内部和相对内部等基本性质, 以及闭包与相对内部运算的性质和凸包的拓扑性质. 接着介绍了两个凸集在不同条件下的超平面分离定理和超平面强分离定理; 另外, 在不同条件下证明了支撑超平面、支撑超平面的极点与面、支撑超平面的配极等重要性质. 最后, 介绍了有关近似凸集和相对近似凸集的性质.

第 3 章介绍了几种凸锥的相关性质和理论. 首先以顶点锥及基本性质为主要内容, 重点介绍锥包和凸锥包的理论性质, 证明了三种情形的凸锥包表示定理. 接着给出了回收锥的定义与基本性质, 还讨论了回收锥的代数运算和拓扑运算的性质, 补充了回收空间及其性质. 而后对共轭锥 (正极锥) 与负极锥的主要性质分别进行了讨论; 引入了集合中点的法锥与切锥的概念, 并证明了它们的一些重要性质; 给出了障碍锥定义及其性质, 重点讨论了多个凸锥在不同条件下的凸锥分离定理. 最后分别证明了多面体、多面锥和多面体集的一些重要结论.

第 4 章根据凸函数的连续性和运算性等分别定义和证明了几种凸函数的一些重要性质. 首先定义了三种基本凸函数: 实值凸函数、可微凸函数和正常与非正常凸函数, 并证明了它们的判定定理等重要性质, 通过不同运算形式定义了几种复合凸函数, 并研究了它们的凸性条件. 然后, 针对半连续凸函数、闭凸函数、连续凸函数和 Lipschitz 连续凸函数, 分别证明了它们在连续性下的凸性. 接着分别研究了三类特殊凸函数: 共轭凸函数、支撑凸函数、规范凸函数及其性质. 随后介绍了严格凸函数、半严格凸函数、显凸函数等基本性质, 以及它们之间的等价关系. 最后证明了强凸函数的几个重要等价定理.

第 5 章重点介绍了几种常见的广义凸函数: 拟凸函数、半严格拟凸函数、显拟凸函数、伪凸函数、二次可微广义凸函数和广义单调性等的理论与性质, 分别证明了它们在半连续性、连续性和各种一点凸性下, 各自的判定定理和它们之间的蕴含关系等结论.

第 6 章是凸函数的微分学基本理论. 首先讨论了几种函数的可微性等价关系, 并证明了凸函数的可微性判定定理, 引入了凸函数的方向导数与次微分, 证明了它们之间的关系. 其次, 给出了凸函数的中值定理和若干运算性质, 其中证明了几种次可加性、复合运算可微性. 最后, 定义了 Dini 方向导数, 证明了拟凸函数与 Dini 方向导数之间的关系.

本书内容全面涵盖了凸分析的基础理论, 全书的许多结果参考了书后所附的文献. 本书适合高年级本科生和研究生在凸分析课程中的学习, 同时也适合最优化和其他领域的专业人员学习. 掌握本书内容对于深入研究凸分析和进一步学习最优化理论具有重要意义. 为了帮助读者学习和提高, 本书每一章都配置了相应的习题.

《重庆师范大学学报 (自然科学版)》编辑部黄颖副编审对全书内容进行了仔细校对和检查, 对她的辛勤细致的工作在此表示感谢!

本书得到国家自然科学基金重大项目 (11991020 和 11991024) 资助, 在此表示感谢.

为了提高全书的可读性, 我们尽量统一了各种符号的表示, 并力求使所有定理和性质的证明简洁明了. 尽管在出版前我们对全书进行了仔细的校对和核查, 但由于书中涉及的符号繁多, 难免会在证明细节或符号使用中出现打印错误, 甚至可能存在描述不尽合理或逻辑上的问题. 如果读者在阅读过程中发现任何不当之处, 恳请不吝赐教, 我们将不胜感激.

杨新民

2024 年 11 月

目　录

第 1 章 预备知识

凸分析往往会涉及数学分析、高等代数、拓扑学和实变函数等领域的一些基本知识. 本章对全书各章节中将用到的有关基本定义、定理和重要性质作一简单介绍, 详细内容可以查阅和参考相关文献.

1.1 符号与运算

1.1.1 基本运算

为了更好地区分向量、常量和集合符号, 书中所有向量用黑斜体表示; 实数分量或常量用白斜体表示; 集合采用大写字母白斜体表示, 其中特殊的数集用黑正体表示, 例如 $\mathbf{R}$ 或 $\mathbf{R}^1$ 表示实数空间, $\mathbf{Z}$ 表示整数空间, $\mathbf{R}_+^n$ 表示其每一个元素的所有分量都不小于 0 构成的空间; 其他具体符号与意义见书后附录 1、附录 2 的符号列表. 本书讨论的范围都限制在 n 维实数空间 $\mathbf{R}^n$ 中, 即欧几里得 (Euclid) 空间 (简称欧氏空间). $\mathbf{R}^n$ 中的点 $\boldsymbol{x}=(x_1,x_2,\cdots,x_n)^{\mathrm{T}}\in\mathbf{R}^n$, 其中, $\boldsymbol{x}$ 是列向量, x_i 表示 $\boldsymbol{x}$ 的第 i 个分量, 上标 T 表示转置; 零向量 $\mathbf{0}=(0,0,\cdots,0)^{\mathrm{T}}\in\mathbf{R}^n$ 表示 $\mathbf{R}^n$ 空间中的坐标原点; $(x_1,x_2,\cdots,x_n)$ 表示行向量, $(x_1,x_2,\cdots,x_n)^{\mathrm{T}}$ 表示列向量.

设 $\mathbf{R}^n$ 空间中任意两点 $\boldsymbol{x}=(x_1,x_2,\cdots,x_n)^{\mathrm{T}}$, $\boldsymbol{y}=(y_1,y_2,\cdots,y_n)^{\mathrm{T}}$, 常数 $\alpha\in\mathbf{R}$. 定义下面的运算规则:

加法: 两个向量相加为对应分量分别相加, 即 $\boldsymbol{x}+\boldsymbol{y}=(x_1+y_1,\cdots,x_n+y_n)^{\mathrm{T}}$.

减法: 两个向量相减为对应分量分别相减, 即 $\boldsymbol{x}-\boldsymbol{y}=(x_1-y_1,\cdots,x_n-y_n)^{\mathrm{T}}$.

数乘: 一个实数与一个向量的数乘为实数分别乘上每个分量: $\alpha\boldsymbol{x}=(\alpha x_1,\alpha x_2,\cdots,\alpha x_n)^{\mathrm{T}}$.

内积: 两个向量的内积为对应分量乘积之和, 即 $\boldsymbol{x}^{\mathrm{T}}\boldsymbol{y}=x_1y_1+x_2y_2+\cdots+x_ny_n=\langle\boldsymbol{x},\boldsymbol{y}\rangle$.

$\langle\boldsymbol{x},\boldsymbol{y}\rangle$ 是表示内积的通用符号, 广泛应用于向量空间, 特别是在抽象空间中 (如 Banach 空间), $\langle\boldsymbol{x},\boldsymbol{y}\rangle$ 表示一个空间点与它的对偶空间的点的乘积. 而 $\boldsymbol{x}^{\mathrm{T}}\boldsymbol{y}$ 则更多地表示 $\mathbf{R}^n$ 空间中两个点之间的乘积. 因此, 在 $\mathbf{R}^n$ 空间中 $\langle\boldsymbol{x},\boldsymbol{y}\rangle$ 与 $\boldsymbol{x}^{\mathrm{T}}\boldsymbol{y}$ 表示意义是一致的.

两个向量进行大小比较时通常有以下几种情况.

小于等于 ($\leqslant$): $\boldsymbol{x} \leqslant \boldsymbol{y} \Leftrightarrow x_i \leqslant y_i, i=1,2,\cdots,n$.

大于等于 ($\geqslant$): $\boldsymbol{x} \geqslant \boldsymbol{y} \Leftrightarrow x_i \geqslant y_i, i=1,2,\cdots,n$.

不等于 ($\neq$): $\boldsymbol{x} \neq \boldsymbol{y} \Leftrightarrow \boldsymbol{x}$ 和 $\boldsymbol{y}$ 中至少一个分量不相等, 即存在某个 i 使得 $x_i \neq y_i$, $i=1,2,\cdots,n$.

等于 ($=$): $\boldsymbol{x} = \boldsymbol{y} \Leftrightarrow x_i = y_i, i=1,2,\cdots,n$.

这里仅介绍了本书常用的基本运算. 在本书中, 还定义了许多其他运算符号, 如 $\oplus$, $\not\subset$, $\exists$, $\in$ 和 $\forall$ 等, 具体含义可参考符号列表.

1.1.2 距离和范数

如果 $\boldsymbol{x} \in \mathbf{R}^n$, 则称

$$||\boldsymbol{x}|| = \langle \boldsymbol{x}, \boldsymbol{x} \rangle^{1/2} = \sqrt{x_1^2 + x_2^2 + \cdots + x_n^2}$$

为 $\boldsymbol{x}$ 的欧氏模 (或长度、范数), 表示向量 $\boldsymbol{x}$ 到原点 $\mathbf{0}$ 的距离或长度. 称

$$|\boldsymbol{x}| = |x_1| + |x_2| + \cdots + |x_n|$$

为 $\boldsymbol{x}$ 的绝对值. 一般有 $||\boldsymbol{x}|| \leqslant |\boldsymbol{x}|$.

在实际应用中, 定义 $||\boldsymbol{x}||_p = \sqrt[p]{|x_1|^p + |x_2|^p + \cdots + |x_n|^p}$ 为 $p\,(>0)$ 范数; 定义 $||\boldsymbol{x}||_0 = \sum\limits_{i=1}^{n} |x_i|_0$ 为 0 范数, 其中 $x_i = \begin{cases} 1, & x \neq 0, \\ 0, & x = 0, \end{cases}$ $i=1,2,\cdots,n$.

定义任意两点 $\boldsymbol{x}, \boldsymbol{y} \in \mathbf{R}^n$ 之间的距离为

$$d\,(\boldsymbol{x}, \boldsymbol{y}) = ||\boldsymbol{x} - \boldsymbol{y}||.$$

定义点 $\boldsymbol{x} \in \mathbf{R}^n$ 到集合 $X \subset \mathbf{R}^n$ 的距离为

$$d\,(\boldsymbol{x}, X) = d_X\,(\boldsymbol{x}) = \inf\left\{d\,(\boldsymbol{x}, \boldsymbol{y}) \mid \boldsymbol{y} \in X\right\}.$$

关于距离和范数我们有下面的性质成立.

性质 1.1.1　设 $\boldsymbol{x}, \boldsymbol{y}, \boldsymbol{z} \in \mathbf{R}^n$, 则下面结论成立:

(1) $d\,(\boldsymbol{x}, \boldsymbol{y}) \geqslant 0$, 当且仅当 $\boldsymbol{x} = \boldsymbol{y}$ 时 $d\,(\boldsymbol{x}, \boldsymbol{y}) = 0$;

(2) $d\,(\boldsymbol{x}, \boldsymbol{z}) \leqslant d\,(\boldsymbol{x}, \boldsymbol{y}) + d\,(\boldsymbol{y}, \boldsymbol{z})$;

(3) $d\,(\boldsymbol{x}, \boldsymbol{y}) = d\,(\boldsymbol{y}, \boldsymbol{x})$;

(4) $\langle \boldsymbol{x}, \boldsymbol{y} \rangle \leqslant ||\boldsymbol{x}||||\boldsymbol{y}||$;

(5) $||\boldsymbol{x} + \boldsymbol{y}|| \leqslant ||\boldsymbol{x}|| + ||\boldsymbol{y}|$.

上面定义了 $\mathbf{R}^n$ 中的范数和距离, 并给出了一些基本性质, 更多的性质可参阅相关文献 [1].

1.2 拓扑概念

本节介绍 $\mathbf{R}^n$ 空间中与集合相对应的一些概念和性质, 如邻域、序列、内点、开集、闭集和紧集等.

1.2.1 邻域与序列

在凸分析的一些性质或定理证明中经常使用到点的邻域这一概念, 它主要用于描述集合中某个点附近的点的性质, 并对于定义和理解内点、内部、相对内部、闭包、函数的连续性和微分等概念具有重要的作用.

定义 1.2.1 假设 $\varepsilon > 0$、点 $\boldsymbol{x} \in \mathbf{R}^n$ 和集合 $C \subset \mathbf{R}^n$, 定义下面一些概念和记号:

(1) 原点半径为 1 的邻域

$$B = \{\boldsymbol{x} \in \mathbf{R}^n \,|\, \|\boldsymbol{x}\| \leqslant 1\};$$

(2) $\boldsymbol{x}$ 的半径为 ε 的闭邻域

$$B(\boldsymbol{x}, \varepsilon) = \{\boldsymbol{y} \in \mathbf{R}^n \,|\, \|\boldsymbol{x} - \boldsymbol{y}\| \leqslant \varepsilon\} = \{\boldsymbol{x} + \boldsymbol{y} | \, \|\boldsymbol{y}\| \leqslant \varepsilon\} = \boldsymbol{x} + \varepsilon B;$$

(3) $\boldsymbol{x}$ 的半径为 ε 的开邻域

$$U(\boldsymbol{x}, \varepsilon) = \{\boldsymbol{y} \in \mathbf{R}^n \,|\, \|\boldsymbol{x} - \boldsymbol{y}\| < \varepsilon\} = \{\boldsymbol{x} + \boldsymbol{y} | \, \|\boldsymbol{y}\| < \varepsilon\};$$

(4) C 的半径为 ε 的闭邻域

$$B(C, \varepsilon) = \{\boldsymbol{y} \,|\, \boldsymbol{x} \in C, \ \|\boldsymbol{x} - \boldsymbol{y}\| \leqslant \varepsilon\} = \cup\{\boldsymbol{x} + \varepsilon B \,|\, \boldsymbol{x} \in C\} = C + \varepsilon B;$$

(5) C 的半径为 ε 的开邻域

$$U(C, \varepsilon) = \{\boldsymbol{y} \,|\, \boldsymbol{x} \in C, \ \|\boldsymbol{x} - \boldsymbol{y}\| < \varepsilon\};$$

(6) $\boldsymbol{x}$ 的半径为 ε 的邻域球面

$$S(\boldsymbol{x}, \varepsilon) = \{\boldsymbol{y} \in \mathbf{R}^n \,|\, \|\boldsymbol{x} - \boldsymbol{y}\| = \varepsilon\}.$$

本书有时也用符号 $N_r(\boldsymbol{x})$ 或 $O_r(\boldsymbol{x})$ 来表示点 $\boldsymbol{x}$ 具有半径 r 的一个闭或开邻域. 显然, 点的邻域与半径有关, 因此, 当我们说存在点 $\boldsymbol{x}$ 的某个邻域时, 是指存在该点某个半径的邻域.

将 $\mathbf{R}^n$ 中的无穷序列集合记为 $\{\boldsymbol{x}_i\}$, 其中 $\boldsymbol{x}_i \in \mathbf{R}^n\,(i = 1, 2, \cdots)$. $\{\boldsymbol{x}_i\}$ 的子序列 $\{\boldsymbol{x}_j\}\,(j = 1, 2, \cdots)$ 是指由 $\{\boldsymbol{x}_i\}$ 中无限个元素构成的集合, 即 $\{\boldsymbol{x}_j\} \subseteq \{\boldsymbol{x}_i\}$. 无穷序列最重要的概念是序列的极限点, 定义如下.

定义 1.2.2 设 $\{\boldsymbol{x}_i\}$ 是 $\mathbf{R}^n$ 中的无穷序列集合, $O_r(\boldsymbol{x}^*)$ 是点 $\boldsymbol{x}^* \in \mathbf{R}^n$ 的任意一个具有半径 r 的开邻域. 如果存在一个常数 N 使得对任意的 $i > N$ 都有 $\boldsymbol{x}_i \in O_r(\boldsymbol{x}^*)$, 则称 $\boldsymbol{x}^*$ 是 $\{\boldsymbol{x}_i\}$ 的极限点, 记为 $\boldsymbol{x}_i \to \boldsymbol{x}^*$ 或 $\lim\limits_{i\to\infty} \boldsymbol{x}_i = \boldsymbol{x}^*$, 且称序列 $\{\boldsymbol{x}_i\}$ 是收敛的. 进一步, 如果序列 $\{\boldsymbol{x}_i\}$ 存在一个子序列收敛到点 $\boldsymbol{x}_0^*$, 则称 $\boldsymbol{x}_0^*$ 是序列 $\{\boldsymbol{x}_i\}$ 的聚点.

序列 $\{\boldsymbol{x}_i\}$ 的聚点可能有许多个. 但是, 如果序列 $\{\boldsymbol{x}_i\}$ 是收敛的, 那么它的任意子序列也收敛到同一个聚点.

性质 1.2.1 设 $\{\boldsymbol{x}_i\}$ 是 $\mathbf{R}^n$ 中的无穷序列集合, $\boldsymbol{x}^* \in \mathbf{R}^n$. 对于任意给定的点 $r > 0$, 若存在一个常数 N 使得对任意的 $i > N$ 满足 $d(\boldsymbol{x}^*, \boldsymbol{x}_i) < r$, 则序列 $\{\boldsymbol{x}_i\}$ 是收敛的.

1.2.2 开集、闭集和紧集

如果 $X \subset \mathbf{R}^n$ 是一个有界集合, 则意味着存在一个常数 $a > 0$ 使得 $X \subset \{\boldsymbol{x} \in \mathbf{R}^n | \, ||\boldsymbol{x}|| < a\}$. 开集是本书中经常使用到的一个概念, 也是刻画一个不含边界集合的概念.

定义 1.2.3 设非空集合 $X \subset \mathbf{R}^n$, $\boldsymbol{x} \in X$, 定义下面几个概念:

(1) 如果存在点 $\boldsymbol{x}$ 的一个邻域 $O_r(\boldsymbol{x}) \subset X$, 则称 $\boldsymbol{x}$ 是 X 的内点;

(2) 如果 X 中的任何一个点都是 X 的内点, 则称 X 是开集;

(3) X 的所有内点构成的集合称为 X 的内部, 记为 $\mathrm{int}X$;

(4) 如果 $\boldsymbol{x}$ 的任何一个邻域都存在不属于 X 的点, 则称 $\boldsymbol{x}$ 是 X 的边界点. X 的所有边界点构成的集合记为 ∂X, 称 ∂X 是 X 的边界.

例 1.2.1 设集合 $X = \{\boldsymbol{x} \in \mathbf{R}^n | \, ||\boldsymbol{x}|| \leqslant a\}$, 其中 $a > 0$, 有

$$\mathrm{int}X = \{\boldsymbol{x} \in \mathbf{R}^n | \, ||\boldsymbol{x}|| < a\}, \quad \partial X = \{\boldsymbol{x} \in \mathbf{R}^n | \, ||\boldsymbol{x}|| = a\}.$$

由定义 1.2.3 可知, 下面的结论是显然的.

性质 1.2.2 设集合 $X \subset \mathbf{R}^n$ 是一个开集, 对于任意给定的 $\boldsymbol{x} \in X$ 都存在一个 $N > 0$, 则当 $r < N$ 时, 对于 $\boldsymbol{x}$ 每一个邻域 $O_r(\boldsymbol{x})$ 都有 $O_r(\boldsymbol{x}) \subset X$.

存在许多开集的例子, 如开区间 (0,1).

例 1.2.2 点 $\boldsymbol{x} \in X$ 的具有半径 r 的邻域 $O_r(\boldsymbol{x})$ 本身就是一个开集.

由定义 1.2.3 知 $\mathrm{int}X$ 是开集. 如果 X 是开集当且仅当 $\mathrm{int}X = X$.

闭集也是凸分析中常见的概念, 它的定义与开集有很大的区别, 闭集是含边界的集合. 也存在既不是开集也不是闭集的集合, 例如区间 $X = (0, 1]$.

定义 1.2.4 设非空集合 $X \subset \mathbf{R}^n$, 定义下面几个概念:

(1) 如果 X 中的任何一个收敛序列的极限点都属于 X, 则称 X 是闭集;

(2) 把 X 所有极限点构成的集合称为 X 的闭包, 记为 $\mathrm{cl}X$;

(3) 如果 Y 是 X 的子集且 $X \subset \mathrm{cl}Y$, 则称 Y 是 X 的稠密子集.

注 闭集 X 表明集合 X 中任意点的任何一个邻域都含有该集合中的点, 而开集 X 中的任何一个点都存在该点的一个邻域使得该邻域中所有点都属于 X. 若 Y 是 X 的稠密子集, 则 X 中每一个点的邻域都包含 Y 中的至少一个点, 且 $X \subset \mathrm{cl}Y$.

性质 1.2.3 设 $X \subset \mathbf{R}^n$ 是闭集, $\boldsymbol{x}^* \in \mathbf{R}^n$ 是序列 $\{\boldsymbol{x}_i\} \subset X$ 的极限点, 有下面结论成立:

(1) $\boldsymbol{x}^* \in X$, $X = \mathrm{cl}X$;

(2) 如果 $\bar{X}$ 是 X 的稠密子集, 则 $X = \mathrm{cl}\bar{X}$;

(3) 如果 $A, B \subset \mathbf{R}^n$, $\mathrm{int}\,(A \cap B) \neq \varnothing$, 则 $\mathrm{cl}\,(A \cap B) = \mathrm{cl}A \cap \mathrm{cl}B$.

证明 (1) 和 (2) 是显然的, 这里证明 (3) 成立即可. 因为 $A \subset \mathrm{cl}A$ 和 $B \subset \mathrm{cl}B$, 所以有

$$A \cap B \subset \mathrm{cl}A \cap \mathrm{cl}B.$$

因此, 有 $\mathrm{cl}\,(A \cap B) \subset \mathrm{cl}\,(\mathrm{cl}A \cap \mathrm{cl}B) = \mathrm{cl}A \cap \mathrm{cl}B$.

反之, 设 $\boldsymbol{x} \in \mathrm{cl}A \cap \mathrm{cl}B$, 则存在点 $\boldsymbol{x}$ 的任何邻域 $O_r(\boldsymbol{x})$ 使得 $A \cap O_r(\boldsymbol{x}) \neq \varnothing$, $B \cap O_r(\boldsymbol{x}) \neq \varnothing$, 这表明 $A \cap B \cap O_r(\boldsymbol{x}) = \varnothing$ 不可能成立, 即有 $\boldsymbol{x} \in \mathrm{cl}\,(A \cap B)$. □

例 1.2.3 设 $X_1 = \{\boldsymbol{x} \in \mathbf{R}^n \,|\, |x_i| < 1, i = 1, 2, \cdots, n\}$ 是一个开集, 它不是一个闭集. 因为存在序列

$$\boldsymbol{x}_i = \left(1 - \frac{1}{i}, 1 - \frac{1}{i}, \cdots, 1 - \frac{1}{i}\right)^{\mathrm{T}} \in X_1 \quad (i = 1, 2, \cdots)$$

满足 $\boldsymbol{x}_i = \left(1 - \frac{1}{i}, 1 - \frac{1}{i}, \cdots, 1 - \frac{1}{i}\right)^{\mathrm{T}} \to (1, 1, \cdots, 1)^{\mathrm{T}}$. 但是 $(1, 1, \cdots, 1)^{\mathrm{T}} \notin X_1$.

例 1.2.4 设 $X_2 = \{\boldsymbol{x} \in \mathbf{R}^n \,|\, |x_i| \leqslant 1, i = 1, 2, \cdots, n\}$ 是一个闭集, 它不是一个开集. 因为存在序列

$$\boldsymbol{x}_i = \left(1 - \frac{1}{i}, 1 - \frac{1}{i}, \cdots, 1 - \frac{1}{i}\right)^{\mathrm{T}} \in X_2 \quad (i = 1, 2, \cdots)$$

满足 $\boldsymbol{x}_i = \left(1 - \frac{1}{i}, 1 - \frac{1}{i}, \cdots, 1 - \frac{1}{i}\right)^{\mathrm{T}} \to (1, 1, \cdots, 1)^{\mathrm{T}}$, 且 $(1, 1, \cdots, 1)^{\mathrm{T}} \in X_2$.

例 1.2.5 设 m 个连续函数 $g_j : \mathbf{R}^n \to \mathbf{R}\,(j = 1, 2, \cdots, m)$, 定义两个集合:

$$X_3 = \{\boldsymbol{x} \in \mathbf{R}^n | g_j(\boldsymbol{x}) < 0, j = 1, 2, \cdots, m\};$$

$$X_4 = \{\boldsymbol{x} \in \mathbf{R}^n | g_j(\boldsymbol{x}) \leqslant 0, j = 1, 2, \cdots, m\}.$$

则 X_3 是一个开集, X_4 是一个闭集.

紧集也是凸分析常见的概念, 主要用于最优解的存在性和收敛性分析. 紧集与闭集有着密切的联系, 它的定义主要是通过序列收敛刻画的.

定义 1.2.5　如果集合 $X \subset \mathbf{R}^n$ 中的任何一个收敛序列都存在收敛的子序列, 且收敛子序列的极限点在 X 中, 则称 X 是紧集.

紧集一定是闭集, 但是闭集不一定是紧集. 我们有下面的定理.

定理 1.2.1　$X \subset \mathbf{R}^n$ 是紧集当且仅当 X 是有界闭集.

在有限维空间中, 紧集的判定只要是有界闭集即可, 如闭区间是紧集. 无界集合一定不是紧集, 例如, 集合 $X_1 = \{x \in \mathbf{R} | x \in [0, +\infty)\}$ 显然不是紧集.

定理 1.2.2　设 $X \subset \mathbf{R}^n$ 是紧集, $f: \mathbf{R}^n \to \mathbf{R}$ 在 X 上是连续函数, 则函数 f 在 X 上达到最大值和最小值.

1.3　连续性与导数

本节主要定义 $\mathbf{R}^n$ 上函数的连续性和导数, 并介绍相关性质.

1.3.1　连续性

凸函数的连续性是本书的一个重要内容, 在后面判定凸函数的凸性时具有重要的作用. 下面给出 $\mathbf{R}^n$ 上函数连续性的定义.

定义 1.3.1　设函数 $f: \mathbf{R}^n \to \mathbf{R}$. 如果对任意序列 $\{\boldsymbol{x}_i\}: \boldsymbol{x}_i \to \boldsymbol{x}, f$ 满足 $f(\boldsymbol{x}_i) \to f(\boldsymbol{x})$, 则称函数 f 在 $\boldsymbol{x}$ 处连续. 如果函数 f 在集合 $X \subset \mathbf{R}^n$ 中的任何点都连续, 则称函数 f 在 X 上是连续的.

性质 1.3.1　函数 f 在 $\boldsymbol{x}$ 处连续当且仅当对于任意给定的 $\varepsilon > 0$ 都存在 $\boldsymbol{x}$ 的一个邻域 $O_r(\boldsymbol{x})$, 对任意的 $\boldsymbol{y} \in O_r(\boldsymbol{x})$ 满足条件 $|f(\boldsymbol{x}) - f(\boldsymbol{y})| \leqslant \varepsilon$.

连续函数在进行加、减、乘、除和复合函数运算后仍然保持连续性.

性质 1.3.2　设函数 $f, g: X \to \mathbf{R}$ 在 $X \subset \mathbf{R}^n$ 上是连续的, 则 $f+g, fg$ 和 $\dfrac{f}{g}$(其中 $g(\boldsymbol{x}) \neq 0$) 在 X 上都是连续的.

性质 1.3.3　设函数 $\phi: \mathbf{R} \to \mathbf{R}$ 在区间 $[a, b]$ 上是连续严格单调的, 则对于 $\phi(a) < \alpha < \phi(b)$ 或 $\phi(a) > \alpha > \phi(b)$, 一定存在某个 $t \in (a, b)$ 使得 $\phi(t) = \alpha$.

1.3.2　导数与微分

可微凸函数是本书最重要的概念之一, 在凸函数的性质研究中发挥了不可替代的重要作用. 下面首先给出函数的导数定义.

定义函数 $f:\mathbf{R}^n\to\mathbf{R}$ 在点 $\boldsymbol{x}$ 处的一阶导数或梯度为

$$\nabla f(\boldsymbol{x})=\left(\frac{\partial f}{\partial x_1},\frac{\partial f}{\partial x_2},\cdots,\frac{\partial f}{\partial x_n}\right)^{\mathrm{T}},$$

其中 f 的第 i 个分量偏导数定义为

$$\frac{\partial f}{\partial x_i}=\lim_{t\to 0}\frac{f(x_1,x_2,\cdots,x_i+t,\cdots,x_n)-f(x_1,x_2,\cdots,x_i,\cdots,x_n)}{t},$$

$$i=1,2,\cdots,n.$$

若 f 的一阶导数或梯度 ∇f 是一个向量, 则当且仅当 f 每一个分量的偏导数都存在时, 它的梯度才存在. 我们一般称 f 在 $X\subset\mathbf{R}^n$ 上是光滑的, 是指 f 在 X 上每一点都存在梯度或微分.

定义函数 $f:\mathbf{R}^n\to\mathbf{R}$ 在点 $\boldsymbol{x}$ 处的方向 $\boldsymbol{d}\in\mathbf{R}^n$ 的方向导数如下:

$$f'(\boldsymbol{x},\boldsymbol{d})=\lim_{t\to 0}\frac{f(\boldsymbol{x}+t\boldsymbol{d})-f(\boldsymbol{x})}{t}.$$

定义函数 $f:\mathbf{R}^n\to\mathbf{R}$ 在点 $\boldsymbol{x}$ 处的二阶导数或梯度为 (称为 f 的 Hessian 矩阵)

$$\nabla^2 f(\boldsymbol{x})=\begin{pmatrix}\dfrac{\partial^2 f}{\partial x_1^2} & \dfrac{\partial^2 f}{\partial x_1\partial x_2} & \cdots & \dfrac{\partial^2 f}{\partial x_1\partial x_n}\\ \dfrac{\partial^2 f}{\partial x_2\partial x_1} & \dfrac{\partial^2 f}{\partial x_2^2} & \cdots & \dfrac{\partial^2 f}{\partial x_2\partial x_n}\\ \vdots & \vdots & & \vdots\\ \dfrac{\partial^2 f}{\partial x_n\partial x_1} & \dfrac{\partial^2 f}{\partial x_n\partial x_2} & \cdots & \dfrac{\partial^2 f}{\partial x_n^2}\end{pmatrix}.$$

性质 1.3.4 设函数 $f:\mathbf{R}^n\to\mathbf{R}$ 在 $X\subset\mathbf{R}^n$ 上是连续可微的, 则对任意的 $\boldsymbol{x}\in X$, $\boldsymbol{d}\in\mathbf{R}^n$, 有

$$f'(\boldsymbol{x},\boldsymbol{d})=\nabla f(\boldsymbol{x})^{\mathrm{T}}\boldsymbol{d},$$

且存在充分小的 $\delta>0$, 使得对任意的 $t\in(0,\delta)$ 有

$$f(\boldsymbol{x}+t\boldsymbol{d})=f(\boldsymbol{x})+t\nabla f(\boldsymbol{x})^{\mathrm{T}}\boldsymbol{d}+o(t),$$

其中 $\lim\limits_{t\to 0}o(t)/t=0$.

性质 1.3.5　设函数 $f:\mathbf{R}^n\to\mathbf{R}$ 在 $X\subset\mathbf{R}^n$ 上是二阶连续可微的, 则对任意的 $\boldsymbol{x}\in X$, $\boldsymbol{d}\in\mathbf{R}^n$, 存在充分小的 $\delta>0$, 使得对任意的 $t\in(0,\delta)$ 有

$$f(\boldsymbol{x}+t\boldsymbol{d})=f(\boldsymbol{x})+t\nabla f(\boldsymbol{x})^{\mathrm{T}}\boldsymbol{d}+\frac{1}{2}t\boldsymbol{d}^{\mathrm{T}}\nabla^2 f(\boldsymbol{x})\boldsymbol{d}+o\left(t^2\right),$$

其中 $\lim\limits_{t\to 0}o\left(t^2\right)/t=0$.

不难看出, 上式也是函数 f 的二阶泰勒展开.

基于上述定义与性质, 下面我们介绍几个相关的定理和推论.

定理 1.3.1 (拉格朗日中值定理)　设函数 $\varphi:[a,b]\to\mathbf{R}$ 是可导的, 则在区间 (a,b) 中存在一点 ξ, 使得 $\varphi(b)-\varphi(a)=\varphi'(\xi)(b-a)$.

推论 1.3.1　设函数 $f:X\to\mathbf{R}$ 是可导的, 且 $X\subset\mathbf{R}^n$ 中任意两点连成的线段

$$[\boldsymbol{x},\boldsymbol{y}]=\{\boldsymbol{z}:=t\boldsymbol{x}+(1-t)\boldsymbol{y}|\forall t\in[0,1]\}$$

都在 X 中, 则在区间 $(0,1)$ 中存在一点 ξ, 使得

$$f(\boldsymbol{x})-f(\boldsymbol{y})=\nabla f(\boldsymbol{u})(\boldsymbol{x}-\boldsymbol{y}),$$

其中 $\boldsymbol{u}=\xi\boldsymbol{x}+(1-\xi)\boldsymbol{y}\in[\boldsymbol{x},\boldsymbol{y}]$.

定理 1.3.2 (二阶拉格朗日中值定理)　设函数 $\varphi:[a,b]\to\mathbf{R}$ 是二阶可导的, 则在区间 (a,b) 中存在一点 ξ, 使得 $\varphi(b)-\varphi(a)=\varphi'(a)(b-a)+\dfrac{1}{2}\varphi''(\xi)(b-a)^2$.

推论 1.3.2　设函数 $f:X\to\mathbf{R}$ 是可导的, 且 $X\subset\mathbf{R}^n$ 中任意两点连成的线段

$$[\boldsymbol{x},\boldsymbol{y}]=\{\boldsymbol{z}:=t\boldsymbol{x}+(1-t)\boldsymbol{y}|\forall t\in[0,1]\}$$

都在 X 中, 则在区间 $(0,1)$ 中存在一点 ξ, 使得

$$f(\boldsymbol{x})-f(\boldsymbol{y})=\nabla f(\boldsymbol{y})(\boldsymbol{x}-\boldsymbol{y})+\frac{1}{2}(\boldsymbol{x}-\boldsymbol{y})^{\mathrm{T}}\nabla^2 f(\boldsymbol{u})(\boldsymbol{x}-\boldsymbol{y}),$$

其中 $\boldsymbol{u}=\xi\boldsymbol{x}+(1-\xi)\boldsymbol{y}\in[\boldsymbol{x},\boldsymbol{y}]$.

定理 1.3.3 (隐函数存在定理)　设二元函数 $F:X\to\mathbf{R}$ 是连续可导的, 且

$$X=\{(x,y)\,|\,|x-x_0|\leqslant a,|y-y_0|\leqslant b\}\subset\mathbf{R}^2$$

是一个闭区间, 其中 $(x_0,y_0)\in X$, 常数 $a,b>0$. 如果

$$F(x_0,y_0)=0,\quad F_y'(x_0,y_0)\neq 0,$$

则存在 x_0 的邻域 $O_r(x_0)$ 上的一个连续可微函数 ϕ, 满足

$$y=\phi(x),\quad y_0=\phi(x_0),\quad F(x,\phi(x))=0.$$

1.4 线性代数

下面介绍本书将涉及的一些线性代数知识, 例如行列式、线性相关、线性无关和矩阵等概念和性质.

1.4.1 行列式

设实数 $a_{ij} \in \mathbf{R}\,(i, j = 1, 2)$, 二阶行列式定义为

$$A_2 = \begin{vmatrix} a_{11} & a_{12} \\ a_{21} & a_{22} \end{vmatrix} = a_{11}a_{22} - a_{21}a_{12};$$

设实数 $a_{ij} \in \mathbf{R}\,(i, j = 1, 2, 3)$, 三阶行列式定义为

$$A_3 = \begin{vmatrix} a_{11} & a_{12} & a_{13} \\ a_{21} & a_{22} & a_{23} \\ a_{31} & a_{32} & a_{33} \end{vmatrix}$$

$$= a_{11}a_{22}a_{33} + a_{21}a_{32}a_{13} + a_{31}a_{12}a_{23} - a_{13}a_{22}a_{31} - a_{11}a_{32}a_{32} - a_{12}a_{21}a_{33};$$

设实数 $a_{ij} \in \mathbf{R}\,(i, j = 1, 2, \cdots, n)$, n 阶行列式定义为

$$A_n = \begin{vmatrix} a_{11} & a_{12} & \cdots & a_{1n} \\ a_{21} & a_{22} & \cdots & a_{2n} \\ \vdots & \vdots & & \vdots \\ a_{n1} & a_{n2} & \cdots & a_{nn} \end{vmatrix}.$$

设 $A_{n-1}\,(a_{ij})$ 是行列式 A_n 划去第 i 行及第 j 列的元素后剩下元素所构成的 $n-1$ 阶子行列式, 因此 n 阶行列式值的计算可以由它的 $n-1$ 阶子行列式计算而得, 具体表述为以下性质.

性质 1.4.1 设实数 $a_{ij} \in \mathbf{R}\,(i, j = 1, 2, \cdots, n)$, 则 n 阶行列式 A_n 的计算公式为

$$A_n = \sum_{i=1}^{n} a_{i1}A_{n-1}\,(a_{i1}) = \sum_{i=1}^{n} a_{i2}A_{n-1}\,(a_{i2}) = \cdots = \sum_{i=1}^{n} a_{in}A_{n-1}\,(a_{in}).$$

定义 1.4.1 设有 k 个向量 $\boldsymbol{a}_i \in \mathbf{R}^n, i = 1, 2, \cdots, k$. 如果存在不全为 0 的 k 个常数 $\alpha_i \in \mathbf{R}$ 使得

$$\alpha_1\boldsymbol{a}_1 + \alpha_2\boldsymbol{a}_2 + \cdots + \alpha_k\boldsymbol{a}_k = \mathbf{0}$$

成立, 则称向量集 $\{\boldsymbol{a}_1, \boldsymbol{a}_2, \cdots, \boldsymbol{a}_k\}$ 是线性相关的, 否则称 $\{\boldsymbol{a}_1, \boldsymbol{a}_2, \cdots, \boldsymbol{a}_k\}$ 是线性无关或线性独立的.

性质 1.4.2 当 $k > n$ 时, 则在 $\mathbf{R}^n$ 中任何给定的向量集 $\{\boldsymbol{a}_1, \boldsymbol{a}_2, \cdots, \boldsymbol{a}_k\}$ 都是线性相关的.

性质 1.4.3 如果向量集 $\{\boldsymbol{a}_1, \boldsymbol{a}_2, \cdots, \boldsymbol{a}_k\}$ 是线性无关的, 则它的任何子集也都是线性无关的.

1.4.2 矩阵

设实数或变量 $a_{ij} \in \mathbf{R}\,(i = 1, 2, \cdots, m; j = 1, 2, \cdots, n)$, 定义 $m \times n$ 的矩阵:

$$\boldsymbol{A} = \begin{pmatrix} a_{11} & a_{12} & \cdots & a_{1n} \\ a_{21} & a_{22} & \cdots & a_{2n} \\ \vdots & \vdots & & \vdots \\ a_{m1} & a_{m2} & \cdots & a_{mn} \end{pmatrix} = (a_{ij})_{m \times n}.$$

矩阵的主要运算有加、减、乘和逆, 等等. 例如, 两个矩阵的加、减可分别定义为

$$(a_{ij})_{m \times n} + (b_{ij})_{m \times n} = (a_{ij} + b_{ij})_{m \times n},$$

$$(a_{ij})_{m \times n} - (b_{ij})_{m \times n} = (a_{ij} - b_{ij})_{m \times n};$$

设 $\alpha \in \mathbf{R}$, 则矩阵的数乘可表示为 $\alpha\,(a_{ij})_{m \times n} = (\alpha a_{ij})_{m \times n}$; 需要注意的是, 两个矩阵的相乘必须是前矩阵的列数与后矩阵的行数相等, 即

$$(a_{ij})_{m \times n} \times (b_{ij})_{n \times s} = \left(\sum_{k=1}^{n} a_{ik} b_{kj}\right)_{m \times s}.$$

设矩阵 $\boldsymbol{A} = (a_{ij})_{m \times n}$ 中向量 $\boldsymbol{a}_i = (a_{i1}, a_{i2}, \cdots, a_{in}), i = 1, 2, \cdots, m$, 向量组 $\boldsymbol{a}_1, \boldsymbol{a}_2, \cdots, \boldsymbol{a}_m$ 的最大线性无关个数称为矩阵 $\boldsymbol{A}$ 的秩.

设 $\boldsymbol{A} = (a_{ij})_{n \times n}$ 是一个 $n \times n$ 的矩阵, 它对应的行列式记为 $\det(\boldsymbol{A}) = |\boldsymbol{A}|$. 记 $\boldsymbol{I}$ 是一个 $n \times n$ 的单位矩阵, 其对角线元素均为 1, 其余元素均为 0.

定义 1.4.2 设 $\boldsymbol{A} = (a_{ij})_{n \times n}$ 是一个 $n \times n$ 的矩阵, 定义下面几个概念:

(1) 如果存在 λ 满足方程 $\det(\boldsymbol{A} - \lambda \boldsymbol{I}) = 0$, 则称 λ 为矩阵 $\boldsymbol{A}$ 的特征值;

(2) 设 λ 是矩阵 $\boldsymbol{A}$ 的特征值. 如果存在向量 $\boldsymbol{d} \in \mathbf{R}^n$ 使得 $\boldsymbol{A}\boldsymbol{d} = \lambda \boldsymbol{d}$ 成立, 则称向量 $\boldsymbol{d}$ 是矩阵 $\boldsymbol{A}$ 的特征向量;

(3) 如果存在矩阵 $\boldsymbol{B} = (b_{ij})_{n \times n}$ 使得 $\boldsymbol{AB} = \boldsymbol{BA} = \boldsymbol{I}$, 则称 $\boldsymbol{B}$ 是 $\boldsymbol{A}$ 的逆矩阵.

定义 1.4.3　设 $\boldsymbol{A}=(a_{ij})_{n\times n}$ 是一个 $n\times n$ 的矩阵. 如果下式

$$\boldsymbol{x}^{\mathrm{T}}\boldsymbol{A}\boldsymbol{x}\geqslant 0,\quad \forall \boldsymbol{x}\in\mathbf{R}^n$$

成立, 则称矩阵 $\boldsymbol{A}$ 是半正定的. 如果下式

$$\boldsymbol{x}^{\mathrm{T}}\boldsymbol{A}\boldsymbol{x}>0,\quad \forall \boldsymbol{x}\in\mathbf{R}^n\backslash\{0\}$$

成立, 则称矩阵 $\boldsymbol{A}$ 是正定的.

定理 1.4.1　已知矩阵 $\boldsymbol{A}=(a_{ij})_{n\times n}$ 中所有元素为实数, 向量 $\boldsymbol{b}\in\mathbf{R}^n$, 变量 $\boldsymbol{x}\in\mathbf{R}^n$. 如果行列式 $\det(\boldsymbol{A})\neq 0$, 则 $\boldsymbol{A}$ 存在逆矩阵 $\boldsymbol{A}^{-1}$, 且线性方程组 $\boldsymbol{A}\boldsymbol{x}=\boldsymbol{b}$ 存在唯一解 $\boldsymbol{x}=\boldsymbol{A}^{-1}\boldsymbol{b}$.

定理 1.4.2　已知矩阵 $\boldsymbol{A}=(a_{ij})_{n\times n}$ 中所有元素为实数, 有下面结论成立:

(1) 如果矩阵 $\boldsymbol{A}$ 的所有特征值均大于或等于 0, 则矩阵 $\boldsymbol{A}$ 是半正定的;

(2) 如果矩阵 $\boldsymbol{A}$ 的所有特征值均大于 0, 则矩阵 $\boldsymbol{A}$ 是正定的.

1.5 习　题

1.5-1　证明性质 1.1.1.

1.5-2　证明性质 1.2.1 和性质 1.2.2.

1.5-3　证明定理 1.2.1 和定理 1.2.2.

1.5-4　证明性质 1.3.1、性质 1.3.2、性质 1.3.3、性质 1.3.4 和性质 1.3.5.

1.5-5　证明定理 1.3.1 和定理 1.3.2.

1.5-6　证明定理 1.3.3.

1.5-7　证明性质 1.4.1.

1.5-8　证明定理 1.4.1 和定理 1.4.2.

第 2 章 凸 集

凸集是凸分析与凸优化中最重要的概念, 是所有凸分析与凸优化论著中最主要的内容 [2-96]. 凸集本质上是从几何角度描述集合的一种拓扑形状, 如非单点凸集中任意两点的连接线段属于该凸集. 因此, 直观地从形状上看, 很容易判定一个平面集合是否为凸集.

本章讨论凸集的各种重要性质, 例如凸集的拓扑: 开凸集、闭凸集、相对内部、凸包和单纯形等, 并介绍以上凸集相互之间的并、交、加与数乘等运算关系的性质, 重点讨论在优化理论中有着重要应用的凸集分离定理, 以及凸集在一些运算下产生的性质, 如极点集、配极和支撑超平面等性质. 除了本章所介绍的凸集知识外, 读者也可参考文献 [25,48,52,63,84-87,94] 所介绍的内容.

2.1 仿 射 集

本节讨论 $\mathbf{R}^n$ 中的仿射集的有关理论. 仿射集是凸集的一种特殊情形, 其中线性子空间和超平面都是特殊的仿射集. 仿射集在刻画凸集的相对内部时具有重要的作用, 第 3 章中凸锥的一些性质也是通过仿射集进行刻画的.

2.1.1 线性子空间

设一组向量 $\boldsymbol{x}_1, \boldsymbol{x}_2, \cdots, \boldsymbol{x}_m \in \mathbf{R}^n$, 对应有 m 个实系数 $\alpha_i\,(i=1,2,\cdots,m)$, 将这组向量与实系数之积线性相加:

$$\alpha_1\boldsymbol{x}_1 + \alpha_2\boldsymbol{x}_2 + \cdots + \alpha_m\boldsymbol{x}_m,$$

称上式为 $\boldsymbol{x}_1, \boldsymbol{x}_2, \cdots, \boldsymbol{x}_m$ 的线性组合. 如果 $\alpha_i \geqslant 0\,(或\alpha_i \leqslant 0), i=1,2,\cdots,m$, 则为非负 (或非正) 线性组合; 如果 $\alpha_i > 0\,(或\alpha_i < 0), i=1,2,\cdots,m$, 则为正 (或负) 线性组合. 下面给出 $\mathbf{R}^n$ 的子空间概念.

定义 2.1.1 设非空集合 $S \subset \mathbf{R}^n$. 如果对于任意的 $\boldsymbol{x}, \boldsymbol{y} \in S$ 和任意的 $\alpha_1, \alpha_2 \in \mathbf{R}$, 均有 $\alpha_1\boldsymbol{x} + \alpha_2\boldsymbol{y} \in S$, 则称 S 是 $\mathbf{R}^n$ 中的子空间 (subspace). 对于子空间 S, 如果存在一个 $\boldsymbol{a} \subset \mathbf{R}^n$ 使得 $L = \boldsymbol{a} + S$, 则称 L 是 $\mathbf{R}^n$ 中的一个平面.

特别地, 当 $n=1$ 时, $\mathbf{R}$ 的子空间有且仅有 $\mathbf{R}$; $\mathbf{R}^2$ 中所有通过原点的直线都是子空间, 且原点也是子空间, 平面则是子空间平行平移所得.

例 2.1.1 设给定点 $c \in \mathbf{R}^n$, 集合 $S = \{\boldsymbol{x} | \boldsymbol{c}^{\mathrm{T}}\boldsymbol{x} = 0\}$ 是 $\mathbf{R}^n$ 的子空间.

根据定义 2.1.1 知, 子空间中的任意两个元素经线性组合运算后, 所得结果仍属于该子空间, 即子空间的线性组合运算具有封闭性.

定义 2.1.2 设 $S \subset \mathbf{R}^n$ 是子空间, 称映射 $L: S \to \mathbf{R}^m$ 是线性变换, 如果满足

$$L(\alpha_1 \boldsymbol{x} + \alpha_2 \boldsymbol{y}) = \alpha_1 L(\boldsymbol{x}) + \alpha_2 L(\boldsymbol{y}), \quad \forall \boldsymbol{x}, \boldsymbol{y} \in S, \quad \forall \alpha_1, \alpha_2 \in \mathbf{R}.$$

若映射 L 是子空间 $S_1 \subset \mathbf{R}^n$ 到子空间 $S_2 \subset \mathbf{R}^m$ 上的一对一线性变换, 则称 L 为线性同构.

显然, 线性变换是一个子空间到另一个子空间的映射, 线性变换保持线性组合运算的封闭性.

例 2.1.2 设映射 $L: S \to \mathbf{R}^m$ 为 $\boldsymbol{y} = L\boldsymbol{x}$. 如果 $\boldsymbol{L}$ 是一个给定的 $m \times n$ 的矩阵, 显然, $\boldsymbol{L}$ 是线性变换.

显然, 根据定义 2.1.1 有下面性质成立.

性质 2.1.1 设 $S \subset \mathbf{R}^n$ 是子空间, 有 m 个任意向量 $\boldsymbol{x}_1, \boldsymbol{x}_2, \cdots, \boldsymbol{x}_m \in S$ 和 m 个不全为 0 的实数 $\alpha_i \, (i = 1, 2, \cdots, m)$, 则线性组合 $\alpha_1 \boldsymbol{x}_1 + \alpha_2 \boldsymbol{x}_2 \cdots + \alpha_m \boldsymbol{x}_m$ 仍属于子空间 S.

性质 2.1.2 设子空间族 $\{S_\gamma \subset \mathbf{R}^n | \gamma \in \varGamma\}$, 其中 $\varGamma$ 是一个集合, 则 $\bigcap_{\gamma \in \varGamma} S_\gamma$ 是子空间.

性质 2.1.3 设非空集合 $S \subset \mathbf{R}^n$, 有下面结论成立:

(1) S 的任意线性组合构成的集合

$$\begin{aligned} \operatorname{span} S = \{ & \alpha_1 \boldsymbol{x}_1 + \alpha_2 \boldsymbol{x}_2 + \cdots + \alpha_m \boldsymbol{x}_m | \boldsymbol{x}_1, \boldsymbol{x}_2, \cdots, \\ & \boldsymbol{x}_m \in S, \alpha_1, \alpha_2, \cdots, \alpha_m \in \mathbf{R}, m = 1, 2, \cdots \} \end{aligned}$$

是 $\mathbf{R}^n$ 的一个子空间;

(2) 若 $\boldsymbol{x}_1, \boldsymbol{x}_2, \cdots, \boldsymbol{x}_m \in S$ 是线性无关的, 则 $\operatorname{span}\{\boldsymbol{x}_1, \boldsymbol{x}_2, \cdots, \boldsymbol{x}_m\}$ 由 $\boldsymbol{x}_1$, $\boldsymbol{x}_2, \cdots, \boldsymbol{x}_m$ 的所有线性组合构成.

上述性质表明, 子空间的任意线性组合仍然在子空间中, 即线性运算具有封闭性; 多个子空间相交仍然是子空间; 任何一个集合由该集合中所有的线性组合生成; 有限个线性无关向量的所有线性组合构成子空间.

注 $\operatorname{span} S$ 是包含 S 的最小子空间.

设 S 是 $\mathbf{R}^n$ 中的一个子空间, 若 $\boldsymbol{x}_1, \boldsymbol{x}_2, \cdots, \boldsymbol{x}_m \in S$ 是线性无关的向量, 则 $\boldsymbol{x}_1, \boldsymbol{x}_2, \cdots, \boldsymbol{x}_m$ 的任何线性组合都在 S 中, 显然有 $\operatorname{span}\{\boldsymbol{x}_1, \boldsymbol{x}_2, \cdots, \boldsymbol{x}_m\} \subseteq S$.

定义 2.1.3 设集合 $S \subset \mathbf{R}^n$, $\boldsymbol{x}_1, \boldsymbol{x}_2, \cdots, \boldsymbol{x}_m \in S$. 如果 $S = \operatorname{span}\{\boldsymbol{x}_1, \boldsymbol{x}_2 \cdots, \boldsymbol{x}_m\}$, 则称 $\boldsymbol{x}_1, \boldsymbol{x}_2, \cdots, \boldsymbol{x}_m$ 是 S 的基. 若 S 的基中元素个数 m 是最大的线性无关向量组中向量的个数, 则称 m 为 S 的维数, 记 $m = \dim(S)$. 当 $m = n$ 时, $S = \mathbf{R}^n$.

性质 2.1.4 如果 $\mathbf{R}^n$ 中两个子空间维数相同, 则这两个子空间的线性变换是线性同构的.

性质 2.1.5 任何非空平面 $L \subset \mathbf{R}^n$ 是由唯一子空间 $S = L - L = \{\boldsymbol{x} - \boldsymbol{y} | \boldsymbol{x}, \boldsymbol{y} \in L\}$ 给出的, 进一步, 满足下面结论:

(1) 对点 $\boldsymbol{a} \in \mathbf{R}^n$, 则 $S = \boldsymbol{a} + L$ 当且仅当 $\boldsymbol{a} \in L$;

(2) 对任意非空子集 $X \subset L$, 有 $L = X + S$;

(3) L 是子空间当且仅当 $\mathbf{0} \in L$.

以上性质说明, 当两个子空间维数相等时, 这两个子空间的线性变换是线性同构的; 两个平行的平面相减得到一个子空间.

本小节得到了子空间与平面的基本性质:

1) 子空间的线性组合运算具有封闭性;

2) 任何集合中线性组合张成的子空间是包含该集合的最小子空间;

3) 子空间等于两个平行平面的差.

2.1.2 $\mathbf{R}^n$ 中的仿射集

$\mathbf{R}^2$ 中所有的子空间都是通过原点的直线, 那么所有不过原点的直线是平面, 但不是子空间, 如果将不过原点的直线平移至通过原点, 则构成子空间. 反过来看, 子空间平移后不过原点, 则不构成子空间了, 但它的线性结构没有变化. 下面讨论集合中任意两点的仿射组合构成的集合的性质.

定义 2.1.4 设集合 $A \subset \mathbf{R}^n$, 给出下面几个定义:

(1) 如果对任意的 $\boldsymbol{x}, \boldsymbol{y} \in A$, 任意的 $\lambda \in \mathbf{R}$, 有 $(1-\lambda)\boldsymbol{x} + \lambda \boldsymbol{y} \in A$, 则称 A 是 $\mathbf{R}^n$ 中的仿射集 (affine set), $(1-\lambda)\boldsymbol{x} + \lambda \boldsymbol{y}$ 是 $\boldsymbol{x}, \boldsymbol{y}$ 的仿射组合.

(2) 设 $\boldsymbol{x}_1, \boldsymbol{x}_2, \cdots, \boldsymbol{x}_m \in \mathbf{R}^n$, $\sum\limits_{i=1}^{m} \alpha_i = 0$, $\alpha_1, \alpha_2, \cdots, \alpha_m \in \mathbf{R}$, 如果

$$\alpha_1 \boldsymbol{x}_1 + \alpha_2 \boldsymbol{x}_2 + \cdots + \alpha_m \boldsymbol{x}_m = \mathbf{0}$$

当且仅当 $\alpha_1 = \alpha_2 = \cdots = \alpha_m = 0$ 时成立, 则称向量 $\boldsymbol{x}_1, \boldsymbol{x}_2, \cdots, \boldsymbol{x}_m$ 是仿射无关的; 否则, 称 $\boldsymbol{x}_1, \boldsymbol{x}_2, \cdots, \boldsymbol{x}_m$ 是仿射相关的.

(3) 称 $\mathrm{aff}X$ 是由 X 张成的仿射包, 它是包含 X 的最小仿射集.

(4) 设 A 是 $\mathbf{R}^n$ 中的仿射集, 如果 $\{\boldsymbol{x}_1, \boldsymbol{x}_2, \cdots, \boldsymbol{x}_m\}$ 是 A 中含元素个数最大的仿射无关集, 那么 $A = \mathrm{aff}\{\boldsymbol{x}_1, \boldsymbol{x}_2, \cdots, \boldsymbol{x}_m\}$, 称 $\boldsymbol{x}_1, \boldsymbol{x}_2, \cdots, \boldsymbol{x}_m$ 是 A 的仿射基, 仿射集的元素个数减 1 称为仿射集 A 的维数, 即 $\dim A = m-1$.

仿射集也称为“仿射流形”“仿射族”“线性流形”等. 仿射集中任意两点间的线段属于该仿射集, 所以仿射集是一种特殊的凸集.

例 2.1.3 设 $\boldsymbol{x}, \boldsymbol{y} \in \mathbf{R}^n$, 且过两点 $\boldsymbol{x}, \boldsymbol{y}$ 的直线集合表示为 $l\langle \boldsymbol{x}, \boldsymbol{y} \rangle = \{t\boldsymbol{x} + (1-t)\boldsymbol{y} | t \in \mathbf{R}\}$, 则该直线是仿射直线.

性质 2.1.6 设 A, A_1, A_2 是 $\mathbf{R}^n$ 中的仿射集, $\boldsymbol{x}_1, \boldsymbol{x}_2, \cdots, \boldsymbol{x}_m \in A$, $\lambda_1, \lambda_2, \cdots, \lambda_m \in \mathbf{R}$ 且 $\sum\limits_{i=1}^{m} \lambda_i = 1$, 则有下面结论成立:

(1) 仿射集 A 中任意直线仍然在 A 中;

(2) 空集 $\varnothing$ 和空间 $\mathbf{R}^n$ 都是仿射集, $\mathbf{R}^n$ 中的点、直线和平面都是仿射集;

(3) $\lambda_1 \boldsymbol{x}_1 + \lambda_2 \boldsymbol{x}_2 + \cdots + \lambda_m \boldsymbol{x}_m \in A$;

(4) $A_1 \cap A_2$ 仍是 $\mathbf{R}^n$ 中的仿射集.

注 称性质 2.1.6(3) 的 $\lambda_1 \boldsymbol{x}_1 + \lambda_2 \boldsymbol{x}_2 + \cdots + \lambda_m \boldsymbol{x}_m$ 是 $\boldsymbol{x}_1, \boldsymbol{x}_2, \cdots, \boldsymbol{x}_m \in A$ 的仿射组合.

性质 2.1.6 说明: 仿射集含自身的任意直线; 直线与平面都是仿射集; 仿射集由其中元素的所有的仿射组合构成; 多个仿射集相交仍然是仿射集.

根据定义 2.1.4 有下面性质成立.

性质 2.1.7 设集合 $X \subset \mathbf{R}^n$, 向量 $\boldsymbol{x}_1, \boldsymbol{x}_2, \cdots, \boldsymbol{x}_m$ 是仿射无关的, 则有下面结论成立:

(1) 当且仅当 $\boldsymbol{x}_1, \boldsymbol{x}_2, \cdots, \boldsymbol{x}_m$ 中任何元素都不是除自身外的其他元素的仿射组合;

(2) 当且仅当对每一个 $i\,(i = 1, 2, \cdots, m)$ 得到 m 个向量组 $\boldsymbol{x}_1 - \boldsymbol{x}_i, \boldsymbol{x}_2 - \boldsymbol{x}_i, \cdots, \boldsymbol{x}_m - \boldsymbol{x}_i$ 且都是线性无关的;

(3) $m \leqslant n$ 时必成立, 即 $\mathbf{R}^n$ 中若有大于 n 个不同向量, 则它们是仿射相关的;

(4) $\operatorname{aff}\{\boldsymbol{x}_1, \boldsymbol{x}_2, \cdots, \boldsymbol{x}_m\}$ 构成包含 $\boldsymbol{x}_1, \boldsymbol{x}_2, \cdots, \boldsymbol{x}_m$ 的最小仿射集;

(5) 如果 $\boldsymbol{x}_1, \boldsymbol{x}_2, \cdots, \boldsymbol{x}_m$ 是 X 中含最大元素个数的仿射无关集, 则 $\operatorname{aff}(X) = \operatorname{aff}\{\boldsymbol{x}_1, \boldsymbol{x}_2, \cdots, \boldsymbol{x}_m\}$, 即 $\operatorname{aff}(X)$ 中的任何元素都可以用 $\boldsymbol{x}_1, \boldsymbol{x}_2, \cdots, \boldsymbol{x}_m$ 的仿射组合表示;

(6) 设 X 是 $\mathbf{R}^n$ 中的任意非空集合, 则 X 中所有元素的仿射组合构成的集合是 $\operatorname{aff}(X)$.

特别地, 空集的维数为 -1, 一个点构成的仿射集维数是 0, 直线的仿射集维数是 1.

定义 2.1.5 设 A 是 $\mathbf{R}^n$ 中的仿射集, 映射 $T: A \to \mathbf{R}^m$. 如果

$$T\left(\sum_{i=1}^{m} \lambda_i \boldsymbol{x}_i\right) = \sum_{i=1}^{m} \lambda_i T(\boldsymbol{x}_i),$$

其中 $\boldsymbol{x}_1, \boldsymbol{x}_2, \cdots, \boldsymbol{x}_m \in A$, $\lambda_1, \lambda_2, \cdots, \lambda_m \in \mathbf{R}$ 且 $\sum\limits_{i=1}^{m} \lambda_i = 1$, 则称 T 是仿射变换. 如果 $m = 1$, 则仿射变换 T 也称为仿射函数.

定义 2.1.6　设 A_1 和 A_2 是仿射集, 如果仿射变换 $T: A_1 \to A_2$ 是一一对应的, 则称仿射变换 T 为仿射同构, A_1 和 A_2 是仿射同构集.

性质 2.1.8　如果 $T: A \to \mathbf{R}^m$ 是仿射变换, 其中 A 是 $\mathbf{R}^n$ 中的仿射集, 则下面结论成立:

(1) $T(A) = \{T\boldsymbol{x} | \boldsymbol{x} \in A\}$ 是 $\mathbf{R}^m$ 中的仿射集;

(2) 仿射变换保持仿射包不变, 即 $\operatorname{aff}(T(X)) = T(\operatorname{aff}(X))$;

(3) 仿射变换将线段变为线段, 且不改变两条平行线段的长度之比.

上述性质表明, 仿射集经过仿射变换后还是仿射集, 线段经过仿射变换后还是线段. 而下面的定理则表明仿射变换是仿射集减去一个仿射集中的固定点的线性变换.

定理 2.1.1　如果 $T: A \to \mathbf{R}^m$ 是仿射变换, 其中 A 是 $\mathbf{R}^n$ 中的仿射集, 令 $L\boldsymbol{x} = T\boldsymbol{x} - T\boldsymbol{a}$, 其中 $\boldsymbol{a}, \boldsymbol{x} \in A$, 则 L 是 $A - \boldsymbol{a} \to \mathbf{R}^m$ 的线性变换.

证明　设 $\forall \boldsymbol{x}, \boldsymbol{y} \in A, S = A - \boldsymbol{a}, \forall \alpha_1 \in \mathbf{R}$ 和 $\boldsymbol{a} \in A$. 我们有

$$\alpha_1(\boldsymbol{x} - \boldsymbol{a}) = \alpha_1(\boldsymbol{x} - \boldsymbol{a}) + (1 - \alpha_1)(\boldsymbol{a} - \boldsymbol{a}) = \alpha_1 \boldsymbol{x} + (1 - \alpha_1)\boldsymbol{a} - \boldsymbol{a} \in S,$$

$$\frac{1}{2}(\boldsymbol{x} - \boldsymbol{a}) + \frac{1}{2}(\boldsymbol{y} - \boldsymbol{a}) = \frac{1}{2}(\boldsymbol{x} - \boldsymbol{a}) + \left(1 - \frac{1}{2}\right)(\boldsymbol{y} - \boldsymbol{a}) = \frac{1}{2}\boldsymbol{x} + \left(1 - \frac{1}{2}\right)\boldsymbol{y} - \boldsymbol{a} \in S.$$

由此可得 $(\boldsymbol{x} - \boldsymbol{a}) + (\boldsymbol{y} - \boldsymbol{a}) \in S$, 即 S 是一个子空间. 由于 TA 是仿射集, 根据上面的证明知 $TA - T\boldsymbol{a}$ 是一个子空间, 因此, L 是一个线性变换. □

反之, 下面定理说明线性变换是子空间减去其中一个固定点的仿射变换.

定理 2.1.2　如果 $L: S \to \mathbf{R}^m$ 是线性变换, 其中 S 是 $\mathbf{R}^n$ 中的子空间, 令 $T\boldsymbol{x} = L\boldsymbol{x} + L\boldsymbol{a}$, 其中 $\boldsymbol{a}, \boldsymbol{x} \in S$, 则 T 是 $S + \boldsymbol{a} \to \mathbf{R}^m$ 的仿射变换.

证明　设 $\forall \boldsymbol{x}, \boldsymbol{y} \in S, A = S + \boldsymbol{a}, \lambda \in \mathbf{R}$ 和 $\boldsymbol{a} \in S$, 有

$$(1 - \lambda)(\boldsymbol{x} + \boldsymbol{a}) + \lambda(\boldsymbol{y} + \boldsymbol{a}) = (1 - \lambda)\boldsymbol{x} + \lambda \boldsymbol{y} + \boldsymbol{a} \in A,$$

则 A 是仿射集, 进一步有

$$\begin{aligned} L[(1 - \lambda)(\boldsymbol{x} + \boldsymbol{a}) + \lambda(\boldsymbol{y} + \boldsymbol{a})] &= L[(1 - \lambda)\boldsymbol{x} + \lambda \boldsymbol{y} + \boldsymbol{a}] \\ &= (1 - \lambda) A\boldsymbol{x} + (1 - \lambda) A\boldsymbol{a} + \lambda A\boldsymbol{y} + \lambda A\boldsymbol{a} \\ &= (1 - \lambda) Lx + \lambda Ly, \end{aligned}$$

所以 T 是仿射变换. □

推论 2.1.1　包含原点的仿射集是子空间, 子空间也是包含原点的仿射集.

定理 2.1.3　$\mathbf{R}^n$ 中的两个仿射无关的集合存在从 $\mathbf{R}^n$ 到自身的一一对应仿射变换 T.

证明 根据性质 2.1.5 和定理 2.1.2 知道, 存在从 $\mathbf{R}^n$ 到自身的一一对应线性变换 A, 即可知本定理结论成立. □

上述定理表明, 当且仅当两个仿射集维数相同时, 它们仿射同构. 任何同构的仿射集也是同胚的 (同胚是指 1 对 1 的连续满射), 保持着相同的仿射结构. 不同维数的仿射集都有一个与之维数相同的仿射集同构. 也就是说, 仿射同构集的点具有相同的仿射结构.

推论 2.1.2 设 A_1 和 A_2 是 $\mathbf{R}^n$ 中具有相同的最大仿射无关数, 则 A_1 和 A_2 存在一一对应仿射变换 T, 使 $T(\mathrm{aff}\,(A_1)) = \mathrm{aff}\,(A_2)$.

下面讨论仿射集平移性质.

定义 2.1.7 设 $X \subset \mathbf{R}^n$, $\boldsymbol{a} \in \mathbf{R}^n$, 则 $X + \boldsymbol{a} = \{\boldsymbol{x} + \boldsymbol{a} | \boldsymbol{x} \in X\}$ 称为 X 平移 $\boldsymbol{a}$ 的集合. 如果 A_1, A_2 都是仿射集, $\boldsymbol{a} \in \mathbf{R}^n$ 且 $A_1 = A_2 + \boldsymbol{a}$, 则称仿射集 A_1 与仿射集 A_2 是平行的.

性质 2.1.9 设 $A \subset \mathbf{R}^n$ 是非空的仿射集, $\boldsymbol{a} \in \mathbf{R}^n$, 则下面结论成立:

(1) 仿射集 A 的平移 $A + \boldsymbol{a}$ 仍然是仿射集, 且相互平行;

(2) 对任意 $\boldsymbol{a}_1, \boldsymbol{a}_2 \in A$, 有

$$A - \boldsymbol{a}_1 = \{\boldsymbol{x} - a_1 | \boldsymbol{x} \in A\} = A - \boldsymbol{a}_2 = \{\boldsymbol{x} - \boldsymbol{a}_2 | \boldsymbol{x} \in A\}.$$

证明 (1) 是显然成立的.

(2) 由定理 2.1.1 知道, $A - \boldsymbol{a}_1$ 和 $A - \boldsymbol{a}_2$ 都是子空间. 设 $\forall \boldsymbol{x} \in A - \boldsymbol{a}_1$, $\exists \boldsymbol{a} \in A$ 使得 $\boldsymbol{x} = \boldsymbol{a} - \boldsymbol{a}_1$. 再由 $\boldsymbol{a}_1 - \boldsymbol{a}_2 \in A - \boldsymbol{a}_2$ 得 $-(\boldsymbol{a}_1 - \boldsymbol{a}_2) \in A - \boldsymbol{a}_2$, 即有

$$\boldsymbol{x} = \boldsymbol{a} - \boldsymbol{a}_2 - (\boldsymbol{a}_1 - \boldsymbol{a}_2) \in A - \boldsymbol{a}_2.$$

因此, $A - \boldsymbol{a}_1 \subseteq A - \boldsymbol{a}_2$, 同理得 $A - \boldsymbol{a}_2 \subseteq A - \boldsymbol{a}_1$.

由性质 2.1.9 知道 A 平行于子空间 $A - A$. □

上述性质说明, 仿射集平移后是平行的仿射集.

定义 2.1.8 设非空集合 $X \subset \mathbf{R}^n$, 则子空间 $\mathrm{dir}X = \mathrm{aff}X - \mathrm{aff}X$ 和 $\mathrm{ort}X = (\mathrm{dir}X)^{\perp}$ 分别称为 X 的方向子空间和垂直子空间, 记 $\mathrm{dir}\varnothing = \varnothing$ 和 $\mathrm{ort}\varnothing = \mathbf{R}^n$.

集合的方向子空间和垂直子空间与仿射集有下面的关系.

性质 2.1.10 设非空集合 $X \subset \mathbf{R}^n$, $\boldsymbol{b} \in \mathrm{aff}X$, 则有下面结论成立:

(1) $\mathrm{dir}X = \mathrm{aff}X - \boldsymbol{b} = \mathrm{aff}(X - \boldsymbol{b}) = \mathrm{span}(X - \boldsymbol{b})$;

(2) $\mathrm{dir}X = \mathrm{aff}X - \mathrm{aff}X = \mathrm{aff}(X - X) = \mathrm{span}(X - X)$;

(3) $\mathrm{dir}X = \mathrm{dir}(\mathrm{cl}X)$, 且 $\mathrm{dir}X = \mathrm{dir}Y$, 其中 Y 是 X 的稠密子集;

(4) $\mathrm{dir}X = \{\alpha_1 \boldsymbol{x}_1 + \cdots + \alpha_k \boldsymbol{x}_k | k \geqslant 1, \boldsymbol{x}_1, \cdots, \boldsymbol{x}_k \in X, \alpha_1 + \cdots + \alpha_k = 0\}$.

证明 (1) 根据性质 2.1.9(1) 得 $\mathrm{aff}(X - \boldsymbol{b}) = \mathrm{aff}(X) - \boldsymbol{b}$. 再根据性质 2.1.9(2) 得

$$\mathrm{aff}X - \mathrm{aff}X = \mathrm{aff}X - \boldsymbol{b}.$$

显然, $X-\boldsymbol{b}$ 的最大线性无关组也是 $\mathrm{aff}(X-\boldsymbol{b})$ 和 $\mathrm{span}(X-\boldsymbol{b})$ 的最大线性无关组. 因此, 结论 (1) 成立.

(2) 在 (1) 的证明过程中包含了结论.

(3) 因为仿射集是闭集, 所以结论 (3) 成立.

(4) 不妨设 $\boldsymbol{x}_1,\cdots,\boldsymbol{x}_k\in X$ 是最大仿射无关组, 根据性质 2.1.7(5) 的结论知本结论成立. □

性质 2.1.10 给出了由一个集合生成的仿射集、子空间和垂直子空间之间的一些等价性. 进一步, 下面的性质 2.1.11 表明两个集合之间生成的子空间与仿射集之间交与并运算后的关系.

性质 2.1.11　设非空子集 $X,Y\subset\mathbf{R}^n$, 有下面结论成立:

(1) $\mathrm{aff}(X\cup Y)=(\mathrm{aff}X+\mathrm{dir}Y)\cup(\mathrm{dir}X+\mathrm{aff}Y)\cup L(X,Y)$, 其中

$$L(X,Y)=\{\alpha\boldsymbol{x}+\beta\boldsymbol{y}|\boldsymbol{x}\in\mathrm{aff}X,\boldsymbol{y}\in\mathrm{aff}Y,\alpha+\beta=1\};$$

(2) 如果 $\mathrm{aff}X\cap\mathrm{aff}Y\neq\varnothing$, $\boldsymbol{c}\in\mathrm{aff}X\cap\mathrm{aff}Y$, 则有 $\mathrm{aff}(X\cup Y)=\mathrm{aff}(X+Y)-\boldsymbol{c}$;

(3) 设 $\mathrm{aff}X\cap\mathrm{aff}Y\neq\varnothing$, $\boldsymbol{c}\in\mathrm{aff}X,\boldsymbol{e}\in\mathrm{aff}Y$, l 是通过点 $\boldsymbol{c}$ 和 $\boldsymbol{e}$ 的直线, 则有

$$\mathrm{aff}(X\cup Y)=\mathrm{aff}(X+Y)+l-\boldsymbol{c}-\boldsymbol{e}.$$

本小节得到了仿射集的一些重要性质:

1) 仿射集包含直线, 任何集合中的所有仿射组合构成一个仿射集, 多个仿射集相交还是仿射集;

2) 有限个仿射无关的点集 X 的仿射组合是包含 X 的最小仿射集;

3) 任何 m 个仿射无关向量减去其中任何一个向量, 构成一组线性无关向量;

4) 仿射集通过仿射变换后得到的集合还是仿射集;

5) 一个集合生成的子空间等于其生成仿射集的差;

6) 两个集合的并集生成的仿射集等于它们各自生成的仿射集加上生成子空间的并集, 再与由两个集合生成的仿射集的并集构成;

7) 若两个集合各自生成的仿射集交包含元素 $\boldsymbol{c}$, 则两个集合的并集生成的仿射集等于两个集合之和生成的仿射集减去元素 $\boldsymbol{c}$.

2.1.3 超平面

本小节讨论在凸分析中一类非常重要的特殊仿射集——超平面. 超平面也是凸集.

定义 2.1.9　$\mathbf{R}^n$ 中任何具有 $n-1$ 维的仿射集称为超平面.

例 2.1.4　超平面 $H=\left\{(x_1,x_2,x_3)^{\mathrm{T}}\in\mathbf{R}^3|x_1+x_2+x_3=1\right\}$, 显然 H 是一个 2 维平面且是一个仿射集.

由此可见, 超平面是平面概念的推广.

设 $\langle \boldsymbol{x}, \boldsymbol{y} \rangle = \boldsymbol{x}^{\mathrm{T}}\boldsymbol{y} = 0$, 表示 $\boldsymbol{x} \perp \boldsymbol{y}$, 即 $\boldsymbol{x}$ 与 $\boldsymbol{y}$ 正交或垂直. 已知 $\mathbf{R}^n$ 的子空间 S, 称 S 的正交集合

$$S^{\perp} = \{\boldsymbol{x} | \langle \boldsymbol{x}, \boldsymbol{y} \rangle = 0, \boldsymbol{y} \in S\}$$

是 S 的正交补空间. 若 S 是一个集合, 记 $S^{\perp} = \mathrm{ort}S$, 表示 S 的正交集合. 有

$$\mathrm{dim}S + \mathrm{dim}S^{\perp} = n$$

成立, 且 $\left(S^{\perp}\right)^{\perp} = S$. 如果 $\boldsymbol{x}_1, \boldsymbol{x}_2, \cdots, \boldsymbol{x}_m$ 是 S 的基, 则 $\boldsymbol{x} \perp S \Leftrightarrow \boldsymbol{x} \perp \{\boldsymbol{x}_1, \boldsymbol{x}_2, \cdots, \boldsymbol{x}_m\}$. 特别地, 当 $m = n-1$ 时, S 是 1 维子空间的正交补空间.

据此, 下面我们有一个重要结论, 即超平面是由一个线性方程构成的.

定理 2.1.4 设任意给定一个非零向量 $\boldsymbol{a} \in \mathbf{R}^n$, 任意 $\alpha \in \mathbf{R}$, 则集合

$$H(\boldsymbol{a}, \alpha) = \{\boldsymbol{x} | \langle \boldsymbol{x}, \boldsymbol{a} \rangle = \alpha\}$$

是 $\mathbf{R}^n$ 的一个超平面.

证明 设 $\forall \boldsymbol{x}, \boldsymbol{y} \in H(\boldsymbol{a}, \alpha), \forall \lambda \in \mathbf{R}$, 有

$$\langle (1-\lambda)\boldsymbol{x} + \lambda \boldsymbol{y}, \boldsymbol{a} \rangle = (1-\lambda)\langle \boldsymbol{x}, \boldsymbol{a} \rangle + \lambda \langle \boldsymbol{y}, \boldsymbol{a} \rangle = \alpha,$$

则 $(1-\lambda)\boldsymbol{x} + \lambda \boldsymbol{y} \in H(\boldsymbol{a}, \alpha)$. 再证明 $H(\boldsymbol{a}, \alpha)$ 是 $\mathbf{R}^n$ 的一个 $n-1$ 维的子空间即可. 设非零向量 $\boldsymbol{b} \in H(\boldsymbol{a}, \alpha)$, 集合 $H(\boldsymbol{a}, \alpha) - \langle \boldsymbol{b}, \boldsymbol{a} \rangle = H(\boldsymbol{a}, \alpha) - \alpha$ 是 $\mathbf{R}^n$ 的一个子空间, 且有 $\boldsymbol{a} \perp (H(\boldsymbol{a}, \alpha) - \langle \boldsymbol{b}, \boldsymbol{a} \rangle)$. 显然

$$\{\lambda \boldsymbol{a}\} = (H(\boldsymbol{a}, \alpha) - \langle \boldsymbol{b}, \boldsymbol{a} \rangle)^{\perp}, \quad \forall \lambda \in \mathbf{R}$$

是 1 维空间, 因此有 $H(\boldsymbol{a}, \alpha)$ 是一个 $n-1$ 维的子空间. □

定理 2.1.4 表明任何一个超平面都可以用唯一的线性方程 $\langle \boldsymbol{x}, \boldsymbol{b} \rangle = \alpha$ 表示. 定理 2.1.4 中的向量 $\boldsymbol{a}$ 称为超平面 $H(\boldsymbol{a}, \alpha)$ 的法线或法向量, $H(\boldsymbol{a}, \alpha)$ 所有的法线是平行的, 且均由 $\lambda \boldsymbol{a}\,(\forall \lambda \in \mathbf{R})$ 构成. 如果 $\alpha_1 \neq \alpha_2$, 则超平面 $H(\boldsymbol{a}, \alpha_1)$ 平行于超平面 $H(\boldsymbol{a}, \alpha_2)$; 如果 $\alpha = 0$, 则 $H(\boldsymbol{a}, 0)$ 是一个包含原点的 $n-1$ 维子空间.

定理 2.1.5 设 $\boldsymbol{B}$ 是一个秩为 m 的 $m \times n$ 的矩阵, 则集合

$$S = \{\boldsymbol{x} \in \mathbf{R}^n | \boldsymbol{B}\boldsymbol{x} = \boldsymbol{0}\}$$

是 $\mathbf{R}^n$ 中的一个 $n-m$ 维子空间, 且可构成任意 $n-m$ 维子空间.

证明 设 $\forall \boldsymbol{x}, \boldsymbol{y} \in S$, $\forall \alpha_1, \alpha_2 \in \mathbf{R}$, 均有 $\boldsymbol{B}(\alpha_1 \boldsymbol{x} + \alpha_2 \boldsymbol{y}) = 0$, 则 S 是子空间. 设矩阵 $\boldsymbol{B} = (\boldsymbol{b}_1, \boldsymbol{b}_2, \cdots, \boldsymbol{b}_m)^{\mathrm{T}}$, 其中 $(\boldsymbol{b}_1, \boldsymbol{b}_2, \cdots, \boldsymbol{b}_m)$ 是线性无关的, 那么

$$\boldsymbol{B}\boldsymbol{x} = \boldsymbol{0} \Leftrightarrow \langle \boldsymbol{x}, \boldsymbol{b}_i \rangle = 0, i = 1, 2, \cdots, m.$$

根据定理 2.1.4 的证明过程知 $\boldsymbol{b}_i \perp S\,(i=1,2,\cdots,m)$. 因此, $\boldsymbol{B}=(\boldsymbol{b}_1,\boldsymbol{b}_2,\cdots,\boldsymbol{b}_m)^{\mathrm{T}} \perp S$, 即 S 是 $\mathbf{R}^n$ 中的一个 $n-m$ 子空间. 任何 m 维子空间可由线性无关向量组 $(\boldsymbol{b}_1,\boldsymbol{b}_2,\cdots,\boldsymbol{b}_m)$ 构成, 则任意 $n-m$ 维子空间都可表示为 $S=\{\boldsymbol{x}\in\mathbf{R}^n|\boldsymbol{B}\boldsymbol{x}=\mathbf{0}\}$. 上述结论显然对 $\mathbf{R}^n$ 或 $\varnothing$ 也成立. □

定理 2.1.5 说明由一个线性方程组构成一个超平面, 若由 m 元线性方程组构成 m 个超平面的交集, 则 S 组成了 $n-m$ 维的仿射集. 特别是 m 元线性方程组左端均等于 0, 则构成一个 $n-m$ 维子空间.

推论 2.1.3　设 $\boldsymbol{B}$ 是一个秩为 m 的 $m\times n$ 的矩阵, 向量 $\boldsymbol{b}\in\mathbf{R}^m$, 则集合

$$S=\{\boldsymbol{x}\in\mathbf{R}^n|\boldsymbol{B}\boldsymbol{x}=\boldsymbol{b}\}$$

是 $\mathbf{R}^n$ 中的 $n-m$ 维仿射集, 且可构成任何 $n-m$ 维仿射集.

例 2.1.5　在 $\mathbf{R}^n$ 中求包含 $\boldsymbol{a}_i\in\mathbf{R}^n\,(i=1,2,\cdots,n)$ 点的超平面 H.

解　根据定理 2.1.4 知, 每个 $\boldsymbol{a}_i\in H$ 必须满足方程组 $\langle\boldsymbol{a}_i,\boldsymbol{a}\rangle=\alpha(i=1,2,\cdots,n)$, 求出一组对应的 $\boldsymbol{a},\alpha$ 即可.

这个例子说明了 $\mathbf{R}^n$ 的子空间可以用 n 元齐次线性方程组的解集表示, 并且 $\mathbf{R}^n$ 的仿射集可以用 n 元线性方程组的解集表示.

本小节得到了超平面的几个重要性质:

1) 一个超平面是由一个线性方程构成的;

2) 左端均等于 0 的 m 元线性方程组构成一个 $n-m$ 维子空间;

3) m 元线性方程组构成一个 $n-m$ 维仿射集.

2.2　凸集及相关性质

凸集是凸分析中最重要的概念, 也是构成凸分析理论的核心概念, 凸集理论在凸优化中是描述最优性、对偶性和全局性等方面的性质不可缺少的部分 [52,85-87,94]. 本节详细地讨论凸集及其相关的基本性质.

2.2.1　基本定义及基本性质

下面首先给出几个基本定义.

定义 2.2.1　设集合 $C\subset\mathbf{R}^n$. 如果对任意的 $\boldsymbol{x},\boldsymbol{y}\in C$ 和任意的 $\lambda\in[0,1]$ 有 $(1-\lambda)\,\boldsymbol{x}+\lambda\boldsymbol{y}\in C$, 则称 C 是凸集.

定义 2.2.2　设非空集合 $C\subset\mathbf{R}^n$, 给定点 $\boldsymbol{x}\in C$. 如果对于任意的 $\boldsymbol{y}\in C$ 和任意的 $\lambda\in[0,1]$, 有

$$(1-\lambda)\,\boldsymbol{x}+\lambda\boldsymbol{y}\in C,$$

则称 C 是关于 $\boldsymbol{x}$ 的星形集.

定义 2.2.3 设非空集合 $C \subset \mathbf{R}^n$, 称集合

$$K = \{\boldsymbol{x} \in C |\ (1-\lambda)\boldsymbol{x} + \lambda \boldsymbol{y} \in C, \forall \boldsymbol{y} \in C\}$$

为 C 的核.

根据上述定义易知, 在 2 维平面中, 图 2.2-1 中圆的边界不构成凸集, 但边界与圆内所有点构成凸集; 图 2.2-2 中除边界外的所有点构成星形集; 图 2.2-3 由于两点之间的连线中存在不属于该集合的点, 因此是非凸集. 由此可以看出, 凸集的几何特征可以表述为其中任意两点间的线段上的点都属于该凸集.

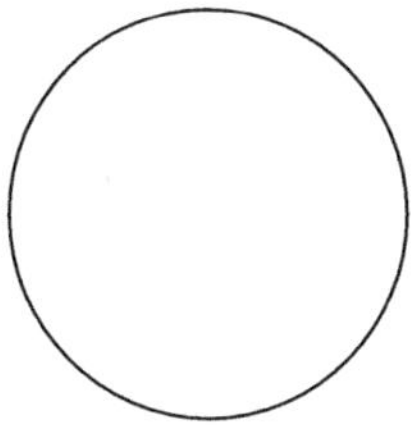

图 2.2-1 凸集

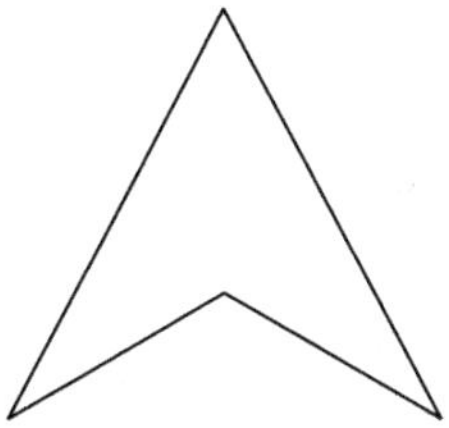

图 2.2-2 星形集

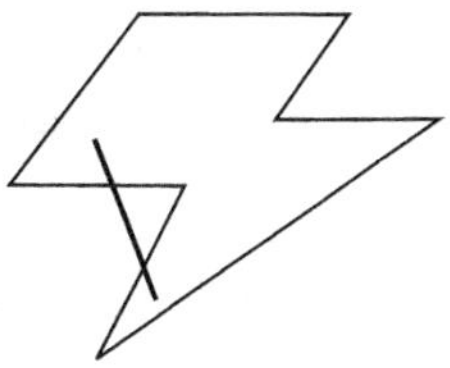

图 2.2-3 非凸集

例 2.2.1 (1) 在 $\mathbf{R}$ 空间中的 $C_1 = (0,1)$ 是一个凸集.

(2) 在 $\mathbf{R}^2$ 空间中的 $C_2 = \left\{(x_1, 0)^{\mathrm{T}} | 0 < x_1 < 1\right\} \cup \left\{(0, x_2)^{\mathrm{T}} | 0 < x_2 < 1\right\}$ 不是凸集, 也不是星形集.

(3) 在 $\mathbf{R}^2$ 空间中的 $C_3 = \left\{(x_1, 0)^{\mathrm{T}} | 0 \leqslant x_1 < 1\right\} \cup \left\{(0, x_2)^{\mathrm{T}} | 0 \leqslant x_2 < 1\right\}$ 是一个关于原点 $\mathbf{0}$ 的星形集.

(4) 在 $\mathbf{R}^2$ 空间中的 $C_4 = \left\{(x_1, x_2)^{\mathrm{T}} | x_1^2 + x_2^2 = 1\right\}$ 由单位圆构成, 不是凸集, 因为圆上任意两点连接的直线都不在圆上.

(5) 在 $\mathbf{R}^2$ 空间中的 $C_5 = \left\{(x_1, x_2)^{\mathrm{T}} | x_1^2 + x_2^2 \leqslant 1\right\}$ 是由圆面构成的凸集, 因为圆中任意两点相连的直线都在圆中.

(6) 设给定两点 $\boldsymbol{x}, \boldsymbol{y} \in \mathbf{R}^n$, 关于点 $\boldsymbol{x}, \boldsymbol{y}$ 的各种线段集合定义如下.

包含两端点线段: $[\boldsymbol{x}, \boldsymbol{y}] = \{t\boldsymbol{x} + (1-t)\boldsymbol{y} | 0 \leqslant t \leqslant 1\}$;

包含右端点闭半线段: $(\boldsymbol{x}, \boldsymbol{y}] = \{(1-t)\boldsymbol{x} + t\boldsymbol{y} | 0 < t \leqslant 1\}$;

包含左端点闭半线段: $[\boldsymbol{x}, \boldsymbol{y}) = \{t\boldsymbol{x} + (1-t)\boldsymbol{y} | 0 < t \leqslant 1\}$;

不包含两端点线段: $(\boldsymbol{x}, \boldsymbol{y}) = \{t\boldsymbol{x} + (1-t)\boldsymbol{y} | 0 < t < 1\}$.

以上线段集合均为凸集.

设两点 $\boldsymbol{x}, \boldsymbol{y} \in \mathbf{R}^n$, 定义过 $\boldsymbol{x}, \boldsymbol{y}$ 的直线集合为

$$l\langle \boldsymbol{x}, \boldsymbol{y} \rangle = \{t\boldsymbol{x} + (1-t)\boldsymbol{y} | t \in \mathbf{R}\};$$

定义 $\boldsymbol{x}, \boldsymbol{y}$ 的闭半直线集合和开半直线集合分别为

$$l[\boldsymbol{x}, \boldsymbol{y}\rangle = \{(1-t)\boldsymbol{x} + t\boldsymbol{y} | t \geqslant 0\}, \quad l\langle \boldsymbol{x}, \boldsymbol{y}] = \{(1-t)\boldsymbol{x} + t\boldsymbol{y} | t \leqslant 1\};$$

$$l(\boldsymbol{x}, \boldsymbol{y}\rangle = \{(1-t)\boldsymbol{x} + t\boldsymbol{y} | t > 0\}, \quad l\langle \boldsymbol{x}, \boldsymbol{y}) = \{(1-t)\boldsymbol{x} + t\boldsymbol{y} | t < 1\}.$$

上述直线、半直线集合均是凸集. $l[\boldsymbol{x}, \boldsymbol{y}\rangle$ 是包含左端点 $\boldsymbol{x}$ 过点 $\boldsymbol{y}$ 的闭半直线, $l\langle \boldsymbol{x}, \boldsymbol{y}]$ 是包含右端点 $\boldsymbol{y}$ 过点 $\boldsymbol{x}$ 的闭半直线, $l(\boldsymbol{x}, \boldsymbol{y}\rangle$ 是不包含左端点 $\boldsymbol{x}$ 过点 $\boldsymbol{y}$ 的开半直线, $l\langle \boldsymbol{x}, \boldsymbol{y})$ 是不包含右端点 $\boldsymbol{y}$ 过点 $\boldsymbol{x}$ 的开半直线. 容易证明连通集合是无解界, 那么该集合存在闭 (开) 半直线. 这些定义的集合在后面的一些性质和定理证明中将会用到.

从凸集定义可知, 凸集的特点是它的任意两个不同点 $\boldsymbol{x}, \boldsymbol{y}$ 连接的线段仍然在凸集中. 因此, 有的文献用线段来定义凸集, 线段可直观地判断凸集, 如果集合中两点的线段上存在不属于集合中的点, 则该集合就不是凸集. 凸集是星形集, 反之却不一定. 非凸集合 C 中存在点不属于它的核, 于是我们有下面性质成立.

性质 2.2.1　设非空集合 $C \subset \mathbf{R}^n$, 则下面结论成立:

(1) $\varnothing$, $\mathbf{R}^n$, 子空间和仿射集都是凸集.

(2) 超平面 $H(\boldsymbol{a}, \alpha) = \{\boldsymbol{x} | \langle \boldsymbol{x}, \boldsymbol{a}\rangle = \alpha\}$ 是凸集, 其中任意非零向量 $\boldsymbol{a} \in \mathbf{R}^n$ 及 $\alpha \in \mathbf{R}$. 对应地, $\{\boldsymbol{x} | \langle \boldsymbol{x}, \boldsymbol{a}\rangle \leqslant \alpha\}$, $\{\boldsymbol{x} | \langle \boldsymbol{x}, \boldsymbol{a}\rangle \geqslant \alpha\}$, $\{\boldsymbol{x} | \langle \boldsymbol{x}, \boldsymbol{a}\rangle < \alpha\}$ 和 $\{\boldsymbol{x} | \langle \boldsymbol{x}, \boldsymbol{a}\rangle > \alpha\}$ 这 4 个集合都是凸集, 前两个集合是闭半空间, 后两个集合是开半空间.

(3) $\mathbf{R}^n$ 中任何点的邻域都是凸集.

(4) $\mathbf{R}^n$ 中集合 C 是凸集当且仅当 C 是自身每一点的星形集.

(5) $\mathbf{R}^n$ 中集合 C 的核 K 是凸集.

(6) $\mathbf{R}^n$ 中凸集 C 的核集合仍是 C.

证明　(1), (2) 和 (3) 可直接由定义 2.2.1 可证.

(4) 根据定义 2.2.1、定义 2.2.2 可得.

(5) 设任意两点 $\boldsymbol{x}, \boldsymbol{y} \in K$, $\forall \lambda \in [0,1]$ 和 $\boldsymbol{u}_\lambda = (1-\lambda)\boldsymbol{x} + \lambda \boldsymbol{y}$. 由定义 2.2.3 可知, 对任意的 $\boldsymbol{y}' \in C$, 任意的 $\alpha \in [0,1]$, 有

$$(1-\alpha)\boldsymbol{x} + \alpha \boldsymbol{y}' \in C, \quad (1-\alpha)\boldsymbol{y} + \alpha \boldsymbol{y}' \in C.$$

因 C 是关于 $\boldsymbol{x}$ 和 $\boldsymbol{y}$ 的星形集, 则 $\boldsymbol{u}_\lambda \in C$. 假设 $\exists \boldsymbol{u}_\lambda \notin K$, $\lambda \in (0,1)$, 则存在某个 $\boldsymbol{y}' \in C$, $\delta \in [0,1]$ 使得 $(1-\delta)\boldsymbol{u}_\lambda + \delta \boldsymbol{y}' \notin C$, 显然 $\delta \in (0,1)$. 我们得到

$$\begin{aligned}(1-\delta)\boldsymbol{u}_\lambda + \delta \boldsymbol{y}' &= (1-\lambda)(1-\delta)\boldsymbol{x} + \lambda(1-\delta)\boldsymbol{y} + \delta \boldsymbol{y}' \\ &= (1-\lambda-\delta+\delta\lambda)\boldsymbol{x} + \lambda(1-\delta)\boldsymbol{y} + \delta \boldsymbol{y}'.\end{aligned}$$

设 $\alpha=\lambda+\delta-\delta\lambda$, $\alpha\boldsymbol{v}=\lambda(1-\delta)\boldsymbol{y}+\delta\boldsymbol{y}'$, 则有 $1>\alpha=\lambda+\delta-\delta\lambda>0$ 和

$$\boldsymbol{v}=\frac{\lambda(1-\delta)}{\alpha}\boldsymbol{y}+\frac{\delta}{\alpha}\boldsymbol{y}'=\left(1-\frac{\delta}{\alpha}\right)\boldsymbol{y}+\frac{\delta}{\alpha}\boldsymbol{y}'\in C.$$

因此, 有 $(1-\delta)\boldsymbol{u}_\lambda+\delta\boldsymbol{y}'=(1-\alpha)\boldsymbol{x}+\alpha\boldsymbol{v}\in C$, 得到矛盾, 所以有 $\boldsymbol{u}_\lambda\in K$.

(6) 根据凸函数定义和核的定义, 结论是显然的. □

综上所述, 可以得到凸集的基本性质:

1) C 是凸集等价于 C 是它所含的每一点的星形集;

2) 任何集合的核都是凸集;

3) 凸集 C 的核集合仍是 C.

2.2.2 凸包

凸集的几何特征是两点之间的凸组合 (线段) 上的点落在该集合内, 这一特征可以推广到多个点的凸组合形式. 换言之, 凸集中的点可以通过其中多个点的凸组合进行表示, 见下面定义.

定义 2.2.4 设 $\boldsymbol{x}_1,\cdots,\boldsymbol{x}_m$ 是 $\mathbf{R}^n$ 中不同的点, $\lambda_i\geqslant 0, i=1,\cdots,m, \sum\limits_{i=1}^m\lambda_i=1$, 则称

$$\boldsymbol{x}=\sum_{i=1}^m\lambda_i\boldsymbol{x}_i$$

为 $\boldsymbol{x}_1,\cdots,\boldsymbol{x}_m$ 的凸组合.

设有限指标集合 $I_n=\{1,2,\cdots,n\}$, 无限指标集合 $I_\infty=\{1,2,\cdots\}$, 则有下面的性质.

性质 2.2.2 设对任意的 $i\in I_n$ 或任意的 $i\in I_\infty$, C_i 是 $\mathbf{R}^n$ 中的凸集, 则下面结论成立:

(1) $\bigcap_{i\in I_n}C_i$ 是凸集;

(2) $\bigcap_{i\in I_\infty}C_i$ 和 $\bigcup_{k=1}^\infty\bigcap_{i=k}^\infty C_i$ 是凸集.

证明 (1) 如果 $\bigcap_{i\in I_n}C_i=\varnothing$ 或仅包含一个点, 结论显然成立. 设 $\boldsymbol{x},\boldsymbol{y}\in\bigcap_{i\in I_n}C_i$, 则对任意的 $\boldsymbol{x},\boldsymbol{y}\in C_i$, 任意的 $i\in I_n$, 有 $(1-\lambda)\boldsymbol{x}+\lambda\boldsymbol{y}\in C_i, \forall\lambda\in[0,1]$ 成立, 故有

$$(1-\lambda)\boldsymbol{x}+\lambda\boldsymbol{y}\in\bigcap_{i\in I_n}C_i,\quad \forall\lambda\in[0,1].$$

所以 $\bigcap_{i\in I_n}C_i$ 是凸集.

(2) $\bigcap_{i\in I_\infty} C_i$ 是凸集的证明与结论 (1) 的证明过程完全类似, 下面来证明 $\bigcup_{k=1}^{\infty}\bigcap_{i=k}^{\infty} C_i$ 是凸集. 设

$$A_k=\bigcap_{i=k}^{\infty} C_i,\quad k\in I_\infty,$$

则 A_k 是凸集, 且 $A_k\subseteq A_{k+1}$. 设

$$\boldsymbol{x},\boldsymbol{y}\in\bigcup_{k=1}^{\infty}\bigcap_{i=k}^{\infty} C_i=\bigcup_{k=1}^{\infty} A_k,$$

可知 $\exists k,\tilde{k}\in I_\infty$ 且 $k\geqslant\tilde{k}$, 使得 $\boldsymbol{x}\in A_k,\boldsymbol{y}\in A_{\tilde{k}}\subseteq A_k$, 则有 $(1-\lambda)\boldsymbol{x}+\lambda\boldsymbol{y}\in A_k,\forall\lambda\in[0,1]$ 成立, 故 $\bigcup_{k=1}^{\infty}\bigcap_{i=k}^{\infty} C_i$ 是凸集. □

例 2.2.2　已知 $\boldsymbol{a}_i\in\mathbf{R}^n,\alpha_i\in\mathbf{R},i\in I_\infty$, 则集合 $\{\boldsymbol{x}\,|\langle\boldsymbol{x},\boldsymbol{b}_i\rangle\leqslant(\text{或}<,\text{或}=)\,\alpha_i,\forall i\in I_\infty\}$ 是凸集, 即联立线性不等式或方程组的解集合是凸集.

由定义 2.2.4 可知, 凸集中任意两点的凸组合仍属于该凸集, 多点的凸组合也仍在该集合中.

定理 2.2.1　设集合 C 由包含它的所有凸组合构成, 则 C 是凸集. 反之, 任何一个凸集都包含了它的所有凸组合.

证明　设 $\sum\limits_{i=1}^{m_1}\lambda_i^1\boldsymbol{x}_i^1,\sum\limits_{j=1}^{m_2}\lambda_j^2\boldsymbol{x}_j^2\in C$, 其中 $\boldsymbol{x}_i^1,\boldsymbol{x}_j^2\in C(i=1,2,\cdots,m_1;j=1,2,\cdots,m_2)$, 且

$$\lambda_i^1\geqslant 0,\sum_{i=1}^{m_1}\lambda_i^1=1,i=1,2,\cdots,m_1;\quad \lambda_j^2\geqslant 0,\sum_{j=1}^{m_2}\lambda_j^2=1,\quad j=1,2,\cdots,m_2.$$

设 $\forall\lambda\in[0,1]$, 有 $\sum\limits_{i=1}^{m_1}(1-\lambda)\lambda_i^1+\sum\limits_{j=1}^{m_2}\lambda\lambda_j^2=1$. 再设 $\lambda_i=(1-\lambda)\lambda_i^1\geqslant 0(i=1,2,\cdots,m_1)$, $\lambda_{j+m_1}=\lambda\lambda_j^2\geqslant 0(j=1,2,\cdots,m_2)$, 有

$$\sum_{i=1}^{m_1}\lambda_i\boldsymbol{x}_i^1+\sum_{j=1}^{m_2}\lambda_{j+m_1}\boldsymbol{x}_j^2\in C,$$

即 C 是凸集.

反过来, 设 $\boldsymbol{x}_1,\cdots,\boldsymbol{x}_m\in C$, $\lambda_i\geqslant 0,\sum\limits_{i=1}^{m}\lambda_i=1$, 下面用归纳法证明 $\sum\limits_{i=1}^{m}\lambda_i\boldsymbol{x}_i\in C$, $i=1,2,\cdots,m$.

显然 $m=2$ 时, 有 $\sum\limits_{i=1}^{2}\lambda_i\boldsymbol{x}_i\in C$ 成立.

假设 $m=k$ 时, 有 $\sum_{i=1}^{k}\lambda_i\boldsymbol{x}_i\in C$; 当 $m=k+1$ 时, 则有

$$\sum_{i=1}^{k+1}\lambda_i'\boldsymbol{x}_i=\sum_{i=1}^{k}\lambda_i'\boldsymbol{x}_i+\lambda_{k+1}'\boldsymbol{x}_{k+1},$$

其中 $\lambda_i'\geqslant 0,\sum_{i=1}^{k+1}\lambda_i'=1$. 设 $\lambda=\sum_{i=1}^{k}\lambda_i'$, 有 $\lambda_{k+1}'=1-\lambda$. 记 $\lambda_i=\dfrac{\lambda_i'}{\lambda}\geqslant 0$, $i=1,2,\cdots,k$, 有 $\sum_{i=1}^{k}\lambda_i=1$, 根据上述所设可得

$$\sum_{i=1}^{k+1}\lambda_i'\boldsymbol{x}_i=\lambda\sum_{i=1}^{k}\lambda_i\boldsymbol{x}_i+(1-\lambda)\,\boldsymbol{x}_{k+1}'\in C.$$

因此, 结论成立. □

定理 2.2.1 说明凸组合由其凸集中任意两点即可构成, 也可以通过任意有限个点的凸组合来构成凸集.

根据性质 2.2.2 知道, 任意多个凸集的交仍是凸集, 所以对于任意给定的集合 X, 存在包含 X 的唯一最小凸集.

定义 2.2.5 设集合 $X\subset\mathbf{R}^n$, 则包含 X 的所有凸集的交集称为 X 的凸包. 用

$$\mathrm{co}X=\left\{\sum_{i=1}^{m}\lambda_i\boldsymbol{x}_i\,\middle|\,\forall\boldsymbol{x}_1,\cdots,\boldsymbol{x}_m\in X,\forall\lambda_i\geqslant 0,i=1,\cdots,m,\right.$$
$$\left.\sum_{i=1}^{m}\lambda_i=1,m=1,2,\cdots\right\}$$

表示, 也可记为 $\mathrm{co}X$ 或 $\mathrm{co}\,(X)$ 或 $\mathrm{co}\,\{X\}$.

例 2.2.3 设集合: $X=\left\{(0,0)^{\mathrm{T}},(1,0)^{\mathrm{T}},(0,1)^{\mathrm{T}},(1,1)^{\mathrm{T}}\right\}$. 它的凸包为

$$\mathrm{co}X=\left\{\sum_{i=1}^{4}\lambda_i\boldsymbol{x}_i\,\middle|\,\forall\boldsymbol{x}_1,\boldsymbol{x}_3,\boldsymbol{x}_3,\boldsymbol{x}_4\in X,\forall\lambda_i\geqslant 0,\sum_{i=1}^{4}\lambda_i=1\right\}$$
$$=\left\{(x_1,x_2)^{\mathrm{T}}\,\middle|\,0\leqslant x_1,x_2\leqslant 1\right\}.$$

由定义 2.2.5 和定理 2.2.1 的证明过程可得到如下性质.

性质 2.2.3 设集合 $X\subset\mathbf{R}^n$, 则下面结论成立:

(1) co$X$ 是 X 中所有点的全部凸组合组成的凸集;

(2) $\mathbf{R}^n$ 的有限子集 $\{\boldsymbol{x}_1, \cdots, \boldsymbol{x}_m\}$ 的凸包由形如 $\sum\limits_{i=1}^{m} \lambda_i \boldsymbol{x}_i$ 的向量构成, 其中 $\lambda_i \geqslant 0, \sum\limits_{i=1}^{m} \lambda_i = 1$.

由定义 2.2.5 知一个集合 X 的凸包是由该集合中任意有限个点的凸组合所组成的集合. 对此, 只需证明一个集合 X 的凸包是由该集合中任意不超过 $n+1$ 个点的凸组合组成的集合.

定理 2.2.2[52](Carathéodory 定理) 设集合 $X \subset \mathbf{R}^n$, 则 X 的凸包是

$$\mathrm{co}X = \left\{ \sum_{i=1}^{m} \lambda_i \boldsymbol{x}_i \middle| \forall \boldsymbol{x}_1, \cdots, \boldsymbol{x}_m \in X, \forall \lambda_1, \cdots, \lambda_m \geqslant 0, \sum_{i=1}^{m} \lambda_i = 1, m \leqslant n+1 \right\}.$$

证明 设 $m > n+1$, $\boldsymbol{x} = \sum\limits_{i=1}^{m} \lambda_i \boldsymbol{x}_i \in X$, 其中 $\boldsymbol{x}_1, \cdots, \boldsymbol{x}_m \in X, \lambda_i \geqslant 0$, $\sum\limits_{i=1}^{m} \lambda_i = 1$. 因为 $\boldsymbol{x}_1, \cdots, \boldsymbol{x}_m$ 的个数大于 $n+1$, 所以向量组 $\boldsymbol{x}_1, \cdots, \boldsymbol{x}_m$ 是线性相关的, 故存在不全为 0 的数 $\phi_i (i = 1, \cdots, m-1)$ 使得 $\sum\limits_{i=1}^{m-1} \phi_i \boldsymbol{x}_i = \boldsymbol{x}_m$. 设 $\phi = \sum\limits_{i=1}^{m-1} \phi_i$, 如果 $\phi \neq 0$, 令 $\beta_m = -\dfrac{1}{\phi}$, $\beta_i = \dfrac{\phi_i}{\phi}$, 那么, 有

$$\sum_{i=1}^{m} \beta_i \boldsymbol{x}_i = \mathbf{0}, \quad \sum_{i=1}^{m} \beta_i = \sum_{i=1}^{m-1} \frac{\phi_i}{\phi} - \frac{1}{\phi} = 0$$

成立; 如果 $\phi = 0$, 令 $\beta_m = -1$, $\beta_i = \dfrac{1}{m-1} + \phi_i$, 有

$$\sum_{i=1}^{m} \beta_i \boldsymbol{x}_i = \mathbf{0}, \quad \sum_{i=1}^{m} \beta_i = 1 + \sum_{i=1}^{m-1} \phi_i - 1 = 0$$

成立.

事实上, β_i 中至少存在一个正数, 设 $\eta = \min\left\{ \dfrac{\lambda_i}{\beta_i} \middle| \beta_i > 0, i = 1, \cdots, m \right\}$, $\overline{\lambda_i} = \lambda_i - \eta\beta_i \geqslant 0$. 则存在某个 $i' \in \{1, 2, \cdots, m\}$ 使得 $\overline{\lambda_{i'}} = 0$, 显然有

$$\sum_{i=1}^{m} \overline{\lambda_i} \boldsymbol{x}_i = \sum_{i=1}^{m} \lambda_i \boldsymbol{x}_i - \eta \sum_{i=1}^{m} \beta_i \boldsymbol{x}_i = \boldsymbol{x},$$

$$\sum_{i=1}^{m}\overline{\lambda_i}=\sum_{i=1}^{m}\lambda_i-\eta\sum_{i=1}^{m}\beta_i=1.$$

上式表明 $\boldsymbol{x}$ 仍是一个凸组合表示, 但减少了至少一项式中的非零项. 我们重复上面的过程直到 $m\leqslant n+1$ 为止. □

定理 2.2.2 说明凸包 co$X$ 中的点是由 X 中有限多个点的凸组合组成的, 只需要 X 中至多 $n+1$ 个点的凸组合就可以表示 co$X$ 中的点. 该定理还表明: 凸包 co$X$ 可以由 X 中的仿射无关元素的全部凸组合构成. 并且为了得到凸包 co$X$, 不一定要取 X 中点的全部凸组合, 而只要构造 X 中所有仿射无关点集的全部凸组合就可以了. 注意, 这里 X 中任何一组固定的仿射无关点集凸包不构成 co$X$. 设 $\boldsymbol{x}_1,\cdots,\boldsymbol{x}_m$ 是仿射集 A 的最大仿射无关集合, 那么 $A=\mathrm{aff}\{\boldsymbol{x}_1,\cdots,\boldsymbol{x}_m\}$, 但是

$$\mathrm{co}\{\boldsymbol{x}_1,\cdots,\boldsymbol{x}_m\}\subset\mathrm{co}A=\mathrm{co}\left(\mathrm{aff}\{\boldsymbol{x}_1,\cdots,\boldsymbol{x}_m\}\right).$$

因此, 我们可以得到下面的推论.

推论 2.2.1 设 $X\subset\mathbf{R}^n$, $\dim(\mathrm{aff}X)=m$, 则 co$X$ 的点可以由 X 的至多 $m+1$ 个点的凸组合构成.

推论 2.2.2 设 $\{X_i\subset\mathbf{R}^n\,|i\in I_m\}$ 是凸集族, I_m 是指标集, $X=\mathrm{co}\left(\bigcup_{i\in I_m}X_i\right)$, 则 X 的任何一点都可以由不同 X_i 中的点, 但至多是 $n+1$ 个仿射无关点的凸组合组成.

下面的定理是 E. Helly 在 1913 年提出并证明的, 最终于 1923 年发表 [97].

定理 2.2.3 (Helly 定理) 设 $C=\{C_i\subset\mathbf{R}^n|i\in I_m\}$, 对每个 $i\in I_m=\{1,2,\cdots,m\}$, 对应的 C_i 都是凸集, $m\geqslant n+1$. 如果 C 中任意 $n+1$ 个集合交集是非空的, 则 $\bigcap_{i=1}^{m}C_i\neq\varnothing$.

证明 当 $m=n+1$ 时, 结论显然成立.

假设 $m=n+2$ 时, C 中任意 $n+1$ 个集合的交集是非空的. 设集合 $B_i=C\backslash\{C_i\}(i=1,2,\cdots,m)$ 表示 C 中去掉一个集合 C_i 后剩余的所有集合, 则每个 B_i 都有 $n+1$ 个集合, 且 $D_i=\bigcap_{C_j\in B_i}C_j\neq\varnothing$, 其中所有 $j\in I_m\,(j\neq i)$, 即遍历 I_m 中除了 i 的指标. 假设 $\bigcap_{i=1}^{m}C_i=\varnothing$, 则对每个 C_i 都有

$$C_i\cap\bigcap_{C_j\in B_i}C_j=\varnothing.$$

设 $\forall\boldsymbol{x}_i\in D_i$ 但 $\boldsymbol{x}_i\notin C_i(i=1,2,\cdots,m)$, 有 $n+1$ 个 $\boldsymbol{x}_i\in C_j\,(j\in I_m\backslash\{i\})$, 即

$$\boldsymbol{x}_1\notin C_1,\boldsymbol{x}_1\in C_2,\boldsymbol{x}_1\in C_3,\cdots,\boldsymbol{x}_1\in C_m,$$

$$\boldsymbol{x}_2\in C_1,\boldsymbol{x}_2\notin C_2,\boldsymbol{x}_2\in C_3,\cdots,\boldsymbol{x}_2\in C_m,$$

$$\cdots\cdots$$

$$\boldsymbol{x}_m \in C_1, \boldsymbol{x}_m \in C_2, \boldsymbol{x}_m \in C_3, \cdots, \boldsymbol{x}_m \notin C_m.$$

根据定理 2.2.2, 由上面第 2 列可以得到 $\boldsymbol{x}_1, \boldsymbol{x}_3, \cdots, \boldsymbol{x}_m \in C_2$ 的凸组合属于 C_2, 于是有

$$\boldsymbol{x}_1 = \alpha_1 \boldsymbol{x}_1 + \alpha_3 \boldsymbol{x}_3 + \cdots + \alpha_m \boldsymbol{x}_m \in C_2,$$

且 $\alpha_1 + \alpha_3 + \cdots + \alpha_m = 1, \alpha_1, \alpha_3, \cdots, \alpha_m \geqslant 0$. 若 $\alpha_1 \neq 1$, 由上式有

$$\boldsymbol{x}_1 = \frac{\alpha_3}{1-\alpha_1}\boldsymbol{x}_3 + \cdots + \frac{\alpha_m}{1-\alpha_1}\boldsymbol{x}_m \in C_2,$$

且 $\dfrac{\alpha_3}{1-\alpha_1} + \cdots + \dfrac{\alpha_m}{1-\alpha_1} = 1$. 再由上面第 1 列得到 $\boldsymbol{x}_1 \in C_1$, 矛盾. 因此, $\bigcap_{i=1}^m C_i \neq \varnothing$.

同理, $m = n+3$ 时也类似可证. 由归纳法得定理结论成立. □

事实上, 容易验证定理 2.2.2 和定理 2.2.3 是等价的, 并且只有对凸集才成立. 容易举例说明, 定理中的假设条件是必要的.

例 2.2.4 在图 2.2-4 中, 4 个凸集 $\{C_1, C_2, C_3, C_4\}$ 中的任意 3 个集合相交存在空集的情况, 不满足任意 3 个集合相交是非空的条件, 所以定理结论不成立.

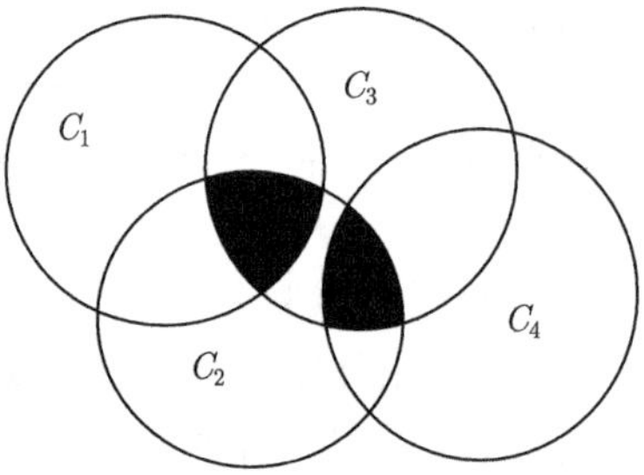

图 2.2-4 4 个凸集中的任意 3 个集合相交存在空集

本小节得到了一个集合张成凸包的几个性质:

1) 任意多个凸集的交还是凸集;

2) 凸集等价于它包含了它的所有凸组合构成的集合;

3) 一个集合 X 凸包是由该集合中任意不超过 $n+1$ 个点的凸组合组成的集合;

4) 一个超过 $n+1$ 个凸集构成的凸集族中, 其任意 $n+1$ 个集合交集是非空的, 则凸集族所有的交非空.

2.2.3 单纯形

在所有的凸集中单纯形是一个重要的概念. 我们知道, 仿射集是一个凸集, 任意仿射无关点构成的凸包包含这些点的最小凸包, 我们称这个最小凸包为单纯形. 单纯形在线性规划的求解算法收敛性分析中起到了重要作用.

定义 2.2.6 如果 $\mathbf{R}^n$ 中任意给定的集合 $\{\boldsymbol{x}_0, \boldsymbol{x}_1, \cdots, \boldsymbol{x}_m\}$ 是仿射无关的, 称 $\operatorname{co}\{\boldsymbol{x}_0, \boldsymbol{x}_1, \cdots, \boldsymbol{x}_m\}$ 为 m 维单纯形, 用 $\operatorname{Sim}(\boldsymbol{x}_0, \boldsymbol{x}_1, \cdots, \boldsymbol{x}_m)$ 或 $\operatorname{Sim}\{\boldsymbol{x}_0, \boldsymbol{x}_1, \cdots, \boldsymbol{x}_m\}$ 表示, 其中 $\boldsymbol{x}_0, \boldsymbol{x}_1, \cdots, \boldsymbol{x}_m$ 称为单纯形的顶点.

当 $m = 0, 1, 2, 3$ 时, 单纯形分别是点、线段、三角形、四面体. 需要注意的是, $\mathbf{R}^2$ 中的平行四边形不是单纯形.

例 2.2.5 设集合 $X = \{(0,0)^{\mathrm{T}}, (1,0)^{\mathrm{T}}, (0,1)^{\mathrm{T}}, (1,1)^{\mathrm{T}}\}$, 它的凸包

$$
\begin{aligned}
\operatorname{co}X &= \left\{ \sum_{i=1}^{4} \lambda_i \boldsymbol{x}_i \middle| \forall \boldsymbol{x}_1, \boldsymbol{x}_3, \boldsymbol{x}_3, \boldsymbol{x}_4 \in X, \forall \lambda_i \geqslant 0, \sum_{i=1}^{4} \lambda_i = 1 \right\} \\
&= \left\{ (x_1, x_2)^{\mathrm{T}} | 0 \leqslant x_1, x_2 \leqslant 1 \right\}
\end{aligned}
$$

不是单纯形, 因为 X 的 4 个点是仿射相关的. 但不难看出, X 中任意 3 点组成的凸包是单纯形, 例如 $\operatorname{co}\left\{(0,0)^{\mathrm{T}}, (1,0)^{\mathrm{T}}, (0,1)^{\mathrm{T}}\right\}$ 是单纯形.

定义 2.2.7 设 $C \subset \mathbf{R}^n$ 是凸集. 如果不存在 C 的线段通过点 $\boldsymbol{x}_0 \in C$, 则称 $\boldsymbol{x}_0$ 为 C 的极点.

凸集的极点可能是有限多个, 例如三角形的顶点; 也可能是无穷多个, 例如圆周上的任意一点. 注意, 顶点的概念是针对单纯形定义的, 极点的概念是针对凸集定义的, 而部分文献常把凸集的极点和顶点视为一致.

性质 2.2.4 设凸集 $C \subset \mathbf{R}^n$, $\{\boldsymbol{x}_0, \boldsymbol{x}_1, \cdots, \boldsymbol{x}_m\}$ 是仿射无关的, 则下面结论成立:

(1) $\bar{\boldsymbol{x}}$ 为 C 的极点等价于对 $\boldsymbol{x}, \boldsymbol{y} \in C$ 和 $\bar{\boldsymbol{x}} \in [\boldsymbol{x}, \boldsymbol{y}]$ 有 $\boldsymbol{x} = \boldsymbol{y} = \bar{\boldsymbol{x}}$;

(2) 单纯形 $\operatorname{Sim}(\boldsymbol{x}_0, \boldsymbol{x}_1, \cdots, \boldsymbol{x}_m)$ 的顶点都是极点;

(3) 单纯形 $\operatorname{Sim}(\boldsymbol{x}_0, \boldsymbol{x}_1, \cdots, \boldsymbol{x}_m)$ 中的每点都可以唯一地表示成它的顶点的凸组合;

(4) $\operatorname{aff}\{\operatorname{Sim}(\boldsymbol{x}_0, \boldsymbol{x}_1, \cdots, \boldsymbol{x}_m)\} = \operatorname{aff}\{\boldsymbol{x}_0, \boldsymbol{x}_1, \cdots, \boldsymbol{x}_m\}$.

证明 (1) 根据定义 2.2.7 知结论显然.

(2) 假设 $\{\boldsymbol{x}_0, \boldsymbol{x}_1, \cdots, \boldsymbol{x}_m\}$ 中存在某个点 $\boldsymbol{x}_k$ 不是极点, 那么存在 $\boldsymbol{x}, \boldsymbol{y} \in \operatorname{Sim}(\boldsymbol{x}_0, \boldsymbol{x}_1, \cdots, \boldsymbol{x}_m)$ 和 $0 < \lambda < 1$ 使得

$$
\boldsymbol{x}_k = (1 - \lambda)\boldsymbol{x} + \lambda \boldsymbol{y},
$$

其中, $\boldsymbol{x}=\sum\limits_{i=0}^{m}\alpha_i\boldsymbol{x}_i,\alpha_i\geqslant 0,\sum\limits_{i=0}^{m}\alpha_i=1;\boldsymbol{y}=\sum\limits_{i=0}^{m}\beta_i\boldsymbol{x}_i,\beta_i\geqslant 0,\sum\limits_{i=0}^{m}\beta_i=1$. 因此, 有

$$\boldsymbol{x}_k=\sum_{i=0}^{m}[(1-\lambda)\,\alpha_i+\lambda\beta_i]\boldsymbol{x}_i.$$

由于 $\sum\limits_{i=0}^{m}[(1-\lambda)\,\alpha_i+\lambda\beta_i]-1=0$, 对于 $k\neq i\,(k,i=0,1,2,\cdots,m)$, 设

$$\bar{\alpha}_i=(1-\lambda)\,\alpha_i+\lambda\beta_i,$$

$$\bar{\alpha}_k=(1-\lambda)\,\alpha_k+\lambda\beta_k-1,$$

有 $\sum\limits_{i=0}^{m}\bar{\alpha}_i\boldsymbol{x}_i=0$. 因为 $\{\boldsymbol{x}_0,\boldsymbol{x}_1,\cdots,\boldsymbol{x}_m\}$ 是仿射无关的, 对于 $k\neq i(k,i=0,1,2,\cdots,m)$, 有

$$(1-\lambda)\,\alpha_i+\lambda\beta_i=0.$$

因为 $\lambda>0,1-\lambda>0,\alpha_i\geqslant 0,\beta_i\geqslant 0$, 故当 $i\neq k$ 时 $\alpha_i=\beta_i=0$, 有 $\alpha_k=\beta_k=1$. 最后得到 $\boldsymbol{x}=\boldsymbol{y}=\boldsymbol{x}_k$, 与假设矛盾.

(3) 由性质 2.2.3(2), 单纯形 $\mathrm{Sim}\,(\boldsymbol{x}_0,\boldsymbol{x}_1,\cdots,\boldsymbol{x}_m)$ 中的点 $\boldsymbol{x}$ 可以表示成其顶点唯一的凸组合

$$\boldsymbol{x}=\sum_{i=0}^{m}\alpha_i\boldsymbol{x}_i,\quad \alpha_i\geqslant 0,\quad \sum_{i=0}^{m}\alpha_i=1.$$

如果该表示不唯一, 则存在另一个凸组合表示:

$$\boldsymbol{x}=\sum_{i=0}^{m}\beta_i\boldsymbol{x}_i,\quad \beta_i\geqslant 0,\quad \sum_{i=0}^{m}\beta_i=1.$$

设 $\lambda_i=\alpha_i-\beta_i,i=0,1,\cdots,m$, 则有 $\sum\limits_{i=0}^{m}\lambda_i\boldsymbol{x}_i=\mathbf{0}$. 因为 $\{\boldsymbol{x}_0,\boldsymbol{x}_1,\cdots,\boldsymbol{x}_m\}$ 是仿射无关的, 故 $\lambda_0=\cdots=\lambda_m=0$, 由此可得 $\alpha_i=\beta_i$. 因此, 点 $\boldsymbol{x}$ 的顶点凸组合表示唯一.

(4) 首先证明 $\boldsymbol{x}\in\mathrm{aff}\,\{\mathrm{Sim}\,(\boldsymbol{x}_0,\boldsymbol{x}_1,\cdots,\boldsymbol{x}_m)\}$, 存在唯一 $\lambda_i\in\mathbf{R}$, $\sum\limits_{i=0}^{m}\lambda_i=1$ 使得 $\boldsymbol{x}=\sum\limits_{i=0}^{m}\lambda_i\boldsymbol{x}_i$. 因为存在

$$\boldsymbol{y}_j\in\mathrm{Sim}\,(\boldsymbol{x}_0,\boldsymbol{x}_1,\cdots,\boldsymbol{x}_m)\,,\quad j=1,\cdots,k,\quad \delta_j\in\mathbf{R},\quad \sum_{j=1}^{k}\delta_j=1,$$

使得 $\boldsymbol{x}=\sum\limits_{j=1}^{k}\delta_j\boldsymbol{y}_j$. 通过 (3) 可知

$$\boldsymbol{y}_j=\sum_{i=0}^{m}\alpha_{ij}\boldsymbol{x}_i,$$

其中 $\alpha_{ij}\geqslant 0$, $\sum\limits_{i=0}^{m}\alpha_{ij}=1$. 故得

$$\boldsymbol{x}=\sum_{j=1}^{k}\delta_j\sum_{i=0}^{m}\alpha_{ij}\boldsymbol{x}_i=\sum_{i=0}^{m}\left(\sum_{j=1}^{k}\delta_j\alpha_{ij}\right)\boldsymbol{x}_i.$$

设 $\lambda_i=\sum\limits_{j=1}^{k}\delta_j\alpha_{ij}$, 则有 $\boldsymbol{x}=\sum\limits_{i=0}^{m}\lambda_i\boldsymbol{x}_i\in\operatorname{aff}\{\boldsymbol{x}_0,\boldsymbol{x}_1,\cdots,\boldsymbol{x}_m\}$, 且

$$\sum_{i=0}^{m}\lambda_i=\sum_{i=0}^{m}\sum_{j=1}^{k}\delta_j\alpha_{ij}=1.$$

反之, 显然. □

性质 2.2.4 说明单纯形中任何点可以由其顶点的凸组合唯一表示, 且单纯形的仿射集合也可以由单纯形顶点的仿射组合表示.

定理 2.2.4 设 $X=\{\boldsymbol{x}_1,\boldsymbol{x}_2,\cdots,\boldsymbol{x}_m\}\subset\mathbf{R}^n$, 其中 $\boldsymbol{x}_1,\boldsymbol{x}_2,\cdots,\boldsymbol{x}_m$ 是仿射相关的, 则可以将集合 X 分解成两个不相交的集合 X_1 和 X_2, 使得 $\operatorname{co}X_1\cap\operatorname{co}X_2\neq\varnothing$.

证明 因为向量组 $\boldsymbol{x}_1,\boldsymbol{x}_2,\cdots,\boldsymbol{x}_m$ 是仿射相关的, 则存在不全为 0 的实数 $\lambda_1,\cdots,\lambda_m$ 使得

$$\sum_{i=1}^{m}\lambda_i\boldsymbol{x}_i=\boldsymbol{0}$$

成立, 其中 $\sum\limits_{i=1}^{m}\lambda_i=0$. 设指标集合 $I_1=\{i\,|\lambda_i>0\,,i\in I_m\}$, $I_2=\{i\,|\lambda_i\leqslant 0\,,i\in I_m\}$, 其中 $I_m=\{1,2,\cdots,m\}$, 取集合 $X_1=\{\boldsymbol{x}_i\,|i\in I_1\}$ 和 $X_2=\{\boldsymbol{x}_i\,|i\in I_2\}$. 显然 $I_1,I_2\neq\varnothing$. 令 $\boldsymbol{x}=\sum\limits_{i\in I_1}\dfrac{\lambda_i}{\lambda}\boldsymbol{x}_i$, 其中 $\lambda=\sum\limits_{i\in I_1}\lambda_i$. 由于 $\sum\limits_{i\in I_1}\dfrac{\lambda_i}{\lambda}=1$, $\lambda_i>0, i\in I_1$, 得 $\boldsymbol{x}\in\operatorname{co}X_1$. 再由 $\dfrac{1}{\lambda}\sum\limits_{i=1}^{m}\lambda_i\boldsymbol{x}_i=\boldsymbol{0}$ 得

$$\sum_{i=1}^{m}\frac{\lambda_i}{\lambda}\boldsymbol{x}_i=\sum_{i\in I_1}\frac{\lambda_i}{\lambda}\boldsymbol{x}_i+\sum_{i\in I_2}\frac{\lambda_i}{\lambda}\boldsymbol{x}_i=\boldsymbol{x}+\sum_{i\in I_2}\frac{\lambda_i}{\lambda}\boldsymbol{x}_i=\boldsymbol{0},\quad \lambda=\sum_{i\in I_2}(-\lambda_i).$$

由上式得 $\boldsymbol{x}=\sum\limits_{i\in I_2}\dfrac{-\lambda_i}{\lambda}\boldsymbol{x}_i$, 显然 $1=\sum\limits_{i\in I_2}\dfrac{-\lambda_i}{\lambda}$, 因此, $\boldsymbol{x}\in \mathrm{co}X_2$, 即

$$\mathrm{co}X_1\cap \mathrm{co}X_2\neq\varnothing. \qquad \square$$

由定理 2.2.4 知一个仿射无关集合可以分解成两个不相交的集合, 但这两个集合凸包相交非空.

例 2.2.6 设集合 $X=\left\{(0,0)^{\mathrm{T}},(1,0)^{\mathrm{T}},(0,1)^{\mathrm{T}},(1,1)^{\mathrm{T}}\right\}$, 令 $X_1=\{(0,0)^{\mathrm{T}},(1,1)^{\mathrm{T}}\}$, $X_2=\left\{(1,0)^{\mathrm{T}},(0,1)^{\mathrm{T}}\right\}$, 显然有 $\mathrm{co}X_1\cap \mathrm{co}X_2\neq\varnothing$.

推论 2.2.3 设 $X=\{\boldsymbol{x}_1,\cdots,\boldsymbol{x}_m\}\subset\mathbf{R}^n, m\geqslant n+2$, 则可以将集合 X 分解成两个不相交的集合 X_1,X_2, 使得 $\mathrm{co}X_1\cap \mathrm{co}X_2\neq\varnothing$.

证明 因为 $m\geqslant n+2$, 所以 $X=\{\boldsymbol{x}_1,\cdots,\boldsymbol{x}_m\}\subset\mathbf{R}^n$ 是仿射相关的. $\square$

定义 2.2.8 设 $X\subset\mathbf{R}^n$, 则 $\mathrm{aff}X$ 的维数等于 X 的维数, 即 $\dim X=\dim(\mathrm{aff}X)$.

根据定义 2.2.8, 任何集合的仿射集的维数都代表了这个集合的维数. 2 维凸集在任何空间中仍然是 2 维的, 与凸集所在的空间维数无关. 因此, 单纯形 $\mathrm{Sim}(\boldsymbol{x}_0,\boldsymbol{x}_1,\cdots,\boldsymbol{x}_m)$ 的维数是它的最大线性无关数 m.

性质 2.2.5 设单纯形 $\mathrm{Sim}(\boldsymbol{x}_0,\boldsymbol{x}_1,\cdots,\boldsymbol{x}_m)$, $X\subset\mathbf{R}^n$, 向量集 $\{\boldsymbol{x}_0,\boldsymbol{x}_1,\cdots,\boldsymbol{x}_m\}$ 是线性无关的, 且 $\boldsymbol{x}_i\in X, i=0,1,\cdots,m$, 则下面结论成立:

(1) $\dim(\mathrm{Sim}(\boldsymbol{x}_0,\boldsymbol{x}_1,\cdots,\boldsymbol{x}_m))=m$;

(2) $\dim X\geqslant m$.

证明 (1) 由性质 2.2.4(4) 有 $\mathrm{aff}\{\mathrm{Sim}(\boldsymbol{x}_0,\boldsymbol{x}_1,\cdots,\boldsymbol{x}_m)\}=\mathrm{aff}\{\boldsymbol{x}_0,\boldsymbol{x}_1,\cdots,\boldsymbol{x}_m\}$, 其中 $\boldsymbol{x}_0,\boldsymbol{x}_1,\cdots,\boldsymbol{x}_m$ 是仿射无关的, 那么 $\boldsymbol{x}_1-\boldsymbol{x}_0,\cdots,\boldsymbol{x}_m-\boldsymbol{x}_0$ 是线性无关的. 由性质 2.1.10 得

$$\mathrm{aff}\{\boldsymbol{x}_0,\boldsymbol{x}_1,\cdots,\boldsymbol{x}_m\}-\boldsymbol{x}_0=\mathrm{span}\{\boldsymbol{x}_1-\boldsymbol{x}_0,\cdots,\boldsymbol{x}_m-\boldsymbol{x}_0\}.$$

因此, $\mathrm{Sim}(\boldsymbol{x}_0,\boldsymbol{x}_1,\cdots,\boldsymbol{x}_m)$ 的维数是 m.

(2) 显然 $\mathrm{aff}\{\boldsymbol{x}_0,\boldsymbol{x}_1,\cdots,\boldsymbol{x}_m\}\subseteq\mathrm{aff}X$, 有 $\dim(\mathrm{aff}\{\boldsymbol{x}_0,\boldsymbol{x}_1,\cdots,\boldsymbol{x}_m\})=m$, 再由定义 2.2.8 可知结论成立. $\square$

下面证明一个凸集 C 的维数等于它的全体单纯形族最大的维数.

定理 2.2.5 设 C 是 $\mathbf{R}^n$ 中的凸集, 定义 C 的全体单纯形族为

$$\varepsilon=\left\{\mathrm{Sim}(\boldsymbol{x}_0,\boldsymbol{x}_1,\cdots,\boldsymbol{x}_m)\,|\,\boldsymbol{x}_0,\boldsymbol{x}_1,\cdots,\boldsymbol{x}_m\text{是}C\text{中任意仿射无关向量组}\right\},$$

则 $\dim C=\max\{\dim(\mathrm{Sim}(\boldsymbol{x}_0,\boldsymbol{x}_1,\cdots,\boldsymbol{x}_m))\,|\,\forall\mathrm{Sim}(\boldsymbol{x}_0,\boldsymbol{x}_1,\cdots,\boldsymbol{x}_m)\in\varepsilon\}$.

证明 如果 $C=\varnothing$, 结论显然成立.

因此, 假设 $C\neq\varnothing$. 根据性质 2.1.7、性质 2.2.5 可知, $\dim(\mathrm{Sim}(\boldsymbol{x}_0,\boldsymbol{x}_1,\cdots,\boldsymbol{x}_m))=m\leqslant n$. 不妨假设 $\dim(\mathrm{Sim}(\boldsymbol{x}_0,\boldsymbol{x}_1,\cdots,\boldsymbol{x}_m))=m$ 是 ε 中最大维数的那

个单纯形的维数, 只要证明 $C \subset \operatorname{aff}\{\operatorname{Sim}(\boldsymbol{x}_0, \boldsymbol{x}_1, \cdots, \boldsymbol{x}_m)\}$, 根据定义 2.2.8 即可知结论成立.

下面假设存在 $\boldsymbol{x}' \in C$, 但 $\boldsymbol{x}' \notin \operatorname{aff}\{\operatorname{Sim}(\boldsymbol{x}_0, \boldsymbol{x}_1, \cdots, \boldsymbol{x}_m)\}$. 因 C 是凸集, 得

$$\operatorname{co}\{\boldsymbol{x}_0, \boldsymbol{x}_1, \cdots, \boldsymbol{x}_m, \boldsymbol{x}'\} \subset C.$$

因为 $\operatorname{co}\{\boldsymbol{x}_0, \boldsymbol{x}_1, \cdots, \boldsymbol{x}_m, \boldsymbol{x}'\}$ 不是单纯形, 否则, $\operatorname{co}\{\boldsymbol{x}_0, \boldsymbol{x}_1, \cdots, \boldsymbol{x}_m, \boldsymbol{x}'\}$ 的维数是 $m+1$, 这与假设矛盾, 所以, 有 $\boldsymbol{x}_1 - \boldsymbol{x}_0, \cdots, \boldsymbol{x}_m - \boldsymbol{x}_0, \boldsymbol{x}' - \boldsymbol{x}_0$ 线性相关, 但又因为 $\boldsymbol{x}_1 - \boldsymbol{x}_0, \cdots, \boldsymbol{x}_m - \boldsymbol{x}_0$ 是线性无关的, 故存在不全为 0 的 $\lambda_1, \cdots, \lambda_m$, 使

$$\boldsymbol{x}' - \boldsymbol{x}_0 = \sum_{i=1}^{m} \lambda_i (\boldsymbol{x}_i - \boldsymbol{x}_0).$$

由上式得 $\boldsymbol{x}' = \sum\limits_{i=1}^{m} \lambda_i \boldsymbol{x}_i + \left(1 - \sum\limits_{i=1}^{m} \lambda_i\right) \boldsymbol{x}_0 \in \operatorname{aff}\{\operatorname{Sim}(\boldsymbol{x}_0, \boldsymbol{x}_1, \cdots, \boldsymbol{x}_m)\}$, 得到矛盾, 所以 $\boldsymbol{x}'$ 不存在. □

例 2.2.7 设在 $\mathbf{R}^2$ 中凸包集合 $\operatorname{co}X = \operatorname{co}\left\{(0,0)^{\mathrm{T}}, (1,0)^{\mathrm{T}}, (0,1)^{\mathrm{T}}, (1,1)^{\mathrm{T}}\right\}$, 有单纯形族:

$$\begin{aligned}\varepsilon = \{&\operatorname{Sim}((0,0)^{\mathrm{T}}, (1,0)^{\mathrm{T}}, (0,1)^{\mathrm{T}}), \operatorname{Sim}((0,0)^{\mathrm{T}}, (1,0)^{\mathrm{T}}), \operatorname{Sim}((0,0)^{\mathrm{T}}, (0,1)^{\mathrm{T}}),\\ &\operatorname{Sim}((0,0)^{\mathrm{T}}, (1,0)^{\mathrm{T}}, (1,1)^{\mathrm{T}}), \operatorname{Sim}((0,0)^{\mathrm{T}}, (1,1)^{\mathrm{T}}), \operatorname{Sim}((1,0)^{\mathrm{T}}, (0,1)^{\mathrm{T}}),\\ &\operatorname{Sim}((0,0)^{\mathrm{T}}, (0,1)^{\mathrm{T}}, (1,1)^{\mathrm{T}}), \operatorname{Sim}((1,0)^{\mathrm{T}}, (1,1)^{\mathrm{T}}), \operatorname{Sim}((0,1)^{\mathrm{T}}, (1,1)^{\mathrm{T}}),\\ &\operatorname{Sim}((1,0)^{\mathrm{T}}, (0,1)^{\mathrm{T}}, (1,1)^{\mathrm{T}}), \operatorname{Sim}((0,1)^{\mathrm{T}}), \operatorname{Sim}((1,0)^{\mathrm{T}}),\\ &\operatorname{Sim}((1,1)^{\mathrm{T}}), \operatorname{Sim}((0,0)^{\mathrm{T}}), \cdots\}.\end{aligned}$$

上面单纯形族的元素是无穷多个 (因为 $\operatorname{co}X$ 存在无数个仿射无关组), 则 $\dim(\operatorname{co}X) = 3$.

本小节得到单纯形的一些重要性质:

1) 单纯形中极点与其顶点等价;

2) 单纯形中任何一点都可由其顶点的凸组合表示;

3) 一个仿射无关集合可以分解成两个不相交的集合, 且这两个集合凸包相交非空;

4) 单纯形 $\operatorname{Sim}(\boldsymbol{x}_0, \boldsymbol{x}_1, \cdots, \boldsymbol{x}_m)$ 的维数等于它的最大线性无关元素个数;

5) 凸集 C 的维数等于它的全体单纯形族最大的维数.

2.2.4 代数运算

凸集的代数运算主要有平移、数乘、加法和直和等, 在后面的章节中将被广泛使用.

定义 2.2.9 设 C, C_1, C_2 是 $\mathbf{R}^n$ 的凸集, 定义以下几个关于凸集的运算形式:

(1) 平移运算: $C+\boldsymbol{a}=\{\boldsymbol{x}+\boldsymbol{a}\,|\,\boldsymbol{x}\in C, \boldsymbol{a}\in\mathbf{R}^n\}$;

(2) 数乘运算: $\lambda C=\{\lambda\boldsymbol{x}\,|\,\boldsymbol{x}\in C\}$, 其中 $\lambda>0$;

(3) 加法运算: $C_1+C_2=\{\boldsymbol{x}_1+\boldsymbol{x}_2\,|\,\boldsymbol{x}_1\in C_1, \boldsymbol{x}_2\in C_2\}$.

根据定义 2.2.9 可以得到下面一些凸集运算性质.

性质 2.2.6 设 C, C_1, C_2, C_3 是 $\mathbf{R}^n$ 中的凸集, $\boldsymbol{a}\in\mathbf{R}^n$ 和 $\lambda\in\mathbf{R}$, 则下面性质成立:

(1) $C+\boldsymbol{a}$, λC 和 C_1+C_2 是凸集;

(2) $(1-\lambda)C+\lambda C\subset C$, 其中 $0\leqslant\lambda\leqslant 1$;

(3) $C_1+C_2=C_2+C_1$;

(4) $(C_1+C_2)+C_3=C_1+(C_2+C_3)$;

(5) $\lambda_1(\lambda_2 C)=\lambda_1\lambda_2 C$, 其中 $\lambda_1>0, \lambda_2>0$;

(6) $\lambda(C_1+C_2)=\lambda C_1+\lambda C_2$, 其中 $\lambda>0$;

(7) $(\lambda_1+\lambda_2)C=\lambda_1 C+\lambda_2 C$, 其中 $\lambda_1>0, \lambda_2>0$.

证明 这里只需要证明 (7) 成立即可, (1)~(6) 均可由 (7) 推导得到.

设 $\forall(\lambda_1+\lambda_2)\boldsymbol{x}\in(\lambda_1+\lambda_2)C$, $\forall\boldsymbol{x}\in C$, 有

$$\lambda_1\boldsymbol{x}+\lambda_2\boldsymbol{x}\in\lambda_1 C+\lambda_2 C,$$

则得 $(\lambda_1+\lambda_2)C\subseteq\lambda_1 C+\lambda_2 C$.

反过来, 设 $\forall\boldsymbol{x}\in\lambda_1 C+\lambda_2 C$, $\exists\boldsymbol{x}_1, \boldsymbol{x}_2\in C$ 使得

$$\boldsymbol{x}=\lambda_1\boldsymbol{x}_1+\lambda_2\boldsymbol{x}_2.$$

由此可得 $\dfrac{1}{\lambda_1+\lambda_2}\boldsymbol{x}=\dfrac{\lambda_1}{\lambda_1+\lambda_2}\boldsymbol{x}_1+\dfrac{\lambda_2}{\lambda_1+\lambda_2}\boldsymbol{x}_2\subset C$, 即 $\boldsymbol{x}\subset(\lambda_1+\lambda_2)C$. □

上述性质说明两个凸集的和是凸集, 凸集的正数乘也是凸集. 由性质 2.2.6(7) 可以推广到不同的凸集情形, 如果 $C_1,\cdots,C_m$ 是凸集, 当 $\lambda_1,\cdots,\lambda_m\geqslant 0$ 时, 那么 $C=\lambda_1 C_1+\cdots+\lambda_m C_m$ 也是凸集. 特别地, 当 $\sum\limits_{i=1}^{m}\lambda_i=1$ 时, 也称 C 为 $C_1,\cdots,C_m$ 的凸组合. 但是, 当 $C_i\,(i=1,2,\cdots,n)$ 不是凸集时, 集合的凸组合是凸集这一结论不成立.

定理 2.2.6 设 $C_1, C_2, \cdots, C_m \subset \mathbf{R}^n$ 是凸集, 则

$$\operatorname{co}\left(\bigcup_{i=1}^{m} C_i\right) = \left\{\sum_{i=1}^{m} \lambda_i \boldsymbol{x}_i \middle| \sum_{i=1}^{m} \lambda_i = 1, \boldsymbol{x}_i \in C_i, \lambda_i \geqslant 0, i = 1, 2, \cdots, m\right\}. \tag{2.2.4-1}$$

证明 显然有

$$\operatorname{co}\left(\bigcup_{i=1}^{m} C_i\right) \supset \left\{\sum_{i=1}^{m} \lambda_i \boldsymbol{x}_i \middle| \sum_{i=1}^{m} \lambda_i = 1, \boldsymbol{x}_i \in C_i, \lambda_i \geqslant 0, i = 1, 2, \cdots, m\right\}.$$

下面证明上式的反向包含关系成立. 设 $\boldsymbol{x} \in \operatorname{co}\left(\bigcup_{i=1}^{m} C_i\right)$, 由定理 2.2.2 得 $\boldsymbol{x}_1, \cdots, \boldsymbol{x}_m \in \bigcup_{i=1}^{m} C_i$ 使得 $\boldsymbol{x} = \sum\limits_{i=1}^{m} \alpha_i \boldsymbol{x}_i$, 其中 $\alpha_i \geqslant 0$, $\sum\limits_{i=1}^{m} \alpha_i = 1, m \leqslant n+1$.

不妨设 $\boldsymbol{x}_1, \cdots, \boldsymbol{x}_k \in C_1, 1 < k < m$, 那么有 $\boldsymbol{x}_1' = \sum\limits_{i=1}^{k} \dfrac{\alpha_i}{\lambda_1} \boldsymbol{x}_i \in C_1$, 其中 $\sum\limits_{i=1}^{k} \alpha_i = \lambda_1$. 于是有

$$\boldsymbol{x} = \sum_{i=1}^{m} \alpha_i \boldsymbol{x}_i = \lambda_1 \boldsymbol{x}_1' + \sum_{i=k+1}^{m} \alpha_i \boldsymbol{x}_i,$$

$$\sum_{i=1}^{m} \alpha_i = \lambda_1 + \sum_{i=k+1}^{m} \alpha_i = 1.$$

重复上述方法可以得到

$$\begin{aligned}\boldsymbol{x} &= \sum_{i=1}^{m} \alpha_i \boldsymbol{x}_i \\ &= \sum_{i=1}^{m'} \lambda_i' \boldsymbol{x}_i' \subset \left\{\sum_{i=1}^{m} \lambda_i \boldsymbol{x}_i \middle| \sum_{i=1}^{m} \lambda_i = 1, \boldsymbol{x}_i \in C_i, \lambda_i \geqslant 0, i = 1, 2, \cdots, m\right\},\end{aligned}$$

其中 $\sum\limits_{i=1}^{m'} \lambda_i' = 1, \lambda_i' \geqslant 0, \boldsymbol{x}_i' \in C_i, m' \leqslant m$. 因此, 由上式可得

$$\operatorname{co}\left(\bigcup_{i=1}^{m} C_i\right) \subset \left\{\sum_{i=1}^{m} \lambda_i \boldsymbol{x}_i \middle| \sum_{i=1}^{m} \lambda_i = 1, \boldsymbol{x}_i \in C_i, \lambda_i \geqslant 0, i = 1, 2, \cdots, m\right\}. \qquad \square$$

(2.2.4-1) 式说明多个凸集并集的凸包等于从每个集合各取一个元素组成的凸组合构成的集合.

定义 2.2.10　设 $C_1 \subset \mathbf{R}^n, C_2 \subset \mathbf{R}^m$, 则 C_1 和 C_2 的直和运算为

$$C_1 \oplus C_2 = \{(\boldsymbol{x}, \boldsymbol{y}) \,|\boldsymbol{x} \in C_1, \boldsymbol{y} \in C_2\}.$$

例 2.2.8　设凸集合 $C_1 = \{(x_1, 0)^{\mathrm{T}} \,|0 \leqslant x_1 \leqslant 1\}$, $C_2 = \{(x_2, 1)^{\mathrm{T}} \,|0 \leqslant x_2 \leqslant 1\}$, 有

$$C_1 + C_2 = \left\{(x_1, 1)^{\mathrm{T}} \,|0 \leqslant x_1 \leqslant 2\right\},$$

$$\mathrm{co}\,(C_1 \cup C_2) = \left\{(x_1, x_2)^{\mathrm{T}} \,|0 \leqslant x_1 \leqslant 1, 0 \leqslant x_2 \leqslant 1\right\},$$

$$C_1 \oplus C_2 = \left\{(x_1, 0, x_2, 1)^{\mathrm{T}} \,|0 \leqslant x_1 \leqslant 1, 0 \leqslant x_2 \leqslant 1\right\}.$$

定理 2.2.7　设 $C_1 \subset \mathbf{R}^n, C_2 \subset \mathbf{R}^m$ 是凸集, 那么 $C_1 \oplus C_2$ 也是凸集.

证明　由凸集定义及直和定义直接可证. □

定理 2.2.8　设 $C_1, C_2 \subset \mathbf{R}^{n+m}$ 是凸集, 则集合

$$C = \{(\boldsymbol{x}, \boldsymbol{y}) \,|\boldsymbol{x} \in \mathbf{R}^n, \boldsymbol{y} \in \mathbf{R}^m, (\boldsymbol{x}, \boldsymbol{y}_1) \in C_1,\ (\boldsymbol{x}, \boldsymbol{y}_2) \in C_2, \boldsymbol{y}_1 + \boldsymbol{y}_2 = \boldsymbol{y}\} \tag{2.2.4-2}$$

是凸集.

证明　设 $\forall\,(\boldsymbol{x}^1, \boldsymbol{y}^1), (\boldsymbol{x}^2, \boldsymbol{y}^2) \in C$, 其中

$$\left(\boldsymbol{x}^1, \boldsymbol{y}_1^1\right) \in C_1, \quad \left(\boldsymbol{x}^1, \boldsymbol{y}_2^1\right) \in C_2, \quad \boldsymbol{y}_1^1 + \boldsymbol{y}_2^1 = \boldsymbol{y}^1,$$

$$\left(\boldsymbol{x}^2, \boldsymbol{y}_1^2\right) \in C_1, \quad \left(\boldsymbol{x}^2, \boldsymbol{y}_2^2\right) \in C_2, \quad \boldsymbol{y}_1^2 + \boldsymbol{y}_2^2 = \boldsymbol{y}^2.$$

设 $0 \leqslant \lambda \leqslant 1$, 根据 C_1, C_2 的凸性定义, 有

$$(1-\lambda)\left(\boldsymbol{x}^1, \boldsymbol{y}_1^1\right) + \lambda\left(\boldsymbol{x}^2, \boldsymbol{y}_1^2\right) = \left((1-\lambda)\,\boldsymbol{x}^1 + \lambda\boldsymbol{x}^2, (1-\lambda)\,\boldsymbol{y}_1^1 + \lambda\boldsymbol{y}_1^2\right) \in C_1,$$

$$(1-\lambda)\left(\boldsymbol{x}^1, \boldsymbol{y}_2^1\right) + \lambda\left(\boldsymbol{x}^2, \boldsymbol{y}_2^2\right) = \left((1-\lambda)\,\boldsymbol{x}^1 + \lambda\boldsymbol{x}^2, (1-\lambda)\,\boldsymbol{y}_2^1 + \lambda\boldsymbol{y}_2^2\right) \in C_2,$$

其中

$$(1-\lambda)\,\boldsymbol{y}^1 + \lambda\boldsymbol{y}^2 = \left[(1-\lambda)\,\boldsymbol{y}_1^1 + \lambda\boldsymbol{y}_1^2\right] + \left[(1-\lambda)\,\boldsymbol{y}_2^1 + \lambda\boldsymbol{y}_2^2\right].$$

从而得

$$(1-\lambda)\left(\boldsymbol{x}^1, \boldsymbol{y}^1\right) + \lambda\left(\boldsymbol{x}^2, \boldsymbol{y}^2\right) = \left((1-\lambda)\,\boldsymbol{x}^1 + \lambda\boldsymbol{x}^2, (1-\lambda)\,\boldsymbol{y}^1 + \lambda\boldsymbol{y}^2\right) \in C,$$

故 C 是凸集. □

定义 2.2.11　设 $L: \mathbf{R}^n \to \mathbf{R}^m$ 是线性变换, $C \subset \mathbf{R}^n$, $D \subset \mathbf{R}^m$, 称 $LC = \{L\boldsymbol{x} \,|\boldsymbol{x} \in C\}$ 是变换 L 关于 C 的象, $L^{-1}D = \{\boldsymbol{x} \,|L\boldsymbol{x} \in D\}$ 是变换 L 关于 D 的逆象.

定理 2.2.9 设 $L:\mathbf{R}^n\to\mathbf{R}^m$ 是线性变换, $C\subset\mathbf{R}^n$ 和 $D\subset\mathbf{R}^m$ 是凸子集, 则 $LC\subset\mathbf{R}^m$ 和 $L^{-1}D\subset\mathbf{R}^n$ 也是凸集.

证明 根据定义 2.2.10 可知, 结论显然成立. □

推论 2.2.4 任何凸集 C 在子空间 L 上的正交投影也是凸集.

证明 根据定理 2.2.6 和正交投影定义可知, 结论显然成立. □

本小节得到了凸集的一些运算性质:

1) 两个凸集之和是凸集, 凸集乘一个正数构成的集合还是凸集;

2) 多个凸集并集的凸包等于从每个集合各取一个元素组成的凸组合构成的集合;

3) 两个凸集的直和也是凸集;

4) 两个凸集根据 (2.2.4-2) 式运算产生的集合是凸集;

5) 凸集进行线性变换后的象是凸集, 进行逆线性变换的逆象是凸集.

2.3 拓 扑 性 质

本节讨论凸集的闭包、内部和相对内部等拓扑性质, 这些性质刻画了凸集内部与外部的结构, 将在后面的分离定理和支撑超平面定理的充分必要条件刻画中起到重要作用, 也是后面的凸锥、(广义) 凸函数及微分理论的重要基础.

2.3.1 闭包、内部与相对内部

本小节首先讨论凸集的闭包和内部等概念与性质. 对定义 1.2.3 中凸集的内部、闭包和边界进行重新定义.

定义 2.3.1 设 $C\subset\mathbf{R}^n$ 是凸集, C 的全体内点构成的集合称为凸集 C 的内部, 用 $\mathrm{int}C$ 或 $\mathrm{int}\,(C)$ 或 $\mathrm{int}\,\{C\}$ 表示.

定义 2.3.2 设 $C\subset\mathbf{R}^n$ 是凸集, C 的全体极限点构成的集合称为 C 的闭包, 用 $\mathrm{cl}C$ 或 $\mathrm{cl}\,\{C\}$ 或 $\mathrm{cl}\,(C)$ 表示, 集合 $\mathrm{bd}C=\mathrm{cl}C\backslash\mathrm{int}C$ 称为凸集 C 的边界集.

集合的内部 $\mathrm{int}C$ 是开集, 集合的闭包 $\mathrm{cl}C$ 是闭集. 注意边界 ∂C 和 $\mathrm{bd}C$ 是不同的集合, 例如, 设集合 $C=(0,1]$, 有 $\partial C=\{1\}$, $\mathrm{bd}C=\{0,1\}$. 因此, $\partial C\subset\mathrm{bd}C$.

例 2.3.1 设集合 $C=\{\boldsymbol{x}\in C\,|0<x_i\leqslant 1, i=1,2,\cdots,n\}$, 显然 C 不是开集也不是闭集, 有

$$\mathrm{int}C=\{\boldsymbol{x}\in C\,|0<x_i<1, i=1,2,\cdots,n\},$$

$$\mathrm{cl}C=\{\boldsymbol{x}\in C\,|0\leqslant x_i\leqslant 1, i=1,2,\cdots,n\}.$$

性质 2.3.1 设集合 $C\subset\mathbf{R}^n$, 则下面性质成立:

(1) $\mathrm{int}C=\{\boldsymbol{x}\in C\,|\exists\varepsilon>0, B(\boldsymbol{x},\varepsilon)\subset C\}$;

(2) $\mathrm{cl}C = \cap\{C + \varepsilon B \,|\, \varepsilon > 0\}$.

下面的一些结论揭示了凸集的一些拓扑性质, 例如, 凸集的闭包和内部仍是凸集. 如果凸集中含有单纯形, 则内部非空; 反之, 如果一个凸集内部为空集, 则它不存在单纯形子集, 但一定存在包含该集合的仿射集, 并且这个仿射集的维数小于空间维度.

性质 2.3.2　设 C 是 $\mathbf{R}^n$ 中的凸集, 则下面性质成立:

(1) $\mathrm{int}C$ 和 $\mathrm{cl}C$ 都是凸集;

(2) 如果存在单纯形 $\mathrm{Sim}\,(\boldsymbol{x}_0, \boldsymbol{x}_1, \cdots, \boldsymbol{x}_n) \subset C$, 那么 $\mathrm{int}C \neq \varnothing$;

(3) 如果 $\mathrm{int}C = \varnothing$, 那么不存在单纯形 $\mathrm{Sim}\,(\boldsymbol{x}_0, \boldsymbol{x}_1, \cdots, \boldsymbol{x}_n) \subset C$, 且存在一个包含 C 的仿射集 A, 且 $\dim A < n$.

证明　(1) 不妨设 $\mathrm{int}C \neq \varnothing$, $\forall \boldsymbol{x}_1, \boldsymbol{x}_2 \in \mathrm{int}C$, 由定义 1.2.3(1) 知, $\exists \varepsilon_1, \varepsilon_2 > 0$, 则有

$$\boldsymbol{x}_1 + \varepsilon_1 \boldsymbol{y} \in C, \quad \boldsymbol{x}_2 + \varepsilon_2 \boldsymbol{y} \in C, \quad \forall \boldsymbol{y} \in B.$$

设 $\lambda \in [0, 1]$, 由 C 是凸集知 $(1-\lambda)\,\boldsymbol{x}_1 + \lambda \boldsymbol{x}_2 \in C$. 由上式得

$$\begin{aligned} &(1-\lambda)\,(\boldsymbol{x}_1 + \varepsilon_1 \boldsymbol{y}) + \lambda\,(\boldsymbol{x}_2 + \varepsilon_2 \boldsymbol{y}) \\ = \;& (1-\lambda)\,\boldsymbol{x}_1 + \lambda \boldsymbol{x}_2 + [(1-\lambda)\,\varepsilon_1 + \lambda \varepsilon_2]\,\boldsymbol{y} \subset C, \quad \forall \boldsymbol{y} \in B, \end{aligned}$$

即 $(1-\lambda)\,\boldsymbol{x}_1 + \lambda \boldsymbol{x}_2 \in \mathrm{int}C$, 所以 $\mathrm{int}C$ 是凸集.

设 $\forall \boldsymbol{x}_1, \boldsymbol{x}_2 \in \mathrm{cl}C$. 由定义 1.2.4(1) 知, 存在点列 $\{\boldsymbol{x}_{1k}\}$ 和 $\{\boldsymbol{x}_{2k}\}$,$\boldsymbol{x}_{ik} \in C, i = 1, 2, k = 1, 2, \cdots$, 使得 $\lim\limits_{k\to\infty} \boldsymbol{x}_{ik} = \boldsymbol{x}_i$. 设 $\lambda \in [0, 1]$, 由 C 是凸集可得 $(1-\lambda)\,\boldsymbol{x}_{1k} + \lambda \boldsymbol{x}_{2k} \in C$. 因此, 有

$$(1-\lambda)\,\boldsymbol{x}_1 + \lambda \boldsymbol{x}_2 = \lim_{k\to\infty} [(1-\lambda)\,\boldsymbol{x}_{1k} + \lambda \boldsymbol{x}_{2k}] \in \mathrm{cl}C,$$

所以 $\mathrm{cl}C$ 是凸集.

(2) 因为 $\mathrm{Sim}\,(\boldsymbol{x}_0, \boldsymbol{x}_1, \cdots, \boldsymbol{x}_n) \subset C$, 所以设 $\boldsymbol{x} \in \mathrm{Sim}\,(\boldsymbol{x}_0, \boldsymbol{x}_1, \cdots, \boldsymbol{x}_n)$, 有

$$\boldsymbol{x} = \sum_{i=0}^{n} \alpha_i \boldsymbol{x}_i, \quad \alpha_i \geqslant 0, \quad \sum_{i=0}^{n} \alpha_i = 1.$$

再设 $\bar{\alpha} = \min\{\alpha_i | \alpha_i > 0, i = 0, 1, 2, \cdots, n\}$, $a = \max\{\|\boldsymbol{x}_i\| \,|\, i = 0, 1, 2, \cdots, n\}$. 取 ε 满足 $0 < \varepsilon \leqslant \dfrac{1}{\bar{\alpha} a}$. 对于 $i = 0, 1, 2, \cdots, n$, 再取 $\lambda_i \in \left[\alpha_i, \alpha_i + \dfrac{1}{n+1}\varepsilon\bar{\alpha}\right]$, 显

然对 $\alpha_i > 0$, 有 $\lambda_i > 0$. 设 $\sum\limits_{i=0}^{n} \lambda_i = \lambda$, 有 $\lambda \in [1, 1+\varepsilon\bar{\alpha}]$ 和

$$\boldsymbol{y} = \sum_{i=0}^{n} \frac{\lambda_i}{\lambda}\boldsymbol{x}_i, \quad \frac{\lambda_i}{\lambda} \geqslant 0, \quad \sum_{i=0}^{n} \frac{\lambda_i}{\lambda} = 1, \quad \frac{\lambda_i}{\lambda} \in \left[\alpha_i - \frac{\varepsilon\bar{\alpha}}{1+\varepsilon\bar{\alpha}}\alpha_i, \alpha_i + \frac{1}{n+1}\varepsilon\bar{\alpha}\right].$$

当 $\lambda_i = \alpha_i$ 时, 有 $\boldsymbol{y} = \boldsymbol{x}$; 当 $\lambda_i \neq \alpha_i$ 时, 有

$$||\boldsymbol{y} - \boldsymbol{x}|| = \left\| \sum_{i=0}^{n} \left(\frac{\lambda_i}{\lambda} - \alpha_i \right) \boldsymbol{x}_i \right\| \leqslant \sum_{i=0}^{n} \left| \frac{\lambda_i}{\lambda} - \alpha_i \right| ||\boldsymbol{x}_i|| \leqslant a\varepsilon\bar{\alpha} \leqslant 1.$$

因此, 满足上述不等式的所有 $\boldsymbol{y}$ 构成 $\boldsymbol{x}$ 的一个邻域, 且有 $\boldsymbol{y} \in \mathrm{Sim}(\boldsymbol{x}_0, \boldsymbol{x}_1, \cdots, \boldsymbol{x}_n)$, 结论得证.

(3) 由上面结论可知, 若 $\mathrm{int}C = \varnothing$, 那么 C 中不包含单纯形 $\mathrm{Sim}(\boldsymbol{x}_0, \boldsymbol{x}_1, \cdots, \boldsymbol{x}_n)$. 因此, 由定理 2.2.2 知 C 中的每个元素由所有凸组合 $\boldsymbol{x}_0, \boldsymbol{x}_1, \cdots, \boldsymbol{x}_k$ 构成, 其中 $\boldsymbol{x}_0, \boldsymbol{x}_1, \cdots, \boldsymbol{x}_k$ 存在最大仿射无关组, 显然 $k < n$. 设 $\boldsymbol{x}_i - \boldsymbol{x}_0\,(i = 1, \cdots, k)$ 扩展的子空间为 S, 显然 $C - \boldsymbol{x}_0 \subset S$, 设 $A = \boldsymbol{x}_0 + S$, 则 $C \subset A$. □

上述性质说明, 如果凸集包含内点, 那么凸集中包含单纯形; 低维空间的凸集在高维空间中不含内点.

例 2.3.2 设凸集 $C = \left\{(x_1, 0)^{\mathrm{T}} \in \mathbf{R}^2 | 0 < x_1 \leqslant 1\right\}$, 显然 $\mathrm{int}C = \varnothing$, 在 C 中 $\mathrm{Sim}\,(\boldsymbol{x}_0, \boldsymbol{x}_1, \boldsymbol{x}_2)$ 不存在.

进一步, 再定义凸集的几个拓扑概念, 如相对内点、相对内部和相对边界等. 这几个概念对于凸集性质刻画具有重要意义, 相对内点和相对内部是凸集都具有的特征, 例如, 存在没有内点却有相对内点凸集. 后面将要讨论的凸集可分离条件是只要有两个凸集的相对内部不相交即可.

定义 2.3.3 设 C 是 $\mathbf{R}^n$ 中的非空凸集, $\boldsymbol{x} \in C$, 定义下面几个概念:

(1) 包含 $C - \boldsymbol{x}$ 的全体子空间的交称为 C 的支撑子空间, 用 $\mathrm{Sup}C$ 表示;

(2) 如果 $\exists\varepsilon > 0$, 使 $(\boldsymbol{x} + \varepsilon B) \cap \mathrm{aff}C \subset C$, 则称 $\boldsymbol{x}$ 是 C 的相对内点;

(3) C 的全体相对内点构成的集合称为 C 的相对内部, 用 $\mathrm{ri}C$ 或 $\mathrm{ri}\,(C)$ 或 $\mathrm{ri}\,\{C\}$ 表示;

(4) 把集合 $\mathrm{cl}C\backslash\mathrm{ri}C$ 称为 C 的相对边界, 用 $\mathrm{rb}C$ 或 $\mathrm{rb}\,(C)$ 或 $\mathrm{rb}\,\{C\}$ 表示, 称 $\mathrm{rb}C$ 中的点为相对边界点;

(5) 若 $\mathrm{ri}C = C$, 则称 C 为相对开集.

根据定义 2.3.3 得到下面的性质.

性质 2.3.3 设 C 是 $\mathbf{R}^n$ 中的非空凸集, 则下面结论成立:

(1) $\mathrm{aff}C = \boldsymbol{x} + \mathrm{Sup}C$, 即 $\mathrm{aff}C$ 与 $\mathrm{Sup}C$ 平行;

(2) $(\boldsymbol{x}+\varepsilon B)\cap \mathrm{aff}C\subset C \Leftrightarrow \boldsymbol{x}+\mathrm{Sup}C\cap\varepsilon B\subset C$;

(3) $\mathrm{ri}C\subset C\subset \mathrm{cl}C$;

(4) $\mathrm{ri}C$ 是凸集;

(5) $\mathrm{cl}\,(\mathrm{cl}C)=\mathrm{cl}C, \mathrm{ri}\,(\mathrm{ri}C)=\mathrm{ri}C$.

证明 由性质 2.2.6 可知结论成立. □

因为 $\mathrm{aff}C$ 是 $\mathbf{R}^n$ 中的闭集, 由性质 2.3.3 知, 当 C 是凸集时, C 的相对闭包就是 C 的闭包; C 的相对边界就是 C 在 $\mathrm{aff}C$ 中的边界; C 相对内点的邻域不一定属于该集合中, 属于它的仿射包这个集合即可; C 内点的邻域中所有点必须在所有集合中.

例 2.3.3 设在 $\mathbf{R}^2$ 中凸集 $C=\{(x_1,0)^{\mathrm{T}}\in\mathbf{R}^2\,|a\leqslant x_1\leqslant b\}$, 显然内部 $\mathrm{int}C=\varnothing$. 由于 C 的仿射包是 $\mathrm{aff}C=\{(x_1,0)^{\mathrm{T}}\in\mathbf{R}^2|x_1\in\mathbf{R}\}$, 所以 $\mathrm{ri}C=\{(x_1,0)^{\mathrm{T}}\in\mathbf{R}^2\,|a<x_1<b\}$. 说明内部空的凸集, 其相对内部不一定非空.

例 2.3.4 设超平面 $C_1=\{\boldsymbol{x}\in\mathbf{R}^2\,|\langle\boldsymbol{a},\boldsymbol{x}\rangle=\alpha\}$, 超平面闭半空间 $C_2=\{\boldsymbol{x}\in\mathbf{R}^2\,|\langle\boldsymbol{a},\boldsymbol{x}\rangle\leqslant\alpha\}$, 其中非零元素 $\boldsymbol{a}\in\mathbf{R}^2$, $\alpha>0$. 显然, $C_1\subset C_2$, $C_1=\mathrm{aff}C_1$ 和 $\mathrm{aff}C_2=\mathbf{R}^2$, 因此有 $\mathrm{cl}C_1\subset\mathrm{cl}C_2$ 和 $\mathrm{int}C_1\subset\mathrm{int}C_2$ 成立, 以及 $\mathrm{ri}C_1=C_1$, $\mathrm{ri}C_2=\{\boldsymbol{x}\in\mathbf{R}^2|\,\langle\boldsymbol{a},\boldsymbol{x}\rangle<\alpha\}$, 所以 $\mathrm{ri}C_1\not\subset\mathrm{ri}C_2$.

上述例子表明, 内点与相对内点是两个不同的概念. 如果一个集合被另一个集合包含, 它的内点集合也被另一个集合的内点集合所包含, 即 $C_1\subset C_2$ 时, 一定有 $\mathrm{int}C_1\subset\mathrm{int}C_2$. 但相对内点就不一定成立了, 例 2.3.4 就说明了, 相对内点不具有运算传递性包含关系. 由此可见, 内点与相对内点这两个概念不尽相同. 下面是保证 $\mathrm{ri}C_1\subset\mathrm{ri}C_2$ 成立的一个充分条件.

定理 2.3.1 设 $C_1,C_2\subset\mathbf{R}^n$ 是凸集, 且 $C_1\subset C_2,\mathrm{aff}C_1=\mathrm{aff}C_2$, 则 $\mathrm{ri}C_1\subset\mathrm{ri}C_2$.

证明 设 $\boldsymbol{x}\in\mathrm{ri}C_1$. 由定义 2.3.3 知, $\exists\varepsilon>0$ 使得

$$(\boldsymbol{x}+\varepsilon B)\cap\mathrm{aff}C_1\subset C_1,$$

结合上式, 由 $\mathrm{aff}C_1=\mathrm{aff}C_2$ 得 $(\boldsymbol{x}+\varepsilon B)\cap\mathrm{aff}C_2\subset C_1\subset C_2$, 故 $\boldsymbol{x}\in\mathrm{ri}C_2$. □

例 2.3.5 设凸集 $C=\left\{(x_1,0)^{\mathrm{T}}\in\mathbf{R}^2|0<x_1\leqslant 1\right\}$, 有

$$\mathrm{int}C=\varnothing,\quad \mathrm{ri}C=\left\{(x_1,0)^{\mathrm{T}}\in\mathbf{R}^2\,|0<x_1<1\right\},$$

$$\mathrm{cl}C=\left\{(x_1,0)^{\mathrm{T}}\in\mathbf{R}^2|0\leqslant x_1\leqslant 1\right\}.$$

显然对任意的 $(x_1,0)^{\mathrm{T}}\in\mathrm{ri}C$, $(1,0)^{\mathrm{T}}\in\mathrm{cl}C$ 和 $0\leqslant\lambda<1$, 有

$$(1-\lambda)\,(x_1,0)^{\mathrm{T}}+\lambda\,(1,0)^{\mathrm{T}}\in\mathrm{ri}C.$$

上例说明任何相对内部的点与闭包的点的凸组合属于相对内部, 下面对此进行证明.

定理 2.3.2 设 C 是 $\mathbf{R}^n$ 中的非空凸集, 对任何 $\boldsymbol{x} \in \mathrm{ri}C$, $\boldsymbol{y} \in \mathrm{cl}C$ 及 $0 \leqslant \lambda < 1$, 有

$$(1-\lambda)\boldsymbol{x} + \lambda\boldsymbol{y} \in \mathrm{ri}C.$$

证明 由定理 2.3.1 知, 若 $\mathrm{int}C = \varnothing$, 结论显然成立.

若 $\mathrm{int}C \neq \varnothing$, 不妨设 $\dim C = n$, 故 $\mathrm{ri}C = \mathrm{int}C$. 由 $\boldsymbol{y} \in \mathrm{cl}C$, 对任意的 $\varepsilon > 0$ 有 $\boldsymbol{y} \in C + \varepsilon B$. 则存在一个 $\boldsymbol{c} \in C$, $\boldsymbol{b} \in B$ 满足 $\boldsymbol{y} = \boldsymbol{c} + \varepsilon\boldsymbol{b}$, 有

$$\begin{aligned}(1-\lambda)\boldsymbol{x} + \lambda\boldsymbol{y} + \varepsilon B &= (1-\lambda)\boldsymbol{x} + \lambda(\boldsymbol{c} + \varepsilon\boldsymbol{b}) + \varepsilon B \\ &= (1-\lambda)\boldsymbol{x} + \lambda\boldsymbol{c} + \lambda\varepsilon\boldsymbol{b} + \varepsilon B \\ &\subset (1-\lambda)\left[\boldsymbol{x} + \varepsilon(1+\lambda)(1-\lambda)^{-1}B\right] + \lambda\boldsymbol{c}.\end{aligned}$$

因为 $\boldsymbol{x} \in \mathrm{int}C$, 所以由上式知, 当 ε 充分小时, 有

$$\boldsymbol{x} + \varepsilon(1+\lambda)(1-\lambda)^{-1}B \subset C.$$

又因为 C 是凸集, 有

$$(1-\lambda)\boldsymbol{x} + \lambda\boldsymbol{y} + \varepsilon B \subset (1-\lambda)C + \lambda C = C,$$

再根据定义 2.3.3 得 $(1-\lambda)\boldsymbol{x} + \lambda\boldsymbol{y} \in \mathrm{ri}C$. □

注 定理 2.3.2 的结论包括了: 若 $\boldsymbol{y} \in C$, 对 $0 \leqslant \lambda < 1$, 有 $(1-\lambda)\boldsymbol{x} + \lambda\boldsymbol{y} \in \mathrm{ri}C$; 对 $0 < \lambda < 1$, 存在 $\boldsymbol{x}_C \in \mathrm{ri}C$ 使得 $\boldsymbol{y} = \dfrac{1}{\lambda}\boldsymbol{x}_C - \left(\dfrac{1}{\lambda} - 1\right)\boldsymbol{x}$.

定理 2.3.3 设 C 是 $\mathbf{R}^n$ 中的非空凸集, 则 $\mathrm{ri}C = \mathrm{int}C$ 的充要条件是 $\mathrm{int}C \neq \varnothing$.

证明 根据性质 2.3.2 知, 如果 $\mathrm{int}C \neq \varnothing$, 有 $\mathrm{aff}C = \mathbf{R}^n$. 再根据定义 2.3.3 可知 $\mathrm{ri}C = \mathrm{int}C$. 反过来, 如果 $\mathrm{ri}C = \mathrm{int}C$, 根据定理 2.3.2 得 $\mathrm{int}C \neq \varnothing$. □

定理 2.3.3 中条件 $\mathrm{int}C \neq \varnothing$ 是不可缺少的, 这可通过前面的例 2.3.3 说明. 一般地, 非空凸集的相对内部一定是非空的. 即, 相对内部是凸集一个非常重要的特征.

性质 2.3.4 设 C 是 $\mathbf{R}^n$ 中的非空凸集, 则下面结论成立:

(1) $\mathrm{cl}C = \mathrm{cl}(\mathrm{cl}C) = \mathrm{cl}(\mathrm{ri}C)$;

(2) $\mathrm{ri}C = \mathrm{ri}(\mathrm{cl}C) = \mathrm{ri}(\mathrm{ri}C)$;

(3) $\mathrm{rb}C = \mathrm{rb}(\mathrm{cl}C) = \mathrm{rb}(\mathrm{ri}C)$;

(4) $\text{aff}C = \text{aff}\,(\text{cl}C) = \text{aff}\,(\text{ri}C)$;

(5) $\dim C = \dim\,(\text{cl}C) = \dim\,(\text{ri}C)$;

(6) $\text{ri}C \neq \varnothing$.

证明 (1) 因为 $\text{ri}C \subset C$, 所以 $\text{cl}\,(\text{ri}C) \subset \text{cl}C$. 反过来, 假设 $\boldsymbol{x}^* \in \text{cl}C$, 则存在一个序列 $\{\boldsymbol{x}_k\} \subset C$ 和 $1 > \lambda_k \to 0, \boldsymbol{x}_k \to \boldsymbol{x}^*, k \to \infty$. 令 $\boldsymbol{y} \in \text{ri}C$, 由定理 2.3.2 知, $(1-\lambda_k)\,\boldsymbol{x}_k + \lambda_k \boldsymbol{y} \in \text{ri}C$. 但 $\lim\limits_{k\to\infty}(1-\lambda_k)\,\boldsymbol{x}_k + \lambda_k \boldsymbol{y} = \boldsymbol{x}^*$, 所以 $\boldsymbol{x}^*$ 是 $\text{ri}C$ 的极限点, 故 $\boldsymbol{x}^* \in \text{cl}\,(\text{ri}C)$. 因此 $\text{cl}C \subset \text{cl}\,(\text{ri}C)$. 综上, 有 $\text{cl}C = \text{cl}\,(\text{cl}C) = \text{cl}\,(\text{ri}C)$.

(2) 显然 $C \subset \text{cl}C$, 由定理 2.3.1 得, $\text{ri}C \subset \text{ri}\,(\text{cl}C)$. 反过来, 设 $\boldsymbol{y} \in \text{ri}\,(\text{cl}C)$ 和 $\boldsymbol{x} \in \text{ri}C$, 且 $\boldsymbol{x} \neq \boldsymbol{y}$. 再设点 $\boldsymbol{y}_\alpha = (1-\alpha)\,\boldsymbol{x} + \alpha\boldsymbol{y}$, 其中 $\alpha > 1$, 当 $\alpha \to 1$ 时有 $\boldsymbol{y}_\alpha \in \text{cl}C$. 最后, 设 $\lambda = \dfrac{1}{\alpha} < 1$, 则由定理 2.3.2 可得

$$\boldsymbol{y} = \lambda\boldsymbol{y}_\alpha + (1-\lambda)\,\boldsymbol{x} \in \text{ri}C,$$

所以 $\text{ri}\,(\text{cl}C) \subset \text{ri}C$.

(3) 根据定义 2.3.3, 因为

$$\text{rb}C = \text{cl}C\backslash\text{ri}C, \quad \text{rb}\,(\text{cl}C) = \text{cl}\,(\text{cl}C)\,\backslash\text{ri}\,(\text{cl}C)\,,$$

$$\text{rb}\,(\text{ri}C) = \text{cl}\,(\text{ri}C)\,\backslash\text{ri}\,(\text{ri}C)\,,$$

则由 (1) 和 (2) 可得 (3) 成立.

(4) 因为 $\text{aff}C$ 是闭集, 所以有 $\text{aff}C = \text{aff}\,(\text{cl}C)$, 将上式中的 C 换为 $\text{ri}C$, 根据 (1) 可知 (4) 成立, 即

$$\text{aff}\,(\text{ri}C) = \text{aff}\,(\text{cl}\,(\text{ri}C)) = \text{aff}\,(\text{cl}C) = \text{aff}C.$$

(5) 结论由 (1) 直接可得.

(6) 根据性质 2.3.2, 不妨设 $\dim C = r$, 即在 C 中存在 $r+1$ 个仿射无关的向量 $\boldsymbol{x}_0, \boldsymbol{x}_1, \cdots, \boldsymbol{x}_r$ 构成一个 r 维单纯形 $S = \text{co}\,\{\boldsymbol{x}_0, \boldsymbol{x}_1, \cdots, \boldsymbol{x}_r\} \subset C$. 由性质 2.3.2(2) 得出 $\text{int}S \neq \varnothing$, 再由定理 2.3.3 得 $\text{ri}S = \text{int}S$. 即有 $\dim\,(\text{aff}S) = r = \dim\,(\text{aff}C)$, 于是得 $\text{aff}S = \text{aff}C$. 由 (4) 有

$$\text{aff}\,(\text{ri}C) = \text{aff}C = \text{aff}S = \text{aff}\,(\text{ri}S)\,.$$

上式说明 C 具有相对于 $\text{aff}C$ 的非空相对内部, 即 $\text{ri}C \neq \varnothing$. □

性质 2.3.4 说明 C, $\text{cl}\,(C)$ 和 $\text{ri}C$ 各自的闭包、相对内部、相对边界、仿射集和维数得到的对应集合分别相等. 在第 3 章一些性质的证明中, 将会反复用到性质 2.3.4(6) 的结论: 任何一个非空凸集的相对内部总是非空的.

推论 2.3.1 设 C_1, C_2 是 $\mathbf{R}^n$ 中的非空凸集, $C_1 \subset \mathrm{rb}C_2$, 则 $\dim C_1 < \dim C_2$.

证明 用反证法. 假设 $\dim C_1 < \dim C_2$ 不成立, 则 $\dim C_1 \geqslant \dim C_2$, 由性质 2.3.4(4), (5) 知 $\dim(\mathrm{ri}C_2) = \dim C_2$, 即有 $\mathrm{ri}C_2 \cap C_1 \neq \varnothing$, 显然与条件 $C_1 \subset \mathrm{rb}C_2$ 矛盾. 故

$$\dim C_1 < \dim C_2.$$

□

下面的定理是一个很有用的结论, 将在后面凸锥的某些性质证明中用到.

定理 2.3.4 设 C 是 $\mathbf{R}^n$ 中的非空凸集, 则 $\boldsymbol{x} \in \mathrm{ri}C$ 当且仅当对于每一个 $\boldsymbol{y} \in C$, $\exists \lambda > 1$, 使

$$(1-\lambda)\boldsymbol{y} + \lambda\boldsymbol{x} \in C.$$

证明 先证明必要性. 设 $\boldsymbol{x} \in \mathrm{ri}C$ 和 $\boldsymbol{y} \in C$, 由相对内点定义 2.3.3 知 $\exists \varepsilon > 0$, 使

$$(\boldsymbol{x} + \varepsilon B) \cap \mathrm{aff}C \subset C,$$

则存在 $0 < \alpha < 1$ 和 $\boldsymbol{c} \in C$ 使得

$$\boldsymbol{x} = (1-\alpha)\boldsymbol{y} + \alpha\boldsymbol{c} \in C.$$

令 $\lambda = \dfrac{1}{\alpha}$, 则

$$\boldsymbol{c} = (1-\lambda)\boldsymbol{y} + \lambda\boldsymbol{x} \in C, \quad \lambda > 1.$$

再证明充分性. 根据性质 2.3.4 知, $\mathrm{ri}C \neq \varnothing$. 对于每一个 $\boldsymbol{y} \in \mathrm{ri}C$, 均存在 $\lambda > 1$, 使

$$(1-\lambda)\boldsymbol{y} + \lambda\boldsymbol{x} = \boldsymbol{c} \in C,$$

即 $\boldsymbol{x} = \dfrac{1}{\lambda}\boldsymbol{c} + \left(1 - \dfrac{1}{\lambda}\right)\boldsymbol{y}$, 由定理 2.3.2 知 $\boldsymbol{x} \in \mathrm{ri}C$. □

本小节得到了凸集 C 的闭包、相对内部等一些非常重要的拓扑性质:

1) 凸集的闭包和内部也是凸集;

2) 如果凸集中含有单纯形, 则内部非空;

3) 如果凸集内部为空集, 则它不存在单纯形子集, 但存在包含该集合的仿射集的维数小于空间维度;

4) 凸集的仿射集与它的支撑子空间平行, 凸集的相对内部也是凸集;

5) 如果两个不相等的凸集对应的仿射集相等, 则它们的相对内部具有包含关系;

6) 凸集的相对内部与闭包的凸组合在该凸集的相对内部中;

7) 如果凸集的相对内部等于它的内部, 则内部非空, 且非空凸集的相对内部非空;

8) C, $\mathrm{cl}\,(C)$ 和 $\mathrm{ri}C$ 各自的闭包、相对内部、相对边界、仿射集和维数得到的对应集合分别相等.

2.3.2　运算性质

下面进一步讨论在凸集族上进行拓扑运算后的闭包与相对内部的一些性质, 例如凸集闭包族、支撑子空间族和相对内部族的交运算性质.

性质 2.3.5　设 $\{C_i\,|i\in I\}$ 是 $\mathbf{R}^n$ 中的凸集族, I 是任意指标集, $\bigcap_{i\in I}\mathrm{ri}C_i\neq\varnothing$, 则下面结论成立:

(1) $\mathrm{cl}\left(\bigcap_{i\in I}C_i\right)=\bigcap_{i\in I}(\mathrm{cl}C_i)$;

(2) $\bigcap_{i\in I}(\mathrm{Sup}C_i)=\mathrm{Sup}\left(\bigcap_{i\in I}C_i\right)$, 当 I 有限时;

(3) $\mathrm{ri}\left(\bigcap_{i\in I}C_i\right)=\bigcap_{i\in I}(\mathrm{ri}C_i)$, 当 I 有限时.

证明　(1) 设 $\boldsymbol{x}\in\bigcap_{i\in I}(\mathrm{ri}C_i)$, $\boldsymbol{y}\in\bigcap_{i\in I}(\mathrm{cl}C_i)$, 由定理 2.3.2 知, 当 $0\leqslant\lambda<1$ 时, 有

$$(1-\lambda)\,\boldsymbol{x}+\lambda\boldsymbol{y}\in\bigcap_{i\in I}(\mathrm{ri}C_i).$$

上式中令 $\lambda\to1$, 则有 $(1-\lambda)\,\boldsymbol{x}+\lambda\boldsymbol{y}\to\boldsymbol{y}$, 即 $\boldsymbol{y}\in\mathrm{cl}\left(\bigcap_{i\in I}(\mathrm{ri}C_i)\right)$. 因此有

$$\bigcap_{i\in I}(\mathrm{cl}C_i)\subset\mathrm{cl}\left(\bigcap_{i\in I}(\mathrm{ri}C_i)\right)\subset\mathrm{cl}\left(\bigcap_{i\in I}C_i\right)\subset\bigcap_{i\in I}(\mathrm{cl}C_i).$$

(2) 不妨设 $I=\{1,2\}$, $\boldsymbol{x}\in\mathrm{ri}C_1\cap\mathrm{ri}C_2$. 根据支撑子空间定义有

$$\mathrm{aff}C_1=\boldsymbol{x}+\mathrm{Sup}C_1,\quad \mathrm{aff}C_2=\boldsymbol{x}+\mathrm{Sup}C_2.$$

设 $\boldsymbol{y}\in\mathrm{Sup}\,(C_1\cap C_2)$, 有

$$\mathrm{aff}\,(C_1\cap C_2)=\boldsymbol{x}+\mathrm{Sup}\,(C_1\cap C_2)\,,$$

$$\begin{aligned}\mathrm{Sup}\,(C_1\cap C_2)&=\mathrm{aff}\,(C_1\cap C_2)-\boldsymbol{x}\\&=(\mathrm{aff}\,(C_1)-\boldsymbol{x})\cap(\mathrm{aff}\,(C_2)-\boldsymbol{x})=\mathrm{Sup}C_1\cap\mathrm{Sup}C_2.\end{aligned}$$

上述证明过程对任意指标集都成立.

(3) 不妨设 $I=\{1,2\}$, $\boldsymbol{x}\in \mathrm{ri}C_1\cap \mathrm{ri}C_2$. 则 $\exists\varepsilon>0$, 使得

$$\boldsymbol{x}+\mathrm{Sup}C_1\cap\varepsilon B\subset C_1,\quad \boldsymbol{x}+\mathrm{Sup}C_2\cap\varepsilon B\subset C_2,$$

即由上式得

$$(\boldsymbol{x}+\mathrm{Sup}C_1\cap\varepsilon B)\cap(\boldsymbol{x}+\mathrm{Sup}C_2\cap\varepsilon B)=\boldsymbol{x}+\mathrm{Sup}\,(C_1\cap C_2)\cap\varepsilon B\subset C_1\cap C_2.$$

所以 $\boldsymbol{x}\in\mathrm{ri}\,(C_1\cap C_2)$, 得到 $\mathrm{ri}C_1\cap\mathrm{ri}C_2\subset\mathrm{ri}\,(C_1\cap C_2)$.

反过来, 设 $\boldsymbol{x}\in\mathrm{ri}\,(C_1\cap C_2)$, $\exists\varepsilon>0$, 使得

$$\boldsymbol{x}+\mathrm{Sup}\,(C_1\cap C_2)\cap\varepsilon B\subset(C_1\cap C_2)\,,$$

即有 $\boldsymbol{x}+(\mathrm{Sup}C_1\cap\mathrm{Sup}C_2)\cap\varepsilon B\subset(C_1\cap C_2)$. 显然有

$$\boldsymbol{x}+\mathrm{Sup}C_1\cap\varepsilon B\subset C_1,\quad \boldsymbol{x}+\mathrm{Sup}C_2\cap\varepsilon B\subset C_2$$

成立, 即 $\boldsymbol{x}\in\mathrm{ri}C_1\cap\mathrm{ri}C_2$, 得 $\mathrm{ri}\,(C_1\cap C_2)\subset\mathrm{ri}C_1\cap\mathrm{ri}C_2$.

同理, 上述证明过程对任意指标集都成立. □

例 2.3.6 在 $\mathbf{R}^2$ 中, 考虑下面两个凸集:

$$C_1=\left\{(x_1,x_2)^{\mathrm{T}}\,|x_1>0,\ x_2>0\right\}\cup\left\{(0,0)^{\mathrm{T}}\right\},\quad C_2=\left\{(x_1,0)^{\mathrm{T}}\,|x_1\in\mathbf{R}\right\}.$$

显然上述两个集合的相对内部相交为空集: $\mathrm{ri}C_1\cap\mathrm{ri}C_2=\varnothing$, 但是它们相交后的集合 $C_1\cap C_2$ 相对内部非空: $\mathrm{ri}\,(C_1\cap C_2)=\{(0,0)\}$. 因此, 我们有 $\mathrm{ri}C_1\cap\mathrm{ri}C_2\neq\mathrm{ri}\,(C_1\cap C_2)$.

另一方面, C_1,C_2 的闭包相交与交集的闭包也不相等:

$$\mathrm{cl}C_1\cap\mathrm{cl}C_2\neq\mathrm{cl}(C_1\cap C_2),$$

因为

$$\mathrm{cl}\,(C_1\cap C_2)=\left\{(0,0)^{\mathrm{T}}\right\},\quad \mathrm{cl}C_1\cap\mathrm{cl}C_2=\left\{(x_1,0)^{\mathrm{T}}\,|0\leqslant x_1<+\infty\right\}.$$

这说明性质 2.3.5 中条件 $\bigcap_{i\in I}\mathrm{ri}C_i\neq\varnothing$ 成立是必要的.

性质 2.3.6 设 C 是 $\mathbf{R}^n$ 中的非空凸集, S 是至少包含 $\mathrm{ri}C$ 的一个点的仿射集, 则

$$\mathrm{ri}\,(S\cap C)=S\cap\mathrm{ri}C,\quad \mathrm{cl}\,(S\cap C)=S\cap\mathrm{cl}C.$$

证明 由于 S 是仿射集, 有 $\mathrm{ri}S=S=\mathrm{cl}S$, 根据性质 2.3.5(1) 和 (3) 可知结论成立. □

性质 2.3.7 设 C_1, C_2 是 $\mathbf{R}^n$ 中的非空凸集, $C_2 \subset \mathrm{cl}C_1$, $C_2 \cap \mathrm{ri}C_1 \neq \varnothing$, 则 $\mathrm{ri}C_2 \subset \mathrm{ri}C_1$.

证明 由性质 2.3.5 知结论显然. □

例 2.3.7 在 $\mathbf{R}^2$ 中, 考虑下面两个凸集:

$$C_1 = \left\{(x_1, 0)^{\mathrm{T}} \,|0 < x_1 < 1\right\}, \quad C_2 = \left\{(x_1, 0)^{\mathrm{T}} \,|0 < x_1 \leqslant 1\right\}.$$

显然有 $C_2 \subset \mathrm{cl}C_1, C_2 \cap \mathrm{ri}C_1 \neq \varnothing$, 且 $\mathrm{ri}C_2 \subset \mathrm{ri}C_1$.

下面是凸集合到另一个空间的线性变换后的拓扑性质: 凸集线性变换后的闭包包含闭包线性变换后的集合, 凸集线性变换后的相对内部等于相对内部线性变换后的集合.

性质 2.3.8 设 C 是 $\mathbf{R}^n$ 中的非空凸集, L 是从 $\mathbf{R}^n$ 到 $\mathbf{R}^m$ 的线性变换, $\lambda \in \mathbf{R}$, 则

(1) $L(\mathrm{cl}C) \subseteq \mathrm{cl}(LC)$;

(2) $L(\mathrm{ri}C) = \mathrm{ri}(LC)$;

(3) $\mathrm{ri}(\lambda C) = \lambda \mathrm{ri}C$.

证明 (1) 设序列 $\{\boldsymbol{x}_k\} \subset C$, $\boldsymbol{x}_k \to \boldsymbol{x}$, 显然 $\{L\boldsymbol{x}_k\} \subset LC$ 和 $L\boldsymbol{x}_k \to L\boldsymbol{x}$ 成立, 因此结论成立.

(2) 根据 (1)、性质 2.3.3(3) 及性质 2.3.4(1) 有

$$L(\mathrm{ri}C) \subseteq LC \subseteq L(\mathrm{cl}C) = L(\mathrm{cl}(\mathrm{ri}C)) \subseteq \mathrm{cl}(L(\mathrm{ri}C)).$$

由上式有 $\mathrm{cl}(LC) = \mathrm{cl}(L(\mathrm{ri}C))$, 所以由性质 2.3.4(2)、性质 2.3.7 得

$$\mathrm{ri}(LC) = \mathrm{ri}(L(\mathrm{ri}C)) \subset L(\mathrm{ri}C).$$

反过来, 设 $\boldsymbol{z} \in L(\mathrm{ri}C)$. 再设 $\forall \boldsymbol{x} \in LC$, 令 $\boldsymbol{z}' \in \mathrm{ri}C, \boldsymbol{x}' \in C$, 使得

$$L\boldsymbol{z}' = \boldsymbol{z}, \quad L\boldsymbol{x}' = \boldsymbol{x}.$$

由定理 2.3.4 可知, $\exists \alpha > 1$ 使 $(1-\alpha)\boldsymbol{x}' + \alpha\boldsymbol{z}' \in C$, 即有

$$L[(1-\alpha)\boldsymbol{x}' + \alpha\boldsymbol{z}'] = (1-\alpha)\boldsymbol{x} + \alpha\boldsymbol{z} \in LC,$$

从而得到 $\boldsymbol{z} \in \mathrm{ri}(LC)$. 所以 $L(\mathrm{ri}C) \subseteq \mathrm{ri}(LC)$, 故 $L(\mathrm{ri}C) = \mathrm{ri}(LC)$ 成立.

(3) 在 (2) 中令 $L\boldsymbol{x} = \lambda\boldsymbol{x}$ 即得. □

性质 2.3.9 设 C_1, C_2 是 $\mathbf{R}^n$ 中的非空凸集, 则下面结论成立:

$$\mathrm{ri}(C_1 + C_2) = \mathrm{ri}C_1 + \mathrm{ri}C_2,$$

$$\mathrm{cl}\,(C_1+C_2)\supset\mathrm{cl}C_1+\mathrm{cl}C_2.$$

证明 根据直和的定义和性质 2.3.4, $\mathbf{R}^{2n}$ 中的直和 $C_1\oplus C_2$ 满足下面的关系:

$$\mathrm{ri}\,(C_1\oplus C_2)=\mathrm{ri}C_1\oplus\mathrm{ri}C_2,$$

$$\mathrm{cl}\,(C_1\oplus C_2)=\mathrm{cl}C_1\oplus\mathrm{cl}C_2.$$

在性质 2.3.4 中设 L 是从 $\mathbf{R}^{2n}$ 到 $\mathbf{R}^n$ 的加法线性变换, 即

$$L\,(\boldsymbol{x}_1,\boldsymbol{x}_2)\to\boldsymbol{x}_1+\boldsymbol{x}_2,\quad \forall\boldsymbol{x}_1,\boldsymbol{x}_2\in\mathbf{R}^n.$$

则 $L\,(C_1\oplus C_2)=C_1+C_2$, 再由性质 2.3.8(1),(2) 得结论成立. □

例 2.3.8 设凸集 $C_1=\{(x_1,0)^{\mathrm{T}}\,|0<x_1\leqslant1\}$ 和 $C_2=\{(x_1,1)^{\mathrm{T}}\,|0\leqslant x_1<1\}$, 有

$$\mathrm{ri}C_1=\left\{(x_1,0)^{\mathrm{T}}\,|0<x_1<1\right\},\ \ \mathrm{ri}C_2=\left\{(x_1,1)^{\mathrm{T}}\,|0<x_1<1\right\},$$

$$C_1+C_2=\left\{(x_1,1)^{\mathrm{T}}\,|0\leqslant x_1<2\right\}.$$

容易计算得

$$\mathrm{ri}\,(C_1+C_2)=\left\{(x_1,1)^{\mathrm{T}}\,|0<x_1<2\right\}=\mathrm{ri}C_1+\mathrm{ri}C_2,$$

$$\mathrm{cl}\,(C_1+C_2)=\left\{(x_1,1)^{\mathrm{T}}\,|0\leqslant x_1\leqslant2\right\}=\mathrm{cl}C_1+\mathrm{cl}C_2.$$

性质 2.3.10 设 C 是 $\mathbf{R}^m$ 中的非空凸集, L 是从 $\mathbf{R}^n$ 到 $\mathbf{R}^m$ 的线性变换, $L^{-1}\,(\mathrm{ri}C)\neq\varnothing$, 则

$$\mathrm{ri}\left(L^{-1}C\right)=L^{-1}\,(\mathrm{ri}C)\,,$$

$$\mathrm{cl}\left(L^{-1}C\right)=L^{-1}\,(\mathrm{cl}C)\,.$$

证明 因为线性变换的逆变换也是线性变换, 由性质 2.3.8(1) 和 (2) 得结论成立. □

本小节得到了凸集的闭包和相对内部的一些运算性质:

1) 如果凸集族相对内部的交集非空, 则凸集族各相对内部的交集等于凸集族交集的相对内部, 凸集族各闭包的交等于凸集族交的闭包;

2) 如果仿射集 S 与 $\mathrm{ri}C$ 相交非空, 则 S 与 C 交的相对内部等于 S 与 $\mathrm{ri}C$ 相交, S 与 C 交的闭包等于 S 与 $\mathrm{cl}C$ 相交;

3) 凸集线性变换后的闭包包含闭包线性变换后的集合, 凸集线性变换后的相对内部等于相对内部线性变换后的集合;

4) 两个非空凸集之和的相对内部等于各自的相对内部之和, 两个非空凸集之和的闭包包含各自的闭包之和.

2.3.3 闭凸包

凸包是包含集合的最小凸集, 凸包有开的与闭的两种情形, 也存在凸包既不是开集也不是闭集的情形, 如凸区间 $C=(0,1]$ 的凸包 $\mathrm{co}C=(0,1]$. 下面引入闭凸包的概念, 它有别于凸包概念, 具体可见下面例子.

例 2.3.9 设集合 $C=\left\{(0,1)^{\mathrm{T}}\right\}\cup\left\{(x_1,0)^{\mathrm{T}}\,|x_1\geqslant 0\right\}\subset\mathbf{R}^2$, 容易验证 C 是闭集, 但是它的凸包

$$\mathrm{co}C=\left\{(x_1,x_2)^{\mathrm{T}}\,|x_1\geqslant 0,0\leqslant x_2\leqslant 1\right\}\cup\left\{(0,1)^{\mathrm{T}}\right\}$$

是一个缺了上边界的集合 $\left\{(x_1,1)^{\mathrm{T}}\,|x_1\geqslant 0\right\}$, 因此, $\mathrm{co}C$ 不是闭集.

定义 2.3.4 设 $C\subset\mathbf{R}^n$, 包含 C 的全体闭凸集的交称为 C 的闭凸包, 用 $\mathrm{clco}C$ 表示.

例 2.3.10 设在 $\mathbf{R}$ 中集合 $C=\{x_1|x_1\in(0,1)\}\cup\{x_1|x_1\in(1,2)\}$ 是开集, 它的凸包

$$\mathrm{co}C=\{x_1|x_1\in(0,2)\}$$

也是一个开集.

性质 2.3.11 设 C 是 $\mathbf{R}^n$ 中的非空开集, 则 $\mathrm{co}C$ 也是开集.

证明 因为 $\mathrm{int}\,(\mathrm{co}C)\subset\mathrm{co}C$ 显然成立, 所以只要证明 $\mathrm{co}C\subset\mathrm{int}\,(\mathrm{co}C)$. 因为 C 是开集, 故

$$C\cap\mathrm{bd}\,(\mathrm{co}C)=\varnothing,$$

其中 $\mathrm{bd}\,(\mathrm{co}C)$ 是 $\mathrm{co}C$ 的边界集. 因为 $C\subset\mathrm{co}C$, 所以有 $C\subset\mathrm{int}\,(\mathrm{co}C)$, 由性质 2.3.2 知道 $\mathrm{int}\,(\mathrm{co}C)$ 是凸集. 因此, 有 $\mathrm{co}C\subset\mathrm{int}\,(\mathrm{co}C)$, 于是得 $\mathrm{co}C=\mathrm{int}\,(\mathrm{co}C)$, 所以 $\mathrm{co}C$ 是开集. □

开集的凸包是开集, 下面证明凸包的闭凸集既是闭包, 又是闭集.

性质 2.3.12 设 $C\subset\mathbf{R}^n$, 则 $\mathrm{clco}C=\mathrm{cl}\,(\mathrm{co}C)$.

证明 根据定义 2.3.4 知, $\mathrm{clco}C$ 是包含 C 的最小闭凸集, 故 $\mathrm{clco}C\subset\mathrm{cl}\,(\mathrm{co}C)$. 但是因为 $\mathrm{co}C$ 是包含 C 的最小凸集, 而 $\mathrm{clco}C$ 是包含 C 的闭凸集, 故 $\mathrm{clco}C\supset\mathrm{co}C$, 即

$$\mathrm{clco}C \supset \mathrm{cl}\,(\mathrm{co}C)\,.$$

因此, 得 $\mathrm{clco}C = \mathrm{cl}\,(\mathrm{co}C)$. □

需要注意的是, 在上述性质中, 闭凸包是凸包的闭集, 而不是闭集的凸包! 因此, 闭凸包与闭集的凸包有时并不是相等的. 例 2.3.8 即说明了闭集的凸包不一定是闭集. 下面有更强的结论: 紧集的凸包是紧集.

性质 2.3.13 设 $C \subset \mathbf{R}^n$ 是非空紧集, 则 $\mathrm{co}C$ 是紧集.

证明 在 $\mathbf{R}^n$ 中的集合为紧集的充要条件是该集合是有界闭集, 因此, 只要证明 $\mathrm{co}C$ 是有界闭集即可. 因为 $C \subset \mathbf{R}^n$ 是紧集, 存在常数 β 使得对任意 $\boldsymbol{y} \in C$ 有 $||\boldsymbol{y}|| < \beta$. 设 $\forall \boldsymbol{x} \in \mathrm{co}C$, 由定理 2.2.2 知, 存在 $n+1$ 个点 $\{\boldsymbol{x}_i\} \subset C, i = 1, \cdots, n+1$, 以及 $\lambda_i \geqslant 0, \sum\limits_{i=1}^{n+1} \lambda_i = 1$ 满足

$$\boldsymbol{x} = \sum_{i=1}^{n+1} \lambda_i \boldsymbol{x}_i.$$

因此, $||\boldsymbol{x}|| \leqslant \sum\limits_{i=1}^{n+1} \lambda_i ||\boldsymbol{x}_i|| < \beta$, 那么 $\mathrm{co}C$ 是有界集.

下面再证明 $\mathrm{co}C$ 是闭集. 设收敛序列 $\left\{\boldsymbol{x}^k\right\} \subset \mathrm{co}C$ 收敛到 $\boldsymbol{x}^*$, 对每一个 $\boldsymbol{x}^k$ 存在 $n+1$ 个点 $\left\{\boldsymbol{x}_i^k\right\} \subset C(i = 1, \cdots, n+1)$, 以及 $\lambda_i^k \geqslant 0, \sum\limits_{i=1}^{n+1} \lambda_i^k = 1$ 使得

$$\boldsymbol{x}^k = \sum_{i=1}^{n+1} \lambda_i^k \boldsymbol{x}_i^k.$$

因为 $C \subset \mathbf{R}^n$ 是紧集, 那么 $n+1$ 个序列 $\left\{\boldsymbol{x}_i^k\right\}$ 都存在收敛子序列, $\left\{\lambda_i^k\right\}$ 也存在收敛子序列. 不失一般性, 对于 $k \to +\infty$, 不妨设 $\lambda_i^k \to \lambda_i$, $\boldsymbol{x}_i^k \to \boldsymbol{x}_i$. 因为 C 是紧集, 有

$$\boldsymbol{x}_i \in C \quad (i = 1, \cdots, n+1)\,,$$

$$\boldsymbol{x}^k \to \boldsymbol{x}^*, \quad \sum_{i=1}^{n+1} \lambda_i^k \boldsymbol{x}_i^k \to \sum_{i=1}^{n+1} \lambda_i \boldsymbol{x}_i,$$

以及 $\lambda_i \geqslant 0, \sum\limits_{i=1}^{n+1} \lambda_i = 1$ 满足 $\boldsymbol{x}^* = \sum\limits_{i=1}^{n+1} \lambda_i \boldsymbol{x}_i$, 故 $\boldsymbol{x}^* \in \mathrm{co}C$. 因此, $\mathrm{co}C$ 是闭集. □

本小节得到了凸包的几个拓扑性质:

1) 开集的凸包是开集;

2) 凸包的闭凸集等于凸包的闭集;

3) 紧集的凸包是紧集.

2.4　凸集分离定理

凸集分离定理在最优化分析中有着广泛的应用, 如证明约束优化最优解的必要性条件 [42,52]. 本节将证明几个重要的凸集分离定理 [52,85]. 后面根据超平面与凸集是否相交的情况, 按分离、严格分离和强分离等 3 种凸集分离定理情形讨论.

设一个超平面 $H(\boldsymbol{a},\alpha)=\{\boldsymbol{x}|\langle\boldsymbol{x},\boldsymbol{a}\rangle=\alpha\}$ 或 $H^{=}(\boldsymbol{a},\alpha)=\{\boldsymbol{x}|\langle\boldsymbol{x},\boldsymbol{a}\rangle=\alpha\}$, 其中非零向量 $\boldsymbol{a}\in\mathbf{R}^n$ 和常数 $\alpha\in\mathbf{R}$, 它对应两个闭半子空间的符号记为

$$H^{\leqslant}(\boldsymbol{a},\alpha)=\{\boldsymbol{x}|\langle\boldsymbol{x},\boldsymbol{a}\rangle\leqslant\alpha\},$$

$$H^{\geqslant}(\boldsymbol{a},\alpha)=\{\boldsymbol{x}|\langle\boldsymbol{x},\boldsymbol{a}\rangle\geqslant\alpha\}.$$

两个开半子空间的符号记为

$$H^{<}(\boldsymbol{a},\alpha)=\{\boldsymbol{x}|\langle\boldsymbol{x},\boldsymbol{a}\rangle<\alpha\},$$

$$H^{>}(\boldsymbol{a},\alpha)=\{\boldsymbol{x}|\langle\boldsymbol{x},\boldsymbol{a}\rangle>\alpha\}.$$

图 2.4-1 是两个凸集 C_1,C_2 的分离情形, C_1,C_2 可能存在相交的部分点属于超平面 $H(\boldsymbol{a},\alpha)$. 图 2.4-2 是两个凸集 C_1,C_2 的严格 (强) 分离情形, C_1,C_2 不存在相交的点, 它们分别属于超平面 $H(\boldsymbol{a},\alpha)$ 的不同半空间. 图 2.4-3 是凸集 C_5 存在支撑超平面情形, 其中 C_5 属于超平面 $H(\boldsymbol{a},\alpha)$ 的半空间, C_5 存在部分点属于超平面.

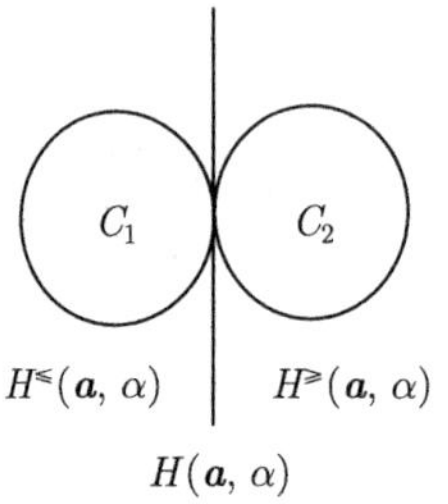

图 2.4-1　凸集接触分离

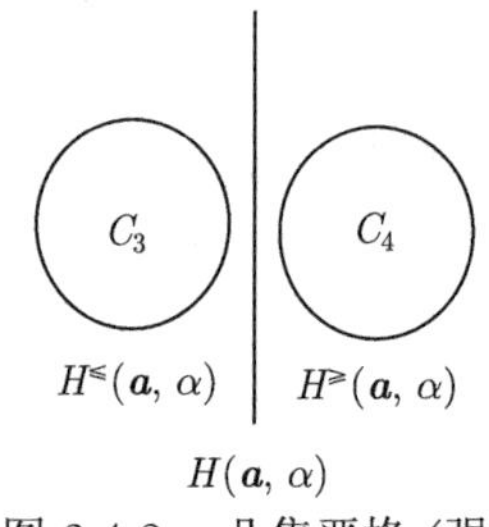

图 2.4-2　凸集严格 (强) 分离

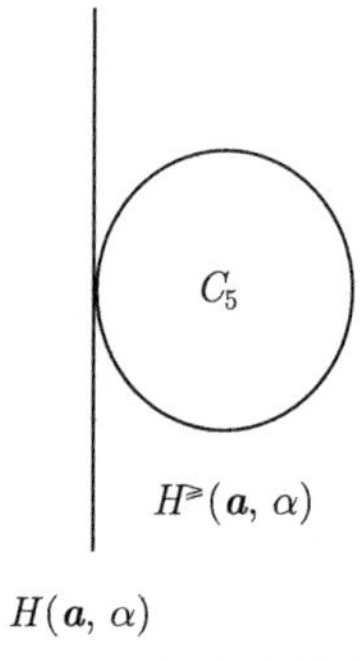

图 2.4-3　凸集支撑分离

2.4.1　超平面分离定理

下面是两个凸集的超平面分离与严格分离定义.

定义 2.4.1　设非空凸集 $C_1,C_2\subset\mathbf{R}^n$. 如果存在点 $\boldsymbol{a}\in\mathbf{R}^n$ 和常数 $\alpha\in\mathbf{R}$ 构成的一个超平面 $H(\boldsymbol{a},\alpha)=\{\boldsymbol{x}|\langle\boldsymbol{x},\boldsymbol{a}\rangle=\alpha\}$ 满足

$$C_1\subset H^{\leqslant}(\boldsymbol{a},\alpha)\quad\text{和}\quad C_2\subset H^{\geqslant}(\boldsymbol{a},\alpha)$$

或者

$$C_2 \subset H^{\leqslant}(\boldsymbol{a},\alpha) \quad 和 \quad C_1 \subset H^{\geqslant}(\boldsymbol{a},\alpha),$$

且 $C_1 \not\subset H(\boldsymbol{a},\alpha)$ 或 $C_2 \not\subset H(\boldsymbol{a},\alpha)$ 之一成立, 则称超平面 $H(\boldsymbol{a},\alpha)$ 分离 C_1 和 C_2.

定义 2.4.2 设非空凸集 $C_1, C_2 \subset \mathbf{R}^n$. 如果存在点 $\boldsymbol{a} \in \mathbf{R}^n$ 和常数 $\alpha \in \mathbf{R}$ 构成的一个超平面 $H(\boldsymbol{a},\alpha) = \{\boldsymbol{x} | \langle \boldsymbol{x}, \boldsymbol{a} \rangle = \alpha\}$ 使以下几组条件其中之一成立:

$$C_1 \subset H^{<}(\boldsymbol{a},\alpha), \quad C_2 \subset H^{\geqslant}(\boldsymbol{a},\alpha).$$

$$C_1 \subset H^{\geqslant}(\boldsymbol{a},\alpha), \quad C_2 \subset H^{<}(\boldsymbol{a},\alpha).$$

$$C_1 \subset H^{\leqslant}(\boldsymbol{a},\alpha), \quad C_2 \subset H^{>}(\boldsymbol{a},\alpha).$$

$$C_1 \subset H^{>}(\boldsymbol{a},\alpha), \quad C_2 \subset H^{\leqslant}(\boldsymbol{a},\alpha),$$

则称超平面 $H(\boldsymbol{a},\alpha)$ 严格分离 C_1, C_2.

注 如果超平面 $H(\boldsymbol{a},\alpha)$ 严格分离 C_1, C_2, 那么超平面 $H(\boldsymbol{a},\alpha)$ 分离 C_1, C_2. 反之不一定成立. 严格分离是两个凸集中至少有一个凸集不与分离超平面相交, 而一般分离两个凸集可能都要与分离超平面相交.

例 2.4.1 (1) 设凸集 $C_1 = \{x | x \in [0,1]\}$ 和 $C_2 = \{x | x \in [1,2]\}$ 是 $\mathbf{R}$ 中的非空凸集. 显然超平面 $H(1,1) = \{x | \langle x, 1 \rangle = 1\}$ 分离 C_1, C_2. 但超平面 $H(1,1) = \{x | \langle x, 1 \rangle = 1\}$ 不严格分离 C_1, C_2.

(2) 设凸集 $C_1 = \{x | x \in (0,1)\}$ 和 $C_2 = \{x | x \in [1,2]\}$ 是 $\mathbf{R}$ 中的非空凸集. 显然超平面 $H(1,1) = \{x | \langle x, 1 \rangle = 1\}$ 严格分离 C_1, C_2.

下面是判定超平面 $H(\boldsymbol{a},\alpha) = \{\boldsymbol{x} | \langle \boldsymbol{x}, \boldsymbol{a} \rangle = \alpha\}$ 分离 C_1, C_2 的充要条件.

定理 2.4.1 设非空凸集 $C_1, C_2 \subset \mathbf{R}^n$, 非零向量 $\boldsymbol{a}^* \in \mathbf{R}^n$, 常数 $\alpha \in \mathbf{R}$, 则超平面

$$H(\boldsymbol{a}^*,\alpha) = \{\boldsymbol{x} | \langle \boldsymbol{x}, \boldsymbol{a}^* \rangle = \alpha\}$$

分离 C_1, C_2 的充要条件是满足

$$\inf\{\langle \boldsymbol{a}, \boldsymbol{a}^* \rangle | \boldsymbol{a} \in C_1\} \geqslant \sup\{\langle \boldsymbol{a}, \boldsymbol{a}^* \rangle | \boldsymbol{a} \in C_2\}, \tag{2.4.1-1}$$

$$\sup\{\langle \boldsymbol{a}, \boldsymbol{a}^* \rangle | \boldsymbol{a} \in C_1\} > \inf\{\langle \boldsymbol{a}, \boldsymbol{a}^* \rangle | \boldsymbol{a} \in C_2\}. \tag{2.4.1-2}$$

证明 先证明充分条件. 假设 $\exists \boldsymbol{a}^* \in \mathbf{R}^n$, $\alpha \in \mathbf{R}$ 使得 (2.4.1-1) 和 (2.4.1-2) 式成立. 不妨设 $\alpha = \sup\{\langle \boldsymbol{a}, \boldsymbol{a}^* \rangle | \boldsymbol{a} \in C_2\}$, 那么 $H(\boldsymbol{a}^*,\alpha) = \{\boldsymbol{x} | \langle \boldsymbol{x}, \boldsymbol{a}^* \rangle = \alpha\}$ 是超平面, 显然有

$$C_1 \subset H^{\geqslant}(\boldsymbol{a}^*,\alpha), \quad C_2 \subset H^{\leqslant}(\boldsymbol{a}^*,\alpha).$$

如果

$$\sup\{\langle \boldsymbol{a}, \boldsymbol{a}^*\rangle \,|\boldsymbol{a}\in C_1\} > \alpha \tag{2.4.1-3}$$

成立, 显然可得 $C_1 \not\subset H(\boldsymbol{a},\alpha)$ 成立. 否则, $C_1 \subset H(\boldsymbol{a}^*,\alpha)$, 则有 $\sup\{\langle \boldsymbol{a}, \boldsymbol{a}^*\rangle \mid \boldsymbol{a}\in C_1\} = \alpha$ 与 (2.4.1-3) 式矛盾. 因此, 根据定义 2.4.1, 超平面 $H(\boldsymbol{a}^*,\alpha)$ 分离 C_1, C_2.

如果 $\sup\{\langle \boldsymbol{a}, \boldsymbol{a}^*\rangle \,|\boldsymbol{a}\in C_1\} = \alpha$, 根据 (2.4.1-1) 式得 $C_1 \subset H(\boldsymbol{a}^*,\alpha)$. 如果 $C_2 \not\subset H(\boldsymbol{a},\alpha)$ 成立, 根据定义 2.4.1, 超平面 $H(\boldsymbol{a}^*,\alpha)$ 分离 C_1, C_2. 因为若 $C_2 \not\subset H(\boldsymbol{a},\alpha)$ 不成立, 则有

$$C_2 \subset H(\boldsymbol{a}^*,\alpha),$$

得 $\inf\{\langle \boldsymbol{a}, \boldsymbol{a}^*\rangle \,|\boldsymbol{a}\in C_2\} = \alpha$ 与 (2.4.1-2) 式矛盾.

下面证明必要条件. 如果超平面 $H(\boldsymbol{a}^*,\alpha)$ 分离 C_1, C_2, 则由定义 2.4.1, 不妨设

$$C_1 \subset H^{\geqslant}(\boldsymbol{a},\alpha), \quad C_2 \subset H^{\leqslant}(\boldsymbol{a},\alpha), \quad C_1 \not\subset H(\boldsymbol{a},\alpha).$$

上述包含关系等价于

$$\langle \boldsymbol{a}, \boldsymbol{a}^*\rangle \geqslant \alpha, \quad \langle \boldsymbol{a}', \boldsymbol{a}^*\rangle \leqslant \alpha, \quad \forall \boldsymbol{a}\in C_1, \quad \forall \boldsymbol{a}'\in C_2,$$

即

$$\inf\{\langle \boldsymbol{a}, \boldsymbol{a}^*\rangle \,|\boldsymbol{a}\in C_1\} \geqslant \alpha, \quad \sup\{\langle \boldsymbol{a}, \boldsymbol{a}^*\rangle \,|\boldsymbol{a}\in C_2\} \leqslant \alpha,$$

即为 (2.4.1-1) 式. 因为 $C_1 \not\subset H(\boldsymbol{a},\alpha)$, 那么 $\exists \boldsymbol{a}_1 \in C_1$ 使得 $\langle \boldsymbol{a}_1, \boldsymbol{a}^*\rangle > \alpha$, 即 (2.4.1-2) 式成立. □

下面的定理说明, 如果一个非空相对开凸集与一个非空仿射集相交为空集, 则存在一个超平面分离这两个集合. 需注意的是, 仿射集也是凸集.

定理 2.4.2 *设非空相对开凸集 $C \subset \mathbf{R}^n$, 非空仿射集 $A \subset \mathbf{R}^n$, 且 $A \cap C = \varnothing$, 则存在超平面 $H(\boldsymbol{a}^*,\alpha) = \{\boldsymbol{x}|\langle \boldsymbol{x}, \boldsymbol{a}^*\rangle = \alpha\} \supset A$, 且 $C \subset H^{\geqslant}(\boldsymbol{a}^*,\alpha)$ 或 $C \subset H^{\leqslant}(\boldsymbol{a}^*,\alpha)$, 其中非零向量 $\boldsymbol{a}^* \in \mathbf{R}^n$ 和常数 $\alpha \in \mathbf{R}$.*

证明 由定义 2.1.9 知道任何一个超平面是一个 $n-1$ 维的仿射集. 不妨设 $\{\boldsymbol{x}_1, \boldsymbol{x}_2, \cdots, \boldsymbol{x}_m\}$ 是 A 中含最大数量的仿射无关集合, 那么 $A = \mathrm{aff}\{\boldsymbol{x}_1, \boldsymbol{x}_2, \cdots, \boldsymbol{x}_m\}$, 其中 $\{\boldsymbol{x}_1, \boldsymbol{x}_2, \cdots, \boldsymbol{x}_m\}$ 是 A 的仿射基, 且 $\dim A = m-1$. 由假设得 $m < n$, 否则 $C \subset \mathrm{aff}\{\boldsymbol{x}_1, \boldsymbol{x}_2, \cdots, \boldsymbol{x}_m\}$, 与 $A \cap C = \varnothing$ 矛盾.

下面用归纳法证明.

如果 $m = n-1$, 那么 A 是一个超平面. 不妨设超平面 $H(\boldsymbol{a}^*,\alpha) = \{\boldsymbol{x}|\langle \boldsymbol{x}, \boldsymbol{a}^*\rangle = \alpha\} = A$, 其中非零向量 $\boldsymbol{a}^* \in \mathbf{R}^n$, 常数 $\alpha \in \mathbf{R}$. 因为 $A \cap C = \varnothing$, 对任意的 $\boldsymbol{a} \in C$, 有

$$\langle \boldsymbol{a}, \boldsymbol{a}^*\rangle > \alpha \quad \text{或} \quad \langle \boldsymbol{a}, \boldsymbol{a}^*\rangle < \alpha.$$

如果存在点 $\boldsymbol{a}_1, \boldsymbol{a}_2 \in C$ 使得

$$\langle \boldsymbol{a}_1, \boldsymbol{a}^* \rangle > \alpha \text{ 和 } \langle \boldsymbol{a}_2, \boldsymbol{a}^* \rangle < \alpha \tag{2.4.1-4}$$

同时成立, 那么对任意的 $\lambda \in [0,1]$ 有

$$\phi(\lambda) = \langle \lambda \boldsymbol{a}_1 + (1-\lambda) \boldsymbol{a}_2, \boldsymbol{a}^* \rangle = \lambda \langle \boldsymbol{a}_1, \boldsymbol{a}^* \rangle + (1-\lambda) \langle \boldsymbol{a}_2, \boldsymbol{a}^* \rangle,$$

其中, $\phi(\lambda)$ 是关于 $\lambda \in [0,1]$ 的单调连续增加函数, 由 (2.4.1-4) 式可得 $\phi(0) < \alpha < \phi(1)$. 因此, $\exists \lambda \in (0,1)$ 使得 $\phi(\lambda) = \langle \lambda \boldsymbol{a}_1 + (1-\lambda) \boldsymbol{a}_2, \boldsymbol{a}^* \rangle = \alpha$, 即有 $\lambda \boldsymbol{a}_1 + (1-\lambda) \boldsymbol{a}_2 \in A$, 这与 $A \cap C = \varnothing$ 矛盾. 所以, 定理结论成立.

如果 $m = n-2$(即 $\dim A = m-1 = n-3$), 那么 A 不是一个超平面. 由推论 2.2.1 得 $\mathrm{co}A$ 的点可以由 A 的至多 m 个点的凸组合构成, 即

$$\mathrm{co}A = \left\{ \sum_{i=1}^{m} \lambda_i \boldsymbol{a}_i \middle| \forall \boldsymbol{a}_1, \cdots, \boldsymbol{a}_m \in A, \forall \lambda_i \geqslant 0, \sum_{i=1}^{m} \lambda_i = 1 \right\}. \tag{2.4.1-5}$$

由 $A = \mathrm{aff}\{\boldsymbol{x}_1, \boldsymbol{x}_2, \cdots, \boldsymbol{x}_m\}$, 其中 $\boldsymbol{x}_1, \boldsymbol{x}_2, \cdots, \boldsymbol{x}_m$ 是 A 的仿射基. 显然有

$$\boldsymbol{a}_i = \sum_{j=1}^{m} \lambda_j^i \boldsymbol{x}_j, \quad \sum_{j=1}^{m} \lambda_j^i = 1, \quad i = 1, 2, \cdots, m. \tag{2.4.1-6}$$

由 (2.4.1-5) 和 (2.4.1-6) 式得

$$\sum_{i=1}^{m} \lambda_i \boldsymbol{a}_i = \sum_{i=1}^{m} \lambda_i \sum_{j=1}^{m} \lambda_j^i \boldsymbol{x}_j = \sum_{j=1}^{m} \sum_{i=1}^{m} \lambda_i \lambda_j^i \boldsymbol{x}_j = \sum_{j=1}^{m} \beta_j \boldsymbol{x}_j, \quad \sum_{j=1}^{m} \beta_j = 1,$$

$$\sum_{j=1}^{m} \beta_j = \sum_{j=1}^{m} \sum_{i=1}^{m} \lambda_i \lambda_j^i = \sum_{i=1}^{m} \lambda_i \sum_{j=1}^{m} \lambda_j^i = 1,$$

其中 $\beta_j = \sum\limits_{i=1}^{m} \lambda_i \lambda_j^i$, 则有 $\mathrm{co}A \subseteq A$. 因此, 得

$$\mathrm{co}A = A = \mathrm{aff}\{\boldsymbol{x}_1, \boldsymbol{x}_2, \cdots, \boldsymbol{x}_m\},$$

又得 $\mathrm{co}A \cap C = \varnothing$. 因为 $C = \mathrm{ri}C$, 任取一个点 $\boldsymbol{a} \in C = \mathrm{ri}C$, 则序列 $\{\boldsymbol{a}, \boldsymbol{x}_1, \boldsymbol{x}_2, \cdots, \boldsymbol{x}_m\}$ 是仿射无关的. 否则, $\boldsymbol{a}$ 可以由 $\boldsymbol{x}_1, \boldsymbol{x}_2, \cdots, \boldsymbol{x}_m$ 表示成一个仿射组合, 那么有

$$\dim(\mathrm{aff}\{\boldsymbol{a}, \boldsymbol{x}_1, \boldsymbol{x}_2, \cdots, \boldsymbol{x}_m\}) = n-1.$$

因此, 得到一个仿射集合 $A_1 = \mathrm{aff}\{\boldsymbol{a}, \boldsymbol{x}_1, \boldsymbol{x}_2, \cdots, \boldsymbol{x}_m\}$, 且 $A_1 \cap C_1 = \varnothing$, 其中 $C_1 = C \backslash \{\boldsymbol{a}\}$, 则 C_1 是凸集. 由前面证明知存在一个超平面

$$H_1(\boldsymbol{a}_1^*, \alpha_1) = \{\boldsymbol{x} | \langle \boldsymbol{x}, \boldsymbol{a}_1^* \rangle = \alpha_1\} = A_1 \supset A,$$

且有 $A_1 \cap C_1 = \varnothing$, $C_1 \subset H_1{}^{>}(\boldsymbol{a}_1^*, \alpha_1)$ 或 $C_1 \subset H_1{}^{<}(\boldsymbol{a}_1^*, \alpha_1)$ 只有一个成立, 其中非零向量 $\boldsymbol{a}_1^* \in \mathbf{R}^n$ 和常数 $\alpha_1 \in \mathbf{R}$. 因此, 对任意的 $\boldsymbol{x} \in A$, $\boldsymbol{a} \in A_1$ 有

$$\langle \boldsymbol{x}, \boldsymbol{a}_1^* \rangle = \langle \boldsymbol{a}, \boldsymbol{a}_1^* \rangle = \alpha_1, \quad \langle \boldsymbol{a}_1, \boldsymbol{a}_1^* \rangle > \alpha_1 \text{ 或 } \langle \boldsymbol{a}_1, \boldsymbol{a}_1^* \rangle < \alpha_1, \quad \forall \boldsymbol{a}_1 \in C_1.$$

不妨取 $\alpha = \langle \boldsymbol{a}, \boldsymbol{a}_1^* \rangle$, 显然有 $\langle \boldsymbol{a}_1, \boldsymbol{a}_1^* \rangle \geqslant \alpha$ 或 $\langle \boldsymbol{a}_1, \boldsymbol{a}_1^* \rangle \leqslant \alpha, \forall \boldsymbol{a}_1 \in C$, 则有

$$C \subset H_1{}^{\geqslant}(\boldsymbol{a}_1^*, \alpha) \text{ 或 } C \subset H_1{}^{\leqslant}(\boldsymbol{a}_1^*, \alpha),$$

这时定理结论成立.

如果 $m = n - 3$, 重复上面的过程, 即可得出定理也成立. □

注　在定理 2.4.2 中集合 A 是单点集时结论也成立.

定理 2.4.3　*设非空凸集 $C_1, C_2 \subset \mathbf{R}^n$, 则存在超平面分离 C_1, C_2 的充要条件是 $\mathrm{ri}C_1 \cap \mathrm{ri}C_2 = \varnothing$.*

证明　先证明充分性. 因为非空凸集 $C_1, C_2 \subset \mathbf{R}^n$, 所以 $C = C_1 - C_2$ 也是凸集, 根据性质 2.3.9 得 $\mathrm{ri}C = \mathrm{ri}C_1 - \mathrm{ri}C_2$. 由 $\mathrm{ri}C_1 \cap \mathrm{ri}C_2 = \varnothing$ 知道 $\mathbf{0} \notin \mathrm{ri}C$. 根据定理 2.4.2, 存在一个超平面:

$$H(\boldsymbol{a}^*, \alpha) = \{\boldsymbol{x} | \langle \boldsymbol{x}, \boldsymbol{a}^* \rangle = \alpha\} \supset \{\mathbf{0}\},$$

且 $\mathrm{ri}C \subset H^{\geqslant}(\boldsymbol{a}^*, \alpha)$ 或 $\mathrm{ri}C \subset H^{\leqslant}(\boldsymbol{a}^*, \alpha)$, 其中非零向量 $\boldsymbol{a}^* \in \mathbf{R}^n$, 常数 $\alpha \in \mathbf{R}$. 因为 $C \subset \mathrm{cl}(\mathrm{ri}C)$, 所以 $C \subset H^{\geqslant}(\boldsymbol{a}^*, \alpha)$ 或 $C \subset H^{\leqslant}(\boldsymbol{a}^*, \alpha)$. 因为 $\mathbf{0} \in H(\boldsymbol{a}^*, \alpha)$, 有 $\alpha = 0$, 不妨设 $C \subset H^{\geqslant}(\boldsymbol{a}^*, \alpha)$, 有

$$\langle \boldsymbol{a}_1 - \boldsymbol{a}_2, \boldsymbol{a}^* \rangle \geqslant 0, \quad \forall \boldsymbol{a}_1 \in C_1, \quad \forall \boldsymbol{a}_2 \in C_2,$$

等价于

$$\inf_{\boldsymbol{a}_1 \in C_1} \langle \boldsymbol{a}_1, \boldsymbol{a}^* \rangle \geqslant \sup_{\boldsymbol{a}_2 \in C_2} \langle \boldsymbol{a}_2, \boldsymbol{a}^* \rangle.$$

根据 $\mathbf{0} \notin \mathrm{ri}C$, 显然存在一点 $\boldsymbol{c} = \boldsymbol{c}_1 - \boldsymbol{c}_2, \boldsymbol{c}_1 \in C_1, \boldsymbol{c}_2 \in C_2$ 使得

$$\langle \boldsymbol{c}_1 - \boldsymbol{c}_2, \boldsymbol{a}^* \rangle > 0.$$

由上式即得

$$\sup_{\boldsymbol{c} \in C} \langle \boldsymbol{c}, \boldsymbol{a}^* \rangle = \sup_{\boldsymbol{c}_1 \in C_1} \langle \boldsymbol{c}_1, \boldsymbol{a}^* \rangle - \inf_{\boldsymbol{c}_2 \in C_2} \langle \boldsymbol{c}_2, \boldsymbol{a}^* \rangle \geqslant \langle \boldsymbol{c}_1 - \boldsymbol{c}_2, \boldsymbol{a}^* \rangle > 0.$$

根据定理 2.4.1, $H(\boldsymbol{a}^*,\alpha)$ 分离 C_1,C_2.

下面证明必要性. 设 C_1,C_2 可以被分离. 根据定理 2.4.1, 存在超平面

$$H(\boldsymbol{a}^*,\alpha)=\{\boldsymbol{x}|\langle\boldsymbol{x},\boldsymbol{a}^*\rangle=\alpha\}$$

分离 C_1,C_2 需满足条件

$$\inf\{\langle\boldsymbol{a},\boldsymbol{a}^*\rangle\,|\boldsymbol{a}\in C_1\}\geqslant\sup\{\langle\boldsymbol{a},\boldsymbol{a}^*\rangle\,|\boldsymbol{a}\in C_2\}.$$

由上式即得 $\langle\boldsymbol{a}_1-\boldsymbol{a}_2,\boldsymbol{a}^*\rangle\geqslant 0,\forall\boldsymbol{a}_1\in C_1,\forall\boldsymbol{a}_2\in C_2$. 再令 $D=\{\boldsymbol{x}\,|\langle\boldsymbol{x},\boldsymbol{a}^*\rangle\geqslant 0\}$, 显然有

$$C_1-C_2\subset D\quad 和\quad \mathbf{0}\notin \mathrm{ri}D,$$

得 $\mathrm{ri}D\cap(C_1-C_2)\neq\varnothing$. 利用定理 2.3.1, $\mathrm{ri}\,(C_1-C_2)\subset\mathrm{ri}D$, 所以 $\mathbf{0}\notin\mathrm{ri}C$. □

下面的非严格分离定理是实际中应用得最多的一种结论, 即如果两个凸集不相交, 则存在超平面分离这两个凸集.

推论 2.4.1 设非空凸集 $C_1,C_2\subset\mathbf{R}^n$, 且 $C_1\cap C_2=\varnothing$, 则存在非零向量 $\boldsymbol{a}^*\in\mathbf{R}^n$ 和常数 $\alpha\in\mathbf{R}$ 构成的一个超平面 $H(\boldsymbol{a}^*,\alpha)=\{\boldsymbol{x}|\langle\boldsymbol{x},\boldsymbol{a}^*\rangle=\alpha\}$ 分离 C_1,C_2, 且

$$\langle\boldsymbol{a}_1,\boldsymbol{a}^*\rangle\leqslant\langle\boldsymbol{a}_2,\boldsymbol{a}^*\rangle,\quad\forall\boldsymbol{a}_1\in C_1,\quad\forall\boldsymbol{a}_2\in C_2$$

或

$$\langle\boldsymbol{a}_1,\boldsymbol{a}^*\rangle\geqslant\langle\boldsymbol{a}_2,\boldsymbol{a}^*\rangle,\quad\forall\boldsymbol{a}_1\in C_1,\quad\forall\boldsymbol{a}_2\in C_2.$$

例 2.4.2 设在 $\mathbf{R}^2$ 中凸集 $C_1=\{(\boldsymbol{z}_1,-1)^{\mathrm{T}}\,|\boldsymbol{z}_1<0\}$ 和 $C_2=\{(\boldsymbol{z}_1,0)^{\mathrm{T}}\,|\boldsymbol{z}_1\geqslant 0\}$, $\mathrm{ri}C_1\cap\mathrm{ri}C_2=\varnothing$. 显然超平面 $H=\{(\boldsymbol{z}_1,0)^{\mathrm{T}}\,|\boldsymbol{z}_1\in\mathbf{R}\}$ 分离 C_1,C_2, 且 $C_2\subset H$.

这个例子再次说明两个凸集分离允许两个集合之一 (而不是两个) 包含在分离超平面之中. 进一步, 存在另一个分离定理: 如果一个凸集与另一个凸集的内部不相交, 则两个凸集可分离, 它们可能边界相交, 见例 2.4.1.

定理 2.4.4 设非空凸集 $C_1,C_2\subset\mathbf{R}^n$, 且 $\mathrm{int}C_1\neq\varnothing$, $C_2\cap\mathrm{int}C_1=\varnothing$, 则存在分离 C_1,C_2 的超平面.

证明 因为 $\mathrm{int}C_1$ 是凸集, 根据推论 2.4.1, 存在非零向量 $\boldsymbol{a}^*\in\mathbf{R}^n$ 和常数 $\alpha\in\mathbf{R}$ 构成的一个超平面 $H(\boldsymbol{a}^*,\alpha)=\{\boldsymbol{x}|\langle\boldsymbol{x},\boldsymbol{a}^*\rangle=\alpha\}$ 分离 $\mathrm{int}C_1$ 和 C_2. 再由定理 2.4.1 得

$$\inf\{\langle\boldsymbol{a},\boldsymbol{a}^*\rangle\,|\boldsymbol{a}\in\mathrm{int}C_1\}\geqslant\sup\{\langle\boldsymbol{a},\boldsymbol{a}^*\rangle\,|\boldsymbol{a}\in C_2\},\tag{2.4.1-7}$$

$$\sup\{\langle\boldsymbol{a},\boldsymbol{a}^*\rangle\,|\boldsymbol{a}\in\mathrm{int}C_1\}>\inf\{\langle\boldsymbol{a},\boldsymbol{a}^*\rangle\,|\boldsymbol{a}\in C_2\}.\tag{2.4.1-8}$$

显然, 有

$$\inf\{\langle\boldsymbol{a},\boldsymbol{a}^*\rangle\,|\boldsymbol{a}\in\mathrm{int}C_1\}=\inf\{\langle\boldsymbol{a},\boldsymbol{a}^*\rangle\,|\boldsymbol{a}\in C_1\},$$

$$\sup\{\langle\boldsymbol{a},\boldsymbol{a}^*\rangle\,|\boldsymbol{a}\in\mathrm{int}C_1\}=\sup\{\langle\boldsymbol{a},\boldsymbol{a}^*\rangle\,|\boldsymbol{a}\in C_1\}.$$

因此, (2.4.1-1) 和 (2.4.1-2) 式成立, 所以超平面 $H(\boldsymbol{a}^*,\alpha)=\{\boldsymbol{x}|\langle\boldsymbol{x},\boldsymbol{a}^*\rangle=\alpha\}$ 还是分离 C_1,C_2. □

下面例子说明定理 2.4.4 的内部非空条件必不可少.

例 2.4.3 设在 $\mathbf{R}^2$ 中两个凸集 $C_1=\{(\boldsymbol{z}_1,0)^{\mathrm{T}}\,|\,|\boldsymbol{z}_1|\leqslant 1\}$, $C_2=\{(\boldsymbol{z}_1,\boldsymbol{z}_2)^{\mathrm{T}}\,|\,|\boldsymbol{z}_2|\leqslant 1\}$, 有 $\mathrm{int}C_1=\varnothing$ 和 $C_2\cap\mathrm{int}C_1=\varnothing$, 因 $\mathrm{ri}C_2\cap\mathrm{ri}C_1\neq\varnothing$, 由推论 2.4.1 知不存在超平面分离 C_1,C_2. 即说明在定理 2.4.4 中条件 $\mathrm{int}C_1\neq\varnothing$ 必不可少.

本小节得到了关于两个非空凸集的几个分离定理:

1) 两个凸集可分离的充要条件是 (2.4.1-1) 和 (2.4.1-2) 式成立;

2) 如果一个仿射集与一个非空相对开集不相交, 则它们可分离;

3) 两个凸集可分离的充要条件是两个凸集的相对内部不相交;

4) 存在超平面分离两个不相交的凸集;

5) 如果一个凸集与另一个凸集的非空内部不相交, 则这两个凸集可分离.

2.4.2 超平面强分离定理

本小节讨论两个凸集的强分离定理. 强分离的两个凸集之间的距离存在大于 0 的空隙, 且强分离一定是严格分离的.

定义 2.4.3 *设非空凸集 $C_1,C_2\subset\mathbf{R}^n$. 如果存在一个超平面 $H(\boldsymbol{a},\alpha)=\{\boldsymbol{x}|\langle\boldsymbol{x},\boldsymbol{a}\rangle=\alpha\}$, $\varepsilon>0$ 满足*

$$C_1+\varepsilon B\subset H^{<}(\boldsymbol{a},\alpha),\quad C_2+\varepsilon B\subset H^{>}(\boldsymbol{a},\alpha)$$

或

$$C_2+\varepsilon B\subset H^{<}(\boldsymbol{a},\alpha),\quad C_1+\varepsilon B\subset H^{>}(\boldsymbol{a},\alpha)$$

成立, 则称超平面 H 强分离 C_1,C_2.

根据定义 2.4.3 和推论 2.4.1 知下面性质是显然的.

性质 2.4.1 *设非空凸集 $C_1,C_2\subset\mathbf{R}^n$, 则 C_1,C_2 是强分离的充要条件是 $\exists\varepsilon>0$, 使 $(C_1+\varepsilon B)\cap(C_2+\varepsilon B)=\varnothing$.*

下面的例子说明了 3 种分离情形的区别.

例 2.4.4 (1) 设 $\boldsymbol{a}=(1,0)^{\mathrm{T}}$, 超平面 $H(\boldsymbol{a},1)=\{\boldsymbol{x}\in\mathbf{R}^2|\langle\boldsymbol{x},\boldsymbol{a}\rangle=1\}$ 分离凸集

$$C_1=\left\{(x_1,x_2)^{\mathrm{T}}\,|\,x_1^2+x_2^2\leqslant 1\right\},\quad C_2=\left\{(x_1,x_2)^{\mathrm{T}}\,|\,(x_1-2)^2+x_2^2\leqslant 1\right\},$$

但因为 $C_1\cap C_2=\left\{(1,0)^{\mathrm{T}}\right\}$, $\mathrm{ri}(C_1)\cap\mathrm{ri}(C_2)=\varnothing$, 所以不是严格分离和强分离.

(2) 设 $\boldsymbol{a}=(1,0)^{\mathrm{T}}$, 超平面 $H(\boldsymbol{a},1)=\{\boldsymbol{x}\in\mathbf{R}^2|\langle\boldsymbol{x},\boldsymbol{a}\rangle=1\}$ 严格分离凸集

$$C_1=\left\{(x_1,x_2)^{\mathrm{T}}\,|\,x_1^2+x_2^2<1\right\},\quad C_2=\left\{(x_1,x_2)^{\mathrm{T}}\,|\,(x_1-2)^2+x_2^2\leqslant 1\right\},$$

但不是强分离. 因为 $C_1 \subset H^{<}(\boldsymbol{a},1) = \{\boldsymbol{x}|\langle \boldsymbol{x},\boldsymbol{a}\rangle < 1\}$, $C_2 \subset H^{\geqslant}(\boldsymbol{a},1) = \{\boldsymbol{x}|\langle \boldsymbol{x},\boldsymbol{a}\rangle \geqslant 1\}$, 且由定理 2.4.5 可得 $\inf\{\langle \boldsymbol{x},\boldsymbol{a}\rangle\,|\boldsymbol{x}\in C_1\} = \sup\{\langle \boldsymbol{x},\boldsymbol{a}\rangle\,|\boldsymbol{x}\in C_2\}$.

(3) 设 $\boldsymbol{a} = (1,0)^{\mathrm{T}}$, 超平面 $H(\boldsymbol{a},1) = \{\boldsymbol{x}\in\mathbf{R}^2|\langle \boldsymbol{x},\boldsymbol{a}\rangle = 2\}$ 强分离凸集

$$C_1 = \left\{(x_1,x_2)^{\mathrm{T}} \mid x_1^2 + x_2^2 \leqslant 1\right\}, \quad C_2 = \left\{(x_1,x_2)^{\mathrm{T}} \mid (x_1-4)^2 + x_2^2 \leqslant 1\right\}.$$

上例 (3) 说明强分离是两个凸集之间存在明显距离间隔 (空隙). 我们有下面结论.

定理 2.4.5 设非空凸集 $C_1, C_2 \subset \mathbf{R}^n$, 则存在超平面强分离 C_1, C_2 的充要条件是 $\exists \boldsymbol{a}^* \in \mathbf{R}^n$ 满足

$$\inf\{\langle \boldsymbol{a},\boldsymbol{a}^*\rangle\,|\boldsymbol{a}\in C_1\} > \sup\{\langle \boldsymbol{a},\boldsymbol{a}^*\rangle\,|\boldsymbol{a}\in C_2\}. \tag{2.4.2-1}$$

证明 先证明必要性. 假设 $\exists \boldsymbol{a}^* \in \mathbf{R}^n$, $\alpha \in \mathbf{R}$ 构成的超平面

$$H(\boldsymbol{a}^*,\alpha) = \{\boldsymbol{a}\,|\langle \boldsymbol{a},\boldsymbol{a}^*\rangle = \alpha\}$$

强分离 C_1, C_2. 不妨设 $C_1 + \varepsilon B \subset \{\boldsymbol{a}\,|\langle \boldsymbol{a},\boldsymbol{a}^*\rangle > \alpha\}$, $C_2 + \varepsilon B \subset \{\boldsymbol{a}\,|\langle \boldsymbol{a},\boldsymbol{a}^*\rangle < \alpha\}$, 其中 $\varepsilon > 0$, 有

$$\alpha \leqslant \inf\{\langle \boldsymbol{a},\boldsymbol{a}^*\rangle + \varepsilon\langle \boldsymbol{x},\boldsymbol{a}^*\rangle\,|\boldsymbol{a}\in C_1, \boldsymbol{x}\in B\} < \inf\{\langle \boldsymbol{a},\boldsymbol{a}^*\rangle\,|\boldsymbol{a}\in C_1\},$$

$$\alpha \geqslant \sup\{\langle \boldsymbol{a},\boldsymbol{a}^*\rangle + \varepsilon\langle \boldsymbol{x},\boldsymbol{a}^*\rangle\,|\boldsymbol{a}\in C_2, \boldsymbol{x}\in B\} > \sup\{\langle \boldsymbol{a},\boldsymbol{a}^*\rangle\,|\boldsymbol{a}\in C_2\}.$$

所以 (2.4.2-1) 式成立.

再证明充分性. 假设 $\exists \boldsymbol{a}^* \in \mathbf{R}^n$ 满足 (2.4.2-1) 式, 显然 $\boldsymbol{a}^* \neq \mathbf{0}$. 那么设

$$\underline{\alpha} = \inf\{\langle \boldsymbol{a},\boldsymbol{a}^*\rangle\,|\boldsymbol{a}\in C_1\}, \quad \bar{\alpha} = \sup\{\langle \boldsymbol{a},\boldsymbol{a}^*\rangle\,|\boldsymbol{a}\in C_2\},$$
$$\alpha = \frac{\underline{\alpha}+\bar{\alpha}}{2}, \quad \varepsilon' = \frac{\underline{\alpha}-\bar{\alpha}}{2},$$

得到

$$\langle \boldsymbol{a}_1,\boldsymbol{a}^*\rangle \geqslant \alpha + \varepsilon', \quad \forall \boldsymbol{a}_1 \in C_1;$$
$$\langle \boldsymbol{a}_2,\boldsymbol{a}^*\rangle \leqslant \alpha - \varepsilon', \quad \forall \boldsymbol{a}_2 \in C_2.$$

对于单位球 B 有界, 取 $0 < \varepsilon < \varepsilon'/\|\boldsymbol{a}^*\|$, 有 $|\langle \boldsymbol{x},\boldsymbol{a}^*\rangle| < \varepsilon, \forall \boldsymbol{x} \in \varepsilon B$. 根据上面不等式得

$$\langle \boldsymbol{a}_1+\boldsymbol{x},\boldsymbol{a}^*\rangle = \langle \boldsymbol{a}_1,\boldsymbol{a}^*\rangle + \langle \boldsymbol{x},\boldsymbol{a}^*\rangle \geqslant \alpha+\varepsilon+\langle \boldsymbol{x},\boldsymbol{a}^*\rangle > \alpha, \quad \forall \boldsymbol{a}_1 \in C_1,$$

$$\langle \boldsymbol{a}_2+\boldsymbol{x},\boldsymbol{a}^*\rangle = \langle \boldsymbol{a}_2,\boldsymbol{a}^*\rangle + \langle \boldsymbol{x},\boldsymbol{a}^*\rangle \leqslant \alpha-\varepsilon+\langle \boldsymbol{x},\boldsymbol{a}^*\rangle < \alpha, \quad \forall \boldsymbol{a}_2 \in C_2.$$

故由上面两个不等式得

$$C_1 + \varepsilon B \subset \{\boldsymbol{a}\,|\langle \boldsymbol{a}, \boldsymbol{a}^* \rangle > \alpha\},$$

$$C_2 + \varepsilon B \subset \{\boldsymbol{a}\,|\langle \boldsymbol{a}, \boldsymbol{a}^* \rangle < \alpha\}.$$

所以根据定义 2.4.3 知超平面 $H(\boldsymbol{a}^*, \alpha) = \{\boldsymbol{a}\,|\langle \boldsymbol{a}, \boldsymbol{a}^* \rangle = \alpha\}$ 强分离 C_1, C_2. □

下面是另一个强分离定理.

定理 2.4.6 设非空凸集 $C_1, C_2 \subset \mathbf{R}^n$, 则存在超平面强分离 C_1, C_2 的充要条件是

$$\mathbf{0} \notin \mathrm{cl}(C_1 - C_2),$$

$$\inf\{|\boldsymbol{a}_1 - \boldsymbol{a}_2|\,|\,\forall \boldsymbol{a}_1 \in C_1, \forall \boldsymbol{a}_2 \in C_2\} > 0.$$

证明 由性质 2.4.1 知 C_1, C_2 是强分离的充要条件是 $\exists \varepsilon > 0$ 使得

$$(C_1 + \varepsilon B) \cap (C_2 + \varepsilon B) = \varnothing,$$

其中 B 是闭单位球. 因此, C_1, C_2 是强分离的等价条件是 $\exists \varepsilon > 0$ 使得

$$\mathbf{0} \notin (C_1 + \varepsilon B) - (C_2 + \varepsilon B) = C_1 - C_2 + 2\varepsilon B.$$

上式等价于 $2\varepsilon B \cap (C_1 - C_2) = \varnothing$, 再等价于 $\mathbf{0} \notin \mathrm{cl}(C_1 - C_2)$. □

事实上, 不相交的两个闭凸集, 不一定是强分离的. 但是, 如果不相交的两个闭凸集, 其中一个是紧集, 则它们是强分离的.

定理 2.4.7 设非空紧凸集 $C_1 \subset \mathbf{R}^n$, 非空闭凸集 $C_2 \subset \mathbf{R}^n$, 则存在超平面强分离 C_1, C_2 的充要条件是 C_1, C_2 不相交.

证明 先证明必要性. 根据性质 2.4.1, 如果超平面 H 强分离 C_1, C_2, $\exists \varepsilon > 0$ 使

$$(C_1 + \varepsilon B) \cap (C_2 + \varepsilon B) = \varnothing.$$

因为 B 包含零点, 所以, $C_1 \cap C_2 = \varnothing$.

再证明充分性. 设 $C_1 \cap C_2 = \varnothing$, 根据推论 2.4.1 知, 存在非零向量 $\boldsymbol{a}^* \in \mathbf{R}^n$ 和常数 $\alpha \in \mathbf{R}$ 构成的一个超平面 $H(\boldsymbol{a}^*, \alpha) = \{\boldsymbol{x}|\langle \boldsymbol{x}, \boldsymbol{a}^* \rangle = \alpha\}$ 分离 C_1, C_2, 且

$$\inf\{\langle \boldsymbol{a}, \boldsymbol{a}^* \rangle\,|\boldsymbol{a} \in C_1\} \geqslant \sup\{\langle \boldsymbol{a}, \boldsymbol{a}^* \rangle\,|\boldsymbol{a} \in C_2\}.$$

因为 C_1 是紧集和 C_2 是闭集, 那么存在 $\bar{\boldsymbol{a}} \in C_1$ 和 $\underline{\boldsymbol{a}} \in C_2$ 使得

$$\langle \bar{\boldsymbol{a}}, \boldsymbol{a}^* \rangle = \inf\{\langle \boldsymbol{a}, \boldsymbol{a}^* \rangle\,|\boldsymbol{a} \in C_1\} \quad \text{和} \quad \langle \underline{\boldsymbol{a}}, \boldsymbol{a}^* \rangle = \sup\{\langle \boldsymbol{a}, \boldsymbol{a}^* \rangle\,|\boldsymbol{a} \in C_2\}.$$

因为 $\bar{\boldsymbol{a}} \neq \underline{\boldsymbol{a}}$, 有 $\delta = \langle \bar{\boldsymbol{a}}, \boldsymbol{a}^* \rangle - \langle \underline{\boldsymbol{a}}, \boldsymbol{a}^* \rangle > 0$. 取 $\varepsilon < \dfrac{\delta}{2}$, 则有 $(C_1 + \varepsilon B) \cap (C_2 + \varepsilon B) = \varnothing$. 否则, 存在 $\tilde{\boldsymbol{a}} + \varepsilon \boldsymbol{x} \in (C_1 + \varepsilon B) \cap (C_2 + \varepsilon B)$, 其中 $\boldsymbol{x} \in B$, 则得

$$\langle \bar{\boldsymbol{a}}, \boldsymbol{a}^* \rangle + \langle \varepsilon \boldsymbol{x}, \boldsymbol{a}^* \rangle \leqslant \langle \tilde{\boldsymbol{a}} + \varepsilon \boldsymbol{x}, \boldsymbol{a}^* \rangle = \langle \tilde{\boldsymbol{a}}, \boldsymbol{a}^* \rangle + \langle \varepsilon \boldsymbol{x}, \boldsymbol{a}^* \rangle \leqslant \langle \underline{\boldsymbol{a}}, \boldsymbol{a}^* \rangle + \langle \varepsilon \boldsymbol{x}, \boldsymbol{a}^* \rangle .$$

由上式得 $\langle \bar{\boldsymbol{a}}, \boldsymbol{a}^* \rangle - \langle \underline{\boldsymbol{a}}, \boldsymbol{a}^* \rangle \leqslant 0$, 产生矛盾. 由性质 2.4.1 知, 超平面 H 强分离 C_1, C_2. □

下面证明两个非空紧集可以强分离的充要条件是两个集合的凸包不相交.

定理 2.4.8 设非空紧集 $C_1, C_2 \subset \mathbf{R}^n$, 则存在超平面强分离 C_1, C_2 的充要条件是 $\mathrm{co}C_1 \cap \mathrm{co}C_2 = \varnothing$.

证明 先证明必要性. 由定理 2.4.5 知, 存在一个超平面 $H(\boldsymbol{a}^*, \alpha) = \{\boldsymbol{x} | \langle \boldsymbol{x}, \boldsymbol{a}^* \rangle = \alpha\}$ 强分离 C_1, C_2, 则有

$$\inf\{\langle \boldsymbol{a}_1, \boldsymbol{a}^* \rangle | \boldsymbol{a}_1 \in C_1\} > \alpha > \sup\{\langle \boldsymbol{a}_2, \boldsymbol{a}^* \rangle | \boldsymbol{a}_2 \in C_2\}. \tag{2.4.2-2}$$

设 $\forall \boldsymbol{x} \in \mathrm{co}C_1$, $\exists \boldsymbol{x}_i \in C_1\,(i = 1, \cdots, m)$ 使得

$$\boldsymbol{x} = \lambda_1 \boldsymbol{x}_1 + \cdots + \lambda_m \boldsymbol{x}_m,$$

其中 $\lambda_i \geqslant 0, \sum\limits_{i=1}^{m} \lambda_i = 1$, 则

$$\langle \boldsymbol{x}, \boldsymbol{a}^* \rangle = \lambda_1 \langle \boldsymbol{x}_1, \boldsymbol{a}^* \rangle + \cdots + \lambda_m \langle \boldsymbol{x}_m, \boldsymbol{a}^* \rangle > \lambda_1 \alpha + \cdots + \lambda_m \alpha = \alpha.$$

类似可得, $\forall \boldsymbol{y} \in \mathrm{co}C_2$ 有 $\langle \boldsymbol{y}, \boldsymbol{a}^* \rangle < \alpha$. 因此, 有

$$\inf\{\langle \boldsymbol{x}, \boldsymbol{a}^* \rangle | \boldsymbol{x} \in \mathrm{co}C_1\} > \sup\{\langle \boldsymbol{y}, \boldsymbol{a}^* \rangle | \boldsymbol{y} \in \mathrm{co}C_2\},$$

所以 $\mathrm{co}C_1$, $\mathrm{co}C_2$ 是强分离的, 即有 $\mathrm{co}C_1 \cap \mathrm{co}C_2 = \varnothing$.

再证明充分性. 假设 $\mathrm{co}C_1 \cap \mathrm{co}C_2 = \varnothing$. 因为 $\mathrm{co}C_1$ 和 $\mathrm{co}C_2$ 是紧集, 由定理 2.4.7 知, 存在超平面 H 强分离 $\mathrm{co}C_1$, $\mathrm{co}C_2$, 即 $\exists \varepsilon > 0$, 有

$$(\mathrm{co}C_1 + \varepsilon B) \cap (\mathrm{co}C_2 + \varepsilon B) = \varnothing.$$

如果 $\boldsymbol{x} \in (C_1 + \varepsilon B) \cap (C_2 + \varepsilon B)$, 因为 B 包含零点, 再由 $C_1 \subset \mathrm{co}C_1$, $C_2 \subset \mathrm{co}C_2$ 成立. 则有

$$\boldsymbol{x} \in (\mathrm{co}C_1 + \varepsilon B) \cap (\mathrm{co}C_2 + \varepsilon B)$$

成立, 得矛盾.

因此, $(C_1 + \varepsilon B) \cap (C_2 + \varepsilon B) = \varnothing$. 由定理 2.4.5 知存在超平面 H 强分离 C_1, C_2. □

注 下面是定理 2.4.8 的 3 种特殊的情形:

(1) 若 $C_1=\{\boldsymbol{a}\}$ 是独点集, 则存在超平面强分离 $\boldsymbol{a}$ 和 C_2 的充要条件是 $\boldsymbol{a}\notin \mathrm{co}C_2$;

(2) 若 $C_1=\{\boldsymbol{a}\}$ 是独点集, 则存在超平面强分离 $\boldsymbol{a}$ 和 C_2 的充要条件是存在超平面强分离 $\boldsymbol{a}$ 和由 C_2 中最多 $n+1$ 个点构成的每一个集合;

(3) 存在超平面强分离 C_1,C_2 的充要条件是存在超平面强分离 C_1 和由 C_2 中最多 $n+1$ 个点构成的每一个集合.

下面是定理 2.4.6、定理 2.4.7 和定理 2.4.8 的反例.

例 2.4.5 设 $\mathbf{R}^2$ 中有两个凸集 $C_1=\{(x_1,0)^{\mathrm{T}}\mid x_1<0\}$, $C_2=\{(x_1,x_2)^{\mathrm{T}}\mid x_1\geqslant 0, x_2\in\mathbf{R}\}$. 显然有

$$\mathrm{ri}\,(C_1)\cap \mathrm{ri}\,(C_2)=\varnothing,$$

设 $\boldsymbol{a}=(1,0)^{\mathrm{T}}$, 超平面 $H\,(\boldsymbol{a},0)=\{\boldsymbol{x}|\langle\boldsymbol{x},\boldsymbol{a}\rangle=0\}$ 分离凸集 C_1,C_2. 但显然不是强分离, 因为 C_1 不是凸紧集, 而且 $\inf\{|\boldsymbol{a}_1-\boldsymbol{a}_2|\mid\forall\boldsymbol{a}_1\in C_1,\forall\boldsymbol{a}_2\in C_2\}=0$.

本小节得到了两个非空凸集强分离的几个性质:

1) 两个非空凸集强分离的充要条件是它们存在不相交的邻域集, 或者是存在一个非零 $\boldsymbol{a}^*$ 满足 (2.4.2-1) 式, 又或者是 $\mathbf{0}$ 不属于两个集合差的闭包;

2) 非空紧凸集与非空闭凸集强分离的充要条件是两个集合不相交;

3) 两个非空紧集强分离的充要条件是两个集合的凸包不相交.

2.5 支撑超平面

本节讨论闭凸集与支撑超平面的关系, 本质上支撑超平面是凸集分离定理的另一种形式. 即总是存在超平面使得一个闭凸集位于该平面的一个半空间中.

2.5.1 基本性质

支撑超平面是由包含它的所有超平面交构成的. 一般情形, 凸集的支撑超平面具有以下特点: 凸集与支撑超平面相交, 且被包含在超平面的半空间中.

定义 2.5.1 设非空凸集 $C\subset\mathbf{R}^n$. 如果存在一个非零向量 $\boldsymbol{x}^*\in\mathbf{R}^n$ 和常数 $\alpha\in\mathbf{R}$ 构成的超平面 $H\,(\boldsymbol{x}^*,\alpha)=\{\boldsymbol{x}\,|\langle\boldsymbol{x},\boldsymbol{x}^*\rangle=\alpha\}$ 满足下面条件:

$$C\subset H^{\leqslant}\,(\boldsymbol{x}^*,\alpha)=\{\boldsymbol{x}\,|\langle\boldsymbol{x},\boldsymbol{x}^*\rangle\leqslant\alpha\}\quad (\text{或}C\subset H^{\geqslant}\,(\boldsymbol{x}^*,\alpha)=\{\boldsymbol{x}\,|\langle\boldsymbol{x},\boldsymbol{x}^*\rangle\geqslant\alpha\}),$$

且 $H\,(\boldsymbol{x}^*,\alpha)\cap C\neq\varnothing$ 成立, 则称 $H\,(\boldsymbol{x}^*,\alpha)$ 为 C 的支撑超平面, 称 $H^{\leqslant}\,(\boldsymbol{x}^*,\alpha)$(或 $H^{\geqslant}\,(\boldsymbol{x}^*,\alpha)$) 为 C 的支撑半空间. 如果 $H\,(\boldsymbol{x}^*,\alpha)$ 是凸集 C 的支撑超平面, 且 $C\subset H\,(\boldsymbol{x}^*,\alpha)$, 则称 $H\,(\boldsymbol{x}^*,\alpha)$ 为非正常支撑超平面. 否则, 如果 $C\subset H\,(\boldsymbol{x}^*,\alpha)$ 不成立, 则称 $H\,(\boldsymbol{x}^*,\alpha)$ 为正常支撑超平面.

例 2.5.1 设 $\boldsymbol{a}=(1,0)^{\mathrm{T}}$, 超平面 $H(\boldsymbol{a},1)=\{\boldsymbol{x}|\langle\boldsymbol{x},\boldsymbol{a}\rangle=1\}$ 是凸集

$$C=\left\{(x_1,x_2)^{\mathrm{T}} \mid x_1^2+x_2^2\leqslant 1\right\}$$

的支撑超平面.

由定义 2.5.1 知有下面的性质成立.

性质 2.5.1 $H(\boldsymbol{x}^*,\alpha)$ 是凸集 C 的支撑超平面等价于存在一个 $\boldsymbol{x}_0\in C$ 和非零 $\boldsymbol{x}^*\in\mathbf{R}^n$ 使得 $\langle\boldsymbol{x}_0,\boldsymbol{x}^*\rangle=\alpha$, 对于任意的 $\boldsymbol{x}\in C$ 有 $\langle\boldsymbol{x},\boldsymbol{x}^*\rangle\leqslant\alpha$(或 $\langle\boldsymbol{x},\boldsymbol{x}^*\rangle\geqslant\alpha$) 成立.

性质 2.5.2 $H(\boldsymbol{x}^*,\alpha)$ 是凸集 C 的正常支撑超平面等价于对任意的 $\boldsymbol{x}\in C$ 有 $\langle\boldsymbol{x},\boldsymbol{x}^*\rangle\leqslant\alpha$(或 $\langle\boldsymbol{x},\boldsymbol{x}^*\rangle\geqslant\alpha$), 且至少存在点 $\boldsymbol{x}',\boldsymbol{x}''\in C$ 使得

$$\langle\boldsymbol{x}',\boldsymbol{x}^*\rangle=\alpha \text{ 和 } \langle\boldsymbol{x}'',\boldsymbol{x}^*\rangle<\alpha \quad (\text{或 } \langle\boldsymbol{x}'',\boldsymbol{x}^*\rangle>\alpha).$$

性质 2.5.3 如果 C 是闭凸集, 则存在一个非零向量 $\boldsymbol{x}^*\in\mathbf{R}^n$ 和常数 $\alpha\in\mathbf{R}$ 构成的超平面 $H(\boldsymbol{x}^*,\alpha)=\{\boldsymbol{x}|\langle\boldsymbol{x},\boldsymbol{x}^*\rangle=\alpha\}$ 是 C 的支撑超平面的充要条件是 $\alpha=\max\{\langle\boldsymbol{x},\boldsymbol{x}^*\rangle|\forall\boldsymbol{x}\in C\}$(或 $\alpha=\min\{\langle\boldsymbol{x},\boldsymbol{x}^*\rangle|\forall\boldsymbol{x}\in C\}$) 成立.

证明 根据性质 2.5.1 可知结论显然成立. □

性质 2.5.4 如果 C 是闭凸集, 则存在一个非零向量 $\boldsymbol{x}^*\in\mathbf{R}^n$ 和常数 $\alpha\in\mathbf{R}$ 构成的超平面 $H(\boldsymbol{x}^*,\alpha)=\{\boldsymbol{x}|\langle\boldsymbol{x},\boldsymbol{x}^*\rangle=\alpha\}$ 是 C 的正常支撑超平面的充要条件是

$$\inf_{\boldsymbol{x}\in C}\langle\boldsymbol{x},\boldsymbol{x}^*\rangle<\max_{\boldsymbol{x}\in C}\langle\boldsymbol{x},\boldsymbol{x}^*\rangle. \tag{2.5.1-1}$$

事实上, 只有闭凸集才存在支撑超平面, 举例如下.

例 2.5.2 $C=\left\{(x_1,x_2)^{\mathrm{T}}\in\mathbf{R}^2|x_1^2+x_2^2<1\right\}$ 是内部非空开凸集, 显然不存在支撑超平面.

在定义 2.5.1 中有一种情形: $H(\boldsymbol{x}^*,\alpha)\cap C=\varnothing$, 且 $C\subset H^{\leqslant}(\boldsymbol{x}^*,\alpha)=\{\boldsymbol{x}|\langle\boldsymbol{x},\boldsymbol{x}^*\rangle\leqslant\alpha\}$(或 $C\subset H^{\geqslant}(\boldsymbol{x}^*,\alpha)=\{\boldsymbol{x}|\langle\boldsymbol{x},\boldsymbol{x}^*\rangle\geqslant\alpha\}$) 成立, 下面的结论是更弱的情形.

定理 2.5.1 设 C 是非空凸集, 且存在一个非零向量 $\boldsymbol{x}^*\in\mathbf{R}^n$ 和常数 $\alpha\in\mathbf{R}$ 构成的超平面 $H(\boldsymbol{x}^*,\alpha)=\{\boldsymbol{x}|\langle\boldsymbol{x},\boldsymbol{x}^*\rangle=\alpha\}$, 则 $H(\boldsymbol{x}^*,\alpha)\cap\mathrm{ri}C=\varnothing$ 成立的充要条件是

$$C\subset H^{\leqslant}(\boldsymbol{x}^*,\alpha)=\{\boldsymbol{x}|\langle\boldsymbol{x},\boldsymbol{x}^*\rangle\leqslant\alpha\} \quad (\text{或}C\subset H^{\geqslant}(\boldsymbol{x}^*,\alpha)=\{\boldsymbol{x}|\langle\boldsymbol{x},\boldsymbol{x}^*\rangle\geqslant\alpha\}),$$

且 $C\subset H(\boldsymbol{x}^*,\alpha)$ 不成立.

证明 先证明必要性. 首先设 $H(\boldsymbol{x}^*,\alpha)\cap \mathrm{ri}C=\varnothing$ 成立. 由于 $\mathrm{ri}C\neq\varnothing$, 显然 $C\subset H(\boldsymbol{x}^*,\alpha)$ 不成立, 即 C 不包含在 H 之中. 假设 $C\subset H^{\leqslant}(\boldsymbol{x}^*,\alpha)=\{\boldsymbol{x}\,|\langle\boldsymbol{x},\boldsymbol{x}^*\rangle\leqslant\alpha\}$ 不成立, 那么 $\mathrm{ri}C\subset H^{\leqslant}(\boldsymbol{x}^*,\alpha)$ 也不成立, 则存在两点 $\boldsymbol{x}'\in\mathrm{ri}C,\boldsymbol{x}''\in C$ 使得

$$\langle\boldsymbol{x}',\boldsymbol{x}^*\rangle<\alpha<\langle\boldsymbol{x}'',\boldsymbol{x}^*\rangle. \tag{2.5.1-2}$$

对于 $\lambda\in[0,1]$, 设函数 $\phi(\lambda)=\langle(1-\lambda)\boldsymbol{x}'+\lambda\boldsymbol{x}'',\boldsymbol{x}^*\rangle$, 显然它关于 λ 是严格单调递增函数. 存在 $\boldsymbol{x}_\lambda=(1-\lambda)\boldsymbol{x}'+\lambda\boldsymbol{x}''$ 使得 $\phi(\lambda)=\langle\boldsymbol{x}_\lambda,\boldsymbol{x}^*\rangle=\alpha$, 故 $\boldsymbol{x}_\lambda\in H(\boldsymbol{x}^*,\alpha)$. 再根据定理 2.3.2 得 $\boldsymbol{x}_\lambda\in\mathrm{ri}C$, 所以 $\boldsymbol{x}_\lambda\in H(\boldsymbol{x}^*,\alpha)\cap\mathrm{ri}C$, 得到矛盾. 若 $C\subset H^{\geqslant}(\boldsymbol{x}^*,\alpha)=\{\boldsymbol{x}\,|\langle\boldsymbol{x},\boldsymbol{x}^*\rangle\geqslant\alpha\}$ 不成立, 同样可以推出矛盾.

下面证明充分性. 假设存在点 $\boldsymbol{x}_1\in H(\boldsymbol{x}^*,\alpha)\cap\mathrm{ri}C$. 不妨设

$$C\subset H^{\leqslant}(\boldsymbol{x}^*,\alpha)=\{\boldsymbol{x}\,|\langle\boldsymbol{x},\boldsymbol{x}^*\rangle\leqslant\alpha\},$$

且 $C\subset H(\boldsymbol{x}^*,\alpha)$ 不成立. 因此, 存在点 $\boldsymbol{x}_2\in C\backslash\{H(\boldsymbol{x}^*,\alpha)\}$ 满足 $\langle\boldsymbol{x}_2,\boldsymbol{x}^*\rangle<\alpha$, 对于 $\lambda\in[0,1]$, 定义

$$\boldsymbol{x}_\lambda=(1-\lambda)\boldsymbol{x}_1+\lambda\boldsymbol{x}_2,$$

那么有

$$\begin{aligned}\langle\boldsymbol{x}_\lambda,\boldsymbol{x}^*\rangle&=\langle(1-\lambda)\boldsymbol{x}_1+\lambda\boldsymbol{x}_2,\boldsymbol{x}^*\rangle\\&=(1-\lambda)\langle\boldsymbol{x}_1,\boldsymbol{x}^*\rangle+\lambda\langle\boldsymbol{x}_2,\boldsymbol{x}^*\rangle<(1-\lambda)\alpha+\lambda\alpha=\alpha.\end{aligned}$$

因为 $\boldsymbol{x}_1\in H(\boldsymbol{x}^*,\alpha)\cap\mathrm{ri}C$, 对于充分小的 λ, 再根据定理 2.3.2 得

$$\boldsymbol{x}_\lambda\in H(\boldsymbol{x}^*,\alpha)\cap\mathrm{ri}C,$$

因此, 有 $\langle\boldsymbol{x}_\lambda,\boldsymbol{x}^*\rangle=\alpha$ 成立, 这与前面不等式矛盾, 所以定理得证. □

下面讨论凸集存在支撑超平面的充要条件.

定理 2.5.2 设凸集 $C,D\subset\mathbf{R}^n$, 且 $D\subset C$, 则 C 存在包含 D 的正常支撑超平面的充要条件是 $D\cap\mathrm{ri}C=\varnothing$.

证明 先证明必要性. 如果 C 存在包含 D 的正常支撑超平面, 即存在一个非零向量 $\boldsymbol{x}^*\in\mathbf{R}^n$ 和常数 $\alpha\in\mathbf{R}$ 构成的包含 D 的超平面 $H(\boldsymbol{x}^*,\alpha)=\{\boldsymbol{x}\,|\langle\boldsymbol{x},\boldsymbol{x}^*\rangle=\alpha\}$, 使得

$$C\subset H^{\leqslant}(\boldsymbol{x}^*,\alpha)=\{\boldsymbol{x}\,|\langle\boldsymbol{x},\boldsymbol{x}^*\rangle\leqslant\alpha\}\quad(\text{或 } C\subset H^{\geqslant}(\boldsymbol{x}^*,\alpha)=\{\boldsymbol{x}\,|\langle\boldsymbol{x},\boldsymbol{x}^*\rangle\geqslant\alpha\}),$$

且 $C\subset H(\boldsymbol{x}^*,\alpha)$ 不成立. 说明 D 和 C 分离, 那么根据分离定理 (定理 2.4.3) 得等价条件

$$\mathrm{ri}D\cap\mathrm{ri}C=\varnothing.$$

因为 $D\subset C$, 那么 $\mathrm{ri}D\subset\mathrm{ri}C$, 可得 $\mathrm{ri}D=\varnothing$, 即 $D\cap\mathrm{ri}C=\varnothing$.

再证明充分性. 如果 $D\cap\mathrm{ri}C=\varnothing$, 显然有 $\mathrm{ri}D\cap\mathrm{ri}C=\varnothing$, 根据分离定理 (定理 2.4.3) 知道 C 存在包含 D 的正常支撑超平面. □

例 2.5.3 在 $\mathbf{R}^2$ 中设 $C=\{(x_1,x_2)^{\mathrm{T}}\mid 0\leqslant x_1,x_2\leqslant 1\}$ 和 $D=\{(x_1,1)^{\mathrm{T}}\mid 0\leqslant x_1\leqslant 1\}$, 显然 $D\subset C$, 且

$$\mathrm{ri}C=\left\{(x_1,x_2)^{\mathrm{T}}|0<x_1,x_2<1\right\},\quad D\cap\mathrm{ri}C=\varnothing.$$

取 $\boldsymbol{a}=(0,1)^{\mathrm{T}}$, 超平面 $H(\boldsymbol{a},1)=\{\boldsymbol{x}|\langle\boldsymbol{x},\boldsymbol{a}\rangle=1\}$ 是 C 的支撑超平面, 且包含 D.

定理 2.5.3 设闭凸集 $C\subset\mathbf{R}^n$, $\forall\boldsymbol{a}\in\mathrm{rb}C$, 则 C 存在过 $\boldsymbol{a}$ 的正常支撑超平面.

证明 由 $\boldsymbol{a}\in\mathrm{rb}C$ 知道 $\boldsymbol{a}$ 是 C 的边界点, 即 $\boldsymbol{a}\notin\mathrm{ri}C$ 和 $\mathrm{rb}C=\mathrm{cl}C\backslash\mathrm{ri}C$. 根据定理 2.4.2 得, 存在一个超平面 $H(\boldsymbol{a}^*,\alpha)=\{\boldsymbol{x}|\langle\boldsymbol{x},\boldsymbol{a}^*\rangle=\alpha\}\supset\boldsymbol{a}$, 且 $\mathrm{ri}C\subset H^{\geqslant}(\boldsymbol{a}^*,\alpha)$ 或 $\mathrm{ri}C\subset H^{\leqslant}(\boldsymbol{a}^*,\alpha)$, 其中非零向量 $\boldsymbol{a}^*\in\mathbf{R}^n$, 常数 $\alpha\in\mathbf{R}$, 即 $H(\boldsymbol{a}^*,\alpha)\cap\mathrm{ri}C=\varnothing$ 成立. 由定理 2.5.1 知结论成立. □

例 2.5.4 在 $\mathbf{R}^2$ 中设 $C=\left\{(x_1,x_2)^{\mathrm{T}}\mid 0\leqslant x_1,x_2\leqslant 1\right\}$, 有

$$\begin{aligned}\mathrm{rb}C=&\left\{(x_1,0)^{\mathrm{T}}|0\leqslant x_1\leqslant 1\right\}\cup\left\{(x_1,1)^{\mathrm{T}}|0\leqslant x_1\leqslant 1\right\}\\&\cup\left\{(0,x_2)^{\mathrm{T}}|0\leqslant x_2\leqslant 1\right\}\cup\left\{(1,x_2)^{\mathrm{T}}|0\leqslant x_2\leqslant 1\right\},\end{aligned}$$

显然取 $\boldsymbol{a}^*=(0,1)^{\mathrm{T}}$, 超平面 $H(\boldsymbol{a}^*,1)=\{\boldsymbol{x}|\langle\boldsymbol{x},\boldsymbol{a}^*\rangle=1\}$ 是 C 的支撑超平面, 过

$$\left\{(x_1,1)^{\mathrm{T}}\mid 0\leqslant x_1\leqslant 1\right\}\subset\mathrm{rb}C.$$

下面的定理说明, 如果一个内部非空的闭集对每一个边界点上都存在支撑超平面, 那么这个集合一定是凸集.

定理 2.5.4 设非空内部的闭集 $C\subset\mathbf{R}^n$, 并且 C 的每一个边界点都存在通过该点的支撑超平面, 则 C 是一定凸集.

证明 显然 $C=\mathbf{R}^n$ 是凸集, 不妨设 $C\neq\mathbf{R}^n$. 首先假设 $\mathrm{int}C$ 不是凸集, 那么一定存在由两个点 $\boldsymbol{x},\boldsymbol{y}\in\mathrm{int}C$ 构成的直线

$$C_{\boldsymbol{x},\boldsymbol{y}}=\{\boldsymbol{x}_\lambda|\boldsymbol{x}_\lambda=(1-\lambda)\boldsymbol{x}+\lambda\boldsymbol{y},\lambda\in[0,1]\}$$

中还存在两点 $\boldsymbol{x}_{\lambda_1},\boldsymbol{x}_{\lambda_2}\in C_{\boldsymbol{x},\boldsymbol{y}}$, 其中 $0<\lambda_1<\lambda_2<1$, 使得产生下面 3 个集合:

$$C^1_{\boldsymbol{x},\boldsymbol{y}}=\{\boldsymbol{x}_\lambda|\boldsymbol{x}_\lambda=(1-\lambda)\boldsymbol{x}+\lambda\boldsymbol{y},\lambda\in[0,\lambda_1]\},$$

$$C_{\boldsymbol{x},\boldsymbol{y}}^{2}=\{\boldsymbol{x}_{\lambda}|\boldsymbol{x}_{\lambda}=(1-\lambda)\,\boldsymbol{x}+\lambda\boldsymbol{y},\lambda\in[\lambda_2,1]\}\,,$$

$$C_{\boldsymbol{x},\boldsymbol{y}}^{3}=\{\boldsymbol{x}_{\lambda}|\boldsymbol{x}_{\lambda}=(1-\lambda)\,\boldsymbol{x}+\lambda\boldsymbol{y},\lambda\in(\lambda_1,\lambda_2)\}\,,$$

其中 $C_{\boldsymbol{x},\boldsymbol{y}}^{3}\not\subset C$, $C_{\boldsymbol{x},\boldsymbol{y}}^{1}\subset C$, $C_{\boldsymbol{x},\boldsymbol{y}}^{2}\subset C$. 显然 $\boldsymbol{x}_{\lambda_1},\boldsymbol{x}_{\lambda_2}$ 是边界点, 根据定理的假设存在一个 $C_{\boldsymbol{x},\boldsymbol{y}}^{1}$ 的支撑超平面 $H(\boldsymbol{a}^*,\alpha)=\{\boldsymbol{x}|\langle\boldsymbol{x},\boldsymbol{a}^*\rangle=\alpha\}$ 使得 $\langle\boldsymbol{x}_{\lambda_1},\boldsymbol{a}^*\rangle=\alpha$. 不妨设 $C\subset H^{\leqslant}(\boldsymbol{a}^*,\alpha)$, 即有

$$\langle\boldsymbol{c},\boldsymbol{a}^*\rangle\leqslant\langle\boldsymbol{x}_{\lambda_1},\boldsymbol{a}^*\rangle=\alpha,\quad\forall\boldsymbol{c}\in C.\tag{2.5.1-3}$$

所以有 $\langle\boldsymbol{x},\boldsymbol{a}^*\rangle\leqslant\alpha$, $\langle\boldsymbol{y},\boldsymbol{a}^*\rangle\leqslant\alpha$. 由 (2.5.1-3) 式必然有 $\langle\boldsymbol{x},\boldsymbol{a}^*\rangle=\alpha$, $\langle\boldsymbol{y},\boldsymbol{a}^*\rangle=\alpha$ 成立. 否则若 $\langle\boldsymbol{x},\boldsymbol{a}^*\rangle<\alpha$, 则有 $\alpha=\langle\boldsymbol{x}_{\lambda_1},\boldsymbol{a}^*\rangle=\langle(1-\lambda_1)\,\boldsymbol{x}+\lambda_1\boldsymbol{y},\boldsymbol{a}^*\rangle<\alpha$, 得到矛盾. 由 (2.5.1-3) 式可得

$$\langle\boldsymbol{c},\boldsymbol{a}^*\rangle=\alpha,\quad\forall\boldsymbol{c}\in C_{\boldsymbol{x},\boldsymbol{y}}.\tag{2.5.1-4}$$

因为 $\boldsymbol{x},\boldsymbol{y}\in\mathrm{int}C$, 那么一定存在一个充分小的 $\varepsilon\in(0,1)$ 使得

$$C_{\boldsymbol{x},\boldsymbol{y}}\cap(\boldsymbol{x}+\varepsilon B)\neq\varnothing.$$

必然有某个 $t\in(0,\varepsilon)$ 和非零向量 $\boldsymbol{b}\in B$ 使得 $\boldsymbol{x}+t\boldsymbol{b}\in C_{\boldsymbol{x},\boldsymbol{y}}\cap(\boldsymbol{x}+\varepsilon B)$, 则由 (2.5.1-4) 式得到

$$\langle\boldsymbol{x}+t\boldsymbol{b},\boldsymbol{a}^*\rangle=\alpha,$$

由此得到 $t=0$ 的矛盾. 所以 $\mathrm{int}C$ 是凸集. 根据性质 2.3.2 可得 C 是凸集. □

下面举例说明定理 2.5.4 中 C 是内部非空条件不能去掉.

例 2.5.5 $C=\{(x_1,x_2)^{\mathrm{T}}\in\mathbf{R}^2|x_1^2+x_2{}^2=1\}$ 是非凸闭集, 但内部是空集.

由定理 2.4.3 和定理 2.5.4 有下面结论成立.

推论 2.5.1 设非空内部的闭集 $C\subset\mathbf{R}^n$, 则 C 是凸集的充要条件是在 C 的每一个边界点均存在 C 的支撑超平面.

定理 2.5.5 设闭凸集 $C\subset\mathbf{R}^n$, 则 C 是它所有支撑超平面的交.

证明 通过推论 2.5.1 知, 若 $\dim C\geqslant1$, 则 $C\neq\mathbf{R}^n$, C 包含在它所有超平面的交中. 因此有

$$C\subset\cap H(\boldsymbol{a}^*,\alpha)\tag{2.5.1-5}$$

成立, 其中超平面 $H(\boldsymbol{a}^*,\alpha)$ 包含 C. 反过来上式也成立, 假设 $\exists\boldsymbol{y}\notin C$ 但 $\boldsymbol{y}\in\cap H(\boldsymbol{a}^*,\alpha)$, 那么由定理 2.4.4 可得, 存在可分离 $\boldsymbol{y}$ 与 C 的超平面 H', 且 $\boldsymbol{y}\notin H'$, 但

$$H'\subset\cap H(\boldsymbol{a}^*,\alpha),$$

这显然矛盾. 当 $\dim C = -1$ 或 0 时, 或 $C = \mathbf{R}^n$ 时, 则 (2.5.1-5) 式是等式, 所以结论成立. □

推论 2.5.2 设 $S \subset \mathbf{R}^n$, 则 $\mathrm{cl}\,(\mathrm{co}S)$ 是 S 的所有支撑超平面的交.

本小节得到支撑超平面判定的一些性质:

1) 凸集 C 的支撑超平面等价于存在一个 $\boldsymbol{x}_0 \in C$ 和非零 $\boldsymbol{x}^* \in \mathbf{R}^n$ 使得 $\langle \boldsymbol{x}_0, \boldsymbol{x}^* \rangle = \alpha$, 对于任意的 $\boldsymbol{x} \in C$ 有 $\langle \boldsymbol{x}, \boldsymbol{x}^* \rangle \leqslant \alpha$(或 $\langle \boldsymbol{x}, \boldsymbol{x}^* \rangle \geqslant \alpha$) 成立;

2) 凸集 C 存在正常支撑超平面的充要条件是 (2.5.1-1) 式成立;

3) 凸集 C 存在包含凸集 D 的正常支撑超平面的充要条件是 D 与 C 的相对内部不相交;

4) 任何闭凸集都存在过其边界点的支撑超平面;

5) 若非空内部的闭集的每个边界点都存在通过该点的支撑超平面, 则该集合是凸集;

6) 闭凸集是所有包含它的支撑超平面的交.

2.5.2 极点与面

本小节专门讨论闭凸集的极点与面的性质, 其中凸集的正常面与支撑超平面有关. 当 C 是凸紧集 (有界闭凸集) 时, 它可以表示为其 0-维面的凸包. 根据极点定义 (定义 2.2.7), 来定义面的概念.

定义 2.5.2 设闭凸集 $C \subset \mathbf{R}^n$, C 的全体极点构成的集合记为 $\mathrm{ext}C$ 或 $\mathrm{ext}\,(C)$ 或 $\mathrm{ext}\,\{C\}$.

定义 2.5.3 设闭凸集 $C \subset \mathbf{R}^n$, 凸集 $F \subset C$. 如果对任意的 $\boldsymbol{x}, \boldsymbol{y} \in C$, 当 $\mathrm{ri}\,[\boldsymbol{x}, \boldsymbol{y}] \cap F \neq \varnothing$ 时, 有 $[\boldsymbol{x}, \boldsymbol{y}] \subset F$ 和 $\dim F = k$ 成立, 则称 F 是 k-维面. 进一步, 若 $\dim F = \dim C - 1$, 则称 F 是超面. 称 C 的极点是 0-维面. 除 $\varnothing$ 和 C 外, C 的真子集构成的面称为正常面.

例 2.5.6 在 $\mathbf{R}^2$ 中设集合

$$C = \left\{ (x_1, x_2)^{\mathrm{T}} \,|x_1^2 + x_2^2 \leqslant 1, x_1, x_2 \geqslant 0 \right\},$$

$$F = \left\{ (x_1, x_2)^{\mathrm{T}} \,|0 \leqslant x_1 \leqslant 1, x_2 = 0 \right\}.$$

设 $\boldsymbol{x} = (x_1, 0)^{\mathrm{T}}\,(0 \leqslant x_1 \leqslant 1)$, $\boldsymbol{y} = (y_1, 0)^{\mathrm{T}}\,(0 \leqslant x_1 < y_1 \leqslant 1)$, 有 $\mathrm{ri}\,[\boldsymbol{x}, \boldsymbol{y}] \cap F \neq \varnothing$, 且 $[\boldsymbol{x}, \boldsymbol{y}] \subset F$, 则 F 是 1-维面. 如果 $F = \left\{ (x_1, x_2)^{\mathrm{T}} \,|0.5 \leqslant x_1 \leqslant 1, x_2 = 0 \right\}$, 则 F 不是 1-维面, 因为如果设 $\boldsymbol{x} = (0.1, 0)^{\mathrm{T}}$ 和 $\boldsymbol{y} = (1, 0)^{\mathrm{T}}$, 虽然有 $\mathrm{ri}\,[\boldsymbol{x}, \boldsymbol{y}] \cap F \neq \varnothing$, 但条件 $[\boldsymbol{x}, \boldsymbol{y}] \subset F$ 却不成立.

正常面上包含了无穷多点, 因此有下面的性质.

性质 2.5.5　有下面结论成立:

(1) $\varnothing$ 和闭凸集 C 是 C 的面;

(2) 极点是一个 0-维面.

定理 2.5.6　设闭凸集 $C \subset \mathbf{R}^n$, $\dim C \geqslant 1$. 如果 $\exists \boldsymbol{x}^* \in \mathbf{R}^n$ 和 $\alpha \in \mathbf{R}$ 使得 $H(\boldsymbol{x}^*, \alpha)$ 是 C 的一个正常支撑超平面, 则 $F = H(\boldsymbol{x}^*, \alpha) \cap C$ 是 C 的正常面.

证明　因为 $H(\boldsymbol{x}^*, \alpha)$ 是 C 的正常支撑超平面, 根据定义 2.5.1, F 一定是 C 的正常非空的凸子集. 假设 $\boldsymbol{x}, \boldsymbol{y} \in C$ 使得 $\mathrm{ri}\,[\boldsymbol{x}, \boldsymbol{y}] \cap F \neq \varnothing$, 则 $\exists \lambda, 0 < \lambda < 1$, 使得

$$(1-\lambda)\,\boldsymbol{x} + \lambda \boldsymbol{y} \in H(\boldsymbol{x}^*, \alpha).$$

不妨设 $C \subset H^{\leqslant}(\boldsymbol{x}^*, \alpha)$, 有

$$\langle \boldsymbol{x}, \boldsymbol{x}^* \rangle \leqslant \alpha, \quad \langle \boldsymbol{y}, \boldsymbol{x}^* \rangle \leqslant \alpha, \quad \langle (1-\lambda)\,\boldsymbol{x} + \lambda \boldsymbol{y}, \boldsymbol{x}^* \rangle = \alpha.$$

从而得 $\langle \boldsymbol{x}, \boldsymbol{x}^* \rangle = \langle \boldsymbol{y}, \boldsymbol{x}^* \rangle = \alpha$. 说明 $\boldsymbol{x}, \boldsymbol{y} \in H(\boldsymbol{x}^*, \alpha)$, 有 $[\boldsymbol{x}, \boldsymbol{y}] \subset F$, 因此, F 是 C 的正常面. □

定理 2.5.6 告诉我们, 闭凸集的正常面只需通过边界点上的支撑超平面寻找即可, 见下面例子.

例 2.5.7　在 $\mathbf{R}^2$ 中设集合

$$C = \left\{ (x_1, x_2)^{\mathrm{T}} \,|\, x_1^2 + x_2^2 \leqslant 1, x_1, x_2 \geqslant 0 \right\},$$

$$F = \left\{ (x_1, x_2)^{\mathrm{T}} \,|\, 0 \leqslant x_1 \leqslant 1, x_2 = 0 \right\}.$$

设支撑超平面 $H(\boldsymbol{x}^*, 0) = \{ \boldsymbol{x} |\, \langle \boldsymbol{x}, \boldsymbol{x}^* \rangle = 0 \}$ 和 $\boldsymbol{x}^* = (0, 1)^{\mathrm{T}}$, 有 $F = H(\boldsymbol{x}^*, \alpha) \cap C$, 则 F 是正常面.

下面证明闭凸集的面的一些性质.

性质 2.5.6　设闭凸集 $C \subset \mathbf{R}^n$, F 是 C 的一个面, 有下面性质成立:

(1) F 是闭集;

(2) $G \subset F$ 是 F 的面当且仅当 G 是 C 的面;

(3) 如果 $F \subset C$, 则 $F \subset \mathrm{rb}C$, 即面一定是在凸集的相对边界上;

(4) 若 $G \subset F$ 是 C 的面且 $G \neq F$, 则 $G \subset \mathrm{rb}F$ 和 $\dim G < \dim F$;

(5) $\boldsymbol{x} \in \mathrm{ri}F$ 当且仅当 F 是包含 $\boldsymbol{x}$ 的最小面;

(6) 对任意的 $\boldsymbol{x} \in C$, 在 C 中存在唯一的面 F 使得 $\boldsymbol{x} \in \mathrm{ri}F$.

证明　当 $\dim C = -1$ 或 0 时, 性质中的结论在通常情况下都成立. 以下均假设 $\dim C \geqslant 1$.

(1) 因为 $\mathrm{ri}F \neq \varnothing$, 设 $\forall \boldsymbol{x} \in \mathrm{cl}F$, $\boldsymbol{x}_0 \in \mathrm{ri}F$, 不妨设 $\boldsymbol{x} \neq \boldsymbol{x}_0$. 由定理 2.3.2 得

$$(1-\lambda)\,\boldsymbol{x}_0 + \lambda\boldsymbol{x} \in \mathrm{ri}F, \quad 0 \leqslant \lambda < 1.$$

当 $0 < \lambda < 1$ 时, $\{(1-\lambda)\,\boldsymbol{x}_0 + \lambda\boldsymbol{x}\,|0 < \lambda < 1\} \cap F \neq \varnothing$. 由面的定义 2.5.3 知, 线段 $[\boldsymbol{x}_0, \boldsymbol{x}]$ 被包含在 F 中, 即 $\boldsymbol{x} \in F$, 因此 F 是闭集.

(2) 由定义 2.5.3 知, 若 G 是 C 的面, 再根据 (1), 显然 G 也是 F 的面. 下面假设 G 是 F 的面, $\exists \boldsymbol{x}, \boldsymbol{y} \in C$ 使得

$$\{(1-\lambda)\,\boldsymbol{x} + \lambda\boldsymbol{y}\,|\,0 < \lambda < 1\} \cap G \neq \varnothing.$$

因此, 由上式得

$$\{(1-\lambda)\,\boldsymbol{x} + \lambda\boldsymbol{y}\,|\,0 < \lambda < 1\} \cap F \neq \varnothing.$$

所以 F 也是 C 的面, 故 $[\boldsymbol{x}, \boldsymbol{y}] \in F$. 但 G 是 F 的面, 又有 $[\boldsymbol{x}, \boldsymbol{y}] \subset G$, 故 G 也是 C 的面.

(3) 因为 $\mathrm{rb}C = \mathrm{cl}C \backslash \mathrm{ri}C$. 假设 $\exists \boldsymbol{x} \in F$ 使得 $\boldsymbol{x} \in \mathrm{ri}C$, 且 $F \subset C$. 那么对 C 中 $\forall \boldsymbol{y} \neq \boldsymbol{x}$ 点, 由定理 2.3.2 知道, 对 $\boldsymbol{x} \in \mathrm{ri}C$, $\boldsymbol{y} \in C$ 及 $0 \leqslant \lambda < 1$, 有

$$(1-\lambda)\,\boldsymbol{x} + \lambda\boldsymbol{y} \in \mathrm{ri}C,$$

即 $\mathrm{ri}\,[\boldsymbol{x}, \boldsymbol{y}] \cap F \neq \varnothing$. 因为 F 是 C 的面, 从而有 $[\boldsymbol{x}, \boldsymbol{y}] \subset F$, 即 $\boldsymbol{y} \in F$, 因此 $F \supset C$. 这与 $F \subset C$ 矛盾, 因此 $F \subset \mathrm{rb}C$.

(4) 由 (2) 和 (3) 即得.

(5) 设 $\boldsymbol{x} \in \mathrm{ri}F$. 假设还存在 C 的面 F' 使得 $\boldsymbol{x} \in F'$, 且 $F' \subset F$, 则 $F' \cap \mathrm{ri}F \neq \varnothing$. 由 (4) 知道 $F' = F$. 所以 F 是包含 $\boldsymbol{x}$ 的最小面. 反过来, 若 $\boldsymbol{x} \in \mathrm{rb}F$, 由 (2) 知存在 F 的面 G 使得 $\boldsymbol{x} \in G$ 和 $G \neq F$, 同时 G 也是 C 的面, 那么 F 不是包含 $\boldsymbol{x}$ 的最小面.

(6) 由 (5) 直接得结论成立. □

上述性质表明, 对闭凸集 C 的任何子集 M 都存在 C 包含 M 的最小面, 这个最小面也是包含 M 的所有面的交. 并且如果 M 包含 $\mathrm{ri}C$ 中的点, 那么包含 M 的最小面就是 C 的本身.

定义 2.5.4 设闭凸集 $C \subset \mathbf{R}^n$. 如果存在 $\boldsymbol{x}^* \in \mathbf{R}^n$ 和 $\alpha \in \mathbf{R}$ 使得 $H(\boldsymbol{x}^*, \alpha)$ 是 C 的一个正常支撑超平面, 则称 $F = H(\boldsymbol{x}^*, \alpha) \cap C$ 是 C 的正常暴露面. 如果 $\{\boldsymbol{x}\}$ 是 C 的暴露面, 则称 $\boldsymbol{x}$ 是暴露点, 所有暴露点构成的集合记为 $\exp C$. 显然有 $\exp C \subset \mathrm{ext}C$.

例 2.5.8 设下面集合

$$C = \left\{(x_1, x_2)^{\mathrm{T}}\,|x_1^2 + x_2^2 \leqslant 1, x_1, x_2 \geqslant 0\right\}, \quad C_1 = \left\{(x_1, x_2)^{\mathrm{T}}\,|0 \leqslant x_1 \leqslant 1, x_2 = 0\right\},$$

$C_2=\left\{(x_1,x_2)^{\mathrm{T}}\,|0\leqslant x_2\leqslant 1, x_1=0\right\}$, $C_3=\left\{(x_1,x_2)^{\mathrm{T}}\,|x_1^2+x_2^2=1, x_1,x_2>0\right\}$.

那么 C_1 和 C_2 都是 C 的暴露面, C_3 中所有点既是 C 的极点, 又是暴露点. 点 $(0,0)^{\mathrm{T}}$, $(0,1)^{\mathrm{T}}$, $(1,0)^{\mathrm{T}}\in C$ 是极点, 但不是暴露点, 说明极点不一定是暴露点.

定理 2.5.7 若 F 是 $\mathbf{R}^n$ 中闭凸集 C 的超面, 则 F 是暴露面.

证明 因为 F 是超面, 由性质 2.5.6(5) 知 $\exists \boldsymbol{x}\in \mathrm{ri}F$, 所以 F 是 C 的包含 $\boldsymbol{x}$ 的最小面. 假设 G 是 C 的暴露面使得 $F\subset G\subset C$, 有 $\boldsymbol{x}\in G$. 因 $G\neq C$, 故由性质 2.5.6(4) 知

$$\mathrm{dim}C-1=\mathrm{dim}F\leqslant \mathrm{dim}G<\mathrm{dim}C.$$

所以 $\mathrm{dim}G=\mathrm{dim}F$, 因此有 $F=G$. □

定理 2.5.7 说明闭凸集的超面都是暴露面. 进一步, 下面证明闭凸集所有正常暴露面的交集也是正常暴露面.

定理 2.5.8 设 $\{F_i\,|i\in I\}$ 是闭凸集 $C\subset\mathbf{R}^n$ 的正常暴露面族, 其中 I 是指标集, 则 $F=\bigcap_{i\in I}F_i$ 是 C 的正常暴露面.

证明 假设 $I=\{1,2\}$. 根据暴露面的定义 (定义 2.5.4), 存在 $\boldsymbol{x}_i^*\in\mathbf{R}^n, \alpha_i\in \mathbf{R}\,(i=1,2)$, 还存在两个支撑超平面 $H\left(\boldsymbol{x}_i^*,\alpha_i\right)=\{\boldsymbol{x}\,|\langle\boldsymbol{x},\boldsymbol{x}_i^*\rangle=\alpha_i\}\,(i=1,2)$ 及闭半空间 $H^{\leqslant}\left(\boldsymbol{x}_i^*,\alpha_i\right)=\{\boldsymbol{x}\,|\langle\boldsymbol{x},\boldsymbol{x}_i^*\rangle\leqslant\alpha_i\}$ 满足

$$\forall \boldsymbol{y}_i\in F_i=H\left(\boldsymbol{x}_i^*,\alpha_i\right)\cap C,\quad \langle\boldsymbol{y}_i,\boldsymbol{x}_i^*\rangle=\alpha_i, \tag{2.5.2-1}$$

$$C\subset H^{\leqslant}\left(\boldsymbol{x}_i^*,\alpha_i\right). \tag{2.5.2-2}$$

设 $\boldsymbol{x}_3^*=\dfrac{1}{2}\left(\boldsymbol{x}_1^*+\boldsymbol{x}_2^*\right), \alpha_3=\dfrac{1}{2}\left(\alpha_1+\alpha_2\right)$. 如果

$$\boldsymbol{y}\in F_1\cap F_2=H\left(\boldsymbol{x}_1^*,\alpha_1\right)\cap H\left(\boldsymbol{x}_2^*,\alpha_2\right)\cap C,$$

那么有 $\langle\boldsymbol{y},\boldsymbol{x}_3^*\rangle=\left\langle\boldsymbol{y},\dfrac{1}{2}\left(\boldsymbol{x}_1^*+\boldsymbol{x}_2^*\right)\right\rangle=\dfrac{1}{2}\left(\alpha_1+\alpha_2\right)=\alpha_3$, 即有

$$F_1\cap F_2\subset H\left(\boldsymbol{x}_3^*,\alpha_3\right)\cap C.$$

反过来, 设 $\boldsymbol{y}\in H\left(\boldsymbol{x}_3^*,\alpha_3\right)\cap C$, 有 $\langle\boldsymbol{y},\boldsymbol{x}_3^*\rangle=\left\langle\boldsymbol{y},\dfrac{1}{2}\left(\boldsymbol{x}_1^*+\boldsymbol{x}_2^*\right)\right\rangle=\alpha_3$. 同时, 由 (2.5.2-2) 式有

$$\langle\boldsymbol{y},\boldsymbol{x}_1^*\rangle\leqslant\alpha_1,\quad \langle\boldsymbol{y},\boldsymbol{x}_2^*\rangle\leqslant\alpha_2.$$

由此可得: $\langle\boldsymbol{y},\boldsymbol{x}_1^*\rangle=\alpha_1, \langle\boldsymbol{y},\boldsymbol{x}_2^*\rangle=\alpha_2$, 则有

$$\boldsymbol{y}\in F_1\cap F_2=H\left(\boldsymbol{x}_1^*,\alpha_1\right)\cap H\left(\boldsymbol{x}_2^*,\alpha_2\right)\cap C.$$

再设对任意的 $\boldsymbol{x} \in C$, 有

$$\langle \boldsymbol{x}, \boldsymbol{x}_3^* \rangle = \left\langle \boldsymbol{x}, \frac{1}{2}(\boldsymbol{x}_1^* + \boldsymbol{x}_2^*) \right\rangle \leqslant \frac{1}{2}(\alpha_1 + \alpha_2) = \alpha_3,$$

即得 $C \subset H^{\leqslant}(\boldsymbol{x}_3^*, \alpha_3)$. 最后得 $F_1 \cap F_2 = H(\boldsymbol{x}_3^*, \alpha_3) \cap C$, 所以 F 是 C 的暴露面.

如果 $I = \{1, 2, 3\}$, 根据上面内容易证 $F_1 \cap F_2$ 是 C 的暴露面, 重复过程可得 $F_1 \cap F_2 \cap F_3$ 是 C 的暴露面. 对于 $I = \{1, 2, \cdots, m\}$, 不断重复可得 $F = \bigcap_{i \in I} F_i$ 是 C 的正常暴露面.

如果 I 是无限指标集. 不妨在 $\{F_i \,|i \in I\}$ 中取 m 个集合 $\{F_1, F_2, \cdots, F_m\}$, 其中 $m > n$, 如果 $F = F_1$, 结论得证. 否则有 $F \subset F_1$, 且 $F \neq F_1$. 如果 $F = F_1 \cap F_2$, 前面证明定理结论也成立, 否则有

$$F \subset F_1 \cap F_2 \subset F_1,$$

$$F_1 \cap F_2 \neq F_1.$$

由性质 2.5.6(4) 得 $\dim(F_1 \cap F_2) < \dim F_1$. 同样如果 $F = F_1 \cap F_2 \cap F_3$, 前面证明定理结论也成立. 否则有

$$F \subset F_1 \cap F_2 \cap F_3 \subset F_1 \cap F_2 \subset F_1,$$

$$F_1 \cap F_2 \cap F_3 \neq F_1 \cap F_2.$$

再由性质 2.5.6(4) 得知

$$\dim(F_1 \cap F_2 \cap F_3) < \dim(F_1 \cap F_2).$$

与上述过程类似地, 可以对 $F_4, F_5, \cdots, F_m$ 继续下去, 这样每经过一步, 维数至少降低一维, 因为不可能降低到 0, 所以在有限步后, 可得 $F = \bigcap_{i \in I} F_i$. □

最后, 讨论凸紧集的极点集性质. 我们知道, 闭半空间和仿射集是没有极点的闭凸集. 对于凸紧集情形就不同了, 下面的定理说明凸紧集是由其极点“张”成的, 即非空凸紧集是由凸集所有的极点集凸包组成的, 这个结论也称为凸集内表示.

定理 2.5.9 设非空凸紧集 $C \subset \mathbf{R}^n$, 则 $C = \mathrm{co}(\mathrm{ext}C)$.

证明 显然 $\mathrm{ext}C \subset C$, 由于 C 是凸集, 那么 $\mathrm{co}(\mathrm{ext}C) \subset C$ 和 $\mathrm{cl}(\mathrm{co}(\mathrm{ext}C)) \subset C$ 成立.

反过来证明 $C \subset \mathrm{co}(\mathrm{ext}C)$ 成立. 设 $\forall \boldsymbol{x} \in C$, $\boldsymbol{x} \notin \mathrm{co}(\mathrm{ext}C)$ 成立, 即 $\boldsymbol{x}$ 不是极点的凸组合. 任取 $\boldsymbol{y} \in \mathrm{ext}C$, 构造直线:

$$C_{\boldsymbol{x},\boldsymbol{y}} = \{\boldsymbol{x}_\lambda | \boldsymbol{x}_\lambda = \boldsymbol{y} + \lambda(\boldsymbol{x} - \boldsymbol{y}), \lambda > 0\}.$$

由于 C 是紧集, 则一定存在一个 $\boldsymbol{x}_\lambda$ 是 C 的边界点. 记 $\boldsymbol{y}_0=\boldsymbol{x}_\lambda, \mu=\dfrac{1}{\lambda}$, 那么得到

$$\boldsymbol{x}=(1-\mu)\,\boldsymbol{y}+\mu\boldsymbol{y}_0,\quad \boldsymbol{x}\in \mathrm{ri}\left\{(1-t)\,\boldsymbol{y}+t\boldsymbol{y}_0\,|0\leqslant t\leqslant 1\right\}.$$

由性质 2.5.6(6) 知, 存在一个支撑超平面 $H\left(\boldsymbol{a}^*,\alpha\right)=\{\boldsymbol{x}|\langle\boldsymbol{x},\boldsymbol{a}^*\rangle=\alpha\}$ 使得

$$\boldsymbol{x}\in \mathrm{ri}F=\mathrm{ri}\left(H\left(\boldsymbol{x}^*,\alpha\right)\cap C\right),\quad \langle\boldsymbol{x},\boldsymbol{x}^*\rangle=\alpha, \tag{2.5.2-3}$$

$$C\subset H^{\leqslant}\left(\boldsymbol{x}^*,\alpha\right). \tag{2.5.2-4}$$

由上两式知 F 是包含了 $\boldsymbol{y}_0,\boldsymbol{y}$ 的 C 的正常暴露面. 但由性质 2.5.6(3) 知 $\boldsymbol{x}\in \mathrm{rb}C$, 由性质 2.3.4(6) 知 $\mathrm{ri}C$ 非空. 因此, $\mathrm{ri}C\subset \mathrm{co}\,(\mathrm{ext}C)$, 再由性质 2.3.4(1) 知

$$C=\mathrm{cl}\,(\mathrm{ri}C)\subset \mathrm{cl}\,(\mathrm{co}\,(\mathrm{ext}C)).$$

由此可得 $\mathrm{cl}\,(\mathrm{co}\,(\mathrm{ext}C))=C$. 再利用性质 2.3.13 得 $\mathrm{co}\,(\mathrm{ext}C)=C$. □

下面证明任何凸紧集等于非空子集的凸包等价于所有极点集属于这个子集.

定理 2.5.10　设非空凸紧集 $C\subset \mathbf{R}^n$, M 是 C 的非空子集, 则 $C=\mathrm{co}M$ 当且仅当 $\mathrm{ext}C\subset M$.

证明　如果 $C=\mathrm{co}M$ 成立, 则 $\mathrm{ext}C\subset M$ 不成立. 假设存在一个极点 $\boldsymbol{x}\notin M$, 由 $M\subset C\backslash\{\boldsymbol{x}\}$ 和极点的定义知 $C\backslash\{\boldsymbol{x}\}$ 还是凸集, 则有 $\mathrm{co}M\subset C\backslash\{\boldsymbol{x}\}$, 得到矛盾. 因此, 有 $\mathrm{ext}C\subset M$.

反过来, 如果 $\mathrm{ext}C\subset M$, 由定理 2.5.9 得 $C=\mathrm{co}\,(\mathrm{ext}C)$, 且

$$C=\mathrm{co}\,(\mathrm{ext}C)\subset \mathrm{co}M\subset C.$$

所以 $C=\mathrm{co}M$. □

推论 2.5.3　设凸紧集 $C\subset \mathbf{R}^n$, 且 $\mathrm{dim}C=n$, 则 C 的每一点均由 C 的至多 $n+1$ 个极点的凸组合组成.

证明　由定理 2.5.9 知道 $\mathrm{ext}C$ 至少存在 n 个线性无关点. 否则, $\dim\,(\mathrm{ext}C)<n$, 即有 $\dim\,(\mathrm{co}\,(\mathrm{ext}C))<n$, 得到矛盾. □

推论 2.5.3 说明, 维数等于 n 的凸紧集是由凸集的至多 $n+1$ 个极点的凸组合组成的.

本小节得到了凸集与极点、面和暴露面之间的一些关系:

1) 如果一个维数大于零的凸集存在一个正常支撑超平面, 那么这个超平面与凸集的交构成一个正常面;

2) 闭凸集 C 的每一个面都是闭集, 且这个面的任何子集也是面等价于子集也是 C 的面;

3) 凸集的真子集属于它的相对边界;

4) 一个面的每个相对内点等价于这个面是包含该点的最小面;

5) 闭凸集的超面都是暴露面, 且闭凸集所有正常暴露面的交集也是正常暴露面;

6) 非空凸紧集是由凸集所有的极点集凸包组成的;

7) 任何凸紧集等于非空子集的凸包等价于所有极点集属于这个子集.

2.5.3 配极

下面讨论支撑超平面理论中集合的配极性质. 集合的配极是由包含集合的超平面半空间交构成的, 即配极由与集合中所有点内积不大于 1 的全体元素构成.

定义 2.5.5 设集合 $S\subset\mathbf{R}^n$, 则称

$$S^\circ=\{\boldsymbol{y}\in\mathbf{R}^n\,|\langle\boldsymbol{x},\boldsymbol{y}\rangle\leqslant 1,\boldsymbol{x}\in S\}=\left\{\boldsymbol{y}\in\mathbf{R}^n\left|\sup_{\boldsymbol{x}\in S}\langle\boldsymbol{x},\boldsymbol{y}\rangle\leqslant 1\right.\right\}\tag{2.5.3-1}$$

是 S 的配极. 若 S° 是 S 的配极, 且 $(S^\circ)^\circ$ 是 S° 的配极, 则称 $(S^\circ)^\circ$ 是 S 的双配极. 记 $S^{\circ\circ}=(S^\circ)^\circ$.

配极是一个非常有意义的概念, 它将非凸集合对应到一个闭凸集上, 这对于研究非凸集理论具有重要意义.

例 2.5.9 设集合 $S=\left\{(x_1,x_2)^{\mathrm{T}}\in\mathbf{R}^2|x_1+x_2=1,x_1,x_2\geqslant 0\right\}$, 考虑下面 3 种情形:

(1) 当 $1\geqslant y_1>y_2$ 时, 有 $\sup\limits_{(x_1,x_2)\in S}\{x_1y_1+x_2y_2\}=y_1\leqslant 1$;

(2) 当 $1\geqslant y_2>y_1$ 时, 有 $\sup\limits_{(x_1,x_2)\in S}\{x_1y_1+x_2y_2\}=y_2\leqslant 1$;

(3) 当 $1\geqslant y_1=y_2$ 时, 有 $\sup\limits_{(x_1,x_2)\in S}\{x_1y_1+x_2y_2\}=y_1\leqslant 1$.

因此, 有

$$\begin{aligned}S^\circ&=\left\{\boldsymbol{y}\in\mathbf{R}^2\,|\sup\{x_1y_1+x_2y_2\}\leqslant 1\text{ s.t. }x_1+x_2=1,x_1,x_2\geqslant 0\right\}\\&=\left\{(y_1,y_2)^{\mathrm{T}}\,|y_1\leqslant y_2\leqslant 1\right\}\cup\left\{(y_1,y_2)^{\mathrm{T}}\,|y_2\leqslant y_1\leqslant 1\right\}\\&=\left\{(y_1,y_2)^{\mathrm{T}}\,|y_1\leqslant 1\right\}\cap\left\{(y_1,y_2)^{\mathrm{T}}\,|y_2\leqslant 1\right\}.\end{aligned}$$

上式表明了 S 的配极 S° 是支撑超平面的交.

事实上, 由定义 2.5.5 可得下式成立:

$$S^\circ=\bigcap_{\boldsymbol{x}\in S}H^{\leqslant}(\boldsymbol{x},1),\tag{2.5.3-2}$$

其中 $H^{\leqslant}(\boldsymbol{x},1)=\{\boldsymbol{y}\in\mathbf{R}^n\,|\langle\boldsymbol{y},\boldsymbol{x}\rangle\leqslant 1\}$. 设 $\forall\boldsymbol{x}\in S$, 如果 $\boldsymbol{y}\in H^{\leqslant}(\boldsymbol{x},1)$, 则有 $\langle\boldsymbol{y},\boldsymbol{x}\rangle\leqslant 1$, 即 $\boldsymbol{y}\in S^{\circ}$, 因此有 $\bigcap_{\boldsymbol{x}\in S}H^{\leqslant}(\boldsymbol{x},1)\subset S^{\circ}$. 反过来, 如果 $\boldsymbol{y}\in S^{\circ}$, 则有 $\langle\boldsymbol{x},\boldsymbol{y}\rangle\leqslant 1,\boldsymbol{x}\in S$, 即 $\boldsymbol{y}\in H^{\leqslant}(\boldsymbol{x},1)$, 得 $\bigcap_{\boldsymbol{x}\in S}H^{\leqslant}(\boldsymbol{x},1)\supset S^{\circ}$. 由配极定义获得下面性质.

性质 2.5.7 设非空集合 $S,S_1,S_2\subset\mathbf{R}^n$, 有下列结论成立:

(1) $(S_1\cup S_2)^{\circ}=S_1^{\circ}\cap S_2^{\circ}$;

(2) 如果 $S_1\subset S_2$, 则 $S_2^{\circ}\subset S_1^{\circ}$;

(3) 如果 $\lambda>0$, 则 $(\lambda S^{\circ})=\dfrac{1}{\lambda}S^{\circ}$;

(4) S° 是包含原点的闭凸集;

(5) 如果 S 有界, 则 $\mathbf{0}$ 是 S° 的内点;

(6) 如果 $\mathbf{0}$ 是 S 的内点, 则 S° 有界.

证明 (1) 由配极的定义 (定义 2.5.5), 设

$$\boldsymbol{y}\in(S_1\cup S_2)^{\circ}=\left\{\boldsymbol{y}\,\middle|\,\sup_{\boldsymbol{x}\in S_1\cup S_2}\langle\boldsymbol{x},\boldsymbol{y}\rangle\leqslant 1,\boldsymbol{x}\in S_1\cup S_2\right\}.$$

显然有 $\sup\limits_{\boldsymbol{x}\in S_1}\langle\boldsymbol{x},\boldsymbol{y}\rangle\leqslant\sup\limits_{\boldsymbol{x}\in S_1\cup S_2}\langle\boldsymbol{x},\boldsymbol{y}\rangle\leqslant 1$ 和 $\sup\limits_{\boldsymbol{x}\in S_2}\langle\boldsymbol{x},\boldsymbol{y}\rangle\leqslant\sup\limits_{\boldsymbol{x}\in S_1\cup S_2}\langle\boldsymbol{x},\boldsymbol{y}\rangle\leqslant 1$ 同时成立, 这等价于同时有 $\boldsymbol{y}\in S_1^{\circ}$ 和 $\boldsymbol{y}\in S_2^{\circ}$. 因此, $(S_1\cup S_2)^{\circ}=S_1^{\circ}\cap S_2^{\circ}$.

(2) 因 $S_1\subset S_2$, 则 $S_2=S_1\cup(S_2\backslash S_1)$, 那么由 (1) 知 $S_2^{\circ}=S_1^{\circ}\cap(S_2\backslash S_1)^{\circ}\subset S_1^{\circ}$.

(3) 设 $\forall\boldsymbol{x}\in(\lambda S)^{\circ}$. 由配极的定义知, 对任意的 $\lambda\boldsymbol{y}\in\lambda S$ 有 $\langle\lambda\boldsymbol{y},\boldsymbol{x}\rangle\leqslant 1$, 等价于 $\langle\boldsymbol{y},\lambda\boldsymbol{x}\rangle\leqslant 1$. 从而, 等价于 $\lambda\boldsymbol{x}\in S^{\circ},\boldsymbol{x}\in\left(\dfrac{1}{\lambda}\right)S^{\circ}$.

(4) 由定义 2.5.5 知显然 $\mathbf{0}\in S^{\circ}$. 由 (2.5.3-2) 式知 S° 是闭半空间族的交, 由性质 2.2.2 知它是闭凸集.

(5) 设 S 有界, 则 $\exists\varepsilon>0$ 使 $S\subset B(\mathbf{0},\varepsilon)$. 由 $\langle\boldsymbol{x},\boldsymbol{y}\rangle\leqslant\|\boldsymbol{x}\|\,\|\boldsymbol{y}\|$, 对任意的 $\boldsymbol{y}\in B(\mathbf{0},\varepsilon^{-1})$ 有

$$\langle\boldsymbol{x},\boldsymbol{y}\rangle\leqslant\|\boldsymbol{x}\|\,\|\boldsymbol{y}\|\leqslant 1,$$

则由 $B(\mathbf{0},\varepsilon^{-1})\subset S^{\circ}$ 可得, $\mathbf{0}$ 是 S° 的内点.

(6) 如果 $\mathbf{0}$ 是 S 的内点, 即 $\exists\varepsilon>0$ 使 $B(\mathbf{0},\varepsilon)\subset S$. 假设 S° 无界, 则 $\exists\boldsymbol{y}\in S^{\circ}$ 使得 $\|\boldsymbol{y}\|>\dfrac{1}{\varepsilon}$, 并且存在充分小的 t 使得 $t\boldsymbol{y}\in B(\mathbf{0},\varepsilon)$ 且 $t\|\boldsymbol{y}\|=\varepsilon$, 则有 $\|\boldsymbol{y}\|=\dfrac{\varepsilon}{t}>\dfrac{1}{\varepsilon}$. 因为 $\boldsymbol{y}\in S^{\circ}$ 使得 $1\geqslant\langle t\boldsymbol{y},\boldsymbol{y}\rangle=t\|\boldsymbol{y}\|^2=\dfrac{\varepsilon^2}{t}>1$, 得到矛盾. 因此, S° 有界. □

性质 2.5.7 说明两个集合的各自配极相交等于它们并集后的配极, 一个集合属于另一个集合, 那么它们各自的配极是反包含关系. 配极是包含原点的闭凸集. 原点是有界集的配极的内点. 如果集合包含原点, 则它的配极是有界的.

下面的结果表明 $S^{\circ\circ}$ 是包含原点和 S 的最小闭凸集.

定理 2.5.11 设集合 $S \subset \mathbf{R}^n$, 则

$$S^{\circ\circ} = \operatorname{clco}(\{\mathbf{0}\} \cup S).$$

证明 根据 (2.5.3-2) 式有

$$S^{\circ\circ} = \bigcap_{\boldsymbol{y} \in S^{\circ}} H^{\leqslant}(\boldsymbol{y}, 1) = \bigcap_{S \subset H^{\leqslant}(\boldsymbol{y},1)} H^{\leqslant}(\boldsymbol{y}, 1). \tag{2.5.3-3}$$

因为 $\{\mathbf{0}\} \cup S \subset H^{\leqslant}(\boldsymbol{y}, 1)$, 则上式说明 $S^{\circ\circ}$ 包含原点和 S 的闭凸集. 因为 $\operatorname{clco}(\{\mathbf{0}\} \cup S)$ 也包含原点和 S 的最小闭凸集, 所以有 $\operatorname{clco}(\{\mathbf{0}\} \cup S) \subset S^{\circ\circ}$.

假设点 $\boldsymbol{x}_0 \in S^{\circ\circ}$, 但 $\boldsymbol{x}_0 \notin \operatorname{clco}(\{\mathbf{0}\} \cup S)$, 得 $\dfrac{1}{2}\boldsymbol{x}_0 \notin \operatorname{clco}(\{\mathbf{0}\} \cup S)$. 根据定理 2.5.5, $H(\boldsymbol{a}^*, \alpha)$ 是包含 $\operatorname{clco}(\{\mathbf{0}\} \cup S)$ 的所有支撑半空间, 且

$$\operatorname{clco}(\{\mathbf{0}\} \cup S) = \cap H(\boldsymbol{a}^*, \alpha).$$

因此, 有 $\dfrac{1}{2}\boldsymbol{x}_0 \notin H(\boldsymbol{a}^*, \alpha)$, 则

$$\sup\{\langle \boldsymbol{x}, \boldsymbol{a}^* \rangle \,|\, \boldsymbol{x} \in \operatorname{clco}(\{\mathbf{0}\} \cup S)\} \leqslant \alpha < \left\langle \frac{1}{2}\boldsymbol{x}_0, \boldsymbol{a}^* \right\rangle,$$

显然 $\alpha \geqslant 0$, 且有上式等价于

$$\sup\left\{\left\langle \boldsymbol{x}, \frac{2}{\langle \boldsymbol{x}_0, \boldsymbol{a}^* \rangle}\boldsymbol{a}^* \right\rangle \middle| \boldsymbol{x} \in \operatorname{clco}(\{\mathbf{0}\} \cup S)\right\} < 1.$$

根据定义 2.5.5, 有

$$\frac{2}{\langle \boldsymbol{x}_0, \boldsymbol{a}^* \rangle}\boldsymbol{a}^* \in \operatorname{clco}(\{\mathbf{0}\} \cup S)^{\circ}.$$

由于 $\boldsymbol{x}_0 \in S^{\circ\circ}$, 根据双配极定义应该有 $2 = \left\langle \dfrac{2}{\langle \boldsymbol{x}_0, \boldsymbol{a}^* \rangle}\boldsymbol{a}^*, \boldsymbol{x}_0 \right\rangle \leqslant 1$. 因此, 推出矛盾, 所以有 $\boldsymbol{x}_0 \in \operatorname{clco}(\{\mathbf{0}\} \cup S)$. □

由上述定理得下面结论成立, 即包含原点的凸紧集的配极也是凸紧集.

推论 2.5.4 设凸紧集 $C \subset \mathbf{R}^n$, $\mathbf{0} \in \operatorname{int}C$, 则 C° 也是凸紧集, 且 $C^{\circ\circ} = C$.

进一步, 下面的结论揭示了凸紧集的配极与支撑超平面的关系: 包含原点为内点的凸紧集的支撑超平面等价于超平面由配极的边界点构成, 该凸紧集配极的支撑超平面 $H(\boldsymbol{x},1)$ 等价于 $\boldsymbol{x}$ 属于边界集.

定理 2.5.12 设凸紧集 $C\subset \mathbf{R}^n$, $\mathbf{0}\in \text{int}C$, 则下面结论成立:

(1) $H(\boldsymbol{y},1)$ 是 C 的支撑超平面当且仅当 $\boldsymbol{y}\in \text{bd}C^\circ$, 其中 $\text{bd}C^\circ$ 表示 C° 的边界;

(2) $H(\boldsymbol{x},1)$ 是 C° 的支撑超平面当且仅当 $\boldsymbol{x}\in \text{bd}C$.

证明 (1) 如果 $H(\boldsymbol{y},1)$ 是 C 的支撑超平面, 根据配极定义知, $\exists \boldsymbol{y}\in C^\circ$ 使得

$$\sup_{\boldsymbol{x}\in C}\langle \boldsymbol{x},\boldsymbol{y}\rangle = 1. \tag{2.5.3-4}$$

如果 $\boldsymbol{y}\in \text{int}C^\circ$, 存在 $\boldsymbol{b}\in B$ 和 $t>0$, 有 $\boldsymbol{y}+t\boldsymbol{b}\in C^\circ$ 和 $\langle \boldsymbol{x},\boldsymbol{b}\rangle>0$. 再由配极定义得

$$\sup_{\boldsymbol{x}\in C}\langle \boldsymbol{x},\boldsymbol{y}+t\boldsymbol{b}\rangle \leqslant 1,$$

根据 (2.5.3-4) 式得到 $0<\sup\limits_{\boldsymbol{x}\in C}\langle \boldsymbol{x},\boldsymbol{b}\rangle \leqslant 0$, 得到矛盾. 故有 $\boldsymbol{y}\in \text{bd}C^\circ$.

反过来, 如果 $\boldsymbol{y}\in \text{bd}C^\circ$, 显然 $\boldsymbol{y}\neq \mathbf{0}$, 由配极的定义知

$$0<\sup_{\boldsymbol{x}\in C}\langle \boldsymbol{x},\boldsymbol{y}\rangle \leqslant 1.$$

上式说明 $\exists t>1$ 使得 $\sup\limits_{\boldsymbol{x}\in C}\langle \boldsymbol{x},t\boldsymbol{y}\rangle = 1$ 成立, 即 $t\boldsymbol{y}\in C^\circ$. 由于 $\mathbf{0}\in \text{int}C^\circ$, $\boldsymbol{y}$ 是线段在 $(\mathbf{0},t\boldsymbol{y})$ 中的一个内点, 即有 $\boldsymbol{y}\in \text{int}C^\circ$, 得到矛盾. 所以有

$$\sup_{\boldsymbol{x}\in C}\langle \boldsymbol{x},\boldsymbol{y}\rangle = 1.$$

显然 $H(\boldsymbol{y},1)$ 是 C 的支撑超平面.

(2) 证明过程与 (1) 类似. □

注 如果凸紧集 $C\subset \mathbf{R}^n$, $\mathbf{0}\in \text{int}C$, 对于任意的 $\boldsymbol{x},\boldsymbol{y}\in \mathbf{R}^n$, 下面 4 个结论等价:

(1) $H(\boldsymbol{y},1)$ 是 C 在 $\boldsymbol{x}$ 的支撑超平面;

(2) $H(\boldsymbol{x},1)$ 是 C° 在 $\boldsymbol{y}$ 的支撑超平面;

(3) $\langle \boldsymbol{x},\boldsymbol{y}\rangle = 1, \boldsymbol{x}\in \text{bd}C, \boldsymbol{y}\in \text{bd}C^\circ$;

(4) $\langle \boldsymbol{x},\boldsymbol{y}\rangle = 1, \boldsymbol{x}\in C, \boldsymbol{y}=C^\circ$.

本小节得到了集合配极对应的一些性质:

1) 两个集合各自的配极相交等于它们并集后的配极;

2) 如果一个集合属于另一个集合, 那么它们各自的配极是反包含关系;

3) 配极是包含原点的闭凸集;

4) 原点是有界集的配极的内点;

5) 如果集合包含原点, 则它的配极是有界的;

6) 集合的双配极等于包含原点和该集合并集的最小闭凸集;

7) 包含原点为内点的凸紧集的配极也是凸紧集, 且其双配极等于该集合;

8) 如果凸紧集包含原点, 则它的支撑超平面等价于超平面由配极的边界点构成, 该凸紧集配极的支撑超平面 $H(\boldsymbol{x},1)$ 等价于 $\boldsymbol{x}$ 属于边界集.

2.6 两种近似凸集

下面讨论与凸集密切相关的近似凸集, 近似凸集可以不是凸集, 可参见文献 [99-102].

2.6.1 近似凸集

近似凸集是非常接近凸集的一种集合, 从下面定义来看近似凸集包含一个凸集, 且被包含在该凸集闭包中.

定义 2.6.1 设非空集合 $E\subset\mathbf{R}^n$. 如果存在一个凸集 C 满足

$$C\subset E\subset \mathrm{cl}C,$$

则称 E 是 $\mathbf{R}^n$ 中的近似凸集.

凸集是近似凸集, 但是近似凸集不一定是凸集.

例 2.6.1 设给定集合 $E\subset\mathbf{R}^2$ 如下:

$$E=\left\{(x_1,x_2)^{\mathrm{T}}\,|\,-1<x_1<1,x_2>0\right\}\cup\left\{(-1,0)^{\mathrm{T}},(1,0)^{\mathrm{T}}\right\},$$

根据定义 2.6.1 容易验证集合 E 是近似凸集. 又因为

$$2E=\left\{(x_1,x_2)^{\mathrm{T}}\,|\,-2<x_1<2,x_2>0\right\}\cup\left\{(-2,0)^{\mathrm{T}},(2,0)^{\mathrm{T}}\right\},$$

$$E+E=2E\cup\left\{(0,0)^{\mathrm{T}}\right\},$$

有 $E+E\neq 2E$, 根据性质 2.2.6(7), 所以说明 E 不是凸集.

例 2.6.2 设给定两个集合 $E_1,E_2\in\mathbf{R}^2$ 如下:

$$E_1=\left\{(x_1,x_2)^{\mathrm{T}}\,|x_1\geqslant 0,x_2\in\mathbf{R}\right\}\Big\backslash\left\{(0,x_2)^{\mathrm{T}}\,||x_2|<1\right\},$$

$$E_2 = \left\{(x_1, x_2)^{\mathrm{T}} \,|x_1 \leqslant 0, x_2 \in \mathbf{R}\right\} \Big\backslash \left\{(0, x_2)^{\mathrm{T}} \,||x_2| < 1\right\}.$$

上式两个集合 E_1, E_2 是近似凸集, 得 $E_1 \cap E_2 = \left\{(0, x_2)^{\mathrm{T}} \,|\, |x_2| \geqslant 1\right\}$ 不是近似凸集. 近似凸集的交不一定是近似凸集. 但是两个凸集的交也是凸集.

因此, 凸集成立的性质, 近似凸集不一定成立.

定义 2.6.2　设非空集合 $C, D \subset \mathbf{R}^n$. 如果下面式子成立:

$$\mathrm{cl}C = \mathrm{cl}D, \quad \mathrm{ri}C = \mathrm{ri}D,$$

则称 C 和 D 近似相等, 记 $C \approx D$.

我们有下面的性质成立.

性质 2.6.1　设 E 是 $\mathbf{R}^n$ 中的近似凸集, 记存在一个凸集 C 满足 $C \subset E \subset \mathrm{cl}C$, 则有

$$E \approx \mathrm{cl}E \approx \mathrm{ri}E \approx \mathrm{co}E \approx \mathrm{ri}\,(\mathrm{co}E) \approx C,$$

且 $\mathrm{cl}E$ 和 $\mathrm{ri}E$ 是凸集, 如果 $E \neq \varnothing$ 有 $\mathrm{ri}E \neq \varnothing$.

(证明作为读者练习.)

性质 2.6.2　设 E_1, E_2 和 E_3 是 $\mathbf{R}^n$ 中的近似凸集, 且满足 $E_1 \approx E_3$ 和 $E_1 \subset E_2 \subset E_3$, 则有

$$E_1 \approx E_2 \approx E_3.$$

(证明作为读者练习.)

性质 2.6.3　设 E 是 $\mathbf{R}^n$ 中的子集, 则下面结论等价:

(1) E 是 $\mathbf{R}^n$ 中的近似凸集;

(2) $E \approx \mathrm{co}E$;

(3) E 近似等于一个凸集;

(4) E 近似等于一个近似凸集;

(5) $\mathrm{ri}\,(\mathrm{co}E) \subset E$;

(6) $\mathrm{ri}E$ 是 $\mathbf{R}^n$ 中的凸集且 $\mathrm{cl}\,(\mathrm{ri}E) = \mathrm{cl}E$;

(7) $\mathrm{cl}E$ 是 $\mathbf{R}^n$ 中的凸集且 $\mathrm{ri}\,(\mathrm{cl}E) \subset \mathrm{ri}E$;

(8) $[\boldsymbol{x}, \boldsymbol{y}) \subset \mathrm{ri}E$, 其中 $\forall \boldsymbol{x} \in \mathrm{ri}E, \forall \boldsymbol{y} \in E$.

(证明作为读者练习.)

性质 2.6.4　设 C 是 $\mathbf{R}^n$ 中的凸集, 则 E 是 $\mathbf{R}^n$ 中的近似凸集的充分必要条件是 $E = C \cup S$, 其中 $S \subset \mathrm{cl}C \backslash \mathrm{ri}C = \mathrm{rb}C$.

证明　证明必要性. 设 E 是 $\mathbf{R}^n$ 中的近似凸集, 令 $E = \mathrm{ri}E \cup (E \backslash \mathrm{ri}E)$, 记 $C = \mathrm{ri}E$ 和 $S = E \backslash \mathrm{ri}E$, 由性质 2.6.1 知 $\mathrm{ri}E$ 是凸集, 有

$$\mathrm{cl}C \backslash \mathrm{ri}C = \mathrm{cl}\,(\mathrm{ri}E) \backslash \mathrm{ri}\,(\mathrm{ri}E) \supset E \backslash \mathrm{ri}\,(E) = S.$$

证明充分性. 如果 $E = C \cup S$, 其中 $S \subset \text{cl}C \backslash \text{ri}C$. 显然有 $\partial S \subset \text{rb}C$, 则得 $\text{ri}E = \text{ri}C$ 且是凸集, 由 $C \subset E$ 和 $\text{cl}C \subset \text{cl}E$, 得 $\text{cl}E = \text{cl}\,(C \cup S) = \text{cl}C \cup \text{cl}S = \text{cl}C$, 故 $E \subset \text{cl}C$. 由定义 2.6.1 知 E 是 $\mathbf{R}^n$ 中的近似凸集. □

注 性质 2.6.4 说明近似凸集是由一个闭凸集去掉它的一部分相对边界构成的.

性质 2.6.5 设 E 是 $\mathbf{R}^n$ 中的近似凸集, 则下面结论成立:

(1) $\text{int}\,(\text{cl}E) = \text{int}E$;

(2) 若 $\text{int}E \neq \varnothing$, 则有 $\text{ri}E = \text{int}E$;

(3) $\text{aff}\,(\text{ri}E) = \text{aff}E = \text{aff}\,(\text{cl}E)$.

(证明作为读者练习.)

定义 2.6.3 设非空集合 $E \subset \mathbf{R}^n$, 如果对任意的 $\boldsymbol{x}, \boldsymbol{y} \in E$, 有 $(\boldsymbol{x}, \boldsymbol{y}) \subset \text{ri}E$, 则称 E 为相对严格凸集.

性质 2.6.6 设 E 是 $\mathbf{R}^n$ 中的近似凸集, 下面结论成立:

(1) 如果 E 是相对严格凸集, 则 E 是凸集;

(2) 如果 E 是开集, 则 E 是凸集;

(3) 如果 E 是闭集, 则 E 是凸集;

(4) 如果对任意的 $\boldsymbol{x}, \boldsymbol{y} \in E \backslash \text{ri}E$, 有 $[\boldsymbol{x}, \boldsymbol{y}] \subset E$, 则 E 是凸集.

证明 (1) 设 $\forall \boldsymbol{x}, \boldsymbol{y} \in E$, 因为 E 是相对严格凸集和近似凸集, 则 $(\boldsymbol{x}, \boldsymbol{y}) \subset \text{ri}E \subset E$, 有 $[\boldsymbol{x}, \boldsymbol{y}] \subset E$, 说明 E 是凸集.

(2) 设 C 是凸集, 因为 E 是开集和近似凸集, 有 $\text{ri}C = \text{ri}E$. 由性质 2.6.5 得 $\text{ri}C = \text{int}E = E$, 所以 E 是凸集.

(3) 根据近似凸集的定义, 存在一个凸集 C 使得 $C \subset E \subset \text{cl}C$. 由 E 是闭集得 $E = \text{cl}E = \text{cl}C$, 则 E 是凸集.

(4) 由 $\forall \boldsymbol{x}, \boldsymbol{y} \in E = \text{ri}E \cup E \backslash \text{ri}E$, 有 $[\boldsymbol{x}, \boldsymbol{y}] \subset E$, 则 E 是凸集. □

2.6.2 相对近似凸集

在文献 [100] 中定义的仿凸集是一种近似凸集, 下面给出仿凸集的另一种名称.

定义 2.6.4 设 E 是 $\mathbf{R}^n$ 中的非空子集, 对任意 $\boldsymbol{x} \in \text{ri}E$, $\boldsymbol{y} \in \text{cl}E$ 及 $0 \leqslant \lambda < 1$, 有

$$(1 - \lambda)\,\boldsymbol{x} + \lambda \boldsymbol{y} \in \text{ri}E,$$

则称 E 是相对近似凸集 (仿凸集).

性质 2.6.7 设 E 是 $\mathbf{R}^n$ 中的相对近似凸集, 下面结论成立:

(1) $\text{ri}E$ 是凸集;

(2) $\text{cl}\,(\text{ri}E) = \text{cl}E = \text{cl}\,(\text{cl}E)$, 且 $\text{cl}E$ 是凸集;

(3) $\text{ri}\,(\text{ri}E) = \text{ri}E = \text{ri}\,(\text{cl}E)$;

(4) E 是近似凸集.

证明 (1) 设任何 $\boldsymbol{x}\in \mathrm{ri}E$, $\boldsymbol{y}\in \mathrm{ri}E$, 由定义 2.6.4 得 $(1-\lambda)\boldsymbol{x}+\lambda\boldsymbol{y}\in \mathrm{ri}E$, 则 $\mathrm{ri}E$ 是凸集.

(2) 先证明 $\mathrm{cl}\,(\mathrm{ri}E)=\mathrm{cl}E$, 显然 $\mathrm{cl}\,(\mathrm{ri}E)\subset \mathrm{cl}E$ 成立. 反过来, 设任何 $\boldsymbol{x}\in \mathrm{cl}E$, 则存在一个序列 $\{\boldsymbol{x}_k\}\subset E\subset \mathrm{cl}E$ 和 $1>\lambda_k\to 0, \boldsymbol{x}_k\to \boldsymbol{x}, k\to\infty$. 令 $\boldsymbol{y}\in \mathrm{ri}E$, 由定义 2.6.4 知

$$(1-\lambda_k)\,\boldsymbol{x}_k+\lambda_k\boldsymbol{y}\in \mathrm{ri}E.$$

但 $\lim\limits_{k\to\infty}(1-\lambda_k)\,\boldsymbol{x}_k+\lambda_k\boldsymbol{y}=\boldsymbol{x}$, 所以 $\boldsymbol{x}$ 是 $\mathrm{ri}E$ 的极限点, 即有 $\mathrm{cl}\,(\mathrm{ri}E)\supset \mathrm{cl}E$. 因此有 $\mathrm{cl}\,(\mathrm{ri}E)=\mathrm{cl}E$. 因为 $\mathrm{ri}E$ 是凸集, 则 $\mathrm{cl}\,(\mathrm{ri}E)$ 是凸集, 说明 $\mathrm{cl}E$ 也是凸集. $\mathrm{cl}E=\mathrm{cl}\,(\mathrm{cl}E)$ 是显然成立的.

(3) 由性质 2.3.3 得 $\mathrm{ri}E$ 是相对开集, 因此, $\mathrm{ri}E=\mathrm{ri}\,(\mathrm{ri}E)$. 由性质 2.3.4 得

$$\mathrm{ri}\,(\mathrm{ri}E)=\mathrm{ri}\,(\mathrm{cl}\,(\mathrm{ri}E))=\mathrm{ri}\,(\mathrm{ri}\,(\mathrm{ri}E))\,,$$

由 (2) 得 $\mathrm{ri}E=\mathrm{ri}\,(\mathrm{cl}E)$

(4) 因为 $\mathrm{ri}E\subset E\subset \mathrm{cl}\,(\mathrm{ri}E)$, 所以 E 是近似凸集. □

注 由性质 2.6.7(4) 知相对近似凸集是近似凸集, 反之, 容易证明近似凸集也是相对近似凸集.

性质 2.6.8 设 E 是 $\mathbf{R}^n$ 中的近似凸集, 则 E 是相对近似凸集.

证明 设任意 $\boldsymbol{x}\in \mathrm{ri}E$, $\boldsymbol{y}\in \mathrm{cl}E$, 则由性质 2.6.7 得 $\mathrm{cl}E$ 是凸集和 $\mathrm{ri}E=\mathrm{ri}\,(\mathrm{cl}E)$. 再由定理 2.3.2 知

$$(1-\lambda)\,\boldsymbol{x}+\lambda\boldsymbol{y}\in \mathrm{ri}E,$$

所以, E 是相对近似凸集. □

本节得到近似凸集的几个重要性质:

1) 近似凸集的相对内部和闭包是凸集, 且它们近似相等.

2) 近似凸集的充分必要条件是它等于一个凸集并上该凸集闭包去掉相对内部的一部分.

3) 如果近似凸集是开集或闭集, 那么它也是凸集.

4) 相对近似凸集等价于近似凸集.

2.7 习 题

2.7-1 证明性质 2.1.1、性质 2.1.2 和性质 2.1.3.

2.7-2 证明性质 2.1.4 和性质 2.1.5.

2.7-3 证明性质 2.1.6 和性质 2.1.7.

2.7-4 证明性质 2.1.8、推论 2.1.1 和推论 2.1.2.

2.7-5 证明性质 2.2.3、推论 2.2.1 和推论 2.2.2.

2.7-6 证明性质 2.3.1 和性质 2.3.3.

2.7-7 已知集合 $X=\{(0,0)^{\mathrm{T}},(1,0)^{\mathrm{T}},(0,1)^{\mathrm{T}},(1,1)^{\mathrm{T}},(-1,-1)^{\mathrm{T}}\}$, 求 $\mathrm{aff}X$, $\mathrm{co}X$, $\dim X$, $\mathrm{dir}X$, $\mathrm{ort}X$ 和 $\mathrm{span}X$, 并求出 X 中的最大单纯形集.

2.7-8 已知 $C=\{(x_1,x_2)^{\mathrm{T}}\,|0<x_1\leqslant 1,0<x_2\leqslant 1\}$, 那么 C 是凸集、开集还是闭集? 求 $\mathrm{co}C$, $\mathrm{int}C$, $\mathrm{cl}C$, $\mathrm{ri}C$, ∂C, $\mathrm{bd}C$, $\mathrm{rb}C$ 和 $\mathrm{clco}C$.

2.7-9 已知 $C_1=\{(x_1,x_2)^{\mathrm{T}}\,|0<x_1\leqslant 1,0<x_2\leqslant 1\}$ 和 $C_2=\{(x_1,x_2)^{\mathrm{T}}\,|1<x_1\leqslant 2,1<x_2\leqslant 2\}$, 求 $\mathrm{co}\,(C_1+C_2)$, $\mathrm{int}\,(C_1+C_2)$, $\mathrm{cl}\,(C_1+C_2)$, $\mathrm{ri}\,(C_1+C_2)$ 和 $\mathrm{clco}\,(C_1+C_2)$, 计算直和 $(C_1\oplus C_2)$ 对应的凸包、内部、相对内部、闭包、相对内部和闭凸包.

2.7-10 证明性质 2.5.1、性质 2.5.2、性质 2.5.3 和性质 2.5.4.

2.7-11 已知 $C_1=\{(x_1,x_2)^{\mathrm{T}}\,|0<x_1\leqslant 1,0<x_2\leqslant 1\}$ 和 $C_2=\{(x_1,x_2)^{\mathrm{T}}\,|1<x_1\leqslant 2,1<x_2\leqslant 2\}$, 求 C_1 和 C_2 的可分离超平面, 它们是严格分离还是强分离.

2.7-12 已知 $C=\{(x_1,x_2)^{\mathrm{T}}\,|0<x_1\leqslant 1,0<x_2\leqslant 1\}$, 求 C 的支撑超平面.

2.7-13 已知集合 $C=\{(x_1,x_2,x_3)^{\mathrm{T}}\,|x_1^2+x_2^2+x_3^2\leqslant 1,x_1,x_2,x_3\geqslant 0\}$, 求 C 的所有极点集、面、正常面、暴露面和暴露点.

2.7-14 已知集合 $C=\{(x_1,x_2,x_3)^{\mathrm{T}}\,|x_1^2+x_2^2+x_3^2\leqslant 1,x_1,x_2,x_3\geqslant 0\}$, 求 $\mathrm{co}\,(\mathrm{ext}C)$.

2.7-15 已知集合 $C=\{(x_1,x_2,x_3)^{\mathrm{T}}\,|x_1^2+x_2^2+x_3^2\leqslant 1,x_1,x_2,x_3\geqslant 0\}$, 求 C 的配极和双配极, 并分别求出 C 和 C° 的一个支撑超平面.

2.7-16 设 $C,D\subset\mathbf{R}^n$, 且 $C\subset D$, 证明 $C\subset\mathrm{co}D$ 和 $\mathrm{co}C\subset\mathrm{co}D$, 且如果 D 是凸集, 则有 $\mathrm{co}C\subset D$.

2.7-17 设 $C,D\subset\mathbf{R}^n$, 证明 $\mathrm{co}\,(C\cup D)\subset\mathrm{co}\,(C)\cup\mathrm{co}D$ 和 $\mathrm{co}\,(C\cap D)\subset\mathrm{co}\,(C)\cap\mathrm{co}D$.

2.7-18 设 $C,D\subset\mathbf{R}^n$ 是凸集, 证明 $\mathrm{cl}\,(\mathrm{ri}C+\mathrm{ri}D)=\mathrm{cl}\,(C+D)$, $\mathrm{ri}(\mathrm{ri}C+\mathrm{ri}D)=\mathrm{ri}\,(C+D)$, $\mathrm{aff}\,(\mathrm{ri}C+\mathrm{ri}D)=\mathrm{aff}\,(C+D)$, $\mathrm{co}\,(\mathrm{cl}C)\subset\mathrm{cl}\,(\mathrm{co}C)$ 和 $\mathrm{co}(\mathrm{cl}C\cup\mathrm{cl}D)\subset\mathrm{cl}(\mathrm{co}\,(C\cup D))$.

2.7-19 设 $C\subset\mathbf{R}^n$ 是不包含原点的非空凸集, 证明存在超平面分离原点.

2.7-20 设 $C\subset\mathbf{R}^n$ 是非空闭凸集, $H^{\geqslant}(\boldsymbol{x}^*,\alpha)$ 和 $H\,(\boldsymbol{x}^*,\alpha)$ 是 C 的支撑半空间和支撑超平面, 证明 $\inf\limits_{\boldsymbol{x}\in C}\langle\boldsymbol{x},\boldsymbol{x}^*\rangle<\max\limits_{\boldsymbol{x}\in C}\langle\boldsymbol{x},\boldsymbol{x}^*\rangle$ 的充分必要条件是 $H\,(\boldsymbol{x}^*,\alpha)$ 是正常支撑超平面.

2.7-21　设 $C \subset \mathbf{R}^n$, 证明 $(C^{\circ\circ})^\circ = C^\circ$.

2.7-22　设 $\{C_i \subset \mathbf{R}^n | i \in I\}$ 是包含原点的闭凸集族, 证明

$$\left(\bigcap_{i \in I} C_i\right)^\circ = \mathrm{clco} \bigcap_{i \in I} C_i^\circ.$$

2.7-23　设 E_1, E_2 是 $\mathbf{R}^n$ 中的近似凸集, 证明下面结论等价:

$$E_1 \approx E_2 \Leftrightarrow \mathrm{ri}E_1 \approx \mathrm{ri}E_2 \Leftrightarrow \mathrm{cl}E_1 \approx \mathrm{cl}E_2.$$

2.7-24　证明性质 2.6.1、性质 2.6.2、性质 2.6.3 和性质 2.6.5.

2.7-25[102]　设函数 $f : [a, b] \to \mathbf{R}$ 的上方图定义:

$$\mathrm{epi}f = \{(x, \alpha) \,|\, f(x) \leqslant \alpha, \forall x \in [a, b]\},$$

证明 epi$f$ 是近似凸集的充分必要条件为: 存在一个 $\lambda \in (0, 1)$ 使得

$$f(\lambda x + (1 - \lambda y)) \leqslant \lambda f(x) + (1 - \lambda) f(y), \quad \forall x, y \in [a, b].$$

2.7-26　如果 E 是 $\mathbf{R}^n$ 中的近似凸集, 证明存在一个 $\lambda \in (0, 1)$, 对于任意的 $\boldsymbol{x}, \boldsymbol{y} \in E$ 满足

$$(1 - \lambda)\boldsymbol{x} + \lambda \boldsymbol{y} \in E.$$

请举例说明, 反之不一定成立.

第 3 章 凸　锥

凸锥是凸分析基础理论中最重要的部分之一. 在几何结构上, 锥的概念不同于子空间与仿射集的概念, 一般情况下, 锥不一定是子空间、仿射集或凸集. 锥的几何特征是含有锥上任何点出发的半直线或直线, 锥不一定含整条直线, 但子空间和仿射集都含有直线. 凸锥指既是凸集又是锥的集合, 因此, 凸锥是一种特殊的凸集. 除凸集的性质对于凸锥都成立外, 凸锥作为一类特殊集合还具有一些重要的特殊性质. 在向量优化理论中, 凸锥是刻画多维空间序的重要工具, 是向量优化研究不可缺少的工具. 在向量空间中几乎所有的序定义都是采用凸锥来刻画的, 通俗地说, 凸锥可以作为“尺度”来比较空间两点之间的“大小”, 若不同两点的差属于凸锥, 则可以定义前者不小于后者.

几乎所有的凸分析著作中都有关于凸锥的介绍, 见文献 [20,25,52,63,85-87,95], 其中 Soltan 的著作中对凸锥的介绍比较深刻和全面 [63]. 为了系统、全面地介绍凸锥的基本知识和性质, 本章内容尽量涵盖了凸锥现有的基本理论与性质, 主要包含顶点锥、锥包、回收锥、共轭锥、极锥、法锥、切锥、障碍锥、凸锥分离、多面体和多面体锥等.

3.1 顶　点　锥

锥是一类特殊集合, 它的最大几何特征是锥内的每一个点乘一个大于 0 的数得到的点仍然在这个集合中, 即意味着锥含有半直线. 后面将说明, 锥的顶点不一定属于锥. 因此, 一个非独点锥是无界集合. 本节主要介绍含顶点的锥和凸锥的一些基本性质.

3.1.1 基本性质

先给出有关顶点锥和顶点凸锥的定义.

定义 3.1.1 设 $C \subset \mathbf{R}^n$ 是非空集合. 如果对于任意的 $\lambda > 0$ 和任意的 $\boldsymbol{x} \in C$ 有 $\lambda \boldsymbol{x} \in C$, 则称 C 是锥. 如果锥 C 还是凸集, 则称 C 是凸锥.

定义 3.1.2 设 $C \subset \mathbf{R}^n$ 是非空集合和点 $\boldsymbol{a} \in \mathbf{R}^n$. 如果对任意的 $\lambda > 0$ 和任意的 $\boldsymbol{x} \in C$ 有

$$l\langle \boldsymbol{a}, \boldsymbol{x}\rangle = \boldsymbol{a} + \lambda(\boldsymbol{x} - \boldsymbol{a}) \in C,$$

则称 C 是关于顶点 $\boldsymbol{a}$ 的锥 (简称 C 是顶点锥), 称 $\boldsymbol{a}$ 是 C 的顶点. 再如果顶点锥 C 还是凸集, 则称 C 是关于顶点 $\boldsymbol{a}$ 的凸锥 (简称 C 是顶点凸锥). 如果 $\boldsymbol{a} \in C$, 则称 C 是关于顶点 $\boldsymbol{a}$ 的正常锥.

注　锥不一定是凸锥, 定义 3.1.1 的锥是指以原点 $\mathbf{0}$ 为顶点的锥 (以下如果没有特别说明锥的顶点, 默认这个顶点为原点 $\mathbf{0}$). 本章给定的锥若不做说明, 则一般不包含顶点.

图 3.1-1 是带有一个顶点的凸锥, 图 3.1-2 是带有一个顶点的非凸锥. 锥的特点是锥中任何一个点连接顶点后沿着大于 0 的方向的射线仍然在锥中, 即有

$$C = \{\boldsymbol{a} + \lambda(\boldsymbol{x} - \boldsymbol{a}) \,|\, \forall \boldsymbol{x} \in C, \forall \lambda > 0\}.$$

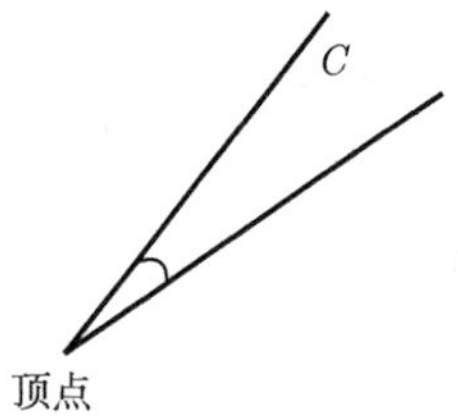

图 3.1-1　凸锥

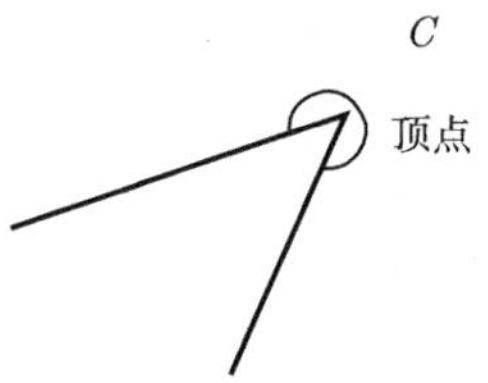

图 3.1-2　非凸锥

下面是锥、凸锥、非凸锥、不含顶点锥的例子.

例 3.1.1　(1) 在 $\mathbf{R}$ 空间中的区间集合 $C = (0, +\infty)$ 是一个锥, 也是关于顶点 $\mathbf{0}$ 的一个锥, 且是一个凸锥, 不含顶点.

(2) 在 $\mathbf{R}^2$ 空间中的集合 $C = \{(x_1, 0)^{\mathrm{T}} \,|\, x_1 \in (0, +\infty)\} \cup \{(0, x_2)^{\mathrm{T}} \,|\, x_2 \in (0, +\infty)\}$ 是由两个半坐标轴构成的, 顶点是原点, 它不含顶点, 且不是凸锥.

(3) 在 $\mathbf{R}^2$ 空间中的集合 $C = \{(x_1, 0)^{\mathrm{T}} \,|\, x_1 \in [0, +\infty)\} \cup \{(0, x_2)^{\mathrm{T}} \,|\, x_2 \in [0, +\infty)\}$ 是由两个半坐标轴构成的, 含有一个顶点且是原点, 但不是凸锥.

(4) 在 $\mathbf{R}$ 空间中的集合 $C = (1, +\infty)$ 是一个锥, 也是关于顶点 1 的一个锥, 且是一个凸锥.

(5) 在 $\mathbf{R}^2$ 空间中的集合 $C_1 = \{(x_1, 0)^{\mathrm{T}} \,|\, x_1 \in (1, +\infty)\}$ 和 $C_2 = \{(0, x_2)^{\mathrm{T}} \,|\, x_2 \in (1, +\infty)\}$ 分别为两个顶点 $(1, 0)^{\mathrm{T}}$ 和 $(0, 1)^{\mathrm{T}}$ 的锥, 它们的并集 $C_1 \cup C_2$ 不是锥.

下面是顶点锥的一些基本性质.

性质 3.1.1　设 $C \subset \mathbf{R}^n$ 是非空集合, 点 $\boldsymbol{a} \in \mathbf{R}^n$, 有下面结论成立:

(1) C 是关于顶点 $\mathbf{0}$ 的锥当且仅当 C 是锥;

(2) 若 C 是锥, 则当 $\lambda > 0$ 时, 有 $\lambda C = C$;

(3) 若 C 是仿射集, $\boldsymbol{a}$ 是 C 的顶点, 则 C 是关于顶点 $\boldsymbol{a}$ 的锥;

(4) 若 C 是凸锥, 当且仅当对 $\lambda, \mu > 0$ 时, 有 $\lambda C + \mu C = C$;

(5) 若 C 是关于顶点 $\boldsymbol{a}$ 的锥, 则 $C \cup \{\boldsymbol{a}\}$ 或 $C \setminus \{\boldsymbol{a}\}$ 是关于顶点 $\boldsymbol{a}$ 的锥.

证明 (1) 从定义 3.1.1 和定义 3.1.2 得知, 定义 3.1.1 所定义的锥已包含以原点 $\mathbf{0}$ 为顶点的锥.

(2) 若 $\forall \boldsymbol{x} \in C$, 由定义 3.1.1 得 $\lambda \boldsymbol{x} \in C$, 即 $\lambda C \subset C$. 反之, $C \subset \lambda C$ 显然成立.

(3) 由仿射集定义可知, $\boldsymbol{a} + \lambda(\boldsymbol{x} - \boldsymbol{a}) = (1-\lambda)\boldsymbol{a} + \lambda \boldsymbol{x} \in C$.

(4) 由 (2) 知只需证明 $C + C \subset C$ 成立. 设 $\forall \boldsymbol{x}, \boldsymbol{y} \in C$, 因为 C 是凸集, 那么对 $\lambda = \dfrac{1}{2}$ 有 $\dfrac{1}{2}\boldsymbol{x} + \dfrac{1}{2}\boldsymbol{y} \in C$ 成立, 根据凸锥定义, 有 $2\left(\dfrac{1}{2}\boldsymbol{x} + \dfrac{1}{2}\boldsymbol{y}\right) = \boldsymbol{x} + \boldsymbol{y} \in C$.

(5) 由定义 3.1.2 知结论显然的. □

性质 3.1.1 揭示了锥与顶点之间的关系, 表明锥的顶点可以属于锥, 也可以不属于锥. 仿射集都是锥, 反之却不一定. 凸锥的结构可以用它含有任意两点的正线性组合表达.

进一步, 性质 3.1.2 表明了顶点锥与凸集的关系: 非空锥张成的仿射集还是锥; 并且顶点锥的凸包还是顶点锥; 若一个顶点锥去掉顶点是一个凸集, 那么该顶点锥具有唯一顶点.

性质 3.1.2 设 $C \subset \mathbf{R}^n$ 是关于顶点 $\boldsymbol{a}$ 的非空锥, 有下面结论成立:

(1) 凸包 $\mathrm{co}C$ 是关于顶点 $\boldsymbol{a}$ 的一个锥;

(2) $\mathrm{aff}C = \mathrm{aff}(\{\boldsymbol{a}\} \cup C)$ 是关于顶点 $\boldsymbol{a}$ 的一个锥;

(3) 如果 C 是凸集, 则 $\{\boldsymbol{a}\} \cup C$ 是凸集;

(4) 如果 $\boldsymbol{a} \in C$ 且 $C \setminus \{\boldsymbol{a}\}$ 是凸集, 则 C 是凸集且 $\boldsymbol{a}$ 是 C 的唯一一个顶点.

证明 (1) 由凸包的定理 (定理 2.2.2) 知, 对于任何点 $\boldsymbol{x} \in \mathrm{co}C$, 存在 $\boldsymbol{x}_1, \cdots, \boldsymbol{x}_m \in C$ 和 $\lambda_i \geqslant 0\left(i = 1, \cdots, m, \sum\limits_{i=1}^{m} \lambda_i = 1\right)$, 使得

$$\boldsymbol{x} = \sum_{i=1}^{m} \lambda_i \boldsymbol{x}_i,$$

那么有

$$\boldsymbol{x} - \boldsymbol{a} = \sum_{i=1}^{m} \lambda_i (\boldsymbol{x}_i - \boldsymbol{a}). \tag{3.1.1-1}$$

设 $\forall \lambda > 0$, 由 (3.1.1-1) 式得

$$\boldsymbol{a} + \lambda(\boldsymbol{x} - \boldsymbol{a}) = \sum_{i=1}^{m} \lambda_i (\boldsymbol{a} + \lambda(\boldsymbol{x}_i - \boldsymbol{a})),$$

由定义 3.1.2 得 $\boldsymbol{a}+\lambda(\boldsymbol{x}_i-\boldsymbol{a})\in C$, 再根据凸包定义知 $\boldsymbol{a}+\lambda(\boldsymbol{x}-\boldsymbol{a})\in \operatorname{co}C$.

(2) 根据性质 2.1.7 知, 对于任意点 $\boldsymbol{x}\in \operatorname{aff}C$, 存在 $\boldsymbol{x}_1,\cdots,\boldsymbol{x}_m\in C$, $\lambda_i\in\mathbf{R}$ $(i=1,\cdots,m)$ 和 $\sum\limits_{i=1}^{m}\lambda_i=1$, 使得 $\boldsymbol{x}=\sum\limits_{i=1}^{m}\lambda_i\boldsymbol{x}_i$. 那么设 $\forall\lambda>0$, 可得

$$\boldsymbol{a}+\lambda(\boldsymbol{x}-\boldsymbol{a})=\sum_{i=1}^{m}\lambda_i(\boldsymbol{a}+\lambda(\boldsymbol{x}_i-\boldsymbol{a})).$$

根据定义 3.1.2 得 $\boldsymbol{a}+\lambda(\boldsymbol{x}_i-\boldsymbol{a})\in C$. 再由仿射集定义知 $\boldsymbol{a}+\lambda(\boldsymbol{x}-\boldsymbol{a})\in\operatorname{aff}C$, 则 $\operatorname{aff}C$ 是关于顶点 $\boldsymbol{a}$ 的一个锥, 故有 $\operatorname{aff}C\subset\operatorname{aff}(\{\boldsymbol{a}\}\cup C)$. 因为仿射集是一个闭集, 当 $\lambda\to 0$ 时有

$$\boldsymbol{a}+\lambda(\boldsymbol{x}-\boldsymbol{a})\to\boldsymbol{a},$$

即有 $\boldsymbol{a}\in\operatorname{aff}C$.

(3) 因 C 是关于顶点 $\boldsymbol{a}$ 的一个非空凸锥, 根据定义 $\boldsymbol{a}+\lambda(\boldsymbol{x}-\boldsymbol{a})\in C$, 则有 $\{\boldsymbol{a}\}\cup C$ 是凸集.

(4) 由 (3) 得 C 是凸集. 设 $C=\{\boldsymbol{a}\}\cup C\setminus\{\boldsymbol{a}\}$, 由性质 3.1.1(5) 得 $C\setminus\{\boldsymbol{a}\}$ 是关于顶点 $\boldsymbol{a}$ 的一个锥. 假设 $\boldsymbol{a}_1$ 是 C 的另一个顶点, 那么有 $\boldsymbol{a}_2=\boldsymbol{a}_1+2(\boldsymbol{a}-\boldsymbol{a}_1)\in C$, 得 $\boldsymbol{a}=\dfrac{1}{2}\boldsymbol{a}_2+\dfrac{1}{2}\boldsymbol{a}_1\in C$. 因为 $\forall\lambda>0$, 有 $\boldsymbol{a}_2=\boldsymbol{a}+\lambda(\boldsymbol{a}_2-\boldsymbol{a})\in C\setminus\{\boldsymbol{a}\}$; 再由 $C\setminus\{\boldsymbol{a}\}$ 是凸集, 于是有

$$\boldsymbol{a}=\boldsymbol{a}_1+\frac{1}{2}(\boldsymbol{a}_2-\boldsymbol{a}_1)\in C\setminus\{\boldsymbol{a}\},$$

得到矛盾. □

以下讨论的结果都是关于顶点锥的. 定理 3.1.1 给出了判断顶点锥的等价关系. 例如, 一个顶点为 $\boldsymbol{a}$ 的锥 C 等价于顶点为原点的锥 $C-\boldsymbol{a}$, 注意其中顶点 $\boldsymbol{a}$ 不一定属于 C.

定理 3.1.1 设集合 $C\subset\mathbf{R}^n$, 点 $\boldsymbol{a}\in\mathbf{R}^n$, 则下面结论等价:

(1) C 是关于顶点 $\boldsymbol{a}$ 的锥;

(2) $\mathbf{R}^n\setminus C$ 是关于顶点 $\boldsymbol{a}$ 的锥;

(3) $C-\boldsymbol{a}$ 是关于顶点 $\mathbf{0}$ 的锥;

(4) $C=\boldsymbol{a}+\lambda(C-\boldsymbol{a})$ 对所有的 $\lambda>0$ 成立.

证明 首先证明 (1)$\Leftrightarrow$ (2). 若 (1) 成立, 设 $\boldsymbol{y}\in\mathbf{R}^n\setminus C$. 如果存在一个 $\lambda>0$ 使得

$$\boldsymbol{a}+\lambda(\boldsymbol{y}-\boldsymbol{a})\notin\mathbf{R}^n\setminus C,$$

则 $\boldsymbol{a}+\lambda(\boldsymbol{y}-\boldsymbol{a})\in C$. 因此, 对任意 $t>0$, 有 $\boldsymbol{a}+t(\boldsymbol{a}+\lambda(\boldsymbol{y}-\boldsymbol{a})-\boldsymbol{a})\in C$. 不妨取 $t=1/\lambda$, 那么有 $\boldsymbol{y}\in C$, 得矛盾. 因此, $\mathbf{R}^n\setminus C$ 是关于顶点 $\boldsymbol{a}$ 的锥. 根据上面的证明可得, 如果 $\mathbf{R}^n\setminus C$ 是关于顶点 $\boldsymbol{a}$ 的锥, 那么 $C=\mathbf{R}^n\setminus(\mathbf{R}^n\setminus C)$ 是关于顶点 $\boldsymbol{a}$ 的锥. 因此, (1) 和 (2) 等价.

再证明 (1)$\Leftrightarrow$ (3). 根据定义 3.1.1 和定义 3.1.2, C 是关于顶点 $\boldsymbol{a}$ 的锥显然等价于 $C-\boldsymbol{a}$ 是关于顶点 $\mathbf{0}$ 的锥.

再通过 (1) 和 (3) 及性质 3.1.1(2) 知, (1) 和 (4) 也等价. □

下面的定理 3.1.2 给出了判断凸集为锥的更弱的条件, 只要存在一个不等于 1 的正数 λ 使得

$$C=\boldsymbol{a}+\lambda(C-\boldsymbol{a})$$

成立, 就可以断定该凸集是一个凸锥.

定理 3.1.2 设 $C\subset\mathbf{R}^n$ 是凸集, 则 C 是关于顶点 $\boldsymbol{a}$ 的凸锥的充要条件是存在一个不等于 1 的正数 λ 使得 $C=\boldsymbol{a}+\lambda(C-\boldsymbol{a})$ 成立.

证明 根据定理 3.1.1(4) 可知必要性成立. 只要证明充分性即可. 假设存在一个不等于 1 的正数 λ 使得 $C=\boldsymbol{a}+\lambda(C-\boldsymbol{a})$ 成立, 那么有

$$C-\boldsymbol{a}=\lambda^{-1}(C-\boldsymbol{a}).$$

如果 $\lambda>1$, 设 $\boldsymbol{x}\in C$, 由 $\lambda(\boldsymbol{x}-\boldsymbol{a})\in C-\boldsymbol{a}$, 得到 $\lambda^k(\boldsymbol{x}-\boldsymbol{a})\in C-\boldsymbol{a}$ 对任意整数 $k\geqslant 1$ 成立. 因为 C 是凸集, 得线段 $\left[\lambda^{k-1}(\boldsymbol{x}-\boldsymbol{a}),\lambda^k(\boldsymbol{x}-\boldsymbol{a})\right]\in C-\boldsymbol{a}$. 因此, 对任意 $t>1$ 有

$$t(C-\boldsymbol{a})\in C-\boldsymbol{a}.$$

如果 $\lambda<1$, 同样有 $\lambda^k(\boldsymbol{x}-\boldsymbol{a})\in C-\boldsymbol{a}$ 对任意整数 $k\geqslant 1$ 成立. 因为 C 是凸集, 得到线段

$$\left[\lambda^{k-1}(\boldsymbol{x}-\boldsymbol{a}),\lambda^k(\boldsymbol{x}-\boldsymbol{a})\right]\in C-\boldsymbol{a}.$$

因此, 对任意 $t<1$ 有 $t(C-\boldsymbol{a})\in C-\boldsymbol{a}$. □

例 3.1.2 设 $\boldsymbol{a}\in\mathbf{R}^n$ 和锥 $C=\{\boldsymbol{x}\in\mathbf{R}^n|\ x_i\geqslant a_i, i=1,2,\cdots,n\}$, 显然它是凸锥, 有

$$C-\boldsymbol{a}=\{\boldsymbol{x}\in\mathbf{R}^n|\ x_i\geqslant 0,\ i=1,2,\cdots,n\},$$

且 $\boldsymbol{a}+\lambda(C-\boldsymbol{a})=C$.

定理 3.1.3 设凸集 $C\subset\mathbf{R}^n$, 则 C 是关于顶点 $\boldsymbol{a}$ 的凸锥的充要条件是

$$\boldsymbol{a}+\lambda(\boldsymbol{x}-\boldsymbol{a})+\mu(\boldsymbol{y}-\boldsymbol{a})\in C,$$

其中 $\forall\lambda,\mu>0$ 且 $\forall\boldsymbol{x},\boldsymbol{y}\in C$.

证明　根据性质 3.1.1(4) 的证明知结论成立. □

定理 3.1.3 中的结果也是凸锥的一种表示形式:

$$C = \boldsymbol{a} + \lambda(C - \boldsymbol{a}) + \mu(C - \boldsymbol{a}), \quad \forall \lambda, \mu > 0.$$

下面推广顶点锥的表示形式. 由锥的表示公式可以看出锥可以通过仿射集来生成, 即通过若干个仿射无关组定义锥.

定义 3.1.3　设集合 $\{\boldsymbol{a}, \boldsymbol{x}_1, \cdots, \boldsymbol{x}_k\}$ 是 $\mathbf{R}^n$ 中的仿射无关集合, 由集合 $\{\boldsymbol{x}_1, \cdots, \boldsymbol{x}_k\}$ 生成的关于顶点 $\boldsymbol{a}$ 的锥如下:

$$\mathrm{Simc}(\boldsymbol{a}, \boldsymbol{x}_1, \cdots, \boldsymbol{x}_k) = \{\boldsymbol{a} + \lambda_1 (\boldsymbol{x}_1 - \boldsymbol{a}) + \cdots + \lambda_k (\boldsymbol{x}_k - \boldsymbol{a}) | \lambda_i \geqslant 0, i = 1, 2, \cdots, k\},$$

称 $\mathrm{Simc}(\boldsymbol{a}, \boldsymbol{x}_1, \cdots, \boldsymbol{x}_k)$ 是关于顶点 $\boldsymbol{a}$ 的 k-简单锥.

下面是一个简单锥的例子.

例 3.1.3　给定集合 $\{\boldsymbol{a}, \boldsymbol{x}_1, \boldsymbol{x}_2\} = \left\{(0,0)^{\mathrm{T}}, (1,0)^{\mathrm{T}}, (0,1)^{\mathrm{T}}\right\}$ 是 $\mathbf{R}^2$ 中的仿射无关集合, 定义

$$\mathrm{Simc}(\boldsymbol{a}, \boldsymbol{x}_1, \boldsymbol{x}_2) = \{(\lambda_1, \lambda_2)^{\mathrm{T}} | \lambda_i \geqslant 0, i = 1, 2\}$$

是一个以原点为顶点的 2-简单锥. 显然有 $\mathrm{aff}\{\boldsymbol{a}, \boldsymbol{x}_1, \boldsymbol{x}_2\} = \mathbf{R}^2$, 且

$$\mathrm{ri}\,(\mathrm{Simc}(\boldsymbol{a}, \boldsymbol{x}_1, \boldsymbol{x}_2)) = \{(\lambda_1, \lambda_2)^{\mathrm{T}} | \lambda_i > 0, i = 1, 2\}.$$

下面的定理表明了简单锥的结构性质, 即简单锥是闭凸锥, 且相对内部是非空的.

定理 3.1.4　如果 $\mathrm{Simc}(\boldsymbol{a}, \boldsymbol{x}_1, \cdots, \boldsymbol{x}_k)$ 是关于顶点 $\boldsymbol{a}$ 的 k-简单锥, 则 $\mathrm{Simc}(\boldsymbol{a}, \boldsymbol{x}_1, \cdots, \boldsymbol{x}_k)$ 是关于顶点 $\boldsymbol{a}$ 的 k-维凸锥, 且顶点 $\boldsymbol{a}$ 是唯一顶点, 并有

$$\mathrm{aff}\,(\mathrm{Simc}(\boldsymbol{a}, \boldsymbol{x}_1, \cdots, \boldsymbol{x}_k)) = \mathrm{aff}\,\{\boldsymbol{a}, \boldsymbol{x}_1, \cdots, \boldsymbol{x}_k\} \tag{3.1.1-2}$$

成立. 进一步, $\mathrm{Simc}(\boldsymbol{a}, \boldsymbol{x}_1, \cdots, \boldsymbol{x}_k)$ 是闭集, 记 $A = \mathrm{Simc}(\boldsymbol{a}, \boldsymbol{x}_1, \cdots, \boldsymbol{x}_k)$ 且有

$$\mathrm{ri}A = \{\boldsymbol{a} + \lambda_1 (\boldsymbol{x}_1 - \boldsymbol{a}) + \cdots + \lambda_k (\boldsymbol{x}_k - \boldsymbol{a}) | \lambda_i > 0, i = 1, 2, \cdots, k\}. \tag{3.1.1-3}$$

证明　设任意 $\boldsymbol{x} \in \mathrm{Simc}(\boldsymbol{a}, \boldsymbol{x}_1, \cdots, \boldsymbol{x}_k)$, 存在 $\lambda_i \geqslant 0, i = 1, 2, \cdots, k$ 使得

$$\boldsymbol{x} = \boldsymbol{a} + \lambda_1 (\boldsymbol{x}_1 - \boldsymbol{a}) + \cdots + \lambda_k (\boldsymbol{x}_k - \boldsymbol{a}). \tag{3.1.1-4}$$

设 $t > 0$, 那么有

$$\boldsymbol{a} + t\,(\boldsymbol{x} - \boldsymbol{a}) = \boldsymbol{a} + t\lambda_1 (\boldsymbol{x}_1 - \boldsymbol{a}) + \cdots + t\lambda_k (\boldsymbol{x}_k - \boldsymbol{a}) \in \mathrm{Simc}(\boldsymbol{a}, \boldsymbol{x}_1, \cdots, \boldsymbol{x}_k),$$

表明 $\operatorname{Simc}(\boldsymbol{a},\boldsymbol{x}_1,\cdots,\boldsymbol{x}_k)$ 是关于顶点 $\boldsymbol{a}$ 的 k-维锥. 由 (3.1.1-4) 式表示知 $\operatorname{Simc}(\boldsymbol{a},\boldsymbol{x}_1,\cdots,\boldsymbol{x}_k)$ 是凸集, 且 $\operatorname{Simc}(\boldsymbol{a},\boldsymbol{x}_1,\cdots,\boldsymbol{x}_k)\setminus\{\boldsymbol{a}\}$ 也是凸集, 由性质 3.1.2(4) 得顶点 $\boldsymbol{a}$ 是唯一顶点.

设 $\boldsymbol{x},\boldsymbol{y}\in\operatorname{Simc}(\boldsymbol{a},\boldsymbol{x}_1,\cdots,\boldsymbol{x}_k)$, 存在 $\lambda_i,\beta_i\geqslant 0, i=1,2,\cdots,k$ 使得

$$\boldsymbol{x}=\boldsymbol{a}+\lambda_1(\boldsymbol{x}_1-\boldsymbol{a})+\cdots+\lambda_k(\boldsymbol{x}_k-\boldsymbol{a}), \tag{3.1.1-5}$$

$$\boldsymbol{y}=\boldsymbol{a}+\beta_1(\boldsymbol{x}_1-\boldsymbol{a})+\cdots+\beta_k(\boldsymbol{x}_k-\boldsymbol{a}). \tag{3.1.1-6}$$

对任意 $t>0$, 通过 (3.1.1-5) 和 (3.1.1-6) 式有

$$\begin{aligned}&(1-t)\boldsymbol{x}+t\boldsymbol{y}\\ =&\,\boldsymbol{a}+((1-t)\lambda_1+t\beta_1)(\boldsymbol{x}_1-\boldsymbol{a})+\cdots+((1-t)\lambda_k+t\beta_k)(\boldsymbol{x}_k-\boldsymbol{a})\\ \in&\,\operatorname{aff}(\boldsymbol{a},\boldsymbol{x}_1,\cdots,\boldsymbol{x}_k).\end{aligned}$$

上式表明 $\operatorname{Simc}(\boldsymbol{a},\boldsymbol{x}_1,\cdots,\boldsymbol{x}_k)\in\operatorname{aff}(\boldsymbol{a},\boldsymbol{x}_1,\cdots,\boldsymbol{x}_k)$, 即有

$$\operatorname{aff}(\operatorname{Simc}(\boldsymbol{a},\boldsymbol{x}_1,\cdots,\boldsymbol{x}_k))\subset\operatorname{aff}(\boldsymbol{a},\boldsymbol{x}_1,\cdots,\boldsymbol{x}_k).$$

因为 $\{\boldsymbol{a},\boldsymbol{x}_1,\cdots,\boldsymbol{x}_k\}$ 是 $\mathbf{R}^n$ 中的仿射无关集合, 由性质 2.1.7(4) 知 $\operatorname{aff}(\boldsymbol{a},\boldsymbol{x}_1,\cdots,\boldsymbol{x}_k)$ 是包含 $\{\boldsymbol{a},\boldsymbol{x}_1,\cdots,\boldsymbol{x}_k\}$ 的最小仿射集. 因此, 有

$$\operatorname{aff}(\operatorname{Simc}(\boldsymbol{a},\boldsymbol{x}_1,\cdots,\boldsymbol{x}_k))=\operatorname{aff}(\boldsymbol{a},\boldsymbol{x}_1,\cdots,\boldsymbol{x}_k).$$

由定义 3.1.3 知, 显然 $\operatorname{Simc}(\boldsymbol{a},\boldsymbol{x}_1,\cdots,\boldsymbol{x}_k)$ 是闭集.

因为 $A-\boldsymbol{a}=\operatorname{Simc}(\boldsymbol{a},\boldsymbol{x}_1,\cdots,\boldsymbol{x}_k)-\boldsymbol{a}$ 是关于顶点为原点的凸锥, 记

$$A_0=\{\lambda_1(\boldsymbol{x}_1-\boldsymbol{a})+\cdots+\lambda_k(\boldsymbol{x}_k-\boldsymbol{a})|\lambda_i>0,i=1,2,\cdots,k\}.$$

显然对任意 $\boldsymbol{x}\in A_0$ 有 (3.1.1-5) 式成立, 且 $\{\boldsymbol{x}_1-\boldsymbol{a},\cdots,\boldsymbol{x}_k-\boldsymbol{a}\}$ 是线性无关向量. 则对任何 $\boldsymbol{y}\in B$, 存在不全为 0 的常数 $\alpha_1,\cdots,\alpha_k$ 使得

$$\boldsymbol{y}=\alpha_1(\boldsymbol{x}_1-\boldsymbol{a})+\cdots+\alpha_k(\boldsymbol{x}_k-\boldsymbol{a}).$$

令 $\bar{\alpha}=\min\{\alpha_1,\cdots,\alpha_k|\ \forall\boldsymbol{y}\in B\}$, 因为 B 是单位球, 所以 $\bar{\alpha}$ 有界且不为 0, 不妨取

$$\varepsilon=0.5\min\{\lambda_1,\cdots,\lambda_k\}/|\bar{\alpha}|.$$

那么有

$$\lambda_i+\varepsilon\alpha_i\geqslant\min\{\lambda_1,\cdots,\lambda_k\}-\varepsilon\bar{\alpha}>0,\quad i=1,2,\cdots,k.$$

我们可得到

$$\boldsymbol{x}+\varepsilon\boldsymbol{y}=(\lambda_1+\varepsilon\alpha_1)(\boldsymbol{x}_1-\boldsymbol{a})+\cdots+(\lambda_k+\varepsilon\alpha_k)(\boldsymbol{x}_k-\boldsymbol{a})\in A_0. \tag{3.1.1-7}$$

由上式得 $A_0\subset \mathrm{ri}A-\boldsymbol{a}$. 另外通过 (3.1.1-7) 式得 $\mathrm{ri}A-\boldsymbol{a}\subset A_0$, 则 (3.1.1-3) 式成立. □

下面是锥族的交、并运算后的性质, 即同一个顶点锥族的交或并后的集合还是同一个顶点锥, 同样, 同一个顶点锥族的并后的凸包还是同一个顶点凸锥.

性质 3.1.3 设 I 是指标集, 集合族 $\{C_i\}(i\in I)$ 中每个集合 C_i 都是关于顶点 $\boldsymbol{a}$ 的锥, 则下面结论均成立:

(1) $\bigcup_{i\in I}C_i$ 和 $\bigcap_{i\in I}C_i$ 均是关于顶点 $\boldsymbol{a}$ 的锥;

(2) $\mathrm{co}(\bigcup_{i\in I}C_i)$ 是关于顶点 $\boldsymbol{a}$ 的凸锥.

证明 (1) 设 $\boldsymbol{x}\in\bigcup_{i\in I}C_i$, 存在某个 C_i 使得 $\boldsymbol{x}\in C_i$. 因此, 对于任意的 $\lambda>0$ 有

$$\boldsymbol{a}+\lambda(\boldsymbol{x}-\boldsymbol{a})\in C_i,\quad i\in I.$$

从而 $\boldsymbol{a}+\lambda(\boldsymbol{x}-\boldsymbol{a})\in\bigcup_{i\in I}C_i$, 故 $\bigcup_{i\in I}C_i$ 是关于顶点 $\boldsymbol{a}$ 的锥. 根据定义 3.1.2 容易知 $\bigcap_{i\in I}C_i$ 是关于顶点 $\boldsymbol{a}$ 的锥.

(2) 根据性质 3.1.2(1) 知结论成立. □

下面的结论表明有限个顶点 (凸) 锥的线性加法运算和还是一个顶点 (凸) 锥.

性质 3.1.4 设指标集 $I=\{1,2,\cdots,k\}$ 和有限集合族 $\{C_i\}(i\in I)$ 中每个集合 C_i 分别是关于顶点 $\boldsymbol{a}_i(i\in I)$ 的锥, $\alpha_i(i\in I)$ 是常数, 有下面结论成立:

(1) $\sum\limits_{i=1}^{k}\alpha_iC_i$ 是关于顶点 $\sum\limits_{i=1}^{k}\alpha_i\boldsymbol{a}_i$ 的锥;

(2) 如果每个集合 $C_i(i\in I)$ 分别是凸集, 则 $\sum\limits_{i=1}^{k}\alpha_iC_i$ 是凸锥.

证明 (1) 对于任意的 $\lambda>0$ 和任意的 $\boldsymbol{x}\in\sum\limits_{i=1}^{k}\alpha_iC_i$, 存在 $\boldsymbol{x}_i\in C_i(i=1,2,\cdots,k)$ 使得 $\boldsymbol{x}=\alpha_1\boldsymbol{x}_1+\cdots+\alpha_k\boldsymbol{x}_k$. 则有

$$\sum_{i=1}^{k}\alpha_i\boldsymbol{a}_i+\lambda\left(\boldsymbol{x}-\sum_{i=1}^{k}\alpha_i\boldsymbol{a}_i\right)=\alpha_i\sum_{i=1}^{k}(\boldsymbol{a}_i+\lambda(\boldsymbol{x}_i-\boldsymbol{a}_i))\in\sum_{i=1}^{k}\alpha_iC_i,$$

由上式表明 $\sum\limits_{i=1}^{k}\alpha_iC_i$ 是关于顶点 $\sum\limits_{i=1}^{k}\alpha_i\boldsymbol{a}_i$ 的锥.

(2) 由 (1) 知, $\sum_{i=1}^{k}\alpha_i C_i$ 是关于顶点 $\sum_{i=1}^{k}\alpha_i \boldsymbol{a}_i$ 的锥. 再根据性质 2.2.6(1) 易知, $\sum_{i=1}^{k}\alpha_i C_i$ 是凸集. 因此, $\sum_{i=1}^{k}\alpha_i C_i$ 是凸锥. □

定义 3.1.4 设 $\mathrm{ap}C$ 是顶点锥 $C\subset\mathbf{R}^n$ 的所有顶点构成的集合, 则称 $\mathrm{ap}C$ 是顶点锥 C 的顶点集. 如果 $C=\varnothing$, 则 $\mathrm{ap}C=\mathbf{R}^n$.

事实上, 存在一些顶点锥具有无限多个顶点, 但在实际应用中仅含一个顶点的锥是使用最多的. 下面的性质 3.1.5 表明, 如果任何锥顶点集 $\mathrm{ap}C$ 是一个平面, 要么 $\mathrm{ap}C$ 在锥 C 中, 要么不在 C 中 ($\mathrm{ap}C$ 和 C 不可能相交), 先看下面两个例子.

例 3.1.4 锥 $C=\{(x_1,x_2)^{\mathrm{T}}|x_1\geqslant 0,x_2\in\mathbf{R}\}$ 的顶点集: $\mathrm{ap}C=\{(0,x_2)^{\mathrm{T}}|x_2\in\mathbf{R}\}$. 说明顶点集可以有无限多个元素, 且 $\mathrm{ap}C\subset C$.

例 3.1.5 锥 $C=\{(x_1,x_2)^{\mathrm{T}}|x_1> 0,x_2\in\mathbf{R}\}$ 的顶点集: $\mathrm{ap}C=\{(0,x_2)^{\mathrm{T}}|x_2\in\mathbf{R}\}$. 说明顶点集可以有无限多个元素, 且 $\mathrm{ap}C\cap C=\varnothing$.

性质 3.1.5 设 $C\subset\mathbf{R}^n$, 则任何锥 C 的顶点集 $\mathrm{ap}C$ 是一个平面, 且有 $\mathrm{ap}C\subset C$ 或者

$$\mathrm{ap}C\cap C=\varnothing$$

成立.

证明 如果 C 是 $\varnothing$ 或单点集, 性质的结论显然成立. 所以假设 C 是非空单点集, 设 $\forall\boldsymbol{a}_1,\boldsymbol{a}_2\in\mathrm{ap}C$, 则 $C-\boldsymbol{a}_1$ 和 $C-\boldsymbol{a}_2$ 都是以原点为顶点的锥. 对于 $\alpha\in\mathbf{R}$, 由性质 3.1.4 得 $(1-\alpha)C+\alpha C$ 是关于顶点 $(1-\alpha)\boldsymbol{a}_1+\alpha\boldsymbol{a}_2$ 的锥. 不妨设 $\alpha>0$ 和 $\boldsymbol{a}=(1-\alpha)\boldsymbol{a}_1+\alpha\boldsymbol{a}_2$, 对任意 $\boldsymbol{x}\in C$ 和 $\alpha>0$ 有

$$C=\boldsymbol{a}_1+(1-\alpha)(C-\boldsymbol{a}_1)=\boldsymbol{a}_2+\alpha(C-\boldsymbol{a}_2). \tag{3.1.1-8}$$

由上式得

$$\alpha C+(1-\alpha)C-\boldsymbol{a}=C-\boldsymbol{a}_1+C-\boldsymbol{a}_2. \tag{3.1.1-9}$$

再由 (3.1.1-8) 式得

$$C-\boldsymbol{a}_2=\alpha(C-\boldsymbol{a}_2)=\boldsymbol{a}_1-\boldsymbol{a}_2+(1-\alpha)(C-\boldsymbol{a}_1).$$

将上式代入 (3.1.1-9) 式得

$$\begin{aligned}\alpha C+(1-\alpha)C-\boldsymbol{a}&=\boldsymbol{a}_1-\boldsymbol{a}_2+(2-\alpha)(C-\boldsymbol{a}_1)\\&=\boldsymbol{a}_1-\boldsymbol{a}_2+C-\boldsymbol{a}_1\\&=C-\boldsymbol{a}_2.\end{aligned}$$

上式说明 $\alpha C+(1-\alpha)C$ 是关于顶点 $\boldsymbol{a}$ 的锥, 等价于 $C-\boldsymbol{a}_2$ 是关于顶点 $\boldsymbol{a}$ 的锥. 因此, 直线 $l\langle \boldsymbol{a}_1,\boldsymbol{a}_2\rangle \subset \mathrm{ap}C$, 可得 $\mathrm{ap}C$ 是一个平面.

假设存在顶点 $\boldsymbol{a}\in C\cap \mathrm{ap}C$. 如果 $\mathrm{ap}C=\{\boldsymbol{a}\}$, 则 $\mathrm{ap}C\subset C$ 是显而易见的.

假设存在 C 的另一个顶点 $\boldsymbol{a}_1$, 由性质 3.1.6(1) 的证明得 $\{(1-\alpha)\boldsymbol{a}_1+\alpha\boldsymbol{a}|\alpha>0\}\subset \mathrm{ri}C$, 由上面证明知 $l\langle \boldsymbol{a}_1,\boldsymbol{a}\rangle\subset C$. 因此, $\mathrm{ap}C\subset C$. □

本小节得到了顶点锥与顶点凸锥的一些基本性质:

1) 锥的顶点可以属于锥, 也可以不属于锥;

2) 仿射集都是锥, 反之不一定;

3) 凸锥等价于它含有任意两个点的正线性组合;

4) 任何一个锥可张成一个仿射集;

5) 顶点锥的凸包仍然是顶点锥;

6) 如果一个含有顶点 $\boldsymbol{a}$ 的非空锥, 去掉该顶点后为凸集, 那么该锥是具有唯一顶点的凸锥;

7) C 是以 $\boldsymbol{a}$ 为顶点的凸锥的充要条件是存在一个不等于 1 的正数 λ 使得 $C=\boldsymbol{a}+\lambda(C-\boldsymbol{a})$ 成立;

8) k-简单锥是一个凸锥;

9) 同一个顶点的锥族的交或并后的集合仍然是同一个顶点的锥, 同一个顶点的锥族的并后的凸包是同一个顶点的凸锥;

10) 有限个顶点 (凸) 锥的线性加法运算和还是顶点 (凸) 锥;

11) 任何锥的顶点集是一个平面, 顶点集要么属于该锥, 要么不属于.

3.1.2　拓扑性质

本小节讨论顶点锥的相对内部、闭包和相对边界之间的拓扑关系. 首先, 下面的性质 3.1.6 揭示了顶点锥中通过内点和锥包的点与顶点构成的半直线关系, 即相对内点与顶点构成的半直线仍然在锥的相对内部中, 锥的闭包中的点与顶点构成的半直线仍然在闭包中.

性质 3.1.6　设 C 是关于顶点 $\boldsymbol{a}$ 的非空锥, 点 $\boldsymbol{x}\in\mathbf{R}^n$, 且 $\boldsymbol{x}\neq\boldsymbol{a}$, 有下面结论成立:

(1) 如果 $\boldsymbol{x}\in\mathrm{ri}C$, 则 $l(\boldsymbol{a},\boldsymbol{x}\rangle=\{(1-\lambda)\boldsymbol{a}+\lambda\boldsymbol{x}|\lambda>0\}\subset\mathrm{ri}C$;

(2) 如果 $\boldsymbol{x}\in\mathrm{cl}C$, 则 $l[\boldsymbol{a},\boldsymbol{x}\rangle=\{(1-\lambda)\boldsymbol{a}+\lambda\boldsymbol{x}|\lambda\geqslant 0\}\subset\mathrm{cl}C$;

(3) 如果 C 不是一个平面且 $\boldsymbol{x}\in\mathrm{rb}C$, 则 $l[\boldsymbol{a},\boldsymbol{x}\rangle=\{(1-\lambda)\boldsymbol{a}+\lambda\boldsymbol{x}|\lambda\geqslant 0\}\subset\mathrm{rb}C$.

证明　(1) 设 $\boldsymbol{x}\in\mathrm{ri}C$ 和 $\lambda>0$, 存在 $\varepsilon>0$ 使得 $(\boldsymbol{x}-\boldsymbol{a}+\varepsilon B)\cap\mathrm{aff}(C-\boldsymbol{a})\subset C-\boldsymbol{a}$. 事实上, 对 $t>0$, 当 $t<\varepsilon$ 时有 $(\boldsymbol{x}-\boldsymbol{a}+tB)\cap\mathrm{aff}(C-\boldsymbol{a})\subset C-\boldsymbol{a}$, 显然有

$$\lambda(\boldsymbol{x}-\boldsymbol{a}+tB)\subset\lambda((\boldsymbol{x}-\boldsymbol{a})+tB)\cap\lambda\mathrm{aff}(C-\boldsymbol{a})\subset\lambda(C-\boldsymbol{a})=C-\boldsymbol{a}.$$

因此, 有 $\lambda(\boldsymbol{x}-\boldsymbol{a}) \in \mathrm{ri}C-\boldsymbol{a}$.

(2) 设 $\boldsymbol{x} \in \mathrm{cl}C$, 对任意的 $\varepsilon>0$ 有 $(\boldsymbol{x}-\boldsymbol{a}+\varepsilon B) \cap (C-\boldsymbol{a}) \neq \varnothing$. 因为 $C-\boldsymbol{a}$ 是以原点为顶点的锥, 所以对于 $\lambda \geqslant 0$, 有

$$\lambda(\boldsymbol{x}-\boldsymbol{a}+\varepsilon B) \cap (C-\boldsymbol{a}) = (\lambda(\boldsymbol{x}-\boldsymbol{a})+\lambda\varepsilon B) \cap \lambda(C-\boldsymbol{a}) \neq \varnothing.$$

上式说明 $\lambda(\boldsymbol{x}-\boldsymbol{a}) \in \mathrm{cl}C-\boldsymbol{a}$.

(3) 设 $\boldsymbol{x} \in \mathrm{rb}C=\mathrm{cl}C \setminus \mathrm{ri}C$, 由 (2) 有 $\{(1-\lambda)\boldsymbol{a}+\lambda\boldsymbol{x} | \lambda \geqslant 0\} \in \mathrm{cl}C$. 如果存在某个 $\lambda_0>0$ 使得 $(1-\lambda_0)\boldsymbol{a}+\lambda_0\boldsymbol{x} \in \mathrm{ri}C$, 其中 $\lambda \neq 1$. 因此, 再由 (1) 有

$$\{(1-\lambda)\boldsymbol{a}+\lambda((1-\lambda_0)\boldsymbol{a}+\lambda_0\boldsymbol{x}) | \lambda>0\} \in \mathrm{ri}C.$$

令 $\lambda\lambda_0=1$, 则由上式知

$$(1-\lambda)\boldsymbol{a}+\lambda((1-\lambda_0)\boldsymbol{a}+\lambda_0\boldsymbol{x}) = (1-\lambda\lambda_0)\boldsymbol{a}+\lambda\lambda_0\boldsymbol{x} = \boldsymbol{x} \in \mathrm{ri}C,$$

这与 $\boldsymbol{x} \in \mathrm{rb}C=\mathrm{cl}C \setminus \mathrm{ri}C$ 矛盾. 因此, $\{(1-\lambda)\boldsymbol{a}+\lambda\boldsymbol{x} | \lambda \geqslant 0\} \in \mathrm{rb}C$. □

例 3.1.6 设 $\boldsymbol{a} \in \mathbf{R}^n$ 和凸锥 $C=\{\boldsymbol{x} \in \mathbf{R}^n |\ x_i \geqslant a_i,\ i=1,2,\cdots,n\}$, 有 $\mathrm{cl}C=C$ 和

$$\mathrm{ri}C=\{\boldsymbol{x} \in \mathbf{R}^n |\ x_i>a_i,\ i=1,2,\cdots,n\},$$

$$\mathrm{rb}C=\{\boldsymbol{x} \in \mathbf{R}^n |\ x_i=a_i,\ i=1,2,\cdots,n\}.$$

显然性质 3.1.6 的结论对例 3.1.6 都成立, 且 $\mathrm{cl}C, \mathrm{ri}C$和 $\mathrm{rb}C$ 是关于顶点 $\boldsymbol{a}$ 的一个非空凸锥, 且知 $\mathrm{ri}C=C+\mathrm{ri}C=\mathrm{cl}C+\mathrm{ri}C$ 也成立.

性质 3.1.7 设 C 是关于顶点 $\boldsymbol{a}$ 的一个非空锥, 有下面结论成立:

(1) $\mathrm{cl}C, \mathrm{ri}C$ 和 $\mathrm{rb}C$ 是关于顶点 $\boldsymbol{a}$ 的非空锥;

(2) 如果 C 是凸集, 则 $\mathrm{cl}C$ 和 $\mathrm{ri}C$ 是凸集, 且有

$$\mathrm{ri}C=C+\mathrm{ri}C=\mathrm{cl}C+\mathrm{ri}C.$$

证明 (1) 由性质 3.1.6 知 $\mathrm{cl}C, \mathrm{ri}C$ 和 $\mathrm{rb}C$ 是关于顶点 $\boldsymbol{a}$ 的一个非空锥.

(2) 由性质 2.3.2 和性质 2.3.3 知 $\mathrm{cl}C$ 和 $\mathrm{ri}C$ 是凸集, 由性质 2.3.9 得

$$\mathrm{ri}C=\mathrm{ri}(2C)=\mathrm{ri}C+\mathrm{ri}C \subset C+\mathrm{ri}C \subset \mathrm{cl}C+\mathrm{ri}C. \tag{3.1.2-1}$$

再由定理 2.3.2, 对于 $0<\lambda<1$ 有

$$(1-\lambda)\,\mathrm{cl}C+\lambda\mathrm{ri}C \subset \mathrm{ri}C, \tag{3.1.2-2}$$

$$\lambda \mathrm{cl} C + (1-\lambda)\,\mathrm{ri} C \subset \mathrm{ri} C. \tag{3.1.2-3}$$

将 (3.1.2-2) 和 (3.1.2-3) 两式相加, 可得

$$\mathrm{cl} C + \mathrm{ri} C \subset 2\mathrm{ri} C = \mathrm{ri} C.$$

结合 (3.1.2-1) 式和上式, 有 $\mathrm{ri} C = \mathrm{ri} C + \mathrm{ri} C = C + \mathrm{ri} C = \mathrm{cl} C + \mathrm{ri} C$. □

性质 3.1.7 表明顶点锥的相对内部、闭包和相对边界仍然构成相同顶点的锥, 顶点凸锥的相对内部和闭包仍然是相同顶点的凸锥.

下面的性质 3.1.8 进一步表明锥与顶点的邻域之间的关系: 顶点锥的凸性和闭性保持了顶点的邻域与顶点锥相交的凸性和闭性一致性, 即顶点邻域与锥相交为凸集等价于该锥是一个凸锥, 并且顶点邻域与锥相交为闭集等价于该锥是一个闭集. 换句话说, 一个锥的拓扑性可以通过其顶点邻域拓扑与锥的交集来判定, 顶点邻域决定锥的凸性或闭性.

性质 3.1.8 设 C 是关于顶点 $\boldsymbol{a}$ 的非空锥, 给定 $\varepsilon > 0$, 则下面结论成立:

(1) C 是凸集当且仅当 $C \cap B(\boldsymbol{a},\varepsilon)$ 和 $C \cap U(\boldsymbol{a},\varepsilon)$ 是凸集;

(2) $\mathrm{cl} C \cap B(\boldsymbol{a},\varepsilon) = \mathrm{cl}(C \cap B(\boldsymbol{a},\varepsilon)) = \mathrm{cl}(C \cap U(\boldsymbol{a},\varepsilon))$;

(3) C 是闭集当且仅当 $C \cap B(\boldsymbol{a},\varepsilon)$ 是闭集;

(4) $\mathrm{cl} C \cap S(\boldsymbol{a},\varepsilon) = \mathrm{cl}(C \cap S(\boldsymbol{a},\varepsilon))$;

(5) $\{\boldsymbol{a}\} \cup C$ 是闭集当且仅当 $C \cap S(\boldsymbol{a},\varepsilon)$ 是闭集.

证明 (1) 由定义 1.2.1 知 $B(\boldsymbol{a},\varepsilon)$ 和 $U(\boldsymbol{a},\varepsilon)$ 是凸集, 所以若 C 是凸集, 则

$$C \cap B(\boldsymbol{a},\varepsilon) \text{ 和 } C \cap U(\boldsymbol{a},\varepsilon)$$

是凸集. 反之, 如果 $C \cap B(\boldsymbol{a},\varepsilon)$ 和 $C \cap U(\boldsymbol{a},\varepsilon)$ 是凸集, 显然可知 $(C-\boldsymbol{a}) \cap (B(\boldsymbol{a},\varepsilon)-\boldsymbol{a})$ 和 $(C-\boldsymbol{a}) \cap (U(\boldsymbol{a},\varepsilon)-\boldsymbol{a})$ 是凸集. 设 $\forall \boldsymbol{x},\boldsymbol{y} \in C$ 和 $\forall \lambda > 0$, 那么有

$$\lambda(\boldsymbol{x}-\boldsymbol{a}), \lambda(\boldsymbol{y}-\boldsymbol{a}) \in C - \boldsymbol{a}.$$

取 $\lambda = \min\{0.5\varepsilon/||\boldsymbol{x}-\boldsymbol{a}||, 0.5\varepsilon/||\boldsymbol{y}-\boldsymbol{a}||\}$, 则得 $||\lambda(\boldsymbol{x}-\boldsymbol{a})|| < \varepsilon, ||\lambda(\boldsymbol{y}-\boldsymbol{a})|| < \varepsilon$. 从而有

$$\lambda(\boldsymbol{x}-\boldsymbol{a}), \lambda(\boldsymbol{y}-\boldsymbol{a}) \in U(\boldsymbol{a},\varepsilon) - \boldsymbol{a} \subset B(\boldsymbol{a},\varepsilon) - \boldsymbol{a}.$$

对任意的 $t \in (0,1)$, 由上式可得: $\lambda(1-t)(\boldsymbol{x}-\boldsymbol{a}) + t\lambda(\boldsymbol{y}-\boldsymbol{a}) \in C - \boldsymbol{a}$. 由 $C-\boldsymbol{a}$ 是一个锥可知 $(1-t)\boldsymbol{x} + t\boldsymbol{y} - \boldsymbol{a} \in C - \boldsymbol{a}$ 成立. 因此, C 是凸集.

(2) 因为有 $C \cap U(\boldsymbol{a},\varepsilon) \subset C \cap B(\boldsymbol{a},\varepsilon) \subset \mathrm{cl} C \cap B(\boldsymbol{a},\varepsilon) = \mathrm{cl}(\mathrm{cl} C \cap B(\boldsymbol{a},\varepsilon))$ 成立, 则有

$$\mathrm{cl}(C \cap U(\boldsymbol{a},\varepsilon)) \subset \mathrm{cl}(C \cap B(\boldsymbol{a},\varepsilon)) \subset \mathrm{cl} C \cap B(\boldsymbol{a},\varepsilon) = \mathrm{cl}(\mathrm{cl} C \cap B(\boldsymbol{a},\varepsilon)).$$

设 $\boldsymbol{x} \in \mathrm{cl}C \cap B(\boldsymbol{a},\varepsilon)$, 由 $\mathrm{cl}C \cap B(\boldsymbol{a},\varepsilon)$ 是闭集, 则存在一个收敛到 $\boldsymbol{x}$ 的序列 $\{\boldsymbol{x}_i\} \subset C$. 若 $\boldsymbol{x}=\boldsymbol{a}$, 显然有 $\boldsymbol{x} \in \mathrm{cl}\,(C \cap U(\boldsymbol{a},\varepsilon))$. 这里假设 $\boldsymbol{x} \neq \boldsymbol{a}$, 设

$$\boldsymbol{y}_i = \boldsymbol{a} + \frac{i||\boldsymbol{x}-\boldsymbol{a}||}{(i+\varepsilon)\,||\boldsymbol{x}_i-\boldsymbol{a}||}\,(\boldsymbol{x}_i-\boldsymbol{a}) \in C, \quad i=1,2,\cdots.$$

由上式得 $||\boldsymbol{y}_i-\boldsymbol{a}|| = \dfrac{i||\boldsymbol{x}-\boldsymbol{a}||}{(i+\varepsilon)\,||\boldsymbol{x}_i-\boldsymbol{a}||}||\boldsymbol{x}_i-\boldsymbol{a}|| < \varepsilon$, 有 $\boldsymbol{y}_i \in (C \cap U(\boldsymbol{a},\varepsilon))$. 令 $i \to +\infty$ 时, 有 $\boldsymbol{y}_i \to \boldsymbol{x}$, 表明 $\boldsymbol{x} \in \mathrm{cl}\,(C \cap U(\boldsymbol{a},\varepsilon))$. 得 $\mathrm{cl}\,(C \cap U(\boldsymbol{a},\varepsilon)) = \mathrm{cl}C \cap B(\boldsymbol{a},\varepsilon)$, 同时有

$$\mathrm{cl}\,(C \cap B(\boldsymbol{a},\varepsilon)) = \mathrm{cl}C \cap B(\boldsymbol{a},\varepsilon).$$

(3) 若 C 是闭集, 显然 $C \cap B(\boldsymbol{a},\varepsilon)$ 是闭集. 反之, 若 $C \cap B(\boldsymbol{a},\varepsilon)$ 是闭集, 设 $\boldsymbol{x} \in \mathrm{cl}C$, 则存在一个收敛到 $\boldsymbol{x}$ 的序列 $\{\boldsymbol{x}_i\} \subset C$. 若 $\boldsymbol{x}=\boldsymbol{a}$, 显然有 $\boldsymbol{x}=\boldsymbol{a} \in C \cap B(\boldsymbol{a},\varepsilon)$. 这里假设 $\boldsymbol{x} \neq \boldsymbol{a}$, 设

$$\boldsymbol{y}_i = \boldsymbol{a} + \frac{\varepsilon}{||\boldsymbol{x}_i-\boldsymbol{a}||}\,(\boldsymbol{x}_i-\boldsymbol{a}) \in C, \quad \boldsymbol{y} = \boldsymbol{a} + \frac{\varepsilon}{||\boldsymbol{x}-\boldsymbol{a}||}\,(\boldsymbol{x}-\boldsymbol{a}), \quad i=1,2,\cdots.$$

由 $||\boldsymbol{y}_i-\boldsymbol{a}||=\varepsilon$ 和 $||\boldsymbol{y}-\boldsymbol{a}||=\varepsilon$, 得 $\boldsymbol{y}_i \in C \cap B(\boldsymbol{a},\varepsilon)$. 令 $i \to +\infty$ 时, 有

$$\boldsymbol{y}_i \to \boldsymbol{y} \in C \cap B(\boldsymbol{a},\varepsilon) \subset C,$$

由顶点锥定义得 $\{(1-\lambda)\boldsymbol{a}+\lambda\boldsymbol{y}|\lambda \geqslant 0\} \subset C$. 取 $\lambda = \dfrac{||\boldsymbol{x}-\boldsymbol{a}||}{\varepsilon}$, 即有

$$(1-\lambda)\boldsymbol{a}+\lambda\left(\boldsymbol{a}+\frac{\varepsilon}{||\boldsymbol{x}-\boldsymbol{a}||}\,(\boldsymbol{x}-\boldsymbol{a})\right) = \boldsymbol{a}+\frac{\varepsilon\lambda}{||\boldsymbol{x}-\boldsymbol{a}||}\,(\boldsymbol{x}-\boldsymbol{a}) = \boldsymbol{x} \in C,$$

说明 C 是闭集.

(4) 和 (5) 的证明与 (2) 和 (3) 的证明类似. □

例 3.1.7 在 $\mathbf{R}^2$ 空间中的锥 $C=\{(x_1,0)^{\mathrm{T}}|x_1 \in (0,+\infty)\} \cup \{(0,x_2)^{\mathrm{T}}|x_2 \in (0,+\infty)\}$ 是由两个半坐标轴构成的, C 的顶点是原点, 但不含该顶点, 且 C 不是凸锥, 也不是闭集. 易知性质 3.1.8 的结论对本例中的 C 都不成立.

顶点凸锥与顶点集存在着密切关系, 下面证明它们之间的等价性: 顶点凸锥是一个平面等价于顶点集与顶点凸锥相等, 或者等价于顶点集属于顶点凸锥的相对内部, 或者等价于顶点集与凸锥的相对内部交非空.

性质 3.1.9 设 C 是关于顶点集 $\mathrm{ap}C$ 的顶点凸锥, 则有下面结论等价:

(1) C 是一个平面;

(2) $\text{ap}C = C$;

(3) $\text{ap}C \subset \text{ri}C$;

(4) $\text{ap}C \cap \text{ri}C \neq \varnothing$.

证明　先证明 (1) 和 (2) 等价. 若 C 是一个平面, 存在点 $\boldsymbol{b}$ 和子空间 S 使得 $C = \boldsymbol{b} + S$. 假设取任意点 $\boldsymbol{a} \in C$, 则有 $\boldsymbol{a} = \boldsymbol{b} + \boldsymbol{s}$, 其中 $\boldsymbol{s} \in S$. 对于任意的点 $\boldsymbol{x} \in C$, 有 $\boldsymbol{x} = \boldsymbol{b} + \boldsymbol{s}'$, 其中 $\boldsymbol{s}' \in S$. 那么对于任何 $\lambda > 0$ 有

$$\boldsymbol{a} + \lambda(\boldsymbol{x} - \boldsymbol{a}) = \boldsymbol{b} + \boldsymbol{s} + \lambda\boldsymbol{s} - \lambda\boldsymbol{s}' \in \boldsymbol{b} + S.$$

上式说明 $\boldsymbol{a} \in \text{ap}C$, 即 $C \subset \text{ap}C$, 再由性质 3.1.5 推出 $\text{ap}C = C$. 反之, 如果 $\text{ap}C = C$, 由性质 3.1.5 即可知 C 是一个平面.

下面证明若 (2) 成立, 则 (3) 也成立. 因为 $\text{ap}C = C$, 且 C 是一个平面和非空顶点凸锥, 根据性质 2.3.4(5) 得 $C = \text{ri}C$. 因此, (3) 成立.

若 (3) 成立, 显然 (4) 也是成立的.

最后证明, 若 (4) 成立, 则 (2) 也成立. 设 $\forall \boldsymbol{x} \in C$ 和 $\boldsymbol{a} \in \text{ap}C \cap \text{ri}C \subset \text{ri}C$, 且 $\boldsymbol{a} \neq \boldsymbol{x}$, 有

$$\{(1-\lambda)\boldsymbol{a} + \lambda\boldsymbol{x} | \lambda > 0\} \in C.$$

由性质 3.1.5 知 $\text{ap}C \subset C$ 且 $\text{ap}C$ 是一个平面, 以及 $(\boldsymbol{a} + \varepsilon B) \cap \text{aff}C \subset C$, 说明直线

$$\{(1-\lambda)\boldsymbol{a} + \lambda\boldsymbol{x} | \lambda > 0\}$$

中存在一个 $\boldsymbol{y} \in \text{ap}C$ 和充分小的 $t > 0$, 使得 $\boldsymbol{y} = (1-t)\boldsymbol{a} + t\boldsymbol{x}$. 当 $0 < \alpha \leqslant t$ 时有

$$(1-\alpha)\boldsymbol{a} + \alpha\boldsymbol{y} \in \text{ap}C.$$

因此, 线段 $l\langle\boldsymbol{a}, \boldsymbol{y}\rangle \subset \{(1-\lambda)\boldsymbol{a} + \lambda\boldsymbol{x} | \lambda > 0\}$, 由于 $\text{ap}C$ 是一个平面, 说明直线 $l\langle\boldsymbol{a}, \boldsymbol{x}\rangle$ 在 $\text{ap}C$ 中, 即有 $\boldsymbol{x} \in \text{ap}C$. □

例 3.1.8　在 $\mathbf{R}^2$ 空间中的锥 $C = \{(x_1, 0)^{\mathrm{T}} | x_1 \in (0, +\infty)\} \cup \{(0, x_2)^{\mathrm{T}} | x_2 \in (0, +\infty)\}$ 是由两个半坐标轴构成的, 容易知锥 C 对性质 3.1.9 的结论都不成立.

例 3.1.9　在 $\mathbf{R}^2$ 空间中的锥 $C = \{(x_1, x_2)^{\mathrm{T}} | x_1, x_2 > 0\}$ 是顶点为原点的凸锥, 但不含原点, 且对性质 3.1.9 的结论均不成立, 因为 C 不是一个平面, $\text{ap}C = \{(0, 0)\}$.

性质 3.1.10　设 C 是关于顶点集 $\text{ap}C$ 的顶点凸锥, 则下面结论等价:

(1) C 不是一个平面;

(2) $\text{ap}C \neq C$;

(3) $\text{ap}C \subset \text{rb}C$;

(4) $\text{ap}C \cap \text{rb}C \neq \varnothing$.

证明 根据性质 3.1.9, 显然结论都成立. □

性质 3.1.11 设 C 是关于顶点集 ap$C$ 的顶点凸锥, 则 cl$C$, ap$C\cup$ri$C$ 和 ap(cl$C$)$\cup$ri$C$ 都是关于顶点集 ap$C$ 的顶点凸锥, 且 ap$C\cup$ri$C\subset C$ 和 ap(cl$C$)$\cup$ri$C\subset$ cl$C$, 以及

$$\mathrm{ap}C=\mathrm{ap}(\mathrm{ap}C\cup\mathrm{ri}C)\subset\mathrm{ap}(\mathrm{cl}C)=\mathrm{ap}\left(\mathrm{ap}(\mathrm{cl}C)\cup\mathrm{ri}C\right).$$

证明 由性质 3.1.7 得 cl$C$ 和 ri$C$ 是凸锥, 下面分两种情形证明后面的结论成立.

(1) 当 C 是一个平面时, 由性质 3.1.8 得 cl$C=C$, ap$C=C$ 和 $C=$ ri$C$. ap$C$ 显然也是顶点凸锥, 由性质 3.1.3 知 ap$C\cup$ri$C$ 和 ap(cl$C$)$\cup$ri$C$ 都是顶点凸锥, 且 ap$C\cup$ri$C=C$, 后面结论均成立.

(2) 当 C 不是一个平面时, 由性质 3.1.9(3) 得 ap$C\cup$ri$C\subset C$ 和 ap(cl$C$)$\cup$ri$C\subset$ cl$C$, 且

$$\mathrm{ap}(\mathrm{ap}C\cup\mathrm{ri}C)\subset\mathrm{ap}C\subset\mathrm{ap}\left(\mathrm{ap}(\mathrm{cl}C)\cup\mathrm{ri}C\right)\subset\mathrm{ap}(\mathrm{cl}C).$$

因为 ap$C$ 和 ri$C$ 是凸集, 设 $\boldsymbol{x}\in$ ap$C$ 和 $\boldsymbol{y}\in$ ri$C$, 由性质 3.1.9 得 $\boldsymbol{x}\in$ ap$C\subset$ rb$C$. 由定理 2.3.2 知, 对任意 $t\in(0,1)$ 有 $(1-t)\boldsymbol{x}+t\boldsymbol{y}\in$ ri$C$, 说明 ap$C\cup$ri$C$ 是凸集. 类似地, 由 cl$C$ 不是一个平面可推出 ap(cl$C$)$\cup$ri$C$ 也是凸集.

下面证明 ap(ap$C\cup$ri$C$) = ap$C$. 设 $\boldsymbol{a}\in$ ap$C$ 和 $\boldsymbol{x}\in$ ap$C\cup$ri$C$, 若 $\boldsymbol{x}\in$ ap$C$, 由性质 3.1.5 得, 对任意的 $\lambda>0$ 有 $(1-\lambda)\boldsymbol{a}+\lambda\boldsymbol{x}\in$ ap$C$. 若 $\boldsymbol{x}\in$ ri$C$, 那么由性质 3.1.6(1) 得

$$\{(1-\lambda)\boldsymbol{a}+\lambda\boldsymbol{x}|\lambda>0\}\in\mathrm{ri}C.$$

因此, 有 $\boldsymbol{a}\in$ ap(ap$C\cup$ri$C$). 类似也可证明 ap(ap(cl$C$)$\cup$ri$C$) = ap(cl$C$). □

容易知性质 3.1.10 和性质 3.1.11 的结论对例 3.1.9 定义的锥 C 成立.

例 3.1.10 在 $\mathbf{R}^2$ 中的顶点锥 $C=\{(x_1,0)^{\mathrm{T}}|x_1\in(0,+\infty)\}\cup\{(0,x_2)^{\mathrm{T}}|x_2\in(0,+\infty)\}$ 是由两个半坐标轴构成的, 不是顶点凸锥, 所以性质 3.1.11 的结论对本例中的顶点锥均不成立.

本小节得到了顶点锥与顶点凸锥的一些拓扑性质:

1) 顶点锥的相对内点与顶点构成的半直线仍然在顶点锥的相对内部中, 顶点锥的闭包中的点与顶点构成的半直线仍然在闭包中;

2) 顶点锥的相对内部、闭包和相对边界都是有相同顶点的锥;

3) 顶点凸锥的相对内部和闭包仍然是顶点凸锥;

4) 顶点锥是一个凸锥等价于该锥的顶点邻域与锥相交为凸集, 并且顶点邻域与顶点锥相交为闭集等价于该顶点锥是一个闭集;

5) 顶点凸锥是一个平面等价于顶点集等于凸锥, 或者等价于顶点集属于凸锥的相对内部, 或者等价于顶点集与凸锥的相对内部交非空;

6) 顶点凸锥不是一个平面等价于顶点集不等于凸锥, 或者等价于顶点集在凸锥的相对边界里, 或者等价于顶点集与凸锥的相对边界交非空;

7) 顶点凸锥的闭包、相对内部与顶点集的并、相对内部与闭包的顶点集的并都是具有相同顶点集的顶点凸锥.

3.1.3　原点凸锥

由定义 3.1.1 知以原点为顶点的凸锥 $C \subset \mathbf{R}^n$ 只含一个顶点. 本小节仅讨论原点凸锥的一些重要性质. 在 $\mathbf{R}^n$ 中存在许多原点凸锥的情形, 例如过原点的闭、开半空间和子空间都是以原点为顶点的凸锥. 为了描述方便, 以下凸锥均是以原点为顶点的凸锥.

在 $\mathbf{R}^n$ 中存在两个重要的凸锥, 定义如下.

非负真锥: $\mathbf{R}^n_+ = \{\boldsymbol{x} = (x_1, \cdots, x_n)\,|x_1 \geqslant 0, \cdots, x_n \geqslant 0\}$,

正真锥: $\operatorname{int} \mathbf{R}^n_+ = \{\boldsymbol{x} = (x_1, \cdots, x_n)\,|x_1 > 0, \cdots, x_n > 0\}$.

在向量优化中, 上述两个锥也称为 Pareto 锥, 可用于刻画 $\mathbf{R}^n$ 中的偏序 (比较大小).

由定理 3.1.3 得下面结论, 凸锥由其中任意两个元素的非负线性组合而成.

定理 3.1.5　设非空集合 $C \subset \mathbf{R}^n$, 则 C 是凸锥的充要条件为

$$C = \lambda C + \mu C, \quad \lambda, \mu > 0.$$

例 3.1.11　设 $b_i \in \mathbf{R}^n\,(i \in I)$, I 是任意指标集, 则闭半空间族的交

$$C = \{\boldsymbol{x} \in \mathbf{R}^n\,|\langle \boldsymbol{x}, b_i \rangle \leqslant 0, i \in I\}$$

是凸锥.

推论 3.1.1　设非空集合 $C \subset \mathbf{R}^n$, 则 C 是凸锥的充要条件为

$$\lambda_1 \boldsymbol{x}_1 + \cdots + \lambda_m \boldsymbol{x}_m \in C, \quad \forall \boldsymbol{x}_i \in C, \quad \forall \lambda_i > 0, \quad i = 1, \cdots, m.$$

由推论 3.1.1 得下面结论.

推论 3.1.2　设 S 为非空集合 $C \subset \mathbf{R}^n$ 的子集, 则集合

$$C = \{\lambda_1 \boldsymbol{x}_1 + \cdots + \lambda_m \boldsymbol{x}_m | \boldsymbol{x}_i \in S, \lambda_i > 0, i = 1, \cdots, m\}$$

是包含 S 的最小凸锥.

推论 3.1.3 设非空凸集 $C \subset \mathbf{R}^n$, 则集合

$$\bar{C} = \{\lambda \boldsymbol{x} \,|\, \lambda > 0, \forall \boldsymbol{x} \in C\} = \bigcup_{\lambda>0} \lambda C$$

是包含 C 的最小凸锥.

证明 设任意的 $\lambda_1 > 0, \lambda_2 > 0, \boldsymbol{x}_1 \in C, \boldsymbol{x}_2 \in C$, 则 $\lambda_1 \boldsymbol{x}_1, \lambda_2 \boldsymbol{x}_2 \in \bar{C}$. 因为 C 是凸集, 且

$$0 < \frac{\lambda_1}{\lambda_1 + \lambda_2} < 1, \quad 0 < \frac{\lambda_2}{\lambda_1 + \lambda_2} < 1,$$

故 $\dfrac{\lambda_1}{\lambda_1 + \lambda_2} \boldsymbol{x}_1 + \dfrac{\lambda_2}{\lambda_1 + \lambda_2} \boldsymbol{x}_2 \in C$. 所以有

$$\lambda_1 \boldsymbol{x}_1 + \lambda_2 \boldsymbol{x}_2 = (\lambda_1 + \lambda_2) \left(\frac{\lambda_1}{\lambda_1 + \lambda_2} \boldsymbol{x}_1 + \frac{\lambda_2}{\lambda_1 + \lambda_2} \boldsymbol{x}_2 \right) \in \bar{C}.$$

根据定理 3.1.5 得 $\bar{C}$ 是凸锥.

上述证明过程容易推广为, 对于任意 $\boldsymbol{x}_1, \cdots, \boldsymbol{x}_m \in C$, $\lambda_i > 0\,(i = 1, \cdots, m)$, 由 C 是凸集可知

$$\begin{aligned} & \lambda_1 \boldsymbol{x}_1 + \cdots + \lambda_m \boldsymbol{x}_m \\ = & (\lambda_1 + \cdots + \lambda_m) \left(\frac{\lambda_1}{\lambda_1 + \cdots + \lambda_m} \boldsymbol{x}_1 + \cdots + \frac{\lambda_m}{\lambda_1 + \cdots + \lambda_m} \boldsymbol{x}_m \right) \in \bar{C}. \end{aligned}$$

这表示 C 的点的正线性组合所成的全体集合包含在 $\bar{C}$ 中, 由推论 2.2.1 可知, $\bar{C}$ 是包含 C 的最小凸锥. □

由性质 3.1.3 可知下面结论成立.

定理 3.1.6 设 I 是指标集, 族 $\{C_i\}(i \in I)$ 中每个集合 C_i 都是凸锥, 则 $\bigcap_{i \in I} C_i$ 是凸锥.

定理 3.1.7 设 C 是凸锥, 则 $C - C = \{\boldsymbol{x} - \boldsymbol{y} \,|\, \boldsymbol{x} \in C, \boldsymbol{y} \in C\} = \mathrm{aff} C$ 是包含 C 的最小子空间, 且 $(-C) \cap C$ 是包含在 C 中的最大子空间.

证明 首先证明 $C - C$ 是子空间. 显然, 设任意的 $\boldsymbol{c}_1, \boldsymbol{c}_2 \in C - C$, 存在 $\boldsymbol{x}_1, \boldsymbol{x}_2 \in C$ 和 $\boldsymbol{y}_1, \boldsymbol{y}_2 \in C$ 使得 $\boldsymbol{c}_1 = \boldsymbol{x}_1 - \boldsymbol{y}_1$ 和 $\boldsymbol{c}_2 = \boldsymbol{x}_2 - \boldsymbol{y}_2$. 对于任意 $\alpha, \beta \in \mathbf{R}$, 若 $\alpha, \beta > 0$, 有

$$\alpha \boldsymbol{c}_1 + \beta \boldsymbol{c}_2 = \alpha \boldsymbol{x}_1 - \alpha \boldsymbol{y}_1 + \beta \boldsymbol{x}_2 - \beta \boldsymbol{y}_2 \in C - C;$$

若 $\alpha > 0, \beta < 0$, 有

$$\alpha \boldsymbol{c}_1 + \beta \boldsymbol{c}_2 = \alpha \boldsymbol{x}_1 - \beta \boldsymbol{y}_2 - \alpha \boldsymbol{y}_1 + \beta \boldsymbol{x}_2 \in C - C;$$

若 $\alpha < 0, \beta > 0$, 有

$$\alpha \boldsymbol{c}_1 + \beta \boldsymbol{c}_2 = -\alpha \boldsymbol{y}_1 + \beta \boldsymbol{x}_2 + \alpha \boldsymbol{x}_1 - \beta \boldsymbol{y}_2 \in C - C;$$

若 $\alpha < 0, \beta < 0$, 有

$$\alpha \boldsymbol{c}_1 + \beta \boldsymbol{c}_2 = -\beta \boldsymbol{y}_2 - \alpha \boldsymbol{y}_1 + \beta \boldsymbol{x}_2 + \alpha \boldsymbol{x}_1 \in C - C.$$

下面证明 $C - C$ 是包含 C 的最小子空间. 因为 $\mathbf{0} \in C$, 所以有 $C \subset C - C$. 假设还存在一个子空间 D, 使 $C \subset D$ 和 $D \subset C - C$, 且 $(C - C) \setminus D \neq \varnothing$. 令 $\boldsymbol{a} \in (C - C) \setminus D$, 则 $\boldsymbol{a} \in (C - C)$, 且 $\boldsymbol{a} \notin D$. 故可设 $\boldsymbol{a} = \boldsymbol{x} - \boldsymbol{y}, \boldsymbol{x} \in C, \boldsymbol{y} \in C$, 有 $\boldsymbol{x} \in D, \boldsymbol{y} \in D$, 因为 D 是子空间, 所以 $\boldsymbol{x} - \boldsymbol{y} \in D$, 得矛盾. 因为包含原点的集合的仿射包是子空间, 所以 $C - C = \text{aff}C$.

下面证明 $(-C) \cap C$ 是包含在 C 中的最大子空间. 若 $\boldsymbol{c}_1, \boldsymbol{c}_2 \in (-C) \cap C$, 对于任意 $\alpha, \beta \in \mathbf{R}$, 有 $\alpha \boldsymbol{c}_1, \beta \boldsymbol{c}_2 \in (-C) \cap C$. 因此, 有 $\alpha \boldsymbol{c}_1 + \beta \boldsymbol{c}_2 \in (-C) \cap C$. 假设还存在一个子空间 D, 满足 $C \supset D$ 和 $D \supset (-C) \cap C$, 且 $D \setminus ((-C) \cap C) \neq \varnothing$. 令 $\boldsymbol{a} \in D \setminus ((-C) \cap C)$, 则有 $\boldsymbol{a} \notin (-C) \cap C$ 和 $\boldsymbol{a} \in D$, 因为 D 是子空间, 所以 $-\boldsymbol{a} \in (-C) \cap C$, 得矛盾. □

例 3.1.12　在例 3.1.11 的闭半空间族的交 C 是凸锥, 由定理 3.1.7 得

$$C - C = \{\boldsymbol{x} - \boldsymbol{y} \in \mathbf{R}^n \,|\langle \boldsymbol{x}, b_i \rangle \leqslant 0, \langle \boldsymbol{y}, b_i \rangle \leqslant 0, i \in I\} = \text{aff}C.$$

因此, 有 $(-C) \cap C = \{\boldsymbol{x} \in \mathbf{R}^n \,|\langle \boldsymbol{x}, b_i \rangle = 0, i \in I\}$.

定理 3.1.8　设 C_1 和 C_2 是 $\mathbf{R}^n$ 凸锥, 则

$$C_1 + C_2 = \text{co}\,(C_1 \cup C_2)\,.$$

证明　因为 C_1 和 C_2 包含原点, 所以有 $C_1, C_2 \subset C_1 + C_2$. 那么有

$$C_1 \cup C_2 \subset (C_1 + C_2) \cup (C_1 + C_2) = C_1 + C_2.$$

由性质 3.1.3 知, $C_1 + C_2$ 和 $\text{co}\,(C_1 \cup C_2)$ 都是凸锥. 因此, $\text{co}\,(C_1 \cup C_2) \subset C_1 + C_2$. 设 $\forall \boldsymbol{c} \in C_1 + C_2$, 存在 $\boldsymbol{c}_1 \in C_1$ 和 $\boldsymbol{c}_2 \in C_2$ 使得 $\boldsymbol{c} = \boldsymbol{c}_1 + \boldsymbol{c}_2$. 显然有 $\boldsymbol{c} = \boldsymbol{c}_1 + \boldsymbol{c}_2 \in \text{co}\,(C_1 \cup C_2)$. 因此, 定理结论成立. □

例 3.1.13　在 $\mathbf{R}^2$ 中 $C_1 = \{(x_1, 0)^{\mathrm{T}} \,|x_1 \in (0, +\infty)\}$ 和 $C_2 = \{(0, x_2)^{\mathrm{T}} \,|x_2 \in (0, +\infty)\}$ 是凸锥, 则 $C_1 + C_2 = C_1 \cup C_2 = \{(x_1, x_2)^{\mathrm{T}} \,|x_1, x_2 \in (0, +\infty)\}$.

本小节得到了原点凸锥 C 的几个性质:

1) 凸锥等价于集合中任两点非负线性组合属于该集合;

2) 凸集乘上任意正数张成包含该集合的最小凸锥;

3) 凸锥族中所有集合的交是凸锥;

4) $C-C$ 是包含凸锥 C 的最小子空间, 且 $(-C)\cap C$ 是包含在凸锥 C 中的最大子空间;

5) 两个凸锥相加等于两个凸锥并的凸包.

3.2 锥　包

本节将讨论两种锥包, 一种是包含集合的所有顶点锥交集构成的锥包 (可能是非凸), 另一种是包含集合的所有顶点凸锥交集构成的凸锥包. 与凸包不同的是, 凸锥包是由锥生成的. 因此, 凸锥包既有凸包的性质又有锥的性质. 同时, 凸锥还包含一些特有的性质.

3.2.1 基本性质

首先给出一个集合的顶点的锥包和凸锥包的定义.

定义 3.2.1 对于给定的集合 $S\subset\mathbf{R}^n$, 所有包含 S 的关于正常顶点 $\boldsymbol{a}\in\mathbf{R}^n$ 的锥交集称为由 S 生成的锥包, 用 $C_{\boldsymbol{a}}(S)$ 表示. 所有包含 S 的关于正常顶点 $\boldsymbol{a}$ 的凸锥交集称为由 S 生成的凸锥包, 用 $\mathrm{cone}_{\boldsymbol{a}}(S)$ 或 $\mathrm{cone}_{\boldsymbol{a}}S$ 表示. 特别地, 当 $\boldsymbol{a}=\mathbf{0}$ 时, 由 S 生成的锥包记为 $C(S)$, 由 S 生成的凸锥包记为 $\mathrm{cone}(S)$ 或 $\mathrm{cone}S$.

从上面的定义知道, 如果 $S\neq\varnothing$, 则 $\mathrm{cone}S$ 由 S 的点的全体非负线性组合所构成的集合组成. 我们规定, $\mathrm{cone}_{\boldsymbol{a}}\varnothing=\{\boldsymbol{a}\}$ 和 $C_{\boldsymbol{a}}\varnothing=\{\boldsymbol{a}\}$. 根据锥包的定义, 有 $\boldsymbol{a}\in\mathrm{cone}_{\boldsymbol{a}}S$ 和 $\boldsymbol{a}\in C_{\boldsymbol{a}}S$.

图 3.2-1 表示集合 S 是平面中的两个线段 S_1 和 S_2 的并, 由 S 生成的锥包 $C_{\boldsymbol{a}}(S)$ 是从顶点 $\boldsymbol{a}$ 出发且包含线段 S_1 和 S_2 的两根射线. 图 3.2-2 表示 $S=S_1\cup S_2$ 生成的凸锥包 $\mathrm{cone}_{\boldsymbol{a}}(S)$ 是由从顶点 $\boldsymbol{a}$ 出发且包含线段 S_1 和 S_2 的两根射线之间所有元素构成.

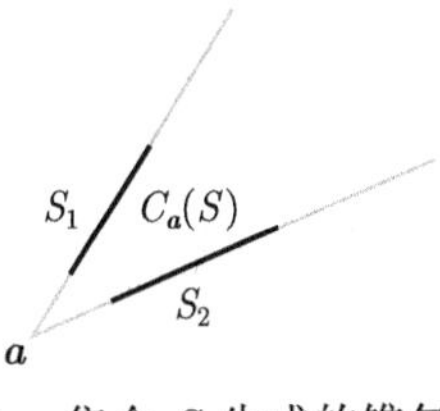

图 3.2-1　集合 S 生成的锥包 $C_{\boldsymbol{a}}(S)$

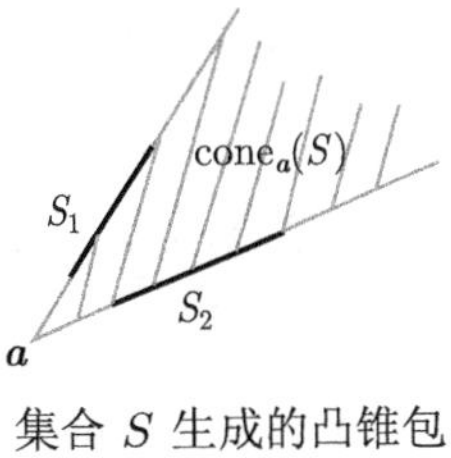

图 3.2-2　集合 S 生成的凸锥包 $\mathrm{cone}_{\boldsymbol{a}}(S)$

由定义 3.2.1 有下面凸锥包的一些基本性质.

性质 3.2.1 设集合 $S,T\subset\mathbf{R}^n$ 和点 $\boldsymbol{a}\in\mathbf{R}^n$, 有下面性质成立:

(1) $S\subset\mathrm{cone}_{\boldsymbol{a}}(S)=\mathrm{cone}_{\boldsymbol{a}}(S\cup\{\boldsymbol{a}\})=\mathrm{cone}_{\boldsymbol{a}}(\mathrm{cone}_{\boldsymbol{a}}(S))$;

(2) $S = \mathrm{cone}_{\boldsymbol{a}}(S)$ 当且仅当 S 是关于正常顶点 $\boldsymbol{a}$ 的凸锥;

(3) $(\boldsymbol{x}-\boldsymbol{a})+\mathrm{cone}_{\boldsymbol{a}}(S)=\mathrm{cone}_{\boldsymbol{x}}(\boldsymbol{x}-\boldsymbol{a}+S), \forall \boldsymbol{x}\in \mathbf{R}^n$;

(4) 如果 $S\subset T$, 则 $\mathrm{cone}_{\boldsymbol{a}}(S)\subset \mathrm{cone}_{\boldsymbol{a}}(T)$;

(5) 如果 $S\subset T\subset \mathrm{cone}_{\boldsymbol{a}}(S)$, 则 $\mathrm{cone}_{\boldsymbol{a}}(S)=\mathrm{cone}_{\boldsymbol{a}}(T)$;

(6) $\mathrm{cone}_{\boldsymbol{a}}(S)=\mathrm{cone}_{\boldsymbol{a}}(S\cup T)$ 当且仅当 $T\subset \mathrm{cone}_{\boldsymbol{a}}(S)$.

证明　(1) 显然 $S\subset \mathrm{cone}_{\boldsymbol{a}}(S)\subset \mathrm{cone}_{\boldsymbol{a}}(S\cup\{\boldsymbol{a}\})$, 根据性质 3.1.3 知 $\mathrm{cone}_{\boldsymbol{a}}(S)$ 是一个关于顶点 $\boldsymbol{a}$ 的凸锥. 再根据性质 3.1.1 和性质 3.1.2 知 $\mathrm{cone}_{\boldsymbol{a}}(S)\cup\{\boldsymbol{a}\}$ 是关于顶点 $\boldsymbol{a}$ 的凸锥. 设任意 $\boldsymbol{x}\in \mathrm{cone}_{\boldsymbol{a}}(S)$, 以及任意 $\boldsymbol{a}\in \mathrm{cone}_{\boldsymbol{a}}(S)$, 则 $\boldsymbol{a}$ 与任意的 $\boldsymbol{s}\in S$ 的凸组合都属于 $\mathrm{cone}_{\boldsymbol{a}}(S)$, 即 $\mathrm{cone}_{\boldsymbol{a}}(S)\supset \mathrm{cone}_{\boldsymbol{a}}(S\cup\{\boldsymbol{a}\})$, 最终得

$$\mathrm{cone}_{\boldsymbol{a}}(S\cup\{\boldsymbol{a}\})=\mathrm{cone}_{\boldsymbol{a}}\left(\mathrm{cone}_{\boldsymbol{a}}(S)\cup\{\boldsymbol{a}\}\right)=\mathrm{cone}_{\boldsymbol{a}}(S).$$

(2) 结论是显然的.

(3) 设 $\forall \boldsymbol{x}\in \mathbf{R}^n$. 因为 $\mathrm{cone}_{\boldsymbol{a}}(S)$ 是关于顶点 $\boldsymbol{a}$ 的凸锥, 那么对任意的 $\boldsymbol{y}\in \mathrm{cone}_{\boldsymbol{a}}(S)$ 和 $\lambda>0$ 有

$$\boldsymbol{x}+\lambda(\boldsymbol{y}+\boldsymbol{x}-\boldsymbol{a}-\boldsymbol{x})=\boldsymbol{x}-\boldsymbol{a}+\boldsymbol{a}+\lambda(\boldsymbol{y}-\boldsymbol{a})\in \boldsymbol{x}-\boldsymbol{a}+\mathrm{cone}_{\boldsymbol{a}}(S). \tag{3.2.1-1}$$

上式说明 $(\boldsymbol{x}-\boldsymbol{a})+\mathrm{cone}_{\boldsymbol{a}}(S)$ 是关于顶点 $\boldsymbol{x}$ 的凸锥, 以及 $\boldsymbol{x}-\boldsymbol{a}+S\subset(\boldsymbol{x}-\boldsymbol{a})+\mathrm{cone}_{\boldsymbol{a}}(S)$, 则有

$$\mathrm{cone}_{\boldsymbol{x}}(\boldsymbol{x}-\boldsymbol{a}+S)\subset(\boldsymbol{x}-\boldsymbol{a})+\mathrm{cone}_{\boldsymbol{a}}(S).$$

设对每一个 $\boldsymbol{y}\in \mathrm{cone}_{\boldsymbol{a}}(S)$ 和 $\lambda>0$, 存在 $\boldsymbol{x}_i\in S$, $\lambda_i\geqslant 0\,(i=1,2,\cdots,k)$ 和 $\sum\limits_{i=1}^{k}\lambda_i=1$ 使得

$$\boldsymbol{y}=\boldsymbol{a}+\lambda_1(\boldsymbol{x}_1-\boldsymbol{a})+\cdots+\lambda_k(\boldsymbol{x}_k-\boldsymbol{a}). \tag{3.2.1-2}$$

因此, 通过 (3.2.1-1) 式和 (3.2.1-2) 式有

$$\boldsymbol{x}+\lambda(\boldsymbol{y}+\boldsymbol{x}-\boldsymbol{a}-\boldsymbol{x})=\sum_{i=1}^{k}\lambda_i\left(\boldsymbol{x}+\lambda(\boldsymbol{x}_i+\boldsymbol{x}-\boldsymbol{a}-\boldsymbol{x})\right).$$

上式说明 $\boldsymbol{y}+\boldsymbol{x}-\boldsymbol{a}\in \mathrm{cone}_{\boldsymbol{x}}(\boldsymbol{x}-\boldsymbol{a}+S)$. □

性质 (4), (5) 和 (6) 都是显然的.

例 3.2.1　(1) 在 $\mathbf{R}$ 空间中的集合 $S=(0,1)$, 它的锥包 $\mathrm{cone}_{\mathbf{0}}S=[0,+\infty)$ 是关于顶点 $\mathbf{0}$ 的一个凸锥, 且是一个凸锥包, 它的凸包和闭包分别为 $\mathrm{co}S=(0,1)$ 和 $\mathrm{cl}S=[0,1]$.

(2) 在 $\mathbf{R}^2$ 空间中的集合 $S=\left\{(x_1,0)^{\mathrm{T}}|x_1\in(0,1)\right\}\cup\left\{(0,x_2)^{\mathrm{T}}|x_2\in(0,2)\right\}$, 它的锥包为

$$C_0(S)=\left\{(x_1,0)^{\mathrm{T}}|x_1\in[0,+\infty)\right\}\cup\left\{(0,x_2)^{\mathrm{T}}|x_2\in[0,+\infty)\right\},$$

凸锥包为 $\mathrm{cone}_{\mathbf{0}}(S)=\left\{(x_1,x_2)^{\mathrm{T}}|x_1\in[0,+\infty),x_2\in[0,+\infty)\right\}$. 设 $\boldsymbol{a}=(a_1,a_2)^{\mathrm{T}}$, 有

$$(a_1,a_2)^{\mathrm{T}}+S=\left\{(a_1+x_1,a_2)^{\mathrm{T}}|x_1\in(0,1)\right\}\cup\left\{(a_1,a_2+x_2)^{\mathrm{T}}|x_2\in(0,2)\right\},$$

则此时锥包为

$$\begin{aligned}&C_{\boldsymbol{a}}\left((a_1,a_2)^{\mathrm{T}}+S\right)\\=&\left\{(a_1+x_1,a_2)^{\mathrm{T}}|x_1\in[0,+\infty)\right\}\cup\left\{(a_1,a_2+x_2)^{\mathrm{T}}|x_2\in[0,+\infty)\right\}\\=&(a_1,a_2)^{\mathrm{T}}+\left\{(x_1,0)^{\mathrm{T}}|x_1\in[0,+\infty)\right\}\cup\left\{(0,x_2)^{\mathrm{T}}|x_2\in[0,+\infty)\right\}\\=&(a_1,a_2)^{\mathrm{T}}+C_{\mathbf{0}}(S).\end{aligned}$$

凸锥包为

$$\begin{aligned}\mathrm{cone}_{\boldsymbol{a}}\left((a_1,a_2)^{\mathrm{T}}+S\right)&=\left\{(a_1+x_1,a_2+x_2)^{\mathrm{T}}|x_1\in[0,+\infty),x_2\in[0,+\infty)\right\}\\&=(a_1,a_2)^{\mathrm{T}}+\left\{(x_1,x_2)^{\mathrm{T}}|x_1\in[0,+\infty),x_2\in[0,+\infty)\right\}\\&=(a_1,a_2)^{\mathrm{T}}+\mathrm{cone}_{\mathbf{0}}(S).\end{aligned}$$

例 3.2.1 说明了凸锥包和锥包都包含顶点, 凸锥包的凸锥包等于原来的凸锥包, 凸锥等于它的凸锥包, 锥包的平移等于顶点的锥包平移. 同时, 也说明性质 3.2.1 中的结论对该例成立.

例 3.2.2　设在 $\mathbf{R}^2$ 空间中的两个集合

$$T=\left\{(x_1,x_2)^{\mathrm{T}}|x_1-x_2\leqslant 0,0\leqslant x_1,x_2\leqslant 1\right\},\quad S=\left\{(x_1,x_2)^{\mathrm{T}}|0\leqslant x_1,x_2\leqslant 1\right\},$$

显然有 $T\subset S$, 且

$$\begin{aligned}C_{\boldsymbol{a}}(T)=\left\{(x_1,x_2)^{\mathrm{T}}|x_1-x_2\leqslant 0,0\leqslant x_1,x_2\leqslant+\infty\right\}\subset C_{\boldsymbol{a}}(S)\\=\left\{(x_1,x_2)^{\mathrm{T}}|x_1,x_2\geqslant 0\right\},\end{aligned}$$

$$\text{cone}_{\boldsymbol{a}}(T)=\left\{(x_1,x_2)^{\mathrm{T}}\,|x_1-x_2\leqslant 0,0\leqslant x_1,x_2\leqslant+\infty\right\}\subset\text{cone}_{\boldsymbol{a}}(S)$$
$$=\left\{(x_1,x_2)^{\mathrm{T}}\,|x_1,x_2\geqslant 0\right\}.$$

上式说明了性质 3.2.1(4) 对该例成立.

例 3.2.3　设在 $\mathbf{R}$ 空间中的两个集合

$$T=\{x_1|0\leqslant x_1\leqslant 1\},\quad S=\{x_1|2\leqslant x_1\leqslant 3\},\quad T\cap S=\varnothing,$$

显然有

$$C_0(T\cap S)=\{0\}\subset C_0(T)\cap C_0(S)=[0,+\infty),$$
$$\text{cone}_0(T\cap S)=\{0\}\subset\text{cone}_0(T)\cap\text{cone}_0(S)=[0,+\infty),$$

以及

$$C_0(T\cup S)=C_0(T)\cup C_0(S)=[0,+\infty),$$
$$\text{cone}_0(T\cup S)=\text{cone}_0(T)\cup\text{cone}_0(S)=[0,+\infty).$$

例 3.2.4　设在 $\mathbf{R}^2$ 空间中的两个集合

$$T=\left\{(x_1,0)^{\mathrm{T}}\,|0<x_1\leqslant 1\right\},\quad S=\left\{(0,x_2)^{\mathrm{T}}\,|0<x_2\leqslant 1\right\},\quad T\cap S=\varnothing,$$

显然有

$$C_0(T)\cup C_0(S)=\left\{(x_1,0)^{\mathrm{T}}\,|x_1\in[0,+\infty)\right\}\cup\left\{(0,x_2)^{\mathrm{T}}\,|x_2\in[0,+\infty)\right\},$$
$$C_0(T\cup S)=\left\{(x_1,0)^{\mathrm{T}}\,|x_1\in[0,+\infty)\right\}\cup\left\{(0,x_2)^{\mathrm{T}}\,|x_2\in[0,+\infty)\right\}.$$

但 $\text{cone}_0(T)\cup\text{cone}_0(S)=\left\{(x_1,0)^{\mathrm{T}}\,|x_1\in[0,+\infty)\right\}\cup\left\{(0,x_2)^{\mathrm{T}}\,|x_2\in[0,+\infty)\right\}\subset$ $\text{cone}_0(T\cup S)$, 其中

$$\text{cone}_0(T\cup S)=\left\{(x_1,x_2)^{\mathrm{T}}\,|x_1\in[0,+\infty),x_2\in[0,+\infty)\right\}.$$

下面是关于凸锥包和锥包的交并运算性质.

性质 3.2.2　设集合族 $\{S_i\}\subset\mathbf{R}^n\,(i\in I)$ 和点 $\boldsymbol{a}\in\mathbf{R}^n$, 下面结论成立:

(1) $\text{cone}_{\boldsymbol{a}}(\bigcap_{i\in I}S_i)\subset\bigcap_{i\in I}\text{cone}_{\boldsymbol{a}}(S_i)$;

(2) $\bigcup_{i\in I}\text{cone}_{\boldsymbol{a}}(S_i)\subset\text{cone}_{\boldsymbol{a}}(\bigcup_{i\in I}S_i)$;

(3) 如果 $\{S_i\}\subset\mathbf{R}^n\,(i\in I)$ 是嵌套的, 那么 $\bigcup_{i\in I}\text{cone}_{\boldsymbol{a}}(S_i)=\text{cone}_{\boldsymbol{a}}(\bigcup_{i\in I}S_i)$;

(4) $\text{cone}_{\boldsymbol{a}}(\bigcup_{i\in I}S_i)=\text{cone}_{\boldsymbol{a}}(\bigcup_{i\in I}\text{co}S_i)=\text{cone}_{\boldsymbol{a}}(\bigcup_{i\in I}\text{cone}_{\boldsymbol{a}}(S_i))$.

上面结果可以根据性质 2.2.2、性质 3.1.3 和性质 3.2.1 直接获得. 前面例 3.2.4 可说明性质 3.2.2(1) 和 (2) 的反包含不成立.

另外, 通过性质 3.1.3 也可以直接得到下面性质.

性质 3.2.3 设集合 $S, T \subset \mathbf{R}^n$ 和点 $\boldsymbol{a} \in \mathbf{R}^n$, 下面结论成立:

(1) $S \subset C_{\boldsymbol{a}}(S) = C_{\boldsymbol{a}}(S \cup \{\boldsymbol{a}\}) = C_{\boldsymbol{a}}(C_{\boldsymbol{a}}(S))$;

(2) $S = C_{\boldsymbol{a}}(S)$ 当且仅当 S 是关于正常顶点 $\boldsymbol{a}$ 的锥;

(3) $(\boldsymbol{x} - \boldsymbol{a}) + C_{\boldsymbol{a}}(S) = C_{\boldsymbol{x}}(\boldsymbol{x} - \boldsymbol{a} + S)$, $\forall \boldsymbol{x} \in \mathbf{R}^n$;

(4) 如果 $S \subset T$, 则 $C_{\boldsymbol{a}}(S) \subset C_{\boldsymbol{a}}(T)$;

(5) 如果 $S \subset T \subset C_{\boldsymbol{a}}(S)$, 则 $C_{\boldsymbol{a}}(S) = C_{\boldsymbol{a}}(T)$;

(6) $C_{\boldsymbol{a}}(S) = C_{\boldsymbol{a}}(S \cup T)$ 当且仅当 $T \subset C_{\boldsymbol{a}}(S)$;

(7) 如果 $S \neq \varnothing$, 则有 $C_{\boldsymbol{a}}(S) = \{\boldsymbol{a} + \lambda(\boldsymbol{x} - \boldsymbol{a}) | \boldsymbol{x} \in S, \lambda \geqslant 0\}$.

性质 3.2.4 设集合族 $\{S_i\} \subset \mathbf{R}^n\,(i \in I)$ 和点 $\boldsymbol{a} \in \mathbf{R}^n$, 则下面结论成立:

(1) $C_{\boldsymbol{a}}(\bigcap_{i \in I} S_i) \subset \bigcap_{i \in I} C_{\boldsymbol{a}}(S_i)$;

(2) $C_{\boldsymbol{a}}(\bigcup_{i \in I} S_i) = C_{\boldsymbol{a}}(\bigcup_{i \in I} C_{\boldsymbol{a}}(S_i)) = \bigcup_{i \in I} C_{\boldsymbol{a}}(S_i)$.

下面, 在证明凸锥包的表示定理之前, 先举例说明 $\operatorname{cone}_{\boldsymbol{a}}(S)$ 中的元素可以表示成 $\boldsymbol{a}, (\boldsymbol{x}_1 - \boldsymbol{a}), \cdots, (\boldsymbol{x}_m - \boldsymbol{a})$ 的非负线性组合.

例 3.2.5 设在 $\mathbf{R}^2$ 空间中的集合

$$S = \left\{(x_1, x_2)^{\mathrm{T}} | 0 \leqslant x_1, x_2 \leqslant 1\right\},$$

显然有 $\operatorname{cone}_0(S) = \{\lambda_1(\boldsymbol{x}_1) + \lambda_2(\boldsymbol{x}_2) | \boldsymbol{x}_i \in S, \lambda_i \geqslant 0, i = 1, 2\}$.

下面证明一般凸锥包表示定理: 集合的凸锥包可以由集合的顶点和集合内有限个点的非负线性组合构成.

定理 3.2.1 设非空集合 $S \subset \mathbf{R}^n$ 和点 $\boldsymbol{a} \in \mathbf{R}^n$, 则

$$\begin{aligned}\operatorname{cone}_{\boldsymbol{a}}(S) = \{&\boldsymbol{a} + \lambda_1(\boldsymbol{x}_1 - \boldsymbol{a}) + \cdots + \lambda_m(\boldsymbol{x}_m - \boldsymbol{a}) | \boldsymbol{x}_i \in S, \lambda_i \geqslant 0,\\ &i = 1, \cdots, m, m \geqslant 1\}.\end{aligned}$$

证明 设集合

$$C = \{\boldsymbol{a} + \lambda_1(\boldsymbol{x}_1 - \boldsymbol{a}) + \cdots + \lambda_m(\boldsymbol{x}_m - \boldsymbol{a}) | \boldsymbol{x}_i \in S, \lambda_i \geqslant 0, i = 1, \cdots, m, m \geqslant 1\}. \tag{3.2.1-3}$$

先证明 C 是关于顶点 $\boldsymbol{a} \in \mathbf{R}^n$ 的凸锥. 设 $\boldsymbol{y} \in C$, 那么有

$$\boldsymbol{y} = \boldsymbol{a} + \lambda_1(\boldsymbol{x}_1 - \boldsymbol{a}) + \cdots + \lambda_m(\boldsymbol{x}_m - \boldsymbol{a}), \quad 其中\ \boldsymbol{x}_i \in S, \lambda_i \geqslant 0, i = 1, \cdots, m. \tag{3.2.1-4}$$

对于 $\lambda > 0$, 由上式得

$$\boldsymbol{a}+\lambda(\boldsymbol{y}-\boldsymbol{a})=\boldsymbol{a}+\lambda\lambda_1(\boldsymbol{x}_1-\boldsymbol{a})+\cdots+\lambda\lambda_m(\boldsymbol{x}_m-\boldsymbol{a})\in C.$$

因此, C 是关于顶点 $\boldsymbol{a}\in\mathbf{R}^n$ 的锥.

下面证明 C 是凸集. 设 $\forall \boldsymbol{y},\boldsymbol{z}\in C$, 其中 $\boldsymbol{y}$ 按 (3.2.1-4) 式表示, $\boldsymbol{z}$ 可表示为

$$\boldsymbol{z}=\boldsymbol{a}+\mu_1(\boldsymbol{z}_1-\boldsymbol{a})+\cdots+\mu_m(\boldsymbol{z}_m-\boldsymbol{a}),\quad 其中\ \boldsymbol{z}_i\in S,\mu_i\geqslant 0,i=1,\cdots,m. \tag{3.2.1-5}$$

设 $\forall t\in(0,1)$, 根据 (3.2.1-4) 式和 (3.2.1-5) 式可得

$$(1-t)\boldsymbol{y}+t\boldsymbol{z}=\boldsymbol{a}+\sum_{i=1}^{m}(1-t)\lambda_i(\boldsymbol{x}_i-\boldsymbol{a})+\sum_{i=1}^{m}t\mu_i(\boldsymbol{z}_i-\boldsymbol{a})\in C.$$

因此, C 是关于顶点 $\boldsymbol{a}\in\mathbf{R}^n$ 的凸锥. 显然 $S\subset C$, $\mathrm{cone}_{\boldsymbol{a}}(S)$ 是包含 S 的最小凸锥. 那么有

$$S\subset\mathrm{cone}_{\boldsymbol{a}}(S)\subset C.$$

根据 C 的定义有 $\mathrm{cone}_{\boldsymbol{a}}(S)\supset C$. □

因此根据定理 3.2.1, 下面推论显然成立, 即有限多个点可扩张成凸锥包.

推论 3.2.1 设集合 $S=\{\boldsymbol{x}_1,\cdots,\boldsymbol{x}_m\}\subset\mathbf{R}^n$ 和点 $\boldsymbol{a}\in\mathbf{R}^n$, 则

$$\mathrm{cone}_{\boldsymbol{a}}(S)=\{\boldsymbol{a}+\lambda_1(\boldsymbol{x}_1-\boldsymbol{a})+\cdots+\lambda_m(\boldsymbol{x}_m-\boldsymbol{a})\,|\,\lambda_i\geqslant 0,i=1,\cdots,m\}.$$

推论 3.2.2 设 $\mathrm{Simc}(\boldsymbol{a},\boldsymbol{x}_1,\cdots,\boldsymbol{x}_k)$ 是关于顶点 $\boldsymbol{a}$ 的 k-简单锥, 则

$$\mathrm{Simc}(\boldsymbol{a},\boldsymbol{x}_1,\cdots,\boldsymbol{x}_k)=\mathrm{cone}_{\boldsymbol{a}}(\{\boldsymbol{x}_1,\cdots,\boldsymbol{x}_k\}).$$

下面证明凸锥包的另一个表示定理, $\mathrm{cone}_{\boldsymbol{a}}(S)$ 中的元素可以表示成 $\boldsymbol{a},(\boldsymbol{x}_1-\boldsymbol{a}),\cdots,(\boldsymbol{x}_m-\boldsymbol{a})$ 的正数线性组合.

定理 3.2.2 设非空集合 $S\subset\mathbf{R}^n$ 和点 $\boldsymbol{a}\in\mathbf{R}^n$, 且 $S\setminus\{\boldsymbol{a}\}\neq\varnothing$, 则

$$\mathrm{cone}_{\boldsymbol{a}}(S)\setminus\{\boldsymbol{a}\}=\{\boldsymbol{a}+\lambda_1(\boldsymbol{x}_1-\boldsymbol{a})+\cdots+\lambda_m(\boldsymbol{x}_m-\boldsymbol{a})\,|\\ \boldsymbol{x}_i\in S,\lambda_i>0,i=1,\cdots,m,m\geqslant 1\}.$$

其中 $\{\boldsymbol{a},\boldsymbol{x}_1,\cdots,\boldsymbol{x}_m\}$ 都是 S 中的仿射无关集合.

证明 设 $\boldsymbol{x}\in\mathrm{cone}_{\boldsymbol{a}}S\setminus\{\boldsymbol{a}\}$, 由定理 3.2.1 知, 存在最少 k 个 $\lambda_i>0(i=1,\cdots,k)$ 和一组向量 $\{\boldsymbol{x}_1,\cdots,\boldsymbol{x}_k\}\subset S$ 使得

$$\boldsymbol{x}=\boldsymbol{a}+\lambda_1(\boldsymbol{x}_1-\boldsymbol{a})+\cdots+\lambda_k(\boldsymbol{x}_k-\boldsymbol{a}). \tag{3.2.1-6}$$

假设 $\{\boldsymbol{x}_1,\cdots,\boldsymbol{x}_k\}\subset S$ 是仿射相关的, 那么存在一个大于 0 的 $\alpha_i(i=1,2,\cdots,k)$ 使得

$$\sum_{i=1}^{k}\alpha_i\boldsymbol{x}_i=\mathbf{0},\quad \sum_{i=1}^{k}\alpha_i=0. \tag{3.2.1-7}$$

设 $\bar{\alpha}=\min\{\lambda_i/\alpha_i|\alpha_i>0,i=1,2,\cdots,k\}$, 由 (3.2.1-6) 式和 (3.2.1-7) 式得

$$\boldsymbol{x}=\boldsymbol{a}+(\lambda_1-\bar{\alpha}\alpha_1)(\boldsymbol{x}_1-\boldsymbol{a})+\cdots+(\lambda_k-\bar{\alpha}\alpha_k)(\boldsymbol{x}_k-\boldsymbol{a}).$$

上式中至少有一个 $\lambda_i-\bar{\alpha}\alpha_i=0$, 但上述表达式与存在最少 k 个 $\lambda_i>0(i=1,\cdots,k)$ 的 (3.2.1-6) 式的假设矛盾. 因此, $\{\boldsymbol{x}_1,\cdots,\boldsymbol{x}_k\}\subset S$ 是仿射无关的.

进一步, 假设 $\{\boldsymbol{a},\boldsymbol{x}_1,\cdots,\boldsymbol{x}_k\}$ 是仿射相关的. 因为 $\{\boldsymbol{x}_1,\cdots,\boldsymbol{x}_k\}\subset S$ 是仿射无关的, 那么存在一个大于 0 的 $\beta_i\,(i=1,2,\cdots,k)$ 使得

$$\sum_{i=1}^{k}\beta_i(\boldsymbol{x}_i-\boldsymbol{a})=\mathbf{0},\quad \sum_{i=1}^{k}\beta_i=1. \tag{3.2.1-8}$$

再设 $\bar{\beta}=\min\{\lambda_i/\beta_i|\beta_i>0,i=1,2,\cdots,k\}$, 由 (3.2.1-7) 式和 (3.2.1-8) 式得

$$\boldsymbol{x}=\boldsymbol{a}+\left(\lambda_1-\bar{\beta}\beta_1\right)(\boldsymbol{x}_1-\boldsymbol{a})+\cdots+\left(\lambda_k-\bar{\beta}\beta_k\right)(\boldsymbol{x}_k-\boldsymbol{a}).$$

上式中至少有一个 $\lambda_i-\bar{\beta}\beta_i=0$, 但上述表达式与存在最少 k 个 $\lambda_i>0(i=1,\cdots,k)$ 的 (3.2.1-6) 式的假设矛盾. 我们已经证明了

$$\begin{aligned}\operatorname{cone}_{\boldsymbol{a}}(S)\setminus\{\boldsymbol{a}\}\subset\{&\boldsymbol{a}+\lambda_1(\boldsymbol{x}_1-\boldsymbol{a})+\cdots+\lambda_m(\boldsymbol{x}_m-\boldsymbol{a})\,|\\ &\lambda_i>0,i=1,\cdots,m,m\geqslant 1\},\end{aligned}$$

但从 $\operatorname{cone}_{\boldsymbol{a}}(S)$ 的定义知道上式是等式. □

进一步, 下面有更弱的凸锥包表示定理.

定理 3.2.3 设非空集合 $S\subset\mathbf{R}^n$ 和点 $\boldsymbol{a}\in\mathbf{R}^n$, 给定 $\boldsymbol{x}_1\in S$, 则

$$\begin{aligned}\operatorname{cone}_{\boldsymbol{a}}(S)=\{&\boldsymbol{a}+\lambda_1(\boldsymbol{x}_1-\boldsymbol{a})+\cdots+\lambda_m(\boldsymbol{x}_m-\boldsymbol{a})\,|\\ &\boldsymbol{x}_i\in S,\lambda_i\geqslant 0,i=1,\cdots,m,m\geqslant 1\},\end{aligned}$$

其中 $\{\boldsymbol{x}_1,\cdots,\boldsymbol{x}_m\}$ 是 S 中的任意仿射无关集合.

证明 如果 $\boldsymbol{x}_1=\boldsymbol{a}$, 根据定理 3.2.2 知结论成立. 假设 $\boldsymbol{x}_1\neq\boldsymbol{a}$, 再由定理 3.2.1 知, 对于 $\boldsymbol{x}\in\operatorname{cone}_{\boldsymbol{a}}(S)$ 存在 k 个 $\lambda_i\geqslant 0\,(i=1,\cdots,k)$ 和一组向量 $\{\boldsymbol{x}_1,\cdots,\boldsymbol{x}_k\}\subset S$ 使得

$$\boldsymbol{x}=\boldsymbol{a}+\lambda_1(\boldsymbol{x}_1-\boldsymbol{a})+\cdots+\lambda_k(\boldsymbol{x}_k-\boldsymbol{a}).$$

当 $\boldsymbol{x}=\boldsymbol{a}$ 时结论显然成立. 当 $\boldsymbol{x}\neq\boldsymbol{a}$ 时, 令 $\beta=\lambda_1+\cdots+\lambda_k$, 有 $\beta>0$, 设

$$\boldsymbol{y}=\beta^{-1}(\boldsymbol{x}-\boldsymbol{a})=\beta^{-1}\lambda_1(\boldsymbol{x}_1-\boldsymbol{a})+\cdots+\beta^{-1}\lambda_k(\boldsymbol{x}_k-\boldsymbol{a}). \tag{3.2.1-9}$$

由上式得 $\boldsymbol{y}\in\mathrm{co}(S-\boldsymbol{a})$, 即有 $\boldsymbol{y}+\boldsymbol{a}\in\mathrm{co}(S)$. 再由推论 2.2.1 知, 存在仿射无关集合 $\{\boldsymbol{x}_1,\cdots,\boldsymbol{x}_m\}$, 使得 $\boldsymbol{y}+\boldsymbol{a}\in\mathrm{co}\{\boldsymbol{x}_1,\cdots,\boldsymbol{x}_m\}$, 即有

$$\boldsymbol{y}+\boldsymbol{a}=\alpha_1\boldsymbol{x}_1+\cdots+\alpha_m\boldsymbol{x}_m,\quad \sum_{i=1}^{k}\alpha_i=1,\alpha_i\geqslant 0,i=1,2,\cdots,m.$$

由 (3.2.1-9) 式得

$$\boldsymbol{x}=\boldsymbol{a}+\beta\alpha_1(\boldsymbol{x}_1-\boldsymbol{a})+\cdots+\beta\alpha_m(\boldsymbol{x}_m-\boldsymbol{a}).$$ □

下面证明凸包、锥包、凸锥包和仿射集之间的关系.

性质 3.2.5　设集合 $S\subset\mathbf{R}^n$ 和点 $\boldsymbol{a}\in\mathbf{R}^n$, 下面结论成立:

(1) $\mathrm{co}S\subset\mathrm{cone}_{\boldsymbol{a}}(S)=\mathrm{co}(\mathrm{cone}_{\boldsymbol{a}}(S))=\mathrm{cone}_{\boldsymbol{a}}(\mathrm{co}S)=\mathrm{co}(C_{\boldsymbol{a}}(S))=C_{\boldsymbol{a}}(\mathrm{co}S)$;

(2) $C_{\boldsymbol{a}}(S)\subset\mathrm{cone}_{\boldsymbol{a}}(S)\subset\mathrm{aff}(\{\boldsymbol{a}\}\cup S)$;

(3) $\mathrm{aff}(C_{\boldsymbol{a}}(S))=\mathrm{aff}(\mathrm{cone}_{\boldsymbol{a}}(S))=\mathrm{aff}(S\cup\{\boldsymbol{a}\})$;

(4) $C_{\boldsymbol{a}}(S)\subset\mathrm{aff}(S)$ 等价于 $\mathrm{cone}_{\boldsymbol{a}}(S)\subset\mathrm{aff}(S)$, 也等价于 $\boldsymbol{a}\in\mathrm{aff}(S)$;

(5) $\mathrm{cone}_{\boldsymbol{a}}(S)$ 是一个平面等价于 $\mathrm{cone}_{\boldsymbol{a}}(S)=\mathrm{aff}(\{\boldsymbol{a}\}\cup S)$;

(6) 如果 $S\neq\varnothing$, 则 $\mathrm{cone}_{\boldsymbol{a}}(S)$ 是一个平面等价于 $\mathrm{cone}_{\boldsymbol{a}}(S)=\mathrm{aff}(S)$;

(7) 如果 $\boldsymbol{a}\notin\mathrm{co}S$, 则 $\mathrm{cone}_{\boldsymbol{a}}(S)\setminus\{\boldsymbol{a}\}$ 是包含 S 关于顶点 $\boldsymbol{a}$ 的最小凸锥;

(8) 如果 $C\subset\mathbf{R}^n$ 是正常顶点 $\boldsymbol{a}$ 的锥使得 $S\cap C\subset\{\boldsymbol{a}\}$, 则 $C_a(S)\cap C=\{\boldsymbol{a}\}$;

(9) 如果 $C\subset\mathbf{R}^n$ 是正常顶点 $\boldsymbol{a}$ 的锥使得 $\mathrm{co}S\cap C\subset\{\boldsymbol{a}\}$, 则

$$\mathrm{cone}_a(S)\cap C=\{\boldsymbol{a}\}.$$

证明　(1) 根据定义 3.2.1, 显然有

$$S\subset\mathrm{co}S\subset\mathrm{cone}_{\boldsymbol{a}}(S)\subset\mathrm{co}(\mathrm{cone}_{\boldsymbol{a}}(S)), \tag{3.2.1-10}$$

$$S\subset C_{\boldsymbol{a}}(S)\subset\mathrm{cone}_{\boldsymbol{a}}(S),\quad C_{\boldsymbol{a}}(S)\subset C_{\boldsymbol{a}}(\mathrm{co}S). \tag{3.2.1-11}$$

根据 (3.2.1-10) 式和性质 3.2.1 得

$$\mathrm{cone}_{\boldsymbol{a}}(S)\subset\mathrm{cone}_{\boldsymbol{a}}(\mathrm{co}S)\subset\mathrm{cone}_{\boldsymbol{a}}(\mathrm{cone}_{\boldsymbol{a}}(S)).$$

由上式即可得到 $\mathrm{cone}_{\boldsymbol{a}}(S)=\mathrm{cone}_{\boldsymbol{a}}(\mathrm{co}S)=\mathrm{cone}_{\boldsymbol{a}}(\mathrm{cone}_{\boldsymbol{a}}(S))$. 因为 $\mathrm{cone}_{\boldsymbol{a}}(S)$ 是凸集, 所以得

$$\mathrm{cone}_{\boldsymbol{a}}(S)=\mathrm{co}(\mathrm{cone}_{\boldsymbol{a}}(S)).$$

再由 (3.2.1-11) 式和性质 3.2.3 得

$$\mathrm{co}\,(S) \subset \mathrm{co}\,(C_{\boldsymbol{a}}(S)) \subset \mathrm{co}\,(\mathrm{cone}_{\boldsymbol{a}}(S)), \quad \mathrm{co}\,(C_{\boldsymbol{a}}(S)) \subset \mathrm{co}\,(C_{\boldsymbol{a}}(\mathrm{co}S)).$$

根据定义 3.2.1 知 $\mathrm{co}\,(C_{\boldsymbol{a}}(S))$ 是包含 S 的凸锥, 所以 $\mathrm{cone}_{\boldsymbol{a}}(S) \subset \mathrm{co}\,(C_{\boldsymbol{a}}(S))$, 由此得到

$$\mathrm{cone}_{\boldsymbol{a}}(S) = \mathrm{co}\,(C_{\boldsymbol{a}}(S)).$$

因为 $C_{\boldsymbol{a}}(\mathrm{co}S)$ 是包含凸集 $\mathrm{co}S$ 的锥, 设 $\boldsymbol{x}, \boldsymbol{y} \in S$, 对 $\lambda > 0$ 和 $t \in (0,1)$, 有

$$\boldsymbol{a} + \lambda\,((1-t)\,\boldsymbol{x} + t\boldsymbol{y} - \boldsymbol{a}) \in C_{\boldsymbol{a}}(\mathrm{co}S),$$

且

$$\begin{aligned} & \boldsymbol{a} + \lambda\,((1-t)\,\boldsymbol{x} + t\boldsymbol{y} - \boldsymbol{a}) \\ = \ & (1-t)\,(\boldsymbol{a} + \lambda\,(\boldsymbol{x} - \boldsymbol{a})) + t\,(\boldsymbol{a} + \lambda\,(\boldsymbol{y} - \boldsymbol{a})) \in \mathrm{co}\,(C_{\boldsymbol{a}}(S)). \end{aligned}$$

上式表明 $\mathrm{co}\,(C_{\boldsymbol{a}}(S)) = C_{\boldsymbol{a}}(\mathrm{co}S)$.

(2) 结论显然.

(3) 根据 (2) 有 $\mathrm{aff}\,(C_{\boldsymbol{a}}(S)) \subset \mathrm{aff}\,(\mathrm{cone}_{\boldsymbol{a}}(S)) \subset \mathrm{aff}(S \cup \{\boldsymbol{a}\})$. 另一方面, 由

$$S \cup \{\boldsymbol{a}\} \subset C_{\boldsymbol{a}}(S) \subset \mathrm{cone}_{\boldsymbol{a}}(S) \subset \mathrm{aff}(S \cup \{\boldsymbol{a}\}),$$

可得

$$\mathrm{aff}(S \cup \{\boldsymbol{a}\}) \subset \mathrm{aff}\,(C_{\boldsymbol{a}}(S)) \subset \mathrm{aff}\,(\mathrm{cone}_{\boldsymbol{a}}(S)) \subset \mathrm{aff}\,(\mathrm{aff}(S \cup \{\boldsymbol{a}\})) = \mathrm{aff}(S \cup \{\boldsymbol{a}\}).$$

(4) 由 (3) 直接得结论成立, 并且 $\mathrm{aff}\,(S) = \mathrm{aff}\,(S \cup \{\boldsymbol{a}\})$ 当且仅当 $\boldsymbol{a} \in \mathrm{aff}\,(S)$.

(5) 如果 $\mathrm{cone}_{\boldsymbol{a}}\,(S)$ 是一个平面, 那么有 $\mathrm{aff}\,(\{\boldsymbol{a}\} \cup S) \subset \mathrm{cone}_{\boldsymbol{a}}(S)$. 反过来, 如果

$$\mathrm{cone}_{\boldsymbol{a}}(S) = \mathrm{aff}\,(\{\boldsymbol{a}\} \cup S),$$

因为 $\mathrm{aff}\,(\{\boldsymbol{a}\} \cup S)$ 是包含 $\{\boldsymbol{a}\} \cup S$ 的平面, 所以结论 (5) 成立.

(6) 由 (4) 和 (5) 可得.

(7) 假设 $\mathrm{cone}_{\boldsymbol{a}}(S) \setminus \{\boldsymbol{a}\}$ 不是包含 S 关于顶点 $\boldsymbol{a} \in \mathbf{R}^n$ 的最小凸锥, 且存在一个凸锥

$$X_{\boldsymbol{a}}\,(S) \subset \mathrm{cone}_{\boldsymbol{a}}(S) \setminus \{\boldsymbol{a}\}.$$

若 $\boldsymbol{y} \in \mathrm{cone}_{\boldsymbol{a}}(S) \setminus \{\boldsymbol{a}\}$, 但 $\boldsymbol{y} \notin X_{\boldsymbol{a}}\,(S)$. 由定理 3.2.2 知, 存在仿射无关的集合 $\{\boldsymbol{a}, \boldsymbol{x}_1, \cdots, \boldsymbol{x}_m\}$ 满足

$$\boldsymbol{y} = \boldsymbol{a} + \lambda_1\,(\boldsymbol{x}_1 - \boldsymbol{a}) + \cdots + \lambda_m\,(\boldsymbol{x}_m - \boldsymbol{a}), \quad \boldsymbol{x}_i \in S, \quad \lambda_i > 0, \quad i = 1, \cdots, m.$$

设 $\lambda = \lambda_1 + \cdots + \lambda_m > 0$, 那么有

$$\lambda^{-1}(\boldsymbol{y} - \boldsymbol{a}) = \lambda^{-1}\lambda_1(\boldsymbol{x}_1 - \boldsymbol{a}) + \cdots + \lambda^{-1}\lambda_m(\boldsymbol{x}_m - \boldsymbol{a}) \in X_{\boldsymbol{a}}(S) - \boldsymbol{a},$$

由上式得 $\boldsymbol{y} \in X_{\boldsymbol{a}}(S)$, 产生矛盾. 故结论 (7) 成立.

(8) 因 C 和 $C_{\boldsymbol{a}}(S)$ 是正常顶点 $\boldsymbol{a}$ 的锥, 那么 $C_{\boldsymbol{a}}(S) \cap C$ 也是正常顶点 $\boldsymbol{a}$ 的锥. 如果存在点 $\boldsymbol{x} \in C_{\boldsymbol{a}}(S) \cap C$, 且 $\boldsymbol{x} \neq \boldsymbol{a}$, 根据性质 3.2.3(7) 知, 存在一个 $t > 0$ 和 $\boldsymbol{x}_0 \in S$ 使得

$$\boldsymbol{x} = (1 - t)\boldsymbol{a} + t\boldsymbol{x}_0.$$

对于任意 $\lambda > 0$, 由上式得

$$\boldsymbol{a} + \lambda(\boldsymbol{x} - \boldsymbol{a}) = \boldsymbol{a} + t\lambda(\boldsymbol{x}_0 - \boldsymbol{a}) \in C.$$

令 $\lambda = t^{-1}$, 得 $\boldsymbol{x}_0 \in C$. 因为 $S \cap C \subset \{\boldsymbol{a}\}$, 得 $\boldsymbol{x}_0 = \boldsymbol{a} = \boldsymbol{x}$, 即有 $C_a(S) \cap C = \{\boldsymbol{a}\}$.

(9) 由 (1) 有 $\text{cone}_{\boldsymbol{a}}(S) = C_{\boldsymbol{a}}(\text{co}S)$, 再根据 (8) 知结论 (9) 成立. □

例 3.2.6　设在 $\mathbf{R}^2$ 空间中的集合

$$S = \left\{(x_1, 0)^{\mathrm{T}} | x_1 \in (0, 1)\right\} \cup \left\{(0, x_2)^{\mathrm{T}} | x_2 \in (0, 1)\right\},$$

显然有 $\text{co}S = \left\{(x_1, x_2)^{\mathrm{T}} | x_1 \in (0, 1), x_2 \in (0, 1), x_1 + x_2 < 1\right\}$, 我们有

$$C_0(S) = \left\{(x_1, 0)^{\mathrm{T}} | x_1 \in [0, +\infty)\right\} \cup \left\{(0, x_2)^{\mathrm{T}} | x_2 \in [0, +\infty)\right\},$$

$$\text{cone}_0(S) = \text{co}(C_0(S)) = C_0(\text{co}S) = \left\{(x_1, x_2)^{\mathrm{T}} | x_1 \in [0, +\infty), x_2 \in [0, +\infty)\right\}.$$

且有 $\text{aff}(\{0\} \cup S) = \mathbf{R}^2$. 性质 3.2.5 中 (5) 和 (6) 的结论在此例的情形下不成立, 因为锥包和凸锥包 $\text{cone}_{\boldsymbol{a}}(S)$ 都不是一个平面.

下面得到最后一个凸锥包表示定理: 非空集合的凸锥包经过线性平移等于顶点和集合的线性平移产生的凸锥包.

定理 3.2.4　设非空集合 $S \subset \mathbf{R}^n$ 和点 $\boldsymbol{a}, \boldsymbol{b} \in \mathbf{R}^n$, 给定常数 α, 则

$$\text{cone}_{\boldsymbol{b}+\alpha\boldsymbol{a}}(\boldsymbol{b} + \alpha S) = \boldsymbol{b} + \alpha\text{cone}_{\boldsymbol{a}}(S). \tag{3.2.1-12}$$

证明　显然 $\boldsymbol{b} + \alpha S \subset \boldsymbol{b} + \alpha\text{cone}_{\boldsymbol{a}}(S)$, 以及通过性质 3.1.4 得 $\boldsymbol{b} + \alpha\text{cone}_{\boldsymbol{a}}(S)$ 是关于顶点 $\boldsymbol{b} + \alpha\boldsymbol{a}$ 的凸锥. 因此, 有

$$\text{cone}_{\boldsymbol{b}+\alpha\boldsymbol{a}}(\boldsymbol{b} + \alpha S) \subset \boldsymbol{b} + \alpha\text{cone}_{\boldsymbol{a}}(S).$$

反过来, 设 $\boldsymbol{y} \subset \boldsymbol{b}+\alpha \operatorname{cone}_{\boldsymbol{a}}(S)$. 因此, 有 $\boldsymbol{y}=\boldsymbol{b}+\alpha \boldsymbol{x}$, 其中 $\boldsymbol{x} \in \operatorname{cone}_{\boldsymbol{a}}(S)$. 根据定理 3.2.1, 存在 $\{\boldsymbol{x}_1, \cdots, \boldsymbol{x}_m\} \subset S$ 使得

$$\boldsymbol{x}=\boldsymbol{a}+\lambda_1(\boldsymbol{x}_1-\boldsymbol{a})+\cdots+\lambda_m(\boldsymbol{x}_m-\boldsymbol{a}), \quad \boldsymbol{x}_i \in S, \quad \lambda_i \geqslant 0, \quad i=1, \cdots, m.$$

那么有

$$\boldsymbol{y}=\boldsymbol{b}+\alpha \boldsymbol{x}=\boldsymbol{b}+\alpha \boldsymbol{a}+\lambda_1(\boldsymbol{b}+\alpha \boldsymbol{x}_1-(\boldsymbol{b}+\alpha \boldsymbol{a}))+\cdots+\lambda_m(\boldsymbol{b}+\alpha \boldsymbol{x}_m-(\boldsymbol{b}+\alpha \boldsymbol{a})),$$

由上式可得 $\boldsymbol{y} \in \operatorname{cone}_{\boldsymbol{b}+\alpha \boldsymbol{a}}(\boldsymbol{b}+\alpha S)$, 因此, (3.2.1-12) 式成立. □

本小节得到了集合的锥包与凸锥包的一些基本性质.

1) 基本性质: 凸锥包和锥包都包含顶点; 凸锥包的凸锥包等于原来的凸锥包; 凸锥等于它的凸锥包; 锥包的平移等于顶点的锥包平移.

2) 交并运算性质: 集合族凸锥包的交集包含集合族交集的凸锥包; 集合族凸锥包的并集属于集合族并集的凸锥包; 如果集合族是嵌套的, 那么集合族凸锥包的并集等于集合族并集的凸锥包; 集合族并集的凸锥包等于集合族中每个集合凸包的并集后再做成的凸锥包, 还等于集合族凸锥包的并集后再做成的凸锥包.

3) 锥包复合后的性质: 集合的锥包等于集合锥包的锥包; 锥等于锥包; 两个集合包含关系对应的锥包有对应的包含关系; 集合的锥包等于顶点与集合的每个点的凸组合.

4) 三个凸锥包表示定理: 集合的凸锥包由集合的顶点和集合内有限个点的非负线性组合构成 (定理 3.2.1); 除去顶点的集合凸锥包由集合的顶点和集合内含顶点与有限多个点组成的仿射无关组的正线性组合构成 (定理 3.2.2); 集合凸锥包由集合内有限多个仿射无关点的非负线性组合构成 (定理 3.2.3).

5) 凸锥包、凸包的锥包、锥包的凸包、锥包的仿射集之间的关系: 凸锥包等于凸锥包的凸包, 等于凸包的凸锥包, 等于锥包的凸包, 等于凸包的锥包; 集合与顶点一起生成的仿射集包含凸锥包; 凸锥包包含锥包; 锥包的仿射集等于凸锥包的仿射集, 还等于集合与顶点一起生成的仿射集; 其余结论见性质 3.2.5.

6) 非空集合的凸锥包经过线性平移后等于顶点与集合经过线性平移生成的凸锥包之和, 见公式 (3.2.1-12).

3.2.2 拓扑性质

本小节讨论锥包和凸锥包的若干拓扑性质, 如闭集的锥包是闭集, 闭集的凸锥包也是闭集, 见下面性质 3.2.6.

性质 3.2.6 设集合 $S \subset \mathbf{R}^n$ 和点 $\boldsymbol{a} \in \mathbf{R}^n$, 下面结论成立:

(1) $C_{\boldsymbol{a}}(\mathrm{cl} S) \subset \mathrm{cl}(C_{\boldsymbol{a}}(S))=\mathrm{cl}(C_{\boldsymbol{a}}(\mathrm{cl} S))$;

(2) 如果 S 是有界的和 $\boldsymbol{a} \notin \mathrm{cl} S$, 则 $C_{\boldsymbol{a}}(\mathrm{cl} S)=\mathrm{cl}(C_{\boldsymbol{a}}(S))$;

(3) $\text{cone}_{\boldsymbol{a}}(\text{cl}S) \subset \text{cl}(\text{cone}_{\boldsymbol{a}} S) = \text{cl}(\text{cone}_{\boldsymbol{a}}(\text{cl}S))$;

(4) 如果 C 是有界的和 $\boldsymbol{a} \notin \text{cl}(\text{co}S)$, 则 $\text{cone}_{\boldsymbol{a}}(\text{cl}S) = \text{cl}(\text{cone}_{\boldsymbol{a}} S)$;

(5) 如果 T 是 S 的一个稠密子集, 则 $\text{cl}(C_{\boldsymbol{a}}(S)) = \text{cl}(C_{\boldsymbol{a}}(T))$, $\text{cl}(\text{cone}_{\boldsymbol{a}} S) = \text{cl}(\text{cone}_{\boldsymbol{a}} T)$.

证明 当 $S = \varnothing$ 时, 性质的结论均显然成立. 下面假设 $S \subset \mathbf{R}^n$ 是非空的.

(1) 因为 $S \subset \text{cl}S$, 有 $C_{\boldsymbol{a}}(S) \subset C_{\boldsymbol{a}}(\text{cl}S)$, 然后得 $\text{cl}(C_{\boldsymbol{a}}(S)) \subset \text{cl}(C_{\boldsymbol{a}}(\text{cl}S))$. 再由性质 3.1.7 知 $\text{cl}(C_{\boldsymbol{a}}(S))$ 和 $\text{cl}(C_{\boldsymbol{a}}(\text{cl}S))$ 都是关于顶点 $\boldsymbol{a} \in \mathbf{R}^n$ 的锥. 设点 $\boldsymbol{x} \in C_{\boldsymbol{a}}(\text{cl}S)$, 则存在一个 $t > 0$ 和 $\boldsymbol{y} \in \text{cl}S$ 使得 $\boldsymbol{x} = \boldsymbol{a} + t(\boldsymbol{y} - \boldsymbol{a})$, 那么存在对任何收敛到 $\boldsymbol{y}$ 的序列

$$\boldsymbol{y}_k \in S \quad (k = 1, 2, \cdots),$$

使得序列 $\boldsymbol{a} + t(\boldsymbol{y}_k - \boldsymbol{a}) \in C_{\boldsymbol{a}}(S)$ 收敛到 $\boldsymbol{a} + t(\boldsymbol{y} - \boldsymbol{a})$, 即得 $\boldsymbol{x} \in \text{cl}(C_{\boldsymbol{a}}(S))$. 因此, 有

$$C_{\boldsymbol{a}}(\text{cl}S) \subset \text{cl}(C_{\boldsymbol{a}}(S))$$

成立. 反过来, 设 $\boldsymbol{x} \in \text{cl}(C_{\boldsymbol{a}}(\text{cl}S))$, 那么存在对任何收敛于 $\boldsymbol{x}$ 的序列

$$\boldsymbol{x}_k \in C_{\boldsymbol{a}}(\text{cl}S) \quad (k = 1, 2, \cdots),$$

由前面证明有 $\boldsymbol{x}_k \in \text{cl}(C_{\boldsymbol{a}}(S))\,(k = 1, 2, \cdots)$, 令 $k \to +\infty$, 即得 $\boldsymbol{x} \in \text{cl}(C_{\boldsymbol{a}}(S))$.

(2) 如果 S 是有界的和 $\boldsymbol{a} \notin \text{cl}S$, 则有 $C_{\boldsymbol{a}}(\text{cl}S) \subset \text{cl}(C_{\boldsymbol{a}}(S)) = \text{cl}(C_{\boldsymbol{a}}(\text{cl}S))$. 只需证明 $C_{\boldsymbol{a}}(\text{cl}S)$ 是闭集即可. 设任何收敛于 $\boldsymbol{x} \in \mathbf{R}^n$ 的序列

$$\boldsymbol{x}_k \in C_{\boldsymbol{a}}(\text{cl}S) \subset \text{cl}(C_{\boldsymbol{a}}(S)) \quad (k = 1, 2, \cdots),$$

上式表明存在一个 $t_k > 0$ 和 $\boldsymbol{y}_k \in S$ 使得 $\boldsymbol{x}_k = \boldsymbol{a} + t_k(\boldsymbol{y}_k - \boldsymbol{a})$. 由假设 S 是有界的和 $\boldsymbol{a} \notin \text{cl}S$, 可知序列 $\{\boldsymbol{y}_k\}$ 存在收敛子序列. 不妨设 $\boldsymbol{y}_k \to \boldsymbol{y} \in \text{cl}(S)\,(k \to +\infty)$, 因为 $||\boldsymbol{x}_k - \boldsymbol{a}|| = t_k||\boldsymbol{y}_k - \boldsymbol{a}||$, 得

$$\lim_{k\to+\infty} t_k = \lim_{k\to+\infty} \frac{||\boldsymbol{x}_k - \boldsymbol{a}||}{||\boldsymbol{y}_k - \boldsymbol{a}||} = \frac{||\boldsymbol{x} - \boldsymbol{a}||}{||\boldsymbol{y} - \boldsymbol{a}||}.$$

当 $k \to +\infty$ 时, 得 $\boldsymbol{x}_k = \boldsymbol{a} + t_k(\boldsymbol{y}_k - \boldsymbol{a}) \to \boldsymbol{x} = \boldsymbol{a} + t(\boldsymbol{y} - \boldsymbol{a}) \in C_{\boldsymbol{a}}(\text{cl}S)$. 说明 $C_{\boldsymbol{a}}(\text{cl}S)$ 是闭集.

(3) 和 (4) 的证明过程完全类似于 (1) 和 (2) 的证明.

(5) 因为 T 是 S 的一个稠密子集, 则由 $T \subset S \subset \text{cl}T$ 得

$$\text{cl}(C_{\boldsymbol{a}}(T)) \subset \text{cl}(C_{\boldsymbol{a}}(S)) \subset \text{cl}(C_{\boldsymbol{a}}(\text{cl}T)).$$

由 (1) 的结论 $\mathrm{cl}(C_{\boldsymbol{a}}(T))=\mathrm{cl}(C_{\boldsymbol{a}}(\mathrm{cl}T))$ 可得: $\mathrm{cl}(C_{\boldsymbol{a}}(S))=\mathrm{cl}(C_{\boldsymbol{a}}(T))$. 由 (3) 得

$$\mathrm{cl}(\mathrm{cone}_{\boldsymbol{a}}S)=\mathrm{cl}(\mathrm{cone}_{\boldsymbol{a}}T).$$ □

例 3.2.7 设在 $\mathbf{R}^2$ 空间中的开集合

$$S=\left\{(x_1,x_2)^{\mathrm{T}}\,|x_1^2+(x_2-1)^2<1\right\}.$$

显然有 $\mathrm{cl}S=\left\{(x_1,x_2)\,|x_1^2+(x_2-1)^2\leqslant 1\right\}$, 进一步有

$$\begin{aligned}C_{\mathbf{0}}(S)=C_{\mathbf{0}}(\mathrm{cl}S)&=\mathrm{cone}_{\mathbf{0}}(S)=\mathrm{cone}_{\mathbf{0}}(\mathrm{cl}S)\\&=\left\{(x_1,x_2)^{\mathrm{T}}\,|x_1\in\mathbf{R},x_2>0\right\}\cup\{\mathbf{0}\},\\ \mathrm{cl}(C_{\mathbf{0}}(S))=\mathrm{cl}(C_{\mathbf{0}}(\mathrm{cl}S))&=\mathrm{cl}(\mathrm{cone}_{\mathbf{0}}(S))\\&=\mathrm{cl}(\mathrm{cone}_{\mathbf{0}}(\mathrm{cl}S))=\left\{(x_1,x_2)^{\mathrm{T}}\,|x_1\in\mathbf{R},x_2\geqslant 0\right\}.\end{aligned}$$

这个例子满足性质 3.2.6 中 (1) 和 (3), 但不满足 (2) 和 (4), 因为 $\mathbf{0}\in\mathrm{cl}S$. 这个例子对下面性质 3.2.7 的结论也不满足.

进一步, 我们可以看到相对内部与锥包之间的关系. 性质 3.2.7 表明, 如果凸锥包是一个平面等价于集合张成的仿射空间, 且等价于顶点属于该集合凸包的相对内部或顶点属于凸锥包的相对内部.

性质 3.2.7 设集合 $S\subset\mathbf{R}^n$ 和点 $\boldsymbol{a}\in\mathbf{R}^n$, 则下面结论等价:

(1) $\boldsymbol{a}\in\mathrm{ri}(\mathrm{co}S)$;

(2) $\boldsymbol{a}\in\mathrm{ri}(\mathrm{cone}_{\boldsymbol{a}}(S))$;

(3) $\mathrm{cone}_{\boldsymbol{a}}(\mathrm{ri}(\mathrm{co}S))=\mathrm{aff}S$;

(4) $\mathrm{cone}_{\boldsymbol{a}}(S)=\mathrm{aff}S$;

(5) $\mathrm{cone}_{\boldsymbol{a}}(S)$ 是一个平面.

证明 首先证明 (1)⇒(2). 如果 $\boldsymbol{a}\in\mathrm{ri}(\mathrm{co}S)$, 因为 $S\subset\mathrm{co}S\subset\mathrm{cone}_{\boldsymbol{a}}(S)$, 由性质 3.2.5(2) 和 (3) 得 $\mathrm{aff}S=\mathrm{aff}(\mathrm{co}S)=\mathrm{aff}(\mathrm{cone}_{\boldsymbol{a}}(S))$. 因此, 再根据定理 2.3.1 有

$$\boldsymbol{a}\in\mathrm{ri}(\mathrm{co}S)\subset\mathrm{ri}(\mathrm{cone}_{\boldsymbol{a}}(S)).$$

再证明 (1)⇒(3). 如果 $\boldsymbol{a}\in\mathrm{ri}(\mathrm{co}S)$, 则有 $\boldsymbol{a}\in\mathrm{aff}(\mathrm{co}S)=\mathrm{aff}S$. 因为

$$\mathrm{ri}(\mathrm{co}S)\subset\mathrm{co}S\subset\mathrm{aff}S,$$

由性质 3.2.5(2) 得 $\mathrm{cone}_{\boldsymbol{a}}(\mathrm{ri}\,(\mathrm{co}S)\,) \subset \mathrm{aff}\,(S)$. 反过来, 设 $\boldsymbol{x} \in \mathrm{aff}\,(S)$, 但 $\boldsymbol{x} \neq \boldsymbol{a} \in \mathrm{ri}(\mathrm{co}S)$, 则存在 $\varepsilon > 0$ 使得

$$(\boldsymbol{a} + \varepsilon B) \cap (\mathrm{aff}\,(\mathrm{co}S)\,) \subset \mathrm{co}S \subset \mathrm{aff}S,$$

上式和定理 2.3.2 表明 $l\,\langle \boldsymbol{a}, \boldsymbol{x}\rangle \cap (\mathrm{co}S) \neq \varnothing$. 因此, 有 $l\,\langle \boldsymbol{a}, \boldsymbol{x}\rangle \subset \mathrm{cone}_{\boldsymbol{a}}(\mathrm{ri}\,(\mathrm{co}S))$, 从而 (3) 成立.

类似地, 运用上面的证明过程, 容易证明 (2)⇒(3) 成立.

证明 (3)⇒(4). 因为 $\mathrm{aff}(\mathrm{ri}\,(\mathrm{co}S)\,) = \mathrm{aff}(\mathrm{co}S) = \mathrm{aff}S$, 所以 (4) 成立.

证明 (4) 和 (5) 等价. 由性质 3.2.5(5) 知结论成立.

证明 (5)⇒(1). 若 $\mathrm{cone}_{\boldsymbol{a}}(S)$ 是一个平面, 由性质 3.1.9 显然 $\boldsymbol{a} \in \mathrm{ri}(\mathrm{co}S)$. □

例 3.2.8 设在 $\mathbf{R}^2$ 空间中的开集合: $S = \left\{(x_1, 0)^{\mathrm{T}}\,|x_1 \in (-1, 1)\right\}$, 显然凸锥包:

$$\mathrm{cone}_{\boldsymbol{a}}S = \left\{(x_1, 0)^{\mathrm{T}}\,|x_1 \in \mathbf{R}\right\}$$

是一个面, 其中 $\boldsymbol{a} = (0, 0)^{\mathrm{T}} \in \mathrm{cone}_{\boldsymbol{a}}S$, 有 $\mathrm{co}S = \left\{(x_1, 0)^{\mathrm{T}}\,|x_1 \in [-1, 1]\right\}$ 和 $\mathrm{aff}S = \left\{(x_1, 0)^{\mathrm{T}}\,|x_1 \in \mathbf{R}\right\}$, 容易检查性质 3.2.7 的结论均成立.

下面结论与性质 3.2.7 不同, 主要描述的是相对内部张成的锥包之间的关系.

性质 3.2.8 设集合 $S \subset \mathbf{R}^n$ 和点 $\boldsymbol{a} \in \mathbf{R}^n$, 下面结论成立:

(1) $\mathrm{cone}_{\boldsymbol{a}}(\mathrm{ri}(\mathrm{co}S)) = \{\boldsymbol{a}\} \cup \mathrm{ri}(\mathrm{cone}_{\boldsymbol{a}}S)$;

(2) $\mathrm{cone}_{\boldsymbol{a}}(\mathrm{ri}(\mathrm{co}S)) = \mathrm{ri}(\mathrm{cone}_{\boldsymbol{a}}S)$ 当且仅当 $\mathrm{cone}_{\boldsymbol{a}}S = \mathrm{aff}(\boldsymbol{a} \cup S)$;

(3) 如果 $S \neq \varnothing$, 则 $\mathrm{cone}_{\boldsymbol{a}}(\mathrm{ri}(\mathrm{co}S)) = \mathrm{ri}(\mathrm{cone}_{\boldsymbol{a}}S)$ 当且仅当 $\mathrm{cone}_{\boldsymbol{a}}S = \mathrm{aff}S$.

证明 当 $\boldsymbol{a} \in \mathrm{ri}(\mathrm{cone}_{\boldsymbol{a}}(S))$ 时, 由性质 3.2.7 知本性质的所有结论成立. 现在设

$$\boldsymbol{a} \notin \mathrm{ri}(\mathrm{cone}_{\boldsymbol{a}}(S)),$$

再进一步证明 (1)~(3) 结论成立.

(1) 因为 $\mathrm{co}S \subset \mathrm{cone}_{\boldsymbol{a}}(S)$, 由性质 3.2.5 得 $\mathrm{aff}\,(\mathrm{co}S) = \mathrm{aff}\,(\mathrm{cone}_{\boldsymbol{a}}(S))$. 因此, 根据定理 2.3.1 有 $\mathrm{ri}(\mathrm{co}S) \subset \mathrm{ri}(\mathrm{cone}_{\boldsymbol{a}}(S))$, 再由性质 3.1.7 知 $\{\boldsymbol{a}\} \cup \mathrm{ri}(\mathrm{cone}_{\boldsymbol{a}}(S))$ 是关于顶点 $\boldsymbol{a}$ 的锥. 因此, 有

$$\mathrm{cone}_{\boldsymbol{a}}(\mathrm{ri}(\mathrm{co}S)) \subset \{\boldsymbol{a}\} \cup \mathrm{ri}(\mathrm{cone}_{\boldsymbol{a}}(S)).$$

反过来, 设 $\boldsymbol{x} \in \{\boldsymbol{a}\} \cup \mathrm{ri}(\mathrm{cone}_{\boldsymbol{a}}(S))$. 只要考虑 $\boldsymbol{x} \in \mathrm{ri}(\mathrm{cone}_{\boldsymbol{a}}(S))$, 存在 $\varepsilon > 0$ 使得

$$(\boldsymbol{x} + \varepsilon B) \cap (\mathrm{aff}\,(\mathrm{cone}_{\boldsymbol{a}}(S))\,) \subset \mathrm{cone}_{\boldsymbol{a}}(S) = \mathrm{cone}_{\boldsymbol{a}}(\mathrm{co}S).$$

由性质 3.2.1 得 $\boldsymbol{a} \in \mathrm{cone}_{\boldsymbol{a}}(S)$, 再由上式和定理 2.3.2 得, 对于任意 $0 \leqslant \lambda < 1$ 有

$$(1-\lambda)\,\boldsymbol{x} + \lambda\boldsymbol{a} \in \mathrm{ri}(\mathrm{cone}_{\boldsymbol{a}}(S)) \subset \mathrm{cone}_{\boldsymbol{a}}(S).$$

上式说明 $[\boldsymbol{a},\boldsymbol{x}) \cap \mathrm{cone}_{\boldsymbol{a}}(S) \neq \varnothing$. 对任意 $t>0$, 有

$$\boldsymbol{a} + t\left((1-\lambda)\,\boldsymbol{x} + \lambda\boldsymbol{a} - \boldsymbol{a}\right) = \boldsymbol{a} + t\,(1-\lambda)\,(\boldsymbol{x}-\boldsymbol{a}) \in \mathrm{cone}_{\boldsymbol{a}}(S).$$

因此, 有 $l\,(\boldsymbol{a},\boldsymbol{x}\rangle \subset \mathrm{cone}_{\boldsymbol{a}}(S) = \mathrm{cone}_{\boldsymbol{a}}(\mathrm{co}S)$. 再由性质 3.2.1(3) 得

$$\boldsymbol{x} \in \boldsymbol{x} - \boldsymbol{a} + \mathrm{cone}_{\boldsymbol{a}}(\mathrm{co}S) = \mathrm{cone}_{\boldsymbol{x}}(\boldsymbol{x} - \boldsymbol{a} + \mathrm{co}S),$$

故 $\boldsymbol{x} \in \mathrm{ri}\,(\mathrm{cone}_{\boldsymbol{x}}(\boldsymbol{x} - \boldsymbol{a} + \mathrm{co}S))$. 再运用性质 3.2.7(1)(2) 和性质 3.2.1(3) 得

$$\mathrm{cone}_{\boldsymbol{x}}(\boldsymbol{x} - \boldsymbol{a} + \mathrm{co}S) = \mathrm{cone}_{\boldsymbol{x}}(\boldsymbol{x} - \boldsymbol{a} + \mathrm{ri}\,(\mathrm{co}S)) = \boldsymbol{x} - \boldsymbol{a} + \mathrm{cone}_{\boldsymbol{a}}(\mathrm{ri}\,(\mathrm{co}S)).$$

因此, 有 $\boldsymbol{x} \in \mathrm{cone}_{\boldsymbol{a}}(\mathrm{ri}\,(\mathrm{co}S))$, 并得到 $\mathrm{cone}_{\boldsymbol{a}}(\mathrm{ri}(\mathrm{co}S)) = \{\boldsymbol{a}\} \cup \mathrm{ri}(\mathrm{cone}_{\boldsymbol{a}}(S))$.

(2) 和 (3) 可由 (1) 直接得到. □

例 3.2.9 设在 $\mathbf{R}^2$ 空间中的开集合

$$S = \left\{(x_1,x_2)^{\mathrm{T}}\,|x_1^2 + (x_2-1)^2 < 1\right\},$$

有 $\mathrm{co}S = \mathrm{ri}\,(\mathrm{co}S) = \left\{(x_1,x_2)^{\mathrm{T}}\,|x_1^2 + (x_2-1)^2 < 1\right\}$, 因为 $\mathbf{0} \notin \mathrm{ri}\,(\mathrm{co}S)$ 和 $\mathbf{0} \notin \mathrm{ri}(\mathrm{cone}_{\boldsymbol{a}}(S))$, 且

$$\mathrm{ri}\,(\mathrm{cone}_{\mathbf{0}}\,(\mathrm{co}S)) = \left\{(x_1,x_2)^{\mathrm{T}}\,|x_1 \in \mathbf{R},\ x_2 > 0\right\} \cup \{\mathbf{0}\},$$

$$\mathrm{cone}_{\mathbf{0}}\,(\mathrm{ri}\,(\mathrm{co}S)) = \left\{(x_1,x_2)^{\mathrm{T}}\,|x_1 \in \mathbf{R},\ x_2 \geqslant 0\right\}.$$

这个例子对性质 3.2.7 的结论均不成立, 对性质 3.2.8 中的结论 (1) 成立, 但对结论 (2) 和 (3) 不成立, 因为条件

$$\mathrm{cone}_{\mathbf{0}}(\mathrm{ri}(\mathrm{co}S)) = \mathrm{ri}(\mathrm{cone}_{\mathbf{0}}S)$$

不成立.

本小节得到锥包和凸锥包的几个拓扑性质.

1) (凸) 锥包的闭包性质: 集合闭包的 (凸) 锥包的闭包等于其 (凸) 锥包的闭包, 且包含其闭包的锥包; 如果集合有界且不包含顶点, 则集合闭包的 (凸) 锥包等于其 (凸) 锥包的闭包; 如果 T 是 S 的一个稠密子集, 则它们的 (凸) 锥包相等.

2) (凸) 锥包与仿射集合相对内部之间的关系: 集合凸锥包是一个平面等价于凸锥包等于该集合的仿射集或等于集合凸包的相对内部生成的凸锥包, 或等价于顶点属于集合凸包的相对内部或属于集合凸锥包的相对内部.

3) 凸包的相对内部生成的凸锥包等于该凸锥包顶点与集合凸锥包的相对内部并集; 凸包的相对内部生成的凸锥包等于集合凸锥包的相对内部等价于凸锥包等于顶点与集合的并集生成的仿射集.

3.2.3 原点锥包

本小节讨论顶点仅为原点的凸锥包一些性质, 用 cone$C$ 表示集合 $C \subset \mathbf{R}^n$ 的凸锥包. 下面的定理说明集合的内点的正半直线仍是这个集合凸锥包的内点.

定理 3.2.5 设 C 是 $\mathbf{R}^n$ 中含原点的凸集. 如果 $\boldsymbol{x}_\lambda = \lambda \boldsymbol{x}, \lambda > 0, \boldsymbol{x} \in \operatorname{int} C$, 则

$$\boldsymbol{x}_\lambda \in \operatorname{int}(\operatorname{cone} C).$$

证明 由定理 2.3.3 得 $\operatorname{int} C = \operatorname{ri} C$, 即有 $\operatorname{int}(\operatorname{cone} C) = \operatorname{ri}(\operatorname{cone} C)$, 由性质 3.2.8 得

$$\boldsymbol{x}_\lambda \in \operatorname{int}(\operatorname{cone} C). \qquad \square$$

性质 3.2.2 说明了凸锥包族的交集关系, 上一节得到多个集合相交的凸锥包与每个集合的凸锥包相交有包含关系, 不一定是等式. 但是如果相交的集合包含相同的顶点, 则集合相交的凸锥包等于每个集合的凸锥包相交. 见定理 3.2.6.

定理 3.2.6 设 $C_i \subset \mathbf{R}^n$ 是含原点的凸集, $i = 1, \cdots, m$, 则

$$\bigcap_{i=1}^{m} \operatorname{cone}(C_i) = \operatorname{cone}\left(\bigcap_{i=1}^{m} C_i\right).$$

证明 设 $\boldsymbol{x} \in \operatorname{cone}\left(\bigcap_{i=1}^{m} C_i\right)$, 则存在 $\lambda > 0$ 和 $\boldsymbol{x}_\lambda \in \bigcap_{i=1}^{m} C_i$, 使得 $\boldsymbol{x} = \lambda \boldsymbol{x}_\lambda$. 对于 $i = 1, \cdots, m$, 有 $\boldsymbol{x}_\lambda \in \operatorname{cone}(C_i)$, 所以 $\boldsymbol{x} \in \bigcap_{i=1}^{m} \operatorname{cone}(C_i)$, 因此有

$$\bigcap_{i=1}^{m} \operatorname{cone}(C_i) \supset \operatorname{cone}\left(\bigcap_{i=1}^{m} C_i\right).$$

反之, 设 $\boldsymbol{x} \in \bigcap_{i=1}^{m} \operatorname{cone}(C_i)$, 有 $\boldsymbol{x} \in \operatorname{cone}(C_i)\ (i = 1, \cdots, m)$, 则存在 $\lambda_i > 0$ 和 $\boldsymbol{x}_i \in C_i$ 使得 $\boldsymbol{x} = \lambda_i \boldsymbol{x}_i$. 因此, 有 $\lambda_i^{-1} \boldsymbol{x} \in C_i$. 设 $t \in (0,1)$, 有

$$t\left(\lambda_i^{-1} \boldsymbol{x}\right) = (1-t)\,\mathbf{0} + t\left(\lambda_i^{-1} \boldsymbol{x}\right) \in C_i, \quad i = 1, \cdots, m.$$

设 $\beta=\min\left\{\lambda_i^{-1}|i=1,2,\cdots,m\right\}$, 有 $0<\beta\lambda_i\leqslant 1\,(i=1,2,\cdots,m)$, 根据上式有

$$\beta\boldsymbol{x}=(\beta\lambda_i)\left(\lambda_i^{-1}\boldsymbol{x}\right)\in C_i,\quad i=1,2,\cdots,m.$$

因此, 得到 $\boldsymbol{x}\in\bigcap_{i=1}^m\operatorname{cone}(C_i)$, 即有 $\bigcap_{i=1}^m\operatorname{cone}(C_i)\subset\operatorname{cone}\left(\bigcap_{i=1}^m C_i\right)$. □

含原点凸集生成的凸锥包有下面的充要条件.

定理 3.2.7 设 C 是 $\mathbf{R}^n$ 中的凸集, $\mathbf{0}\in C$, 则 $\boldsymbol{x}\in\operatorname{cone}C$ 的充要条件是对于充分小的 $\lambda>0$ 有 $\lambda\boldsymbol{x}\in C$.

证明 设 $\boldsymbol{x}\in\operatorname{cone}C$, 存在 $\lambda>0$ 和 $\boldsymbol{x}_\lambda\in C$ 使得 $\boldsymbol{x}=\lambda\boldsymbol{x}_\lambda$. 当 $\beta\leqslant\dfrac{1}{\lambda}$, 即 $\lambda\beta\leqslant 1$ 时, 有

$$\beta\boldsymbol{x}=\lambda\beta\boldsymbol{x}_\lambda=(1-\lambda\beta)\,\mathbf{0}+\lambda\beta\boldsymbol{x}_\lambda\in C.$$

反过来, 如果对 $\lambda>0$ 有 $\lambda\boldsymbol{x}=\boldsymbol{x}_\lambda\in C$, 则有 $\boldsymbol{x}=\lambda^{-1}\boldsymbol{x}_\lambda\in\operatorname{cone}C$. □

本小节得到了原点凸锥包的三个性质:

1) 集合的内点的正半直线仍是这个集合凸锥包的内点;

2) 集合相交的凸锥包等于每个集合的凸锥包相交;

3) $\boldsymbol{x}\in\operatorname{cone}C$ 的充要条件是对于充分小的 $\lambda>0$ 有 $\lambda\boldsymbol{x}\in C$.

3.3 回 收 锥

回收锥是集合生成的一种特殊锥, 由集合内任何点为起点的正半直线构成. 回收锥与 (凸) 锥包是一个完全不同的概念.

3.3.1 基本性质

定义 3.3.1 设非空集合 $X\subset\mathbf{R}^n$, 则集合

$$\operatorname{rec}X=\{\boldsymbol{e}\in\mathbf{R}^n\mid\lambda\boldsymbol{e}+\boldsymbol{x}\in X,\forall\lambda\geqslant 0,\forall\boldsymbol{x}\in X\}\tag{3.3.1-1}$$

称为 X 的回收锥. 定义空集的回收锥仍是空集.

由上面定义, 除了单点集合, 回收锥包含了原集合中任何点到原点所有直线延伸, 见下面例子.

例 3.3.1 (1) 单点集合 $X=\{\boldsymbol{a}\}$ 的回收锥 $\operatorname{rec}X=\{\mathbf{0}\}$.

(2) 有界集合 $X=[1,2]$, 则 X 的回收锥为 $\operatorname{rec}X=\{0\}$.

(3) 集合 $X=[1,+\infty)\subset\mathbf{R}$, X 的回收锥为 $\operatorname{rec}X=[0,+\infty)\subset\mathbf{R}$.

图 3.3-1 表示的集合是一个有界集, 它的回收锥是一个原点. 图 3.3-2 表示的集合 X 是一个无限集合, 它由集合 $X\subset\mathbf{R}^n$ 延伸到原点直线构成, 补充了不构成锥的部分集合 $\operatorname{rex}X\setminus X$.

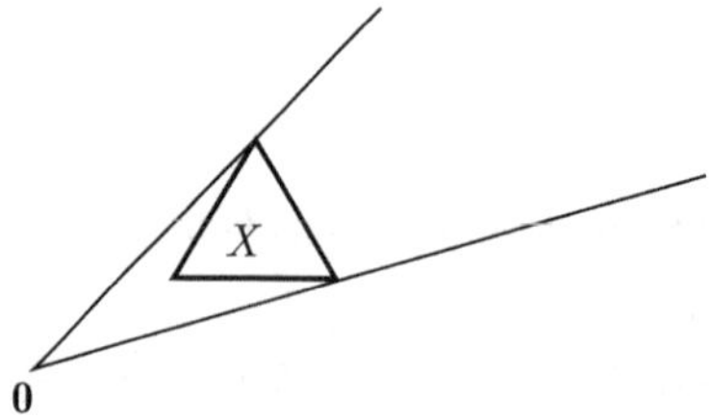

图 3.3-1　有界集生成的回收锥

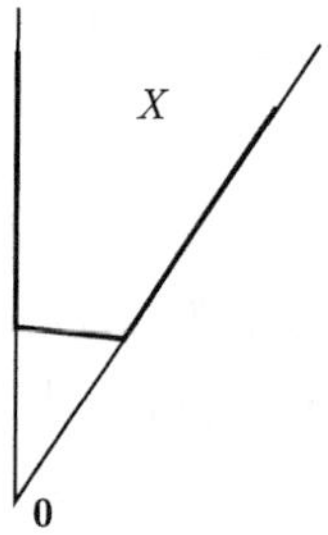

图 3.3-2　无界集生成的回收锥

下面的性质 3.3.1 揭示了回收锥结构几个最基本的性质: 集合的回收锥内任意点乘以非负数加上集合的任何点等于该集合; 集合的回收锥是含原点凸锥的最大凸锥等价于凸锥加上该集合等于该集合; 如果集合有界, 则回收锥仅含原点; 如果给定点 $\boldsymbol{a} \in X$, 则回收锥属于集合 $\cap\{\lambda(X-\boldsymbol{a}) \mid \lambda > 0\}$; 集合的回收锥属于其方向子空间等价于集合是一个平面.

性质 3.3.1　设 $X \subset \mathbf{R}^n$ 是非空集合, 下面结论成立:

(1) $\mathrm{rec}X$ 是满足条件 $\lambda\boldsymbol{e}+X \subset X$ $(\forall\lambda \geqslant 0)$ 的所有 $\boldsymbol{e} \in \mathbf{R}^n$ 构成的最大集合, 等价于 $\lambda\boldsymbol{e}+X = X, \forall\lambda \geqslant 0$;

(2) 如果 $C+X \subset X$, 则 $\mathrm{rec}X$ 是包含以原点为顶点凸锥 $C \subset \mathbf{R}^n$ 的最大凸锥, 且等价于 $C+X=X$;

(3) 如果 X 是有界的, 则 $\mathrm{rec}X = \{\mathbf{0}\}$;

(4) 如果给定点 $\boldsymbol{a} \in X$, 则 $\mathrm{rec}X \subset \cap\{\lambda(X-\boldsymbol{a}) \mid \lambda > 0\}$;

(5) $\mathrm{rec}X \subset \mathrm{dir}X$, $\mathrm{rec}X = \mathrm{dir}X$ 当且仅当 X 是一个平面.

证明　(1) 根据定义 3.1.1, 当 $\lambda \geqslant 0$ 时, 有 $\lambda\boldsymbol{e}+X \subset X$. 另一方面, 由于 $\mathrm{rec}X$ 包含原点, 因此, $X \subset \lambda\boldsymbol{e}+X$.

(2) 因为 $C \subset \mathbf{R}^n$ 包含以原点为顶点凸锥, 那么有 $C+X \supset X$. 因此, $C+X = X$. 根据 (3.3.1-1) 式, 设 $\boldsymbol{e}_1, \boldsymbol{e}_2 \in \mathrm{rec}X$ 和 $\alpha, \beta \geqslant 0$, 有

$$\alpha\boldsymbol{e}_1+\beta\boldsymbol{e}_2+\boldsymbol{x} \in \alpha\boldsymbol{e}_1+X \subset X.$$

因此, $\alpha\boldsymbol{e}_1+\beta\boldsymbol{e}_2 \in \mathrm{rec}X$, 说明 $\mathrm{rec}X$ 是包含以原点为顶点的凸锥. 设 $\boldsymbol{e} \in C$, 对于 $\boldsymbol{x} \in X$ 有

$$\alpha\boldsymbol{e}+\boldsymbol{x} \in C+X \subset X,$$

说明 $\boldsymbol{e} \in C \subset \mathrm{rec}X$.

(3) 如果 X 是有界的, 则存在常数 A 使得对于任意 $\boldsymbol{x} \in X$ 有 $\|\boldsymbol{x}\| < A$. 若 $\boldsymbol{e} \in \mathrm{rec}X$, 则对于任意 $\lambda \geqslant 0$ 有 $\lambda\boldsymbol{e}+\boldsymbol{x} \in X$, 即有 $\|\lambda\boldsymbol{e}+\boldsymbol{x}\| < A$. 然而, 假设 $\boldsymbol{e} \neq \mathbf{0}$, 有

$$\lambda\|\boldsymbol{e}\| = \|\lambda\boldsymbol{e}+\boldsymbol{x}-\boldsymbol{x}\| \leqslant \|\lambda\boldsymbol{e}+\boldsymbol{x}\| + \|-\boldsymbol{x}\| < 2A.$$

因为 $\lambda \geqslant 0$ 可以任意大, 与上式矛盾. 因此结论成立.

(4) 设 $\boldsymbol{e} \in \mathrm{rec}X$, 那么对于任意 $\lambda \geqslant 0$ 有 $\lambda\boldsymbol{e}+\boldsymbol{x} \in X$, 则 $\lambda\boldsymbol{e}+\boldsymbol{x}-\boldsymbol{a} \in X-\boldsymbol{a}$. 当 $\boldsymbol{x}=\boldsymbol{a}$ 时, 有 $\lambda\boldsymbol{e} \in X-\boldsymbol{a}$, 即对任意 $t>0$ 有 $t\lambda\boldsymbol{e} \in t(X-\boldsymbol{a})$. 因为 λ 的任意性, 有

$$\mathrm{rec}X \subset \cap\{\lambda(X-\boldsymbol{a}) \mid \lambda > 0\}.$$

(5) 设 $\boldsymbol{e} \in \mathrm{rec}X$, 则对于任意 $\lambda \geqslant 0$ 有 $\lambda\boldsymbol{e}+\boldsymbol{x} \in X$, 由定义 2.1.8 和性质 2.1.10(1) 得 $\mathrm{dir}X = \mathrm{aff}X - \boldsymbol{x}$, 其中 $\boldsymbol{x} \in \mathrm{aff}X$. 显然有 $\lambda\boldsymbol{e} \in X-\boldsymbol{x}$, 即 $\mathrm{rec}X \subset \mathrm{dir}X$. 如果 X 是一个平面, 存在 $\boldsymbol{a} \in \mathbf{R}^n$ 和子空间 S 使得 $X=\boldsymbol{a}+S$, 即有 $X-\boldsymbol{a}=S$, 等价于

$$\mathrm{dir}X = \mathrm{aff}X - \boldsymbol{x} - \boldsymbol{a} + \boldsymbol{a} = S - \boldsymbol{x} + \boldsymbol{a} = X - \boldsymbol{x}.$$

根据回收锥定义, 由上式有 $\mathrm{rec}X = \mathrm{dir}X$. 反之, 若 $\mathrm{rec}X = \mathrm{dir}X$, 那么由性质 2.1.10(2) 有 $\mathrm{rec}X = \mathrm{aff}X - \mathrm{aff}X$, 即说明 $\mathrm{rec}X$ 是子空间. 事实上, 再根据回收锥定义, 那么对于任意 $\lambda \geqslant 0$ 和 $\boldsymbol{x} \in X$ 有

$$\lambda\mathrm{rec}X + \boldsymbol{x} = \lambda(\mathrm{aff}X - \boldsymbol{y}) + \boldsymbol{x} \subset X, \quad \forall \boldsymbol{y} \in \mathrm{aff}X.$$

在上式中取 $\lambda = 1$ 和 $\boldsymbol{y}=\boldsymbol{x}$, 得 $X \subset \mathrm{aff}X \subset X$, 表明 X 是一个仿射集. 由于 $\mathrm{rec}X = X - \boldsymbol{x}$, 因此, X 是一个平面. □

下面的定理揭示了顶点锥的回收锥与顶点集的关系, 即如果集合是一个顶点锥, 则顶点集合减去顶点包含于回收锥, 且回收锥包含于顶点集与该集合的并集减去顶点. 当顶点锥是凸锥时, 回收锥等于顶点集与该集合的并集减去顶点.

定理 3.3.1 如果 $C \subset \mathbf{R}^n$ 是关于顶点 $\boldsymbol{a}$ 的非空锥, 则

$$\mathrm{ap}C - \boldsymbol{a} \subset \mathrm{rec}C \subset (\mathrm{ap}C \cup C) - \boldsymbol{a},$$

进一步, 有 $\mathrm{rec}C = (\mathrm{ap}C \cup C) - \boldsymbol{a}$ 当且仅当 C 是凸集.

证明 设任意的 $\boldsymbol{e} \in \mathrm{ap}C - \boldsymbol{a}$, 则存在 $\boldsymbol{b} \in \mathrm{ap}C$ 使得 $\boldsymbol{e} = \boldsymbol{b} - \boldsymbol{a}$. 同时, 也可得 C 是关于顶点 $\boldsymbol{b}$ 的非空锥, 那么对于任意 $\lambda \geqslant 0$ 有

$$\boldsymbol{b} + \lambda(C - \boldsymbol{b}) = C, \tag{3.3.1-2}$$

$$\boldsymbol{a} + \lambda(C - \boldsymbol{a}) = C. \tag{3.3.1-3}$$

由 (3.3.1-2) 式与 (3.3.1-3) 式相减得 $C = (\lambda - 1)\boldsymbol{e} + C$, 根据性质 3.3.1 得 $\boldsymbol{e} \in \mathrm{rec}C$.

因为 $\mathbf{0} \in (\mathrm{ap}C \cup C) - \boldsymbol{a}$, 所以由性质 3.3.1 有

$$\begin{aligned}\mathrm{rec}C &= \mathrm{rec}C + \mathbf{0} \subset \mathrm{rec}C + ((\mathrm{ap}C \cup C) - \boldsymbol{a}) \\ &= (\mathrm{rec}(\mathrm{ap}C \cup C) + (\mathrm{ap}C \cup C)) - \boldsymbol{a} \subset (\mathrm{ap}C \cup C) - \boldsymbol{a}.\end{aligned} \tag{3.3.1-4}$$

如果 C 是凸锥, 由于 $\mathrm{ap}C$ 是平面, 那么 $\mathrm{ap}C \cup C$ 也是凸锥. 于是有

$$((\mathrm{ap}C \cup C) - \boldsymbol{a}) + (\mathrm{ap}C \cup C) = (\mathrm{ap}C \cup C) - \boldsymbol{a},$$

由上式和 (3.3.1-4) 式得 $(\mathrm{ap}C \cup C) - \boldsymbol{a} \subset \mathrm{rec}(\mathrm{ap}C \cup C)$. 因此, $\mathrm{rec}C = (\mathrm{ap}C \cup C) - \boldsymbol{a}$.

反过来, 若 $\mathrm{rec}C = (\mathrm{ap}C \cup C) - \boldsymbol{a}$ 成立, 由性质 3.3.1 有

$$((\mathrm{ap}C \cup C) - \boldsymbol{a}) + C = C.$$

上式说明 $(\mathrm{ap}C \cup C) - \boldsymbol{a}$ 是包含 C 的最大凸锥. 根据性质 3.1.5 知, 若 $\mathrm{ap}C \subset C$, 则 C 是凸锥. 若 $\mathrm{ap}C \cap C = \varnothing$, 那么对任意 $\boldsymbol{b}_1, \boldsymbol{b}_2 \in (\mathrm{ap}C \cup C)$, 下面分两种情形讨论. 若 $\boldsymbol{b}_1, \boldsymbol{b}_2 \in \mathrm{ap}C$, 因 $\mathrm{ap}C$ 是平面, 则对任意 $\lambda \in (0, 1)$ 有

$$\lambda \boldsymbol{b}_1 + (1 - \lambda)\, \boldsymbol{b}_2 \in \mathrm{ap}C;$$

若 $\boldsymbol{b}_1, \boldsymbol{b}_2$ 至少一个属于 $\mathrm{ap}C$, 根据 C 是锥, 那么也有 $\lambda \boldsymbol{b}_1 + (1 - \lambda)\, \boldsymbol{b}_2 \in C$; 若 $\boldsymbol{b}_1, \boldsymbol{b}_2 \in C$, 则 $\lambda \boldsymbol{b}_1 + (1 - \lambda)\, \boldsymbol{b}_2 \in C$. □

例 3.3.2 (1) 设在 $\mathbf{R}^2$ 中的集合 $C = \{(x_1, 0)^{\mathrm{T}} | x_1 \in (0, +\infty)\} \cup \{(0, x_2)^{\mathrm{T}} | x_2 \in (0, +\infty)\}$, 它的回收锥 $\mathrm{rec}C = \{(x_1, 0)^{\mathrm{T}} \mid x_1 \in [0, +\infty)\} \cup \{(0, x_2)^{\mathrm{T}} \mid x_2 \in [0, +\infty)\}$, $\mathrm{ap}C = \{\mathbf{0}\}$, 有

$$\mathrm{ap}C \subset \mathrm{rec}C \subset (\mathrm{ap}C \cup C).$$

(2) 设在 $\mathbf{R}^2$ 中的凸集 $C = \{(x_1, x_2)^{\mathrm{T}} | x_1 + x_2 \geqslant 1, x_1, x_2 \geqslant 0\}$, 它的回收锥为

$$\mathrm{rec}C = \left\{(x_1, x_2)^{\mathrm{T}} | x_1, x_2 \geqslant 0\right\},$$

有 $\mathrm{ap}C \subset \mathrm{rec}C \subset (\mathrm{ap}C \cup C)$, 但 C 不是凸锥.

本小节得到了回收锥的一些基本性质:

1) 集合的回收锥中的点乘以非负数加上集合的任何点等于该集合.

2) 集合的回收锥是含原点凸锥的最大凸锥等价于凸锥加上该集合等于该集合.

3) 如果集合有界, 则回收锥仅含原点.

4) 如果给定点 $\boldsymbol{a} \in X$, 则回收锥属于集合 $\cap\{\lambda(X - \boldsymbol{a}) \mid \lambda > 0\}$.

5) 集合的回收锥属于其方向子空间等价于集合是一个平面.

6) 如果集合是一个顶点锥, 则顶点集合减去顶点属于回收锥, 回收锥又属于顶点集与该集合的并集减去顶点. C 是凸集等价于 $\mathrm{rec}C = (\mathrm{ap}C \cup C) - \boldsymbol{a}$.

3.3.2 代数性质

本小节讨论集合族并与交后的回收锥性质. 假设集合族中所有集合的交是非空的, 下面结论成立: 集合族中所有集合交的回收锥包含所有集合回收锥的交, 集合族中所有集合并的回收锥包含所有集合回收锥的交.

性质 3.3.2 如果在 $\mathbf{R}^n$ 中一个集合族 $\{X_i\}(i \in I)$ 满足 $\bigcap_{i\in I} X_i \neq \varnothing$, 则

$$\bigcap_{i\in I} \mathrm{rec}X_i \subset \mathrm{rec}\left(\bigcap_{i\in I} X_i\right), \quad \bigcap_{i\in I} \mathrm{rec}X_i \subset \mathrm{rec}\left(\bigcup_{i\in I} X_i\right). \tag{3.3.2-1}$$

证明 设 $\boldsymbol{e} \in \bigcap_{i\in I} \mathrm{rec}X_i$, 那么对任意 $\lambda \geqslant 0$ 有 $\lambda\boldsymbol{e} + \boldsymbol{x} \in X_i$, 其中 $\boldsymbol{x} \in \bigcap_{i\in I} X_i$, 由定义 3.3.1 得到 $\boldsymbol{e} \in \mathrm{rec}\left(\bigcap_{i\in I} X_i\right)$. 第 2 个结论显然也成立. □

性质 3.3.2 中的条件 $\bigcap_{i\in I} X_i \neq \varnothing$ 不能缺少, 见下面例子.

例 3.3.3 在 $\mathbf{R}^2$ 空间中 $C_1 = \{(x_1, 0)^{\mathrm{T}} \,|x_1 \in (1, +\infty)\}$ 和 $C_2 = \{(0, x_2)^{\mathrm{T}} \,| x_2 \in (1, +\infty)\}$ 分别是由顶点 $(1, 0)^{\mathrm{T}}$ 和 $(0, 1)^{\mathrm{T}}$ 构成的凸锥, 它们的并集 $C_1 \cup C_2$ 并不构成凸锥. 它们的交集 $C_1 \cap C_2 = \varnothing$, 且有

$$\mathrm{rec}C_1 = \left\{(x_1, 0)^{\mathrm{T}} \,|x_1 \in [0, +\infty)\right\}, \quad \mathrm{rec}C_2 = \left\{(0, x_2)^{\mathrm{T}} \,|x_2 \in [0, +\infty)\right\}.$$

并有 $\mathrm{rec}\,(C_1 \cap C_2) = \varnothing$, $\mathrm{rec}C_1 \cap \mathrm{rec}C_2 = \{\mathbf{0}\}$, 进一步有

$$\mathrm{rec}\,(C_1 \cup C_2) = \left\{(x_1, 0)^{\mathrm{T}} \,|x_1 \in [0, +\infty)\right\} \cup \left\{(0, x_2)^{\mathrm{T}} \,|x_2 \in [0, +\infty)\right\}.$$

如果集合族的交是空集, 性质 3.3.2 的结论不一定成立. 下面的定理表明对于给定集合作线性运算平移的回收锥与原集合回收锥的关系.

定理 3.3.2 设给定集合 $X \subset \mathbf{R}^n$, $\boldsymbol{a} \in \mathbf{R}^n$ 和常数 t, 则

$$\mathrm{rec}(\boldsymbol{a} + tX) = t\ \mathrm{rec}X = \mathrm{sgn}(t)\mathrm{rec}X. \tag{3.3.2-2}$$

证明 当 $t = 0$ 时结论显然成立. 假设 $t \neq 0$, 设 $\lambda \geqslant 0$, 由性质 3.3.1(1) 得

$$\lambda\boldsymbol{e} + \boldsymbol{a} + tX = \boldsymbol{a} + tX \Leftrightarrow \text{等价于}\lambda\boldsymbol{e} + tX = tX,$$

还等价于若 $t > 0$ 有 $\lambda'\boldsymbol{e} + X = X$, 其中 $\lambda' = \dfrac{\lambda}{t} \geqslant 0$; 若 $t < 0$ 有 $\lambda'\boldsymbol{e} - X = -X$, 其中 $\lambda' = -\dfrac{\lambda}{t} \geqslant 0$. 则 (3.3.2-2) 式成立. □

定理 3.3.2 说明回收锥具有线性平移性, 即集合 $\boldsymbol{a} + tX$ 的回收锥等于回收锥乘以常数 t, 与平移无关. 定理 3.3.2 可以推广为以下结论, 即多个集合线性组合的回收锥包含各回收锥的线性组合 (见公式 (3.3.2-3)).

定理 3.3.3 设集合 $X_1, X_2, \cdots, X_m \subset \mathbf{R}^n$ 和常数 $t_1, t_2, \cdots, t_m \in \mathbf{R}$, 则

$$t_1 \mathrm{rec} X_1 + \cdots + t_m \mathrm{rec} X_m \subset \mathrm{rec}(t_1 X_1 + \cdots + t_m X_m). \tag{3.3.2-3}$$

进一步, 如果所有的集合 $X_1, X_2, \cdots, X_m$ 非空, 且对应的 $\mathrm{aff} X_1, \mathrm{aff} X_2, \cdots, \mathrm{aff} X_m$ 是仿射无关的, 则

$$t_1 \mathrm{rec} X_1 + \cdots + t_m \mathrm{rec} X_m = \mathrm{rec}(t_1 X_1 + \cdots + t_m X_m). \tag{3.3.2-4}$$

证明 当 $m = 1$ 时, 由定理 3.3.2 得结论 (3.3.2-3) 式和 (3.3.2-4) 式都成立. 证明当 $m = 2$ 时, 设任意 $\boldsymbol{e}_1 \in \mathrm{rec} X_1$ 和 $\boldsymbol{e}_2 \in \mathrm{rec} X_2$, 对任意的 $\lambda \geqslant 0$ 有

$$\lambda \boldsymbol{e}_1 + X_1 = X_1, \quad \lambda \boldsymbol{e}_2 + X_2 = X_2.$$

得 $\lambda t_1 \boldsymbol{e}_1 + \lambda t_2 \boldsymbol{e}_2 + t_1 X_1 + t_2 X_2 = t_1 X_1 + t_2 X_2$, 有 $t_1 \boldsymbol{e}_1 + t_2 \boldsymbol{e}_2 \in \mathrm{rec}\,(t_1 X_1 + t_2 X_2)$.

假设 $\mathrm{aff} X_1$ 和 $\mathrm{aff} X_2$ 是仿射无关的, 设一个 $\boldsymbol{e} \in \mathrm{rec}\,(t_1 X_1 + t_2 X_2)$, 那么对任意的 $\lambda \geqslant 0$ 有

$$\lambda \boldsymbol{e} + t_1 X_1 + t_2 X_2 = t_1 X_1 + t_2 X_2. \tag{3.3.2-5}$$

设 $\boldsymbol{x}_1 \in X_1$ 和 $\boldsymbol{x}_2 \in X_2$, 由性质 2.1.10 得两个子空间

$$\mathrm{dir} X_1 = \mathrm{aff} X_1 - \boldsymbol{x}_1, \quad \mathrm{dir} X_2 = \mathrm{aff} X_2 - \boldsymbol{x}_2. \tag{3.3.2-6}$$

因为 $\mathrm{aff} X_1$ 和 $\mathrm{aff} X_2$ 是仿射无关的, 故 $\mathrm{dir} X_1$ 和 $\mathrm{dir} X_2$ 是独立的子空间, 由定理 3.3.2 得

$$\mathrm{rec} X_1 = \mathrm{rec}\,(X_1 - \boldsymbol{x}_1), \quad \mathrm{rec} X_2 = \mathrm{rec}\,(X_2 - \boldsymbol{x}_2).$$

再根据性质 3.3.1(5) 得

$$\mathrm{rec} X_1 = \mathrm{rec}\,(X_1 - \boldsymbol{x}_1) \subset \mathrm{aff} X_1 - \boldsymbol{x}_1, \quad \mathrm{rec} X_2 = \mathrm{rec}\,(X_2 - \boldsymbol{x}_2) \subset \mathrm{aff} X_2 - \boldsymbol{x}_2.$$

根据上式得

$$\mathrm{rec}\,(t_1 (X_1 - \boldsymbol{x}_1) + t_2 (X_2 - \boldsymbol{x}_2)) = \mathrm{rec}\,(t_1 X_1 + t_2 X_2), \tag{3.3.2-7}$$

由 (3.3.2-5) 式有 $\boldsymbol{e} \in \mathrm{rec}\,(t_1 (X_1 - \boldsymbol{x}_1) + t_2 (X_2 - \boldsymbol{x}_2))$. 再由性质 3.3.1(5) 得

$$\boldsymbol{e} \in \mathrm{dir}\,(t_1 (X_1 - \boldsymbol{x}_1) + t_2 (X_2 - \boldsymbol{x}_2)) = \mathrm{dir}\,(t_1 (X_1 - \boldsymbol{x}_1)) + \mathrm{dir}\,(t_2 (X_2 - \boldsymbol{x}_2)).$$

由 (3.3.2-6) 式和上式得 $\boldsymbol{e}_1 \in \mathrm{dir}\,(X_1 - \boldsymbol{x}_1), \boldsymbol{e}_2 \in \mathrm{dir}\,(X_2 - \boldsymbol{x}_2)$, 其中 $\boldsymbol{e} = \boldsymbol{e}_1 + \boldsymbol{e}_2$. 那么对任意的 $\lambda \geqslant 0$ 由 (3.3.2-7) 式得

$$\lambda (\boldsymbol{e}_1 + \boldsymbol{e}_2) + t_1 (X_1 - \boldsymbol{x}_1) + t_2 (X_2 - \boldsymbol{x}_2) = t_1 (X_1 - \boldsymbol{x}_1) + t_2 (X_2 - \boldsymbol{x}_2).$$

因为 $\operatorname{dir}X_1$ 和 $\operatorname{dir}X_2$ 是独立的子空间, 那么由上式得

$$\lambda(\boldsymbol{e}_1)+t_1(X_1-\boldsymbol{x}_1)\in \operatorname{aff}(X_1)-\boldsymbol{x}_1,\quad \lambda(\boldsymbol{e}_2)+t_2(X_2-\boldsymbol{x}_2)\in \operatorname{aff}(X_2)-\boldsymbol{x}_2.$$

即有 $\boldsymbol{e}_1\in t_1\operatorname{rec}(X_1-\boldsymbol{x}_1)$ 和 $\boldsymbol{e}_2\in t_2\operatorname{rec}(X_2-\boldsymbol{x}_2)$ 成立, 由 (3.3.2-2) 式得

$$\boldsymbol{e}\in t_1\operatorname{rec}(X_1-\boldsymbol{x}_1)+t_2\operatorname{rec}(X_2-\boldsymbol{x}_2)=t_1\operatorname{rec}(X_1)+t_2\operatorname{rec}(X_2).$$

当 $m>2$ 时, 由归纳法可知结论成立. □

例 3.3.4 在 $\mathbf{R}^2$ 空间中 $C_1=\{(x_1,0)^{\mathrm{T}}\,|x_1\in(1,+\infty)\}$ 和 $C_2=\{(0,x_2)^{\mathrm{T}}\,|x_2\in(1,+\infty)\}$ 分别是由两个顶点 $(1,0)^{\mathrm{T}}$ 和 $(0,1)^{\mathrm{T}}$ 构成的凸锥, 且有

$$\operatorname{rec}C_1=\left\{(x_1,0)^{\mathrm{T}}\,|x_1\in[0,+\infty)\right\},\quad \operatorname{rec}C_2=\left\{(0,x_2)^{\mathrm{T}}\,|x_2\in[0,+\infty)\right\}.$$

设 $t_1,t_2>0$, 有

$$t_1\operatorname{rec}C_1+t_2\operatorname{rec}C_2=\left\{(x_1,x_2)^{\mathrm{T}}\,|x_1,x_2\in[0,+\infty)\right\}.$$

从 $t_1C_1+t_2C_2=\left\{(x_1,x_2)^{\mathrm{T}}\,|x_1\in(t_1,+\infty),x_2\in(t_2,+\infty)\right\}$ 得其回收锥为

$$\operatorname{rec}(t_1C_1+t_2C_2)=\left\{(x_1,x_2)^{\mathrm{T}}\,|x_1\in[0,+\infty),x_2\in[0,+\infty)\right\},$$

当 $t_1,t_2>0$ 时, 定理 3.3.3 的结论对于上述例子成立.

下面的定理是集合在线性变换后回收锥的性质.

定理 3.3.4 设线性变换 $L:\mathbf{R}^n\to\mathbf{R}^m$, 集合 $X\subset\mathbf{R}^n$ 和 $Y\subset\mathbf{R}^m$, $\operatorname{rng}L$ 是变换 L 的值域, 则

$$L(\operatorname{rec}X)\subset\operatorname{rec}L(X),\quad \operatorname{rec}L^{-1}(Y)=L^{-1}(\operatorname{rec}(Y\cap\operatorname{rng}L)).$$

进一步, 如果 L 是一对一映射, 则 $L(\operatorname{rec}X)=\operatorname{rec}L(X)$.

证明 设任何 $\boldsymbol{e}\in\operatorname{rec}X$ 和 $\lambda\geqslant0$, 由性质 3.3.1(1) 得 $\lambda\boldsymbol{e}+X=X$. 经线性变换得

$$L(\lambda\boldsymbol{e}+X)=\lambda L\boldsymbol{e}+L(X)=L(X).$$

因此, $L\boldsymbol{e}\in\operatorname{rec}L(X)$. 设 $X=L^{-1}(Y)$, $\boldsymbol{e}\in\operatorname{rec}X$ 和 $\lambda\geqslant0$. 再由性质 3.3.1(1) 得 $\lambda\boldsymbol{e}+L^{-1}(Y)=L^{-1}(Y)$, 那么有

$$L(\lambda\boldsymbol{e})+L\left(L^{-1}(Y)\right)=L\left(L^{-1}(Y)\right),$$

即有

$$\lambda L(\boldsymbol{e}) + Y \cap \text{rng}L = Y \cap \text{rng}L,$$

得 $L(\boldsymbol{e}) \in \text{rec}(Y \cap \text{rng}L)$, 则 $\boldsymbol{e} \in L^{-1}(\text{rec}(Y \cap \text{rng}L))$.

反过来, 设 $\boldsymbol{e} \in L^{-1}(\text{rec}(Y \cap \text{rng}L))$, 那么有 $L(\boldsymbol{e}) \in \text{rec}(Y \cap \text{rng}L)$. 从而有

$$L(\lambda \boldsymbol{e}) + L\left(L^{-1}(Y)\right) = L\left(L^{-1}(Y)\right),$$

即

$$L\left(\lambda \boldsymbol{e} + L^{-1}(Y)\right) = L\left(L^{-1}(Y)\right).$$

由上式得 $\lambda \boldsymbol{e} + L^{-1}(Y) \subset L^{-1}(Y)$, 即有 $\boldsymbol{e} \in \text{rec}X$. 如果 L 是一对一映射, 则 $L(\text{rec}X) = \text{rec}L(X)$ 是显然的. □

本小节得到了回收锥几个代数运算性质:

1) 如果集合族中所有集合交是非空的, 则集合族中所有集合交的回收锥包含所有集合回收锥的交, 集合族所有集合并的回收锥包含所有集合回收锥的交;

2) 集合 $\boldsymbol{a} + tX$ 的回收锥等于 X 的回收锥乘以常数 t;

3) 多个集合线性组合的回收锥包含各回收锥的线性组合;

4) 集合 X 的线性变换后的回收锥包含 X 的回收锥后的线性变换;

5) 集合 Y 逆线性变换后的回收锥等于线性变换的值域与集合 Y 的交集的回收锥的逆线性变换.

3.3.3 拓扑性质

本小节进一步讨论集合的回收锥与它的闭包和相对内部的拓扑关系. 下面的性质讨论了集合闭包和相对内部的点与回收锥的关系, 即属于集合闭包的点加上回收锥仍然属于该集合的闭包, 属于集合相对内部的点加上回收锥仍然属于该集合的相对内部.

性质 3.3.3 设 $X \subset \mathbf{R}^n$ 是非空集合, 下面结论成立:

(1) 如果 $\boldsymbol{x} \in \text{cl}X$, 则 $\boldsymbol{x} + \text{rec}X \subset \text{cl}X$;

(2) 如果 $\text{ri}X \neq \varnothing$ 和 $\boldsymbol{x} \in \text{ri}X$, 则 $\boldsymbol{x} + \text{rec}X \subset \text{ri}X$.

证明 (1) 设序列 $\boldsymbol{e}_i \in \text{rec}X$ 收敛到 $\boldsymbol{e}$, 对于任意 $\lambda \geqslant 0$ 有 $\lambda \boldsymbol{e}_i + \boldsymbol{x} \in X$, 显然当 $i \to +\infty$ 时有 $\lambda \boldsymbol{e}_i + \boldsymbol{x} \to \lambda \boldsymbol{e} + \boldsymbol{x} \in \text{cl}X$.

(2) 设 $\boldsymbol{e} \in \text{rec}X$, 对于任意 $\lambda \geqslant 0$ 有 $\lambda \boldsymbol{e} + \boldsymbol{x} \in X$. 因为 $\boldsymbol{x} \in \text{ri}X$, 所以存在一个 $\varepsilon > 0$ 有

$$(\boldsymbol{x} + \varepsilon B) \cap \text{aff}X \subset X.$$

因此, 存在一个 $\lambda' > 0$, 当 $\lambda \in (0, \lambda')$ 时有 $\lambda \boldsymbol{e} + \boldsymbol{x} \in (\boldsymbol{x} + \varepsilon B) \cap \text{aff}X \subset X$. 设 $\boldsymbol{a} = \boldsymbol{x} + \boldsymbol{e}$, 令

$$\boldsymbol{u} \in (\boldsymbol{a} + \lambda' B) \cap \text{aff}X.$$

得 $\|\boldsymbol{u}-\boldsymbol{e}-\boldsymbol{x}\|=\|\boldsymbol{u}-\boldsymbol{a}\|\leqslant\lambda'$, 则 $\boldsymbol{u}-\boldsymbol{e}\in(\boldsymbol{x}+\lambda'B)\cap\text{aff}X\subset X$, 再由性质 3.3.1(5) 知, 对于任意 $\boldsymbol{e}\in\text{dir}X=\text{aff}X-\boldsymbol{e}=\text{aff}X$ 有

$$\boldsymbol{u}=\boldsymbol{u}-\boldsymbol{e}+\boldsymbol{e}\in(\boldsymbol{x}+\lambda'B)\cap\text{aff}X+\boldsymbol{e}\subset X+\boldsymbol{e}\subset X.$$

因此, $(\boldsymbol{x}+\boldsymbol{e}+\lambda'B)\cap\text{aff}X\subset X$, 即 $\boldsymbol{x}+\text{rec}X\subset\text{ri}X$. □

例 3.3.5 在 $\mathbf{R}^2$ 空间中 $X=\{(x_1,0)\,|x_1\in(1,+\infty)\}$ 的回收锥为

$$\text{rec}X=\{(x_1,0)\,|x_1\in[0,+\infty)\},$$

有 $\text{cl}X=\{(x_1,0)\,|x_1\in[1,+\infty)\}$. 如果 $\boldsymbol{x}\in\text{cl}X$, 有 $\boldsymbol{x}+\text{rec}X\subset\text{cl}X$; 如果 $\boldsymbol{x}\in\text{ri}X$, 显然有 $\boldsymbol{x}+\text{rec}X\subset\text{ri}X$. 这个例子对于性质 3.3.3 成立.

例 3.3.6 在 $\mathbf{R}^2$ 空间中, $X=\{(x_1,0)^{\text{T}}\,|x_1\in\mathbf{R}\}\cup\{(0,1)\}$, 且有 X 和 $\text{cl}X$ 的回收锥为

$$\text{rec}X=\text{rec}\,(\text{cl}X)=\{\mathbf{0}\}.$$

则有 $\text{aff}X=\mathbf{R}^2$ 和 $\text{ri}X=\varnothing$, 其回收锥 $\text{rec}\,(\text{ri}X)=\varnothing$, 性质 3.3.3(2) 对此例不成立.

性质 3.3.4 设 $X\subset\mathbf{R}^n$, 下面结论成立:

(1) $\text{rec}(\text{cl}X)$ 和 $\text{rec}(\text{ri}X)$ 是闭集;

(2) $\text{cl}(\text{rec}X)\subset\text{rec}(\text{cl}X)$;

(3) 如果 $\text{ri}X\neq\varnothing$, 则 $\text{cl}(\text{rec}X)\subset\text{rec}(\text{ri}X)$.

证明 (1) 设序列 $\boldsymbol{e}_i\in\text{rec}\,(\text{cl}X)$ 收敛到 $\boldsymbol{e}$, 对于任意 $\lambda\geqslant0$ 有 $\lambda\boldsymbol{e}_i+\boldsymbol{x}\in\text{cl}X$, 其中 $\boldsymbol{x}\in\text{cl}X$. 显然当 $i\to+\infty$ 时有 $\lambda\boldsymbol{e}_i+\boldsymbol{x}\to\lambda\boldsymbol{e}+\boldsymbol{x}\in\text{cl}X$.

设序列 $\boldsymbol{e}_i\in\text{rec}\,(\text{ri}X)$ 收敛到 $\boldsymbol{e}$. 对于每一个 $\lambda\geqslant0$ 有 $\lambda\boldsymbol{e}_i+\boldsymbol{x}\subset\text{ri}X$, 其中 $\boldsymbol{x}\in\text{ri}X$. 则存在 $\varepsilon>0$ 有

$$(\boldsymbol{x}+\lambda\boldsymbol{e}_i+\varepsilon B)\cap\text{aff}X\subset X\quad\text{和}\quad\|\lambda\boldsymbol{e}_i-\lambda\boldsymbol{e}\|<0.5\varepsilon.$$

设 $\boldsymbol{u}\in(\boldsymbol{x}+\lambda\boldsymbol{e}_i+0.5\varepsilon B)\cap\text{aff}X$, 我们有

$$\|\boldsymbol{u}-\boldsymbol{x}-\lambda\boldsymbol{e}\|=\|\boldsymbol{u}-\boldsymbol{x}-\lambda\boldsymbol{e}_i+\lambda\boldsymbol{e}_i-\lambda\boldsymbol{e}\|<0.5\varepsilon+0.5\varepsilon.$$

因此, $\boldsymbol{u}\in(\boldsymbol{x}+\lambda\boldsymbol{e}+\varepsilon B)\cap\text{aff}X\subset X$, 即 $\boldsymbol{x}+\lambda\boldsymbol{e}\in\text{ri}X$, 得 $\boldsymbol{e}\in\text{rec}\,(\text{ri}X)$.

(2) 设序列 $\boldsymbol{e}_i\in\text{rec}\,(X)$ 收敛到 $\boldsymbol{e}$. 对于任意 $\lambda\geqslant0$ 有 $\lambda\boldsymbol{e}_i+\boldsymbol{x}\in X$, 其中 $\boldsymbol{x}\in X$. 显然令 $i\to+\infty$ 有 $\lambda\boldsymbol{e}_i+\boldsymbol{x}\to\lambda\boldsymbol{e}+\boldsymbol{x}\in\text{cl}X$, 即 $\boldsymbol{e}\in\text{cl}\,(\text{rec}\,(X))$.

(3) 设序列 $\boldsymbol{e}_i\in\text{rec}\,(X)$ 收敛到 $\boldsymbol{e}$, 由性质 3.3.1(2) 有 $\boldsymbol{e}_i+\boldsymbol{x}\subset\text{ri}X$, 其中 $\boldsymbol{x}\in\text{ri}X$. 则存在 $\varepsilon>0$ 有

$$(\boldsymbol{x}+\boldsymbol{e}_i+\varepsilon B)\cap\text{aff}X\subset X\quad\text{和}\quad\|\boldsymbol{e}_i-\boldsymbol{e}\|<0.5\varepsilon.$$

再设 $\boldsymbol{u} \in (\boldsymbol{x} + \boldsymbol{e}_i + 0.5\varepsilon B) \cap \mathrm{aff}X$, 我们有

$$\|\boldsymbol{u} - \boldsymbol{x} - \boldsymbol{e}\| = \|\boldsymbol{u} - \boldsymbol{x} - \boldsymbol{e}_i + \boldsymbol{e}_i - \boldsymbol{e}\| < 0.5\varepsilon + 0.5\varepsilon.$$

因此, $\boldsymbol{u} \in (\boldsymbol{x} + \boldsymbol{e} + \varepsilon B) \cap \mathrm{aff}X \subset X$, 即 $\boldsymbol{x} + \boldsymbol{e} \in \mathrm{ri}X$, 得 $\boldsymbol{e} \in \mathrm{rec}\,(\mathrm{ri}X)$. □

性质 3.3.4 表明了集合的闭包与相对内部构成的回收锥之间的拓扑性质, 即集合闭包的回收锥和集合相对内部的回收锥均是闭集, 集合闭包的回收锥包含集合回收锥的闭包, 集合相对内部的回收锥包含集合回收锥的闭包.

另一方面, 下面的性质 3.3.5 表明凸集的闭包与相对内部与回收锥之间的拓扑关系, 即如果集合是凸集, 则集合的相对内部的回收锥等于集合闭包的回收锥, 集合是无界的等价于集合闭包的回收锥含非零点, 闭包回收锥可以用 (3.3.3-1) 式表达, 相对内部回收锥可以用 (3.3.3-2) 式表达.

性质 3.3.5 设 $C \subset \mathbf{R}^n$ 是凸集, 下面结论成立:

(1) $\mathrm{rec}(\mathrm{ri}C) = \mathrm{rec}(\mathrm{cl}C)$;

(2) C 是无界的等价于 $\mathrm{rec}(\mathrm{cl}C)$ 含非零点, 且等价于 $\mathrm{rec}(\mathrm{ri}C)$ 含非零点;

(3) 如果 $C \neq \varnothing$ 和 $\boldsymbol{a} \in \mathrm{cl}C$, 则

$$\mathrm{rec}(\mathrm{cl}C) = \bigcap_{\lambda>0} \lambda(\mathrm{cl}C - \boldsymbol{a}); \tag{3.3.3-1}$$

(4) 如果 $C \neq \varnothing$ 和 $\boldsymbol{a} \in \mathrm{ri}C$, 则

$$\mathrm{rec}(\mathrm{ri}C) = \bigcap_{\lambda>0} \lambda(\mathrm{ri}C - \boldsymbol{a}) = \bigcap_{\lambda>0} \lambda(C - \boldsymbol{a}); \tag{3.3.3-2}$$

(5) 如果 $M \subset \mathbf{R}^n$ 是包含 C 的凸集, 则 $\mathrm{rec}(\mathrm{cl}C) \subset \mathrm{rec}(\mathrm{cl}M)$.

证明 (1) 设序列 $\boldsymbol{e} \in \mathrm{rec}\,(\mathrm{ri}C)$. 若 $\boldsymbol{e} = \boldsymbol{0}$, 显然, $\boldsymbol{0} \in \mathrm{rec}\,(\mathrm{cl}C)$. 假设 $\boldsymbol{e} \neq \boldsymbol{0}$, 则对于任意 $\lambda \geqslant 0$ 有 $\lambda\boldsymbol{e} + \boldsymbol{x} \in \mathrm{ri}C$, 其中 $\boldsymbol{x} \in \mathrm{ri}C$. 设 $\boldsymbol{y} \in \mathrm{cl}C$, 由定理 2.3.2 得

$$(1-t)\,(\lambda\boldsymbol{e} + \boldsymbol{x}) + t\boldsymbol{y} = (1-t)\,\lambda\boldsymbol{e} + (1-t)\,\boldsymbol{x} + t\boldsymbol{y} \in \mathrm{cl}C, \quad 0 \leqslant t < 1.$$

由上式得 $(1-t)\,\lambda\boldsymbol{e} + t\,(\boldsymbol{y} - \boldsymbol{x}) \in \mathrm{cl}C - \boldsymbol{x}$. 因为也有 $(1-t)\,\boldsymbol{x} + t\boldsymbol{y} \in \mathrm{cl}C$, 即 $t\,(\boldsymbol{y} - \boldsymbol{x}) \in \mathrm{cl}C - \boldsymbol{x}$. 因此, 由定义 3.3.1 可得 $\boldsymbol{e} \in \mathrm{rec}\,(\mathrm{cl}C)$.

反过来, 设 $\boldsymbol{e} \in \mathrm{rec}\,(\mathrm{cl}C)$. 若 $\boldsymbol{e} = \boldsymbol{0}$, 显然, 有 $\boldsymbol{0} \in \mathrm{rec}\,(\mathrm{ri}C)$. 假设 $\boldsymbol{e} \neq \boldsymbol{0}$, 那么对于任意 $\lambda \geqslant 0$ 有 $\lambda\boldsymbol{e} + \boldsymbol{x} \in \mathrm{cl}C$, 其中 $\boldsymbol{x} \in \mathrm{cl}C$. 设 $\boldsymbol{y} \in \mathrm{ri}C$, 由定理 2.3.2 得

$$t\,(\lambda\boldsymbol{e} + \boldsymbol{x}) + (1-t)\,\boldsymbol{y} = t\lambda\boldsymbol{e} + t\boldsymbol{x} + (1-t)\,\boldsymbol{y} \in \mathrm{ri}C,$$

其中 $0 \leqslant t < 1$. 又得 $t\lambda\boldsymbol{e} + (1-t)\,(\boldsymbol{y} - \boldsymbol{x}) \in \mathrm{cl}C - \boldsymbol{x}$, 因为也有 $t\boldsymbol{x} + (1-t)\,\boldsymbol{y} \in \mathrm{ri}C$, 即有

$$(1-t)\,(\boldsymbol{y} - \boldsymbol{x}) \in \mathrm{ri}C - \boldsymbol{x}.$$

由性质 3.3.4(3) 可得 $\boldsymbol{e} \in \mathrm{rec}\,(\mathrm{ri}C - \boldsymbol{x}) = \mathrm{rec}\,(\mathrm{ri}C)$.

(2) 若 $\mathrm{rec}(\mathrm{cl}C)$ 含非零点, 那么 $\mathrm{cl}C$ 是无界的, 否则由性质 3.3.1 知, 若 $\mathrm{cl}C$ 是有界的, 有 $\mathrm{rec}(\mathrm{cl}C)$ 仅含一个零点, 则 C 是无界的. 反过来, 如果 C 是无界的, 假设 $\mathrm{rec}(\mathrm{cl}C)$ 不含非零元素. 事实上 $\mathrm{rec}(\mathrm{cl}C)$ 仅含一个零点, 因为对于任意的 $\boldsymbol{x} \in C$, 存在非零点 $\boldsymbol{e} \in C$, 对于任意 $\lambda \geqslant 0$ 有 $\lambda(\boldsymbol{e} - \boldsymbol{x}) + \boldsymbol{x} \in C$. 如果它不成立, 对于每一个 $\boldsymbol{x} \in C$, 存在一个常数 $\delta > 1$, 当 $0 \leqslant \lambda \leqslant \delta$ 时有

$$\lambda(\boldsymbol{e} - \boldsymbol{x}) + \boldsymbol{x} \in \mathrm{cl}C, \quad \delta(\boldsymbol{e} - \boldsymbol{x}) + \boldsymbol{x} \in \mathrm{bd}C.$$

而当 $\lambda > \delta$ 时有 $\lambda(\boldsymbol{e} - \boldsymbol{x}) + \boldsymbol{x} \notin \mathrm{cl}C$, 表明 $\mathrm{cl}C$ 都是由有界线段 $[\boldsymbol{x}, \delta(\boldsymbol{e} - \boldsymbol{x}) + \boldsymbol{x}]$ 构成的, 它与 C 是无界的矛盾. 因此, 存在 $\boldsymbol{x} \in \mathrm{cl}C$ 和非零点 $\boldsymbol{x} \neq \boldsymbol{e} \in C$, 对于任意 $\lambda \geqslant 0$ 有 $\lambda(\boldsymbol{e} - \boldsymbol{x}) + \boldsymbol{x} \in C$, 故 $\mathrm{rec}(\mathrm{cl}C)$ 含非零点.

(3) 设 $\boldsymbol{e} \in \mathrm{rec}\,(\mathrm{cl}C)$, 那么对于任意 $\lambda \geqslant 0$ 有 $\lambda\boldsymbol{e} + \boldsymbol{x} \in \mathrm{cl}C$, 得 $\lambda\boldsymbol{e} + \boldsymbol{x} - \boldsymbol{a} \in \mathrm{cl}C - \boldsymbol{a}$. 当 $\boldsymbol{x} = \boldsymbol{a}$ 时, 有 $\lambda\boldsymbol{e} \in \mathrm{cl}C - \boldsymbol{a}$, 即有对任意 $t > 0$ 有 $t\lambda\boldsymbol{e} \in t\,(\mathrm{cl}C - \boldsymbol{a})$. 因为 t 的任意性, 有

$$\mathrm{rec}\,(\mathrm{cl}C) \subset \cap\{t(\mathrm{cl}C - \boldsymbol{a}) \mid t > 0\}.$$

设 $\forall \boldsymbol{e} \in D = \cap\{t(\mathrm{cl}C - \boldsymbol{a}) \mid t > 0\}$. 对于 $t > 0$ 有 $\boldsymbol{e} \in t(\mathrm{cl}C - \boldsymbol{a})$, 即 $\dfrac{1}{t}\boldsymbol{e} \in (\mathrm{cl}C - \boldsymbol{a})$, 有 $\dfrac{1}{t}\boldsymbol{e} + \boldsymbol{a} \in \mathrm{cl}C$, 根据回收锥定义, 有 $\boldsymbol{e} \in \mathrm{rec}\,(\mathrm{cl}C)$.

(4) 证明过程与 (3) 类似.

(5) 如果 $C \subset M \subset \mathbf{R}^n$, 有 $\mathrm{cl}C \subset \mathrm{cl}M$, 设 $\boldsymbol{a} \in \mathrm{cl}C \subset \mathrm{cl}M$. 由性质 3.3.1 得

$$\mathrm{rec}\,(\mathrm{cl}C) \subset \cap\{t(\mathrm{cl}C - \boldsymbol{a}) \mid t > 0\} \subset \cap\{t(\mathrm{cl}M - \boldsymbol{a}) \mid t > 0\},$$

根据 (3) 结论成立. □

下面性质说明 $\mathrm{rec}(\mathrm{cl}C)$ 是包含 C 的最大的凸锥.

性质 3.3.6 设非空凸集 $C \subset \mathbf{R}^n$, 下面结论成立:

(1) 如果 $\boldsymbol{a} \in \mathrm{cl}C$ 和 $\boldsymbol{a} + C \subset \mathrm{cl}C$, 则 $\mathrm{rec}(\mathrm{cl}C)$ 是包含 C 的最大的凸锥;

(2) 如果 $\boldsymbol{a} \in \mathrm{ri}C$ 和 $\boldsymbol{a} + C \subset \mathrm{ri}C$, 则 $\mathrm{rec}(\mathrm{cl}C)$ 是包含 C 的最大的凸锥.

证明 (1) 假设存在一个凸锥 D 使得 $C \subset \mathrm{rec}(\mathrm{cl}C) \subset D$, 且有 $D \backslash \mathrm{rec}(\mathrm{cl}C) \neq \varnothing$. 设 $\boldsymbol{y} \in D \setminus \mathrm{rec}(\mathrm{cl}C)$, 但 $\boldsymbol{y} \notin \mathrm{rec}(\mathrm{cl}C)$. 根据 D 是凸锥知, 对于任意 $\lambda \geqslant 0$ 有 $\lambda\boldsymbol{y} \in D$, 因为 $\lambda \to 0$ 时, 有 $\lambda\boldsymbol{y} \to \mathbf{0}$, 由假设知 $\mathbf{0} \in \mathrm{rec}(\mathrm{cl}C)$, 得对充分小的 $\lambda \geqslant 0$ 有 $\lambda\boldsymbol{y} \in \mathrm{rec}(\mathrm{cl}C)$, 得矛盾.

(2) 证明同上. □

例 3.3.7 设 $\mathbf{R}^2$ 空间中的集合 $C=\{(x_1,0)^{\mathrm{T}}|x_1\in(1,+\infty)\}\cup\{(0,x_2)^{\mathrm{T}}|x_2\in(1,+\infty)\}$, 它的回收锥 $\mathrm{rec}C=\{(x_1,0)^{\mathrm{T}}|x_1\in[0,+\infty)\}\cup\{(0,x_2)^{\mathrm{T}}|x_2\in[0,+\infty)\}$, 由 $C=\mathrm{ri}C$ 和

$$\mathrm{cl}C=\left\{(x_1,0)^{\mathrm{T}}|x_1\in[1,+\infty)\right\}\cup\left\{(0,x_2)^{\mathrm{T}}|x_2\in[1,+\infty)\right\},$$

得 $\mathrm{rec}(\mathrm{ri}C)=\mathrm{rec}(\mathrm{cl}C)=\mathrm{rec}(C)$. $\mathrm{rec}(\mathrm{cl}C)$ 和 $\mathrm{rec}(\mathrm{ri}C)$ 都含非零点, 集合 C 是非凸集, 因此 $\mathrm{rec}(\mathrm{cl}C)$ 和 $\mathrm{rec}(\mathrm{ri}C)$ 不是凸锥. 性质 3.3.6 的结论对此例不成立.

该例说明上面性质成立的条件"C 是凸集"是充分的.

性质 3.3.7 设 $C\subset\mathbf{R}^n$ 是非空凸集, 则

$$\mathrm{rec}(\mathrm{cl}C)=\{\boldsymbol{e}\in\mathbf{R}^n|\exists\boldsymbol{x}\in C\text{ 使得 }t\boldsymbol{e}+\boldsymbol{x}\in C,\forall t\geqslant 0\}.\tag{3.3.3-3}$$

证明 由性质 3.3.3(1) 得 $\mathrm{rec}(\mathrm{cl}C)\subset\mathrm{cl}C-\boldsymbol{x}$. 反过来, 根据定义 3.3.1 知, 若 $\boldsymbol{e}\in\mathbf{R}^n$, 存在 $\boldsymbol{x}\in C$, 使得对于任何 $t\geqslant 0$ 有 $t\boldsymbol{e}+\boldsymbol{x}\in C$, 那么由回收锥定义有 $\boldsymbol{e}\in\mathrm{rec}(\mathrm{cl}C)$, 即结论成立. □

性质 3.3.7 和下面定理 3.3.5 都描述了凸集的回收锥的拓扑结构性质, 表明了凸集的回收锥, 只要存在其中一点加上该凸集属于该集合闭包的回收锥, 或者该集合闭包的回收锥由集合中序列乘以收敛到 $\mathbf{0}$ 的收敛序列的极限点构成.

定理 3.3.5 设 $C\subset\mathbf{R}^n$ 是一个非空凸集, 则 $\mathrm{rec}(\mathrm{cl}C)$ 是所有收敛序列 $\{t_k\boldsymbol{x}_k\}$ 极限点构成的集合, 其中 $\{\boldsymbol{x}_k\}\subset C$ 和正数序列 $t_k\to 0(k\to+\infty)$.

证明 设 $\forall\boldsymbol{e}\in\mathrm{rec}(\mathrm{cl}C)$, 如果 $\boldsymbol{e}=\{\mathbf{0}\}$ 结论显然成立. 因此, 假设 $\boldsymbol{e}\neq\{\mathbf{0}\}$, 再设集合

$$D=\{\boldsymbol{e}|t_k\boldsymbol{x}_k\to\boldsymbol{e},t_k\to 0,k\to+\infty,\boldsymbol{x}_k\in C,t_k>0,k=1,2,\cdots\},$$

那么有 $k\boldsymbol{e}+\boldsymbol{x}=\boldsymbol{x}_k\in C\,(k=1,2,\cdots)$, 其中 $\boldsymbol{x}\in C$, 令 $k\to+\infty$, 有 $\boldsymbol{e}+\dfrac{1}{k}\boldsymbol{x}=\dfrac{1}{k}\boldsymbol{x}_k\to\boldsymbol{e}$, 得 $\boldsymbol{e}\in D$.

反过来, 设 $\boldsymbol{e}=\lim\limits_{k\to+\infty}\lambda_k\boldsymbol{x}_k\in D$, 其中 $\boldsymbol{x}_k\in C$ 和正数列 $\lambda_k\to 0(k=1,2,\cdots)$. 若 $\boldsymbol{e}=\mathbf{0}$, 则 $\mathbf{0}\in\mathrm{rec}(\mathrm{cl}C)$. 设 $\boldsymbol{e}\neq\mathbf{0}$, 不妨设 $\boldsymbol{x}_k$ 均是非零向量, 令 $\boldsymbol{y}_k=\dfrac{\boldsymbol{x}_k}{||\boldsymbol{x}_k||}\,(k=1,2,\cdots)$ 和 $\boldsymbol{a}=\dfrac{\boldsymbol{e}}{||\boldsymbol{e}||}$, 再令 $k\to+\infty$, 显然有

$$\boldsymbol{y}_k=\frac{t_k\boldsymbol{x}_k}{||t_k\boldsymbol{x}_k||}\to\boldsymbol{a}.$$

由上式得 $[\boldsymbol{e},\boldsymbol{e}+\boldsymbol{a})\subset\mathrm{cl}C$ 对于 $\boldsymbol{e}\in\mathrm{cl}C$ 成立. 因此, $\boldsymbol{e}\in l\,[0,\boldsymbol{a})\subset\mathrm{rec}(\mathrm{cl}C)$, 即 $D=\mathrm{rec}(\mathrm{cl}C)$. □

定理 3.3.6 设 $C \subset \mathbf{R}^n$ 是一个非空凸集, 且不是一个平面. 如果 X 是 $\mathrm{rb}C$ 的稠密集, 则

(1) $\mathrm{cl}C = \bigcap_{\boldsymbol{a}\in X}(C_{\boldsymbol{a}}(\mathrm{cl}C))$; (3.3.3-4)

(2) $\mathrm{rec}(\mathrm{cl}C) = \bigcap_{\boldsymbol{a}\in X}(C_{\boldsymbol{a}}(\mathrm{cl}C) - \boldsymbol{a})$. (3.3.3-5)

证明 (1) 显然 $\mathrm{cl}C \subset \bigcap_{\boldsymbol{a}\in X}(C_{\boldsymbol{a}}(\mathrm{cl}C))$. 反过来, 设 $\boldsymbol{x} \in \mathbf{R}^n \setminus \mathrm{cl}C$, 下面证明存在点 $\boldsymbol{a} \in X$ 使得 $\boldsymbol{x} \notin C_{\boldsymbol{a}}(\mathrm{cl}C)$. 因为 $\mathbf{R}^n \setminus \mathrm{cl}C = (\mathbf{R}^n \setminus \mathrm{aff}C) \cup (\mathrm{aff}C \setminus \mathrm{cl}C)$, 则

$$\boldsymbol{x} \in \mathbf{R}^n \setminus \mathrm{aff}C \quad \text{或者} \quad \boldsymbol{x} \in \mathrm{aff}C \setminus \mathrm{cl}C.$$

若 $\boldsymbol{x} \in \mathbf{R}^n \setminus \mathrm{aff}C$, 由性质 3.2.5 可知, 对于任意点 $\boldsymbol{a} \in X$ 有 $C_{\boldsymbol{a}}(\mathrm{cl}C) \subset \mathrm{aff}C$. 因此, $\boldsymbol{x} \notin C_{\boldsymbol{a}}(\mathrm{cl}C)$. 若 $\boldsymbol{x} \in \mathrm{aff}C \setminus \mathrm{cl}C$, 对于 $\boldsymbol{y} \in \mathrm{ri}C$, 则存在常数 $\varepsilon > 0$ 有 $\mathrm{aff}C \cap U(\boldsymbol{y},\varepsilon) \subset \mathrm{ri}C$. 由定理 2.3.2 知 $(\boldsymbol{x},\boldsymbol{y}) = \{t\boldsymbol{x} + (1-t)\boldsymbol{y} | t \in (0,1)\}$ 与 $\mathrm{rb}C$ 有相交点 $\boldsymbol{z}$, 使得

$$\boldsymbol{z} = (1-t)\boldsymbol{x} + t\boldsymbol{y}, \quad 0 < t < 1,$$

$\boldsymbol{y} = \left(1 - \dfrac{1}{t}\right)\boldsymbol{x} + \dfrac{1}{t}\boldsymbol{z}$. 则取 β 满足 $\beta < t\varepsilon$, 因为 X 是 $\mathrm{rb}C$ 的稠密子集, 存在点 $\boldsymbol{c} \in X$ 使得 $\|\boldsymbol{c} - \boldsymbol{z}\| < \beta$. 令 $\boldsymbol{u} = \left(1 - \dfrac{1}{t}\right)\boldsymbol{x} + \dfrac{1}{t}\boldsymbol{c}$, 有 $\boldsymbol{u} \in \mathrm{aff}C$ 和

$$\|\boldsymbol{y} - \boldsymbol{u}\| = \frac{1}{t}\|\boldsymbol{z} - \boldsymbol{c}\| < \frac{1}{t}\beta < \varepsilon.$$

上式表明 $\boldsymbol{u} \in \mathrm{aff}C \cap U(\boldsymbol{y},\varepsilon) \subset \mathrm{ri}C$. 设 $\boldsymbol{x} \in C_{\boldsymbol{a}}(\mathrm{cl}C)$, 因为 $C_{\boldsymbol{a}}(\mathrm{cl}C) \neq \mathrm{aff}C$ 和 $\boldsymbol{u} \in \mathrm{ri}C \subset \mathrm{ri}C_{\boldsymbol{a}}(\mathrm{cl}C)$, 若有 $\boldsymbol{a} \in \mathrm{rb}C_{\boldsymbol{a}}(C)$, 可得

$$\boldsymbol{a} \in (\boldsymbol{x},\boldsymbol{u}) \subset \mathrm{ri}C_{\boldsymbol{a}}(\mathrm{cl}C),$$

显然上式与 $\boldsymbol{a} \in \mathrm{rb}C_{\boldsymbol{a}}(\mathrm{cl}C)$ 矛盾. 因此, 有 $\boldsymbol{x} \notin C_{\boldsymbol{a}}(\mathrm{cl}C)$.

(2) 由性质 3.3.3(1) 得 $\mathrm{rec}(\mathrm{cl}C) \subset \mathrm{cl}C - \boldsymbol{a}$, 其中 $\forall \boldsymbol{a} \in X \subset \mathrm{rb}C \subset \mathrm{cl}C$, 即有

$$\mathrm{rec}(\mathrm{cl}C) \subset \bigcap_{\boldsymbol{a}\in X}(C_{\boldsymbol{a}}(\mathrm{cl}C) - \boldsymbol{a}).$$

反过来, 设开半直线 $l\langle\boldsymbol{0},\boldsymbol{d}\rangle \not\subset \mathrm{rec}(\mathrm{cl}C)$, 点 $\boldsymbol{u} \in \mathrm{ri}C$, 则存在 $\rho > 0$ 使得

$$\mathrm{aff}C \cap U(\boldsymbol{u},\rho) \subset \mathrm{ri}C.$$

由回收锥定义知 $l\langle\boldsymbol{u},\boldsymbol{d}+\boldsymbol{u}\rangle \not\subset \mathrm{cl}C$. 但直线 $l\langle\boldsymbol{u},\boldsymbol{d}+\boldsymbol{u}\rangle$ 与边界集 $\mathrm{rb}C$ 一定存在相交, 则存在点 $\boldsymbol{x} \in \mathrm{rb}C$ 使得 $l\langle\boldsymbol{x},\boldsymbol{d}+\boldsymbol{x}\rangle \not\subset \mathrm{cl}C$. 又因为 X 是 $\mathrm{rb}C$ 的稠密子集, 则存在点 $\boldsymbol{a} \in X$ 使得 $\|\boldsymbol{x} - \boldsymbol{a}\| < \rho$. 令 $\boldsymbol{y} = \boldsymbol{u} + (\boldsymbol{x} - \boldsymbol{a})$, 得 $l\langle\boldsymbol{y},\boldsymbol{d}+\boldsymbol{y}\rangle$

与 rb$C$ 相交, 导致 $l\langle \boldsymbol{a}, \boldsymbol{d}+\boldsymbol{a}\rangle \not\subset \mathrm{cl}C$ 成立, 故 $l\langle \boldsymbol{a}, \boldsymbol{d}+\boldsymbol{a}\rangle \not\subset C_{\boldsymbol{a}}(\mathrm{cl}C)$. 因此, 有 $l\langle \boldsymbol{a}, \boldsymbol{d}+\boldsymbol{a}\rangle - \boldsymbol{a} \not\subset C_{\boldsymbol{a}}(\mathrm{cl}C) - \boldsymbol{a}$, 得

$$\mathrm{rec}(\mathrm{cl}C) \supset \bigcap_{\boldsymbol{a}\in X} (C_{\boldsymbol{a}}(\mathrm{cl}C) - \boldsymbol{a}). \qquad \square$$

本小节得到回收锥的一些拓扑性质, 这些性质揭示了集合和凸集的闭包和相对内部的回收锥的关系:

1) 属于集合闭包的点加上回收锥仍然属于该集合的闭包, 属于集合相对内部的点加上回收锥仍然属于该集合的相对内部.

2) 集合闭包的回收锥和集合相对内部的回收锥均是闭集, 集合闭包的回收锥包含集合回收锥的闭包, 集合相对内部的回收锥包含集合回收锥的闭包.

3) 如果集合是凸集, 则集合的相对内部的回收锥等于集合闭包的回收锥, 集合是无界的等价于集合闭包的回收锥含非零点, 闭包回收锥有 (3.3.3-1) 式成立, 相对内部回收锥有 (3.3.3-2) 式成立.

4) 如果凸集的闭包中一个点加上该集合属于该集合的闭包, 则闭包的回收锥是包含该集合的最大凸锥; 如果凸集的相对内部中一个点加上该集合属于该集合的相对内部, 则闭包的回收锥是包含该集合的最大凸锥.

5) 存在凸集中一点加上该凸集属于该集合就可以得到该集合闭包的回收锥.

6) 凸集闭包的回收锥可由该集合中序列乘以收敛到 $\mathbf{0}$ 的收敛序列的极限点构成.

7) 设 C 是非平面凸集, 如果 X 是 rb$C$ 的稠密集, 则 C 的闭包可由 X 中所有点作为顶点的锥包交构成, C 的闭包回收锥可由 X 中所有点作为顶点的锥包减顶点的交构成.

3.3.4　回收空间

集合回收空间是回收锥的推广, 由集合的正向回收锥与负向回收锥交构成, 见下面定义.

定义 3.3.2　设集合 $X \subset \mathbf{R}^n$, 则称

$$\mathrm{lin}X = \{\boldsymbol{a} \in \mathbf{R}^n | \lambda\boldsymbol{a} + \boldsymbol{x} \in X, \quad \forall\lambda \in \mathbf{R}, \quad \forall\boldsymbol{x} \in X\}$$

是 X 的回收空间 (或线性空间). 定义 lin$\varnothing = \varnothing$.

例 3.3.8　(1) 单点集合 $X = \{\boldsymbol{a}\}$ 的回收空间 lin$X = \{\mathbf{0}\}$.

(2) 有界集合 $X = [1, 2]$, X 的回收空间为 lin$X = \{0\}$.

(3) 集合 $X = [1, +\infty) \subset \mathbf{R}$, X 的回收空间为 lin$X = \{0\}$.

显然集合的回收锥包含它的回收空间. 下面的性质说明了回收空间与回收锥之间的关系: 回收空间等于回收锥的回收空间, 也等于回收锥的顶点集. 换句话说, 回收空间由回收锥的顶点集组成, 回收空间属于集合的方向集.

性质 3.3.8 设非空集合 $X \subset \mathbf{R}^n$, 下面结论成立:

(1) $\text{lin}X = \text{rec}X \cap (-\text{rec}X) = \text{lin}(\text{rec}X) = \text{ap}(\text{rec}X)$;

(2) $\text{lin}X$ 是所有满足条件 $\lambda\boldsymbol{a} + X \subset X$ 的点 $\boldsymbol{a} \in \mathbf{R}^n$ 构成的最大集合, 其中 $\lambda \in \mathbf{R}$;

(3) $\text{lin}X$ 是满足条件 $S + X \subset X$ 的最大子空间, 其中子空间 $S \subset \mathbf{R}^n$;

(4) $\text{lin}X \subset \bigcap_{\lambda \neq 0} \lambda(X - \boldsymbol{x})$, 其中 $\boldsymbol{x} \in X$ 是给定点;

(5) $\text{lin}X \subset \text{dir}X$, 进一步, $\text{lin}X = \text{dir}X$ 当且仅当 X 是一个平面.

证明 (1) 由回收锥和回收空间定义得 $\text{lin}X = \text{rec}X \cap (-\text{rec}X)$. 由性质 3.3.1 知 $\text{rec}X$ 是以原点为顶点的凸锥, 得 $\text{ap}(\text{rec}X) = \text{rec}X \cap (-\text{rec}X)$, 故有

$$\text{lin}(\text{rec}X) = \text{rec}(\text{rec}X) \cap (-\text{rec}(\text{rec}X)) = \text{rec}X \cap (-\text{rec}X) = \text{lin}X.$$

(2) 根据回收空间定义, 对于任意 $\lambda \in \mathbf{R}$, 任意 $\boldsymbol{a} \in \text{lin}X$, 显然有 $\lambda\boldsymbol{a} + X \subset X$. 由于对于任意 $\boldsymbol{x} \in X$ 都有 $\boldsymbol{x} = \boldsymbol{0} + \boldsymbol{x}$, 可得 $X \subset \lambda\boldsymbol{a} + X$.

(3) 由 (1) 和 (2) 得 $\text{lin}X = \text{ap}(\text{rec}X)$ 和 $\text{lin}X + X \subset X$. 设 $S \subset \mathbf{R}^n$ 满足 $S + X \subset X$, 那么设 $\boldsymbol{a} \in S$, $\lambda \in \mathbf{R}$ 和 $\boldsymbol{x} \in X$, 有 $\lambda\boldsymbol{a} + \boldsymbol{x} \in S + X \subset X$, 得 $\boldsymbol{a} \in \text{lin}X$. 因此, $S \subset \text{lin}X$, 即 $\text{lin}X$ 是满足条件 $S + X \subset X$ 的最大子空间, 由 $\boldsymbol{x} = \boldsymbol{0} + \boldsymbol{x}\,(\forall \boldsymbol{x} \in X)$ 有 $X \subset \text{lin}X + X$.

(4) 由 (3) 得 $\text{lin}X$ 是子空间且 $\text{lin}X + \boldsymbol{x} \subset X$, 得 $\text{lin}X \subset X - \boldsymbol{x}$.

(5) 设 $\boldsymbol{a} \in \text{lin}X$ 和 $\boldsymbol{x} \in X$, 则 $\boldsymbol{a} + \boldsymbol{x} \in X$. 由性质 2.1.10 有

$$\boldsymbol{a} \in X - \boldsymbol{x} \subset \text{aff}X - \boldsymbol{x} = \text{span}(X - \boldsymbol{x}) = \text{dir}X,$$

得 $\text{lin}X \subset \text{dir}X$. 如果 X 是一个平面, 则 $\text{lin}X = \text{dir}X$. 反过来, 若 $\text{lin}X = \text{dir}X$, 由 (3) 得 $X = \text{lin}X + X = \text{aff}X$, 说明 X 是一个平面. □

定理 3.3.7 如果 $C \subset \mathbf{R}^n$ 是关于顶点 $\boldsymbol{a}$ 的凸锥, 则 $\text{lin}C = \text{ap}C - \boldsymbol{a}$.

证明 由性质 3.1.5 知 $\text{ap}C$ 是过点 $\boldsymbol{a}$ 的平面, 则 $\text{ap}C - \boldsymbol{a}$ 是一个子空间. 设 $\forall \boldsymbol{c} \in C$, 得 $\boldsymbol{c} + (\text{ap}C - \boldsymbol{a}) \subset C$, 由性质 3.3.8 得 $\text{ap}C - \boldsymbol{a} \subset \text{lin}C$.

反过来, 如果 $\text{lin}C = \{\boldsymbol{0}\}$, 有 $\text{lin}C = \{\boldsymbol{0}\} = \{\boldsymbol{a}\} - \boldsymbol{a} \subset \text{ap}C - \boldsymbol{a}$. 设 $\text{lin}C \neq \{\boldsymbol{0}\}$, 那么 $C \neq \{\boldsymbol{a}\}$. 令 $\boldsymbol{b} \in \boldsymbol{a} + \text{lin}C$, 证明 $\boldsymbol{b} \in \text{ap}C$. 设点 $\boldsymbol{x} \in C \backslash \{\boldsymbol{b}\}$, 如果 $\boldsymbol{x} \in \boldsymbol{a} + \text{lin}C$, 则 $\boldsymbol{x} + \text{lin}C \subset C$, 有

$$l\,(\boldsymbol{b}, \boldsymbol{x}\rangle \subset \boldsymbol{a} + \text{lin}C \subset C.$$

再设 $\boldsymbol{x} \in C \setminus \{\boldsymbol{a} + \text{lin}C\}$, 由 $\boldsymbol{a} \in \text{ap}C$, 则 $l\,(\boldsymbol{a}, \boldsymbol{x}\rangle \subset C$. 设 $\forall \boldsymbol{y} \in l\,(\boldsymbol{a}, \boldsymbol{x}\rangle \subset C$, 由前面证明得 $\boldsymbol{y} + \text{lin}C \subset C$, 所以有

$$(l\,(\boldsymbol{a},\boldsymbol{x}\rangle-\boldsymbol{a})+\boldsymbol{a}+\mathrm{lin}C=l\,(\boldsymbol{a},\boldsymbol{x}\rangle+\mathrm{lin}C=\bigcup_{\boldsymbol{y}\in l(\boldsymbol{a},\boldsymbol{x}\rangle}\boldsymbol{y}+\mathrm{lin}C\subset C,$$

由上式得 $l\,(\boldsymbol{b},\boldsymbol{x}\rangle\subset\bigcup_{\boldsymbol{y}\in l(\boldsymbol{a},\boldsymbol{x}\rangle}\boldsymbol{y}+\mathrm{lin}C\subset C$. 因此, $\boldsymbol{b}\in\mathrm{ap}C$, $\boldsymbol{a}+\mathrm{lin}C\subset\mathrm{ap}C$. □

定理 3.3.7 说明顶点 $\boldsymbol{a}$ 的凸锥的回收空间等于凸锥顶点集减去顶点, 举例如下.

例 3.3.9 凸锥 $C=(-\infty,+\infty)$, C 的回收空间为 $\mathrm{lin}C=\mathrm{ap}C=\mathbf{R}$.

根据性质 3.3.8 和定理 3.3.2 得下面推论成立.

推论 3.3.1 设非空集合 $X\subset\mathbf{R}^n$, $\boldsymbol{a}\in\mathbf{R}^n$ 和 $t\in\mathbf{R}$, 则

$$\mathrm{lin}(\boldsymbol{a}+tX)=t\mathrm{lin}X=\mathrm{sgn}(t)\mathrm{lin}X.$$

定义 3.3.3 如果一个凸集 $C\subset\mathbf{R}^n$ 不含直线, 则称 C 是不含直线的 (line-free). 显然, 空集是不含直线的.

相反地, 例 3.3.9 中的集合是含直线的. 下面的性质 3.3.9 说明了回收空间与直线的等价关系: 凸集不含直线等价于凸集的闭包和相对内部不含直线, 也等价于凸集的闭包回收锥和相对内部回收锥均不含直线, 也等价于凸集的闭包回收锥和相对内部回收锥仅含原点.

性质 3.3.9 设凸集 $C\subset\mathbf{R}^n$, 则下面结论等价:

(1) C, $\mathrm{cl}C$ 和 $\mathrm{ri}C$ 均不含直线;

(2) $\mathrm{rec}(\mathrm{cl}C)$ 和 $\mathrm{rec}(\mathrm{ri}C)$ 不含直线;

(3) $\mathrm{lin}(\mathrm{cl}C)=\{\mathbf{0}\}$ (等价于 $\mathrm{lin}(\mathrm{ri}C)=\{\mathbf{0}\}$).

证明 因为 C 不含直线, 则 C 是有界的, $\mathrm{cl}C$ 也是有界的. 则等价于 $\mathrm{rec}(\mathrm{cl}C)$ 仅含原点, 由性质 3.3.8 等价于 $\mathrm{lin}(\mathrm{cl}C)=\{\mathbf{0}\}$. □

由性质 3.3.9 和性质 3.3.5 得下面推论.

推论 3.3.2 凸集 $C\subset\mathbf{R}^n$ 不含直线当且仅当由顶点 $\boldsymbol{a}\in\mathbf{R}^n\setminus\mathrm{cl}C$ 生成的锥 $C_{\boldsymbol{a}}(C)$ 不含直线, 同时 $C_{\boldsymbol{a}}(\mathrm{cl}C)$ 和 $C_{\boldsymbol{a}}(\mathrm{ri}C)$ 均不含直线.

进一步, 有两个凸集的和与不含直线的关系, 即如果两个凸集不含直线, 那么它们的和、闭包的和、并集的凸包、闭包并集的凸包、闭包回收锥的和均不含直线, 一个凸集的回收锥与另一个负向凸集的回收锥的交集仅含一个原点.

性质 3.3.10 设两个非空凸集 C_1 和 C_2 不含直线, 则下面结论等价:

(1) C_1+C_2 不含直线;

(2) $\mathrm{cl}C_1+\mathrm{cl}C_2$ 不含直线;

(3) $\mathrm{co}(C_1\cup C_2)$ 不含直线;

(4) $\mathrm{co}(\mathrm{cl}C_1\cup\mathrm{cl}C_2)$ 不含直线;

(5) $\mathrm{rec}(\mathrm{cl}C_1)+\mathrm{rec}(\mathrm{cl}C_2)$ 不含直线;

(6) $\mathrm{rec}(\mathrm{cl}C_1)\cap(-\mathrm{rec}(\mathrm{cl}C_2))=\{\mathbf{0}\}$.

证明 我们先证明 (1) 和 (2) 等价、(3) 和 (4) 等价、(5) 和 (6) 等价.

由于 $C_1 + C_2 \subset \mathrm{cl}C_1 + \mathrm{cl}C_2 \subset \mathrm{cl}(C_1 + C_2)$, 根据性质 3.3.9 可得 (1) 和 (2) 等价.

通过

$$\mathrm{co}(C_1 \cup C_2) \subset \mathrm{co}(\mathrm{cl}C_1 \cup \mathrm{cl}C_2) \subset \mathrm{cl}(\mathrm{co}(C_1 \cup C_2)),$$

再根据性质 3.3.9 可得 (3) 和 (4) 等价.

(5) 和 (6) 的证明可通过下面 (2) 和 (6) 等价的证明过程得到.

下面证明 (2) 和 (6) 等价. 若 (2) 成立, 设非零 $\boldsymbol{e} \in \mathrm{rec}(\mathrm{cl}C_1) \cap (-\mathrm{rec}(\mathrm{cl}C_2))$, $\boldsymbol{x}_1 \in \mathrm{cl}C_1$ 和 $\boldsymbol{x}_2 \in \mathrm{cl}C_2$, 令半直线

$$l\,[\boldsymbol{x}_1, \boldsymbol{e}\rangle = \boldsymbol{x}_1 + \{\lambda\boldsymbol{e} : \lambda \geqslant 0\} \subset \mathrm{cl}C_1, \quad l\,[\boldsymbol{x}_2, -\boldsymbol{e}\rangle = \boldsymbol{x}_2 + \{\lambda(-\boldsymbol{e}) : \lambda \geqslant 0\} \subset \mathrm{cl}C_2.$$

显然有直线

$$l\,[\boldsymbol{x}_1, \boldsymbol{e}\rangle + l\,[\boldsymbol{x}_2, -\boldsymbol{e}\rangle = (\boldsymbol{x}_1 + \boldsymbol{x}_2) + \{\lambda\boldsymbol{e} : \lambda \in \mathbf{R}\} \in \mathrm{cl}C_1 + \mathrm{cl}C_2,$$

与 (2) 矛盾, 所以 (6) 成立. 反之, 若 (6) 成立, 有 $\mathrm{lin}(\mathrm{cl}C_1) = \mathrm{lin}(\mathrm{cl}C_2) = \{\mathbf{0}\}$, 由于

$$\mathrm{lin}(\mathrm{cl}(C_1 + C_2)) = \mathrm{lin}(\mathrm{cl}C_1) + \mathrm{lin}(\mathrm{cl}C_2) = \{\mathbf{0}\},$$

由性质 3.3.9 知 (2) 成立.

最后证明 (3) 和 (6) 等价. 假设 $\mathrm{co}(C_1 \cup C_2)$ 不含直线, 则得到 $\mathrm{rec}(\mathrm{cl}(\mathrm{co}(C_1 \cup C_2)))$ 是不含直线的, 再由

$$\mathrm{rec}(\mathrm{cl}C_1) \cup \mathrm{rec}(\mathrm{cl}C_2) \subset \mathrm{rec}(\mathrm{cl}(\mathrm{co}(C_1 \cup C_2)))$$

导出

$$\mathrm{rec}(\mathrm{cl}C_1) + \mathrm{rec}(\mathrm{cl}C_2) = \mathrm{co}(\mathrm{rec}(\mathrm{cl}C_1) \cup \mathrm{rec}(\mathrm{cl}C_2)) \subset \mathrm{rec}(\mathrm{cl}(\mathrm{co}(C_1 \cup C_2))).$$

因此, 得到 $\mathrm{rec}(\mathrm{cl}C_1) + \mathrm{rec}(\mathrm{cl}C_2)$ 不含直线. 反之, 结论显然. □

本小节得到了回收空间的一些基本性质:

1) 回收空间等于回收锥的回收空间, 也等于回收锥的顶点集. 换句话说, 回收空间由回收锥的顶点集组成, 回收空间属于集合的方向集.

2) 顶点凸锥的回收空间等于凸锥顶点集减去顶点.

3) 凸集不含直线等价于凸集的闭包和相对内部不含直线, 也等价于凸集的闭包回收锥和相对内部回收锥均不含直线, 也等价于凸集的闭包回收锥和相对内部回收锥仅含原点.

4) 如果两个凸集不含直线, 那么它们的和、闭包的和、并集的凸包、闭包并集的凸包、闭包回收锥的和均不含直线, 一个凸集的回收锥与另一个负向凸集的回收锥的交集仅含一个原点.

3.4 共 轭 锥

以下讨论包含原点凸锥的对偶锥, 称之为共轭锥或正极锥.

定义 3.4.1 设凸锥 $C \subset \mathbf{R}^n$, 称

$$C^{+}=\{\boldsymbol{x}^{*}\in\mathbf{R}^{n}\,|\langle\boldsymbol{x},\boldsymbol{x}^{*}\rangle\geqslant 0,\forall\boldsymbol{x}\in C\}$$

是 C 的共轭锥或正极锥. 称 $C^{++}=(C^{+})^{+}$ 是 C 的双共轭锥. 并有 $(\mathbf{R}^{n})^{+}=\{\mathbf{0}\}$.

例 3.4.1 设凸锥 $C=\{x|x\geqslant 0\}\subset\mathbf{R}$, 显然 $C^{++}=C^{+}=C$.

性质 3.4.1 设凸锥 $C\subset\mathbf{R}^n$, 则下面结论成立:

(1) C^{+} 是闭凸锥;

(2) $C^{+}=(\mathrm{cl}C)^{+}$.

证明 (1) 显然 C^{+} 是凸锥, 只要证明它是闭的. 设序列 $\{\boldsymbol{x}_i^{*}\}\subset C^{+}(i=1,2,\cdots)$ 收敛于 $\boldsymbol{x}^{*}$, 根据定义 3.4.1, 对任意的 $\boldsymbol{x}\in C$, 有 $\langle\boldsymbol{x},\boldsymbol{x}_i^{*}\rangle\geqslant 0$. 令 $i\to\infty$ 时取极限, 得 $\langle\boldsymbol{x},\boldsymbol{x}^{*}\rangle\geqslant 0$, 即 $\boldsymbol{x}^{*}\in C^{+}$, 故 C^{+} 是闭集.

(2) 由于 C 是 $\mathbf{R}^n$ 中的凸锥, 则 $\mathrm{cl}C$ 是 $\mathbf{R}^n$ 中的凸锥. 设 $\boldsymbol{x}^{*}\in(\mathrm{cl}C)^{+}$, 对任意的 $\boldsymbol{x}\in\mathrm{cl}C$, 有 $\langle\boldsymbol{x},\boldsymbol{x}^{*}\rangle\geqslant 0$. 因此, 对任意的 $\boldsymbol{x}\in C$ 也有 $\langle\boldsymbol{x},\boldsymbol{x}^{*}\rangle\geqslant 0$ 成立, 说明 $\boldsymbol{x}^{*}\in C^{+}$. 反过来, 若 $\boldsymbol{x}^{*}\in C^{+}$, 设序列 $\{\boldsymbol{x}_i\}\subset C\,(i=1,2,\cdots)$ 收敛于 $\boldsymbol{x}\in\mathrm{cl}C$, 根据定义 3.4.1, 有 $\langle\boldsymbol{x}_i,\boldsymbol{x}^{*}\rangle\geqslant 0$. 令 $i\to\infty$ 时取极限, 得 $\langle\boldsymbol{x},\boldsymbol{x}^{*}\rangle\geqslant 0$, 表明 $\boldsymbol{x}^{*}\in(\mathrm{cl}C)^{+}$ 成立. □

性质 3.4.1 表明共轭锥是闭凸锥, 且等于闭包的共轭锥. 下面证明闭凸锥的共轭锥的共轭锥就是原来的锥.

性质 3.4.2 设闭凸锥 $C\subset\mathbf{R}^n$, 如果存在 $\boldsymbol{x}_0$ 使得 $\langle\boldsymbol{x}_0,\boldsymbol{x}^{*}\rangle\geqslant 0\,(\forall\boldsymbol{x}^{*}\in C^{+})$ 成立, 则

(1) $\boldsymbol{x}_0\in C$;

(2) $C^{++}=C$.

证明 (1) 假设 $\boldsymbol{x}_0\notin C$, 根据凸集分离定理 (定理 2.4.7), 存在一个非零 $\boldsymbol{y}^{*}$ 满足

$$\langle\boldsymbol{x}_0,\boldsymbol{y}^{*}\rangle<\langle\boldsymbol{x},\boldsymbol{y}^{*}\rangle,\quad\forall\boldsymbol{x}\in C.\tag{3.4.1-1}$$

因为 C 是凸锥, 有 $\lambda\boldsymbol{x}\,(\lambda>0)$, 令 $\lambda\to 0$, 知 $\lambda\boldsymbol{x}$ 充分接近于零点, 由于 C 是闭集, 故 $\mathbf{0}\in C$. 于是有 $\langle\boldsymbol{x}_0,\boldsymbol{y}^{*}\rangle<0$. 假设存在 $\boldsymbol{x}_1\in C$, 使 $\langle\boldsymbol{x}_1,\boldsymbol{y}^{*}\rangle<0$. 对于任意 $\lambda>0$, 有 $\lambda\boldsymbol{x}_1\in C$, 令 $\lambda\to+\infty$, 则 $\langle\lambda\boldsymbol{x}_1,\boldsymbol{y}^{*}\rangle\to-\infty$, 与 (3.4.1-1) 式矛盾. 因此, $\boldsymbol{y}^{*}\in C^{+}$, 这又与已知条件矛盾.

(2) 因为对任意 $\boldsymbol{x}\in C$, 任意 $\boldsymbol{x}^{*}\in C^{+}$ 有 $\langle\boldsymbol{x},\boldsymbol{x}^{*}\rangle\geqslant 0$, 再由定义 3.4.1 知

$$C^{++}=\left\{\boldsymbol{x}\,|\langle\boldsymbol{x},\boldsymbol{x}^{*}\rangle\geqslant 0,\boldsymbol{x}^{*}\in C^{+}\right\},$$

故 $\boldsymbol{x} \in C^{++}$, 所以 $C \subset C^{++}$. 反之, 设 $\boldsymbol{x} \in C^{++}$, 因 C 是闭集, 则对任意的 $\boldsymbol{x}^* \in C^+$, 有

$$\langle \boldsymbol{x}, \boldsymbol{x}^* \rangle \geqslant 0.$$

从 (1) 的证明可知 $\boldsymbol{x} \in C$, 即 $C^{++} \subset C$. 故 $C^{++} = C$. □

性质 3.4.2 中 C 是闭集的条件是必要的. 如果没有 C 是闭集的条件, 只能得到

$$C^{++} = \mathrm{cl}C.$$

例 3.4.2 设凸锥 $C = \{x|x > 0\} \subset \mathbf{R}$, 显然 $C^+ = \{x|x \geqslant 0\}$, 而 $C^{++} = \{x|x \geqslant 0\}$.

下面证明两个凸锥和的共轭锥等于各自共轭锥的交, 两个凸锥闭包交的共轭锥等于各自共轭锥的和的闭包.

性质 3.4.3 设 C_1 和 C_2 是 $\mathbf{R}^n$ 中的凸锥, 则 $C_1 + C_2$ 是凸锥, 且有

$$(C_1 + C_2)^+ = C_1^+ \cap C_2^+,$$

$$(\mathrm{cl}C_1 \cap \mathrm{cl}C_2)^+ = \mathrm{cl}\left(C_1^+ + C_2^+\right).$$

证明 由定理 3.1.8 知, $C_1 + C_2$ 是凸锥, 且 $C_1 + C_2 = \mathrm{co}(C_1 \cup C_2)$. 设 $\forall \boldsymbol{x} \in C_1 + C_2$, 则存在 $\boldsymbol{x}_1 \in C_1$ 和 $\boldsymbol{x}_2 \in C_2$, 满足 $\boldsymbol{x} = \boldsymbol{x}_1 + \boldsymbol{x}_2$. 由定义 3.4.1 知, 对任意的 $\boldsymbol{x}_1 \in C_1, \boldsymbol{x}_2 \in C_2$, 有

$$\langle \boldsymbol{x}_1 + \boldsymbol{x}_2, \boldsymbol{x}^* \rangle \geqslant 0.$$

令 $\boldsymbol{x}_1 = \mathbf{0}$ 或 $\boldsymbol{x}_2 = \mathbf{0}$, 上式等价于

$$\langle \boldsymbol{x}_1, \boldsymbol{x}^* \rangle \geqslant 0, \boldsymbol{x}_1 \in C_1 \quad \text{和} \quad \langle \boldsymbol{x}_2, \boldsymbol{x}^* \rangle \geqslant 0, \boldsymbol{x}_2 \in C_2,$$

这两个式子同时成立, 表明 $\boldsymbol{x}^* \in (C_1 + C_2)^+$ 等价于 $\boldsymbol{x}^* \in C_1^+ \cap C_2^+$.

根据性质 3.4.2 知

$$C_1^{++} = \mathrm{cl}C_1, \quad C_2^{++} = \mathrm{cl}C_2,$$

有

$$\begin{aligned}(\mathrm{cl}C_1 \cap \mathrm{cl}C_2)^+ &= \left(C_1^{++} \cap C_2^{++}\right)^+ = \left[\left(C_1^+ + C_2^+\right)^+\right]^+ \\ &= (C_1^+ + C_2^+)^{++} = \mathrm{cl}\left(C_1^+ + C_2^+\right).\end{aligned}$$

□

例 3.4.3 设凸锥 $C_1=\{x|x>0\}\subset\mathbf{R}$ 和 $C_2=\{x|x<0\}\subset\mathbf{R}$, 显然

$$C_1^+=\{x|x\geqslant 0\},\quad C_2^+=\{x|x\leqslant 0\},$$

因此有 $(C_1+C_2)^+=C_1^+\cap C_2^+=\{0\}$.

定理 3.4.1 设凸锥 $C\subset\mathbf{R}^n$, 且 $\inf\limits_{\boldsymbol{x}\in C}\langle\boldsymbol{x},\boldsymbol{x}^*\rangle>-\infty$, 则 $\boldsymbol{x}^*\in C^+$. 进一步, 如果 $\boldsymbol{x}\in\mathrm{int}C$, 则对任意 $\boldsymbol{x}^*\in C^+$ 和任意 $\boldsymbol{x}^*\neq\mathbf{0}$ 有 $\langle\boldsymbol{x},\boldsymbol{x}^*\rangle>0$.

证明 假设 $\boldsymbol{x}^*\notin C^+$, 根据凸集分离定理 (定理 2.4.7), 存在一个非零 $\boldsymbol{y}\in\mathbf{R}^n$ 满足

$$\langle\boldsymbol{x}^*,\boldsymbol{y}\rangle<\langle\boldsymbol{y}^*,\boldsymbol{y}\rangle,\quad\forall\boldsymbol{y}^*\in C^+.\tag{3.4.1-2}$$

因为 C^+ 是凸锥, 故 $\mathbf{0}\in C^+$, 则有 $\langle\boldsymbol{x}^*,\boldsymbol{y}\rangle<0$. 如果存在 $\boldsymbol{y}_1^*\in C^+$, 使 $\langle\boldsymbol{y}_1^*,\boldsymbol{y}\rangle<0$, 对于任意 $\lambda>0$, 有 $\lambda\boldsymbol{y}_1^*\in C^+$, 令 $\lambda\to+\infty$, 则 $\langle\lambda\boldsymbol{y}_1^*,\boldsymbol{y}\rangle\to-\infty$, 与 (3.4.1-2) 式矛盾. 因此, 对任意的 $\boldsymbol{y}^*\in C^+$, 有 $\langle\boldsymbol{y}^*,\boldsymbol{y}\rangle\geqslant 0$, 故 $\boldsymbol{y}\in C^{++}$. 由性质 3.4.2 得 $C^{++}=\mathrm{cl}C$ (因为 $\mathrm{cl}C=(\mathrm{cl}C)^{++}=\left((C)^+\right)^+=(C)^{++}$). 由于

$$\inf_{\boldsymbol{x}\in C}\langle\boldsymbol{x},\boldsymbol{x}^*\rangle=\inf_{\boldsymbol{x}\in\mathrm{cl}C}\langle\boldsymbol{x},\boldsymbol{x}^*\rangle,$$

得 $\langle\boldsymbol{y},\boldsymbol{x}^*\rangle\geqslant\inf\limits_{\boldsymbol{x}\in\mathrm{cl}C}\langle\boldsymbol{x},\boldsymbol{x}^*\rangle>-\infty$. 由于 C 是 $\mathbf{R}^n$ 中的凸锥, 那么有

$$\langle\boldsymbol{y},\boldsymbol{x}^*\rangle\geqslant\inf_{\boldsymbol{x}\in\mathrm{cl}C}\langle\boldsymbol{x},\boldsymbol{x}^*\rangle\geqslant 0.$$

因此, $\boldsymbol{x}^*\in C^+$, 得矛盾.

如果 $\boldsymbol{x}\in\mathrm{int}C$, 则存在 $\varepsilon>0$, 使 $\boldsymbol{x}+\varepsilon B\subset C$, 其中 B 是单位球, 由 $\forall\boldsymbol{x}^*\in C^+$ 及 $\forall\boldsymbol{b}\in B$ 有

$$\langle\boldsymbol{x}+\varepsilon\boldsymbol{b},\boldsymbol{x}^*\rangle\geqslant 0.$$

当 $\boldsymbol{x}^*\neq\mathbf{0}$ 时, 上式为

$$\langle\boldsymbol{x},\boldsymbol{x}^*\rangle\geqslant\varepsilon\sup_{\boldsymbol{b}\in B}\langle-\boldsymbol{b},\boldsymbol{x}^*\rangle\geqslant\varepsilon\left\langle\frac{\boldsymbol{x}^*}{\|\boldsymbol{x}^*\|},\boldsymbol{x}^*\right\rangle=\varepsilon\|\boldsymbol{x}^*\|>0.\qquad\square$$

本节得到凸锥共轭锥的一些重要性质:

1) 共轭锥是闭凸锥, 也等于闭包的共轭锥;

2) 任何一点与闭凸锥的共轭锥中的一点内积不小于 0, 则这点属于这个闭凸锥;

3) 两个凸锥之和的共轭锥等于各自共轭锥的交, 两个凸锥闭包交集的共轭锥等于各自共轭锥之和的闭包;

4) 若凸锥与某给定点的内积有下界, 则给定的点属于该凸锥的共轭锥, 且凸锥的内点与共轭锥非零点的内积大于 0.

3.5 极　锥

本节讨论与共轭锥对应的另一种对偶锥形式——极锥, 极锥是负共轭锥. 因此, 本节极锥的性质对共轭锥也成立, 由于极锥 C^- 是负共轭锥 $-C^+$, 将 $C^- = -C^+$ 代入极锥成立的性质中即可得到共轭锥成立的性质. 所以, 本节许多性质没有写出对应的共轭锥成立的形式, 而侧重介绍了极锥许多重要的理论性质, 其他一些性质也可参见文献 [63] 中有关极锥的部分.

3.5.1 基本性质

下面首先介绍极锥的定义.

定义 3.5.1 设非空集合 $C \subset \mathbf{R}^n$, 则称集合

$$C^- = \{\boldsymbol{x}^* \in \mathbf{R}^n |\ \langle \boldsymbol{x}^*, \boldsymbol{x} \rangle \leqslant 0, \forall \boldsymbol{x} \in C\} \tag{3.5.1-1}$$

是 C 的极锥或负极锥, 空集的极锥为 $\varnothing^- = \mathbf{R}^n$.

例 3.5.1 设凸锥 $C = \{x | x \geqslant 0\} \subset \mathbf{R}$, 显然 $C^- = -C$.

根据正极锥的定义显然有: $C^- = -C^+$. 因此, 共轭锥是极锥的负集合. 下面的有关极锥性质和定理对于正极锥 (共轭锥) 也成立.

性质 3.5.1 设两个集合 $C, S \in \mathbf{R}^n$, 下面结论成立:

(1) 如果 $C \subset S$, 则 $S^- \subset C^-$;

(2) $(C \cup S)^- = C^- \cap S^-$;

(3) $C^- = (\mathrm{co}C)^- = (C_{\mathbf{0}}(C))^- = (\mathrm{cone}_{\mathbf{0}}C)^-$;

(4) $C^{\perp} \subset C^- = (\mathrm{cl}C)^-$ 和 $(\mu C)^- = \mu C^-$, 其中 $\mu \neq 0$;

(5) $(C + S)^- = (\mathrm{cl}C + \mathrm{cl}S)^-$, 进一步如果 C 和 S 是闭集, 则 $(C + S)^- = C^- \cap S^-$;

(6) C^- 是关于原点的闭凸锥, 且 $\mathrm{cl}C \cap C^- \subset \{\mathbf{0}\}$;

(7) $(C^-)^- = \mathrm{cl}(\mathrm{cone}_{\mathbf{0}}C)$;

(8) 如果 C 是凸锥, 则 $(C^-)^- = \mathrm{cl}C$;

(9) $C^- \neq \mathbf{R}^n$ 当且仅当 $C \not\subset \{\mathbf{0}\}$;

(10) $C^- \not\subset \{\mathbf{0}\}$ 当且仅当 $\mathrm{cone}_{\mathbf{0}}C \neq \mathbf{R}^n$.

证明 (1) 设 $\boldsymbol{x}^* \in S^-$, 则对于任意 $\boldsymbol{x} \in C \subset S$ 有 $\langle \boldsymbol{x}^*, \boldsymbol{x} \rangle \leqslant 0$, 即 $\boldsymbol{x}^* \in C^-$.

(2) 如果 $\boldsymbol{x}^* \in (C \cup S)^-$, 那么对于任意 $\boldsymbol{x} \in C \cup S$ 有 $\langle \boldsymbol{x}^*, \boldsymbol{x} \rangle \leqslant 0$. 即 $\forall \boldsymbol{x} \in C, \forall \boldsymbol{x} \in S$ 有 $\langle \boldsymbol{x}^*, \boldsymbol{x} \rangle \leqslant 0$, 说明 $\boldsymbol{x}^* \in C^- \cap S^-$. 反过来, 当 $\boldsymbol{x}^* \in C^- \cap S^-$ 时, 对于任意 $\boldsymbol{x} \in C$ 和任意 $\boldsymbol{y} \in S$ 有 $\langle \boldsymbol{x}^*, \boldsymbol{x} \rangle \leqslant 0$ 及 $\langle \boldsymbol{x}^*, \boldsymbol{y} \rangle \leqslant 0$, 表明 $\boldsymbol{x}^* \in (C \cup S)^-$.

(3) 显然有 $C \subset \mathrm{co}C \subset \mathrm{cone}_{\mathbf{0}}C$ 和 $C \subset C_{\mathbf{0}}(C)$, 则有 $C^- \supset (\mathrm{co}C)^- \supset (\mathrm{cone}_{\mathbf{0}}C)^-$ 和 $C^- \supset (C_{\mathbf{0}}(C))^-$. 下面设 $\boldsymbol{x}^* \in C^-$, 那么 $\langle \boldsymbol{x}^*, \boldsymbol{x} \rangle \leqslant 0, \forall \boldsymbol{x} \in C$. 对于

任意 $\lambda \geqslant 0$ 有

$$\langle \boldsymbol{x}^*, \lambda \boldsymbol{x} \rangle \leqslant 0, \quad \forall \lambda \boldsymbol{x} \in \mathrm{cone}_{\mathbf{0}} C, \quad \forall \lambda \boldsymbol{x} \in C_{\mathbf{0}}(C).$$

上式说明 $\boldsymbol{x}^* \in (\mathrm{cone}_{\mathbf{0}} C)^-$ 和 $\boldsymbol{x}^* \in (C_{\mathbf{0}}(C))^-$, 即有 $C^- \subset (\mathrm{co}C)^- \subset (\mathrm{cone}_{\mathbf{0}} C)^-$ 和 $C^- \subset (C_{\mathbf{0}}(C))^-$.

(4)~(6) 的结论根据定义 3.5.1 显然成立.

(7) 设 $\boldsymbol{x} \in (C^-)^- = \{\boldsymbol{x} \in \mathbf{R}^n | \langle \boldsymbol{x}^*, \boldsymbol{x} \rangle \leqslant 0, \forall \boldsymbol{x}^* \in C^-\}$. 因为对 $\boldsymbol{x}^* \in C^-$, 有 $\langle \boldsymbol{x}^*, \boldsymbol{x} \rangle \leqslant 0$, 说明 $\boldsymbol{x} \in C \subset \mathrm{cl}(\mathrm{cone}_{\mathbf{0}} C)$. 反过来, 设 $\boldsymbol{x} \in \mathrm{cl}(\mathrm{cone}_{\mathbf{0}} C)$, 由 (3) 和 (6) 的结论知 $C^- = (\mathrm{cone}_{\mathbf{0}} C)^-$ 是关于原点的闭凸锥, 则 $\mathrm{cl}(\mathrm{cone}_{\mathbf{0}} C)^{-} = (\mathrm{cone}_{\mathbf{0}} C)^{-} = C^-$. 因此, $\boldsymbol{x} \in (C^-)^-$.

(8) 由 (7) 直接可得.

(9) 若 $C^- = \mathbf{R}^n$, 对任意的 $\boldsymbol{x} \in C$ 有 $\langle \boldsymbol{x}, \boldsymbol{x} \rangle \leqslant 0$ 和 $\langle -\boldsymbol{x}, \boldsymbol{x} \rangle \leqslant 0$, 由此可得 $\langle \boldsymbol{x}, \boldsymbol{x} \rangle = 0$. 因此, $C = \{\mathbf{0}\}$. 反过来, $C = \{\mathbf{0}\}$, 显然 $C^- = \mathbf{R}^n$.

(10) 由 (7) 和 (9) 的结论直接可得. □

性质 3.5.2 设 $\{C_i\}(i \in I)$ 是 $\mathbf{R}^n$ 中的一个凸锥族, 其中 I 是指标集合, 则

$$\left(\bigcap_{i \in I} \mathrm{cl} C_i\right)^- = \mathrm{cl}\left(\mathrm{cone}_{\mathbf{0}}\left(\bigcup_{i \in I} C_i^-\right)\right) = \mathrm{cl}\left(\mathrm{co}\left(\bigcup_{i \in I} C_i^-\right)\right). \tag{3.5.1-2}$$

进一步, 如果指标集合是有限个元素 $I = \{1, 2, \cdots, m\}$, 则

$$(C_1 + \cdots + C_m)^- = (C_1 \cup \cdots \cup C_m)^- = C_1^- \cap \cdots \cap C_m^-, \tag{3.5.1-3}$$

$$(\mathrm{cl} C_1 \cap \cdots \cap \mathrm{cl} C_m)^- = \mathrm{cl}(\mathrm{co}(C_1^- \cup \cdots \cup C_m^-)) = \mathrm{cl}(C_1^- + \cdots + C_m^-). \tag{3.5.1-4}$$

证明 设 $\boldsymbol{x}^* \in \left(\bigcap_{i \in I} \mathrm{cl} C_i\right)^-$, 即有

$$\langle \boldsymbol{x}^*, \boldsymbol{x} \rangle \leqslant 0, \forall \boldsymbol{x} \in \bigcap_{i \in I} \mathrm{cl} C_i \Leftrightarrow \langle \boldsymbol{x}^*, \boldsymbol{x} \rangle \leqslant 0, \forall \boldsymbol{x} \in \mathrm{cl} C_i, \ i \in I,$$

等价地, 有 $\boldsymbol{x}^* \in \bigcup_{i \in I} (\mathrm{cl} C_i)^- = \bigcup_{i \in I} (C_i)^-$, 即有 $\boldsymbol{x}^* \in \mathrm{cl}\left(\mathrm{co}\left(\bigcup_{i \in I} C_i^-\right)\right) \subset \mathrm{cl}\left(\mathrm{cone}_{\mathbf{0}}\left(\bigcup_{i \in I} C_i^-\right)\right)$. 反过来, 设 $\boldsymbol{x}^* \in \mathrm{cl}\left(\mathrm{cone}_{\mathbf{0}}\left(\bigcup_{i \in I} C_i^-\right)\right)$, 对每一个 $\varepsilon > 0$, 有 $(\boldsymbol{x}^* + \varepsilon B) \cap \mathrm{cone}_{\mathbf{0}}\left(\bigcup_{i \in I} C_i^-\right) \neq \varnothing$. 由定理 3.2.1 得, 存在 $\boldsymbol{x}_k^* \in \bigcup_{i \in I} C_i^-, \lambda_k \geqslant 0\,(k = 1, \cdots, m, m \geqslant 1)$ 和 $\boldsymbol{b} \in B$ 使得

$$\boldsymbol{x}^* + \varepsilon \boldsymbol{b} = \lambda_1 \boldsymbol{x}_1^* + \cdots + \lambda_m \boldsymbol{x}_m^*. \tag{3.5.1-5}$$

对于 $\boldsymbol{x}_k^* \in \bigcup_{i\in I} C_i^-$ $(k=1,\cdots,m)$, 存在 C_i, 对于任意的 $\boldsymbol{x}\in C_i$ 有 $\langle \boldsymbol{x}_k^*, \boldsymbol{x}\rangle \leqslant 0$, 等价于对任意的 $\boldsymbol{x}\in \bigcap_{i\in I} \mathrm{cl}C_i$ 有 $\langle \boldsymbol{x}_k^*, \boldsymbol{x}\rangle \leqslant 0$, 由 (3.5.1-5) 式得到 $\langle \boldsymbol{x}^* + \varepsilon \boldsymbol{b}, \boldsymbol{x}\rangle \leqslant 0$. 令 $\varepsilon \to 0$, 有 $\langle \boldsymbol{x}^*, \boldsymbol{x}\rangle \leqslant 0$, 则 $\boldsymbol{x}^* \in \left(\bigcap_{i\in I} \mathrm{cl}C_i\right)^-$. 再由性质 3.5.1-1(5) 得 (3.5.1-3) 式和 (3.5.1-4) 式. □

性质 3.4.3 是性质 3.5.2 的特殊情形.

本小节得到了极锥的一些基本性质:

1) 若两个集合是包含关系, 则对应的极锥是反包含关系;

2) 两个集合并集的极锥等于各自极锥的交;

3) 集合的闭包、锥包和凸锥包对应的极锥均相等;

4) 集合的正交集合属于集合的极锥;

5) 两个集合之和的极锥等于各自极锥的交;

6) 极锥是包含原点的闭凸锥;

7) 极锥的极锥等于原点凸锥包的闭包;

8) 凸锥的极锥的极锥等于凸锥的闭包;

9) 若极锥不是整个空间, 极锥等价于该集合是非独点集;

10) 对于给定的一个凸锥族, 族中所有闭包交的极锥等于它们各自极锥并集的原点锥的闭包, 也等于各自极锥并集的凸包的闭包.

3.5.2 拓扑性质

本小节讨论极锥的一些拓扑性质. 这些性质揭示了凸锥的正交集合、凸锥的极锥、凸锥的顶点集、极锥的顶点集、凸锥的生成子空间、极锥的生成子空间、凸锥的闭包顶点集等集合之间存在的蕴含关系.

性质 3.5.3 设 $C\subset \mathbf{R}^n$ 是关于零点的非空凸锥, 则下面结论成立:

(1) $C^\perp = (\mathrm{cl}C)^\perp = \mathrm{ap}C^-$ 和 $\mathrm{ap}C \subset \mathrm{ap}(\mathrm{cl}C) = (C^-)^\perp$;

(2) $\mathrm{span}C = (\mathrm{ap}C^-)^\perp$ 和 $\mathrm{span}C^- = (\mathrm{ap}(\mathrm{cl}C))^\perp$, 且 $\mathrm{ap}C^-$ 和 $\mathrm{ap}(\mathrm{cl}C)$ 是正交子空间;

(3) $\dim(\mathrm{ap}C^-) = n - \dim C$ 和 $\dim C^- = n - \dim(\mathrm{ap}(\mathrm{cl}C))$;

(4) $\dim C = n$ 当且仅当 $\mathrm{ap}C^- = \{\mathbf{0}\}$ (等价于 $\mathrm{int}C \neq \varnothing$ 当且仅当 C^- 不含直线);

(5) $\dim C^- = n$ 当且仅当 $\mathrm{ap}(\mathrm{cl}C) = \{\mathbf{0}\}$ (等价于 $\mathrm{int}C^- \neq \varnothing$ 当且仅当 C 不含直线);

(6) C 是子空间当且仅当 C^- 是子空间;

(7) $C^- = C^\perp$ 当且仅当 C 是子空间;

(8) $C \neq \mathrm{ap}C$ 当且仅当 $C^- \neq \mathrm{ap}C^-$.

证明 (1) 由性质 3.1.9 和性质 3.1.10 知, $\mathrm{ap}C \subset \mathrm{rb}C \subset \mathrm{cl}C$ 或者 $\mathrm{ap}C \subset C$, 由性质 3.5.1 得 $(\mathrm{ap}C)^- \supset (\mathrm{cl}C)^- = C^-$. 由定义知 $C^\perp = (\mathrm{cl}C)^\perp$, 若 $\boldsymbol{x}^* \in C^\perp$, 对任意的 $\boldsymbol{x} \in C$ 有 $\langle \boldsymbol{x}^*, \boldsymbol{x} \rangle = 0$. 则 $\boldsymbol{x}^* \in (\mathrm{ap}C)^-$, 得 $C^\perp = (\mathrm{cl}C)^\perp \subset \mathrm{ap}C^-$.

反之, $\boldsymbol{x}^* \in (\mathrm{ap}C)^-$, 对任意的 $\boldsymbol{a} \in \mathrm{ap}C$ 有 $\langle \boldsymbol{x}^*, \boldsymbol{a} \rangle \leqslant 0$. 由性质 3.1.5 知 $\mathrm{ap}C$ 是一个平面, 且 $\mathrm{ap}C$ 是包含零点的平面, 则 $\mathrm{ap}C$ 是子空间. 因此, 得 $-\boldsymbol{a} \in \mathrm{ap}C$ 和 $\langle \boldsymbol{x}^*, -\boldsymbol{a} \rangle \leqslant 0$. 所以, 对任意的 $\boldsymbol{a} \in \mathrm{ap}C$ 有 $\langle \boldsymbol{x}^*, \boldsymbol{a} \rangle = 0$, 表明了 $(\mathrm{ap}C)^-$ 是子空间, 又因为 $\mathbf{0} \in \mathrm{ap}C \subset \mathrm{cl}C$, 有

$$(\mathrm{ap}C)^- \supset (\mathrm{cl}C)^- = C^-.$$

表明 $C^- \subset (\mathrm{ap}C)^-$, 也有 $-C^- \subset (\mathrm{ap}C)^-$. 那么对任意的 $\boldsymbol{x} \in C$ 有 $\langle \boldsymbol{x}^*, \boldsymbol{x} \rangle \leqslant 0$, 也有 $\langle -\boldsymbol{x}^*, \boldsymbol{x} \rangle \leqslant 0$, 说明 $\boldsymbol{x}^* \in C^\perp$, 即 $C^\perp = (\mathrm{cl}C)^\perp \supset \mathrm{ap}C^-$.

最后, 由性质 3.5.1 得 $(C^-)^- = \mathrm{cl}C$, 有 $\mathrm{ap}(\mathrm{cl}C) = \mathrm{ap}(C^-)^- = (C^-)^\perp$.

(2) 由性质 2.1.3 知, 包含 C 的最小子空间为

$$\begin{aligned}\mathrm{span}C = \{&\alpha_1\boldsymbol{x}_1 + \alpha_2\boldsymbol{x}_2 + \cdots + \alpha_m\boldsymbol{x}_m | \boldsymbol{x}_1, \boldsymbol{x}_2, \cdots, \boldsymbol{x}_m \in C;\\ &\alpha_1, \alpha_2, \cdots, \alpha_m \in \mathbf{R}, m = 1, 2, \cdots\},\end{aligned}$$

即有 $C \subset \mathrm{span}C$. 由 (1) 知 $\mathrm{ap}(\mathrm{cl}C) = (C^-)^\perp$ 和 $C^\perp = (\mathrm{cl}C)^\perp = \mathrm{ap}C^-$. 设

$$\boldsymbol{x} = \alpha_1\boldsymbol{x}_1 + \alpha_2\boldsymbol{x}_2 + \cdots + \alpha_m\boldsymbol{x}_m \in \mathrm{span}C,$$

其中 $\boldsymbol{x}_1, \boldsymbol{x}_2, \cdots, \boldsymbol{x}_m \in C, \alpha_1, \alpha_2, \cdots, \alpha_m \in \mathbf{R}$, 对任意的 $\boldsymbol{y} \in C^\perp = \mathrm{ap}C^-$, 有 $\langle \boldsymbol{x}_k, \boldsymbol{y} \rangle = 0\,(k = 1, 2, \cdots, m)$ 等价于 $\langle \boldsymbol{x}, \boldsymbol{y} \rangle = 0$, 表明 $\boldsymbol{x} \in (\mathrm{ap}C^-)^\perp$. 反之显然. 再由 (1) 得 $\mathrm{ap}C^-$ 和 $\mathrm{ap}(\mathrm{cl}C)$ 是正交子空间.

(3) 可由 (1) 和 (2) 直接获得.

(4) 和 (5) 由 (3) 直接得到.

(6) C 是子空间等价于

$$\langle \boldsymbol{x}^*, \alpha\boldsymbol{x} \rangle = \langle \alpha\boldsymbol{x}^*, \boldsymbol{x} \rangle \leqslant 0, \quad \forall \boldsymbol{x} \in C, \quad \forall \alpha \in \mathbf{R},$$

即 $\alpha\boldsymbol{x}^* \in C^-$, 等价于 C^- 是子空间.

(7) $C^- = C^\perp$ 等价于对于任意 $\boldsymbol{x} \in C$ 和任意 $\alpha \in \mathbf{R}$ 有 $\langle \boldsymbol{x}^*, \alpha\boldsymbol{x} \rangle = \langle \alpha\boldsymbol{x}^*, \boldsymbol{x} \rangle = 0$, 当且仅当 C 是子空间.

(8) 若 $C^- = \mathrm{ap}C^-$, 那么有 $C^\perp = C^-$, 即 C 是子空间. 由性质 3.1.9 得 $C = \mathrm{ap}C$. 反过来, 由性质 3.1.9 知, 若 $C = \mathrm{ap}C$, 则 C 是子空间, 即有 $C^- = \mathrm{ap}C^-$. □

例 3.5.2 设凸锥 $C=\{x|x\geqslant 0\}\subset\mathbf{R}$, 显然 $C^-=-C$, 则有 $C^\perp=\{0\}$, $\mathrm{ap}C^-=\{0\}$, $\mathrm{ap}C=\{0\}$, $(\mathrm{ap}C^-)^\perp=\mathbf{R}$ 和 $(C^-)^\perp=\{0\}$. 这些结果表明性质 3.5.3 的结论对本例均成立.

性质 3.5.3 揭示了凸锥的极锥及顶点集与相应的其他集合的关系. 下面的结论描述了凸锥闭包的分解性质, 表明了凸锥闭包由凸锥的闭包与它的仿射集中顶点集的正交补加上它的顶点集构成. 我们称这个结论为原点凸锥的闭包分解定理, 其中一个重要结论是 $\mathbf{R}^n$ 可以分解成凸锥闭包的顶点集、凸锥的极锥顶点集以及凸锥生成子空间中闭包顶点集的正交补集等三个集合相加.

性质 3.5.4 (凸锥闭包分解定理) 设 $C\subset\mathbf{R}^n$ 是顶点为 $\mathbf{0}$ 的非空凸锥, S 是在 $\mathrm{span}C$ 中 $\mathrm{ap}\,(\mathrm{cl}C)$ 的正交补, 则 S 是在 $\mathrm{span}C^-$ 中 $\mathrm{ap}C^-$ 的正交补, $\mathbf{R}^n$ 是成对正交子空间的和:

$$\mathbf{R}^n=\mathrm{ap}(\mathrm{cl}C)+\mathrm{ap}C^-+S,$$

即

$$\mathrm{cl}C=(\mathrm{cl}C\cap S)+\mathrm{ap}(\mathrm{cl}C),\quad C^-=(C^-\cap S)+\mathrm{ap}C^-. \tag{3.5.2-1}$$

进一步, $\mathrm{cl}C\cap S$ 和 $C^-\cap S$ 不含直线, 且

$$\mathrm{span}(\mathrm{cl}C\cap S)=\mathrm{span}(C^-\cap S)=S, \tag{3.5.2-2}$$

$$\mathrm{ri}C=(\mathrm{ri}C\cap S)+\mathrm{ap}(\mathrm{cl}C)=\mathrm{ri}(C\cap S)+\mathrm{ap}(\mathrm{cl}C), \tag{3.5.2-3}$$

$$\mathrm{ri}C^-=(\mathrm{ri}C^-\cap S)\mathrm{ap}C^-=\mathrm{ri}(C^-\cap S)+\mathrm{ap}C^-. \tag{3.5.2-4}$$

证明 由性质 3.5.3(2) 知 $\mathrm{ap}C^-$ 和 $\mathrm{ap}(\mathrm{cl}C)$ 是正交子空间, 且

$$\mathrm{span}C=(\mathrm{ap}C^-)^\perp,\quad \mathrm{span}C^-=(\mathrm{ap}(\mathrm{cl}C))^\perp,$$

即有

$$\mathbf{R}^n=\mathrm{ap}C^-+\mathrm{span}C=\mathrm{ap}C^-+\mathrm{ap}(\mathrm{cl}C)+S$$

和

$$\mathrm{span}C^-=(\mathrm{ap}(\mathrm{cl}C))^\perp=\mathrm{ap}C^-+S.$$

因此, S 是在 $\mathrm{span}C^-$ 中 $\mathrm{ap}C^-$ 的正交补. 再由 $\mathrm{ap}(\mathrm{cl}C)=\mathrm{lin}(\mathrm{cl}C)$ 可得

$$S=(\mathrm{ap}(\mathrm{cl}C))^\perp\cap\mathrm{span}(\mathrm{cl}C)=\mathrm{span}C^-\cap(\mathrm{ap}C^-)^\perp. \tag{3.5.2-5}$$

根据上式得

$$\mathrm{cl}C=(\mathrm{cl}C\cap S)+\mathrm{ap}(\mathrm{cl}C),\quad \mathrm{span}(\mathrm{cl}C\cap S)=S,$$

$$C^- = (C^- \cap S) + \mathrm{ap}C^-, \quad \mathrm{span}(C^- \cap S) = S.$$

这说明 $\mathrm{cl}C \cap S$ 不含直线, 得

$$\mathrm{ri}C = \mathrm{ri}(\mathrm{cl}C) = (\mathrm{ri}(\mathrm{cl}C) \cap S) + \mathrm{ap}(\mathrm{cl}C) = (\mathrm{ri}C \cap S) + \mathrm{ap}(\mathrm{cl}C),$$

$$\mathrm{ri}C^- = (\mathrm{ri}C^- \cap S) + \mathrm{ap}C^- = \mathrm{ri}(C^- \cap S) + \mathrm{ap}C^-. \qquad \square$$

例 3.5.3　设凸锥 $C = \left\{ (x_1, 0)^{\mathrm{T}} | x_1 > 0 \right\} \subset \mathbf{R}^2$, 有 $C^- = \left\{ (x_1, x_2)^{\mathrm{T}} | x_1 \leqslant 0, x_2 \in \mathbf{R} \right\}$. 我们还有

$$\mathrm{ap}\,(\mathrm{cl}C) = \{\mathbf{0}\}, \quad (\mathrm{ap}\,(\mathrm{cl}C))^{\perp} = \mathbf{R}^2,$$

$$\mathrm{ap}C^- = \{\mathbf{0}\}, \quad \left(\mathrm{ap}C^-\right)^{\perp} = \mathbf{R}^2, \quad \mathrm{ap}C = \{\mathbf{0}\}, \quad \left(C^-\right)^{\perp} = \{\mathbf{0}\},$$

$$\mathrm{span}C^- = \mathbf{R}^2, \quad \mathrm{span}C = (\mathrm{ap}C^-)^{\perp}, \quad \mathrm{ri}C = \{(x_1, 0)\,|x_1 \in \mathbf{R}\},$$

将这些结果代入 (3.5.2-1)~(3.5.2-4) 式均成立. 容易验证, 性质 3.5.5 对本例也成立.

性质 3.5.5　设 $C \subset \mathbf{R}^n$ 是顶点为原点的非空凸锥, 且不是子空间, 下面结论成立:

(1) $\boldsymbol{a} \in \mathrm{ri}C^-$ 当且仅当 $\langle \boldsymbol{x}, \boldsymbol{a} \rangle < 0, \forall \boldsymbol{x} \in \mathrm{cl}C \setminus \mathrm{ap}(\mathrm{cl}C)$;

(2) $\boldsymbol{a} \in \mathrm{ri}C$ 当且仅当 $\langle \boldsymbol{x}, \boldsymbol{a} \rangle < 0, \forall \boldsymbol{x} \in C^- \setminus \mathrm{ap}C^-$.

证明　(1) 若 $\boldsymbol{a} \in \mathrm{ri}C^-$, 对任意的 $\boldsymbol{x} \in \mathrm{cl}C$, 有 $\langle \boldsymbol{a}, \boldsymbol{x} \rangle \leqslant 0$. 那么 $\mathrm{ri}C^-$ 的正交集合为

$$(\mathrm{ri}C^-)^{\perp} = \left\{ \boldsymbol{x} | \langle \boldsymbol{a}, \boldsymbol{x} \rangle = 0, \boldsymbol{a} \in \mathrm{ri}C^- \right\},$$

由性质 3.5.3 得 $\mathrm{ap}(\mathrm{cl}C) = (C^-)^{\perp}$, 即 $\forall \boldsymbol{x} \in \mathrm{cl}C \setminus \mathrm{ap}(\mathrm{cl}C) = \mathrm{cl}C \setminus (C^-)^{\perp}$, 又即 $\langle \boldsymbol{a}, \boldsymbol{x} \rangle < 0$.

反过来, 若 $\langle \boldsymbol{x}, \boldsymbol{a} \rangle < 0, \forall \boldsymbol{x} \in \mathrm{cl}C \setminus \mathrm{ap}(\mathrm{cl}C)$, 有 $\mathrm{cl}C = \left(\mathrm{cl}C \setminus (C^-)^{\perp}\right) \cup (C^-)^{\perp}$. 当

$$\boldsymbol{x} \in (C^-)^{\perp} = (\mathrm{cl}C^-)^{\perp} \subset C,$$

即有 $\langle \boldsymbol{a}, \boldsymbol{x} \rangle = 0\,(\boldsymbol{a} \in \mathrm{cl}C^-)$. 因此, $\boldsymbol{a} \in \mathrm{cl}C^-$.

下面证明 $\boldsymbol{a} \in \mathrm{ri}C^-$, 采用反证法. 假设 $\boldsymbol{a} \notin \mathrm{ri}C^-$, 有 $\left(\boldsymbol{a} + (C^-)^{\perp}\right) \cap \mathrm{ri}C^- = \varnothing$. 因为如果有

$$\left(\boldsymbol{a} + (C^-)^{\perp}\right) \cap \mathrm{ri}C^- \neq \varnothing,$$

则存在一个 $\boldsymbol{c} \in (C^-)^{\perp}$ 使得 $\boldsymbol{a} + \boldsymbol{c} \in \mathrm{ri}C^-$, 有 $\langle \boldsymbol{c}, \boldsymbol{a} + \boldsymbol{c} \rangle = 0$, 得 $\boldsymbol{c} = 0$ 和 $\boldsymbol{a} \in \mathrm{ri}C^-$, 故得到矛盾. 又因为 $\mathrm{ri}C^-$ 和 $\boldsymbol{a} + (C^-)^{\perp}$ 是凸集, 由凸集分离定理 (定

理 2.4.3) 得, 存在非零向量 $\boldsymbol{a}^* \in \mathbf{R}^n$ 和常数 $\alpha \in \mathbf{R}$ 构成的一个超平面 $H(\boldsymbol{a}^*, \alpha) = \{\boldsymbol{x} \mid \langle \boldsymbol{x}, \boldsymbol{a}^* \rangle = \alpha\}$ 分离 $\boldsymbol{a} + (C^-)^\perp$ 和 C^-, 且

$$\langle \boldsymbol{a}_1, \boldsymbol{a}^* \rangle \leqslant \alpha \leqslant \langle \boldsymbol{a} + \boldsymbol{c}, \boldsymbol{a}^* \rangle, \quad \forall \boldsymbol{a}_1 \in C^-, \quad \boldsymbol{c} \in (C^-)^\perp. \tag{3.5.2-6}$$

因为 C^- 是闭锥且含零点, 则 $\sup\limits_{\forall \boldsymbol{a}_1 \in C^-} \langle \boldsymbol{a}_1, \boldsymbol{a}^* \rangle = 0$, 得 $\boldsymbol{a}^* \in \mathrm{cl}C$. 因为 $\boldsymbol{a} \in \mathrm{cl}C^-$, 则 $\langle \boldsymbol{a}, \boldsymbol{a}^* \rangle = 0$, 那么 $\boldsymbol{a}^* \in (C^-)^\perp = \mathrm{ap}(\mathrm{cl}C) \subset \mathrm{rb}(\mathrm{cl}C)$. 因此, $\boldsymbol{a} \in \mathrm{span}C^- = (\mathrm{ap}(\mathrm{cl}C))^\perp$, 但由性质 3.5.4 得

$$\begin{aligned} \mathrm{span}C^- &= (\mathrm{ap}(\mathrm{cl}C))^\perp = \mathrm{ap}C^- + \mathrm{span}C^- \cap \mathrm{span}C \\ &= (\mathrm{cl}C)^\perp + (\mathrm{ap}(\mathrm{cl}C))^\perp \cap (\mathrm{ap}C^-)^\perp. \end{aligned}$$

对任意的 $\boldsymbol{x} \in \mathrm{cl}C \setminus \mathrm{ap}(\mathrm{cl}C)$ 有 $\langle \boldsymbol{x}, \boldsymbol{a} \rangle < 0$, 由 (3.5.2-1) 式和 (3.5.2-5) 式得

$$\mathrm{cl}C = (\mathrm{cl}C \cap (\mathrm{ap}(\mathrm{cl}C))^\perp \cap (\mathrm{ap}C^-)^\perp) + \mathrm{ap}(\mathrm{cl}C).$$

根据上式得 $\boldsymbol{x} \in (\mathrm{ap}(\mathrm{cl}C))^\perp \cap (\mathrm{ap}C^-)^\perp$. 因此, 有

$$\langle \boldsymbol{x}, \boldsymbol{a}^* \rangle = 0 \quad (\boldsymbol{x} \in \mathrm{cl}C \setminus \mathrm{ap}(\mathrm{cl}C)).$$

对任意的 $\boldsymbol{x} \in \mathrm{ap}(\mathrm{cl}C) = (C^-)^\perp$, 有

$$\langle \boldsymbol{x}, \boldsymbol{a}^* \rangle = \langle \boldsymbol{x} + \boldsymbol{a}, \boldsymbol{a}^* \rangle \geqslant 0.$$

由定义 3.5.1 得 $-\boldsymbol{a}^* \in C^-$, 有 $\langle -\boldsymbol{a}^*, \boldsymbol{a}^* \rangle = 0$. 因此, $\boldsymbol{a}^* = \mathbf{0}$, 这是一个矛盾.

(2) 根据性质 3.5.1、性质 3.5.3 和 (1) 的证明可得. □

性质 3.5.5 说明凸锥与极锥之间点的内积关系, 即如果凸锥不是子空间, 则一个点属于该凸锥的极锥相对内部等价于该点与凸锥不含其闭包顶点集合中点的内积均小于 0; 一个点属于该凸锥的相对内部等价于该点与凸锥不含其极锥顶点的极锥中点的内积均小于 0.

性质 3.5.6 设 $C \subset \mathbf{R}^n$ 是顶点为原点的非空真凸锥. 如果 $S \subset \mathbf{R}^n$ 是 $(n-1)$ 维子空间, 则 $S \cap \mathrm{cl}C = \mathrm{ap}(\mathrm{cl}C)$ 当且仅当 $\left(S^\perp \setminus \{\mathbf{0}\}\right) \cap \mathrm{ri}C^- \neq \varnothing$. 此外, 如果 C 不是子空间, 则 $S \cap \mathrm{ri}C = \varnothing$.

证明 如果 $S \cap \mathrm{cl}C = \mathrm{ap}(\mathrm{cl}C)$ 成立, 假设 $\left(S^\perp \setminus \{\mathbf{0}\}\right) \cap \mathrm{ri}C^- = \varnothing$, 那么有

$$\left(S^\perp \setminus \{\mathbf{0}\} + (C^-)^\perp\right) \cap \mathrm{ri}C^- = \varnothing.$$

若上式不成立, 那么有 $\boldsymbol{a} + \boldsymbol{c} \in \mathrm{ri}C^-$, 其中 $\boldsymbol{c} \in (C^-)^\perp = \mathrm{ap}(\mathrm{cl}C)$ 和 $\boldsymbol{a} \in S^\perp \setminus \{\mathbf{0}\}$. 则有 $\langle \boldsymbol{c}, \boldsymbol{a} + \boldsymbol{c} \rangle \leqslant 0$, 可得 $\boldsymbol{c} = \mathbf{0}$, 有 $\boldsymbol{a} \in \mathrm{ri}C^-$, 得矛盾.

又因为 $\mathrm{ri}C^-$ 和 $\left(S^\perp\setminus\{\mathbf{0}\}+(C^-)^\perp\right)$ 是凸集, 由凸集分离定理 (定理 2.4.3) 知, 存在非零向量 $\boldsymbol{a}^*\in\mathbf{R}^n$ 和常数 $\alpha\in\mathbf{R}$ 构成的一个超平面

$$H(\boldsymbol{a}^*,\alpha)=\{\boldsymbol{x}\mid\langle\boldsymbol{x},\boldsymbol{a}^*\rangle=\alpha\}$$

分离 $\left(S^\perp\setminus\{\mathbf{0}\}+(C^-)^\perp\right)$ 和 C^-, 即有

$$\langle\boldsymbol{a}_1,\boldsymbol{a}^*\rangle\leqslant\alpha\leqslant\langle\boldsymbol{a},\boldsymbol{a}^*\rangle,\quad\forall\boldsymbol{a}_1\in C^-,\quad\boldsymbol{a}\in\left(S^\perp\setminus\{\mathbf{0}\}+(C^-)^\perp\right).\tag{3.5.2-7}$$

因为 C^- 是闭锥且含零点, 则 $\sup\limits_{\forall\boldsymbol{a}_1\in C^-}\langle\boldsymbol{a}_1,\boldsymbol{a}^*\rangle=0$. 因为 $S^\perp$ 是子空间, 由 (3.5.2-7) 式得 $\alpha=0$, 有 $\boldsymbol{a}^*\in\mathrm{cl}C$. 因 $S\subset\mathbf{R}^n$ 是 $(n-1)$ 维子空间, 则 $S^\perp$ 是 1 维子空间. 所以, $S^\perp$ 可以表示为

$$S^\perp=\{\lambda\boldsymbol{a}|\lambda\in\mathbf{R},\boldsymbol{a}\neq\mathbf{0}\}.$$

根据 (3.5.2-7) 式得 $\langle\boldsymbol{a},\boldsymbol{a}^*\rangle=0$. 因此 $\boldsymbol{a}^*\in S$, 又有 $\boldsymbol{a}^*\in(C^-)^\perp=\mathrm{ap}(\mathrm{cl}C)$. 故得

$$\boldsymbol{a}\in\mathrm{span}C^-=(\mathrm{ap}(\mathrm{cl}C))^\perp.$$

对任意的 $\boldsymbol{x}\in\mathrm{cl}C\setminus\mathrm{ap}(\mathrm{cl}C)$, 由 (3.5.2-1) 式和 (3.5.2-5) 式得

$$\mathrm{cl}C=(\mathrm{cl}C\cap(\mathrm{ap}(\mathrm{cl}C))^\perp\cap(\mathrm{ap}C^-)^\perp)+\mathrm{ap}(\mathrm{cl}C).$$

根据上式得 $\boldsymbol{x}\in(\mathrm{ap}(\mathrm{cl}C))^\perp\cap(\mathrm{ap}C^-)^\perp$, 因此, 有 $\langle\boldsymbol{x},\boldsymbol{a}^*\rangle=0\,(\boldsymbol{x}\in\mathrm{cl}C\setminus\mathrm{ap}(\mathrm{cl}C))$. 对任意的

$$\boldsymbol{x}\in\mathrm{ap}(\mathrm{cl}C)=(C^-)^\perp,$$

有 $\langle\boldsymbol{x},\boldsymbol{a}^*\rangle=\langle\boldsymbol{x}+\boldsymbol{a},\boldsymbol{a}^*\rangle\geqslant0$, 由定义 3.5.1 得 $-\boldsymbol{a}^*\in C^-$, 有 $\langle-\boldsymbol{a}^*,\boldsymbol{a}^*\rangle=0$, 因此, $\boldsymbol{a}^*=\mathbf{0}$, 这是一个矛盾.

反过来, 如果 $\left(S^\perp\setminus\{\mathbf{0}\}\right)\cap\mathrm{ri}C^-\neq\varnothing$, 设 $\boldsymbol{a}^*\in\left(S^\perp\setminus\{\mathbf{0}\}\right)\cap\mathrm{ri}C^-$, 则有

$$\langle\boldsymbol{x},\boldsymbol{a}^*\rangle=0,\quad\forall\boldsymbol{x}\in S\cap\mathrm{cl}C,$$

上式等价于 $\boldsymbol{x}\in(C^-)^\perp=\mathrm{ap}(\mathrm{cl}C)$, 即有 $S\cap\mathrm{cl}C=\mathrm{ap}\,(\mathrm{cl}C)$.

接着再证明, 如果 C 不是子空间, 则 $S\cap\mathrm{ri}C=\varnothing$. 采用反证法, 设 $\boldsymbol{a}\in S\cap\mathrm{ri}C$. 由性质 3.5.5 得

$$\langle\boldsymbol{x},\boldsymbol{a}\rangle<0,\quad\forall\boldsymbol{x}\in C^-\setminus\ \mathrm{ap}C^-.$$

再由性质 3.5.4 得

$$C^-=(C^-\cap(\mathrm{ap}(\mathrm{cl}C))^\perp\cap(\mathrm{ap}C^-)^\perp)+\mathrm{ap}C^-.$$

对任意的 $\boldsymbol{x} \in C^- \setminus \mathrm{ap}C^-$, 由上式有 $\boldsymbol{x} \in (\mathrm{ap}(\mathrm{cl}C))^{\perp}$. 根据假设得 $S \cap \mathrm{cl}C = \mathrm{ap}\,(\mathrm{cl}C)$, 如果 $\boldsymbol{a} \in S$, 有 $\langle \boldsymbol{x}, \boldsymbol{a} \rangle = 0$, 得出矛盾. 因此, 有 $S \cap \mathrm{ri}C = \varnothing$. □

例 3.5.4 设凸锥 $C = \{ (x_1, 0)^{\mathrm{T}} | x_1 > 0 \} \subset \mathbf{R}^2$, 有 $C^- = \{ (x_1, x_2)^{\mathrm{T}} | x_1 \leqslant 0, x_2 \in \mathbf{R} \}$. 设

$$S = \left\{ (x_1, 0)^{\mathrm{T}} | x_1 \in \mathbf{R} \right\} \subset \mathbf{R}^2,$$

则有 $S \cap \mathrm{cl}C = \mathrm{ap}\,(\mathrm{cl}C) = \{\mathbf{0}\}$, $\mathrm{ri}C^- = \left\{ (x_1, x_2)^{\mathrm{T}} | x_1 < 0, x_2 \in \mathbf{R} \right\}$ 和 $S^{\perp} = \left\{ (0, x_2)^{\mathrm{T}} | x_2 \in \mathbf{R} \right\}$, 得

$$\left(S^{\perp} \setminus \{\mathbf{0}\} \right) \cap \mathrm{ri}C^- \neq \varnothing, \quad S \cap \mathrm{ri}C \neq \varnothing.$$

上述例子说明性质 3.5.6 结论成立.

由性质 3.5.6 和性质 3.3.9 获得下面推论.

推论 3.5.1 设 $C \subset \mathbf{R}^n$ 是不含直线的凸锥, 则有下面结论成立:

(1) 存在一个 $(n-1)$ 维子空间 $S \subset \mathbf{R}^n$ 满足条件 $S \cap \mathrm{cl}C = \{\mathbf{0}\}$;

(2) 存在 $(n-1)$ 维子空间 $S \subset \mathbf{R}^n$ 满足条件 $S \cap \mathrm{cl}C = \{\mathbf{0}\}$ 当且仅当

$$(S^{\perp} \{\mathbf{0}\}) \cap \mathrm{int}C^- \neq \varnothing.$$

推论 3.5.2 设 $C \subset \mathbf{R}^n$ 是顶点为原点的非空凸锥, S 是在 $\mathrm{span}C$ 内 $\mathrm{ap}\,(\mathrm{cl}C)$ 的正交补, 则

$$(\mathrm{cl}C \cap S)^- = C^- + \mathrm{ap}(\mathrm{cl}C), \quad (C^- \cap S)^- = \mathrm{cl}C + \mathrm{ap}C^-.$$

证明 由性质 3.5.3 和性质 3.5.4 有 $S^- = S^{\perp}$, $(\mathrm{cl}C)^- = C^-$ 和

$$(\mathrm{cl}C \cap S)^- = \mathrm{cl}((\mathrm{cl}C)^- + S^-) = \mathrm{cl}(C^- + S^{\perp}) = \mathrm{cl}(C^- + \mathrm{ap}(\mathrm{cl}C) + \mathrm{ap}C^-).$$

因为 $C^- + \mathrm{ap}C^- = C^-$, 所以 C^- 的正交子空间是 $\mathrm{ap}(\mathrm{cl}C)$, 可得

$$\mathrm{cl}(C^- + \mathrm{ap}(\mathrm{cl}C) + \mathrm{ap}C^-) = \mathrm{cl}(C^- + \mathrm{ap}(\mathrm{cl}C)) = C^- + \mathrm{ap}(\mathrm{cl}C)$$

和

$$\begin{aligned}(C^- \cap S)^- &= \mathrm{cl}((C^-)^- + S^-) = \mathrm{cl}(\mathrm{cl}C + S^{\perp}) \\ &= \mathrm{cl}(\mathrm{cl}C + \mathrm{ap}(\mathrm{cl}C) + \mathrm{ap}C^-) \\ &= \mathrm{cl}(\mathrm{cl}C + \mathrm{ap}C^-) = \mathrm{cl}C + \mathrm{ap}C^-.\end{aligned}$$

□

下面的性质 3.5.7 进一步揭示了凸锥的极锥与其生成子空间、正极锥和相对内部之间的密切联系. 即凸锥的闭包与正极锥相交是不含直线的闭凸锥, 凸锥的极锥等于闭包加上闭包与正极锥交的极锥, 闭包与正极锥交的相对内部等于凸锥的相对内部与正极锥的相对内部的交集.

性质 3.5.7 设 $C \subset \mathbf{R}^n$ 是顶点为原点的非空凸锥, S 是在 $\mathrm{span}C$ 内 $\mathrm{ap}\,(\mathrm{cl}C)$ 的正交补 ($S = (\mathrm{ap}(\mathrm{cl}C))^{\perp} \cap \mathrm{span}(\mathrm{cl}C) = \mathrm{span}C^- \cap (\mathrm{ap}C^-)^{\perp}$), 则有下面结论成立:

(1) $\mathrm{cl}C \cap C^+$ 是不含直线的闭凸锥;

(2) $\mathrm{span}(\mathrm{cl}C \cap C^+) = S$, $\dim(\mathrm{cl}C \cap C^+) = \dim S$ 和 $(\mathrm{cl}C \cap C^+)^- = C^- - \mathrm{cl}C$;

(3) $\mathrm{ri}(\mathrm{cl}C \cap C^+) = \mathrm{ri}C \cap (\mathrm{ri}C^+) = \mathrm{ri}(C \cap S) \cap (\mathrm{ri}(C^+ \cap S))$.

证明 分两种情形证明. 当 C 是子空间时, 有 $C^- = C^{\perp}$ 和 $\mathrm{cl}C \cap C^+ = S = \{\mathbf{0}\}$, 定理的所有结论显然成立. 下面只要证明在 C 不是子空间的情形下, 定理各个结论也成立.

(1) 由性质 3.5.1 知 $\mathrm{cl}C \cap C^+$ 是闭凸锥, 再由性质 3.5.4 的 (3.5.2-1) 式可得

$$\begin{aligned}\mathrm{cl}C \cap C^+ &= ((\mathrm{cl}C \cap S) + \mathrm{ap}(\mathrm{cl}C)) \cap ((C^+ \cap S) + \mathrm{ap}C^+)\\ &= (\mathrm{cl}C \cap S) \cap (C^+ \cap S)\\ &= \left(\mathrm{cl}C \cap C^+\right) \cap S \subset S.\end{aligned}$$

注意 $\mathrm{ap}\,(\mathrm{cl}C)$ 与 $\mathrm{ap}C^+$ 是正交子空间, 因此上式说明结论成立.

(2) 因为

$$(\mathrm{cl}C \cap S) \cap (C^- \cap S) = (\mathrm{cl}C \cap C^-) \cap S = \{\mathbf{0}\},$$

由性质 3.3.10 可知 $F = (\mathrm{cl}C \cap S) + (C^+ \cap S)$ 不含直线. 再利用性质 3.5.3 和推论 3.5.2 得 $\dim F^- = n$ 和

$$\begin{aligned}F^- &= ((\mathrm{cl}C \cap S) + (C^+ \cap S))^-\\ &= (\mathrm{cl}C \cap S)^- \cap (C^+ \cap S)^-\\ &= (C^- + \mathrm{ap}(\mathrm{cl}C)) \cap (-\mathrm{cl}C + \mathrm{ap}C^-)\\ &= ((C^- \cap S) + \mathrm{ap}(\mathrm{cl}C) + \mathrm{ap}C^-) \cap (-(\mathrm{cl}C \cap S) + \mathrm{ap}(\mathrm{cl}C) + \mathrm{ap}C^-)\\ &= ((C^- \cap S) + S^{\perp}) \cap (-(\mathrm{cl}C \cap S) + S^{\perp})\\ &= (C^- \cap S) \cap (-(\mathrm{cl}C \cap S)) + S^{\perp}\\ &= ((\mathrm{cl}C \cap C^+) \cap S) + S^{\perp}.\end{aligned}$$

由上式得

$$\begin{aligned}\dim(\mathrm{cl}C\cap C^+) &= \dim((\mathrm{cl}C\cap C^+)\cap S)\\ &= \dim F^- - \dim S^\perp\\ &= n - \dim S^\perp = \dim S.\end{aligned}$$

因此, 得 $\mathrm{cl}C\cap C^+\subset S$ 和 $\mathrm{span}(\mathrm{cl}C\cap C^+) = S$, 根据性质 3.3.9 和性质 3.5.1 得

$$\begin{aligned}(\mathrm{cl}C\cap C^+)^- &= ((\mathrm{cl}C\cap C^+))^-\\ &= \mathrm{cl}((\mathrm{cl}C)^- + \left(-C^-\right)^-)\\ &= \mathrm{cl}(C^- + (-\mathrm{cl}C))\\ &= C^- - \mathrm{cl}C.\end{aligned}$$

(3) 因 $\mathbf{0}\in\mathrm{cl}C\cap C^+$, 有 $\mathrm{span}C = \mathrm{span}C^- = \mathrm{span}(\mathrm{cl}C\cap C^+)$, 得到

$$\mathrm{span}(\mathrm{cl}C\cap C^+) = \mathrm{span}(\mathrm{cl}C\cap S) = \mathrm{span}(C^-\cap S) = S.$$

再由性质 3.5.4 得

$$\begin{aligned}\mathrm{ri}(\mathrm{cl}C\cap C^+) &= \mathrm{ri}(\mathrm{cl}C\cap S)\cap(\mathrm{ri}(C^+\cap S))\\ &= \mathrm{ri}(C\cap S)\cap(\mathrm{ri}(C^+\cap S))\end{aligned}$$

和

$$\begin{aligned}\mathrm{ri}C\cap(\mathrm{ri}C^+) &= (\mathrm{ri}(C\cap S) + \mathrm{ap}(\mathrm{cl}C))\cap(\mathrm{ri}(C^+\cap S) + \mathrm{ap}C^-)\\ &= \mathrm{ri}(C\cap S)\cap(\mathrm{ri}(C^+\cap S)).\end{aligned}$$

□

性质 3.5.8 设 $C\subset\mathbf{R}^n$ 是顶点为原点的非空凸锥, 则 C 是子空间当且仅当 $\mathrm{cl}C\cap C^+ = \{\mathbf{0}\}$ 或 $\mathrm{ri}C\cap(\mathrm{ri}C^+) = \{\mathbf{0}\}$.

证明 由性质 3.5.4 和性质 3.5.8 可得. □

本小节得到了顶点为原点的非空凸锥下的极锥与该凸锥的闭包、生成子空间、相对内部和正交集之间的关系:

1) (闭) 凸锥的正交集等于凸锥极锥的顶点集, 凸锥的顶点集属于凸锥闭包的顶点集, 凸锥闭包的顶点集等于凸锥极锥的正交集.

2) 凸锥生成仿射子空间等于凸锥极锥的顶点集的正交集, 凸锥的极锥的生成仿射子空间等于凸锥闭包的顶点集的正交集.

3) 极锥的顶点集维度等于 n 减去凸锥的维度, 凸锥极锥的维度等于 n 减去凸锥的闭包的顶点集的维度.

4) 凸锥是子空间等价于凸锥的极锥是子空间, 凸锥是子空间等价于凸锥的极锥等于凸锥的正交集, 凸锥不等于其顶点集等价于凸锥的极锥不等于极锥的顶点集.

5) (原点凸锥的闭包分解定理) $\mathbf{R}^n$ 分解成凸锥闭包的顶点集、凸锥极锥顶点集以及凸锥生成子空间中闭包顶点集的正交补集等三个集合相加; 凸锥的闭包等于闭包与凸锥生成子空间中闭包顶点集的正交补集交再加上凸锥闭包的顶点集.

6) 如果凸锥不是子空间, 则一个点属于凸锥的极锥相对内部等价于该点与凸锥的不含其闭包顶点集中点的内积均小于 0, 一个点属于凸锥的相对内部等价于该点与凸锥不含其极锥顶点集的极锥中点的内积均小于 0.

7) 在性质 3.5.7 条件下, 凸锥的闭包与正极锥相交是不含直线的闭凸集; 凸锥的极锥等于凸锥闭包加上闭包与正极锥交的极锥; 闭包与正极锥交的相对内部等于凸锥的相对内部与正极锥的相对内部的交集.

8) 凸锥是子空间等价于凸锥的闭包与正极锥的交集仅含原点, 或等价于凸锥的相对内部与正极锥相对内部的交集仅含原点.

3.6　法锥与切锥

本节介绍在凸分析中常见的两种特殊锥: 法锥与切锥. 法锥与切锥都是由锥包生成的, 可用来描述集合边界点上的几何方向, 它们在非线性函数映射中常用于刻画方向导数与梯度.

3.6.1　度量投影

下面首先给出一个点到给定集合最近点的定义, 即点与集合之间的距离函数.

定义 3.6.1　*设集合 $C\subset\mathbf{R}^n$, 给定点 $\boldsymbol{a}\in\mathbf{R}^n$ 和函数 $\phi_{\boldsymbol{a}}(\boldsymbol{x})=\|\boldsymbol{x}-\boldsymbol{a}\|$ $(\forall\boldsymbol{x}\in C)$. 如果存在点 $\boldsymbol{c}\in C$ 满足*

$$\|\boldsymbol{a}-\boldsymbol{c}\|=\inf\{\phi_{\boldsymbol{a}}(\boldsymbol{x})\mid\boldsymbol{x}\in C\},\tag{3.6.1-1}$$

则称 $\boldsymbol{c}$ 是 $\boldsymbol{a}$ 到 C 的最近点, 如图 3.6-1. 记 $\phi_{\boldsymbol{a}}(C)=\|\boldsymbol{c}-\boldsymbol{a}\|$, 称 $\phi_{\boldsymbol{a}}(C)=\|\boldsymbol{c}-\boldsymbol{a}\|$ 是 $\boldsymbol{a}$ 到 C 的距离.

注　如果定义 3.6.1 中的集合 C 是闭的, 则 $\mathbf{R}^n$ 中的任何点 $\boldsymbol{a}$ 到 C 上至少有一个最近的点. 实际上, 连续函数 $\phi_{\boldsymbol{a}}(\boldsymbol{x})=\|\boldsymbol{x}-\boldsymbol{a}\|$ 在闭集 C 上达到最小值, 即有

$$\|\boldsymbol{a}-\boldsymbol{c}\|=\phi_{\boldsymbol{a}}(C)=\min_{\boldsymbol{x}\in C}\phi_{\boldsymbol{a}}(\boldsymbol{x}).$$

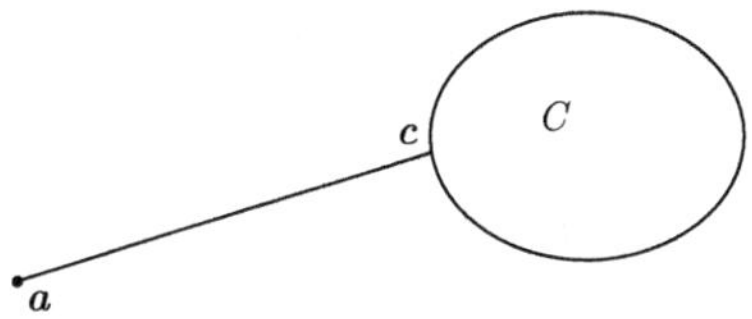

图 3.6-1 点 $\boldsymbol{a}$ 到集合 C 的最近点

例 3.6.1 设集合 $B=\{\boldsymbol{x}\mid\|\boldsymbol{x}\|<\varepsilon\}$, 其中 $\varepsilon>0$. 设 $\boldsymbol{a}\in\mathbf{R}^n$, 那么有

$$\|\boldsymbol{a}-\boldsymbol{c}\|=\phi_{\boldsymbol{a}}(B)=\inf\{\|\boldsymbol{a}-\boldsymbol{x}\|:\boldsymbol{x}\in B\},$$

当 $\boldsymbol{a}\in B$ 时, 有 $\boldsymbol{c}=\boldsymbol{a}$ 是 $\boldsymbol{a}$ 到 B 的最近点.

当 C 是凸集时, 因为 $\phi_{\boldsymbol{a}}(\boldsymbol{x})=\|\boldsymbol{x}-\boldsymbol{a}\|$ 在 C 上是连续的, 点 $\boldsymbol{a}$ 到 C 的闭包上有唯一最小点.

定理 3.6.1 设 $C\subset\mathbf{R}^n$ 是非空凸集和点 $\boldsymbol{a}\in\mathbf{R}^n$, 则 $\mathrm{cl}C$ 包含唯一的最近点 $\boldsymbol{c}$.

证明 设 $\boldsymbol{a}\in\mathrm{cl}C$, 根据定义 3.6.1 知 $\boldsymbol{c}=\boldsymbol{a}$ 是到 $\boldsymbol{a}\in\mathbf{R}^n$ 的最近点. 设 $\boldsymbol{a}\notin\mathrm{cl}C$, 那么函数

$$\inf\{\|\boldsymbol{a}-\boldsymbol{x}\|:\boldsymbol{x}\in\mathrm{cl}C\}\Leftrightarrow\min\left\{\|\boldsymbol{a}-\boldsymbol{x}\|^2:\boldsymbol{x}\in\mathrm{cl}C\right\}.$$

由于上述问题中目标函数是一个严格凸函数, 所以结论成立.

还可以用反证法进行证明.

若 $\mathrm{cl}C$ 包含两个不同的最近的点 $\boldsymbol{c}_1$ 和 $\boldsymbol{c}_2$, 则对于任何 $\boldsymbol{x}\in\mathrm{cl}C$ 有

$$\|\boldsymbol{c}_1-\boldsymbol{a}\|=\|\boldsymbol{c}_2-\boldsymbol{a}\|\leqslant\|\boldsymbol{a}-\boldsymbol{x}\|.$$

取 $\boldsymbol{c}_3=\dfrac{\boldsymbol{c}_1+\boldsymbol{c}_2}{2}$, 利用上式有

$$\begin{aligned}\|\boldsymbol{c}_3-\boldsymbol{a}\|^2&=\left\|\frac{\boldsymbol{c}_1+\boldsymbol{c}_2}{2}-\boldsymbol{a}\right\|^2=\frac{1}{4}(\boldsymbol{c}_1+\boldsymbol{c}_2-2\boldsymbol{a})^{\mathrm{T}}(\boldsymbol{c}_1+\boldsymbol{c}_2-2\boldsymbol{a})\\&=\frac{1}{4}\left[2(\boldsymbol{c}_1-\boldsymbol{a})^{\mathrm{T}}(\boldsymbol{c}_1-\boldsymbol{a})+2(\boldsymbol{c}_2-\boldsymbol{a})^{\mathrm{T}}(\boldsymbol{c}_2-\boldsymbol{a})-(\boldsymbol{c}_2-\boldsymbol{c}_1)^{\mathrm{T}}(\boldsymbol{c}_2-\boldsymbol{c}_1)\right]\\&=\frac{1}{4}\left[4(\boldsymbol{c}_1-\boldsymbol{a})^{\mathrm{T}}(\boldsymbol{c}_1-\boldsymbol{a})-(\boldsymbol{c}_2-\boldsymbol{c}_1)^{\mathrm{T}}(\boldsymbol{c}_2-\boldsymbol{c}_1)\right]\\&=\|\boldsymbol{c}_1-\boldsymbol{a}\|^2-\frac{1}{4}\|\boldsymbol{c}_2-\boldsymbol{c}_1\|^2<\|\boldsymbol{c}_1-\boldsymbol{a}\|^2,\end{aligned}$$

说明 $\mathrm{cl}C$ 包含最近的点 $\boldsymbol{c}_3$, 比 $\boldsymbol{c}_1$ 和 $\boldsymbol{c}_2$ 更好, 显然矛盾. □

由定理 3.6.1 可定义 $\mathrm{cl}C$ 上的度量投影.

定义 3.6.2　设 $C \subset \mathbf{R}^n$ 是非空凸集, 且映射 $p_C:\ \mathbf{R}^n \to \mathrm{cl}C$ 对任何 $\boldsymbol{a} \in \mathbf{R}^n$ 有 $p_C(\boldsymbol{a}) = \boldsymbol{c}$, 其中 $\boldsymbol{c}$ 是 $\mathrm{cl}C$ 中到 $\boldsymbol{a}$ 的唯一最近点, 则称 p_C 在 C 上是度量投影.

度量投射是将空间 $\mathbf{R}^n$ 中的点映射到闭集 $\mathrm{cl}C$ 上, 根据 (3.6.1-1) 式和定理 3.6.1 可定义

$$p_C(\boldsymbol{a}) = \{\boldsymbol{c} \mid \|\boldsymbol{c}-\boldsymbol{a}\| = \inf\{\|\boldsymbol{x}-\boldsymbol{a}\| : \boldsymbol{x} \in C\}\}, \tag{3.6.1-2}$$

则 p_C 在 C 上是度量投影.

下面的定理给出了度量投影判定的充要条件.

定理 3.6.2　设 $C \subset \mathbf{R}^n$ 是非空凸集, $\boldsymbol{c} \in \mathrm{cl}C$, $\boldsymbol{a} \in \mathbf{R}^n \setminus \mathrm{cl}C$ 和 $p_C(\boldsymbol{a}) = \boldsymbol{c}$, 则 p_C 在 C 上是度量投影的充要条件是

$$\mathrm{cl}C \subset V = \{\boldsymbol{x} \in \mathbf{R}^n | \langle \boldsymbol{c}-\boldsymbol{x}, \boldsymbol{c}-\boldsymbol{a}\rangle \leqslant 0\} \tag{3.6.1-3}$$

成立.

证明　如果 p_C 在 C 上是度量投影, 当 $\boldsymbol{x} = \boldsymbol{c} \in \mathrm{cl}C$ 时, 则有 $p_C(\boldsymbol{c}) = \boldsymbol{c} \in V$. 反过来, 设 (3.6.1-3) 式成立, 对任意的 $\boldsymbol{x} \in C$, 得

$$\begin{aligned}
\langle \boldsymbol{c}-\boldsymbol{x}, \boldsymbol{c}-\boldsymbol{a}\rangle \leqslant 0 \Rightarrow \langle \boldsymbol{c}-\boldsymbol{a}, \boldsymbol{c}-\boldsymbol{a}\rangle &\leqslant \langle \boldsymbol{x}-\boldsymbol{a}, \boldsymbol{c}-\boldsymbol{a}\rangle \\
&= \langle \boldsymbol{x}-\boldsymbol{a}, \boldsymbol{x}-\boldsymbol{a}\rangle + \langle \boldsymbol{x}-\boldsymbol{a}, \boldsymbol{c}-\boldsymbol{x}\rangle \\
&= \langle \boldsymbol{x}-\boldsymbol{a}, \boldsymbol{x}-\boldsymbol{a}\rangle + \langle \boldsymbol{c}-\boldsymbol{x}, \boldsymbol{c}-\boldsymbol{a}\rangle + \langle \boldsymbol{c}-\boldsymbol{x}, \boldsymbol{x}-\boldsymbol{c}\rangle \\
&\leqslant \langle \boldsymbol{x}-\boldsymbol{a}, \boldsymbol{x}-\boldsymbol{a}\rangle,
\end{aligned}$$

上式说明 p_C 在 C 上是度量投影. □

下面证明度量投影映射是满足 Lipschitz 条件的.

定理 3.6.3　设 $C \subset \mathbf{R}^n$ 是非空凸集, 则在 C 上的度量投影 p_C 满足 Lipschitz 条件:

$$\|p_C(\boldsymbol{x}) - p_C(\boldsymbol{y})\| \leqslant \|\boldsymbol{x}-\boldsymbol{y}\|, \quad \forall \boldsymbol{x}, \boldsymbol{y} \in \mathbf{R}^n, \tag{3.6.1-4}$$

且 p_C 是连续映射.

证明　设 $\forall \boldsymbol{x}, \boldsymbol{y} \in \mathbf{R}^n$, 由定理 3.6.2 得, 存在 $p_C(\boldsymbol{x}), p_C(\boldsymbol{y}) \in \mathrm{cl}C$ 使得

$$\langle p_C(\boldsymbol{y}) - p_C(\boldsymbol{x}),\ \boldsymbol{x} - p_C(\boldsymbol{x})\rangle \leqslant 0 \quad 和 \quad \langle p_C(\boldsymbol{x}) - p_C(\boldsymbol{y}),\ \boldsymbol{y} - p_C(\boldsymbol{y})\rangle \leqslant 0$$

成立. 令 $\boldsymbol{u} = \boldsymbol{x} - p_C(\boldsymbol{x})$ 和 $\boldsymbol{v} = \boldsymbol{y} - p_C(\boldsymbol{y})$, 那么有

$$(p_C(\boldsymbol{x}) - p_C(\boldsymbol{y}))^{\mathrm{T}}\boldsymbol{u} \geqslant 0 \quad 和 \quad (p_C(\boldsymbol{x}) - p_C(\boldsymbol{y}))^{\mathrm{T}}\boldsymbol{v} \leqslant 0.$$

所以有 $(p_C(\boldsymbol{x})-p_C(\boldsymbol{y}))^{\mathrm{T}}(\boldsymbol{u}-\boldsymbol{v})\geqslant 0$, 并由此得

$$\begin{aligned}\|\boldsymbol{x}-\boldsymbol{y}\|^2&=\|(p_C(\boldsymbol{x})-p_C(\boldsymbol{y}))+(\boldsymbol{u}-\boldsymbol{v})\|^2\\&=\|p_C(\boldsymbol{x})-p_C(\boldsymbol{y})\|^2+2(p_C(\boldsymbol{x})-p_C(\boldsymbol{y}))\cdot(\boldsymbol{u}-\boldsymbol{v})+\|\boldsymbol{u}-\boldsymbol{v}\|^2\\&\geqslant\|p_C(\boldsymbol{x})-p_C(\boldsymbol{y})\|^2,\end{aligned}$$

即 $\|p_C(\boldsymbol{x})-p_C(\boldsymbol{y})\|\leqslant\|\boldsymbol{x}-\boldsymbol{y}\|$. □

本小节得到度量投影的几个性质:

1) 在凸集 C 上的度量投影的充要条件是 (3.6.1-3) 式成立;

2) 在凸集 C 上的度量投影映射满足 Lipschitz 条件.

3.6.2 基本性质

为了刻画凸集外点到凸集上线段的性质, 下面通过 $\mathbf{R}^n$ 上凸集 C 度量投影映射和其顶点凸锥包来定义 C 的切锥和法锥的概念.

图 3.6-2 表示 C 的锥包的闭包为支撑锥, 图 3.6-3 表示 C 的切锥是支撑锥平移 $-\boldsymbol{a}$.

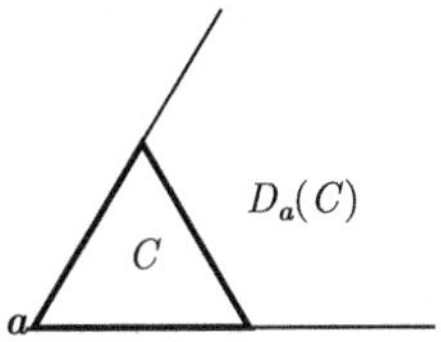

图 3.6-2 C 的支撑锥

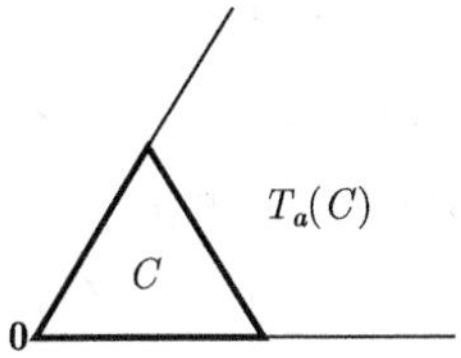

图 3.6-3 C 的切锥

定义 3.6.3 设 $C\subset\mathbf{R}^n$ 是非空凸集, $\boldsymbol{a}\in\mathrm{cl}C$, 则锥

$$D_{\boldsymbol{a}}(C)=\mathrm{cl}C_{\boldsymbol{a}}(C)\quad 和\quad T_{\boldsymbol{a}}(C)=D_{\boldsymbol{a}}(C)-\boldsymbol{a}\tag{3.6.2-1}$$

分别称为 C 在点 $\boldsymbol{a}$ 处的支撑锥和切锥. 集合

$$Q_{\boldsymbol{a}}(C)=\{\boldsymbol{x}\in\mathbf{R}^n|p_C(\boldsymbol{x})=\boldsymbol{a}\}\quad 和\quad N_{\boldsymbol{a}}(C)=Q_{\boldsymbol{a}}(C)-\boldsymbol{a}\tag{3.6.2-2}$$

分别称为 C 在点 $\boldsymbol{a}$ 处的法集和法锥. 如果对任意的 $\boldsymbol{x}\in C,\boldsymbol{a}\in C$ 和 $\boldsymbol{x}^*\in\mathbf{R}^n$, 有

$$\langle\boldsymbol{x}-\boldsymbol{a},\boldsymbol{x}^*\rangle\leqslant 0,$$

使得 $\boldsymbol{x}^*\in N_{\boldsymbol{a}}(C)$, 则称 $\boldsymbol{x}^*$ 是凸集 C 在点 $\boldsymbol{a}$ 的法线.

上述定义表明法锥是 $C-\boldsymbol{a}$ 的极锥. 因此, 法锥可以使用极锥来定义.

例 3.6.2 设超平面 $H(\boldsymbol{b},\alpha)=\{\boldsymbol{x}\in\mathbf{R}^n\,|\langle\boldsymbol{x},\boldsymbol{b}\rangle=\alpha\}$, $H^{\leqslant}(\boldsymbol{b},\alpha)=\{\boldsymbol{x}\in\mathbf{R}^n\,|\langle\boldsymbol{x},\boldsymbol{b}\rangle\leqslant\alpha\}$ 是闭半空间. 则对 $\boldsymbol{x}\in H^{\leqslant}(\boldsymbol{b},\alpha)$ 和 $\boldsymbol{a}\in H(\boldsymbol{b},\alpha)$, 满足 $\langle\boldsymbol{x}-\boldsymbol{a},\boldsymbol{b}\rangle\leqslant 0$, 所以 $\boldsymbol{b}$ 是 $H^{\leqslant}(\boldsymbol{b},\alpha)$ 在 $\boldsymbol{a}$ 的法线. 对于 $\lambda>0$, 形如 $\lambda\boldsymbol{b}$ 的向量的全体构成 $H^{\leqslant}(\boldsymbol{b},\alpha)$ 在点 $\boldsymbol{a}$ 处的法锥.

例 3.6.3 设凸集 $C=\{x_1\in(1,2)\}$, $D_{\boldsymbol{a}}(C)=\mathrm{cl}C_{\boldsymbol{a}}(C)=\{x_1|x_1\in[1,+\infty)\}$ 是 C 在点 $a=1$ 的支撑锥, $T_{\boldsymbol{a}}(C)=\{x_1|x_1\in[0,+\infty)\}$ 是 C 在点 $a=1$ 的切锥, C 在点 $a=1$ 的法锥为

$$N_{\boldsymbol{a}}(C)=\{x_1|x_1\in[0,-\infty)\}.$$

容易验证, 性质 3.6.1 的所有结论对例 3.6.3 的切锥与法锥都满足.

性质 3.6.1 设 $C\subset\mathbf{R}^n$ 是非空凸集和 $\boldsymbol{a}\in\mathrm{cl}C$, 则下面结论成立:

(1) $T_{\boldsymbol{a}}(C)=D_{\boldsymbol{0}}(C-\boldsymbol{a})$;

(2) $N_{\boldsymbol{a}}(C)=Q_{\boldsymbol{0}}(C-\boldsymbol{a})$;

(3) $N_{\boldsymbol{a}}(C)=(T_{\boldsymbol{a}}(C))^-=(D_{\boldsymbol{0}}(C-\boldsymbol{a}))^-=(C_{\boldsymbol{0}}(C-\boldsymbol{a}))^-=(C-\boldsymbol{a})^-$;

(4) $T_{\boldsymbol{a}}(C)=(N_{\boldsymbol{a}}(C))^-$.

证明 (1) 根据定义 3.6.3 和性质 3.2.3(7) 得

$$D_{\boldsymbol{a}}(C)=\mathrm{cl}C_{\boldsymbol{a}}(C)=\mathrm{cl}\{\boldsymbol{a}+\lambda(\boldsymbol{x}-\boldsymbol{a})|\boldsymbol{x}\in C,\lambda\geqslant 0\},$$

等价于 $D_{\boldsymbol{0}}(C-\boldsymbol{a})=\mathrm{cl}C_{\boldsymbol{a}}(C-\boldsymbol{a})=\mathrm{cl}C_{\boldsymbol{a}}(C)-\boldsymbol{a}=D_{\boldsymbol{a}}(C)$.

(2) 根据定义 3.6.3, 结论显然.

(3) 设 $\boldsymbol{c}\in N_{\boldsymbol{a}}(C)=Q_{\boldsymbol{a}}(C)-\boldsymbol{a}$, 由定理 3.6.2 得

$$\langle\boldsymbol{x}-\boldsymbol{a},\boldsymbol{c}-\boldsymbol{a}\rangle\leqslant 0,\quad\forall\boldsymbol{x}\in D_{\boldsymbol{a}}(C),$$

由上式得 $\boldsymbol{c}\in(T_{\boldsymbol{a}}(C))^-$. 显然, 上面的过程反之也成立. 其余结论由 (1), (2) 和性质 3.5.1 得.

(4) 因为 $T_{\boldsymbol{a}}(C)$ 是闭集, 由性质 3.5.1(8) 可得结论. □

性质 3.6.1 揭示了凸集的支撑锥与切锥、法集与法锥、切锥与法锥之间的蕴含关系, 即切锥等价于 $C-\boldsymbol{a}$ 的零点锥包的闭包、切锥等价于法锥的极锥、法锥等价于切锥的极锥. 同时还表明了凸集的法锥与切锥互为极锥的关系.

下面的性质描述了给定多个凸集产生的切锥与法锥之间的运算关系. 多个集合的线性组合的切锥等于各自切锥线性组合的闭包, 多个凸集的线性组合的法锥等于各自法锥的交集.

性质 3.6.2 设 $C_1,\cdots,C_m\subset\mathbf{R}^n$ 是非空凸集, $\boldsymbol{a}_i\in\mathrm{cl}C_i(i=1,2,\cdots,m)$ 和 m 个非零常数 $t_1,\cdots,t_m\in\mathbf{R}$, 则

$$N_{t_1\boldsymbol{a}_1+\cdots+t_m\boldsymbol{a}_m}(t_1C_1+\cdots+t_mC_m)$$

$$= t_1 N_{\boldsymbol{a}_1}(C_1) \cap \cdots \cap t_m N_{\boldsymbol{a}_m}(C_m), \tag{3.6.2-3}$$

$$T_{t_1\boldsymbol{a}_1+\cdots+t_m\boldsymbol{a}_m}(t_1C_1+\cdots+t_mC_m)$$

$$= \mathrm{cl}(t_1T_{\boldsymbol{a}_1}(C_1)+\cdots+t_mT_{\boldsymbol{a}_m}(C_m)). \tag{3.6.2-4}$$

证明 只需证明 $m=2$ 情形. 由性质 3.5.2 和性质 3.6.1 可得

$$\begin{aligned} N_{t_1\boldsymbol{a}_1+t_2\boldsymbol{a}_2}(t_1C_1+t_2C_2) &= (t_1C_1+t_2C_2-t_1\boldsymbol{a}_1-t_2\boldsymbol{a}_2)^- \\ &= (t_1C_1-t_1\boldsymbol{a}_1)^- \cap (t_2C_2-t_2\boldsymbol{a}_2)^- \\ &= t_1N_{\boldsymbol{a}_1}(C_1)\cap t_2N_{\boldsymbol{a}_2}(C_2), \end{aligned}$$

以及 $T_{t_1\boldsymbol{a}_1+t_2\boldsymbol{a}_2}(t_1C_1+t_2C_2) = (N_{t_1\boldsymbol{a}_1+t_2\boldsymbol{a}_2}(t_1C_1+t_2C_2))^-$. 再由上式和性质 3.5.2 得

$$\begin{aligned} T_{t_1\boldsymbol{a}_1+t_2\boldsymbol{a}_2}(t_1C_1+t_2C_2) &= (t_1N_{\boldsymbol{a}_1}(C_1)\cap t_2N_{\boldsymbol{a}_2}(C_2))^- \\ &= \mathrm{cl}\left(t_1N_{\boldsymbol{a}_1}(C_1)^- + t_2N_{\boldsymbol{a}_2}(C_2)^-\right) \\ &= \mathrm{cl}\left(t_1T_{\boldsymbol{a}_1}(C_1)+t_2T_{\boldsymbol{a}_2}(C_2)\right). \end{aligned}$$

类似可证: 一般情形下, 上面证明过程中的 (3.6.2-3) 式和 (3.6.2-4) 式也成立. □

进一步, 有下面有限个集合的交的法锥与切锥性质.

性质 3.6.3 设 $C_1,\cdots,C_m\subset\mathbf{R}^n$ 是非空凸集, $\boldsymbol{a}\in\bigcap_{i=1}^m \mathrm{cl}C_i$, 则

$$N_{\boldsymbol{a}}(C_1)+\cdots+N_{\boldsymbol{a}}(C_m)\subset N_{\boldsymbol{a}}(C_1\cap\cdots\cap C_m), \tag{3.6.2-5}$$

$$\mathrm{cl}C_{\boldsymbol{a}}(C_1\cap\cdots\cap C_m)\subset \mathrm{cl}C_{\boldsymbol{a}}(C_1)\cap\cdots\cap\mathrm{cl}C_{\boldsymbol{a}}(C_m), \tag{3.6.2-6}$$

$$T_{\boldsymbol{a}}(C_1\cap\cdots\cap C_m)\subset T_{\boldsymbol{a}}(C_1)\cap\cdots\cap T_{\boldsymbol{a}}(C_m), \tag{3.6.2-7}$$

其中上述式子中如果有一个为等式成立, 另两个也为等式.

证明 设任意的 $\boldsymbol{x}_i\in N_{\boldsymbol{a}}(C_i)$, $1\leqslant i\leqslant m$, 由性质 3.6.1 得 $\boldsymbol{x}_i\in(C_i-\boldsymbol{a})^-$, 有

$$\langle \boldsymbol{x}-\boldsymbol{a},\boldsymbol{x}_i\rangle\leqslant 0,\quad \forall\boldsymbol{x}\in C_i,\quad i=1,2,\cdots,m.$$

因此, 对于 $\boldsymbol{x}\in C_1\cap\cdots\cap C_m$, 有

$$\langle \boldsymbol{x}-\boldsymbol{a},\boldsymbol{x}_1+\cdots+\boldsymbol{x}_m\rangle = \langle \boldsymbol{x}-\boldsymbol{a},\boldsymbol{x}_1\rangle+\cdots+\langle \boldsymbol{x}-\boldsymbol{a},\boldsymbol{x}_m\rangle\leqslant 0.$$

由上式得 $\boldsymbol{x}_1+\cdots+\boldsymbol{x}_m\in N_{\boldsymbol{a}}(C_1\cap\cdots\cap C_m)$. 因为 $C_{\boldsymbol{a}}(C_i)=\mathrm{cone}_{\boldsymbol{a}}(C_i)$, 由性质 3.2.4 得

$$\mathrm{cl}C_{\boldsymbol{a}}(C_1\cap\cdots\cap C_m)=\mathrm{cl}(C_{\boldsymbol{a}}(C_1)\cap\cdots\cap C_{\boldsymbol{a}}(C_m))\subset\mathrm{cl}C_{\boldsymbol{a}}(C_1)\cap\cdots\cap\mathrm{cl}C_{\boldsymbol{a}}(C_m)$$

和

$$\begin{aligned}T_{\boldsymbol{a}}(C_1\cap\cdots\cap C_m)&=\mathrm{cl}C_{\boldsymbol{a}}(C_1\cap\cdots\cap C_m)-\boldsymbol{a}\\&\subset\mathrm{cl}C_{\boldsymbol{a}}(C_1)\cap\cdots\cap\mathrm{cl}C_{\boldsymbol{a}}(C_m)-\boldsymbol{a}\\&=(\mathrm{cl}C_{\boldsymbol{a}}(C_1)-\boldsymbol{a})\cap\cdots\cap(\mathrm{cl}C_{\boldsymbol{a}}(C_m)-\boldsymbol{a})\\&=T_{\boldsymbol{a}}(C_1)\cap\cdots\cap T_{\boldsymbol{a}}(C_m).\end{aligned}$$

假设 (3.6.2-5) 式是等式情形, 再由性质 3.6.2 得

$$\begin{aligned}\mathrm{cl}C_{\boldsymbol{a}}(C_1\cap\cdots\cap C_m)&=\boldsymbol{a}+(N_{\boldsymbol{a}}(C_1\cap\cdots\cap C_m))^-,\\\mathrm{cl}C_{\boldsymbol{a}}(C_1)\cap\cdots\cap\mathrm{cl}C_{\boldsymbol{a}}(C_m)&=\boldsymbol{a}+(N_{\boldsymbol{a}}(C_1))^-\cap\cdots\cap(N_{\boldsymbol{a}}(C_m))^-\\&=\boldsymbol{a}+(N_{\boldsymbol{a}}(C_1)+\cdots+N_{\boldsymbol{a}}(C_m))^-.\end{aligned}$$

综上所述, (3.6.2-6) 式和 (3.6.2-7) 式也为等式. □

例 3.6.4　设凸集 $C_1=\{x_1\in(0,1)\}$ 和 $C_2=\{x_1\in(-1,0)\}$, C_1 和 C_2 在点 $a=0$ 的支撑锥和切锥为

$$\begin{aligned}D_0(C_1)=T_0(C_1)=\mathrm{cl}C_0(C_1)=\{x_1|x_1\in[0,+\infty)\},\\D_0(C_2)=T_0(C_2)=\mathrm{cl}C_0(C_2)=\{x_1|x_1\in(-\infty,0]\}.\end{aligned}$$

C_1 和 C_2 在点 $a=0$ 的法锥为

$$N_0(C_1)=\{x_1|x_1\in(-\infty,0]\}\quad 和\quad N_0(C_2)=\{x_1|x_1\in[0,+\infty)\}.$$

有 $C_1+C_2=\{x_1\in(-1,1)\}$, C_1+C_2 在点 $a=0$ 的支撑锥和切锥为

$$D_0(C_1+C_2)=T_0(C_1+C_2)=\mathrm{cl}C_{\boldsymbol{0}}(C_1+C_2)=\{x_1|x_1\in\mathbf{R}\}.$$

C_1+C_2 在点 $a=0$ 的法锥为

$$N_0(C_1+C_2)=\{0\}.$$

因此, 可得 $N_0(C_1+C_2)=N_0(C_1)\cap N_0(C_2)$ 和 $T_0(C_1+C_2)=T_0(C_1)+T_0(C_2)$.

另一方面, 我们有

$$C_1 \cap C_2 = \varnothing, \quad C_{\boldsymbol{a}}\left(C_1 \cap C_2\right) = \{0\},$$

$$N_{\boldsymbol{a}}\left(C_1 \cap C_2\right) = \mathbf{R}, \quad N_{\boldsymbol{a}}(C_1) + N_{\boldsymbol{a}}(C_2) = \mathbf{R}.$$

但我们又有 $\mathrm{ri}C_1 \cap \mathrm{ri}C_2 = \varnothing$, $\mathrm{cl}C_{\boldsymbol{a}}(C_1) \cap \mathrm{cl}C_{\boldsymbol{a}}(C_2) = \{0\}$. 这说明性质 3.6.2、性质 3.6.3 及性质 3.6.4 的结论对该例成立, 且性质 3.6.3 和性质 3.6.4 的条件必不可少.

性质 3.6.4 设 $C_1, \cdots, C_m \subset \mathbf{R}^n$ 是非空凸集, $\boldsymbol{a} \in \bigcap_{i=1}^m \mathrm{cl}C_i$. 如果 $\bigcap_{i=1}^m \mathrm{ri}C_i \neq \varnothing$, 则

$$\mathrm{cl}C_{\boldsymbol{a}}\left(\bigcap_{i=1}^m C_i\right) = \bigcap_{i=1}^m \mathrm{cl}C_{\boldsymbol{a}}\left(C_i\right). \tag{3.6.2-8}$$

证明 因为由性质 2.3.4 得 $\mathrm{ri}C_i = \mathrm{ri}(\mathrm{cl}C_i)$, $1 \leqslant i \leqslant m$, 再由假设 $\bigcap_{i=1}^m \mathrm{ri}C_i \neq \varnothing$ 得

$$\bigcap_{i=1}^m \mathrm{ri}\left(\mathrm{cl}C_i\right) \neq \varnothing.$$

则由性质 3.2.8 有 $\bigcap_{i=1}^m \mathrm{ri}C_{\boldsymbol{a}}\left(\mathrm{cl}C_i\right) \neq \varnothing$, 再由性质 2.3.5 得

$$\mathrm{cl}\left(\bigcap_{i=1}^m C_i\right) = \bigcap_{i=1}^m \left(\mathrm{cl}C_i\right), \quad \mathrm{cl}\left(\bigcap_{i=1}^m C_{\boldsymbol{a}}\left(\mathrm{cl}C_i\right)\right) = \bigcap_{i=1}^m \mathrm{cl}C_{\boldsymbol{a}}\left(\mathrm{cl}C_i\right). \tag{3.6.2-9}$$

因为 $\boldsymbol{a} \in \bigcap_{i=1}^m \mathrm{cl}C_i$, 得 $\mathrm{cl}\left(\bigcap_{i=1}^m C_{\boldsymbol{a}}\left(\mathrm{cl}C_i\right)\right) = \bigcap_{i=1}^m \mathrm{cl}C_{\boldsymbol{a}}\left(C_i\right)$. 最后由性质 3.2.6 和 (3.6.2-9) 式得

$$\begin{aligned}
\mathrm{cl}C_{\boldsymbol{a}}\left(\bigcap_{i=1}^m C_i\right) &= \mathrm{cl}C_{\boldsymbol{a}}\left(\mathrm{cl}\left(\bigcap_{i=1}^m C_i\right)\right) = \mathrm{cl}C_{\boldsymbol{a}}\left(\bigcap_{i=1}^m \mathrm{cl}C_i\right) \\
&= \mathrm{cl}\left(\bigcap_{i=1}^m C_{\boldsymbol{a}}(\mathrm{cl}C_i)\right) = \bigcap_{i=1}^m \mathrm{cl}C_{\boldsymbol{a}}(\mathrm{cl}C_i) = \bigcap_{i=1}^m \mathrm{cl}C_{\boldsymbol{a}}(C_i).
\end{aligned}$$ □

性质 3.6.3 和性质 3.6.4 给出了切锥与法锥的和与交的运算关系. 定义 3.6.2 给出的是集合中某一点的法锥概念, 下面给出在整个集合上的法锥概念.

定义 3.6.4 设凸集 $C \subset \mathbf{R}^n$, 则称

$$\mathrm{nor}C = \bigcup_{\boldsymbol{a} \in \mathrm{cl}C} N_{\boldsymbol{a}}(C) \tag{3.6.2-10}$$

是集合 C 的法锥, 每一个向量 $\boldsymbol{e} \in \mathrm{nor}C$ 称为 C 的法向量. 定义 $\mathrm{nor}\varnothing = \mathbf{R}^n$.

例 3.6.5　设凸集 $C = \{x_1 \in (1,2)\}$, 可得 C 在点 $a \in \mathrm{cl}C = [1,2]$ 的支撑锥为

$$D_a(C) = \mathrm{cl}C_a(C) = \{x_1 | x_1 \in [a, +\infty)\};$$

C 在点 $a = 1$ 的切锥为

$$T_a(C) = \{x_1 | x_1 \in [0, +\infty)\};$$

C 在点 $a = 1$ 的法锥为 $N_a(C) = \{x_1 | x_1 \in (-\infty, 0]\}$; 假设 $a = 2$, 则 C 在点 $a = 2$ 的法锥为

$$(C-2)^- = N_a(C) = \{x_1 | x_1 \in [0, +\infty)\};$$

C 的法锥 $\mathrm{nor}C = \mathbf{R}$.

下面证明集合的法锥的一些基本性质, 这些性质揭示了法锥与集合的有界性、正交性以及与平面之间的关系. 如集合是整个空间等价于法锥仅含原点; 法锥为整个空间等价于集合有界; 法锥是集合正交集的正常锥; 法锥等于集合正交集的正常锥当且仅当该集合是一个平面, 否则, 若该集合不是一个平面则集合上的法锥是集合所有相对边界上点的法锥并.

性质 3.6.5　设凸集 $C \subset \mathbf{R}^n$, 有下面结论成立:

(1) $\mathrm{nor}C = \{\mathbf{0}\}$ 当且仅当 $C = \mathbf{R}^n$;

(2) $\mathrm{nor}C = \mathbf{R}^n$ 当且仅当 C 是有界的;

(3) $\mathrm{nor}C$ 是以集合 $\mathrm{ort}C$ 为顶点集的正常锥;

(4) $\mathrm{nor}C = \mathrm{ort}C$ 当且仅当 C 是一个平面;

(5) 如果 C 不是平面, 则 $\mathrm{nor}C = \bigcup_{\boldsymbol{a} \in \mathrm{rb}C} N_{\boldsymbol{a}}(C)$;

(6) $\mathrm{nor}C = \mathrm{nor}(\mathrm{cl}C) = \mathrm{nor}(\mathrm{ri}C)$.

证明　(1) 若 $C = \mathbf{R}^n$, 则 $\mathrm{nor}C = \{\mathbf{0}\}$ 显然成立. 反之, 如果 $C \neq \mathbf{R}^n$, 则 $\mathrm{cl}C \neq \mathbf{R}^n$. 设点 $\boldsymbol{a} \in \mathbf{R}^n \setminus \mathrm{cl}C$. 由定理 3.6.1 知 $\mathrm{cl}C$ 存在离点 $\boldsymbol{a} \in \mathbf{R}^n \setminus \mathrm{cl}C$ 最近的点 $\boldsymbol{c}$, 由定理 3.6.2 有 $\boldsymbol{a} \in Q_{\boldsymbol{c}}(C)$ 和

$$\mathbf{0} \neq \boldsymbol{a} - \boldsymbol{c} \in N_{\boldsymbol{c}}(C) \subset \mathrm{nor}C.$$

(2) 先假设集合 C 是有界的. 设 $\boldsymbol{a} \in \mathbf{R}^n$ 是非零点, 那么优化 $\max\{\langle \boldsymbol{x}, \boldsymbol{a} \rangle | \boldsymbol{x} \in \mathrm{cl}C\}$ 存在一个最优解 $\boldsymbol{c}$ (见定理 1.2.2) , 则得

$$\mathrm{cl}C \subset D = \{\boldsymbol{x} \in \mathbf{R}^n | \langle \boldsymbol{x}, \boldsymbol{a} \rangle \leqslant \langle \boldsymbol{c}, \boldsymbol{a} \rangle\}.$$

由定理 3.6.2 知, $\boldsymbol{c}$ 是 $\mathrm{cl}C$ 到点 $\boldsymbol{a} + \boldsymbol{c}$ 最近的点, 即有 $\boldsymbol{a} + \boldsymbol{c} \in Q_{\boldsymbol{c}}(C)$. 这表明 $\boldsymbol{a} \in N_{\boldsymbol{c}}(C) \subset \mathrm{nor}C$, 即 $C = \mathbf{R}^n$.

反过来, 假设 C 是无界的, 那么 C 包含闭半直线

$$l[\boldsymbol{a},\boldsymbol{b}\rangle=\{(1-t)\boldsymbol{a}+t\boldsymbol{b}|t\geqslant 0\}=\boldsymbol{a}+\{t\boldsymbol{c}\ :t\geqslant 0\}\subset C,$$

其中 $\boldsymbol{a},\boldsymbol{b}\in C$, $\boldsymbol{c}=\boldsymbol{b}-\boldsymbol{a}\neq\boldsymbol{0}$. 因为 $C=\mathbf{R}^n$, 所以结论显然成立, 否则 $\mathrm{nor}C=\{\boldsymbol{0}\}\neq\mathbf{R}^n$. 所以假设 $C\neq\mathbf{R}^n$, 因此有 $\mathrm{nor}C\neq\{\boldsymbol{0}\}$. 设非零点 $\boldsymbol{e}\in\mathrm{nor}C$, 那么存在点 $\boldsymbol{x}\in\mathrm{cl}C$ 使得 $\boldsymbol{x}$ 是 $\mathrm{cl}C$ 到 $\boldsymbol{e}+\boldsymbol{x}$ 的最近点. 因此, 由定理 3.6.2 得

$$\mathrm{cl}C\subset D=\{\boldsymbol{y}\in\mathbf{R}^n:\langle\boldsymbol{y}-\boldsymbol{x},\boldsymbol{e}\rangle\leqslant 0\}.$$

所以, 由 $l[\boldsymbol{a},\boldsymbol{b}\rangle\subset D$ 和上式得

$$\langle\boldsymbol{a}+t\boldsymbol{c}-\boldsymbol{x},\boldsymbol{e}\rangle\leqslant 0,\quad \forall t\geqslant 0.$$

上式表明 $\boldsymbol{e}\in\mathrm{nor}C$ 都被包含在子空间 $\{\boldsymbol{x}\in\mathbf{R}^n:\langle\boldsymbol{x},\boldsymbol{e}\rangle\leqslant 0\}$ 中, 即 $\mathrm{nor}C\neq\mathbf{R}^n$.

(3) 设 $\boldsymbol{a}\in\mathrm{cl}C$, 根据法锥定义和性质 3.6.1, 得 $N_{\boldsymbol{a}}(C)$ 是关于顶点集合 $\boldsymbol{a}+\mathrm{ort}C$ 的闭凸锥. 因此, $\mathrm{nor}C$ 是以集合 $\mathrm{ort}C$ 为顶点的正常锥.

(4) 由性质 3.1.9 和 (3) 知 $\mathrm{nor}C=\mathrm{ort}C$ 当且仅当 $N_{\boldsymbol{a}}(C)=\boldsymbol{a}+\mathrm{ort}C\,(\boldsymbol{a}\in\mathrm{cl}C)$ 成立, 且 $\boldsymbol{a}\in\mathrm{ri}C$. 因此, 得 $\mathrm{nor}C=\mathrm{ort}C$ 当且仅当 $\mathrm{rb}C=\varnothing$, 即当且仅当 C 是一个平面.

(5) 由性质 3.1.10 知, C 不是一个平面, 有 $\mathrm{rb}C\neq\varnothing$. 再根据法锥定义, 对于 $\boldsymbol{a}\in\mathrm{cl}C$ 有 $N_{\boldsymbol{a}}(C)\subset\mathrm{ort}C$. 而 $\boldsymbol{a}\in\mathrm{ri}C$, 有 $N_{\boldsymbol{a}}(C)=\mathrm{ort}C$, 所以结论成立.

(6) 因为 $\mathrm{cl}(\mathrm{cl}C)=\mathrm{cl}C=\mathrm{cl}(\mathrm{ri}C)$ 成立, 所以结论成立. □

本小节得到了凸集 C 中的点的切锥与法锥和 C 的法锥的一些重要性质.

1) 凸集 C 中的点的切锥与法锥的关系: 切锥等价于 $C-\boldsymbol{a}$ 的零点锥包的闭包; 切锥等价于法锥的极锥; 法锥等价于切锥的极锥.

2) 多个凸集的点的切锥与法锥的线性运算性质: 多个凸集的线性组合的切锥等于各自切锥线性组合的闭包; 多个凸集的线性组合的法锥等于各自法锥的交集.

3) 多个凸集的点的切锥与法锥的交运算性质: 当点属于多个凸集闭包的交集时, 所有在该点的法锥交集属于所有集合交集在该点的法锥; 所有凸集的交集在该点上的切锥属于各自切锥的交集. 反之, 以上性质不一定成立.

4) 凸集 C 上的法锥性质: 凸集 C 是整个空间等价于法锥仅含原点; 法锥为整个空间等价于集合 C 有界; 法锥是凸集正交集的正常锥; 法锥等于凸集正交集的正常锥当且仅当 C 是一个平面. 否则, C 不是一个平面则凸集上的法锥是集合所有相对边界上点的法锥并.

3.7 障 碍 锥

本节讨论障碍锥的性质. 障碍锥可以看成法锥的一种推广, 根据前面的讨论, 法锥可以用极锥描述. 因此, 障碍锥的许多性质也是由极锥推导而来的.

3.7.1 基本性质

定义 3.7.1 设非空集合 $C \subset \mathbf{R}^n$. 对于任意的 $\boldsymbol{x} \in C$, 如果存在 $\delta = \delta(\boldsymbol{b})$ 使得 $\langle \boldsymbol{x}, \boldsymbol{b} \rangle \leqslant \delta$ 成立, 则所有的 $\boldsymbol{b}$ 构成的集合称为集合 C 的障碍锥, 记为

$$\mathrm{bar}C = \{\boldsymbol{b} \in \mathbf{R}^n | \langle \boldsymbol{x}, \boldsymbol{b} \rangle \leqslant \delta, \forall \boldsymbol{x} \in C\}, \tag{3.7.1-1}$$

空集的障碍锥定义为: $\mathrm{bar}\varnothing = \mathbf{R}^n$.

根据定义 3.7.1, 有 $\mathbf{0} \in \mathrm{bar}C$ 和 $\mathrm{bar}C = \mathrm{bar}(\mathrm{co}C)$ 成立.

例 3.7.1 (1) 对于给定的 $\boldsymbol{a} \in \mathbf{R}^n$ 和常数 $\alpha \in \mathbf{R}$, 设超平面 $C = \{\boldsymbol{x} \in \mathbf{R}^n | \langle \boldsymbol{x}, \boldsymbol{a} \rangle = \alpha\}$, 那么 C 的障碍锥是

$$\mathrm{bar}C = \{\boldsymbol{b} \in \mathbf{R}^n | \langle \boldsymbol{x}, \boldsymbol{b} \rangle = 0, \forall \boldsymbol{x} \in C\} = \mathrm{ort}C.$$

(2) 对于给定的非零 $\boldsymbol{a} \in \mathbf{R}^n$ 和常数 $\alpha \in \mathbf{R}$, 设集合 $C = \{\boldsymbol{x} \in \mathbf{R}^n | \langle \boldsymbol{x}, \boldsymbol{a} \rangle \leqslant \alpha\}$, 那么 C 的障碍锥是 $\mathrm{bar}C = \{\lambda \boldsymbol{a} \in \mathbf{R}^n | \lambda \geqslant 0\}$.

通过下面的性质可知障碍锥是法锥的推广, 凸集的法锥属于它的障碍锥, 整个空间 $\mathbf{R}^n$ 的障碍锥仅含零点一个元素. 如果一个凸集的障碍锥是整个空间 $\mathbf{R}^n$, 那么这个凸集是有界的. 凸集是平面的等价于它的障碍锥是该凸集的正交集.

性质 3.7.1 设非空凸集 $C \subset \mathbf{R}^n$, 则下面结论成立:

(1) $\mathrm{nor}C \subset \mathrm{bar}C$;

(2) $\mathrm{bar}C = \{\mathbf{0}\}$ 当且仅当 $C = \mathbf{R}^n$;

(3) $\mathrm{bar}C = \mathbf{R}^n$ 当且仅当 C 是有界的;

(4) $\mathrm{bar}C$ 是包含正常点集 $\mathrm{ort}C$ 的凸锥;

(5) $\mathrm{bar}C = \mathrm{ort}C$ 当且仅当 C 是平面;

(6) $\mathrm{bar}C = \mathrm{bar}(\mathrm{cl}C) = \mathrm{bar}(\mathrm{ri}C)$.

证明 (1) 设 $\boldsymbol{b} \in \mathrm{nor}C$, 当 $\boldsymbol{b} = \mathbf{0}$ 时, 显然有 $\mathbf{0} \in \mathrm{bar}C$. 假设 $\boldsymbol{b} \neq \mathbf{0}$, 存在 $\boldsymbol{a} \in \mathrm{cl}C$ 使得

$$\boldsymbol{b} \in N_{\boldsymbol{a}}(C) = Q_{\boldsymbol{a}}(C) - \boldsymbol{a}.$$

因此, 有 $p_C(\boldsymbol{b} + \boldsymbol{a}) = \boldsymbol{a}$. 根据定理 3.6.2 得

$$C \subset D = \{\boldsymbol{x} \in \mathbf{R}^n | \langle \boldsymbol{x}, \boldsymbol{b} \rangle \leqslant \langle \boldsymbol{a}, \boldsymbol{b} \rangle\},$$

再由定义 3.7.1 得 $\boldsymbol{b}\in \mathrm{bar}C$.

(2) 若 $C=\mathbf{R}^n$, 则按性质 3.6.4 和 (1) 得 $\mathrm{bar}C=\{\mathbf{0}\}$. 如果 $C\neq\mathbf{R}^n$, 同样有 $\mathbf{0}\neq \mathrm{nor}C\subset \mathrm{bar}C$.

(3) 若 C 是有界的, 由性质 3.6.4 和 (1) 得 $\mathrm{nor}C=\mathbf{R}^n=\mathrm{bar}C$. 反之, 若 C 是无界的, 那么 C 包含闭半直线:

$$l\,[\boldsymbol{a},\boldsymbol{b})=\{(1-t)\boldsymbol{a}+t\boldsymbol{b}|t\geqslant 0\}=\boldsymbol{b}+\{t\boldsymbol{c}|t\geqslant 0\}\subset C,$$

其中 $\boldsymbol{a},\boldsymbol{b}\in C$, $\boldsymbol{c}=\boldsymbol{a}-\boldsymbol{b}\neq\mathbf{0}$. 按不等式 $t>(\delta-\langle\boldsymbol{a},\boldsymbol{c}\rangle)/\langle\boldsymbol{c},\boldsymbol{c}\rangle$ 选择 δ, 得 $\langle\boldsymbol{a}+t\boldsymbol{c},\boldsymbol{c}\rangle>\delta$, 于是有 $\boldsymbol{a}+t\boldsymbol{c}\in l\,[\boldsymbol{a},\boldsymbol{b})\subset C$. 根据定义 3.7.1 得 $\boldsymbol{c}\in\mathrm{bar}C$ 和 $\mathrm{bar}C\neq\mathbf{R}^n$.

(4) 设 $\boldsymbol{b}\in\mathrm{bar}C$, 则对每一个 $\boldsymbol{x}\in C$ 和 δ 有 $\langle\boldsymbol{x},\boldsymbol{b}\rangle\leqslant\delta$ 成立. 若 $\delta\geqslant 0$, 则由 $\langle\boldsymbol{x},\delta\boldsymbol{b}\rangle\leqslant\delta^2$ 有 $\delta\boldsymbol{b}\in\mathrm{bar}C$. 因此, $\mathrm{bar}C$ 是关于顶点为零点的锥. 设 $\boldsymbol{b}_1,\boldsymbol{b}_2\in\mathrm{bar}C$, $t\in[0,1]$ 和 δ_i 使得 $\langle\boldsymbol{x},\boldsymbol{b}_1\rangle\leqslant\delta_1,\langle\boldsymbol{x},\boldsymbol{b}_2\rangle\leqslant\delta_2$, 则当 $\boldsymbol{x}\in C$ 时有

$$\langle\boldsymbol{x},(1-t)\boldsymbol{b}_1+t\boldsymbol{b}_2\rangle=(1-t)\langle\boldsymbol{x},\boldsymbol{b}_1\rangle+t\langle\boldsymbol{x},\boldsymbol{b}_2\rangle\leqslant(1-t)\delta_1+t\delta_2.$$

由上式得 $(1-t)\boldsymbol{b}_1+t\boldsymbol{b}_2\in\mathrm{bar}C$, 表明 $\mathrm{bar}C$ 是凸锥.

下面证明 $\mathrm{ort}C\subset\mathrm{bar}C$. 如果 $\mathrm{ort}C=\{\mathbf{0}\}$, 结论成立. 假设 $\mathrm{ort}C\neq\{\mathbf{0}\}$, 选择 $\boldsymbol{b}\in\mathrm{ort}C,\boldsymbol{c}\in C$, 则有

$$\mathrm{dir}C\subset S=\{\boldsymbol{x}\in\mathbf{R}^n|\langle\boldsymbol{x},\boldsymbol{b}\rangle=0\}.$$

由性质 2.1.10(1) 得 $\mathrm{aff}C\subset H=\boldsymbol{c}+S$. 再设 $\delta=\langle\boldsymbol{b},\boldsymbol{c}\rangle$, 得

$$C\subset\mathrm{aff}C\subset\boldsymbol{c}+S=\{\boldsymbol{x}\in\mathbf{R}^n|\langle\boldsymbol{x},\boldsymbol{b}\rangle=\delta\},$$

说明 $C\subset\{\boldsymbol{x}\in\mathbf{R}^n|\langle\boldsymbol{x},\boldsymbol{b}\rangle\leqslant\delta\}$ 成立. 据此, 有 $\boldsymbol{b}\in\mathrm{bar}C$, 所以有 $\mathrm{ort}C\subset\mathrm{bar}C$. 因为 $\mathrm{ort}C$ 是子空间, 所以 $\mathrm{ort}C\subset\mathrm{ap}(\mathrm{bar}C)$.

(5) 如果 C 是一个平面, 由性质 3.6.4 得 $\mathrm{bar}C=\mathrm{nor}C=\mathrm{ort}C$. 反之, 如果 $\mathrm{bar}C=\mathrm{ort}C$, 由 $\mathrm{ort}C\subset\mathrm{nor}C\subset\mathrm{bar}C=\mathrm{ort}C$, 得 $\mathrm{ort}C=\mathrm{nor}C$, 再由性质 3.6.4 知 C 是一个平面.

(6) 由障碍锥定义知: $\boldsymbol{b}\in\mathrm{bar}C$ 当且仅当 $C\subset D=\{\boldsymbol{x}\in\mathbf{R}^n|\langle\boldsymbol{x},\boldsymbol{b}\rangle\leqslant\delta\}$. 由于 D 是闭集, 且 $\mathrm{cl}C=\mathrm{cl}(\mathrm{ri}C)$, 那么当且仅当对于 $\mathrm{cl}C,\mathrm{ri}C\subset D$, 有 $\boldsymbol{b}\in\mathrm{bar}C$ 当且仅当 $\boldsymbol{b}\in\mathrm{bar}(\mathrm{cl}C)$ 和 $\boldsymbol{b}\in\mathrm{bar}(\mathrm{ri}C)$ 成立. □

注意, 集合的障碍锥不一定是闭集, 且障碍锥不一定等于它的法锥, 见下面的例子.

例 3.7.2 设集合 $C=\{(x_1,x_2)^{\mathrm{T}}|x_2\geqslant x_1^2\}$, 容易得到

$$\mathrm{nor}C=\mathrm{bar}C=\{\mathbf{0}\}\cup\{(0,x_2)^{\mathrm{T}}|x_2<0\}.$$

下面证明当凸集给定时, 集合线性平移后的障碍锥和法锥不变.

定理 3.7.1　设 $C \subset \mathbf{R}^n$ 是凸集, 点 $\boldsymbol{a} \in \mathbf{R}^n$ 和 ρ 是常数, 则

$$\mathrm{bar}(\boldsymbol{a}+\rho C)=\rho\mathrm{bar}C, \quad \mathrm{nor}(\boldsymbol{a}+\rho C)=\rho\mathrm{nor}C. \tag{3.7.1-2}$$

证明　根据 ρ 的取值分三种情形证明.

(1) 当 $\rho=0$ 时, 结论显然成立, 此时 $\mathrm{bar}(\boldsymbol{a})=\{\mathbf{0}\}$, $\mathrm{nor}(\boldsymbol{a})=\{\mathbf{0}\}$.

(2) 当 $\rho>0$ 时, 设非零 $\boldsymbol{b} \in \mathrm{bar}(\boldsymbol{a}+\rho C)$. 由障碍锥定义知, 存在 δ 使得

$$\langle \boldsymbol{a}+\rho\boldsymbol{x}, \boldsymbol{b}\rangle \leqslant \delta, \quad \forall \boldsymbol{x} \in C.$$

由上式可得, 对任意的 $\boldsymbol{x} \in C$ 有 $\langle \boldsymbol{x}, \boldsymbol{b}\rangle \leqslant (\delta-\langle \boldsymbol{a}, \boldsymbol{b}\rangle)/\rho$. 因此, $\boldsymbol{b} \in \mathrm{bar}C$.

(3) 当 $\rho<0$ 时, 同样有 $\langle \boldsymbol{x}, -\boldsymbol{b}\rangle \leqslant (\delta-\langle \boldsymbol{a}, \boldsymbol{b}\rangle)/(-\rho)$ $(\boldsymbol{x} \in C)$, 有 $-\boldsymbol{b} \in \mathrm{bar}C$. 因此得

$$\mathrm{bar}(\boldsymbol{a}+\rho C) \subset \rho\mathrm{bar}C.$$

反过来, 容易证明 $\mathrm{bar}(\boldsymbol{a}+\rho C) \supset \rho\mathrm{bar}C$.

接下来, 设 $\boldsymbol{x} \in \mathrm{cl}(\boldsymbol{a}+\rho C)=\boldsymbol{a}+\rho\mathrm{cl}C$, 则存在 $\boldsymbol{y} \in \mathrm{cl}C$ 使得 $\boldsymbol{x}=\boldsymbol{a}+\rho\boldsymbol{y}$, 由性质 3.6.5 得

$$\begin{aligned}
\mathrm{nor}(\boldsymbol{a}+\rho C) &= \bigcap_{\boldsymbol{x}\in\mathrm{cl}(\boldsymbol{a}+\rho C)} N_{\boldsymbol{x}}(\boldsymbol{a}+\rho C)=\bigcap_{\boldsymbol{x}\in\mathrm{cl}(\boldsymbol{a}+\rho C)} (\boldsymbol{a}+\rho C-\boldsymbol{x})^- \\
&= \bigcap_{\rho\boldsymbol{y}\in\rho\mathrm{cl}C} (\rho(C-\boldsymbol{y}))^- = \bigcap_{\boldsymbol{y}\in\mathrm{cl}C} (\rho(C-\boldsymbol{y}))^- \\
&= \rho\left(\bigcap_{\boldsymbol{y}\in\mathrm{cl}C} N_{\boldsymbol{x}}(C)\right)=\rho\mathrm{nor}C.
\end{aligned}$$

□

进一步, 下面的定理说明了多个集合障碍锥交集的关系: 多个凸集线性组合的障碍锥等于各自的障碍锥乘以线性系数后的交集, 多个凸集线性组合的法锥包含各自的法锥乘以线性系数后的交集.

定理 3.7.2　设非空凸集 $C_1, C_2, \cdots, C_m \subset \mathbf{R}^n$ 和非零常数 $\rho_1, \rho_2, \cdots, \rho_m$, 则

$$\bigcap_{i=1}^m \rho_i\mathrm{bar}C_i=\mathrm{bar}\left(\sum_{i=1}^m \rho_i C_i\right), \tag{3.7.1-3}$$

$$\bigcap_{i=1}^m \rho_i\mathrm{nor}C_i \subset \mathrm{nor}\left(\sum_{i=1}^m \rho_i C_i\right). \tag{3.7.1-4}$$

证明 首先证明 $\bigcap_{i=1}^m \text{bar}\,(\rho_i C_i) \subset \text{bar}\left(\sum\limits_{i=1}^m \rho_i C_i\right)$. 设 $\boldsymbol{b} \in \bigcap_{i=1}^m \text{bar}\,(\rho_i C_i)$, 若 $\boldsymbol{b}=\boldsymbol{0}$, 则结论显然. 假设 $\boldsymbol{b} \neq \boldsymbol{0}$, 存在 $\delta_i\,(i=1,2,\cdots,m)$ 使得

$$\rho_i\langle \boldsymbol{x}, \boldsymbol{b}\rangle \leqslant \delta_i, \quad \forall \boldsymbol{x} \in \rho_i C_i, \quad i=1,2,\cdots,m.$$

不妨设 $\boldsymbol{y}=\rho_1\boldsymbol{y}_1+\rho_2\boldsymbol{y}_2+\cdots+\rho_m\boldsymbol{y}_m$, 其中 $\boldsymbol{y}_i \in C_i(1 \leqslant i \leqslant m)$, 根据上式得

$$\langle \boldsymbol{y}, \boldsymbol{b}\rangle = \langle \rho_1\boldsymbol{y}_1+\rho_2\boldsymbol{y}_2+\cdots+\rho_m\boldsymbol{y}_m, \boldsymbol{b}\rangle \leqslant \delta_1+\cdots+\delta_m,$$

由此可得 $\boldsymbol{b} \in \text{bar}\left(\sum\limits_{i=1}^m \rho_i C_i\right)$.

下面证明 $\bigcap_{i=1}^m \text{bar}\,(\rho_i C_i) \supset \text{bar}\left(\sum\limits_{i=1}^m \rho_i C_i\right)$. 设 $\boldsymbol{b} \in \text{bar}\left(\sum\limits_{i=1}^m \rho_i C_i\right)$, 若 $\boldsymbol{b}=\boldsymbol{0}$ 结论显然. 假设 $\boldsymbol{b} \neq \boldsymbol{0}$, 存在常数 δ 使得 $\langle \boldsymbol{x}, \boldsymbol{b}\rangle \leqslant \delta$ 对所有 $\boldsymbol{x} \in \rho_1 C_1+\rho_2 C_2+\cdots+\rho_m C_m$ 成立, 如果对于

$$\boldsymbol{x}=\rho_1\boldsymbol{x}_1+\rho_2\boldsymbol{x}_2+\cdots+\rho_m\boldsymbol{x}_m, \quad \boldsymbol{x}_i \in C_i, \quad i=1,2,\cdots,m,$$

依次固定下标 $i=2,\cdots,m$ 对应的变量 $\boldsymbol{x}_i \in C_i, 2 \leqslant i \leqslant m$, 那么对任意的 $\boldsymbol{x}_1 \in C_1$, 有

$$\langle \rho_1\boldsymbol{x}_1+\rho_2\boldsymbol{x}_2+\cdots+\rho_m\boldsymbol{x}_m, \boldsymbol{b}\rangle \leqslant \delta.$$

上式可写为 $\langle \rho_1\boldsymbol{x}_1, \boldsymbol{b}\rangle \leqslant \delta-\langle \rho_2\boldsymbol{x}_2+\cdots+\rho_m\boldsymbol{x}_m, \boldsymbol{b}\rangle$, 得 $\boldsymbol{b} \in \text{bar}(\rho_1 C_1)$. 这样依次固定 $m-1$ 个变量得 $\boldsymbol{b} \in \text{bar}(\rho_i C_i)(2 \leqslant i \leqslant m)$, 因此, 有

$$\boldsymbol{b} \in \text{bar}(\rho_1 C_1) \cap \cdots \cap \text{bar}(\rho_m C_m).$$

最后由定理 3.7.1 得 $\rho_i \text{bar} C_i = \text{bar}(\rho_i C_i)\,(1 \leqslant i \leqslant m)$, 因此有 (3.7.1-3) 式成立.

由定理 3.7.1 知, 只需要证明 $\bigcap_{i=1}^m \text{nor}\,(\rho_i C_i) \subset \text{nor}\left(\sum\limits_{i=1}^m \rho_i C_i\right)$. 设 $\boldsymbol{b} \in \bigcap_{i=1}^m \text{nor}\,(\rho_i C_i)$, 存在点

$$\boldsymbol{x}_i \in \text{cl}(\rho_i C_i) \quad (i=1,\cdots,m)$$

使得 $\boldsymbol{b} \in N_{\boldsymbol{x}_i}(\rho_i C_i)$. 令 $\boldsymbol{x}=\boldsymbol{x}_1+\cdots\cdots+\boldsymbol{x}_m$, 由性质 2.3.9 得

$$\boldsymbol{x} \in \text{cl}(\rho_1 C_1)+\cdots+\text{cl}(\rho_i C_m) \subset \text{cl}(\rho_1 C_1+\cdots+\rho_m C_m).$$

根据上式和性质 3.6.2 得

$$\boldsymbol{b} \in \bigcap_{i=1}^m N_{\boldsymbol{x}_i}(\rho_i C_i) = N_{\boldsymbol{x}_1+\cdots+\boldsymbol{x}_m}\left(\bigcap_{i=1}^m \rho_i C_i\right) \subset \text{nor}\left(\bigcap_{i=1}^m \rho_i C_i\right). \qquad \square$$

本小节得到了凸集的障碍锥与法锥的一些基本性质, 这些性质揭示了凸集的障碍锥与法锥之间的密切关系:

1) 障碍锥包含法锥;

2) 障碍锥仅含原点等价于凸集是整个空间, 障碍锥是整个空间等价于凸集有界;

3) 障碍锥是包含正交集合的正常点构成的凸锥;

4) 凸集是平面等价于障碍锥等于凸集的正交集;

5) 凸集闭包和相对内部产生的障碍锥与原集合障碍锥相等;

6) 凸集线性平移后的障碍锥和法锥不变;

7) 多个凸集线性组合的障碍锥等于各自障碍锥乘以线性系数后的交集;

8) 多个凸集线性组合的法锥包含各自法锥乘以线性系数后的交集.

3.7.2 拓扑性质

本小节讨论障碍锥和法锥之间的拓扑性质. 下面的定理说明一个非空闭凸集的回收锥等于它的法锥或障碍锥的极锥.

定理 3.7.3 设非空凸集 $C \subset \mathbf{R}^n$, 则

$$\mathrm{ri}(\mathrm{rec}(\mathrm{cl}C))^- \subset \mathrm{nor}C \subset \mathrm{bar}C \subset (\mathrm{rec}(\mathrm{cl}C))^-, \tag{3.7.2-1}$$

且 $\mathrm{rec}(\mathrm{cl}C) = (\mathrm{nor}C)^- = (\mathrm{bar}C)^-$.

证明 根据性质 3.7.1 和性质 3.6.4 得 $\mathrm{nor}C = \mathrm{nor}(\mathrm{cl}C)$ 和 $\mathrm{bar}C = \mathrm{bar}(\mathrm{cl}C)$, 那么只需要证明 $\mathrm{ri}(\mathrm{rec}\,(\mathrm{cl}C))^- \subset \mathrm{nor}\,(\mathrm{cl}C) \subset \mathrm{bar}\,(\mathrm{cl}C) \subset (\mathrm{rec}\,(\mathrm{cl}C))^-$ 即可.

如果 $\mathbf{0} \in \mathrm{ri}(\mathrm{rec}\,(\mathrm{cl}C))^-$ 成立, 由 (3.7.2-1) 式有 $\mathbf{0} \in \mathrm{nor}\,(\mathrm{cl}C)$. 设 $\boldsymbol{e} \in \mathrm{ri}(\mathrm{rec}\,(\mathrm{cl}C))^-$ 是一个非零点, 并假设 $\boldsymbol{e} \notin \mathrm{nor}\,(\mathrm{cl}C)$. 根据性质 3.6.1 和法锥定义得 $\boldsymbol{e} \notin (\mathrm{cl}C - \boldsymbol{a})^-$, 其中 $\boldsymbol{a} \in \mathrm{cl}C$. 再根据凸集分离定理得, 存在一个非零点 $\boldsymbol{a}^* \in \mathbf{R}^n$ 和常数 $\alpha \in \mathbf{R}$ 构成的超平面

$$H\,(\boldsymbol{a}^*, \alpha) = \{\boldsymbol{x} \mid \langle \boldsymbol{x}, \boldsymbol{a}^* \rangle = \alpha\}$$

分离 $\boldsymbol{e}$ 和 $(\mathrm{cl}C - \boldsymbol{a})^-$, 且

$$\langle \boldsymbol{c}, \boldsymbol{a}^* \rangle \leqslant \alpha \leqslant \langle \boldsymbol{e}, \boldsymbol{a}^* \rangle, \quad \forall \boldsymbol{c} \in (\mathrm{cl}C - \boldsymbol{a})^-. \tag{3.7.2-2}$$

因为 $(\mathrm{cl}C - \boldsymbol{a})^-$ 是闭凸锥且含零点, 则 $\sup\limits_{\boldsymbol{c} \in (\mathrm{cl}C - \boldsymbol{a})^-} \langle \boldsymbol{c}, \boldsymbol{a}^* \rangle = 0$. 由 (3.7.2-2) 式可得, 当 $\lambda > 0$ 时, 有

$$\lambda \boldsymbol{a}^* \in \left((\mathrm{cl}C - \boldsymbol{a})^-\right)^- = \mathrm{cl}\,(\mathrm{cl}C - \boldsymbol{a}) = \mathrm{cl}C - \boldsymbol{a}. \tag{3.7.2-3}$$

由 $\boldsymbol{e} \notin (\mathrm{cl}C - \boldsymbol{a})^-$ 和极锥定义得 $\langle \boldsymbol{e}, \lambda \boldsymbol{a}^* \rangle > 0$. 再由 $\boldsymbol{e} \in \mathrm{ri}(\mathrm{rec}\,(\mathrm{cl}C))^-$ 和定理 3.3.1 得

$$\langle \boldsymbol{e}, \boldsymbol{x} \rangle \leqslant 0, \quad \forall \boldsymbol{x} \in \mathrm{rec}\,(\mathrm{cl}C) = \bigcap_{\lambda>0} \lambda(\mathrm{cl}C - \boldsymbol{a}).$$

根据上式和 (3.7.2-3) 式, 有 $\langle \boldsymbol{e}, \lambda \boldsymbol{a}^* \rangle \leqslant 0$, 这是一个矛盾. 所以

$$\mathrm{ri}(\mathrm{rec}\,(\mathrm{cl}C))^- \subset \mathrm{nor}\,(\mathrm{cl}C) \subset \mathrm{bar}\,(\mathrm{cl}C)$$

成立.

下面设 $\boldsymbol{b} \in \mathrm{bar}\,(\mathrm{cl}C)$. 根据障碍锥定义, 存在 $\delta = \delta(\boldsymbol{b}) \in \mathbf{R}$ 使得

$$\langle \boldsymbol{x}, \boldsymbol{b} \rangle \leqslant \delta, \quad \forall \boldsymbol{x} \in \mathrm{cl}C. \tag{3.7.2-4}$$

于是有 $\boldsymbol{a} = \underset{\boldsymbol{x}\in\mathrm{cl}C}{\arg\max}\,\{\langle \boldsymbol{x}, \boldsymbol{b} \rangle\} \in \mathrm{cl}C$ 使得 $\langle \boldsymbol{a}, \boldsymbol{b} \rangle = \delta$. 假设 $\boldsymbol{b} \notin (\mathrm{rec}\,(\mathrm{cl}C))^-$, 根据凸集分离定理得, 存在一个非零点 $\boldsymbol{a}^* \in \mathbf{R}^n$ 和常数 $\alpha \in \mathbf{R}$ 构成的超平面 $H\,(\boldsymbol{a}^*, \alpha) = \{\boldsymbol{x} \mid \langle \boldsymbol{x}, \boldsymbol{a}^* \rangle = \alpha\}$ 分离 $\boldsymbol{b}$ 和 $(\mathrm{rec}\,(\mathrm{cl}C))^-$, 且

$$\langle \boldsymbol{c}, \boldsymbol{a}^* \rangle \leqslant \alpha \leqslant \langle \boldsymbol{b}, \boldsymbol{a}^* \rangle, \quad \forall \boldsymbol{c} \in (\mathrm{rec}\,(\mathrm{cl}C))^-. \tag{3.7.2-5}$$

因为 $(\mathrm{rec}\,(\mathrm{cl}C))^-$ 是闭凸锥且含零点, 则 $\sup\limits_{\boldsymbol{c}\in(\mathrm{rec}(\mathrm{cl}C))^-} \langle \boldsymbol{c}, \boldsymbol{a}^* \rangle = 0$. 对任意的 $t > 0$, 由 (3.7.2-5) 式得

$$t\boldsymbol{a}^* \in \left((\mathrm{rec}\,(\mathrm{cl}C))^-\right)^- = \mathrm{cl}(\mathrm{rec}\,(\mathrm{cl}C)) \subset \mathrm{rec}\,(\mathrm{cl}C) = \bigcap_{\lambda>0} \lambda(\mathrm{cl}C - \boldsymbol{a}).$$

再由上式得 $\langle \boldsymbol{b}, t\boldsymbol{a}^* \rangle > 0$, 存在 $\lambda > 0$ 和 $\boldsymbol{c} \in \mathrm{cl}C$ 使得 $t\boldsymbol{a}^* = \lambda(\boldsymbol{c} - \boldsymbol{a})$, 且有 $\boldsymbol{a} \in \mathrm{cl}C$. 但 $\boldsymbol{b} \in \mathrm{bar}\,(\mathrm{cl}C)$, 则结合 (3.7.2-4) 式得

$$\left\langle \frac{t}{\lambda}\boldsymbol{a}^* + \boldsymbol{a}, \boldsymbol{b} \right\rangle \leqslant \delta,$$

即有 $\left\langle \dfrac{t}{\lambda}\boldsymbol{a}^*, \boldsymbol{b} \right\rangle \leqslant \delta - \langle \boldsymbol{a}, \boldsymbol{b} \rangle = 0$. 因此, 有 $\langle \boldsymbol{b}, t\boldsymbol{a}^* \rangle \leqslant 0$, 得到一个矛盾. 所以有 $\mathrm{bar}\,(\mathrm{cl}C) \subset (\mathrm{rec}\,(\mathrm{cl}C))^-$.

由性质 3.5.1 和 (3.7.2-5) 式得

$$\left(\mathrm{ri}(\mathrm{rec}(\mathrm{cl}C))^-\right)^- \supset (\mathrm{nor}C)^- \supset (\mathrm{bar}C)^- \supset \left((\mathrm{rec}(\mathrm{cl}C))^-\right)^- = \mathrm{cl}(\mathrm{rec}(\mathrm{cl}C)).$$

设 $\boldsymbol{a} \in \left(\mathrm{ri}(\mathrm{rec}(\mathrm{cl}C))^-\right)^-$, 有 $\langle \boldsymbol{c}, \boldsymbol{a} \rangle \leqslant 0, \forall \boldsymbol{c} \in \mathrm{ri}(\mathrm{rec}\,(\mathrm{cl}C))^-$, 则有

$$\langle \boldsymbol{c}, \boldsymbol{x} \rangle \leqslant 0, \quad \forall \boldsymbol{x} \in (\mathrm{rec}\,(\mathrm{cl}C)).$$

上式说明 $\boldsymbol{a} \in \mathrm{cl}(\mathrm{rec}\,(\mathrm{cl}C))$, 即有 $\mathrm{rec}(\mathrm{cl}C) = (\mathrm{nor}C)^- = (\mathrm{bar}C)^-$. □

推论 3.7.1 设非空凸集 $C \subset \mathbf{R}^n$, 则

$$\mathrm{cl}(\mathrm{bar}C) = \mathrm{cl}(\mathrm{nor}C) = (\mathrm{rec}(\mathrm{cl}C))^-, \quad \mathrm{ri}(\mathrm{bar}C) = \mathrm{ri}(\mathrm{nor}C) = \mathrm{ri}(\mathrm{rec}(\mathrm{cl}C))^-,$$

且 $\mathrm{cl}(\mathrm{nor}C)$ 和 $\mathrm{ri}(\mathrm{nor}C)$ 是凸锥.

定理 3.7.4 设 C 是关于顶点 $\boldsymbol{a}$ 的非空凸锥, 则 $\mathrm{bar}C = \mathrm{nor}C = (C-\boldsymbol{a})^-$, 且 $\mathrm{bar}C$ 和 $\mathrm{nor}C$ 是闭锥. 进一步, 对任何非空平面 $L \in \mathbf{R}^n$ 有 $\mathrm{bar}L = \mathrm{nor}L = \mathrm{ort}L$, 对任何半平面 $F \subset L$ 有 $\mathrm{bar}F = \mathrm{nor}F$.

证明 根据性质 3.7.1 和性质 3.6.4 得 $\mathrm{nor}C = \mathrm{nor}(\mathrm{cl}C)$ 和 $\mathrm{bar}C = \mathrm{bar}(\mathrm{cl}C)$, 根据定理假设和定理 3.3.1 得

$$\mathrm{rec}(\mathrm{cl}C) = \mathrm{cl}C - \boldsymbol{a}, \quad (\mathrm{cl}C - \boldsymbol{a})^- = (C - \boldsymbol{a})^-.$$

再根据推论 3.7.1 得

$$\mathrm{cl}(\mathrm{bar}C) = \mathrm{cl}(\mathrm{nor}C) = (\mathrm{rec}(\mathrm{cl}C))^- = (\mathrm{cl}C - \boldsymbol{a})^- = (C - \boldsymbol{a})^-.$$

由于 $\mathrm{nor}(\mathrm{cl}C) = \mathrm{nor}C \subset \mathrm{bar}C \subset \mathrm{cl}\,(\mathrm{bar}C) = (\mathrm{cl}C - \boldsymbol{a})^-$, 根据法锥定义 3.6.4 和性质 3.6.1 有

$$N_{\boldsymbol{a}}(\mathrm{cl}C) = (\mathrm{cl}C - \boldsymbol{a})^- \subset \mathrm{nor}\,(\mathrm{cl}C)\,.$$

进一步的结论由 $(L - a)^- = (\mathrm{dir}L)^- = \mathrm{ort}L$ 可得. □

定理 3.7.4 表明一个非空的凸锥的障碍锥与法锥是等价的, 且是它的极锥. 因此, 求一个凸锥的法锥等价于求它的极锥.

本小节得到了凸集闭包与凸集回收锥、法锥、障碍锥和它们的极锥之间的关系:

1) 凸集闭包的回收锥的极锥相对内部属于凸集的法锥, 法锥属于障碍锥, 障碍锥属于凸集闭包的回收锥的极锥;

2) 凸集闭包的回收锥等于法锥的极锥, 还等于障碍锥的极锥;

3) 非空顶点凸锥的障碍锥等于它的法锥, 也等于凸锥减去顶点后的极锥, 法锥、切锥与障碍锥均是闭凸锥;

4) 平面的障碍锥与它的法锥及正交集均相等.

3.8 凸锥分离定理

凸锥是含有从顶点出发半直线的无界凸集. 凸锥在优化中有两个主要应用: 一是用于刻画函数的梯度方向; 二是刻画决策者的偏好方向 [23,25,83,88]. 其中凸集

分离定理和凸锥分离定理是这些应用的重要工具, 所以凸集与凸锥分离定理是凸分析中重要的内容.

本节主要建立多个凸锥 (而不仅是两个凸锥) 分离定理, 凸锥分离与凸集分离不同, 不再是寻找分离的超平面, 而是寻找凸锥对应的共轭锥的非零变量. 从下面的凸锥分离定理看到, 如果多个不相交的凸锥成立, 则存在各自共轭锥中的非零变量之和等于 0, 使得多个凸锥可分离，并称这些非零变量为可分离变量. 本节讨论的可分离凸锥均是顶点为原点的凸锥.

定理 3.8.1 设 C_1 和 C_2 是 $\mathbf{R}^n$ 中的凸锥, $C_1 \cap C_2 = \varnothing$, 则存在非零的 $\boldsymbol{x}_1^* \in C_1^+$ 和 $\boldsymbol{x}_2^* \in C_2^+$ 使得 $\boldsymbol{x}_1^* + \boldsymbol{x}_2^* = \mathbf{0}$.

证明 设变量 $\boldsymbol{x} = (\boldsymbol{x}_1, \boldsymbol{x}_2)$ 是构成 $2n$ 维空间 $\mathbf{R}^{2n}$ 的元素, 其中 $\boldsymbol{x}_1, \boldsymbol{x}_2 \in \mathbf{R}^n$. 再设对偶空间变量为 $\boldsymbol{x}^* = (\boldsymbol{x}_1^*, \boldsymbol{x}_2^*) \in \mathbf{R}^{2n}$, 定义 $\mathbf{R}^{2n}$ 的内积为

$$\langle \boldsymbol{x}, \boldsymbol{x}^* \rangle = \langle \boldsymbol{x}_1, \boldsymbol{x}_1^* \rangle + \langle \boldsymbol{x}_2, \boldsymbol{x}_2^* \rangle.$$

设集合 $D = \{(\boldsymbol{x}_1, \boldsymbol{x}_2) \,|\, \boldsymbol{x}_1 \in C_1, \boldsymbol{x}_2 \in C_2\}$ 和 $E = \{(\boldsymbol{x}, \boldsymbol{x}) \,|\, \boldsymbol{x} \in \mathbf{R}^n\}$, 显然 D 和 E 都是凸锥. 由假设 $C_1 \cap C_2 = \varnothing$ 知 $D \cap E = \varnothing$, 由推论 2.4.1 知, 存在不为零的点 $(\boldsymbol{x}_1^*, \boldsymbol{x}_2^*)$, 使得对任意 $(\boldsymbol{x}, \boldsymbol{x}) \in E, \boldsymbol{x}_1 \in C_1$ 和 $\boldsymbol{x}_2 \in C_2$, 满足

$$\langle \boldsymbol{x}, \boldsymbol{x}_1^* \rangle + \langle \boldsymbol{x}, \boldsymbol{x}_2^* \rangle \leqslant \langle \boldsymbol{x}_1, \boldsymbol{x}_1^* \rangle + \langle \boldsymbol{x}_2, \boldsymbol{x}_2^* \rangle. \tag{3.8.1-1}$$

由上式知, $\langle \boldsymbol{x}_1, \boldsymbol{x}_1^* \rangle + \langle \boldsymbol{x}_2, \boldsymbol{x}_2^* \rangle$ 在 D 上有下界, 由定理 3.4.1 得 $(\boldsymbol{x}_1^*, \boldsymbol{x}_2^*) \in D^+$. 另一方面, $\langle \boldsymbol{x}, \boldsymbol{x}_1^* + \boldsymbol{x}_2^* \rangle$ 在 $\mathbf{R}^n$ 中有上界, 故 $\boldsymbol{x}_1^* + \boldsymbol{x}_2^* = \mathbf{0}$. □

上述结论说明, 如果凸锥 C_1 和 C_2 在 $\mathbf{R}^n$ 中不相交, 那么存在非零的 $\boldsymbol{x}_1^* \in C_1^+$ 和 $\boldsymbol{x}_2^* \in C_2^+$ 使得 $\boldsymbol{x}_1^* + \boldsymbol{x}_2^* = \mathbf{0}$. 由 (3.8.1-1) 式说明

$$\langle \boldsymbol{x}_2, \boldsymbol{x}_1^* \rangle \leqslant \langle \boldsymbol{x}_1, \boldsymbol{x}_1^* \rangle \quad \text{或} \quad \langle \boldsymbol{x}_1, \boldsymbol{x}_2^* \rangle \leqslant \langle \boldsymbol{x}_2, \boldsymbol{x}_2^* \rangle, \quad \boldsymbol{x}_1 \in C_1, \boldsymbol{x}_2 \in C_2,$$

即 $\boldsymbol{x}_1^* \in C_1^+$ 和 $\boldsymbol{x}_2^* \in C_2^+$ 可分离 C_1 和 C_2. 即, 如果两个凸锥不相交, 则在它们各自的共轭锥中存在对应非零点使得两点的和为零. 下面例 3.8.1 说明它们都存在超平面分离两个凸锥.

例 3.8.1 设在 $\mathbf{R}^2$ 空间中 $C_1 = \left\{(x_1, 0)^{\mathrm{T}} \,|\, x_1 \in (0, +\infty)\right\}$ 和 $C_2 = \left\{(0, x_2)^{\mathrm{T}} \,|\, x_2 \in (0, +\infty)\right\}$ 是两个凸锥, 有 $C_1 \cap C_2 = \varnothing$, 它们的共轭锥分别为

$$C_1^+ = \left\{(x_1^*, x_2^*)^{\mathrm{T}} \,|\, x_1^* \in [0, +\infty), x_2^* \in \mathbf{R}\right\}$$

和

$$C_2^+ = \left\{(x_1^*, x_2^*)^{\mathrm{T}} \,|\, x_1^* \in \mathbf{R}, x_2^* \in [0, +\infty)\right\},$$

取非零的 $(x_1^*, -x_2^*)^{\mathrm{T}} \in C_1^+$ 和 $(-x_1^*, x_2^*)^{\mathrm{T}} \in C_2^+$ 时, 有定理 3.8.1 的结论成立.

由归纳法可以得到下面推论成立.

推论 3.8.1　设 $C_1, \cdots, C_m$ 是 $\mathbf{R}^n$ 中的凸锥, $\bigcap_{i=1}^m C_i = \varnothing$, 则存在不全为 $\mathbf{0}$ 的 $\boldsymbol{x}_i^* \in C_i^+ \ (i=1,\cdots,m)$ 使得

$$\boldsymbol{x}_1^* + \cdots + \boldsymbol{x}_m^* = \mathbf{0}. \tag{3.8.1-2}$$

进一步, 当一个凸锥与另一个凸锥内部相交非空时, 则两个凸锥交集的共轭锥等于各自共轭锥的和.

定理 3.8.2　设凸锥 $C_1, C_2 \subset \mathbf{R}^n$, 且 $C_1 \cap \mathrm{int} C_2 \neq \varnothing$, 则

$$(C_1 \cap C_2)^+ = C_1^+ + C_2^+.$$

证明　根据共轭锥的定义 (定义 3.4.1), 设 $\boldsymbol{x}_1^* \in C_1^+$ 和 $\boldsymbol{x}_2^* \in C_2^+$, 对于任意 $\boldsymbol{x} \in C_1 \cap C_2$ 有

$$\langle \boldsymbol{x}, \boldsymbol{x}_1^* \rangle \geqslant 0, \quad \langle \boldsymbol{x}, \boldsymbol{x}_2^* \rangle \geqslant 0,$$

得 $\langle \boldsymbol{x}, \boldsymbol{x}_1^* + \boldsymbol{x}_2^* \rangle \geqslant 0$. 因此

$$C_1^+ + C_2^+ \subset (C_1 \cap C_2)^+.$$

反过来, 设 $\boldsymbol{x}^* \in (C_1 \cap C_2)^+$, 且 $\boldsymbol{x}^* \neq \mathbf{0}, C_0 = \{\boldsymbol{x} \,|\, \langle \boldsymbol{x}, \boldsymbol{x}^* \rangle < 0\}$, 则有

$$C_0 \cap (C_1 \cap C_2) = \varnothing. \tag{3.8.1-3}$$

设 $\forall \boldsymbol{y}^* \in C_0^+ \setminus \{\mathbf{0}\}$ 和 $\boldsymbol{x} \in C_0$, 因为 $\langle \boldsymbol{x}, \boldsymbol{x}^* \rangle < 0$ 和 $\langle \boldsymbol{x}, \boldsymbol{y}^* \rangle \geqslant 0$ 同时成立, 可知 $\boldsymbol{x}^*, \boldsymbol{y}^*$ 线性相关, 故存在不全为 0 的 α_1, α_2, 满足 $\alpha_1 \boldsymbol{x}^* - \alpha_2 \boldsymbol{y}^* = \mathbf{0}$, 显然 $\alpha_2 \neq 0$. 因此 $\boldsymbol{y}^* = \dfrac{\alpha_1}{\alpha_2} \boldsymbol{x}^*$, 对于 $\boldsymbol{x} \in C_0$, 有

$$0 \leqslant \langle \boldsymbol{x}, \boldsymbol{y}^* \rangle = \frac{\alpha_1}{\alpha_2} \langle \boldsymbol{x}, \boldsymbol{x}^* \rangle,$$

所以 $\dfrac{\alpha_1}{\alpha_2} \leqslant 0$. 于是得到

$$C_0^+ = \left\{ \boldsymbol{y}^* \,\middle|\, \boldsymbol{y}^* = \frac{\alpha_1}{\alpha_2} \boldsymbol{x}^*, \frac{\alpha_1}{\alpha_2} \leqslant 0 \right\}. \tag{3.8.1-4}$$

由 (3.8.1-3) 式和推论 3.8.1 知, 存在不全为零的 $\boldsymbol{y}^* \in C_0^+, \boldsymbol{x}_1^* \in C_1^+$ 和 $\boldsymbol{x}_2^* \in C_2^+$ 使得 $\boldsymbol{y}^* + \boldsymbol{x}_1^* + \boldsymbol{x}_2^* = \mathbf{0}$. 根据 (3.8.3-4) 式, 上式即为

$$-\frac{\alpha_1}{\alpha_2} \boldsymbol{x}^* = \boldsymbol{x}_1^* + \boldsymbol{x}_2^*. \tag{3.8.1-5}$$

下面证明 $\frac{\alpha_1}{\alpha_2} \neq 0$. 假设 $\frac{\alpha_1}{\alpha_2} = 0$, 由 (3.8.1-5) 式得 $\boldsymbol{x}_1^* + \boldsymbol{x}_2^* = \boldsymbol{0}$, 因为 $\boldsymbol{x}_1^*$ 和 $\boldsymbol{x}_2^*$ 至少一个不为 0, 则可得到 $\boldsymbol{x}_1^*$ 和 $\boldsymbol{x}_2^*$ 均不为 0. 设 $\boldsymbol{x}_0 \in C_1 \cap \text{int}C_2$, 则由定理 3.4.1 知

$$\langle \boldsymbol{x}_0, \boldsymbol{x}_1^* \rangle \geqslant 0, \langle \boldsymbol{x}_0, \boldsymbol{x}_2^* \rangle > 0 \Rightarrow \langle \boldsymbol{x}_0, \boldsymbol{x}_1^* + \boldsymbol{x}_2^* \rangle > 0,$$

有上式与 $\boldsymbol{x}_1^* + \boldsymbol{x}_2^* = \boldsymbol{0}$ 矛盾, 所以 $\frac{\alpha_1}{\alpha_2} < 0$. 再由 (3.8.1-5) 式得

$$\boldsymbol{x}^* = -\frac{\alpha_2}{\alpha_1}\boldsymbol{x}_1^* - \frac{\alpha_2}{\alpha_1}\boldsymbol{x}_2^* \in C_1^+ + C_2^+,$$

即有 $C^+ \subset C_1^* + C_2^*$. □

由归纳法得下面结论.

推论 3.8.2 设 $C_1, \cdots, C_m$ 是 $\mathbf{R}^n$ 中的凸锥, $\forall C_i \in \{C_1, \cdots, C_m\}$ 且 $C_i \cap \text{int}C_2 \cap \cdots \cap \text{int}C_m \neq \varnothing$, 则

$$(C_i \cap \cdots \cap C_m)^+ = C_i^+ + \cdots + C_m^+.$$

推论的结论说明, 在多个凸锥中, 其中一个与其余凸锥内部交非空时, 所有凸锥相交的共轭锥等于各自凸锥的共轭锥之和.

例 3.8.2 设在 $\mathbf{R}^2$ 空间中的两个凸锥 $C_1 = \left\{(x_1, x_2)^{\mathrm{T}} | x_1 \geqslant 0\right\}$ 和 $C_2 = \left\{(x_1, x_2)^{\mathrm{T}} | x_2 \geqslant 0\right\}$, 它们的共轭锥分别为

$$C_1^+ = \left\{(x_1^*, 0)^{\mathrm{T}} \in \mathbf{R}^2 | x_1^* \geqslant 0\right\} \quad 和 \quad C_2^+ = \left\{(0, x_2^*)^{\mathrm{T}} \in \mathbf{R}^2 | x_2^* \geqslant 0\right\},$$

且有 $C_1 \cap C_2 = \{(x_1, x_2) | x_1 \geqslant 0, x_2 \geqslant 0\}$, 易得 $(C_1 \cap C_2)^+ = C_1^+ + C_2^+$. 显然, 不存在非零的 $\boldsymbol{x}_1^* \in C_1^+$ 和 $\boldsymbol{x}_2^* \in C_2^+$ 使得 $\boldsymbol{x}_1^* + \boldsymbol{x}_2^* = \boldsymbol{0}$, 说明这两个凸锥不可分离.

由定理 3.8.1 和定理 3.8.2 得下面结论及推论成立.

定理 3.8.3 设 C_1 和 C_2 是 $\mathbf{R}^n$ 中的凸锥, 则 $(C_1 \cap C_2)^+ = C_1^+ + C_2^+$, 或存在非零的 $\boldsymbol{x}_1^* \in C_1^+$ 和 $\boldsymbol{x}_2^* \in C_2^+$ 使得 $\boldsymbol{x}_1^* + \boldsymbol{x}_2^* = \boldsymbol{0}$.

推论 3.8.3 设 $C_1, \cdots, C_m$ 是 $\mathbf{R}^n$ 中的凸锥, 则

$$(C_1 \cap \cdots \cap C_m)^+ = C_1^+ + \cdots + C_m^+$$

成立, 或存在不全为 $\boldsymbol{0}$ 的点 $\boldsymbol{x}_i^* \in C_i^+ \ (i = 1, \cdots, m)$ 使

$$\boldsymbol{x}_1^* + \cdots + \boldsymbol{x}_m^* = \boldsymbol{0}.$$

根据上面的几个结论, 下面给出凸锥分离定义及定理, 并举例说明它是凸集分离定理的一种推广.

定义 3.8.1 设 $C_1, \cdots, C_m$ 是 $\mathbf{R}^n$ 中的凸锥. 如果存在不全为 $\mathbf{0}$ 的点 $\boldsymbol{x}_i^* \in C_i^+$ $(i = 1, \cdots, m)$ 满足

$$\boldsymbol{x}_1^* + \cdots + \boldsymbol{x}_m^* = \mathbf{0}, \tag{3.8.1-6}$$

则称凸锥 $C_1, \cdots, C_m$ 可以分离.

(3.8.1-6) 式是凸锥分离的必要条件. 下面证明两个凸锥可分离的充要条件.

定理 3.8.4 设 C_1 和 C_2 是 $\mathbf{R}^n$ 中的凸锥, 则 C_1 和 C_2 不能分离的充要条件是

$$\mathbf{0} \in \operatorname{int}(C_1 - C_2). \tag{3.8.1-7}$$

证明 采用反证法. 如果 C_1 和 C_2 可以分离, 则根据定义 3.8.1, 可以找到不为 $\mathbf{0}$ 的 $\boldsymbol{x}_1^* \in C_1^+$ 和 $\boldsymbol{x}_2^* \in C_2^+$ 使得 $\boldsymbol{x}_1^* = -\boldsymbol{x}_2^*$, 且有

$$\langle \boldsymbol{x}_1 - \boldsymbol{x}_2, \boldsymbol{x}_1^* \rangle = \langle \boldsymbol{x}_1, \boldsymbol{x}_1^* \rangle + \langle \boldsymbol{x}_2, \boldsymbol{x}_2^* \rangle \geqslant 0, \quad \forall \boldsymbol{x}_1 \in C_1, \quad \forall \boldsymbol{x}_2 \in C_2. \tag{3.8.1-8}$$

上式说明 $\mathbf{0} \notin \operatorname{int}(C_1 - C_2)$. 否则, 有 $\mathbf{0} \in \operatorname{int}(C_1 - C_2)$, 且存在 $\varepsilon > 0$ 使得 $-\varepsilon \boldsymbol{x}_1^* \in C_1 - C_2$, 即表明分别存在 $\bar{\boldsymbol{x}}_1 \in C_1$ 和 $\bar{\boldsymbol{x}}_2 \in C_2$, 使得 $-\varepsilon \boldsymbol{x}_1^* = \bar{\boldsymbol{x}}_1 - \bar{\boldsymbol{x}}_2$. 则由 (3.8.1-8) 式有

$$\langle \bar{\boldsymbol{x}}_1 - \bar{\boldsymbol{x}}_2, \boldsymbol{x}_1^* \rangle = -\varepsilon \left| \boldsymbol{x}_1^* \right|^2 < 0,$$

这与 (3.8.1-8) 式矛盾.

反之, 如果 $\mathbf{0} \notin \operatorname{int}(C_1 - C_2)$, 表明 $\mathbf{0}$ 和 $\operatorname{int}(C_1 - C_2)$ 可以分离. 则由定理 2.4.4 知存在 $\boldsymbol{x}^* \neq \mathbf{0}$, 当 $\boldsymbol{x}_1 \in C_1, \boldsymbol{x}_2 \in C_2$, 且 $\boldsymbol{x}_1 - \boldsymbol{x}_2 \in \operatorname{int}(C_1 - C_2)$ 时有

$$\langle \boldsymbol{x}_1 - \boldsymbol{x}_2, \boldsymbol{x}^* \rangle \geqslant 0.$$

在上式中先假设 $\boldsymbol{x}_1^* = \boldsymbol{x}^*$, 令 $\boldsymbol{x}_2 \to \mathbf{0}$ 得 $\boldsymbol{x}_1^* \in C_1^+$. 然后再设 $\boldsymbol{x}_2^* = -\boldsymbol{x}^*$, 令 $\boldsymbol{x}_1 \to \mathbf{0}$ 得 $\boldsymbol{x}_2^* \in C_2^*$, 这样得到了 $\boldsymbol{x}_1^* = -\boldsymbol{x}_2^*$. 那么由定义 3.8.1 知, C_1 和 C_2 可以分离. □

在定理 2.4.6 中得到了存在强分离 C_1 和 C_2 的超平面的充要条件是 $\mathbf{0} \notin \operatorname{cl}(C_1 - C_2)$, 定理 3.8.4 的结论对凸锥更弱些. 因此, 将定理 3.8.4 重写为下面的结论.

推论 3.8.4 设 C_1 和 C_2 是 $\mathbf{R}^n$ 中的凸锥, 则 C_1 和 C_2 可分离的充要条件是

$$\mathbf{0} \notin \operatorname{int}(C_1 - C_2).$$

例 3.8.3 设在 $\mathbf{R}^2$ 空间中的两个凸锥:

$$C_1 = \left\{ (x_1, x_2)^{\mathrm{T}} \middle| x_1 \geqslant 0, x_2 \geqslant 0, x_1 - x_2 > 0 \right\},$$

$$C_2=\left\{(x_1,x_2)^{\mathrm{T}}\,|x_1\geqslant 0,x_2\geqslant 0,x_2-x_1>0\right\},$$

有 $C_1\cap C_2=\varnothing$, 它们共轭锥分别为

$$C_1^+=\left\{(x_1^*,x_2^*)^{\mathrm{T}}\in\mathbf{R}^2|x_1^*+x_2^*\geqslant 0\right\},$$

$$C_2^+=\left\{(x_1^*,x_2^*)^{\mathrm{T}}\in\mathbf{R}^2|x_1^*+x_2^*\geqslant 0\right\}.$$

取非零的 $(x_1^*,-x_1^*)\in C_1^+$ 和 $(-x_1^*,x_1^*)\in C_2^+$, C_1 和 C_2 可以分离, 则 (3.8.1-7) 式不成立. 事实上, 存在超平面 $H\left((-1,1),0\right)=\{(x_1,x_2)\,|x_2-x_1=0\}$ 分离 C_1 和 C_2. 因此, 该例说明两个凸锥分离定理本质上与两个凸集分离定理是等价的.

例 3.8.4 设在 $\mathbf{R}^2$ 空间中的两个凸锥:

$$C_1=\left\{(x_1,x_2)^{\mathrm{T}}\,|x_1\geqslant 0\right\},\quad C_2=\left\{(x_1,x_2)^{\mathrm{T}}\,|x_2\geqslant 0\right\},$$

有 $\mathbf{0}\in\operatorname{int}(C_1-C_2)$, 它们共轭锥分别为

$$C_1^+=\left\{(x_1^*,0)^{\mathrm{T}}\in\mathbf{R}^2|x_1^*\geqslant 0\right\},\quad C_2^+=\left\{(0,x_2^*)^{\mathrm{T}}\in\mathbf{R}^2|x_2^*\geqslant 0\right\},$$

显然 C_1 和 C_2 不可以分离.

在定理 2.4.3 中 $\mathrm{ri}C_1\cap\mathrm{ri}C_2=\varnothing$ 是两个凸集可分离的充要条件, 对于凸锥也有一样的类似结果. 对于凸锥不能分离的情形, 下面的定理 3.8.5 中条件 $\mathrm{ri}C_1\cap\mathrm{ri}C_2\neq\varnothing$ 也是充要条件.

定理 3.8.5 设 C_1 和 C_2 是 $\mathbf{R}^n$ 中的凸锥, 则 C_1 和 C_2 不能分离的充要条件是

$$\mathrm{ri}C_1\cap\mathrm{ri}C_2\neq\varnothing\quad \text{和}\quad \mathrm{Sup}C_1+\mathrm{Sup}C_2=\mathbf{R}^n$$

成立.

证明 先证明必要性, 采用反证法. 若 C_1 和 C_2 不能分离, 且 $\mathrm{ri}C_1\cap\mathrm{ri}C_2=\varnothing$, 则由定理 2.4.3 知存在 $\boldsymbol{x}^*\neq\mathbf{0}$ 满足

$$\langle\boldsymbol{x}_1,\boldsymbol{x}^*\rangle\geqslant\langle\boldsymbol{x}_2,\boldsymbol{x}^*\rangle,\quad\forall\boldsymbol{x}_1\in C_1,\quad\forall\boldsymbol{x}_2\in C_2.\tag{3.8.1-9}$$

在 (3.8.1-9) 式中先设 $\boldsymbol{x}_1^*=\boldsymbol{x}^*$, 令 $\boldsymbol{x}_2\to\mathbf{0}$ 得 $\boldsymbol{x}_1^*\in C_1^+$. 然后再设 $\boldsymbol{x}_2^*=-\boldsymbol{x}^*$, 令 $\boldsymbol{x}_1\to\mathbf{0}$ 得 $\boldsymbol{x}_2^*\in C_2^+$. 得到 $\boldsymbol{x}_1^*=-\boldsymbol{x}_2^*$, 由于 $\mathrm{cl}C_i=\mathrm{cl}\left(\mathrm{ri}C_i\right),C_i^+=(\mathrm{cl}C_i)^+=(\mathrm{ri}C_i)^+\,(i=1,2)$, 那么 C_1 和 C_2 可分离, 得矛盾.

再假设 $\mathrm{Sup}C_1+\mathrm{Sup}C_2\neq\mathbf{R}^n$, 因为凸锥的闭凸锥包含原点, 且

$$\mathrm{Sup}\left(\mathrm{cl}C_1\right)=\mathrm{Sup}C_1\quad\text{和}\quad\mathrm{Sup}\left(\mathrm{cl}C_2\right)=\mathrm{Sup}C_2,$$

有 $C_i \subset \mathrm{Sup}C_i\ (i=1,2)$. 因此,

$$C_1 - C_2 \subset \mathrm{Sup}C_1 - \mathrm{Sup}C_2 = \mathrm{Sup}C_1 + \mathrm{Sup}C_2.$$

因为 $\mathrm{Sup}C_1 + \mathrm{Sup}C_2$ 是 $\mathbf{R}^n$ 的真子空间, 则有 $(\mathrm{Sup}C_1 + \mathrm{Sup}C_2)^{\perp} \neq \varnothing$. 设非零向量

$$\boldsymbol{x}^* \in (\mathrm{Sup}C_1 + \mathrm{Sup}C_2)^{\perp},$$

则对任意的 $\boldsymbol{x}_1 \in C_1, \boldsymbol{x}_2 \in C_2$, 有 $\langle \boldsymbol{x}_1 - \boldsymbol{x}_2, \boldsymbol{x}^* \rangle = 0$. 先设 $\boldsymbol{x}_1^* = \boldsymbol{x}^*$, 令 $\boldsymbol{x}_2 \to \mathbf{0}$ 得 $\boldsymbol{x}_1^* \in C_1^+$. 然后再设 $\boldsymbol{x}_2^* = -\boldsymbol{x}^*$, 令 $\boldsymbol{x}_1 \to \mathbf{0}$ 得 $\boldsymbol{x}_2^* \in C_2^+$, 得到 $\boldsymbol{x}_1^* = -\boldsymbol{x}_2^*$, 那么 C_1 和 C_2 可分离, 得矛盾.

再证明充分性. 假设 $\mathrm{ri}C_1 \cap \mathrm{ri}C_2 \neq \varnothing$ 和 $\mathrm{Sup}C_1 + \mathrm{Sup}C_2 = \mathbf{R}^n$ 成立, 令单位向量

$$\boldsymbol{a}_1 = (1,0,\cdots,0)^{\mathrm{T}}, \boldsymbol{a}_2 = (0,1,0,\cdots,0)^{\mathrm{T}}, \cdots, \boldsymbol{a}_n = (0,\cdots,0,1)^{\mathrm{T}},$$

$\boldsymbol{a}_k \in \mathrm{Sup}C_1 + \mathrm{Sup}C_2, k = 1,\cdots,n$. 则存在 $\boldsymbol{b}_k \in \mathrm{Sup}C_1$ 和 $\boldsymbol{c}_k \in \mathrm{Sup}C_2$, 使 $\boldsymbol{a}_k = \boldsymbol{b}_k - \boldsymbol{c}_k$. 设

$$\boldsymbol{b} = \boldsymbol{b}_1 + \cdots + \boldsymbol{b}_n, \quad \boldsymbol{c} = \boldsymbol{c}_1 + \cdots + \boldsymbol{c}_n,$$

显然 $\boldsymbol{b} \in \mathrm{Sup}C_1, \boldsymbol{c} \in \mathrm{Sup}C_2$. 设 $\boldsymbol{x}_0 \in \mathrm{ri}C_1 \cap \mathrm{ri}C_2$, 选择一个充分小的 $\varepsilon > 0$ 使

$$\boldsymbol{x}_0 + \varepsilon\left(\boldsymbol{b}_k - \frac{1}{n+1}\boldsymbol{b}\right) \in C_1, \quad \boldsymbol{x}_0 + \varepsilon\left(\boldsymbol{c}_k - \frac{1}{n+1}\boldsymbol{c}\right) \in C_2, \quad k = 1,\cdots,n,$$

$$\boldsymbol{x}_0 - \frac{\varepsilon}{n+1}\boldsymbol{b} \in C_1, \quad \boldsymbol{x}_0 - \frac{\varepsilon}{n+1}\boldsymbol{c} \in C_2.$$

设 $\boldsymbol{d} = \boldsymbol{b} - \boldsymbol{c} = \boldsymbol{d}_1 + \cdots + \boldsymbol{d}_n$, 则

$$\begin{aligned}
\boldsymbol{y}_k &= \left[\boldsymbol{x}_0 + \varepsilon\left(\boldsymbol{b}_k - \frac{1}{n+1}\boldsymbol{b}\right)\right] - \left[\boldsymbol{x}_0 + \varepsilon\left(\boldsymbol{c}_k - \frac{1}{n+1}\boldsymbol{c}\right)\right] \\
&= \varepsilon\left(\boldsymbol{d}_k - \frac{1}{n+1}\boldsymbol{d}\right) \in C_1 - C_2, \\
\boldsymbol{y}_0 &= \left(\boldsymbol{x}_0 - \frac{\varepsilon}{n+1}\boldsymbol{b}\right) - \left(\boldsymbol{x}_0 - \frac{\varepsilon}{n+1}\boldsymbol{c}\right) = \frac{-\varepsilon}{n+1}\boldsymbol{d} \in C_1 - C_2,
\end{aligned}$$

其中 $k = 1,\cdots,n$, 因为 $\boldsymbol{y}_0, \boldsymbol{y}_1, \cdots, \boldsymbol{y}_n$ 仿射无关, 故单纯形 $\mathrm{Sim}(\boldsymbol{y}_0, \boldsymbol{y}_1, \cdots, \boldsymbol{y}_n) \subset C_1 - C_2$. 但 $\mathbf{0} \in \mathrm{intSim}(\boldsymbol{y}_0, \boldsymbol{y}_1, \cdots, \boldsymbol{y}_n)$, 所以 $\mathbf{0} \in \mathrm{int}(C_1 - C_2)$. 由定理 3.8.4 知充分性成立. □

用归纳法可由定理 3.8.5 得下面的结果.

定理 3.8.6 设 $C_1,\cdots,C_m$ 是 $\mathbf{R}^n$ 中的凸锥, 则 $C_1,\cdots,C_m$ 不能分离的充要条件是:

(1) $\mathrm{ri}C_1\cap\mathrm{ri}C_2\cap\cdots\cap\mathrm{ri}C_m\neq\varnothing$;

(2) $\mathrm{Sup}C_1\cap\cdots\cap\mathrm{Sup}C_{i-1}+\mathrm{Sup}C_i=\mathbf{R}^n,\ i=2,\cdots,m$.

定理 3.8.7 设 C 是 $\mathbf{R}^n$ 中的凸集, 则 $\boldsymbol{x}^*\in(\mathrm{cone}C)^+$ 的充要条件是对于任意的 $\boldsymbol{x}\in C$ 有

$$\langle\boldsymbol{x},\boldsymbol{x}^*\rangle\geqslant 0.$$

证明 若 $\boldsymbol{x}^*\in(\mathrm{cone}C)^+$, 则对任意的 $\boldsymbol{x}\in\mathrm{cone}C$, 有 $\langle\boldsymbol{x},\boldsymbol{x}^*\rangle\geqslant 0$. 显然对于任意的 $\boldsymbol{x}\in C$ 也有

$$\langle\boldsymbol{x},\boldsymbol{x}^*\rangle\geqslant 0.$$

反过来, 对任意的 $\boldsymbol{x}\in C$ 有 $\langle\boldsymbol{x},\boldsymbol{x}^*\rangle\geqslant 0$. 显然, $\langle\lambda\boldsymbol{x},\boldsymbol{x}^*\rangle\geqslant 0$ 对任意 $\lambda\geqslant 0$ 成立, 这说明 $\boldsymbol{x}^*\in(\mathrm{cone}C)^+$. □

本节得到了凸锥的一些分离定理, 它与 2.4 节的凸集分离定理不同的是, 超过两个以上的凸锥分离不能通过超平面来定义, 而是通过各自的共轭锥或负极锥中存在的非零点之和为零来定义. 但是在两个凸锥情形下, 两个凸锥分离等价于存在超平面可分离它们.

本节得到凸锥分离定理的如下性质.

1) 如果有限个凸锥交为空集, 则它们各自对应的共轭锥中存在不全为零的点使得这些点全部加起来等于 0.

2) 如果在有限个凸锥中有一个凸锥与其余的凸锥内部交为非空, 则它们交集的共轭锥等于各自对应的共轭锥之和.

3) 对于给定的有限个凸锥, 下面仅有一种情形成立: 它们交集的共轭锥等于各自对应的共轭锥之和, 或者它们各自对应的共轭锥中存在不全为 0 的点使得这些点全部加起来等于 0.

4) 两个凸锥可分离的充要条件是原点不属于两个凸锥差的内部.

5) 多个凸锥可分离的充要条件是所有凸锥的相对内部交为空集, 或它们的支撑集中存在一个支撑集加上其余支撑集的交不为整个空间.

6) 凸集与它的凸锥包的共轭锥可分离的充要条件是存在凸锥包的共轭锥的一点与凸集中的任何点的内积不小于 0.

3.9 多面体和多面锥

多面体 (锥) 是一类重要的凸集, 区别在于多面体是用有限多个闭半空间的交构成的有界闭集, 而多面锥是用有限多个闭半空间的交构成的无限集. 多面体集

又是多面体的推广, 因此, 本节先介绍多面体的一些性质, 然后再讨论多面锥的一些性质, 最后讨论多面体集的性质.

3.9.1 多面体

在第 2 章曾经讨论过有限多个点生成的仿射集和单纯形等内容, 从下面多面体定义中可以看出, 多面体是单纯形的一种推广, 单纯形是多面体的特殊情形.

定义 3.9.1 设 $S=\{\boldsymbol{x}_1,\cdots,\boldsymbol{x}_m\}$ 是 $\mathbf{R}^n$ 中的 m 个不同点. 如果由这 m 个不同点生成的凸包 $\mathrm{co}S$ 含有 k 个线性无关点, 即满足 $\dim(\mathrm{co}S)=k$, 则称这个凸包为 k -多面体.

事实上, k -多面体 $\mathrm{co}S$ 等价于由 $k+1$ 个仿射无关点构成的凸集, 它的任何一个由 $k+2$ 个点组成的子集一定是仿射相关的. k 维单纯形一定是 k-多面体, 但 k -多面体不一定是 k 维单纯形, 因为 k -多面体中可以含有多个单纯形. 见下面的例子.

例 3.9.1 k 维单纯形 $\mathrm{Sim}(\boldsymbol{x}_0,\boldsymbol{x}_1,\cdots,\boldsymbol{x}_k)$ 在 $\mathbf{R}^n$ 中含有 k 个线性无关的点构成的凸包, 因此单纯形是一个多面体.

例 3.9.2 设 $S=\left\{(1,0)^{\mathrm{T}},(0,1)^{\mathrm{T}},(0,0)^{\mathrm{T}},(1,1)^{\mathrm{T}}\right\}$ 是 $\mathbf{R}^2$ 中的 4 个点, S 的凸包为

$$\mathrm{co}S=\left\{(x_1,x_2)^{\mathrm{T}}\,|0\leqslant x_1\leqslant 1,0\leqslant x_2\leqslant 1\right\},$$

构成了一个 2-多面体, $\mathrm{co}S$ 含有 4 个不同的单纯形:

$$\mathrm{Sim}\left\{(1,0)^{\mathrm{T}},(0,1)^{\mathrm{T}},(0,0)^{\mathrm{T}}\right\},\quad \mathrm{Sim}\left\{(1,0)^{\mathrm{T}},(0,1)^{\mathrm{T}},(1,1)^{\mathrm{T}}\right\},$$
$$\mathrm{Sim}\left\{(1,0)^{\mathrm{T}},(0,0)^{\mathrm{T}},(1,1)^{\mathrm{T}}\right\},\quad \mathrm{Sim}\left\{(0,1)^{\mathrm{T}},(0,0)^{\mathrm{T}},(1,1)^{\mathrm{T}}\right\}.$$

下面的定理给出了多面体的判定结果.

定理 3.9.1 设 $S=\{\boldsymbol{x}_1,\cdots,\boldsymbol{x}_m\}$ 是 $\mathbf{R}^n$ 中的 m 个不同点, 则 $\mathrm{co}S$ 是多面体当且仅当 $\mathrm{co}S$ 是具有有限个极点的凸紧集.

证明 如果 S 是多面体, 由性质 2.3.13 得 S 是凸紧集, 再由定理 2.5.10 知 $\mathrm{ext}S$ 是有限点集.

反之, $\mathrm{co}S$ 是具有有限个极点的凸紧集. 则由定理 2.5.9, S 是其极点集 $\mathrm{ext}S$ 的凸包, 而 $\mathrm{ext}S$ 是有限点集, 故由定义 3.9.1 知 S 是多面体. □

定理 3.9.1 给出了多面体结构性质, 即多面体是具有有限个极点或顶点的凸紧集. 这些极点的全体构成了多面体, 构成了生成多面体的最小集合. 但全体的极点集可能是仿射相关的, 除非 $\dim(\mathrm{co}S)=n$. 下面的结论是多面体的一些基本性质.

性质 3.9.1 如果 S 是 $\mathbf{R}^n$ 中的多面体, 则下面结论均成立:

(1) $S = \mathrm{co}\,(\mathrm{ext}S)$;

(2) 对于 S 中的任意有限子集 $\{\boldsymbol{x}_1, \cdots, \boldsymbol{x}_m\}$, 有

$$S = \mathrm{co}\,\{\boldsymbol{x}_1, \cdots, \boldsymbol{x}_m\} \quad \text{当且仅当} \quad \mathrm{ext}S \subset \{\boldsymbol{x}_1, \cdots, \boldsymbol{x}_m\};$$

(3) 如果 F 是 S 的正常面, 则 F 是多面体且 $\mathrm{ext}F = F \cap \mathrm{ext}S$;

(4) S 的面数量具有有限个;

(5) S 的所有的面都是暴露面;

(6) S 的每一个暴露面都是多面体, 且仅有有限个不同的暴露面.

证明 (1) 结论可由定理 2.5.9 直接获得. 以后称 $\mathrm{ext}S$ 是多面体 S 的最小表示.

(2) 由定理 2.5.10 易得.

(3) 由性质 2.5.5 和定理 3.9.1 知 S 和 F 都是紧集, 且 F 的极点集合是 S 的位于 F 中的极点集合, 即 $\mathrm{ext}F = F \cap \mathrm{ext}S$, 因此 F 是多面体.

(4) 由定理 3.9.1 知 S 的极点数是有限的, S 的每一个面是由它的极点构成的, 所以它的面的个数是有限个.

(5) 和 (6) 容易证明 (作为读者练习). □

例 3.9.3 设 $S = \mathrm{Sim}\left\{(1,0)^{\mathrm{T}}, (0,1)^{\mathrm{T}}, (0,0)^{\mathrm{T}}\right\}$ 是 $\mathbf{R}^2$ 中的 2 维单纯形, 显然它的正常面都是单纯形.

下面结论说明单纯形的正常面还是单纯形.

定理 3.9.2 设 S 是 $\mathbf{R}^n$ 中的 k 维单纯形, 且 F 是 S 的正常面, 则 F 是单纯形.

证明 由性质 3.9.1 知 F 的顶点由 S 的顶点组成. 因为仿射无关集的任何子集都是仿射无关的, 所以 F 的顶点集合也是仿射无关的, 所以 F 是单纯形. □

定理 3.9.3 设 S 是 $\mathbf{R}^n$ 中的 k 维单纯形, $C \subset \mathrm{ext}S$, 则 $\mathrm{co}C$ 是 S 的面, 且 $\mathrm{ext}\,(\mathrm{co}C) = C$.

证明 设 $\mathrm{ext}S = \{\boldsymbol{x}_1, \cdots, \boldsymbol{x}_{k+1}\}$ 和 $C = \{\boldsymbol{x}_1, \cdots, \boldsymbol{x}_m\}\,(k \geqslant m)$. 因为 S 是单纯形, 对任意的 $\boldsymbol{x} \in S$, 存在一组唯一的 $t_i \geqslant 0 \left(i = 1, \cdots, k+1, \sum\limits_{i=1}^{k+1} t_i = 1\right)$ 使得 $\boldsymbol{x} = t_1\boldsymbol{x}_1 + \cdots + t_{k+1}\boldsymbol{x}_{k+1}$. 因为 $\mathrm{co}C \subset S$, 设 $\boldsymbol{c}_1, \boldsymbol{c}_2 \in S$, 存在 $t_{1i} \geqslant 0$ $\left(i = 1, \cdots, k+1, \sum\limits_{i=1}^{k+1} t_{1i} = 1\right)$ 和 $t_{2i} \geqslant 0 \left(i = 1, \cdots, k+1, \sum\limits_{i=1}^{k+1} t_{2i} = 1\right)$ 使得

$$\boldsymbol{c}_1 = \sum_{i=1}^{k+1} t_{1i}\boldsymbol{x}_i \quad \text{和} \quad \boldsymbol{c}_2 = \sum_{i=1}^{k+1} t_{1i}\boldsymbol{x}_i.$$

假设 $t\in(0,1)$ 和 $\boldsymbol{c}=(1-t)\,c_1+tc_2=\sum\limits_{i=1}^{k+1}[(1-t)\,t_{1i}+tt_{2i}]\boldsymbol{x}_i\in\mathrm{co}C$, 则有

$$(1-t)\,t_{1i}+tt_{2i}=0,\quad i=m+1,\cdots,k+1.$$

所以 $t_{0i}=t_{1i}=0, i=m+1,\cdots,k+1$, 即有 $\boldsymbol{c}_1,\boldsymbol{c}_2\in\mathrm{co}C$ 属于 F, 说明 $\mathrm{co}C$ 是 S 的面. 根据性质 3.9.1 知 $\mathrm{ext}\,(\mathrm{co}C)\subset\{\boldsymbol{x}_1,\cdots,\boldsymbol{x}_m\}$, 即 $\mathrm{ext}\,(\mathrm{co}C)=C$. □

定理 3.9.3 说明如果集合是单纯形中若干极点集组成的, 那么这个集合的凸包是单纯形的面, 显然不是极点构成的, 不可能成为它的面, 见下例.

例 3.9.4 设 $S=\mathrm{Sim}\left\{(1,0)^{\mathrm{T}},(0,1)^{\mathrm{T}},(0,0)^{\mathrm{T}}\right\}$ 是 $\mathbf{R}^2$ 中的 2 维单纯形, 则它的正常面都是单纯形, 显然这三个极点不可能形成面.

通过性质 3.9.1 可知, 多面体可以看成有限多个闭半空间的交构成的有界闭集.

定理 3.9.4 设指标集 $I=\{1,\cdots,m\}$, 给定 m 个点 $\boldsymbol{x}_i^*\in\mathbf{R}^n$ 和常数 $\alpha_i\in\mathbf{R}$, 凸集

$$C=\{\boldsymbol{x}\,|\langle\boldsymbol{x},\boldsymbol{x}_i^*\rangle\leqslant\alpha_i,i\in I\}. \tag{3.9.1-1}$$

设 $\boldsymbol{x}_0\in C$ 和对应的指标集

$$I\,(\boldsymbol{x}_0)=\{i\,|\langle\boldsymbol{x}_0,\boldsymbol{x}_i^*\rangle=\alpha_i,i\in I\}, \tag{3.9.1-2}$$

则 $\boldsymbol{x}_0\in\mathrm{ext}C$ 的充要条件是集合 $\{\boldsymbol{x}_i^*\,|i\in I\,(\boldsymbol{x}_0)\}$ 中含有 n 个线性无关的向量. 进一步, 如果 C 是有界集, 则 C 是多面体.

证明 设 $\boldsymbol{x}_0\in\mathrm{ext}C$. 假设 $\{\boldsymbol{x}_i^*\,|i\in I\,(\boldsymbol{x}_0)\}$ 中线性无关的元素个数小于 n, 则变量 $\boldsymbol{x}$ 的线性方程组

$$\langle\boldsymbol{x},\boldsymbol{x}_i^*\rangle=0,\quad i\in I\,(\boldsymbol{x}_0) \tag{3.9.1-3}$$

存在一个非零解 $\boldsymbol{x}_1$. 因此, 存在一个充分小的 $t>0$ 使得

$$\langle\boldsymbol{x}_0-t\boldsymbol{x}_1,\boldsymbol{x}_i^*\rangle=\langle\boldsymbol{x}_0+t\boldsymbol{x}_1,\boldsymbol{x}_i^*\rangle=\alpha_i,\quad i\in I\,(\boldsymbol{x}_0),$$

$$\langle\boldsymbol{x}_0-t\boldsymbol{x}_1,\boldsymbol{x}_i^*\rangle<\alpha_i,\quad \langle\boldsymbol{x}_0+t\boldsymbol{x}_1,\boldsymbol{x}_i^*\rangle<\alpha_i,\quad i\in I\setminus I\,(\boldsymbol{x}_0).$$

因此, 由于 C 是凸集, 存在 $\boldsymbol{x}_0-t\boldsymbol{x}_1\in C$ 和 $\boldsymbol{x}_0+t\boldsymbol{x}_1\in C$ 使得

$$\boldsymbol{x}_0=\frac{1}{2}\,(\boldsymbol{x}_0+t\boldsymbol{x}_1)+\frac{1}{2}\,(\boldsymbol{x}_0-t\boldsymbol{x}_1)\in C.$$

上式说明 $\boldsymbol{x}_0$ 不是 C 的极点, 得到矛盾. 所以 $\{\boldsymbol{x}_i^*\,|i\in I\,(\boldsymbol{x}_0)\}$ 含 n 个线性无关的向量.

反过来, 假设 $\{\boldsymbol{x}_i^* | i \in I(\boldsymbol{x}_0)\}$ 中线性无关的元素个数是 n. 由 (3.9.1-1) 式, 将定义 C 的不等式组分成下面形式

$$\langle \boldsymbol{x}_0, \boldsymbol{x}_i^* \rangle = \alpha_i, \quad i \in I(\boldsymbol{x}_0), \tag{3.9.1-4}$$

$$\langle \boldsymbol{x}_0, \boldsymbol{x}_i^* \rangle \leqslant \alpha_i, \quad i \in I \setminus I(\boldsymbol{x}_0). \tag{3.9.1-5}$$

假设 $\boldsymbol{x}_0$ 不是 C 的极点, 则存在两个不同点 $\boldsymbol{x}_1, \boldsymbol{x}_2 \in C$ 以及 $t\,(0 < t < 1)$ 使得

$$\boldsymbol{x}_0 = (1-t)\,\boldsymbol{x}_1 + t\boldsymbol{x}_2 \in C.$$

通过 (3.9.1-4) 式必然有

$$\langle \boldsymbol{x}_1, \boldsymbol{x}_i^* \rangle = \alpha_i, \quad \langle \boldsymbol{x}_2, \boldsymbol{x}_i^* \rangle = \alpha_i, \quad i \in I(\boldsymbol{x}_0). \tag{3.9.1-6}$$

设 I_0 是 $I(\boldsymbol{x}_0)$ 中 n 个线性无关元素 $\boldsymbol{x}_i^*$ 对应的指标集, 则 (3.9.1-6) 式是关于 $\boldsymbol{x}_i^*$ 的一个线性方程组

$$\langle \boldsymbol{x}, \boldsymbol{x}_i^* \rangle = \alpha_i, \quad i \in I_0,$$

线性无关的系数向量 $\{\boldsymbol{x}_i^* | i \in I(\boldsymbol{x}_0)\}$ 是 n 个. 因此, 它具有唯一解. 根据 (3.9.1-6) 式, 存在两个不同的解, 得到矛盾. 因此, $\boldsymbol{x}_0$ 是 C 的极点.

如果 C 有界, 根据 C 的定义, 它是凸紧集. 再由定理 2.5.9, 它由极点凸包组成. 根据上面的证明, C 的每一个极点是形如具有 n 个变量的 n 个线性方程的方程组 (3.9.1-3) 的唯一解. 因为 I 是有限的, 所以极点个数也是有限的, 根据定理 3.9.1 知 C 是多面体. □

上述定理说明有限个闭半空间的交形成的有界集构成多面体. 但是, 多面体是由无穷多个支撑超平面构成的, 定理 3.9.4 说明实际上只需要有限多个支撑超平面组成多面体即可.

定义 3.9.2 设 $\boldsymbol{x}_1, \cdots, \boldsymbol{x}_m$ 是 $\mathbf{R}^n$ 中的不同点, $C = \mathrm{co}\{\boldsymbol{x}_1, \cdots, \boldsymbol{x}_m\}$, 称函数

$$\delta_C(\boldsymbol{x}^*) = \sup_{\boldsymbol{x} \in C} \langle \boldsymbol{x}, \boldsymbol{x}^* \rangle = \max_{1 \leqslant i \leqslant m} \langle \boldsymbol{x}_i, \boldsymbol{x}^* \rangle \tag{3.9.1-7}$$

是 C 的支撑函数, C 的支撑超平面定义为

$$H_C(\boldsymbol{x}^*) = \{\boldsymbol{x} | \langle \boldsymbol{x}, \boldsymbol{x}^* \rangle = \delta_C(\boldsymbol{x}^*)\}. \tag{3.9.1-8}$$

设指标集 $I(\boldsymbol{x}^*) = \{i | \langle \boldsymbol{x}_i, \boldsymbol{x}^* \rangle = \delta_C(\boldsymbol{x}^*), i = 1, \cdots, m\}$. 若给定一个 $j \in I(\boldsymbol{x}^*)$, 如果当任意 $i \in I(\boldsymbol{x}^*)$ 时有形如 $\boldsymbol{x}_i - \boldsymbol{x}_j$ 的 $n-1$ 个元素线性无关, 则称 $H_C(\boldsymbol{x}^*)$ 是 C 的极支撑超平面.

例 3.9.5　设 $S=\left\{(1,0)^{\mathrm{T}},(0,1)^{\mathrm{T}},(0,0)^{\mathrm{T}},(1,1)^{\mathrm{T}}\right\}$ 是 $\mathbf{R}^2$ 中的 4 个点, S 的凸包为

$$C=\mathrm{co}S=\left\{(x_1,x_2)^{\mathrm{T}}\,|0\leqslant x_1\leqslant 1,0\leqslant x_2\leqslant 1\right\},$$

$$\delta_C\left(\boldsymbol{x}^*\right)=\sup_{\boldsymbol{x}\in C}\langle\boldsymbol{x},\boldsymbol{x}^*\rangle=\max_{1\leqslant i\leqslant m}\langle\boldsymbol{x}_i,\boldsymbol{x}^*\rangle=\max\left\{x_1^*,x_2^*,0,x_1^*+x_2^*\right\}.$$

取 $\boldsymbol{x}^*=(1,0)^{\mathrm{T}}$, 得支撑超平面 $H_C\left(\boldsymbol{x}^*\right)=\{\boldsymbol{x}\,|x_1=1\}$, $I\left(\boldsymbol{x}^*\right)=\{1,4\}$, 显然 $(1,0)^{\mathrm{T}}-(1,1)^{\mathrm{T}}$ 是 $n-1=1$ 个元素线性无关, 则 $H_C\left(\boldsymbol{x}^*\right)$ 是 C 的极支撑超平面. 取 $\boldsymbol{x}_0=(2,2)\notin C$, 显然有

$$\langle\boldsymbol{x},\boldsymbol{x}^*\rangle<\langle\boldsymbol{x}_0,\boldsymbol{x}^*\rangle,$$

对任意的 $\boldsymbol{x}\in C$ 成立.

例 3.9.5 说明了多面体存在极支撑超平面. 这一结论即为下面定理的.

定理 3.9.5　设 $C=\mathrm{co}\left\{\boldsymbol{x}_1,\cdots,\boldsymbol{x}_m\right\}$ 是 $\mathbf{R}^n$ 中的内部非空的多面体. 如果 $\boldsymbol{x}_0\notin C$, 则存在一个 $\boldsymbol{x}^*$ 构成的极支撑超平面 $H_C\left(\boldsymbol{x}^*\right)$ 使得 $\langle\boldsymbol{x},\boldsymbol{x}^*\rangle<\langle\boldsymbol{x}_0,\boldsymbol{x}^*\rangle$ 对任意的 $\boldsymbol{x}\in C$ 成立.

证明　因多面体 C 是凸紧集和 $\boldsymbol{x}_0\notin C$, 则由定理 2.4.7, 存在 $\boldsymbol{x}^*\in\mathbf{R}^n$ 和 $\alpha\in\mathbf{R}$ 使得超平面

$$H_C\left(\boldsymbol{x}^*,\alpha\right)=\left\{\boldsymbol{x}\,|\langle\boldsymbol{x},\boldsymbol{x}^*\rangle=\alpha\right\}$$

满足下面分离条件

$$\langle\boldsymbol{x},\boldsymbol{x}^*\rangle\leqslant\alpha<\langle\boldsymbol{x}_0,\boldsymbol{x}^*\rangle,\quad\forall\boldsymbol{x}\in C.\tag{3.9.1-9}$$

根据定义 3.9.2 得

$$\delta_C\left(\boldsymbol{x}^*\right)\leqslant\alpha<\langle\boldsymbol{x}_0,\boldsymbol{x}^*\rangle.$$

因为 C 是内部非空点, 不妨设 $\mathbf{0}\in\mathrm{int}C$, 有 $\delta_C\left(\boldsymbol{x}^*\right)=\alpha>0$. 如果支撑超平面 $H_C\left(\boldsymbol{x}^*,\alpha\right)$ 是极支撑超平面, 则定理得证. 如果 $\boldsymbol{x}^*$ 不是极支撑超平面, 则由定义 3.9.2 知向量集 $\{\boldsymbol{x}_i-\boldsymbol{x}_j\}$ 中 (其中 $\forall i\in I\left(\boldsymbol{x}^*\right)$, j 是 $I\left(\boldsymbol{x}^*\right)$ 中一个给定元素) 线性无关的向量不超过 $n-2$ 个, 而向量集 $\left\{\boldsymbol{x}_0-\boldsymbol{x}_j,\boldsymbol{x}_i-\boldsymbol{x}_j,i\in I\left(\boldsymbol{x}^*\right)\right\}$ 中线性无关的向量不超过 $n-1$ 个. 所以存在一个 $\boldsymbol{x}_1^*$ 使得

$$\langle\boldsymbol{x}_0,\boldsymbol{x}_1^*\rangle\leqslant 0,\quad\langle\boldsymbol{x}_i-\boldsymbol{x}_j,\boldsymbol{x}_1^*\rangle=\langle\boldsymbol{x}_0-\boldsymbol{x}_j,\boldsymbol{x}_1^*\rangle=0,\quad\forall i\in I\left(\boldsymbol{x}^*\right).\tag{3.9.1-10}$$

因为在 $\{\boldsymbol{x}_1,\cdots,\boldsymbol{x}_m\}$ 中存在一个元素 $\boldsymbol{x}_k$ 使得 $\langle\boldsymbol{x}_k,\boldsymbol{x}_1^*\rangle>0$, 由 (3.9.1-10) 式知, 当 $i\in I\left(\boldsymbol{x}^*\right)$ 时, $\langle\boldsymbol{x}_i,\bar{\boldsymbol{x}}^*\rangle\leqslant 0$, 所以 $k\notin I\left(\boldsymbol{x}^*\right)$. 设

$$\alpha_k=\frac{\delta_C\left(\boldsymbol{x}^*\right)-\langle\boldsymbol{x}_k,\boldsymbol{x}^*\rangle}{\langle\boldsymbol{x}_k,\boldsymbol{x}_1^*\rangle-\langle\boldsymbol{x}_0,\boldsymbol{x}_1^*\rangle}.\tag{3.9.1-11}$$

由 (3.9.1-10) 式知 $\alpha_k > 0$. 这样有任意 $k \notin I(\boldsymbol{x}^*)$, 得到一个 $\alpha_k > 0$, 不妨设

$$\alpha_k = \min\{\alpha_l | l \in \{1, \cdots, m\}, l \notin I(\boldsymbol{x}^*)\}.$$

因此有 $\langle \boldsymbol{x}_k, \boldsymbol{x}^* \rangle < \delta_C(\boldsymbol{x}^*)$, 由 (3.9.1-11) 式得

$$\langle \boldsymbol{x}_k, \boldsymbol{x}^* + \alpha_k \boldsymbol{x}_1^* \rangle = \delta_C(\boldsymbol{x}^*) + \alpha_k \langle \boldsymbol{x}_0, \boldsymbol{x}_1^* \rangle. \tag{3.9.1-12}$$

当 $i \in I(\boldsymbol{x}^*)$ 时, 由 (3.9.1-10) 式及 $I(\boldsymbol{x}^*)$ 的定义, 对任意的 $\alpha \in \mathbf{R}$, 均有

$$\langle \boldsymbol{x}_i, \boldsymbol{x}^* + \alpha \boldsymbol{x}_1^* \rangle = \delta_C(\boldsymbol{x}^*) + \alpha \langle \boldsymbol{x}_0, \boldsymbol{x}_1^* \rangle. \tag{3.9.1-13}$$

再由 (3.9.1-12) 式和 (3.9.1-13) 式得

$$\delta_C(\boldsymbol{x}^* + \alpha_k \boldsymbol{x}_1^*) = \delta_C(\boldsymbol{x}^*) + \alpha_k \langle \boldsymbol{x}_0, \boldsymbol{x}_1^* \rangle, \tag{3.9.1-14}$$

$$I(\boldsymbol{x}^* + \alpha_k \boldsymbol{x}_1^*) \supset I(\boldsymbol{x}^*) \cup \{k\}. \tag{3.9.1-15}$$

因为 $\alpha_k > 0$, 由 (3.9.1-10) 式得

$$\langle \boldsymbol{x}_k - \boldsymbol{x}_i, \boldsymbol{x}_1^* \rangle = \langle \boldsymbol{x}_k, \boldsymbol{x}_1^* \rangle - \langle \boldsymbol{x}_0, \boldsymbol{x}_1^* \rangle > 0,$$

所以向量集 $\{\boldsymbol{x}_k - \boldsymbol{x}_j, \boldsymbol{x}_i - \boldsymbol{x}_j, i \in I(\boldsymbol{x}^*)\}$ 线性无关, 即向量集

$$\{\boldsymbol{x}_i - \boldsymbol{x}_j, i \in I(\boldsymbol{x}^* + \alpha_k \boldsymbol{x}_1^*)\}$$

中线性无关的向量个数至少 $n-1$ 个. 因此, 由 (3.9.1-14) 式, 对任意的 $\boldsymbol{x} \in C$, 有

$$\langle \boldsymbol{x}, \boldsymbol{x}^* + \alpha_k \boldsymbol{x}_1^* \rangle \leqslant \delta_C(\boldsymbol{x}^* + \alpha_k \boldsymbol{x}_1^*) < \langle \boldsymbol{x}_0, \boldsymbol{x}^* + \alpha_k \boldsymbol{x}_1^* \rangle.$$

上式表明了 $\boldsymbol{x}^* + \alpha_k \boldsymbol{x}_1^*$ 满足定理的结论. 因为 $\{1, \cdots, m\}$ 有限, $\mathrm{int}C \neq \varnothing$, 说明重复上述的过程可以在有限步骤内作出分离 $\boldsymbol{x}_0$ 和 C 的极支撑超平面. □

最后, 证明 $\mathbf{R}^n$ 中的多面体的结构是由有限个不等式组构成的.

定理 3.9.6 C 在 $\mathbf{R}^n$ 中的任何多面体都可以用有限个线性不等式组构成.

证明 如果 $\mathrm{int}C \neq \varnothing$, 则由定理 3.9.5 知 C 的极支撑超平面的个数有限, 设

$$\boldsymbol{x}_i^* \quad (i = 1, \cdots, m)$$

是极支撑超平面对应的向量. 于是不等式组

$$\langle \boldsymbol{x}, \boldsymbol{x}_i^* \rangle \leqslant \delta_C(\boldsymbol{x}_i^*), \quad i = 1, \cdots, m \tag{3.9.1-16}$$

定义了 C 的支撑超平面.

如果 $\text{int}C=\varnothing$, 取 $\boldsymbol{x}_0\in C$, 那么支撑子空间 $\text{Sup}C$ 中集合 $C-\boldsymbol{x}_0$ 应包含内点, 通过 $\text{Sup}C$ 定义有限的不等式组确定 $C-\boldsymbol{x}_0$. □

本小节得到了多面体的一些重要性质:

1) 有限多个点的凸包是多面体等价于该凸包是具有有限个极点的凸紧集.

2) 多面体由它的极点集凸包组成; 多面体由其有限点构成的凸包组成等价于多面体的极点集属于这个有限点集合; 多面体的一个正常面还是多面体等价于该正常面的极点恰好等于该正常面与多面体极点的交; 多面体的面的个数是有限个, 且都是暴露面与多面体.

3) 有限个闭半空间的交形成的有界集构成多面体.

4) (多面体分离定理) 若 C 是内部非空的多面体和 $\boldsymbol{x}_0\notin C$, 则存在一个 $\boldsymbol{x}^*$ 构成的极支撑超平面 $H_C(\boldsymbol{x}^*)$ 使得 $\langle\boldsymbol{x},\boldsymbol{x}^*\rangle<\langle\boldsymbol{x}_0,\boldsymbol{x}^*\rangle$ 对任意的 $\boldsymbol{x}\in C$ 成立.

5) 多面体是由有限个线性不等式组构成的.

3.9.2 多面锥

多面锥是多面体的另一种情形, 它由有限多个点的非负线性组合的点组成, 有限个点集组成的凸包是多面体, 显然多面体不是多面锥. 本节讨论这种特殊多面体的一些性质.

定义 3.9.3 设 C 是 $\mathbf{R}^n$ 中的非空集合, 如果存在 $\{\boldsymbol{x}_1,\cdots,\boldsymbol{x}_m\}$ 使得

$$C=\left\{\sum_{i=1}^{m}\lambda_i\boldsymbol{x}_i\,\middle|\,\lambda_i\geqslant 0,i=1,\cdots,m\right\},\tag{3.9.2-1}$$

则称 C 是多面锥或多面体锥.

根据多面锥定义, 多面锥是多面体的推广, 因此多面体 $\text{co}\{\boldsymbol{x}_1,\cdots,\boldsymbol{x}_m\}\subset C$.

例 3.9.6 设 $S=\left\{(1,0)^{\mathrm{T}},(0,1)^{\mathrm{T}},(0,0)^{\mathrm{T}},(1,1)^{\mathrm{T}}\right\}$ 是 $\mathbf{R}^2$ 中的 4 个点, S 的多面锥

$$C=\left\{(\lambda_1+\lambda_4,\lambda_2+\lambda_4)^{\mathrm{T}}\,|\lambda_i\geqslant 0,i=1,2,3,4\right\}$$

等价于 $C=\left\{(\lambda_1,\lambda_2)^{\mathrm{T}}\,|\lambda_i\geqslant 0,i=1,2\right\}$. 因此, 有

$$\text{co}S=\left\{(\lambda_1,\lambda_2)^{\mathrm{T}}\,|1\geqslant\lambda_i\geqslant 0,i=1,2\right\}.$$

取 $\boldsymbol{x}_1^*=(1,0)^{\mathrm{T}},\boldsymbol{x}_2^*=(0,1)^{\mathrm{T}}$, 下面的定理 3.9.7 与性质 3.9.2 的结论对应成立.

下面证明多面体锥的共轭锥还是多面体锥.

定理 3.9.7 设 $\mathbf{R}^n$ 中的凸锥 C 由 $C=\{\boldsymbol{x}\,|\langle\boldsymbol{x},\boldsymbol{x}_k^*\rangle\geqslant 0,k=1,\cdots,p\}$ 定义，则 C 的共轭锥 C^+ 是多面锥，且

$$C^+=\left\{\boldsymbol{x}^*\left|\boldsymbol{x}^*=\sum_{k=1}^{m}\lambda_k\boldsymbol{x}_k^*,\lambda_k\geqslant 0,k=1,\cdots,p\right.\right\}. \tag{3.9.2-2}$$

证明 设 $\{\boldsymbol{x}_1^*,\cdots,\boldsymbol{x}_p^*\}\subset\mathbf{R}^n$ 使

$$C=\{\boldsymbol{x}\,|\langle\boldsymbol{x},\boldsymbol{x}_k^*\rangle\geqslant 0,k=1,\cdots,p\}$$

成立，由性质 3.9.2 和共轭锥定义得

$$\langle\boldsymbol{x},\boldsymbol{x}_k^*\rangle\geqslant 0,\quad k=1,\cdots,p.$$

说明 $\{\boldsymbol{x}_1^*,\cdots,\boldsymbol{x}_p^*\}\subset C^+$. 设 $\forall\boldsymbol{x}\in C$，对于 $\lambda_k\geqslant 0,k=1,\cdots,p$，由上式有

$$\left\langle\boldsymbol{x},\sum_{k=1}^{p}\lambda_k\boldsymbol{x}_k^*\right\rangle\geqslant 0.$$

即说明 (3.9.2-2) 式成立，根据表示形式由定义 3.9.3 知 C^+ 是多面锥. □

以上结论说明了多面体与共轭锥具有对偶性. 下面证明多面锥是闭凸锥，且由有限多个点构成的线性不等式组的交组成.

性质 3.9.2 设 C 是多面锥，则 C 是闭凸锥，且存在 $\{\boldsymbol{x}_1^*,\cdots,\boldsymbol{x}_p^*\}\subset\mathbf{R}^n$ 使得

$$C=\{\boldsymbol{x}\,|\langle\boldsymbol{x},\boldsymbol{x}_k^*\rangle\geqslant 0,k=1,\cdots,p\}. \tag{3.9.2-3}$$

证明 设 $\forall t>0$ 和 $\boldsymbol{x}=\sum\limits_{i=1}^{m}\lambda_i\boldsymbol{x}_i\in C$，其中 $\lambda_i\geqslant 0\,(i=1,\cdots,m)$，所以

$$t\boldsymbol{x}=\sum_{i=1}^{m}t\lambda_i\boldsymbol{x}_i\in C,\quad t\lambda_i\geqslant 0\quad(i=1,\cdots,m).$$

再设 $\boldsymbol{x}'=\sum\limits_{i=1}^{m}\lambda_i'\boldsymbol{x}_i\in C$，其中 $\lambda_i'\geqslant 0\,(i=1,\cdots,m)$，显然有

$$\boldsymbol{x}+\boldsymbol{x}'=\sum_{i=1}^{m}(\lambda_i\boldsymbol{x}_i+\lambda_i'\boldsymbol{x}_i)\in C,\quad \lambda_i+\lambda_i'\geqslant 0,\quad i=1,2,\cdots,m,$$

即 C 是凸锥，闭性显然. 设

$$D=\left\{\sum_{i=1}^{m}\lambda_i\boldsymbol{x}_i\left|1\geqslant\lambda_i\geqslant 0,i=1,\cdots,m\right.\right\},$$

有 $\operatorname{co}\{\boldsymbol{x}_1,\cdots,\boldsymbol{x}_m\}=D\subset C$. 由定理 3.9.7 知存在有限个 $\boldsymbol{x}_k^*\in\mathbf{R}^n\,(k=1,2,\cdots,p)$ 使得

$$D=\{\boldsymbol{x}|\langle\boldsymbol{x},\boldsymbol{x}_i^*\rangle\geqslant\alpha_i,i=1,2,\cdots,p\}.$$

因为 $\mathbf{0}\in D$, 则有 $\alpha_i=0\,(i=1,2,\cdots,p)$, 得 $D=\{\boldsymbol{x}|\langle\boldsymbol{x},\boldsymbol{x}_i^*\rangle\geqslant 0,i=1,2,\cdots,p\}$, 由此得 $C\subset D$. □

下面证明两个多面锥之和是多面锥, 两个多面锥交集还是多面锥, 这也说明了多面锥的加和交运算具有不变性.

性质 3.9.3　设 C_1 和 C_2 是 $\mathbf{R}^n$ 中的多面锥, 则 C_1+C_2 和 $C_1\cap C_2$ 是多面锥.

证明　设 C_1 和 C_2 是 $\mathbf{R}^n$ 中的多面锥, 如果存在 $\{\boldsymbol{x}_1^1,\boldsymbol{x}_2^1,\cdots,\boldsymbol{x}_m^1\}$ 和 $\{\boldsymbol{x}_1^2,\boldsymbol{x}_2^2,\cdots,\boldsymbol{x}_m^2\}$ 使得

$$C_1=\left\{\sum_{i=1}^m\lambda_i\boldsymbol{x}_i^1\middle|\lambda_i\geqslant 0,i=1,\cdots,m\right\},$$

$$C_2=\left\{\sum_{i=1}^m\mu_i\boldsymbol{x}_i^2\middle|\mu_i\geqslant 0,i=1,\cdots,m\right\}.$$

那么有

$$C_1+C_2=\left\{\sum_{i=1}^m\left(\lambda_i\boldsymbol{x}_i^1+\mu_i\boldsymbol{x}_i^2\right)\middle|\lambda_i,\mu_2\geqslant 0,i=1,\cdots,m\right\},$$

$$C_1\cap C_2=\left\{\sum_{i=1}^m\lambda_i\boldsymbol{x}_i^1\middle|\boldsymbol{x}_i^1=\boldsymbol{x}_i^2,\lambda_i=\mu_i,\lambda_i\geqslant 0,i=1,\cdots,m\right\}.$$

由定义 3.9.3 得 C_1+C_2 和 $C_1\cap C_2$ 是多面锥. □

最后得到下面的性质, 即多个多面锥的交的共轭锥等于每个多面锥的共轭锥之和.

性质 3.9.4　设 $C_1,\cdots,C_m$ 是 $\mathbf{R}^n$ 中的多面锥, 则

$$(C_1\cap\cdots\cap C_m)^+=C_1^++\cdots+C_m^+.$$

证明　由性质 3.4.3 和定理 3.9.7 得结论成立. □

本小节得到多面锥的几个性质:

1) 多面锥是闭凸锥, 且由有限多个点构成的线性不等式交组成, 多面体锥的共轭锥还是多面体锥;

2) 两个多面锥之和是多面锥, 两个多面锥交集是多面锥, 多个多面锥的交的共轭锥等于每个多面锥的共轭锥之和.

3.9.3 多面体集

根据定理 3.9.6 可知, 多面体是由有限个闭半空间的交构成的有界闭凸集, 而多面锥是由有限个过原点的闭半空间的交构成的无界凸锥. 换句话说, 有限个闭半空间的交构成的集合是多面体或多面锥. 事实上, 有界的多面体集就是多面体, 而多面锥则是多面体的无界情形. 下面进一步推广多面体到多面体集的概念.

定义 3.9.4 设 $C \subset \mathbf{R}^n$ 是非空集合. 如果 $\boldsymbol{x}_i^* \in \mathbf{R}^n, \alpha_i \in \mathbf{R}\,(i=1,\cdots,m)$ 使得

$$C=\bigcap_{i=1}^{m} C^{\geqslant}(\boldsymbol{x}_i^*,\alpha_i), \tag{3.9.3-1}$$

其中 $C^{\geqslant}(\boldsymbol{x}_i^*,\alpha_i)=\{\boldsymbol{x}\,|\langle \boldsymbol{x},\boldsymbol{x}_i^*\rangle \geqslant \alpha_i\}\,(i=1,\cdots,m)$, 则称 C 是多面体集.

多面体集与多面体、多面锥存在着区别, 多面体集可以是多面锥, 不一定是多面体, 也可能既不是多面体也不是多面锥, 见下面例子.

例 3.9.7 设 $C=\left\{(1,0)^{\mathrm{T}},(0,1)^{\mathrm{T}},(1,1)^{\mathrm{T}}\right\}$ 是 $\mathbf{R}^2$ 中 3 个点构成的集合, 且

$$C^{\geqslant}\left((1,0)^{\mathrm{T}},0\right)=\left\{(x_1,x_2)^{\mathrm{T}}\,|x_1\geqslant 0, x_2\in\mathbf{R}\right\}, \tag{3.9.3-2}$$

$$C^{\geqslant}\left((0,1)^{\mathrm{T}},0\right)=\left\{(x_1,x_2)^{\mathrm{T}}\,|x_2\geqslant 0, x_1\in\mathbf{R}\right\}, \tag{3.9.3-3}$$

$$C^{\geqslant}\left((1,1)^{\mathrm{T}},0\right)=\left\{(x_1,x_2)^{\mathrm{T}}\,|x_1+x_2\geqslant 0\right\}. \tag{3.9.3-4}$$

则有 $C=C^{\geqslant}\left((1,0)^{\mathrm{T}},0\right)\cap C^{\geqslant}\left((0,1)^{\mathrm{T}},0\right)\cap C^{\geqslant}\left((1,1)^{\mathrm{T}},0\right)=\{(x_1,x_2)^{\mathrm{T}}\,|x_1\geqslant 0, x_2\geqslant 0\}$ 是多面体集, 也是多面锥, 但不是多面体.

设 $S=\left\{(1,0)^{\mathrm{T}},(0,1)^{\mathrm{T}},(-1,-1)^{\mathrm{T}}\right\}$ 是 $\mathbf{R}^2$ 中 3 个点构成的集合, 将 (3.9.3-4) 式改为

$$S^{\geqslant}\left((-1,-1)^{\mathrm{T}},-1\right)=\left\{(x_1,x_2)^{\mathrm{T}}\,|x_1+x_2\leqslant 1\right\}, \tag{3.9.3-5}$$

则有

$$\begin{aligned}
S&=S^{\geqslant}\left((1,0)^{\mathrm{T}},0\right)\cap S^{\geqslant}\left((0,1)^{\mathrm{T}},0\right)\cap S^{\geqslant}\left((-1,-1)^{\mathrm{T}},-1\right)\\
&=\left\{(x_1,x_2)^{\mathrm{T}}\,|x_1\geqslant 0, x_2\geqslant 0, x_1+x_2\leqslant 1\right\}
\end{aligned}$$

是多面体集, 但不是多面体也不是多面锥.

多面体集具有下面的几个基本性质.

性质 3.9.5 设 $C\subset\mathbf{R}^n$ 是多面体集, 下面结论成立:

(1) C 是闭凸集;

(2) 有限个多面体集的交仍是多面体集;

(3) 多面体集平移后得到的集合也是多面体集;

(4) $\mathbf{R}^n$ 中的多面体集等价于多面体与多面锥的和.

证明 (1)~(3) 由定义 3.9.4 直接得到. 下面只需要证明结论 (4) 即可.

设 C 是多面体集, 令 $\forall \boldsymbol{x} \in C$, 有 $\boldsymbol{x}_i^* \in \mathbf{R}^n, \alpha_i \in \mathbf{R}\,(i=1,\cdots,m)$ 使得

$$\langle \boldsymbol{x}, \boldsymbol{x}_i^* \rangle \geqslant \alpha_i, \quad i=1,\cdots,m.$$

设 $\boldsymbol{y}=(\boldsymbol{x},1)^{\mathrm{T}} \in \mathbf{R}^{n+1}$, $\boldsymbol{y}_i^*=(\boldsymbol{x}_i^*,-\alpha_i)^{\mathrm{T}} \in \mathbf{R}^{n+1}$, 则

$$\langle \boldsymbol{y}, \boldsymbol{y}_i^* \rangle \geqslant 0, \quad i=1,\cdots,m$$

是 $\mathbf{R}^{n+1}$ 中的齐次线性不等式组. 任取 $t>0$, 显然有 $\langle t\boldsymbol{y}, \boldsymbol{y}_i^* \rangle \geqslant 0\,(\,i=1,\cdots,m)$ 成立. 由性质 3.9.2 得 C 定义了一个在 $\mathbf{R}^{n+1}$ 中的多面锥, 存在一组元素 $\{\boldsymbol{y}_1, \boldsymbol{y}_2,\cdots,\boldsymbol{y}_p\}$ 使得

$$\boldsymbol{y}=\sum_{j=1}^{p}\lambda_j\boldsymbol{y}_j, \quad \lambda_j \geqslant 0, \quad j=1,2,\cdots,p,$$

其中 $\boldsymbol{y}_j=\left(\boldsymbol{x}_j, x_j^{n+1}\right)^{\mathrm{T}} \in \mathbf{R}^{n+1}\,(j=1,2,\cdots,p)$, $\boldsymbol{x}=\sum\limits_{j=1}^{p}\lambda_j\boldsymbol{x}_j$ 和 $1=\sum\limits_{j=1}^{p}\lambda_j x_j^{n+1}$.

设指标集:

$$I^0=\left\{j\,\middle|\,x_j^{n+1}=0, j=1,\cdots,p\right\},$$

$$I^+=\left\{j\,\middle|\,x_j^{n+1}>0, j=1,\cdots,p\right\}.$$

因此, 我们有

$$\boldsymbol{x}=\sum_{j=1}^{p}\lambda_j\boldsymbol{x}_j=\sum_{j\in I^0}\lambda_j\boldsymbol{x}_j+\sum_{j\in I^+}\lambda_j x_j^{n+1}\left(\boldsymbol{x}_j/x_j^{n+1}\right).$$

在 $j\in I^+$ 时, 令 $\boldsymbol{x}_j'=\boldsymbol{x}_j/x_j^{n+1}, \mu_j=\lambda_j x_j^{n+1}$, 则上式为

$$\boldsymbol{x}=\sum_{j\in I^0}\lambda_j\boldsymbol{x}_j+\sum_{j\in I^+}\mu_j\boldsymbol{x}_j', \quad \lambda_j\geqslant 0, \quad j\in I^0, \quad \mu_j\geqslant 0, \quad j\in I^+, \quad \sum_{j\in I^+}\mu_j=1,$$

上式右边第一部分求和式是多面锥的点, 第二部分求和式是多面体的点, 所以多面体集是多面体与多面锥的和. 反之亦然, 即多面体与多面锥的和是多面体集. □

下面的性质说明多面体集的边界由多面体集的有限多个超平面半空间边界组成, 这些边界与多面体的交集构成多面体集的超平面, 多面体集的每个面还是多面体集, 且面的个数是有限的. 这些结论说明了多面体集与其支持的超平面有密切关系.

性质 3.9.6 设 C 是 $\mathbf{R}^n$ 中的 n 维多面体集. 如果 $\boldsymbol{x}_i^*\in\mathbf{R}^n,\alpha_i\in\mathbf{R}(i=1,\cdots,m)$ 使得

$$C=\bigcap_{i=1}^{m}H^{\geqslant}(\boldsymbol{x}_i^*,\alpha_i),$$

其中 $H^{\geqslant}(\boldsymbol{x}_i^*,\alpha_i)=\{\boldsymbol{x}\,|\langle\boldsymbol{x},\boldsymbol{x}_i^*\rangle\geqslant\alpha_i\}\,(i=1,\cdots,m)$, 则下面的结果成立:

(1) $\mathrm{bd}C=\bigcup_{i=1}^{m}\mathrm{bd}\left(H^{\geqslant}(\boldsymbol{x}_i^*,\alpha_i)\right)\cap C$;

(2) C 的每一个超平面的形式为 $H(\boldsymbol{x}_i^*,\alpha_i)\cap C$;

(3) 如果 F 是 C 的正常面, 则在 C 中存在包含 F 的超平面 G;

(4) C 的每一个面都是多面体集;

(5) 多面体集 C 的面的个数有限.

证明 (1) 因为

$$\mathrm{int}C=\mathrm{int}\bigcap_{i=1}^{m}H^{\geqslant}(\boldsymbol{x}_i^*,\alpha_i)=\bigcap_{i=1}^{m}\mathrm{int}H^{\geqslant}(\boldsymbol{x}_i^*,\alpha_i)=\bigcap_{i=1}^{m}H^{\geqslant}(\boldsymbol{x}_i^*,\alpha_i)\setminus H(\boldsymbol{x}_i^*,\alpha_i),$$

所以 (1) 成立.

(2) 设 F 是 C 的超平面, $\boldsymbol{x}\in\mathrm{ri}F$. 则由性质 2.5.5 知 F 是 C 包含 $\boldsymbol{x}$ 的最小面. 由 (1) 知存在一个超平面满足 $\boldsymbol{x}\in H\left(\boldsymbol{x}_j^*,\alpha_j\right)\cap C$, 得 $F\subset H\left(\boldsymbol{x}_j^*,\alpha_j\right)\cap C$. 因为 F 与 $H\left(\boldsymbol{x}_j^*,\alpha_j\right)\cap C$ 都是 C 的面, 且 $\mathrm{dim}F=n-1$. 因此, 有

$$F=H\left(\boldsymbol{x}_j^*,\alpha_j\right)\cap C.$$

(3) 根据 (2), 设 $\boldsymbol{x}\in\mathrm{ri}F$, 存在一个超平面使得 $\boldsymbol{x}\in H\left(\boldsymbol{x}_j^*,\alpha_j\right)\cap C$. 由性质 2.5.5, 设 F 是包含 $\boldsymbol{x}$ 的最小面, 知 $H\left(\boldsymbol{x}_j^*,\alpha_j\right)\cap C$ 是包含 $\boldsymbol{x}$ 的超平面, 再令 $G=H\left(\boldsymbol{x}_j^*,\alpha_j\right)\cap C$, 则得 G 是包含 F 的超平面.

(4) 由 (3) 的结论, C 的每一个正常面都是它的某一个超平面的面, 故 C 的超平面也是多面体集, 由此结论成立.

(5) 由性质 3.9.5 和性质 3.9.1 可知, 多面体集可以等价表示成多面体与多面锥的和. 因此, 多面体集 C 的面的个数有限. □

最后, 讨论多面体和多面体集的配极之间的关系. 下面的定理 3.9.8 说明了多面体的配极等价于一个多面体集, 多面体集的配极等价于一个包含原点的多面体, 多面体集与多面体可通过配极等价刻画.

定理 3.9.8 设 $\boldsymbol{x}_1,\cdots,\boldsymbol{x}_m$ 是 $\mathbf{R}^n$ 中两两互不相同的点, $m\geqslant 1$. 令

$$C=\operatorname{co}\{\boldsymbol{x}_1,\cdots,\boldsymbol{x}_m\},\quad D=\bigcap_{i=1}^{m}C^{\leqslant}(\boldsymbol{x}_i,1),$$

其中 $C^{\leqslant}(\boldsymbol{x}_i,1)=\{\boldsymbol{x}\,|\langle\boldsymbol{x},\boldsymbol{x}_i\rangle\leqslant 1\}$, 则下面的结论成立:

(1) $C^{\circ}=D$;

(2) $D^{\circ}=\operatorname{co}\{\mathbf{0},\boldsymbol{x}_1,\cdots,\boldsymbol{x}_m\}$;

(3) $\mathbf{0}\in C$ 当且仅当 C 和 D 两两配极;

(4) $\mathbf{0}\in\operatorname{int}C$ 当且仅当 C 和 D 两两配极, 且 D 有界.

证明 (1) 由定义 2.5.5, 对任意的 $\boldsymbol{x}_1,\cdots,\boldsymbol{x}_m$, 由 $\langle\boldsymbol{x},\boldsymbol{x}_i\rangle\leqslant 1(i=1,2,\cdots,m)$ 得

$$\left\langle\boldsymbol{x},\sum_{i=1}^{m}\lambda_i\boldsymbol{x}_i\right\rangle\leqslant 1,\quad \lambda_i\geqslant 1\,(i=1,2,\cdots,m),\quad \sum_{i=1}^{m}\lambda_i=1,$$

上式说明 $C^{\circ}=D$.

(2) 由定理 2.5.10 和 C 是凸紧集, 得

$$D^{\circ}=C^{\circ\circ}=\operatorname{clco}\{\mathbf{0},\boldsymbol{x}_1,\cdots,\boldsymbol{x}_m\}=\operatorname{co}\{\mathbf{0},\boldsymbol{x}_1,\cdots,\boldsymbol{x}_m\}.$$

(3) 直接由 (1) 和 (2) 得到.

(4) 由性质 2.5.6 和 (3) 得到结论. □

本小节得到了多面体集的一些重要性质:

1) 多面体集是闭凸集, 它平移后的集合仍是多面体集, 它可以等价于多面体与多面锥的和, 有限个多面体集的交仍是多面体集;

2) 多面体集的边界由其有限多个超平面半空间边界组成, 这些边界与多面体交集构成多面体集的超平面, 多面体集的每个面还是多面体集, 且面的个数是有限的;

3) 多面体的配极等价于一个多面体集, 多面体集的配极等价于一个包含原点的多面体, 多面体集与多面体可通过配极等价刻画.

3.10 习 题

3.10-1 设集合 $X=\left\{(-1,-1)^{\mathrm{T}},(1,0)^{\mathrm{T}},(0,1)^{\mathrm{T}}\right\}$, 求它的一个简单锥, 并求简单锥的凸包.

3.10-2 设 $C=\left\{(x_1,x_2)^{\mathrm{T}}\,|x_1>0,x_2\geqslant 0\right\}$, C 和 $\operatorname{co}C$ 是否为一个凸锥? 求它的顶点集.

3.10-3 设 $C=\left\{(x_1,x_2)^{\mathrm{T}}\,|x_1\geqslant 1,x_2\geqslant 1\right\}$, 求 $\mathrm{ri}C$, $\mathrm{cl}C$ 和 $\mathrm{ap}C$.

3.10-4 设 $C=\left\{(x_1,x_2)^{\mathrm{T}}\,|x_1>1,x_2=0\right\}$, 求 $\mathrm{ri}C$, $\mathrm{cl}C$ 和 $\mathrm{ap}C$.

3.10-5 设 $C=\left\{(x_1,x_2)^{\mathrm{T}}\,|x_1\geqslant 0,x_2\geqslant 0\right\}$, 求 $C-C$, $(-C)\cap C$.

3.10-6 设 $C_1=\left\{(x_1,x_2)^{\mathrm{T}}\,|x_1\geqslant 1,x_2\geqslant 1\right\}$ 和 $C_2=\left\{(x_1,x_2)^{\mathrm{T}}\,|x_1\geqslant -1,x_2\geqslant -1\right\}$, 求 C_1+C_2.

3.10-7 证明推论 3.1.1、推论 3.1.2 和定理 3.1.5.

3.10-8 证明性质 3.2.2、性质 3.2.3 和性质 3.2.4.

3.10-9 在 $\mathbf{R}$ 空间中的开区间集合 $S=(-1,1)$, 求 $\mathrm{cone}_{\boldsymbol{a}}S$, $C_{\boldsymbol{a}}S$ 和顶点 $\boldsymbol{a}$.

3.10-10 在 $\mathbf{R}^2$ 空间中的集合 $S=\left\{(x_1,0)^{\mathrm{T}}\,|x_1\in(0,1)\right\}\cup\left\{(0,x_2)^{\mathrm{T}}\,|x_2\in(2,3)\right\}$, 求 $\mathrm{cone}_{\boldsymbol{a}}S, C_{\boldsymbol{a}}S$ 和顶点 $\boldsymbol{a}$.

3.10-11 设在 $\mathbf{R}^2$ 空间中的两个集合

$$T=\left\{(x_1,x_2)^{\mathrm{T}}\,|x_1-x_2\leqslant 1,0\leqslant x_1,x_2\leqslant 5\right\}\subset S=\left\{(x_1,x_2)^{\mathrm{T}}\,|0\leqslant x_1,x_2\leqslant 2\right\},$$

令 $\boldsymbol{a}=(0,0)^{\mathrm{T}}$ 或 $(1,1)^{\mathrm{T}}$ 或 $(-1,-1)^{\mathrm{T}}$, 分别求 $\mathrm{cone}_{\boldsymbol{a}}T$, $C_{\boldsymbol{a}}T$, $\mathrm{cone}_{\boldsymbol{a}}S$, $C_{\boldsymbol{a}}S$.

3.10-12 设在 $\mathbf{R}$ 空间中的两个集合

$$T=\{x_1|0\leqslant x_1\leqslant 2\},\quad S=\{x_1|3\leqslant x_1\leqslant 4\},\quad T\cap S=\varnothing,$$

求 $C_0(T\cap S), C_0(T)\cap C_0(S)$ 和 $\mathrm{cone}_0(T\cap S), \mathrm{cone}_0(T)\cap\mathrm{cone}_0(S)$.

3.10-13 设在 $\mathbf{R}^2$ 空间中的集合 $S=\left\{(x_1,x_2)^{\mathrm{T}}\,|0\leqslant x_1,x_2\leqslant 2\right\}$, 根据凸锥包表示定理求 $\mathrm{cone}_0(S)$.

3.10-14 设在 $\mathbf{R}^2$ 空间中的集合 $S=\left\{(x_1,x_2)^{\mathrm{T}}\,|-1\leqslant x_1,x_2\leqslant 2\right\}$, 令 $\boldsymbol{a}=(0,0)^{\mathrm{T}}$ 或 $(-1,-1)^{\mathrm{T}}$, 根据凸锥包表示定理, 求 $\mathrm{cone}_{\boldsymbol{a}}(S)$.

3.10-15 设在 $\mathbf{R}^2$ 空间中的集合 $S=\left\{(x_1,0)^{\mathrm{T}}\,|x_1\in(0,2)\right\}\cup\left\{(0,x_2)^{\mathrm{T}}\,|x_2\in(0,2)\right\}$, 取顶点: $\boldsymbol{a}=(0,0)^{\mathrm{T}}$ 和 $\boldsymbol{a}=(-1,-1)^{\mathrm{T}}$, 分别求下面集合, 并做比较分析:

(1) $\mathrm{co}S, \mathrm{cone}_{\boldsymbol{a}}(S), \mathrm{co}\left(\mathrm{cone}_{\boldsymbol{a}}(S)\right), \mathrm{cone}_{\boldsymbol{a}}(\mathrm{co}S), \mathrm{co}(C_{\boldsymbol{a}}(S)), C_{\boldsymbol{a}}(\mathrm{co}S)$;

(2) $C_{\boldsymbol{a}}(S), \mathrm{cone}_{\boldsymbol{a}}(S), \mathrm{aff}\left(\{\boldsymbol{a}\}\cup S\right), \mathrm{aff}\,(S)$;

(3) $\mathrm{aff}\left(C_{\boldsymbol{a}}(S)\right), \mathrm{aff}\left(\mathrm{cone}_{\boldsymbol{a}}(S)\right), \mathrm{aff}(S\cup\{\boldsymbol{a}\})$.

3.10-16 设在 $\mathbf{R}^2$ 空间中的集合 $S=\left\{(x_1,0)^{\mathrm{T}}\,|x_1\in(0,2)\right\}\cup\left\{(0,x_2)^{\mathrm{T}}\,|x_2\in(0,2)\right\}$, 取顶点: $\boldsymbol{a}=(0,0)^{\mathrm{T}}$ 和 $\boldsymbol{a}=(-1,-1)^{\mathrm{T}}$, 分别求下面集合, 并做比较分析:

(1) $C_{\boldsymbol{a}}(\mathrm{cl}S), \mathrm{cl}(C_{\boldsymbol{a}}(S)),\ \mathrm{cl}(C_{\boldsymbol{a}}(\mathrm{cl}S))$;

(2) $\mathrm{cone}_{\boldsymbol{a}}(\mathrm{cl}S), \mathrm{cl}(\mathrm{cone}_{\boldsymbol{a}}S), \mathrm{cl}(\mathrm{cone}_{\boldsymbol{a}}(\mathrm{cl}S))$;

(3) $\mathrm{ri}(\mathrm{co}S)$, $\mathrm{ri}(\mathrm{cone}_{\boldsymbol{a}}(S))$, $\mathrm{cone}_{\boldsymbol{a}}(\mathrm{ri}(\mathrm{co}S))$, $\mathrm{aff}S$;

(4) $\mathrm{cone}_{\boldsymbol{a}}(\mathrm{ri}(\mathrm{co}S)), \{\boldsymbol{a}\}\cup\mathrm{ri}(\mathrm{cone}_{\boldsymbol{a}}S)$, $\mathrm{cone}_{\boldsymbol{a}}S$, $\mathrm{aff}(\boldsymbol{a}\cup S)$.

3.10-17　设在 $\mathbf{R}^2$ 空间中的集合 $S=\left\{(x_1,0)^{\mathrm{T}}\,|x_1\in(0,2)\right\}\cup\left\{(0,x_2)^{\mathrm{T}}\,|x_2\in(0,2)\right\}$, 求 $\mathrm{rec}S$ 和 $\mathrm{lin}S$.

3.10-18　在 $\mathbf{R}^2$ 空间中的集合 $C=\left\{(x_1,0)^{\mathrm{T}}\,|x_1\in(-1,+\infty)\right\}\cup\left\{(0,x_2)^{\mathrm{T}}\,|\,x_2\in(-1,+\infty)\right\}$, 求它的回收锥 $\mathrm{rec}C$ 和顶点集 $\mathrm{ap}C$, 验证: $\mathrm{ap}C\subset\mathrm{rec}C\subset(\mathrm{ap}C\cup C)$. 进一步, 求 $\mathrm{cl}C$, $\mathrm{ri}C$, $\mathrm{rec}\,(\mathrm{co}C)$ 和 $\mathrm{rec}\,(\mathrm{ri}C)$; 并求 $\mathrm{lin}\,(\mathrm{co}C)$.

3.10-19　在 $\mathbf{R}^2$ 空间中的集合 $C_1=\left\{(x_1,x_2)^{\mathrm{T}}\,|x_1>x_2\geqslant 1\right\}$ 和 $C_2=\left\{(x_1,x_2)^{\mathrm{T}}\,|x_2>x_1\geqslant 1\right\}$, 求 $\mathrm{rec}\,(C_1+C_2)$, $\mathrm{rec}\,(\mathrm{co}C_1\cap\mathrm{co}C_2)$, $\mathrm{rec}\,(C_1\cup C_2)$.

3.10-20　设 $C=\left\{(x_1,x_2)^{\mathrm{T}}\,|x_1>0,x_2\geqslant 0\right\}$, 求 C^+, C^{++}, $(\mathrm{cl}C)^+$, $(\mathrm{cl}C)^{++}$.

3.10-21　写出性质 3.5.1 和性质 3.5.2 的正极锥形式, 并给出证明.

3.10-22　设 $C_1=\left\{(x_1,x_2)^{\mathrm{T}}\,|x_1>x_2\geqslant 1\right\}$ 和 $C_2=\left\{(x_1,x_2)^{\mathrm{T}}\,|x_2>x_1\geqslant 1\right\}$, 求 C_1^-, C_1^{--}, $(\mathrm{cl}C_1)$, $(\mathrm{cl}C_1)^{--}$, $(C_1\cup C_2)^-=C_1^-\cap C_2^-$, $(C_1+C_2)^-=(\mathrm{cl}C_1+\mathrm{cl}C_2)^-$, $\mathrm{cl}(\mathrm{cone}_{\mathbf{0}}C_1)$.

3.10-23　证明推论 3.5.1.

3.10-24　设凸锥 $C=\left\{(x_1,0)^{\mathrm{T}}\,|x_1>0\right\}\subset\mathbf{R}^2$, 验证性质 3.5.5 ~ 性质 3.5.8 的结论是否成立.

3.10-25　设集合 $B=\{\boldsymbol{x}\mid\|\boldsymbol{x}\|<\varepsilon\}$, 求它的投影度量映射, 并验证定理 3.6.2 是否成立.

3.10-26　设凸集 $C=\left\{(x_1,x_2)^{\mathrm{T}}\mid x_1\in(1,2),x_2\in(1,2)\right\}$, 求点 $\boldsymbol{a}\in\mathrm{cl}C$ 的切锥和法锥.

3.10-27　设凸集 $C_1=\{x_1\in(0,1)\}$ 和 $C_2=\{x_1\in(1,2)\}$, 求 C_1+C_2 的切锥和法锥, 验证性质 3.6.2、性质 3.6.3 和性质 3.6.4 的结论是否成立.

3.10-28　设凸集 $C=\left\{(x_1,x_2)^{\mathrm{T}}\mid x_1\in(1,2),x_2\in(1,2)\right\}$, 求凸集的法锥和正交集.

3.10-29　设集合 $C=\left\{(x_1,x_2)^{\mathrm{T}}\mid x_1\geqslant x_2\geqslant 0\right\}$, 求凸集 C 的障碍锥、正交集和法锥, 验证定理 3.7.3 的结论是否成立.

3.10-30　设在 $\mathbf{R}^2$ 空间中的三个凸锥:

$$C_1=\left\{(x_1,x_2)^{\mathrm{T}}\,|x_1\geqslant 0,x_2\geqslant 0,x_1-x_2>0\right\},$$

$$C_2=\left\{(x_1,x_2)^{\mathrm{T}}\left|x_1\geqslant 0,x_2\geqslant 0,x_2-x_1>0\right.\right\},$$

$$C_3=\left\{(x_1,x_2)^{\mathrm{T}}\left|x_1\geqslant 0,x_2\geqslant 0,x_2=x_1\right.\right\},$$

这三个凸锥是否可分离?

3.10-31　如果 2 个凸锥可分离, 那么它们是否存在分离超平面?

3.10-32　证明性质 3.9.1(5)~(6).

3.10-33　设 $S=\left\{(1,0,0)^{\mathrm{T}},(0,0,1)^{\mathrm{T}},(0,0,0)^{\mathrm{T}},(1,1,1)^{\mathrm{T}},(0,1,0)^{\mathrm{T}}\right\}$ 是 $\mathbf{R}^2$ 中的 5 个点, 求它构成的多面体的线性不等式组. 如果 $\boldsymbol{x}_0=(-1,-1,-1)^{\mathrm{T}}$, 则寻找一个 $\boldsymbol{x}^*$ 构成的极支撑超平面 $H_C(\boldsymbol{x}^*)$ 使得 $\langle\boldsymbol{x},\boldsymbol{x}^*\rangle<\langle\boldsymbol{x}_0,\boldsymbol{x}^*\rangle$ 对任意的 $\boldsymbol{x}\in C$ 成立.

3.10-34　设 $S=\left\{(1,0,0)^{\mathrm{T}},(0,0,1)^{\mathrm{T}},(0,0,0)^{\mathrm{T}},(1,1,1)^{\mathrm{T}},(0,1,0)^{\mathrm{T}}\right\}$ 是 $\mathbf{R}^2$ 中的 5 个点, 求 S 构成的多面锥, 以及 S 的共轭锥, 试通过 S 构造一个多面体集.

第 4 章 凸 函 数

凸函数理论在凸分析中有着极其重要的地位. 凸函数最早应用于经济学研究中, 由于它具有全局最优的特性, 所以在经济学领域得到了广泛应用 [2,4,16,22,24,29]. 随着凸优化的应用范围越来越广, 凸函数理论也变得越来越重要, 近几十年来凸分析专著更是不断推陈出新 [52,63,84-87], 关于凸函数的一些新的理论例如半严格凸函数被提出. 因此, 本章除了介绍凸函数的经典基本性质外, 还补充介绍了近些年来凸函数的最新研究成果 [71,81,89-92].

本章首先介绍实值凸函数、可微凸函数和广义凸函数等性质, 然后讨论一些复合凸函数的性质, 侧重介绍半连续、闭凸和连续凸函数的性质, 接着介绍共轭凸函数、支撑凸函数和规范凸函数等三种特殊凸函数的性质, 以及严格凸函数和半严格凸函数的一些性质与判定定理, 最后介绍强凸函数的概念和性质.

4.1 三种基本凸函数

本书仅讨论在 $\mathbf{R}^n$ 中的凸函数性质. 本节介绍实值凸函数、可微凸函数、正常和非正常凸函数 (广义凸函数) 等性质, 其中实值凸函数是指取值不含无穷大值的凸函数, 可微凸函数是指取值不含无穷大值的凸可微函数, 正常和非正常凸函数是指取值含无穷大值的凸函数.

4.1.1 实值凸函数

下面在 $\mathbf{R}^n$ 的凸集上定义一个凸函数, 其值不取无穷大. 后面的讨论中, 除特别强调外, $\mathbf{R}^n$ 中的实值函数均不取无穷大值. 这种凸函数的定义可以用不等式来描述, 见下面定义.

定义 4.1.1 设 $C \subset \mathbf{R}^n$ 是凸集和函数 $f: C \to \mathbf{R}$. 如果对于任意的 $\boldsymbol{x}, \boldsymbol{y} \in C$ 和任意的 $\lambda \in (0,1)$, 有

$$f\left[(1-\lambda)\boldsymbol{x}+\lambda\boldsymbol{y}\right] \leqslant (1-\lambda) f(\boldsymbol{x})+\lambda f(\boldsymbol{y}), \tag{4.1.1-1}$$

则称 f 在 C 上是凸函数. 对于任意的 $\boldsymbol{x}, \boldsymbol{y} \in C, \boldsymbol{x} \neq \boldsymbol{y}$ 和任意的 $\lambda \in (0,1)$, 有

$$f\left[(1-\lambda)\boldsymbol{x}+\lambda\boldsymbol{y}\right] < (1-\lambda) f(\boldsymbol{x})+\lambda f(\boldsymbol{y}), \tag{4.1.1-2}$$

则称 f 在 C 上是严格凸函数. 如果 $-f$ 在 C 上是凸函数, 则称 f 在 C 上是凹函数; 如果 $-f$ 在 C 上是严格凸函数, 则称 f 在 C 上是严格凹函数.

如果对任意的 $\boldsymbol{x},\boldsymbol{y}\in C$, $\exists\lambda\in(0,1)$, 有 (4.1.1-1) 或 (4.1.1-2) 式成立, 则称函数 f 在 C 上关于 $\lambda\in(0,1)$ 满足一点凸性或一点严格凸性.

注 上述定义的实值凸 (凹) 函数后面都简称为凸 (凹) 函数. 如果不强调说明, 后面讨论的凸函数都是有限实值的.

一般凸函数不一定是严格凸函数, 例如函数 $f(\boldsymbol{x})=\boldsymbol{x}$ 是凸函数, 但不是严格凸函数.

例 4.1.1 函数 $f(x_1,x_2)=x_1^2+x_2^2$ 是 $\mathbf{R}^2$ 上的凸函数, 且是严格凸函数.

凸函数可能在函数的某个定义域内是凸的, 而在更大的区域内或整个空间上并不是凸函数, 见下面的例子.

例 4.1.2 函数 $f(x_1)=\left(x_1^2-x_1\right)^2$ 在 $\mathbf{R}^2$ 上不是凸函数. 该函数的一阶导数、二阶导数分别为

$$f'(x_1)=2\left(x_1^2-x_1\right)(2x_1-1)=0,\quad f''(x_1)=2\left(6x_1^2-6x_1+1\right)=0,$$

由上两式可求得 $a_1=\dfrac{3-\sqrt{3}}{6}$, $a_2=\dfrac{3+\sqrt{3}}{6}$. 因此, $f(x_1)$ 在 $(-\infty,a_1)$ 上是凸函数, 在 (a_1,a_2) 上是凹函数, 在 $(a_2,+\infty)$ 上是凸函数, 见图 4.1-1.

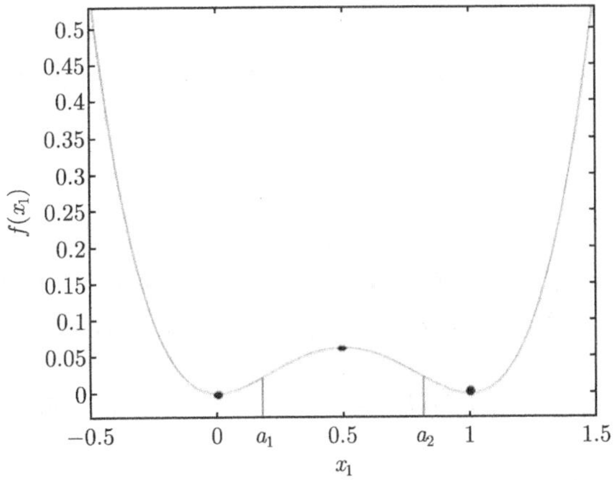

图 4.1-1 非凸函数 $f(x_1)=\left(x_1^2-x_1\right)^2$

这里我们引入了一点 (严格) 凸性概念, 它是判定 (严格) 凸函数的充分或必要条件, 许多非凸函数大多不满足一点凸性. 例如 $f(x)=x^3$ 在 $\mathbf{R}$ 上不是 (严格) 凸函数, 不满足一点 (严格) 凸性. 但是, 后面将证明的一些性质表明, 如果函数同时满足一点凸性和连续性, 可以保证函数的凸性.

定义 4.1.1 的 (4.1.1-1) 式可以推广到 n 个点凸组合形式, 见下面的定理 4.1.1.

定理 4.1.1 设非空凸集 $C\subset\mathbf{R}^n$, 函数 $f:C\to\mathbf{R}$, 则 f 在 C 上是凸函数的充要条件是对任意的 $\boldsymbol{x}_i\in C$ 和 $\lambda_i\geqslant 0\,(i=1,\cdots,m)$, 有

$$f(\lambda_1\boldsymbol{x}_1+\cdots+\lambda_m\boldsymbol{x}_m)\leqslant\lambda_1 f(\boldsymbol{x}_1)+\cdots+\lambda_m f(\boldsymbol{x}_m) \tag{4.1.1-3}$$

成立, 其中 $\sum\limits_{i=1}^{m}\lambda_i=1$.

证明 先用归纳法证明必要性. 若 $f(\boldsymbol{x})$ 是在 C 上的凸函数, 显然 $m=2$ 时有 (4.1.1-3) 式成立. 假设 $m-1$ 时 (4.1.1-3) 式成立. 设 $\forall\boldsymbol{x}_i\in C$ 和 $\lambda_i\geqslant 0$ $(i=1,\cdots,m)$, 其中 $\sum\limits_{i=1}^{m}\lambda_i=1$. 令 $\bar\lambda_2=\sum\limits_{i=2}^{m}\lambda_i$, $\lambda_2\boldsymbol{x}_2+\cdots+\lambda_m\boldsymbol{x}_m=(1-\lambda_1)\bar{\boldsymbol{x}}_2$, 有 $\lambda_1+\bar\lambda_2=1$, $1=\sum\limits_{i=2}^{m}\dfrac{\lambda_i}{1-\lambda_1}$, 则有 $\bar{\boldsymbol{x}}_2\in C$ 和

$$\lambda_1\boldsymbol{x}_1+\lambda_2\boldsymbol{x}_2+\cdots+\lambda_m\boldsymbol{x}_m=\lambda_1\boldsymbol{x}_1+(1-\lambda_1)\bar{\boldsymbol{x}}_2.$$

由归纳法假设得

$$\begin{aligned}f(\lambda_1\boldsymbol{x}_1+\cdots+\lambda_m\boldsymbol{x}_m)&=f(\lambda_1\boldsymbol{x}_1+(1-\lambda_1)\bar{\boldsymbol{x}}_2)\leqslant\lambda_1 f(\boldsymbol{x}_1)+(1-\lambda_1)f(\bar{\boldsymbol{x}}_2)\\&\leqslant\lambda_1 f(\boldsymbol{x}_1)+(1-\lambda_1)f\left(\frac{\lambda_2}{1-\lambda_1}\boldsymbol{x}_2+\cdots+\frac{\lambda_m}{1-\lambda_1}\boldsymbol{x}_m\right)\\&\leqslant\lambda_1 f(\boldsymbol{x}_1)+\lambda_2 f(\boldsymbol{x}_2)+\cdots+\lambda_m f(\boldsymbol{x}_m).\end{aligned}$$

上式说明必要性成立.

充分性显然. □

类似地, 对严格凸函数有下面结论.

定理 4.1.2 设非空凸集 $C\subset\mathbf{R}^n$, 函数 $f:C\to\mathbf{R}$, 则 f 在 C 上是严格凸函数的充要条件是对任意的 $\boldsymbol{x}_i\in C$ 和 $\lambda_i>0\,(i=1,\cdots,m)$ 有下式

$$f(\lambda_1\boldsymbol{x}_1+\cdots+\lambda_m\boldsymbol{x}_m)<\lambda_1 f(\boldsymbol{x}_1)+\cdots+\lambda_m f(\boldsymbol{x}_m) \tag{4.1.1-4}$$

成立, 其中 $\sum\limits_{i=1}^{m}\lambda_i=1$.

下面是关于凸函数及其限制凸函数的等价性, 后面一些凸函数的性质可利用它的限制函数来证明. 我们有两种限制函数的定义结果.

定理 4.1.3 设非空凸集 $C\subset\mathbf{R}^n$, 函数 $f:C\to\mathbf{R}$, 则 f 在 C 上是凸函数当且仅当对任意的 $\boldsymbol{x}\in C$ 及任意的 $\boldsymbol{d}\in\mathbf{R}^n$, 当 $\boldsymbol{x}$ 和 $\boldsymbol{d}$ 固定时, 限制函数 $\varphi_{\boldsymbol{x},\boldsymbol{d}}(t)$ 在 $T(\boldsymbol{x},\boldsymbol{d})$ 上是凸函数, 其中

$$\varphi_{\boldsymbol{x},\boldsymbol{d}}(t)=f(\boldsymbol{x}+t\boldsymbol{d}),\quad T(\boldsymbol{x},\boldsymbol{d})=\{t\in\mathbf{R}|\boldsymbol{x}+t\boldsymbol{d}\in C\}.$$

或者对任意的 $\boldsymbol{x},\boldsymbol{y}\in C$, $\varphi_{\boldsymbol{x},\boldsymbol{y}}(t)=f(\boldsymbol{x}+t(\boldsymbol{y}-\boldsymbol{x}))$ 是 $t\in[0,1]$ 上的凸函数.

证明 设 f 在 C 上是凸函数, $t_1,t_2\in T(\boldsymbol{x},\boldsymbol{d})$, 当 $0\leqslant\lambda\leqslant 1$ 时, 有

$$\begin{aligned}\varphi_{\boldsymbol{x},\boldsymbol{d}}[(1-\lambda)t_1+\lambda t_2]&=f(\boldsymbol{x}+[(1-\lambda)t_1+\lambda t_2]\boldsymbol{d})\\&=f([(1-\lambda)(\boldsymbol{x}+t_1\boldsymbol{d})+\lambda(\boldsymbol{x}+t_2\boldsymbol{d})])\\&\leqslant(1-\lambda)f(\boldsymbol{x}+t_1\boldsymbol{d})+\lambda f(\boldsymbol{x}+t_2\boldsymbol{d})\\&=(1-\lambda)\varphi_{\boldsymbol{x},\boldsymbol{d}}(t_1)+\lambda\varphi_{\boldsymbol{x},\boldsymbol{d}}(t_2),\end{aligned}$$

故 $\varphi_{\boldsymbol{x},\boldsymbol{d}}(t)$ 是 $t\in T(\boldsymbol{x},\boldsymbol{d})$ 的凸函数.

反之, 对任意的 $\boldsymbol{x}\in C$, 任意的 $\boldsymbol{d}\in\mathbf{R}^n$, 设 $\varphi_{\boldsymbol{x},\boldsymbol{d}}(t)$ 是 $t\in T(\boldsymbol{x},\boldsymbol{d})$ 的凸函数. 对任意的 $\boldsymbol{x}_1,\boldsymbol{x}_2\in C$, $0\leqslant\lambda\leqslant 1$, 设 $T(\boldsymbol{x}_1,\boldsymbol{x}_2-\boldsymbol{x}_1)=\{t\in\mathbf{R}|\boldsymbol{x}_1+t(\boldsymbol{x}_2-\boldsymbol{x}_1)\in C\}$, 则有

$$\begin{aligned}f[(1-\lambda)\boldsymbol{x}_1+\lambda\boldsymbol{x}_2]&=f(\boldsymbol{x}_1+[(1-\lambda)\cdot 1+\lambda\cdot 0](\boldsymbol{x}_2-\boldsymbol{x}_1))\\&=\varphi_{\boldsymbol{x}_1,\boldsymbol{x}_2-\boldsymbol{x}_1}((1-\lambda)\cdot 1+\lambda\cdot 0)\\&\leqslant(1-\lambda)\varphi_{\boldsymbol{x}_1,\boldsymbol{x}_2-\boldsymbol{x}_1}(1)+\lambda\varphi_{\boldsymbol{x}_1,\boldsymbol{x}_2-\boldsymbol{x}_1}(0)\\&=(1-\lambda)f(\boldsymbol{x}_2)+\lambda f(\boldsymbol{x}_1),\end{aligned}$$

故 f 是在 C 上是凸函数.

在上述证明过程中, 若将 $T(\boldsymbol{x},\boldsymbol{d})$ 换成 $[0,1]$, 对 $\varphi_{\boldsymbol{x},\boldsymbol{y}}(t)$ 也成立. □

对于严格凸函数, 有下面对应的结论.

定理 4.1.4 设非空凸集 $C\subset\mathbf{R}^n$, 函数 $f:C\to\mathbf{R}$, 则 f 在 C 上是严格凸函数当且仅当对任意的 $\boldsymbol{x}\in C$ 及任意的 $\boldsymbol{d}\in\mathbf{R}^n$, 当 $\boldsymbol{x}$ 和 $\boldsymbol{d}$ 固定时, 限制函数 $\varphi_{\boldsymbol{x},\boldsymbol{d}}(t)$ 在 $T(\boldsymbol{x},\boldsymbol{d})$ 上是严格凸函数, 其中

$$\varphi_{\boldsymbol{x},\boldsymbol{d}}(t)=f(\boldsymbol{x}+t\boldsymbol{d})\quad 和\quad T(\boldsymbol{x},\boldsymbol{d})=\{t\in\mathbf{R}|\boldsymbol{x}+t\boldsymbol{d}\in C\}.$$

或者对任意的 $\boldsymbol{x},\boldsymbol{y}\in C$, $\varphi_{\boldsymbol{x},\boldsymbol{y}}(t)=f(\boldsymbol{x}+t(\boldsymbol{y}-\boldsymbol{x}))$ 是 $t\in[0,1]$ 上的严格凸函数.

例 4.1.3 $f(x_1,x_2)=x_1^2+x_2^2$ 是 $\mathbf{R}^2$ 上的凸函数, 令 $\boldsymbol{d}\in\mathbf{R}^2$, 显然限制函数

$$\begin{aligned}\varphi_{\boldsymbol{x},\boldsymbol{d}}(t)&=f(\boldsymbol{x}+t\boldsymbol{d})=(x_1+td_1)^2+(x_2+td_2)^2\\&=\left(d_1^2+d_2^2\right)t^2+2(d_1x_1+d_2x_2)t+\left(x_1^2+x_2^2\right)\end{aligned}$$

是关于 t 的凸函数.

本小节得到凸集 C 上的 (严格) 凸函数 f 的几个基本结论:

1) f 在 C 上是凸函数等价于 (4.1.1-3) 式成立;

2) f 在 C 上是严格凸函数等价于 (4.1.1-4) 式成立;

3) f 在 C 上是 (严格) 凸函数等价于它的单变量限制函数是 (严格) 凸函数.

4.1.2　可微凸函数

可微凸函数指函数既是凸函数又是可微函数, 具体定义如下.

定义 4.1.2　设非空凸集 $C \subset \mathbf{R}^n$, 可微函数 $f: C \to \mathbf{R}$. 如果 f 在 C 上是凸函数, 则称 f 在 C 上是可微凸函数; 如果 f 在 C 上是严格凸函数, 则称 f 在 C 上是可微严格凸函数; 如果 $-f$ 在 C 上是凸函数, 则称 f 在 C 上是可微凹函数; 如果 f 在 C 上是严格凸函数, 则称 f 是在 C 上的可微严格凹函数.

下面的例子说明存在凸函数是不可微函数.

例 4.1.4　$f(x_1) = |x_1|$ 是 $\mathbf{R}$ 上的凸函数, 但不是可微凸函数.

性质 4.1.1　设非空开凸集 $C \subset \mathbf{R}^n$, 可微函数 $f: C \to \mathbf{R}$, 则以下结论等价:

(1) f 在 C 上是凸函数;

(2) $f(\boldsymbol{x}_2) - f(\boldsymbol{x}_1) \geqslant \langle \boldsymbol{x}_2 - \boldsymbol{x}_1, \nabla f(\boldsymbol{x}_1) \rangle, \forall \boldsymbol{x}_1, \boldsymbol{x}_2 \in C$;

(3) 对任意的 $\boldsymbol{d} \in \mathbf{R}^n$ 和 $\boldsymbol{x} \in C$, $\phi(t) = \langle \boldsymbol{d}, \nabla f(\boldsymbol{x} + t\boldsymbol{d}) \rangle$ 是 $t \in T(\boldsymbol{x}, \boldsymbol{d})$ 的不减函数, 其中

$$T(\boldsymbol{x}, \boldsymbol{d}) = \{t \in \mathbf{R} | \boldsymbol{x} + t\boldsymbol{d} \in C\};$$

(4) 如果 f 在 C 上是二阶连续可微, 那么 $\nabla^2 f(\boldsymbol{x})$ 在 C 上是半正定的.

证明　(1) $\Rightarrow$ (2): 如果 $f(\boldsymbol{x})$ 在 C 上是凸函数, 那么对任意的 $\boldsymbol{x}_1, \boldsymbol{x}_2 \in C$, 有

$$(1-\lambda) f(\boldsymbol{x}_1) + \lambda f(\boldsymbol{x}_2) \geqslant f[(1-\lambda)\boldsymbol{x}_1 + \lambda \boldsymbol{x}_2], \quad 0 < \lambda < 1.$$

由上式得到

$$f(\boldsymbol{x}_2) - f(\boldsymbol{x}_1) \geqslant \frac{f[\boldsymbol{x}_1 + \lambda(\boldsymbol{x}_2 - \boldsymbol{x}_1)] - f(\boldsymbol{x}_1)}{\lambda}.$$

令 $\lambda \to 0$ 取极限, 得 $f(\boldsymbol{x}_2) - f(\boldsymbol{x}_1) \geqslant \langle \boldsymbol{x}_2 - \boldsymbol{x}_1, \nabla f(\boldsymbol{x}_1) \rangle$.

(2) $\Rightarrow$(1): 设 $\forall \boldsymbol{x}_1, \boldsymbol{x}_2 \in C$, $\boldsymbol{x} = (1-\lambda)\boldsymbol{x}_1 + \lambda \boldsymbol{x}_2 \, (0 < \lambda < 1)$, 则有

$$f(\boldsymbol{x}_1) - f(\boldsymbol{x}) \geqslant \langle \boldsymbol{x}_1 - \boldsymbol{x}, \nabla f(\boldsymbol{x}) \rangle,$$

$$f(\boldsymbol{x}_2) - f(\boldsymbol{x}) \geqslant \langle \boldsymbol{x}_2 - \boldsymbol{x}, \nabla f(\boldsymbol{x}) \rangle.$$

将前式两边分别乘 $1-\lambda$, 后式两边分别乘 λ, 然后两式两边分别相加得

$$(1-\lambda) f(\boldsymbol{x}_1) + \lambda f(\boldsymbol{x}_2) \geqslant f[(1-\lambda)\boldsymbol{x}_1 + \lambda \boldsymbol{x}_2], \quad 0 < \lambda < 1.$$

(2)⇒(3): 设 $\boldsymbol{x}_1=\boldsymbol{x}+t_1\boldsymbol{d}\in C, \boldsymbol{x}_2=\boldsymbol{x}+t_2\boldsymbol{d}\in C\,(t_2>t_1)$, 那么有

$$f(\boldsymbol{x}+t_2\boldsymbol{d})-f(\boldsymbol{x}+t_1\boldsymbol{d})\geqslant\langle(t_2-t_1)\,\boldsymbol{d},\nabla f(\boldsymbol{x}+t_1\boldsymbol{d})\rangle,$$

$$f(\boldsymbol{x}+t_1\boldsymbol{d})-f(\boldsymbol{x}+t_2\boldsymbol{d})\geqslant\langle(t_1-t_2)\,\boldsymbol{d},\nabla f(\boldsymbol{x}+t_2\boldsymbol{d})\rangle.$$

将两式两边相加得 $\phi(t_2)\geqslant\phi(t_1)$.

(3)⇒(4): 若 (3) 成立, 对任意的 $\boldsymbol{d}\in\mathbf{R}^n$ 和 $\boldsymbol{x}\in C$, 则 $\exists t'>0$, 对任意的 $t\in(0,t')$ 满足 $\boldsymbol{x}+t\boldsymbol{d}\in C$, 有

$$\langle\boldsymbol{d},\nabla f(\boldsymbol{x}+t\boldsymbol{d})\rangle-\langle\boldsymbol{d},\nabla f(\boldsymbol{x})\rangle\geqslant 0.$$

由上式知, 存在充分小的 $t\in(0,t')$ 使得 $\langle\boldsymbol{d},\nabla f(\boldsymbol{x}+t\boldsymbol{d})-\nabla f(\boldsymbol{x})\rangle=t\boldsymbol{d}^{\mathrm{T}}\nabla^2 f(\boldsymbol{x})\,\boldsymbol{d}+o(t^2)\geqslant 0$, 即有

$$\boldsymbol{d}^{\mathrm{T}}\nabla^2 f(\boldsymbol{x})\,\boldsymbol{d}+o(t)\geqslant 0.$$

令 $t\to 0$, 得 $\nabla^2 f(\boldsymbol{x})$ 在 C 上是半正定的.

(4)⇒(2): 若 (4) 成立, 设 $\forall\boldsymbol{x}_1,\boldsymbol{x}_2\in C$ 和 $\boldsymbol{x}=(1-\lambda)\,\boldsymbol{x}_1+\lambda\boldsymbol{x}_2\,(0<\lambda<1)$, 由泰勒展开存在 λ 使得

$$f(\boldsymbol{x}_2)=f(\boldsymbol{x}_1)+\langle\boldsymbol{x}_2-\boldsymbol{x}_1,\nabla f(\boldsymbol{x}_1)\rangle+(\boldsymbol{x}_2-\boldsymbol{x}_1)^{\mathrm{T}}\,\nabla^2 f(\boldsymbol{x})\,(\boldsymbol{x}_2-\boldsymbol{x}_1)\,.$$

由上式显然 (2) 成立. □

对于严格可微凸函数, 有下面性质.

性质 4.1.2 设非空开凸集 $C\subset\mathbf{R}^n$, 可微函数 $f:X\to\mathbf{R}$, 则以下结论等价:

(1) f 在 C 上是严格凸函数;

(2) $f(\boldsymbol{x}_2)-f(\boldsymbol{x}_1)>\langle\boldsymbol{x}_2-\boldsymbol{x}_1,\nabla f(\boldsymbol{x}_1)\rangle,\forall\boldsymbol{x}_1,\boldsymbol{x}_2\in C$;

(3) 对于任意的 $\boldsymbol{d}\in\mathbf{R}^n$ 和任意的 $\boldsymbol{x}\in C$, 则 $\phi(t)=\langle\boldsymbol{d},\nabla f(\boldsymbol{x}+t\boldsymbol{d})\rangle$ 是 $t\in T(\boldsymbol{x},\boldsymbol{d})$ 的严格增加函数, 其中 $T(\boldsymbol{x},\boldsymbol{d})=\{t\in\mathbf{R}|\boldsymbol{x}+t\boldsymbol{d}\in C\}$.

(4) 当 f 在 C 上是二阶连续可微时, $\nabla^2 f(\boldsymbol{x})$ 在 C 上是正定的.

注 当 $C=\mathbf{R}^n$ 时, 性质 4.1.1 和性质 4.1.2 均成立.

本小节定义了可微凸函数与严格可微凸函数的概念, 得到了判定可微凸函数的几个性质:

1) 凸函数等价于 (4.1.2-1) 式成立;

2) 凸函数等价于其可微函数的限制函数是单调增加的;

3) 如果 f 是二阶连续可微的, 则凸函数等价于它的 Hessian 矩阵是正定的.

对于严格凸性也有类似结果 (见性质 4.1.2).

4.1.3 正常与非正常凸函数

本小节讨论的凸函数是可取值为 $-\infty$ 或 $+\infty$ 的广义函数, 下面是可能出现的包含 $-\infty$ 和 $+\infty$ 的计算规则:

(1) 当 $-\infty < \alpha \leqslant +\infty$ 时, $\alpha + \infty = \infty + \alpha = +\infty$;

(2) 当 $-\infty \leqslant \alpha < +\infty$ 时, $\alpha - \infty = -\infty + \alpha = -\infty$;

(3) 当 $0 < \alpha \leqslant +\infty$ 时, $\alpha \cdot \infty = \infty \cdot \alpha = \infty$;

(4) 当 $-\infty \leqslant \alpha < 0$ 时, $\alpha \cdot \infty = \infty \cdot \alpha = -\infty$, $\alpha \cdot (-\infty) = (-\infty) \cdot \alpha = +\infty$;

(5) $0 \cdot \infty = \infty \cdot 0 = 0 \cdot (-\infty) = (-\infty) \cdot 0 = 0$, $-(-\infty) = +\infty$, $\inf \varnothing = +\infty, \sup \varnothing = -\infty$;

(6) $\infty - \infty$ 和 $-\infty + \infty$ 没有意义.

与定义 4.1.1 中的实值凸函数不同, 含无穷值的凸函数将通过函数上方图的凸集进行定义.

定义 4.1.3 设非空凸集 $C \subset \mathbf{R}^n$, 广义函数 $f: C \to [-\infty, +\infty]$(也称为广义实值函数) , 使 f 只取有限值或 $-\infty$ 的 $\boldsymbol{x}$ 的集合, 称为 f 的有效定义域, 用 $\operatorname{dom} f$ 表示, 即

$$\operatorname{dom} f = \{\boldsymbol{x} \in C \,|\, f(\boldsymbol{x}) < +\infty\}. \tag{4.1.3-1}$$

集合 $\operatorname{epi} f = \{(\boldsymbol{x}, t) \,|\, \boldsymbol{x} \in C \subset \mathbf{R}^n,\ t \in \mathbf{R}, t \geqslant f(\boldsymbol{x})\}$ 称为 f 的上方图 (上图像).

从定义 4.1.3 知, 只有当 $\boldsymbol{x} \in \operatorname{dom} f$ 时, 才存在有限值 $t \geqslant f(\boldsymbol{x})$ 使得 $(\boldsymbol{x}, t) \in \operatorname{epi} f$ 成立.

一个函数的上方图也确定了这个函数本身, 下面结论显然成立.

性质 4.1.3 设非空凸集 $C \subset \mathbf{R}^n$, 广义函数 $f: C \to [-\infty, +\infty]$, 则 f 的上方图等价于

$$f(\boldsymbol{x}) = \inf\{t \,|\, (\boldsymbol{x}, t) \in \operatorname{epi} f\}.$$

从性质 4.1.3 知, 函数的上方图包含函数值上面的相同形状. 因此, 上方图可以用来定义凸函数.

定义 4.1.4 设非空凸集 $C \subset \mathbf{R}^n$, 广义函数 $f: C \to [-\infty, +\infty]$, 有

(1) 如果 $\operatorname{epi} f$ 是凸集, 且 $f(\boldsymbol{x}) \neq -\infty\ (\forall \boldsymbol{x} \in C)$ 和 $C = \operatorname{dom} f$ 成立, 则称 f 在 C 上是正常凸函数;

(2) 如果 $-f$ 在 C 上是正常凸函数, 则称 f 在 C 上是正常凹函数;

(3) 如果 f 在 C 上是正常凸函数也是正常凹函数, 则称 f 在 C 上是仿射函数;

(4) 如果 $\operatorname{epi} f$ 是凸集, 且 $f(\boldsymbol{x}) \neq -\infty\ (\forall \boldsymbol{x} \in C)$ 或 $C = \operatorname{dom} f$ 不成立, 则称 f 是非正常凸函数;

(5) 如果 $-f$ 在 C 上是非正常凸函数, 则称 f 在 C 上是非正常凹函数.

注 定义 4.1.1 的凸函数与正常凸函数限制在凸集 C 上定义本质上是一致的, 区别在于正常凸函数值在有效定义域外有取无穷值的点. 非正常凸函数的函数值一定包含了 $-\infty$ 或 $+\infty$. f 是正常凸函数的等价条件是 $\mathrm{epi}f$ 非空且不包含垂直直线, 或 $\mathrm{dom}f \neq \varnothing$ 和 $\boldsymbol{x} \in \mathrm{dom}f$ 时 f 有限.

例 4.1.5 设函数 $f: \mathbf{R} \to \mathbf{R}$,

$$f(x) = \begin{cases} -\infty, & |x| < 1, \\ 0, & |x| = 1, \\ +\infty, & |x| > 1 \end{cases}$$

是非正常凸函数, 该函数的上方图为

$$\begin{aligned} \mathrm{epi}f = & \left\{ (x, +\infty)^{\mathrm{T}} \,\middle|\, |x| > 1 \right\} \cup \left\{ (x, t)^{\mathrm{T}} \,\middle|\, |x| < 1, t \in \mathbf{R} \cup \{-\infty\} \right\} \\ & \cup \left\{ (x, t)^{\mathrm{T}} \,\middle|\, |x| = 1, t \geqslant 0 \right\}. \end{aligned}$$

但需要注意的是, 如果 $f(x) = \begin{cases} 0, & |x| \leqslant 1, \\ +\infty, & |x| > 1, \end{cases}$ 则 f 是正常凸函数.

因为非正常凸函数的上方图中有效定义域内的值都是有限的, 所以有下面结论成立.

性质 4.1.4 设 f 在凸集 $C \subset \mathbf{R}^n$ 上是非正常凸函数, 则 $\mathrm{dom}f$ 是凸集.

证明 设 $\boldsymbol{x}_1, \boldsymbol{x}_2 \in \mathrm{dom}f = \{\boldsymbol{x} \in C \,|\, f(\boldsymbol{x}) < +\infty\}$, 则 $\exists t_1, t_2 \in \mathbf{R}$ 使得

$$(\boldsymbol{x}_1, t_1), (\boldsymbol{x}_2, t_2) \in \mathrm{epi}f.$$

因 $\mathrm{epi}f$ 是凸集, 对于 $0 < \lambda < 1$, 有

$$(1-\lambda)(\boldsymbol{x}_1, t_1) + \lambda(\boldsymbol{x}_2, t_2) \in \mathrm{epi}f,$$

即

$$f[(1-\lambda)\boldsymbol{x}_1 + \lambda\boldsymbol{x}_2] \leqslant (1-\lambda)t_1 + \lambda t_2.$$

有 $f[(1-\lambda)\boldsymbol{x}_1 + \lambda\boldsymbol{x}_2] \neq +\infty$, 故 $(1-\lambda)\boldsymbol{x}_1 + \lambda\boldsymbol{x}_2 \in \mathrm{dom}f$. 则 $\mathrm{dom}f$ 是凸集. □

但是, 非广义正常凸函数 (不含无穷值) 等价于上方图是凸集.

定理 4.1.5 设非空凸集 $C \subset \mathbf{R}^n$, 函数 $f: C \to \mathbf{R}$, 则 f 在 C 上是正常凸函数的充要条件是 $\mathrm{epi}f$ 是凸集.

证明 如果 $\mathrm{epi}f$ 是凸集. 设 $\forall \boldsymbol{x}_1, \boldsymbol{x}_2 \in C$, 那么 $(\boldsymbol{x}_1, f(\boldsymbol{x}_1)), (\boldsymbol{x}_2, f(\boldsymbol{x}_2)) \in \mathrm{epi}f$, 对 $0 < \lambda < 1$, 有

$$(1-\lambda)(\boldsymbol{x}_1, f(\boldsymbol{x}_1)) + \lambda(\boldsymbol{x}_2, f(\boldsymbol{x}_2)) \in \mathrm{epi}f.$$

由此可得

$$(1-\lambda) f(\boldsymbol{x}_1)+\lambda f(\boldsymbol{x}_2) \geqslant f[(1-\lambda) \boldsymbol{x}_1+\lambda \boldsymbol{x}_2],$$

所以由定义 4.1.4 得 f 是正常凸函数.

反之, 若 f 是正常凸函数. 设 $\forall \boldsymbol{x}_1, \boldsymbol{x}_2 \in C$, $(\boldsymbol{x}_1, t_1),(\boldsymbol{x}_2, t_2) \in \operatorname{epi} f$, 对 $0<\lambda<1$, 有

$$(1-\lambda) t_1+\lambda t_2 \geqslant(1-\lambda) f(\boldsymbol{x}_1)+\lambda f(\boldsymbol{x}_2) \geqslant f[(1-\lambda) \boldsymbol{x}_1+\lambda \boldsymbol{x}_2].$$

所以 $((1-\lambda) \boldsymbol{x}_1+\lambda \boldsymbol{x}_2,(1-\lambda) t_1+\lambda t_2) \in \operatorname{epi} f$, 那么 $\operatorname{epi} f$ 是凸集. □

定理 4.1.6 设非空凸集 $C \subset \mathbf{R}^n$, 广义函数 $f: C \rightarrow[-\infty,+\infty]$, 则 f 在 C 上是非正常凸函数的充要条件是当 $f(\boldsymbol{x}_1)<t_1$ 和 $f(\boldsymbol{x}_2)<t_2$ 时, 有不等式

$$(1-\lambda) t_1+\lambda t_2>f[(1-\lambda) \boldsymbol{x}_1+\lambda \boldsymbol{x}_2], \quad 0<\lambda<1, \quad \forall \boldsymbol{x}_1, \boldsymbol{x}_2 \in C \tag{4.1.3-2}$$

成立.

证明过程与定理 4.1.5 的证明完全一致.

事实上, 根据定理 4.1.5 的证明可有下面结论.

推论 4.1.1 若 f 在凸集 $C \subset \mathbf{R}^n$ 上是非正常凸函数, 则对任意的 $\boldsymbol{x}_1, \boldsymbol{x}_2 \in C$, $0<\lambda<1$, 有

$$(1-\lambda) f(\boldsymbol{x}_1)+\lambda f(\boldsymbol{x}_2) \geqslant f[(1-\lambda) \boldsymbol{x}_1+\lambda \boldsymbol{x}_2].$$

下面定义几个特殊的凸函数, 它们在凸分析中经常被使用.

定义 4.1.5 设非空集合 $C \subset \mathbf{R}^n$, 定义以下几个特殊函数:

(1) 示性函数

$$\theta_C(\boldsymbol{x})=\begin{cases}0, & \boldsymbol{x} \in C, \\ +\infty, & \boldsymbol{x} \notin C.\end{cases}$$

显然, C 是凸集的等价条件是 $\theta_C(\boldsymbol{x})$ 是 $\mathbf{R}^n$ 中的凸函数.

(2) 支撑函数

$$\delta_C(\boldsymbol{x})=\sup \{\langle\boldsymbol{x}, \boldsymbol{y}\rangle \mid \boldsymbol{y} \in C\}.$$

(3) 规范函数 (Minkowski 函数)

$$v_C(\boldsymbol{x})=\inf \{\lambda \mid \lambda \geqslant 0, \boldsymbol{x} \in \lambda C\}.$$

(4) 距离函数

$$d_C(\boldsymbol{x})=\inf (\|\boldsymbol{x}-\boldsymbol{y}\| \mid \boldsymbol{y} \in C).$$

性质 4.1.5 如果非空凸集 $C \subset \mathbf{R}^n$, 则 $\theta_C(\boldsymbol{x})$, $\delta_C(\boldsymbol{x})$, $\upsilon_C(\boldsymbol{x})$ 和 $d_C(\boldsymbol{x})$ 都是正常凸函数.

与函数的上方图不同, 凸函数的水平集是考虑函数值基于小于、等于、大于、大于等于或小于等于等关系在可行域上的限制集合. 函数的水平集与函数的凸性判定有着密切联系, 后面在刻画凸函数性质时会经常使用, 定义如下.

定义 4.1.6 设非空凸集 $C \subset \mathbf{R}^n$, 广义函数 $f: C \to [-\infty, +\infty]$, $\alpha \in [-\infty, +\infty]$, 集合

$$C_{f\leqslant\alpha} = \{\boldsymbol{x} \in C \,|\, f(\boldsymbol{x}) \leqslant \alpha\}, \quad C_{f<\alpha} = \{\boldsymbol{x} \in C \,|\, f(\boldsymbol{x}) < \alpha\},$$

$$C_{f\geqslant\alpha} = \{\boldsymbol{x} \in C \,|\, f(\boldsymbol{x}) \geqslant \alpha\}, \quad C_{f>\alpha} = \{\boldsymbol{x} \in C \,|\, f(\boldsymbol{x}) > \alpha\},$$

$$C_{f=\alpha} = \{\boldsymbol{x} \in C \,|\, f(\boldsymbol{x}) = \alpha\}$$

均称为 f 的水平集.

定理 4.1.7 设 f 在非空凸集 $C \subset \mathbf{R}^n$ 上是正常凸函数, 则水平集 $C_{f\leqslant\alpha}$ 和 $C_{f<\alpha}$ 是凸集.

证明 设 $\boldsymbol{x}_1, \boldsymbol{x}_2 \in C_{f\leqslant\alpha}$, 当 $0 < \lambda < 1$ 时, 有

$$f[(1-\lambda)\boldsymbol{x}_1 + \lambda\boldsymbol{x}_2] \leqslant (1-\lambda) f(\boldsymbol{x}_1) + \lambda f(\boldsymbol{x}_2) \leqslant (1-\lambda)\alpha + \lambda\alpha = \alpha.$$

故 $(1-\lambda)\boldsymbol{x}_1 + \lambda\boldsymbol{x}_2 \in C_{f\leqslant\alpha}$, 那么 $C_{f\leqslant\alpha}$ 是凸集.

$C_{f<\alpha}$ 的情形类似可证. □

定理 4.1.8 设函数族 $f_i(\boldsymbol{x})\,(i \in I)$ 在非空凸集 $C \subset \mathbf{R}^n$ 上是非正常凸函数, 其中 I 是任意指标集. 设 $\alpha_i \in [-\infty, +\infty]\,(i \in I)$, 则 $\{\boldsymbol{x} \in C \,|\, f_i(\boldsymbol{x}) \leqslant \alpha_i, \forall i \in I\}$ 是凸集.

证明过程与定理 4.1.6 的证明类似.

后面还会经常讨论的一类特殊函数: 设 f 是 $\mathbf{R}^n$ 上的实值函数, 如果对任意的 $\boldsymbol{x} \in \mathbf{R}^n$, $0 < \lambda < +\infty$, 有 $f(\lambda\boldsymbol{x}) = \lambda f(\boldsymbol{x})$ 成立, 则称 f 是正齐次函数.

定理 4.1.9 如果 $f: \mathbf{R}^n \to \mathbf{R}$ 是正齐次函数, 则 f 是凸函数充要条件是对任意的 $\boldsymbol{x}_1, \boldsymbol{x}_2 \in \mathbf{R}^n$ 有

$$f(\boldsymbol{x}_1 + \boldsymbol{x}_2) \leqslant f(\boldsymbol{x}_1) + f(\boldsymbol{x}_2).$$

例 4.1.6 容易验证范数函数 $f(\boldsymbol{x}) = \|\boldsymbol{x}\|$ 是正齐次函数, 且是凸函数.

本小节得到了正常与非正常凸函数的一些基本性质:

1) 非正常凸函数的有效定义域是凸集;

2) f 的上方图等价于 $f(\boldsymbol{x}) = \inf\{t \,|\, (\boldsymbol{x}, t) \in \mathrm{epi} f\}$;

3) 非广义正常凸函数等价于上方图是凸集;

4) 非正常凸函数等价于对于任意 $\boldsymbol{x}_1, \boldsymbol{x}_2 \in C$ 和 $f(\boldsymbol{x}_1) < t_1, f(\boldsymbol{x}_2) < t_2$ 有 (4.1.3-2) 式成立;

5) 正常凸函数的水平集 $C_{f\leqslant\alpha}$ 和 $C_{f<\alpha}$ 是凸集.

4.2 复合凸函数

本节将讨论凸函数的判定, 这是一个非常重要的问题. 例如, 判定若干个凸函数的相加、取上确界或下确界函数、几个凸函数复合映射后得到的函数是否还是凸函数.

我们把凸函数经运算后所得的新的凸函数称为复合凸函数, 下面将证明这些复合函数仍保持凸性.

4.2.1 线性复合凸函数

下面证明多个凸函数的线性组合后仍是凸函数的性质.

性质 4.2.1 设非空凸集 $C \subset \mathbf{R}^n$, $\lambda \geqslant 0$. 若 f 在 C 上是 (非) 正常凸函数, 则 λf 在 C 上是 (非) 正常凸函数.

证明 根据定义 4.1.1 和定义 4.1.4, 结论显然. □

推论 4.2.1 凸函数 $f: \mathbf{R}^n \to \mathbf{R}$ 是正齐次函数的充要条件是对任意的 $t>0$, 有 $f_t(\boldsymbol{x}) = tf(\boldsymbol{x})$.

性质 4.2.2 设非空凸集 $C \subset \mathbf{R}^n$, 函数 $f_1, f_2\colon C \to [-\infty, +\infty]$, 有

(1) 如果 f_1 和 f_2 在 C 上是非正常凸函数, 则 $f_1 + f_2$ 在 C 上是非正常凸函数;

(2) 如果 f_1 在 C 上是非正常凸函数和 f_2 在 C 上是正常凸函数, 则 $f_1 + f_2$ 是非正常凸函数;

(3) 如果 f_1 在 C 上是正常凸函数和 f_2 在 C 上是正常凸函数, 则 $f_1 + f_2$ 在 C 上是正常凸函数.

证明 这里只需对情形 (1) 进行证明, (2) 和 (3) 的情形类似可证.

(1) 设 $\forall \boldsymbol{x}_1, \boldsymbol{x}_2 \in C, 0<\lambda<1, t_1 > f_1(\boldsymbol{x}_1) + f_2(\boldsymbol{x}_1)$ 和 $t_2 > f_1(\boldsymbol{x}_2) + f_2(\boldsymbol{x}_2)$, 有

$$(1-\lambda)t_1 + \lambda t_2 > (1-\lambda) f_1(\boldsymbol{x}_1) + (1-\lambda) f_2(\boldsymbol{x}_1) + \lambda f_1(\boldsymbol{x}_2) + \lambda f_2(\boldsymbol{x}_2). \quad (4.2.1\text{-}1)$$

因 $f_1(\boldsymbol{x})$ 和 $f_2(\boldsymbol{x})$ 在 C 上是非正常凸函数, 则由推论 4.1.1 有

$$(1-\lambda) f_1(\boldsymbol{x}_1) + \lambda f_1(\boldsymbol{x}_2) \geqslant f_1[(1-\lambda)\boldsymbol{x}_1 + \lambda \boldsymbol{x}_2], \quad (4.2.1\text{-}2)$$

$$(1-\lambda) f_2(\boldsymbol{x}_1)+\lambda f_2(\boldsymbol{x}_2) \geqslant f_2[(1-\lambda) \boldsymbol{x}_1+\lambda \boldsymbol{x}_2]. \tag{4.2.1-3}$$

由 (4.2.1-1), (4.2.1-2) 和 (4.2.1-3) 式得

$$(1-\lambda) t_1+\lambda t_2>f_1[(1-\lambda) \boldsymbol{x}_1+\lambda \boldsymbol{x}_2]+f_2[(1-\lambda) \boldsymbol{x}_1+\lambda \boldsymbol{x}_2].$$

根据定理 4.1.6 得 f_1+f_2 在 C 上是非正常凸函数. □

由性质 4.2.1、性质 4.2.2 可得下面性质的结论成立.

性质 4.2.3 设非空凸集 $C \subset \mathbf{R}^n$, 函数 $f_1, f_2: C \rightarrow[-\infty,+\infty]$, $\alpha, \beta \geqslant 0$, 有

(1) 如果 f_1 和 f_2 在 C 上是非正常凸函数, 则 $\alpha f_1+\beta f_2$ 在 C 上是非正常凸函数;

(2) 如果 f_1 在 C 上是非正常凸函数和 f_2 在 C 上是正常凸函数, 则 $\alpha f_1+\beta f_2$ 是非正常凸函数;

(3) 如果 f_1 在 C 上是正常凸函数和 f_2 在 C 上是正常凸函数, 则 $\alpha f_1+\beta f_2$ 在 C 上是正常凸函数.

例 4.2.1 $f_1(x)=x^2$, $f_2(x)=\mathrm{e}^x$ 在 $\mathbf{R}$ 上是凸函数, 则 $f_1(x)+f_2(x)=x^2+\mathrm{e}^x$ 在 $\mathbf{R}$ 上是凸函数.

性质 4.2.4 设 f 是非空凸集 $C \subset \mathbf{R}^n$ 上的正常凸函数, $g: \mathbf{R} \rightarrow \mathbf{R}$ 是单调递增正常凸函数, 则复合函数 $h(\boldsymbol{x})=g[f(\boldsymbol{x})]$ 在 C 上是正常凸函数.

证明 设 $\forall \boldsymbol{x}_1, \boldsymbol{x}_2 \in C$, $0<\lambda<1$ 有

$$f[(1-\lambda) \boldsymbol{x}_1+\lambda \boldsymbol{x}_2] \leqslant(1-\lambda) f(\boldsymbol{x}_1)+\lambda f(\boldsymbol{x}_2).$$

由 $g: \mathbf{R} \rightarrow \mathbf{R}$ 是单调递增的正常凸函数及上式得

$$\begin{aligned} g(f[(1-\lambda) \boldsymbol{x}_1+\lambda \boldsymbol{x}_2]) & \leqslant g((1-\lambda) f(\boldsymbol{x}_1)+\lambda f(\boldsymbol{x}_2)) \\ & \leqslant(1-\lambda) g(f(\boldsymbol{x}_1))+\lambda g(f(\boldsymbol{x}_2)), \end{aligned}$$

即

$$h[(1-\lambda) \boldsymbol{x}_1+\lambda \boldsymbol{x}_2] \leqslant(1-\lambda) h(\boldsymbol{x}_1)+\lambda h(\boldsymbol{x}_2).$$ □

下面是一个复合映射保持凸性的例子.

例 4.2.2 $f_1(x)=x^2$ 和 $f_2(x)=\mathrm{e}^x$ 在 $\mathbf{R}$ 上是凸函数, 则 $f_2(f_1(x))=\mathrm{e}^{x^2}$ 在 $\mathbf{R}$ 上是凸函数.

本小节得到如下结论:

1) 若干个凸函数经正线性组合后还是凸函数;

2) 单调递增的正常凸函数的复合函数还是正常凸函数.

4.2.2 逐点下确界复合凸函数

下面定义凸集上的逐点下确界函数, 把它看成一类复合函数.

性质 4.2.5 设凸集 $C \subset \mathbf{R}^{n+1}$, 则 C 的逐点下确界函数

$$f(\boldsymbol{x}) = \inf\{t \,|(\boldsymbol{x}, t) \in C\} \tag{4.2.2-1}$$

在 $\mathbf{R}^n$ 上是非正常凸函数.

证明 设 $\forall \boldsymbol{x}_1, \boldsymbol{x}_2$, $0 < \lambda < 1$, $\exists t_1, t_2 \in \mathbf{R}$, 满足 $f(\boldsymbol{x}_1) < t_1$ 和 $f(\boldsymbol{x}_2) < t_2$. 由 f 的定义, $\exists t, t' \in \mathbf{R}$ 使得 $t < t_1, t' < t_2$ 满足 $(\boldsymbol{x}_1, t), (\boldsymbol{x}_2, t') \in C$. 因 C 是凸集, 有

$$(1-\lambda)(\boldsymbol{x}_1, t) + \lambda(\boldsymbol{x}_2, t') = ((1-\lambda)\boldsymbol{x}_1 + \lambda\boldsymbol{x}_2, (1-\lambda)t + \lambda t') \in C.$$

则得

$$f[(1-\lambda)\boldsymbol{x}_1 + \lambda\boldsymbol{x}_2] \leqslant (1-\lambda)t + \lambda t' < (1-\lambda)t_1 + \lambda t_2.$$

由定理 4.1.5 知, f 在 $\mathbf{R}^n$ 上是非正常凸函数. □

性质 4.2.5 说明凸集的每个分量点的下确界式 (4.2.2-1) 构成一个非正常凸函数. 例如, 凸函数 f 的上方图 epi$f$ 是一个凸集, 上方图关于函数值的分量构成的下确界函数等于 f, 见性质 4.1.3. 下面的结论是性质 4.2.5 的推广.

性质 4.2.6 设 f 在 $\mathbf{R}^n$ 上是 (非) 正常凸函数, $s > 0$, 则函数 $f_s(\boldsymbol{x}) = \inf\{t \,|(\boldsymbol{x}, t) \in s\text{epi}f\}$ 是 (非) 正常凸函数.

证明 因 f 在 $\mathbf{R}^n$ 上是 (非) 正常凸函数, $s > 0$, 则 $s$epi$f$ 是凸集, 由性质 4.2.5 知 f_s 是 (非) 正常凸函数. □

进一步推广性质 4.2.6 的结果, 它是一种多个凸函数通过变量加法运算后的复合函数.

性质 4.2.7 设 $f_i\,(i = 1, \cdots, m)$ 在 $\mathbf{R}^n$ 上是非正常凸函数, 称函数

$$\phi(\boldsymbol{x}) = \inf\{f_1(\boldsymbol{x}_1) + \cdots + f_m(\boldsymbol{x}_m) \,|\boldsymbol{x}_i \in \mathbf{R}^n, \boldsymbol{x}_1 + \cdots + \boldsymbol{x}_m = \boldsymbol{x}\} \tag{4.2.2-2}$$

是 $f_1, \cdots, f_m$ 的下卷积复合函数, 则 ϕ 是非正常凸函数.

证明 设集合

$$C = \{(\boldsymbol{x}, t) \,|t \geqslant f_1(\boldsymbol{x}_1) + \cdots + f_m(\boldsymbol{x}_m), \boldsymbol{x} = \boldsymbol{x}_1 + \cdots + \boldsymbol{x}_m\} \subset \mathbf{R}^{mn+1}.$$

由 $f_i(\boldsymbol{x})\,(i = 1, \cdots, m)$ 在 $\mathbf{R}^n$ 上是非正常凸函数, 设 $(\boldsymbol{x}^1, t^1), (\boldsymbol{x}^2, t^2) \in C$, $0 < \lambda < 1$ 满足

$$\boldsymbol{x}^1 = \boldsymbol{x}_1^1 + \cdots + \boldsymbol{x}_m^1, \quad \boldsymbol{x}^2 = \boldsymbol{x}_1^2 + \cdots + \boldsymbol{x}_m^2, \tag{4.2.2-3}$$

$$t^1 \geqslant f_1\left(\boldsymbol{x}_1^1\right)+\cdots+f_m\left(\boldsymbol{x}_m^1\right), \quad t^2 \geqslant f_1\left(\boldsymbol{x}_1^2\right)+\cdots+f_m\left(\boldsymbol{x}_m^2\right). \qquad (4.2.2\text{-}4)$$

由 (4.2.2-3) 式和 (4.2.2-4) 式得

$$(1-\lambda)\boldsymbol{x}^1+\lambda\boldsymbol{x}^2=(1-\lambda)\boldsymbol{x}_1^1+\lambda\boldsymbol{x}_1^2+\cdots+(1-\lambda)\boldsymbol{x}_m^1+\lambda\boldsymbol{x}_m^2.$$

由 (4.2.2-4) 式和推论 4.1.1 得

$$\begin{aligned}(1-\lambda)t^1+\lambda t^2 &\geqslant (1-\lambda)f\left(\boldsymbol{x}_1^1\right)+\lambda f\left(\boldsymbol{x}_1^2\right)+\cdots+(1-\lambda)f\left(\boldsymbol{x}_m^1\right)+\lambda f\left(\boldsymbol{x}_m^2\right)\\ &\geqslant f\left[(1-\lambda)\boldsymbol{x}_1^1+\lambda\boldsymbol{x}_1^2\right]+\cdots+f\left[(1-\lambda)\boldsymbol{x}_m^1+\lambda\boldsymbol{x}_m^2\right].\end{aligned}$$

则 C 是凸集, 由性质 4.2.5 知 ϕ 是非正常凸函数. □

例 4.2.3 范数函数 $f_1(\boldsymbol{x})=\|\boldsymbol{x}\|$ 和示性函数 $f_2(\boldsymbol{x})=\theta_C(\boldsymbol{x})$ 在 $\mathbf{R}^n$ 上是凸函数, 得

$$\begin{aligned}\phi(\boldsymbol{x}) &= \inf\{\|\boldsymbol{x}_1\|+\theta_C(\boldsymbol{x}_2)\,|\boldsymbol{x}_1,\boldsymbol{x}_2\in\mathbf{R}^n,\boldsymbol{x}_1+\boldsymbol{x}_2=\boldsymbol{x}\}\\ &= \inf\{\|\boldsymbol{x}-\boldsymbol{x}_2\|+\theta_C(\boldsymbol{x}_2)\,|\boldsymbol{x}_1,\boldsymbol{x}_2\in\mathbf{R}^n,\boldsymbol{x}_1+\boldsymbol{x}_2=\boldsymbol{x}\}\\ &= \inf\{\|\boldsymbol{x}-\boldsymbol{x}_2\|\,|\boldsymbol{x}_2\in C\}=d_C(\boldsymbol{x}).\end{aligned}$$

本例说明范数加上示性函数的卷积复合后是一个距离函数.

再由性质 4.2.5 知下面两个性质显然成立.

性质 4.2.8 设 f 在 $\mathbf{R}^n$ 上是凸函数, 则 $h(\boldsymbol{x})=\inf\{u|\,(\boldsymbol{x},u)\in\operatorname{cone}(\operatorname{epi}f)\}$ 是 f 生成的正齐次凸函数.

性质 4.2.9 设 f 是 $\mathbf{R}^n$ 上的实函数, 则 $\phi(\boldsymbol{x})=\inf\{u\,|(\boldsymbol{x},u)\in\operatorname{co}(\operatorname{epi}f)\}$ 在 $\mathbf{R}^n$ 上是非正常凸函数.

例 4.2.4 容易验证凸集 $C\subset\mathbf{R}^n$ 的规范函数 $v_C(\boldsymbol{x})$ 可以由示性函数族 $\{\theta_{tC}(\boldsymbol{x})+t|t\geqslant 0\}$ 的下确界函数构成, 即 $v_C(\boldsymbol{x})=\inf\{\theta_{tC}(\boldsymbol{x})+t|t\geqslant 0\}$.

再由性质 4.2.9 知下面推论成立.

推论 4.2.2 设 $\{f_i\,|i\in I\}$ 是 $\mathbf{R}^n$ 上的任意函数族, I 是任意指标集. 定义函数

$$f(\boldsymbol{x})=\inf\{f_i(\boldsymbol{x})\,|i\in I\},$$

则

$$\phi(\boldsymbol{x})=\inf\{t\,|(\boldsymbol{x},t)\in\operatorname{co}(\operatorname{epi}f)\}$$

是凸函数, 它是这个函数族的凸包, 用 $\operatorname{co}\{f_i(\boldsymbol{x})\,|i\in I\}$ 表示.

推论 4.2.2 说明, 将给定若干凸函数的上方图并集后得到的凸包设为: $C=\operatorname{co}\bigcup_{i\in I}\operatorname{epi}f_i$. 记

$$\operatorname{co}\{f_i(\boldsymbol{x})\,|i\in I\}=\inf\{t\,|(\boldsymbol{x},t)\in C\}.$$

显然, 对任意的 $\boldsymbol{x} \in \mathbf{R}^n$, 任意的 $i \in I, f_i(\boldsymbol{x})$ 的下确界函数上方图的凸包的逐点下确界函数是凸函数. 进一步, 我们有下面的性质.

性质 4.2.10 设 $\{f_i(\boldsymbol{x})\,|i \in I\}$ 是 $\mathbf{R}^n$ 上的正常凸函数族, 则

$$\begin{aligned}\phi(\boldsymbol{x}) &= \inf\{t\,|(\boldsymbol{x},t) \in \operatorname{co}(\operatorname{epi}f)\}\\ &= \inf\left\{\sum_{i\in I}\lambda_i f_i(\boldsymbol{x}_i)\,\middle|\,\boldsymbol{x} = \sum_{i\in I}\lambda_i\boldsymbol{x}_i, \sum_{i\in I}\lambda_i = 1, \lambda_i \geqslant 0, i \in I\right\},\end{aligned}$$

其中下确界是在 $\boldsymbol{x}$ 作为元素 $\boldsymbol{x}_i$ 的凸组合时的全部表示式上取的, 只有有限个 λ_i 不为 0.

证明 根据定理 2.2.2, $(\boldsymbol{x},t) \in \operatorname{co}(\operatorname{epi}f)$ 的等价条件是

$$(\boldsymbol{x},t) = \sum_{i\in I}\lambda_i(\boldsymbol{x}_i,t_i) = \left(\sum_{i\in I}\lambda_i\boldsymbol{x}_i, \sum_{i\in I}\lambda_i t_i,\right),$$

其中 $(\boldsymbol{x}_i,t_i) \in \operatorname{epi}f_i, \lambda_i \geqslant 0, i \in I, \sum\limits_{i\in I}\lambda_i = 1$, 且只有有限个 λ_i. 根据 $\{f_i(\boldsymbol{x})\,|i \in I\}$ 是 $\mathbf{R}^n$ 上的正常凸函数族, 由推论 4.2.2 得

$$\begin{aligned}f(\boldsymbol{x}) &= \inf\left\{\sum_{i\in I}\lambda_i t_i\,\middle|\,\boldsymbol{x} = \sum_{i\in I}\lambda_i\boldsymbol{x}_i, f_i(\boldsymbol{x}_i) \leqslant t_i, \lambda_i \geqslant 0, i \in I, \sum_{i\in I}\lambda_i = 1\right\}\\ &= \inf\left\{\sum_{i\in I}\lambda_i f_i(\boldsymbol{x}_i)\,\middle|\,\boldsymbol{x} = \sum_{i\in I}\lambda_i\boldsymbol{x}_i, \lambda_i \geqslant 0, i \in I, \sum_{i\in I}\lambda_i = 1\right\}.\end{aligned}$$

□

性质 4.2.10 说明, $f(\boldsymbol{x}) = \inf\{f_i(\boldsymbol{x})\,|i \in I\}$ 的上方图凸包的下确界函数, 可以由这个函数族 $\{f_i(\boldsymbol{x})\,|i \in I\}$ 中有限多个凸组合复合函数的下确界构成.

例 4.2.5 设 f 在凸集合 $C \subset \mathbf{R}^n$ 上是正常凸函数, 则示性函数 $\theta_C(\boldsymbol{x})$ 加上 $f(\boldsymbol{x})$ 得到限制函数, 即为 $\theta_C(\boldsymbol{x}) + f(\boldsymbol{x}) = \begin{cases} f(\boldsymbol{x}), & \boldsymbol{x} \in C, \\ +\infty, & \boldsymbol{x} \notin C. \end{cases}$

进一步, 线性变换与凸函数复合后的下确界函数仍然是凸函数; 线性变换与下确界函数复合保持凸性. 见下面性质.

性质 4.2.11 设 A 是从 $\mathbf{R}^n$ 到 $\mathbf{R}^m$ 的线性变换, g 是 $\mathbf{R}^m$ 上的凸函数, h 是 $\mathbf{R}^n$ 上的凸函数. 设 $(gA)(\boldsymbol{x}) = g(A\boldsymbol{x})$ 是在 A 中 g 的逆象, $(Ah)(\boldsymbol{y}) = \inf\{h(\boldsymbol{x})\,|A\boldsymbol{x} = \boldsymbol{y}\}$ 是在 A 中 h 的象, 则 $(gA)(\boldsymbol{x})$ 是 $\mathbf{R}^n$ 上的凸函数, $(Ah)(\boldsymbol{y})$ 是 $\mathbf{R}^m$ 上的凸函数.

证明 设 $\forall \boldsymbol{x}_1, \boldsymbol{x}_2 \in \mathbf{R}^n$, $0 < \lambda < 1$, 有 $\boldsymbol{y}_1 = A\boldsymbol{x}_1$, $\boldsymbol{y}_2 = A\boldsymbol{x}_2$, 得

$$\begin{aligned}(gA)((1-\lambda)\boldsymbol{x}_1 + \lambda\boldsymbol{x}_2) &= g(A((1-\lambda)\boldsymbol{x}_1 + \lambda\boldsymbol{x}_2))\\ &= g((1-\lambda)\boldsymbol{y}_1 + \lambda\boldsymbol{y}_2)\\ &\geqslant (1-\lambda)g(\boldsymbol{y}_1) + \lambda g(\boldsymbol{y}_2)\\ &= (1-\lambda)(gA)(\boldsymbol{x}_1) + \lambda(gA)(\boldsymbol{x}_2),\end{aligned}$$

$(gA)(\boldsymbol{x})$ 是 $\mathbf{R}^n$ 上的凸函数. 定义集合 $C = \{(\boldsymbol{y}, t)\,|t \geqslant h(\boldsymbol{x}), A\boldsymbol{x} = \boldsymbol{y}, \boldsymbol{x} \in \mathbf{R}^n\}$, 由 h 是 $\mathbf{R}^n$ 上的凸函数, 知 C 是凸集, 那么由性质 4.2.5 得 $(Ah)(\boldsymbol{y})$ 是 $\mathbf{R}^m$ 上的凸函数. □

本小节得到了逐点下确界复合函数的几个性质:

1) 凸集的每个分量点的下确界 (4.2.2-1) 式是非正常凸函数;

2) 由 $f_1, \cdots, f_m$ 构成的下卷积复合函数是非正常凸函数;

3) 非正常凸函数 $f(\boldsymbol{x}) = \inf\{f_i(\boldsymbol{x})\,|i \in I\}$ 的上方图的凸包的下确界复合函数是非正常凸函数, 且可以由这个函数族 $\{f_i(\boldsymbol{x})\,|i \in I\}$ 中有限多个凸组合复合函数的下确界构成.

4.2.3 凸函数族上确界复合凸函数

对应的逐点上确界函数的凸性判定见下面性质.

性质 4.2.12 设 $f_i\,(i \in I)$ 在非空凸集 $C \subset \mathbf{R}^n$ 上是正常凸函数族, 其中 I 是任意指标集, 则凸函数族上确界复合函数

$$f(\boldsymbol{x}) = \sup\{f_i(\boldsymbol{x})\,|\boldsymbol{x} \in C, i \in I\}$$

是正常凸函数.

证明 因 $f_i\,(i \in I)$ 在 C 上是正常凸函数族, 则 $\mathrm{epi} f_i$ 是凸集, 得 $\bigcap_{i\in I}\mathrm{epi} f_i$ 是凸集. 有

$$\begin{aligned}\mathrm{epi} f &= \left\{(\boldsymbol{x}, t)\,\middle|\,\boldsymbol{x} \in C, t \in \mathbf{R}, t \geqslant \sup_{i\in I} f_i(\boldsymbol{x})\right\}\\ &= \{(\boldsymbol{x}, t)\,|\boldsymbol{x} \in C, t \in \mathbf{R}, t \geqslant f_i(\boldsymbol{x}), i \in I\}\\ &= \bigcap_{i\in I} \mathrm{epi} f_i,\end{aligned}$$

故 $\mathrm{epi} f$ 是凸集, 于是 f 在 C 上是正常凸函数. □

例 4.2.6 设凸集合 $C \subset \mathbf{R}^n$, 支撑函数 $\delta_C(\boldsymbol{x}) = \sup\{\langle \boldsymbol{x}, \boldsymbol{y}\rangle | \boldsymbol{y} \in C\}$ 是集合 C 上逐点构成的线性函数 $\langle \boldsymbol{x}, \boldsymbol{y}\rangle$ 的上确界, 在 C 上是凸函数.

综上可得, 逐点上确界复合函数的凸性判定性质: 正常凸函数 $\{f_i(\boldsymbol{x})|i\in I\}$ 的上确界复合函数是正常凸函数.

注意, 通过上面推导过程知, 性质 4.2.12 对于非正常凸函数也成立.

4.3 四种连续凸函数

本节讨论具有连续性的凸函数, 当函数具有上半连续、下半连续或连续时, 同时满足一点凸性, 则它是凸函数.

4.3.1 半连续凸函数

本小节讨论半连续凸函数的性质, 先看下面的定义.

定义 4.3.1 设 f 是 $X\subset\mathbf{R}^n$ 上的广义实值函数, $\boldsymbol{x}\in X$, 有

(1) 对于收敛到 $\boldsymbol{x}$ 的任意序列 $\{\boldsymbol{x}_i\}\subset X\,(i=1,2,\cdots)$, 如果有 $\{\boldsymbol{x}_i\}$ 的 $\lim\limits_{i\to\infty}f(\boldsymbol{x}_i)$ 存在, 且 $f(\boldsymbol{x})\leqslant\lim\limits_{i\to\infty}f(\boldsymbol{x}_i)$ 成立, 则称 f 在 $\boldsymbol{x}$ 处是下半连续的;

(2) 如果 $-f$ 在 $\boldsymbol{x}$ 处是下半连续的, 称 f 在 $\boldsymbol{x}$ 处是上半连续的;

(3) 如果 f 在 X 的每一点都是下半连续的, 则称 f 在 X 上是下半连续函数;

(4) 如果 f 在 X 上既是下半连续又是上半连续的, 则称 f 在 X 上是连续函数.

定义 4.3.2 设 f 是在凸集 $C\subset\mathbf{R}^n$ 上的广义实值凸函数, 有

(1) 如果 f 在 C 上是下半连续的, 则称 f 在 C 上是下半连续凸函数;

(2) 如果 f 在 C 上是上半连续的, 则称 f 在 C 上是上半连续凸函数;

(3) 如果 f 在 C 上是连续的, 则称 f 在 C 上是连续凸函数.

下面是下半连续函数的等价性.

性质 4.3.1 设 f 是在 $X\subset\mathbf{R}^n$ 上的广义实值函数, $\boldsymbol{x}\in X$, 则下面结论等价:

(1) f 在 $\boldsymbol{x}$ 处是下半连续的;

(2) 对任意的 $\varepsilon>0$, $f(\boldsymbol{x})\leqslant\lim\limits_{\varepsilon\to0}\inf\limits_{\boldsymbol{x}_\varepsilon\in B(\boldsymbol{x},\varepsilon)}f(\boldsymbol{x}_\varepsilon)$, 其中 $B(\boldsymbol{x},\varepsilon)=\big\{\boldsymbol{y}\in\mathbf{R}^n\big|\,\|\boldsymbol{x}-\boldsymbol{y}\|\leqslant\varepsilon\big\}$;

(3) 对任意的 $\varepsilon>0$, $\exists t>0$, 当 $\forall\boldsymbol{y}\in B(\boldsymbol{x},t)$ 时, 有 $f(\boldsymbol{x})-\varepsilon\leqslant f(\boldsymbol{y})$;

(4) 对任意的 $\alpha\in\mathbf{R}$, $X_{f>\alpha}=\{\boldsymbol{x}\,|f(\boldsymbol{x})>\alpha\}$ 是开集;

(5) 对任意的 $\alpha\in\mathbf{R}$, $X_{f\leqslant\alpha}=\{\boldsymbol{x}\,|f(\boldsymbol{x})\leqslant\alpha\}$ 是闭集;

(6) $\mathrm{epi}f$ 是 $\mathbf{R}^{n+1}$ 中的闭集.

证明 (1) $\Rightarrow$ (2): 设 f 在 $\boldsymbol{x}$ 处是下半连续的, 则对 X 中每个序列 $\{\boldsymbol{x}_i\}$ 有 $\boldsymbol{x}_i\to\boldsymbol{x}\in X$, $\lim\limits_{i\to\infty}f(\boldsymbol{x}_i)$ 存在, 且使得 $f(\boldsymbol{x})\leqslant\lim\limits_{i\to\infty}f(\boldsymbol{x}_i)$ 成立. 即对任意的

$\varepsilon > 0$, $\exists I$ 使得当 $i > I$ 时有 $\boldsymbol{x}_i \in B(\boldsymbol{x}, \varepsilon)$, 满足

$$f(\boldsymbol{x}_i) - \varepsilon \leqslant \inf_{\boldsymbol{x}_\varepsilon \in B(\boldsymbol{x},\varepsilon)} f(\boldsymbol{x}_\varepsilon).$$

再令 $\varepsilon \to 0$, $\boldsymbol{x}_i \to \boldsymbol{x}$, 有

$$f(\boldsymbol{x}) - \varepsilon \leqslant \lim_{\varepsilon \to 0} f(\boldsymbol{x}_i) - \varepsilon \leqslant \lim_{\varepsilon \to 0} \inf_{\boldsymbol{x}_\varepsilon \in B(\boldsymbol{x},\varepsilon)} f(\boldsymbol{x}_\varepsilon),$$

所以结论 (2) 成立.

(2) $\Rightarrow$ (3): 设 (2) 成立, 对任意 $\varepsilon > 0$, 存在充分小 $t > 0$, 当 $\forall \boldsymbol{y} \in B(\boldsymbol{x}, t)$ 时有

$$f(\boldsymbol{x}) \leqslant \inf_{\boldsymbol{y} \in B(\boldsymbol{x},t)} f(\boldsymbol{y}) + \varepsilon \leqslant f(\boldsymbol{y}) + \varepsilon.$$

(3) $\Rightarrow$ (4): 对任意 $\alpha \in \mathbf{R}$, 设 $\boldsymbol{x} \in X_{f>\alpha}$, 即有 $f(\boldsymbol{x}) > \alpha$. 令 $\varepsilon = \frac{1}{2}(f(\boldsymbol{x}) - \alpha) > 0$. 由 (3) 知存在充分小的 $t > 0$, 当任意 $\boldsymbol{y} \in B(\boldsymbol{x}, t)$ 时有

$$f(\boldsymbol{y}) \geqslant f(\boldsymbol{x}) - \varepsilon = \frac{1}{2}(f(\boldsymbol{x}) + \alpha) > \alpha,$$

即得 $\boldsymbol{y} \in X_{f>\alpha}$. 因此, $X_{f>\alpha} = \{\boldsymbol{x} \,|\, f(\boldsymbol{x}) > \alpha\}$ 是开集.

(4) $\Rightarrow$ (5): 对任意 $\alpha \in \mathbf{R}$, 设序列 $\{\boldsymbol{x}_i\} \subset X_{f \leqslant \alpha}\ (i = 1, 2, \cdots)$. 假设 $\boldsymbol{x}_i \xrightarrow{i \to +\infty} \boldsymbol{x}$, 下证 $\boldsymbol{x} \in X_{f \leqslant \alpha}$. 如果 $\boldsymbol{x} \in X_{f>\alpha}$, 由 (4) 成立知存在充分小的 $t > 0$, 当 $\forall \boldsymbol{y} \in B(\boldsymbol{x}, t)$ 时有 $f(\boldsymbol{y}) > \alpha$. 由 $\boldsymbol{x}_i \xrightarrow{i \to +\infty} \boldsymbol{x}$ 知存在充分大的 i 使得 $\boldsymbol{x}_i \in B(\boldsymbol{x}, t)$, $f(\boldsymbol{x}_i) > \alpha$, 因 $f(\boldsymbol{x}_i) \leqslant \alpha$, 显然得到矛盾.

(5) $\Rightarrow$ (6): 设 $\mathrm{epi} f = \{(\boldsymbol{x}, \alpha) \,|\, \boldsymbol{x} \in X \subset \mathbf{R}^n,\ f(\boldsymbol{x}) \leqslant \alpha, \alpha \in \mathbf{R}\}$, 序列

$$(\boldsymbol{x}_i, \alpha_i) \in \mathrm{epi} f \quad (i = 1, 2, \cdots),$$

其中 $(\boldsymbol{x}_i, \alpha_i) \xrightarrow{i \to \infty} (\boldsymbol{x}, \alpha)$, $f(\boldsymbol{x}_i) \leqslant \alpha_i$. 对任意 $\varepsilon > 0$, $\exists N > 0$, 使得当 $i > N$ 时有 $\alpha_i < \alpha + \varepsilon$, 得 $f(\boldsymbol{x}_i) \leqslant \alpha_i < \alpha + \varepsilon$. 根据 (5) 得到 $f(\boldsymbol{x}) < \alpha + \varepsilon$, 令 $\varepsilon \to 0$, 有 $f(\boldsymbol{x}) \leqslant \alpha$. 故 $(\boldsymbol{x}, \alpha) \in \mathrm{epi} f$, $\mathrm{epi} f$ 是闭集.

(6) $\Rightarrow$ (1): 设收敛到 $\boldsymbol{x} \in X$ 的任何序列 $\{\boldsymbol{x}_i\} \subset X\,(i = 1, 2, \cdots)$ 使得 $\lim\limits_{i \to \infty} f(\boldsymbol{x}_i)$ 存在, 令

$$\alpha = \lim_{i \to \infty} f(\boldsymbol{x}_i), \quad \alpha_i = f(\boldsymbol{x}_i),$$

则序列 $(\boldsymbol{x}_i, \alpha_i) \in \mathrm{epi} f$ 成立. 由 $\mathrm{epi} f$ 是闭集得

$$(\boldsymbol{x}_i, \alpha_i) \xrightarrow{i \to +\infty} (\boldsymbol{x}, \alpha),$$

因此, $(\boldsymbol{x},\alpha)\in \mathrm{epi}f$, 即有 $f(\boldsymbol{x})\leqslant\alpha$, 由定义 4.3.1 知 f 在 $\boldsymbol{x}$ 处是下半连续的.

上述证明推导了 (1) ⇒(2) ⇒(3) ⇒(4) ⇒(5) ⇒(6) ⇒(1) 成立, 表明性质的各个结论之间是等价的. □

下面是下半连续函数族的逐点上确界函数对应的下半连续性的性质.

定理 4.3.1 设 $\{f_i(\boldsymbol{x})\,|i\in I\}$ 在 $\boldsymbol{x}$ 处是下半连续的实值函数族, 其中 I 是任意指标集, 则函数族的逐点上确界 $f(\boldsymbol{x})=\sup\{f_i(\boldsymbol{x})\,|i\in I\}$ 在 $\boldsymbol{x}$ 处是下半连续的.

证明 设收敛到 $\boldsymbol{x}\in X$ 的任何序列 $\{\boldsymbol{x}_j\}\subset X\,(j=1,2,\cdots)$ 使得 $\lim\limits_{j\to\infty}f(\boldsymbol{x}_j)$ 存在, 有

$$f(\boldsymbol{x}_j)=\sup\{f_i(\boldsymbol{x}_j)\,|i\in I\}.$$

设 $\alpha_j=f(\boldsymbol{x}_j)$, $\alpha=\lim\limits_{j\to\infty}f(\boldsymbol{x}_j)$, 则有 $f_i(\boldsymbol{x}_j)\leqslant\alpha_j$, $i\in I$. 得到序列

$$(\boldsymbol{x}_j,\alpha_j)\xrightarrow{j\to\infty}(\boldsymbol{x},\alpha),$$

且 $(\boldsymbol{x}_j,\alpha_j)\in\mathrm{epi}f_j$ 成立. 根据性质 4.3.1 得 $(\boldsymbol{x},\alpha)\in\mathrm{epi}f_j$, 即有 $f_i(\boldsymbol{x})\leqslant\alpha$, 对任意 $\varepsilon>0$, 存在一个数 $N>0$, 使得当 $i>N$ 时有 $f(\boldsymbol{x})-\varepsilon\leqslant f_i(\boldsymbol{x})\leqslant\alpha$, 令 $\varepsilon\to0$ 有 $f(\boldsymbol{x})\leqslant\alpha$. □

定义 4.3.3 设非空集合 $X\subset\mathbf{R}^n$, 如果 $\exists\lambda\in(0,1)$, 对任意 $\boldsymbol{x},\boldsymbol{y}\in X$ 有

$$(1-\lambda)\,\boldsymbol{x}+\lambda\boldsymbol{y}\in X,$$

则称 X 关于 $\lambda\in(0,1)$ 满足一点凸性, 或称 X 是弱近似凸集.

实际上近似凸集是弱近似凸集, 反之不一定成立.

显然任何凸集满足一点凸性, 凸函数的一点凸性等价于它的上方图满足一点凸性. 因为, 如果 f 在 C 上关于 $\lambda\in(0,1)$ 满足一点凸性, 设 $\forall(\boldsymbol{x},\alpha),(\boldsymbol{y},\beta)\in\mathrm{epi}f$, 有

$$f(\boldsymbol{x})\leqslant\alpha,\quad f(\boldsymbol{y})\leqslant\beta,\quad \boldsymbol{x},\boldsymbol{y}\in C.$$

根据假设有 $f[(1-\lambda)\,\boldsymbol{x}+\lambda\boldsymbol{y}]\leqslant(1-\lambda)\,f(\boldsymbol{x})+\lambda f(\boldsymbol{y})\leqslant(1-\lambda)\,\alpha+\lambda\beta$, 得

$$((1-\lambda)\,\boldsymbol{x}+\lambda\boldsymbol{y},(1-\lambda)\,\alpha+\lambda\beta)\in\mathrm{epi}f,$$

则 $\mathrm{epi}f$ 关于 $\lambda\in(0,1)$ 满足一点凸性. 反之, 显然.

例 4.3.1 设单变量函数 $f(x)=\begin{cases}1, & x\neq0,\\ 0, & x=0,\end{cases}$ f 不满足一点凸性. 令 $x_1=0$, $x_2=1$, 有

$$f((1-\lambda)\,x_1+\lambda x_2)=f(\lambda)=1>(1-\lambda)\,f(x_1)+\lambda f(x_2)=\lambda,\quad \forall\lambda\in(0,1),$$

f 的上方图 $\mathrm{epi}f = \{(x,t)\,|t \geqslant 1, x \neq 0\} \cup \{(x,t)\,|t \geqslant 0, x = 0\}$ 不是凸集, 且不满足一点凸性.

下面引理告诉我们, 如果一个闭集满足一点凸性, 那么它是凸集.

引理 4.3.1 设非空闭集 $X \subset \mathbf{R}^n$, 如果 $\exists \lambda \in (0,1)$, 对任意 $\boldsymbol{x}, \boldsymbol{y} \in X$ 有

$$(1-\lambda)\,\boldsymbol{x} + \lambda \boldsymbol{y} \in X,$$

则 X 是一个凸集.

证明 设 $\boldsymbol{x}_t = (1-t)\,\boldsymbol{x} + t\boldsymbol{y}, \forall t \in (0,1)$. 如果 X 不是一个凸集, 则 $\exists \boldsymbol{x}, \boldsymbol{y} \in X$, $t \in (0,1)$, 使得 $\boldsymbol{x}_t = (1-t)\,\boldsymbol{x} + t\boldsymbol{y} \notin X$. 令

$$t_1 = \inf\{t \in (0,1)\,|\,(1-t)\,\boldsymbol{x} + t\boldsymbol{y} \in X\}, \quad t_2 = \sup\{t \in (0,1)\,|\,(1-t)\,\boldsymbol{x} + t\boldsymbol{y} \in X\}.$$

显然 $0 < t_1 \leqslant \lambda \leqslant t_2 < 1$, 因为 X 是一个非空闭集, 那么 $\boldsymbol{x}_{t_1}, \boldsymbol{x}_{t_2} \in X$. 因此有任意的 $t \in (0, t_1)$ 和 $t' \in (t_2, 1)$ 满足 $\boldsymbol{x}_t, \boldsymbol{x}_{t'} \notin X$.

如果 $\boldsymbol{x}_t = (1-t)\,\boldsymbol{x} + t\boldsymbol{y} \notin X, \forall t \in (0, t_1)$, 但是有

$$\begin{aligned}(1-\lambda)\,\boldsymbol{x} + \lambda \boldsymbol{x}_{t_1} &= (1-\lambda)\,\boldsymbol{x} + \lambda\left((1-t_1)\,\boldsymbol{x} + t_1\boldsymbol{y}\right) \\ &= (1-\lambda t_1)\,\boldsymbol{x} + \lambda t_1 \boldsymbol{y} \in X,\end{aligned}$$

其中 $\lambda t_1 \in (0, t_1)$, 有 $(1-\lambda)\,\boldsymbol{x} + \lambda \boldsymbol{x}_{t_1} \notin X$, 与上式矛盾.

如果 $\boldsymbol{x}_{t'} = (1-t')\,\boldsymbol{x} + t'\boldsymbol{y} \notin X, \forall t' \in (t_2, 1)$, 但是有

$$\begin{aligned}(1-\lambda)\,\boldsymbol{x}_{t_2} + \lambda \boldsymbol{y} &= (1-\lambda)\left((1-t_2)\,\boldsymbol{x} + t_2\boldsymbol{y}\right) + \lambda \boldsymbol{y} \\ &= \left(1 - \left(t_2 + \lambda\,(1-t_2)\right)\right)\boldsymbol{x} + \left(t_2 + \lambda\,(1-t_2)\right)\boldsymbol{y} \in X,\end{aligned}$$

其中 $t_2 + \lambda\,(1-t_2) \in (t_2, 1)$, 得 $(1-\lambda)\,\boldsymbol{x}_{t_2} + \lambda \boldsymbol{y} \notin X$, 与上式矛盾. □

由引理 4.3.1 知, 把满足一点凸性的集合称为近似凸集, 对于刻画凸函数具有重要意义. 下面结果说明, 如果一个下半连续函数满足一点 (严格) 凸性, 那么它是凸函数 [90]. 本小节中讨论的凸函数均指正常凸函数.

定理 4.3.2 设非空闭凸集 $C \subset \mathbf{R}^n$, f 在 C 上是下半连续函数. 如果 f 是在 C 上关于 $\lambda \in (0,1)$ 满足一点凸性, 则 f 在 C 上是凸函数.

证明 因 $f: C \to \mathbf{R}$ 是下半连续函数, 根据性质 4.3.1 知 $\mathrm{epi}f$ 是闭集. 设 $\forall (\boldsymbol{x}, \alpha), (\boldsymbol{y}, \beta) \in \mathrm{epi}f$ 有

$$f(\boldsymbol{x}) \leqslant \alpha, \quad f(\boldsymbol{y}) \leqslant \beta, \quad \boldsymbol{x}, \boldsymbol{y} \in C.$$

根据假设有 $f[(1-\lambda)\,\boldsymbol{x} + \lambda \boldsymbol{y}] \leqslant (1-\lambda)\,f(\boldsymbol{x}) + \lambda f(\boldsymbol{y}) \leqslant (1-\lambda)\,\alpha + \lambda\beta$, 得

$$((1-\lambda)\,\boldsymbol{x} + \lambda \boldsymbol{y}, (1-\lambda)\,\alpha + \lambda\beta) \in \mathrm{epi}f.$$

所以 epi$f$ 是凸集, 再由引理 3.4.1 得 $f:C\to\mathbf{R}$ 是在 C 上的凸函数. □

推论 4.3.1 设非空闭凸集 $C\subset\mathbf{R}^n$, $f:C\to\mathbf{R}$ 在 C 上是下半连续函数. 如果 f 在 C 上关于 $\lambda\in(0,1)$ 满足一点严格凸性, 则 f 在 C 上是凸函数.

证明过程完全类似于定理 4.3.2 的证明.

定理 4.3.3 设非空闭凸集 $C\subset\mathbf{R}^n$, $f:C\to\mathbf{R}$ 是在 C 上的下半连续函数. 如果 f 在 C 上关于 $\lambda\in(0,1)$ 满足一点半严格凸性, 则 f 在 C 上是凸函数.

证明 根据假设, 由定义 4.7.2 知, $\exists\lambda\in(0,1)$, 对任意 $\boldsymbol{x},\boldsymbol{y}\in C$, $f(\boldsymbol{x})\neq f(\boldsymbol{y})$, 满足

$$f[(1-\lambda)\boldsymbol{x}+\lambda\boldsymbol{y}]<(1-\lambda)f(\boldsymbol{x})+\lambda f(\boldsymbol{y}).$$

根据定理 4.3.2, 只要证明 $\exists\lambda\in(0,1)$, 对任意 $\boldsymbol{x},\boldsymbol{y}\in C$ 有

$$f[(1-\lambda)\boldsymbol{x}+\lambda\boldsymbol{y}]\leqslant(1-\lambda)f(\boldsymbol{x})+\lambda f(\boldsymbol{y}),$$

则 f 是 C 上的凸函数. 否则, $\exists\boldsymbol{x},\boldsymbol{y}\in C$, 对任意 $t\in(0,1)$ 有

$$f[(1-t)\boldsymbol{x}+t\boldsymbol{y}]>(1-t)f(\boldsymbol{x})+tf(\boldsymbol{y}).$$

如果 $f(\boldsymbol{x})\neq f(\boldsymbol{y})$, 与定理假设矛盾. 因此有 $f(\boldsymbol{x})=f(\boldsymbol{y})$, 即得

$$f[(1-t)\boldsymbol{x}+t\boldsymbol{y}]>f(\boldsymbol{x})=f(\boldsymbol{y}),\quad\forall t\in(0,1).\tag{4.3.1-1}$$

用 $\lambda t\in(0,1)$ 替换上式中的 t 后, 式子显然也成立:

$$f(\boldsymbol{x})<f[(1-\lambda t)\boldsymbol{x}+\lambda t\boldsymbol{y}]=f[(1-\lambda)\boldsymbol{x}+\lambda((1-t)\boldsymbol{x}+t\boldsymbol{y})].\tag{4.3.1-2}$$

再由定理的假设、(4.3.1-1) 和 (4.3.1-2) 式得

$$\begin{aligned}f(\boldsymbol{x})&<f[(1-\lambda)\boldsymbol{x}+\lambda((1-t)\boldsymbol{x}+t\boldsymbol{y})]\\&<(1-\lambda)f(\boldsymbol{x})+\lambda f((1-t)\boldsymbol{x}+t\boldsymbol{y})\\&<f((1-t)\boldsymbol{x}+t\boldsymbol{y}),\end{aligned}$$

令 $t=\lambda$, 得 $f(\boldsymbol{x})<f((1-\lambda)\boldsymbol{x}+\lambda\boldsymbol{y})<f(\boldsymbol{x})$, 矛盾. □

下面的引理 4.3.2 是满足一点凸性的另一个重要性质, 即如果 f 在 C 上满足一点凸性, 则 f 在 C 上几乎就是凸函数 (近似凸函数).

引理 4.3.2 设非空凸集 $C\subset\mathbf{R}^n$, $f:C\to\mathbf{R}$ 是实值函数, 且 f 在 C 上关于 $\alpha\in(0,1)$ 满足一点凸性, 则集合

$$A=\{\lambda\in[0,1]\,|f[\lambda\boldsymbol{x}+(1-\lambda)\boldsymbol{y}]\leqslant\lambda f(\boldsymbol{x})+(1-\lambda)f(\boldsymbol{y}),\forall\boldsymbol{x},\boldsymbol{y}\in C\}$$

在 $[0,1]$ 中稠密.

证明 显然 $0,1\in A$, 即集合 A 不是单点集. 下面假设 A 在 $[0,1]$ 中不稠密, 则 $\exists\lambda_0\in(0,1)$ 和它的一个邻域 $U(\lambda_0)$ 满足 $U(\lambda_0)\cap A=\varnothing$. 据此, 令

$$\lambda_1=\inf\{\lambda\in A|\lambda\geqslant\lambda_0\},\quad \lambda_2=\sup\{\lambda\in A|\lambda\leqslant\lambda_0\}, \tag{4.3.1-3}$$

则有 $0\leqslant\lambda_2<\lambda_1\leqslant 1$. 因 $\alpha\in(0,1)$, 我们可选 $t_1,t_2\in A$ 满足 $t_1\geqslant\lambda_1$ 和 $t_2\leqslant\lambda_2$, 且

$$\max\{\alpha(t_1-t_2),(1-\alpha)(t_1-t_2)\}<\lambda_1-\lambda_2. \tag{4.3.1-4}$$

记 $\bar\lambda=\alpha t_1+(1-\alpha)t_2$, 因为对任意的 $\boldsymbol{x},\boldsymbol{y}\in C$, 有

$$\bar\lambda\boldsymbol{x}+\left(1-\bar\lambda\right)\boldsymbol{y}=\alpha(t_1\boldsymbol{x}+(1-t_1)\boldsymbol{y})+(1-\alpha)(t_2\boldsymbol{x}+(1-t_2)\boldsymbol{y}).$$

由 A 的定义得

$$\begin{aligned}f\left(\bar\lambda\boldsymbol{x}+\left(1-\bar\lambda\right)\boldsymbol{y}\right)&=f(\alpha(t_1\boldsymbol{x}+(1-t_1)\boldsymbol{y})+(1-\alpha)(t_2\boldsymbol{x}+(1-t_2)\boldsymbol{y}))\\&\leqslant\alpha f(t_1\boldsymbol{x}+(1-t_1)\boldsymbol{y})+(1-\alpha)f(t_2\boldsymbol{x}+(1-t_2)\boldsymbol{y})\\&\leqslant\alpha f(\boldsymbol{x})+(1-\alpha)f(\boldsymbol{y}),\end{aligned}$$

即得 $\bar\lambda\in A$. 如果 $\bar\lambda\geqslant\lambda_0$, 则由 (4.3.1-3) 式有 $\lambda_1\leqslant\bar\lambda$. 另外, 再由 (4.3.1-4) 式又有

$$\bar\lambda-t_2=\alpha(t_1-t_2)<\lambda_1-\lambda_2.$$

于是得 $\lambda_1>\bar\lambda-t_2+\lambda_2\geqslant\bar\lambda-\lambda_2+\lambda_2=\bar\lambda$, 因此得矛盾. 如果 $\bar\lambda\leqslant\lambda_0$, 类似地, 再由 (4.3.1-3) 和 (4.3.1-4) 式导出矛盾. 因此, 引理成立. □

下面进一步证明, 满足一点凸性的函数只要再加上上半连续性就是凸函数.

定理 4.3.4 *设非空开凸集 $C\subset\mathbf{R}^n$, $f:C\to\mathbf{R}$ 在 C 上是上半连续函数. 如果 f 在 C 上关于 $\alpha\in(0,1)$ 满足一点凸性, 则 f 在 C 上是凸函数.*

证明 由定理的假设和引理 4.3.2 得, 集合 A 在 $[0,1]$ 中稠密. $\forall\lambda\in(0,1)$, 如果 $\lambda\in A$, 设 $\forall\boldsymbol{x},\boldsymbol{y}\in C$, 那么有

$$f[\lambda\boldsymbol{x}+(1-\lambda)\boldsymbol{y}]\leqslant\lambda f(\boldsymbol{x})+(1-\lambda)f(\boldsymbol{y}).$$

若 $\lambda\notin A$, 由 A 在 $[0,1]$ 中稠密性可得, 存在收敛到 λ 的两个序列 $\{\alpha_i\},\{\beta_i\}\subset A$, 满足 $\alpha_i<\lambda<\beta_i$. 显然有 $0<\lambda-\alpha_i<1$, $\dfrac{\alpha_i}{\beta_i}\to 1$. 当 i 充分大时, 有 $0<\lambda-\alpha_i+\dfrac{\alpha_i}{\beta_i}<1$. 令

$$\boldsymbol{y}_i=\boldsymbol{y}+(\lambda-\alpha_i)(\boldsymbol{x}-\boldsymbol{y}),\quad \boldsymbol{x}_i=\boldsymbol{y}+\left(\lambda-\alpha_i+\frac{\alpha_i}{\beta_i}\right)(\boldsymbol{x}-\boldsymbol{y}),$$

则有 $\boldsymbol{y}_i \to \boldsymbol{y}, \boldsymbol{x}_i \to \boldsymbol{x}$, 得 $\beta_i \boldsymbol{x}_i + (1-\beta_i)\boldsymbol{y}_i = \lambda \boldsymbol{x} + (1-\lambda)\boldsymbol{y}$. 由 $\{\beta_i\} \subset A$ 有

$$f[\lambda \boldsymbol{x} + (1-\lambda)\boldsymbol{y}] = f(\beta_i \boldsymbol{x}_i + (1-\beta_i)\boldsymbol{y}_i) \leqslant \beta_i f(\boldsymbol{x}_i) + (1-\beta_i) f(\boldsymbol{y}_i).$$

再由 f 是 C 上的上半连续函数, 在上式中令 $i \to +\infty$ 得

$$f[\lambda \boldsymbol{x} + (1-\lambda)\boldsymbol{y}] \leqslant \lambda f(\boldsymbol{x}) + (1-\lambda) f(\boldsymbol{y}).$$

综上, 定理结论成立. □

推论 4.3.2 设非空开凸集 $C \subset \mathbf{R}^n$, $f: C \to \mathbf{R}$ 在 C 上是上半连续函数. 如果 f 在 C 上关于 $\alpha \in (0,1)$ 满足一点严格凸性, 则 f 在 C 上是严格凸函数.

例 4.3.2 单变量函数 $f(x) = \begin{cases} 1, & x \neq 0, \\ 0, & x = 0 \end{cases}$ 不满足上半连续性和下半连续性, 也不满足一点凸性, 因此, f 在 C 上不是凸函数.

事实上, 凸函数需要保证一点凸性和上、下半连续性是非常必要的. 从定理 4.3.2 和定理 4.3.4 知, 如果一个函数满足一点凸性但不是凸函数, 则一定是不连续函数.

本小节得到下半连续函数与凸性的几个重要性质:

1) 下半连续函数等价于它的上方图是闭集;

2) 下半连续函数族的逐点上确界复合函数也是下半连续函数;

3) 满足一点凸性的闭集是凸集;

4) 满足一点凸性的函数几乎是凸函数, 即该函数在 $[0,1]$ 上几乎处处满足一点凸性;

5) 上半连续或下半连续且满足一点 (严格) 凸性的函数是 (严格) 凸函数.

4.3.2 闭凸函数

闭集的性质是集合的极限点均在该集合内. 下面通过函数上方图的闭性来定义闭凸函数, 并证明正常闭凸函数是下半连续凸函数.

定义 4.3.4 设 $f: \mathbf{R}^n \to [-\infty, +\infty]$ 是广义凸函数. 如果 f 的上方图是闭集, 即

$$\mathrm{epi}(f) = \mathrm{cl}(\mathrm{epi} f),$$

则称 f 在 $\mathbf{R}^n$ 上是闭凸函数. f 是 $\mathbf{R}^n$ 上不取 $-\infty$ 的闭凸函数, 用 $\mathrm{cl} f$ 表示, 即

$$\mathrm{cl} f(\boldsymbol{x}) = \inf\{t \,|\, (\boldsymbol{x}, t) \in \mathrm{cl}(\mathrm{epi} f)\}.$$

若有

$$\mathrm{epi}(\mathrm{cl} f) = \mathrm{cl}(\mathrm{epi} f), \tag{4.3.2-1}$$

特别地, 对某一个 $\boldsymbol{x}$, 如果 $f(\boldsymbol{x})=-\infty$, 定义 $\mathrm{cl}f(\boldsymbol{x})=-\infty$, 且 (4.3.2-1) 式成立, 则称 f 在 $\mathbf{R}^n$ 上是广义闭凸函数.

例 4.3.3　设单变量函数 $f(x)=\begin{cases}x^2, & |x|\leqslant 1,\\ +\infty, & |x|>1\end{cases}$ 是正常凸函数, 且有

$$\mathrm{epi}(f)=\left\{(x,t)^{\mathrm{T}}\,|x^2\leqslant t,|x|\leqslant 1\right\}\cup\left\{(x,+\infty)^{\mathrm{T}}\mid |x|>1\right\},$$

得 $\mathrm{cl}f=f$, 则 f 在 $\mathbf{R}^n$ 上是闭凸函数. 需要说明的是, $(x,+\infty)^{\mathrm{T}}$ 是两维变量, 而不是开区间.

性质 4.3.2　设 f 是 $\mathbf{R}^n$ 上的非正常凸函数, 下面结论成立:

(1) 对任意的 $\boldsymbol{x}\in\mathrm{ri}(\mathrm{dom}f)$, 有 $f(\boldsymbol{x})=-\infty$;

(2) 若 f 在 $\mathbf{R}^n$ 上是下半连续的, 则 $f(\boldsymbol{x})=-\infty$, 其中 $\forall\boldsymbol{x}\in\mathrm{dom}f$;

(3) $\mathrm{cl}f(\boldsymbol{x})$ 是闭非正常凸函数, 且当 $\boldsymbol{x}\in\mathrm{ri}(\mathrm{dom}f)$ 时, $\mathrm{cl}f(\boldsymbol{x})=f(\boldsymbol{x})$.

证明　(1) 如果 $f(\boldsymbol{x})\equiv+\infty$, 则 $\mathrm{dom}f$ 为空集. 否则, 因为 $f(\boldsymbol{x})$ 是 $\mathbf{R}^n$ 上的非正常凸函数, $\exists\boldsymbol{x}'\in\mathrm{dom}f$ 使得 $f(\boldsymbol{x}')=-\infty$. 根据性质 4.1.4 知 $\mathrm{dom}f$ 是凸集, 对任意的 $\boldsymbol{x}\in\mathrm{ri}(\mathrm{dom}f)$, 再由定理 2.3.4 知, $\exists\lambda>1$, 使

$$\boldsymbol{y}=(1-\lambda)\boldsymbol{x}'+\lambda\boldsymbol{x}\in\mathrm{dom}f,$$

有 $\dfrac{1}{\lambda}\boldsymbol{y}+\left(1-\dfrac{1}{\lambda}\right)\boldsymbol{x}'=\boldsymbol{x}$. 再根据定理 4.1.6 得当 $f(\boldsymbol{x}')<t_1, f(\boldsymbol{y})<t_2$ 时, 有不等式

$$f(\boldsymbol{x})=f\left[\left(1-\frac{1}{\lambda}\right)\boldsymbol{x}'+\frac{1}{\lambda}\boldsymbol{y}\right]<\left(1-\frac{1}{\lambda}\right)t_1+\frac{1}{\lambda}t_2.$$

令 $t_1\to-\infty$, 得 $f(\boldsymbol{x})=-\infty$.

(2) 因为 $\boldsymbol{x}\notin\mathrm{dom}f$, 有 $f(\boldsymbol{x})=+\infty$, 则根据 (1) 的结论得 $\mathrm{cl}(\mathrm{ri}(\mathrm{dom}f))=\mathrm{cl}(\mathrm{dom}f)$. 因此, 对于任意的 $\boldsymbol{x}\in\mathrm{dom}f$ 有 $f(\boldsymbol{x})=-\infty$.

(3) 由定义 4.3.2, 以及 (1) 和 (2) 知结论 (3) 成立. □

例 4.3.4　设单变量函数 $f(x)=\begin{cases}-\infty, & |x|<1,\\ 0, & |x|=1,\\ +\infty, & |x|>1\end{cases}$ 是非正常凸函数, 且有

$$\mathrm{epi}(f)=\left\{(x,t)^{\mathrm{T}}\,|-\infty\leqslant t,|x|<1\right\}\cup\left\{(x,t)^{\mathrm{T}}\,|0\leqslant t,|x|=1\right\}$$
$$\cup\left\{(x,+\infty)^{\mathrm{T}}\,\||x|>1\right\}.$$

但是由

$$\mathrm{cl}\,(\mathrm{epi}\,(f))=\left\{(x,t)^{\mathrm{T}}\,|-\infty\leqslant t,|x|\leqslant 1\right\}\cup\left\{(x,+\infty)^{\mathrm{T}}\,\|x|>1\right\},$$

可得 $\mathrm{cl}f\,(x)=-\infty$, 其中 $x\in\mathrm{dom}f=[-1,1]$, 显然性质 4.3.2(1) 和 (2) 的结论都成立.

因为正常凸函数的下半连续性等价于它的上方图是闭集, 所以有下面的结论成立.

定理 4.3.5 设 f 在 $\mathbf{R}^n$ 上是正常凸函数, 则 f 在 $\mathbf{R}^n$ 上为闭凸函数的充要条件是 f 在 $\mathbf{R}^n$ 上是下半连续的.

证明 设 f 为闭凸函数, 由定义 4.3.4 得 $\mathrm{epi}\,(f)=\mathrm{cl}\,(\mathrm{epi}f)$, 再由性质 4.3.1 知 f 是下半连续的. 反过来, 假设 f 是下半连续的, 由性质 4.3.1 知由 $\mathrm{epi}\,(f)=\mathrm{cl}\,(\mathrm{epi}f)$. 即 f 是下半连续的, 故 $\mathrm{cl}f=f$. □

注 闭正常凸函数是下半连续凸函数.

性质 4.3.3 设 f 是 $\mathbf{R}^n$ 上的正常凸函数, 有下面结论成立:

(1) $\mathrm{ri}\,(\mathrm{epi}f)=\{(\boldsymbol{x},t)\,|\boldsymbol{x}\in\mathrm{ri}\,(\mathrm{dom}f)\,,f\,(\boldsymbol{x})<t<+\infty\}$;

(2) 如果存在 $\boldsymbol{x}$ 和 $a\in\mathbf{R}$ 满足 $f\,(\boldsymbol{x})<a$, 则存在 $\bar{\boldsymbol{x}}\in\mathrm{ri}\,(\mathrm{dom}f)$ 使得 $f\,(\bar{\boldsymbol{x}})<a$;

(3) 如果凸集 C 满足 $\mathrm{ri}C\subset\mathrm{dom}f$, 且存在 $\boldsymbol{x}\in\mathrm{cl}C$, $a\in\mathbf{R}$ 满足 $f\,(\boldsymbol{x})<a$, 则存在 $\bar{\boldsymbol{x}}\in\mathrm{ri}C$ 使得 $f\,(\bar{\boldsymbol{x}})<a$;

(4) 如果 f 在凸集 C 上有限, 对任意的 $\boldsymbol{x}\in C$ 有 $f\,(\boldsymbol{x})\geqslant a$, 则对任意的 $\boldsymbol{x}\in\mathrm{cl}C$ 有 $f\,(\boldsymbol{x})\geqslant a$;

(5) $\mathrm{cl}f$ 是闭的正常凸函数, 且当 $\boldsymbol{x}\notin\mathrm{cl}\,(\mathrm{dom}f)\setminus\mathrm{ri}(\mathrm{dom}f)$ 时有 $\mathrm{cl}f\,(\boldsymbol{x})=f\,(\boldsymbol{x})$;

(6) 如果 $\mathrm{dom}f$ 是仿射集, 则 f 是闭凸函数.

证明 (1) 设 $\mathrm{epi}f=\{(\boldsymbol{x},t)\,|\boldsymbol{x}\in\mathbf{R}^n,\ t\in\mathbf{R},t\geqslant f\,(\boldsymbol{x})\}$ 和

$$A=\{(\boldsymbol{x},t)\,|\boldsymbol{x}\in\mathrm{ri}\,(\mathrm{dom}f)\,,f\,(\boldsymbol{x})<t<+\infty\}. \tag{4.3.2-2}$$

对任意 $(\boldsymbol{x},t)\in A$, 有 $(\boldsymbol{x},t)\in\mathrm{epi}f\subset\mathbf{R}^{n+1}$, 下面证明 $(\boldsymbol{x},t)\in\mathrm{ri}\,(\mathrm{epi}f)$. 由于 $\boldsymbol{x}\in\mathrm{ri}\,(\mathrm{dom}f)$ 且 $f\,(\boldsymbol{x})<t$, 那么 $\exists\varepsilon>0$, 使 $(\boldsymbol{x}+\varepsilon B)\cap\mathrm{aff}\,(\mathrm{dom}f)\subset\mathrm{dom}f$, 其中 B 是 $\mathbf{R}^n$ 的单位球. 根据定理 4.3.11 知 f 在 $\boldsymbol{x}\in\mathrm{ri}\,(\mathrm{dom}f)$ 处是连续函数. 令 $\varepsilon_1=\min\left\{\dfrac{1}{2}\,(t-f\,(\boldsymbol{x}))\,,\varepsilon\right\}$, $\exists\varepsilon_2>0$, 使得对任意的 $\boldsymbol{y}\in\boldsymbol{x}+\varepsilon_2B$ 有

$$f\,(\boldsymbol{y})\leqslant f\,(\boldsymbol{x})+\varepsilon_1<t-\varepsilon_1. \tag{4.3.2-3}$$

令 $\varepsilon_3 = \min\{\varepsilon_2, \varepsilon_1\}$, 有

$$(\boldsymbol{x} + \varepsilon_3 B) \cap \text{aff}(\text{dom} f) \subset (\boldsymbol{x} + \varepsilon B) \cap \text{aff}(\text{dom} f) \subset \text{dom} f. \tag{4.3.2-4}$$

设 B_1 是 $\mathbf{R}^{n+1}$ 的单位球, 显然 $B_1 \subset B \times [-1,1]$, 由上式得

$$((\boldsymbol{x}, t) + \varepsilon_3 B_1) \subset ((\boldsymbol{x}, t) + \varepsilon_3 B \times [-1,1]) = (\boldsymbol{x} + \varepsilon_3 B, t + \varepsilon_3 [-1,1]). \tag{4.3.2-5}$$

由 (4.3.2-3) 和 (4.3.2-5) 式得

$$f(\boldsymbol{y}) < t - \varepsilon_1 < t - \varepsilon_3, \quad \forall \boldsymbol{y} \in \boldsymbol{x} + \varepsilon_3 B.$$

由上式和 (4.3.2-4) 式得 $(\boldsymbol{x}, t) \in \text{ri}(\text{epi} f)$, 即 $A \subset \text{ri}(\text{epi} f)$. 反过来, 设 $(\boldsymbol{x}, t) \in \text{ri}(\text{epi} f)$, 则有 $f(\boldsymbol{x}) \leqslant t$. 设 B_1 是 $\mathbf{R}^{n+1}$ 的单位球和 B 是 $\mathbf{R}^n$ 的单位球, 那么存在一个 $\varepsilon > 0$, 使

$$((\boldsymbol{x}, t) + \varepsilon B_1) \cap \text{aff}(\text{epi} f) \subset \text{epi} f. \tag{4.3.2-6}$$

显然存在一个充分小 $\varepsilon_1 > 0$ 使得 $\varepsilon_1 (B \times [-1,1]) \subset \varepsilon B_1$, 并且有 $\text{aff}(\text{epi} f) = \text{aff}(\text{dom} f) \times \mathbf{R}$, 则根据 (4.3.2-6) 式得

$$(\boldsymbol{x} + \varepsilon_1 B, t + \varepsilon_1 [-1,1]) \cap \text{aff}(\text{epi} f) \subset ((\boldsymbol{x}, t) + \varepsilon B_1) \cap \text{aff}(\text{epi} f) \subset \text{epi} f. \tag{4.3.2-7}$$

上式即得 $(\boldsymbol{x} + \varepsilon_1 B) \cap \text{aff}(\text{dom} f) \subset \text{dom} f$, $f(\boldsymbol{x}) \leqslant t - \varepsilon_1 < t$, 说明 $(\boldsymbol{x}, t) \in A$.

(2) 由已知条件得 $(\boldsymbol{x}, a) \in \text{epi} f$, 说明 $\text{ri}(\text{epi} f) \neq \varnothing$. 否则有 (4.3.2-7) 式在 $\text{ri}(\text{epi} f) = \varnothing$ 和 $t = a$ 时不可能成立, 即 $f(\boldsymbol{x}) < a$ 不成立. 因此, 根据 (1) 的结论得 (2) 成立.

(3) 令函数 $h(\boldsymbol{x}) = \begin{cases} f(\boldsymbol{x}), & \boldsymbol{x} \in \text{cl} C, \\ +\infty, & \boldsymbol{x} \notin \text{cl} C, \end{cases}$ 有 $\text{ri}(\text{dom} f) = \text{ri} C$, 根据 (1) 的结论得 (3) 成立.

(4) 假设对任意 $\boldsymbol{x} \in \text{cl} C$ 有 $f(\boldsymbol{x}) \geqslant a$ 不成立, 则 $\exists \boldsymbol{x} \in \text{cl} C$ 和 $a \in \mathbf{R}$ 满足 $f(\boldsymbol{x}) < a$. 由 (3) 知, $\exists \bar{\boldsymbol{x}} \in \text{ri} C$ 使得 $f(\bar{\boldsymbol{x}}) < a$, 这显然与 (4) 的已知矛盾.

(5) 设 $g(\boldsymbol{x}) = \text{cl} f(\boldsymbol{x})$, 根据定义 4.3.4 得 $\text{cl}(\text{epi} f) = \text{epi} g$. 因为 $\text{epi} f$ 是凸集, 则 $\text{cl}(\text{epi} f)$ 也是凸集. 再根据定义 4.3.4 知 $g(\boldsymbol{x}) = -\infty$ 不成立, 说明 $\text{cl} f$ 是闭的正常凸函数.

假设 $\boldsymbol{x} \notin \text{cl}(\text{dom} f) \backslash \text{ri}(\text{dom} f)$ 时. 若 $\boldsymbol{x} \in \text{ri}(\text{dom} f)$, 因 $\text{epi}(\text{cl} f) = \text{cl}(\text{epi} f)$, 由性质 2.3.4 有

$$\text{ri}(\text{epi}(\text{cl} f)) = \text{ri}(\text{cl}(\text{epi} f)) = \text{ri}(\text{epi} f),$$

即得 $\text{cl} f(\boldsymbol{x}) = f(\boldsymbol{x})$. 若 $\boldsymbol{x} \notin \text{cl}(\text{dom} f)$, 则有 $\text{cl} f(\boldsymbol{x}) = f(\boldsymbol{x}) = +\infty$.

(6) 因 $\mathrm{dom}f$ 是仿射集, 则有 $\mathrm{cl}\,(\mathrm{dom}f)\setminus\mathrm{ri}(\mathrm{dom}f)=\varnothing$. 根据 (5) 得 $\mathrm{cl}f\,(\boldsymbol{x})=f\,(\boldsymbol{x})$ 在 $\mathbf{R}^n$ 上成立. 因此, f 在 $\mathbf{R}^n$ 上是闭凸函数. □

性质 4.3.3 给出了正常凸函数与有效定义域相对内部之间的关系, 由此, 得到正常凸函数的有效定义域相对内部点与闭凸函数之间存在收敛关系, 见下面的性质 4.3.4.

性质 4.3.4 设 f 在 $\mathbf{R}^n$ 上是正常凸函数和 $\boldsymbol{x}\in\mathrm{ri}\,(\mathrm{dom}f)$, 则对任意的 $\boldsymbol{y}\in\mathbf{R}^n$, 有

$$\mathrm{cl}f\,(\boldsymbol{y})=\lim_{t\uparrow 1}f\,[(1-t)\,\boldsymbol{x}+t\boldsymbol{y}]\,. \tag{4.3.2-8}$$

当 f 是非正常凸函数, 且 $\boldsymbol{y}\in\mathrm{cl}\,(\mathrm{dom}f)$ 时, 上式也成立.

证明 设序列 $\{t_i\}\subset[0,1]\,(i=1,2,\cdots)$ 和 $t_i\xrightarrow{i\to+\infty}1$, 因为 $\mathrm{cl}f$ 是下半连续的和 $\mathrm{cl}f\,(\boldsymbol{x})\leqslant f\,(\boldsymbol{x})$, 有

$$\mathrm{cl}f\,(\boldsymbol{y})\leqslant\lim_{i\to+\infty}\mathrm{cl}f\,[(1-t_i)\,\boldsymbol{x}+t_i\boldsymbol{y}]\leqslant\lim_{i\to+\infty}f\,[(1-t_i)\,\boldsymbol{x}+t_i\boldsymbol{y}]\,. \tag{4.3.2-9}$$

设 $\mathrm{cl}f\,(\boldsymbol{y})\leqslant a, f\,(\boldsymbol{x})\leqslant b$, 由性质 4.3.2 得

$$(\boldsymbol{y},a)\in\mathrm{epi}\,(\mathrm{cl}f)=\mathrm{cl}\,(\mathrm{epi}f)\,,\quad(\boldsymbol{x},b)\in\mathrm{ri}\,(\mathrm{epi}f)\,.$$

根据定理 2.3.2, 当 $0\leqslant t<1$ 时, 有

$$(1-t)\,(\boldsymbol{x},b)+t\,(\boldsymbol{y},a)\in\mathrm{ri}\,(\mathrm{epi}f)\,.$$

因为 $f\,(\boldsymbol{x})$ 是凸函数, 故当 $0\leqslant t<1$ 时, 有

$$f\,[(1-t)\,\boldsymbol{x}+t\boldsymbol{y}]\leqslant(1-t)\,b+ta.$$

由上式得

$$\lim_{i\to+\infty}f\,[(1-t_i)\,\boldsymbol{x}+t_i\boldsymbol{y}]\leqslant\lim_{i\to+\infty}\,[(1-t_i)\,b+t_ia]=a.$$

由 a 的任意性和 (4.3.2-9) 式, 得 (4.3.2-8) 式成立.

当 f 是非正常且 $\boldsymbol{y}\in\mathrm{cl}\,(\mathrm{dom}f)$ 时, 因为 $0\leqslant t<1$, 根据性质 4.3.2 知

$$f\,[(1-t)\,\boldsymbol{x}+t\boldsymbol{y}]=-\infty.$$

故 (4.3.2-8) 式成立. □

对于正常凸函数, 性质 4.3.4 给出了一个闭凸函数的表达式. 通过这个表达式能够求出闭凸函数形式. 例如在例 4.3.3 中, 对于 $x\in\mathrm{ri}\,(\mathrm{dom}f)=(-1,1)$, 取

$x=0$, 由 (4.3.2-8) 式得

$$\mathrm{cl}f(y)=\lim_{t\uparrow 1} f[ty]=\lim_{t\uparrow 1}(ty)^2=\begin{cases} y^2, & |y|\leqslant 1,\\ +\infty, & |y|>1.\end{cases}$$

在例 4.3.4 中, 对于 $y\in \mathrm{cl}(\mathrm{dom}f)=[-1,1]$, 取 $x=0$, 由 (4.3.2-8) 式得

$$\mathrm{cl}f(y)=\lim_{t\uparrow 1} f[ty]=-\infty.$$

对于 $y\in \mathrm{ri}(\mathrm{dom}f)$, 有 $\mathrm{cl}f(y)=f(y)$.

推论 4.3.3 设 f 是闭正常凸函数, 则对任意的 $\boldsymbol{x}\in \mathrm{dom}f$ 和 $\boldsymbol{y}\in \mathbf{R}^n$, 有

$$f(\boldsymbol{y})=\lim_{t\uparrow 1} f[(1-t)\boldsymbol{x}+t\boldsymbol{y}].$$

证明 根据性质 4.3.4 直接可得. □

正常凸函数的水平集的闭包和相对内部分别有等价性质, 见下面结论.

性质 4.3.5 设 f 在 $\mathbf{R}^n$ 上是正常凸函数. 如果 $a\in\mathbf{R}$ 满足 $a>\inf\limits_{\boldsymbol{x}\in\mathbf{R}^n} f(\boldsymbol{x})$, 则

(1) $\mathrm{cl}\{\boldsymbol{x}\,|f(\boldsymbol{x})\leqslant a\}=\mathrm{cl}\{\boldsymbol{x}\,|f(\boldsymbol{x})<a\}=\{\boldsymbol{x}\,|\mathrm{cl}f(\boldsymbol{x})\leqslant a\}$;

(2) $\mathrm{ri}\{\boldsymbol{x}\,|f(\boldsymbol{x})\leqslant a\}=\mathrm{ri}\{\boldsymbol{x}\,|f(\boldsymbol{x})<a\}=\{\boldsymbol{x}\in \mathrm{ri}(\mathrm{dom}f)\,|f(\boldsymbol{x})<a\}$;

(3) 在 (1) 和 (2) 中各集合维数与 $\mathrm{dom}f$ 的维数相同.

证明 对于固定点 $a\in\mathbf{R}$, 设集合 $H(a)=\{(\boldsymbol{x},a)\,|\boldsymbol{x}\in\mathbf{R}^n\}$. 显然 $H(a)$ 是仿射集, 根据性质 4.3.3 知 $H(a)\cap \mathrm{ri}(\mathrm{epi}f)\neq\varnothing$, 有 $\{(\boldsymbol{x},a)\,|f(\boldsymbol{x})\leqslant a\}=H(a)\cap \mathrm{epi}f$. 再由性质 2.3.6 得

$$\begin{aligned}\mathrm{cl}\{(\boldsymbol{x},a)\,|f(\boldsymbol{x})\leqslant a\}&=H(a)\cap \mathrm{cl}(\mathrm{epi}f)=H(a)\cap \mathrm{epi}(\mathrm{cl}f)\\&=\{(\boldsymbol{x},a)\,|\boldsymbol{x}\in\mathbf{R}^n\}\cap\{(\boldsymbol{x},a)\,|\mathrm{cl}f(\boldsymbol{x})\leqslant a\}\\&=\{(\boldsymbol{x},a)\,|\mathrm{cl}f(\boldsymbol{x})\leqslant a\},\end{aligned}$$

并且

$$\begin{aligned}\mathrm{ri}\{(\boldsymbol{x},a)\,|f(\boldsymbol{x})\leqslant a\}&=H(a)\cap \mathrm{ri}(\mathrm{epi}f)\\&=\{(\boldsymbol{x},a)\,|\boldsymbol{x}\in\mathbf{R}^n\}\cap\{(\boldsymbol{x},a)\,|\mathrm{ri}(\mathrm{dom}f),f(\boldsymbol{x})<a\}.\\&=\{(\boldsymbol{x},a)\,|\mathrm{ri}(\mathrm{dom}f),f(\boldsymbol{x})<a\}.\end{aligned}$$

下面结论显然:

$$\mathrm{ri}\{\boldsymbol{x}\,|f(\boldsymbol{x})\leqslant a\}=\mathrm{ri}\{\boldsymbol{x}\,|f(\boldsymbol{x})<a\},$$

由于凸集的闭包和相对内部具有相同的仿射包, 故 (3) 成立. □

推论 4.3.4 设 f 是 $\mathbf{R}^n$ 上的闭正常凸函数, $\mathrm{dom}f$ 是相对开集. 如果 $a \in \mathbf{R}$ 满足 $a > \inf\limits_{\boldsymbol{x}\in\mathbf{R}^n} f(\boldsymbol{x})$, 则

$$\mathrm{ri}\,\{\boldsymbol{x}\,|f(\boldsymbol{x}) \leqslant a\} = \{\boldsymbol{x}\,|f(\boldsymbol{x}) < a\},$$

$$\mathrm{cl}\,\{\boldsymbol{x}\,|f(\boldsymbol{x}) < a\} = \{\boldsymbol{x}\,|f(\boldsymbol{x}) \leqslant a\}.$$

证明 因为 f 是 $\mathbf{R}^n$ 上的闭正常凸函数, $\mathrm{dom}f$ 是相对开集, 根据定义 4.3.4 得

$$\mathrm{cl}f(\boldsymbol{x}) = f(\boldsymbol{x}) \quad 和 \quad \mathrm{ri}\,(\mathrm{dom}f) = \mathrm{dom}f,$$

由性质 4.3.5 知结论成立. □

性质 4.3.5 说明正常凸函数的水平集的闭包等于它的闭凸函数的水平集闭包. 下面给出了多个闭凸函数之和的运算性质. 定理 4.3.6 表明多个闭正常凸函数的和也是闭正常凸函数.

定理 4.3.6 设 $f_1, \cdots, f_m$ 在 $\mathbf{R}^n$ 上是正常凸函数, 令 $g(\boldsymbol{x}) = \sum\limits_{i=1}^{m} f_i(\boldsymbol{x})$, 有下面结论成立:

(1) 如果 $\mathrm{cl}f_i(\boldsymbol{x}) = f_i(\boldsymbol{x})\,(i = 1, \cdots, m)$ 和 $g(\boldsymbol{x}) \not\equiv +\infty$, 则 g 是闭正常凸函数;

(2) 如果 $f_1, \cdots, f_m$ 不全是闭凸函数, 但 $\bigcap_{i=1}^{m} \mathrm{ri}\,(\mathrm{dom}f_i) \neq \varnothing$, 则

$$\mathrm{cl}g(\boldsymbol{x}) = \sum_{i=1}^{m} \mathrm{cl}f_i(\boldsymbol{x}).$$

证明 (1) 根据性质 4.2.2(3) 得, g 是正常凸函数. 下面只需证明它的闭性. 设

$$\boldsymbol{x} \in \mathrm{ri}\,(\mathrm{dom}g) = \mathrm{ri}\left(\bigcap_{i=1}^{m} \mathrm{dom}f_i\right).$$

对于任意的 $\boldsymbol{y} \in \mathbf{R}^n$, 由性质 4.3.4 有

$$\begin{aligned}\mathrm{cl}g(\boldsymbol{y}) &= \lim_{\lambda\uparrow 1} g\,[(1-\lambda)\,\boldsymbol{x} + \lambda\boldsymbol{y}] \\ &= \lim_{\lambda\uparrow 1} \sum_{i=1}^{m} f_i\,[(1-\lambda)\,\boldsymbol{x} + \lambda\boldsymbol{y}] \\ &= \sum_{i=1}^{m} \mathrm{cl}f_i(\boldsymbol{y}) = \sum_{i=1}^{m} f_i(\boldsymbol{y}) = g(\boldsymbol{y}),\end{aligned}$$

上式说明 g 是闭凸函数.

(2) 由性质 2.3.4 得 $\bigcap_{i=1}^{m}\mathrm{ri}\,(\mathrm{dom}f_i)=\mathrm{ri}\,(\mathrm{dom}g)\neq\varnothing$, 再利用性质 4.3.4 得结论成立. □

下面定理说明正常凸函数族上确界函数的闭凸性成立.

定理 4.3.7 设 $f_i\,(i\in I)$ 是在 $\mathbf{R}^n$ 上的正常凸函数族, 其中 I 是任意指标集, 令 $f\,(\boldsymbol{x})=\sup\limits_{i\in I}f_i\,(\boldsymbol{x})$, 有

(1) 如果 f 在某些点有限且满足 $\mathrm{cl}f_i\,(\boldsymbol{x})=f_i\,(\boldsymbol{x})\,(i\in I)$, 则 f 是闭正常凸函数;

(2) 如果 $f_i\,(i\in I)$ 不全是闭凸函数, 且存在 $\bar{\boldsymbol{x}}\in\bigcap\limits_{i\in I}\mathrm{ri}\,(\mathrm{dom}f_i)$ 和 $f\,(\bar{\boldsymbol{x}})<+\infty$, 则

$$\mathrm{cl}f\,(\boldsymbol{x})=\sup_{i\in I}\,\mathrm{cl}f_i\,(\boldsymbol{x})\,.$$

证明 (1) 由性质 4.2.12 知 f 是正常凸函数, 因为 $\mathrm{epi}f_i$ 是闭集, $\mathrm{epi}f=\bigcap\limits_{i\in I}\mathrm{epi}f_i$ 也是闭集, 所以 f 是闭正常凸函数.

(2) 由性质 4.2.12 得

$$\begin{aligned}\mathrm{epi}\,(\mathrm{cl}f)=\mathrm{cl}\,(\mathrm{epi}f)=\mathrm{cl}\left(\bigcap_{i\in I}\mathrm{epi}f_i\right)&=\bigcap_{i\in I}\mathrm{cl}\,(\mathrm{epi}f_i)\\&=\bigcap_{i\in I}\mathrm{epi}\,(\mathrm{cl}f_i)=\mathrm{epi}\,(\sup\{\mathrm{cl}f_i\,|i\in I\})\,,\end{aligned}$$

故 $\mathrm{cl}f\,(\boldsymbol{x})=\sup\{\mathrm{cl}f_i\,(\boldsymbol{x})\,|i\in I\}$. □

定理 4.3.7 说明闭正常凸函数族的上确界复合函数也是闭正常凸函数; 下面的定理则说明闭正常凸函数的线性变换复合函数也是闭正常凸函数.

定理 4.3.8 设线性变换 $A:\mathbf{R}^n\to\mathbf{R}^m$, $g\,(\boldsymbol{y})$ 在 $\mathbf{R}^m$ 上使 $(gA)\,(\boldsymbol{x})\not\equiv+\infty$ 是正常凸函数, 有

(1) 如果 $\mathrm{cl}g\,(\boldsymbol{y})=g\,(\boldsymbol{y})$, 则 $\mathrm{cl}\,(gA)\,(\boldsymbol{x})=(gA)\,(\boldsymbol{x})$;

(2) 如果 g 不是闭凸函数, 且 $A\boldsymbol{x}\in\mathrm{ri}\,(\mathrm{dom}g)$, 则 $\mathrm{cl}\,(gA)\,(\boldsymbol{x})=((\mathrm{cl}g)\,A)\,(\boldsymbol{x})$.

证明 (1) 由性质 4.2.11 知 $(gA)\,(\boldsymbol{x})$ 是正常凸函数. 由于 $\mathrm{epi}\,(gA)$ 是在线性变换 $B:(\boldsymbol{x},t)\to(A\boldsymbol{x},t)$ 下 $\mathrm{epi}g$ 的逆象, 故当 $\mathrm{cl}g\,(\boldsymbol{y})=g\,(\boldsymbol{y})$ 时, $\mathrm{cl}\,(gA)\,(\boldsymbol{x})=(gA)\,(\boldsymbol{x})$.

(2) 由性质 2.3.11 得

$$\mathrm{epi}\,(\mathrm{cl}\,(gA))=\mathrm{cl}\,(\mathrm{epi}\,(gA))=\mathrm{cl}\left(B^{-1}\mathrm{epi}g\right)=B^{-1}\,(\mathrm{cl}\,(\mathrm{epi}g))$$

$$= B^{-1}\mathrm{epi}\,(\mathrm{cl}g) = \mathrm{epi}\,(\mathrm{cl}g)\,A,$$

故 $\mathrm{cl}\,(gA)\,(\boldsymbol{x}) = ((\mathrm{cl}g)\,A)\,(\boldsymbol{x})$. □

本小节得到了闭 (非) 正常凸函数的一些重要性质:

1) 非正常凸函数值在其有效定义域中相对内点对应的闭凸函数值是 $-\infty$;

2) 正常凸函数的下半连续性等价于它的上方图是闭集;

3) (非) 正常凸函数的闭凸函数可由有限定义域内部点对应的 (4.3.2-8) 式确定;

4) 正常凸函数的水平集的闭包等于它的闭凸函数的水平集闭包;

5) 多个闭正常凸函数的和也是闭正常凸函数;

6) 闭正常凸函数族的上确界复合函数也是闭正常凸函数;

7) 闭正常凸函数的线性变换复合函数也是闭正常凸函数.

4.3.3 连续凸函数

本小节讨论的凸函数均具有连续性, 下面证明后面常用的两个不等式.

引理 4.3.3 设 $\phi(t)$ 是 $\mathbf{R}$ 上的凸函数, $a_0 < a_1 < a_2$, 且 $a_0, a_1, a_2 \in \mathrm{dom}\phi$, 则

$$\varphi\,(a_2, a_0) \geqslant \varphi\,(a_1, a_0)\,, \quad \varphi\,(a_2, a_1) \geqslant \varphi\,(a_0, a_1)\,, \tag{4.3.3-1}$$

其中记二元函数 $\varphi\,(t_1, t_2) = \dfrac{\phi\,(t_1) - \phi\,(t_2)}{t_1 - t_2}, \forall t_1, t_2 \in \mathbf{R}$.

证明 设 $t = \dfrac{a_1 - a_0}{a_2 - a_0}$, 则 $1 - t = \dfrac{a_2 - a_1}{a_2 - a_0}$. 由于 $a_0 < a_1 < a_2$, 得 $0 < t < 1$, 且

$$ta_2 + (1 - t)\,a_0 = a_1.$$

由 $\phi(t)$ 是凸函数, 得

$$\frac{a_1 - a_0}{a_2 - a_0}\phi\,(a_2) + \frac{a_2 - a_1}{a_2 - a_0}\phi\,(a_0) \geqslant \phi\,[ta_2 + (1 - t)\,a_0] = \phi\,(a_1)\,,$$

移项得 $\varphi\,(a_2, a_0) \geqslant \varphi\,(a_1, a_0)$. 而上式也可改写为

$$\frac{a_1 - a_0}{a_2 - a_0}\,[\phi\,(a_2) - \phi\,(a_1)] \geqslant \frac{a_2 - a_1}{a_2 - a_0}\,[\phi\,(a_1) - \phi\,(a_0)]\,,$$

即 $\varphi\,(a_2, a_1) \geqslant \varphi\,(a_0, a_1)$. □

同理, 根据上述推导, 若 $0 < t_1 < t_2$, 且 $0, t_1, t_2 \in \mathrm{dom}\phi$, 则有

$$\frac{\phi\,(t_2) - \phi\,(0)}{t_2} \geqslant \frac{\phi\,(t_1) - \phi\,(0)}{t_1}.$$

当 $\phi(0)=0$ 时, $\dfrac{\phi(t)}{t}$ 是单调递增函数.

下面证明在 $\mathbf{R}^n$ 上的凸函数在有效定义域相对内部中的任何紧子集上是有界的.

定理 4.3.9 设 f 在 $\mathbf{R}^n$ 上是凸函数, C 是集合 $\mathrm{ri}(\mathrm{dom}f)$ 的任何有界闭子集 (紧子集), 则 f 在 C 上是有上界的.

证明 由于 $\mathrm{ri}(\mathrm{dom}f)$ 是开集, 设 $\{\mathrm{Sim}(\boldsymbol{x}_1,\cdots,\boldsymbol{x}_{r+1})\}$ 是 $\mathrm{ri}(\mathrm{dom}f)$ 中的一个单纯形族, 其中 $\boldsymbol{x}_1,\cdots,\boldsymbol{x}_{r+1}\in\mathrm{ri}(\mathrm{dom}f)$. 根据有限覆盖定理知, C 可以被 $\{\mathrm{Sim}(\boldsymbol{x}_1,\cdots,\boldsymbol{x}_{r+1})\}$ 的有限个单纯形覆盖. 这样只要证明 f 在 $\mathrm{ri}(\mathrm{dom}f)$ 中的每一个单纯形上是有上界的. 如果 f 是非正常的, 结论显然成立.

下面设 f 是正常凸函数, 按单纯形的维数 r 进行归纳证明.

当 $r=1$ 时, 对于任意的 $\boldsymbol{x}\in\mathrm{Sim}(\boldsymbol{x}_1,\boldsymbol{x}_2)$ 有

$$f[t\boldsymbol{x}_1+(1-t)\boldsymbol{x}_2]\leqslant tf(\boldsymbol{x}_1)+(1-t)f(\boldsymbol{x}_2)\leqslant\max\{f(\boldsymbol{x}_1),f(\boldsymbol{x}_2)\},\quad 0<t<1.$$

说明 f 在一维单纯形 $\mathrm{Sim}(\boldsymbol{x}_1,\boldsymbol{x}_2)$ 上有上界.

假设 f 在 $\mathrm{dom}f$ 内任意 $(r-1)$ 维单纯形 $\{\mathrm{Sim}(\boldsymbol{x}_1,\cdots,\boldsymbol{x}_r)\}$ 上有上界.

对于 r 维单纯形 $\mathrm{Sim}(\boldsymbol{x}_1,\cdots,\boldsymbol{x}_{r+1})$, 设 $\boldsymbol{x}\in\mathrm{Sim}(\boldsymbol{x}_1,\cdots,\boldsymbol{x}_{r+1})$, 且 $\boldsymbol{x}\neq\boldsymbol{x}_{r+1}$, 有

$$\boldsymbol{x}=t_1\boldsymbol{x}_1+\cdots+t_r\boldsymbol{x}_r+t_{r+1}\boldsymbol{x}_{r+1},\quad t_i\geqslant 0,\quad i=1,\cdots,r+1,\quad \sum_{i=1}^{r+1}t_i=1,\quad t_{r+1}<1,$$

上式可记为 $\boldsymbol{x}=(1-t_{r+1})\boldsymbol{u}+t_{r+1}\boldsymbol{x}_{r+1}$, 其中 $\boldsymbol{u}=\dfrac{1}{1-t_{r+1}}(t_1\boldsymbol{x}_1+\cdots+t_r\boldsymbol{x}_r)$. 则 $\boldsymbol{x}_1,\cdots,\boldsymbol{x}_r$ 仿射无关, 所以 $\boldsymbol{u}$ 是 $\mathrm{dom}f$ 中 $r-1$ 维单纯形的点. 根据归纳假设, $f(\boldsymbol{u})$ 是有上界的, 且

$$f(\boldsymbol{x})\leqslant(1-t_{r+1})f(\boldsymbol{u})+t_{r+1}f(\boldsymbol{x}_{r+1})\leqslant\max[f(\boldsymbol{u}),f(\boldsymbol{x}_{r+1})].$$

因此, f 在 $\mathrm{Sim}(\boldsymbol{x}_1,\cdots,\boldsymbol{x}_{r+1})$ 上也有上界. 故定理结论成立. □

接着证明正常凸函数的连续性. 如果一个正常凸函数在有效定义域内一个点的邻域内具有上界, 则这个正常凸函数在该点处连续.

定理 4.3.10 设 f 是 $\mathbf{R}^n$ 上的正常凸函数, 且在 $\boldsymbol{x}_0\in\mathrm{dom}f$ 的一个邻域 $U(\boldsymbol{x}_0,\delta)$ 内有上界, 则 f 在 $\boldsymbol{x}_0\in\mathrm{dom}f$ 处连续.

证明 假设 f 在 $\boldsymbol{x}_0=\boldsymbol{0}$ 的一个邻域 $U(\boldsymbol{x}_0,\varepsilon)=\{\boldsymbol{x}\mid\|\boldsymbol{x}\|<\varepsilon\}$ 上的上界是 A, 其中 ε 可以是任意给定的小于 δ 的正数. 设 $\forall\boldsymbol{x}\in U(\boldsymbol{x}_0,\varepsilon)$, 定义函数

$$\phi(a)=f(a\boldsymbol{x}),\quad\forall a\in\mathbf{R}.$$

在引理 4.3.3 中, 取 $a_0 = 0, 1 > a_1 = a > 0, a_2 = 1$, 得

$$\frac{\phi(a) - \phi(0)}{a} \leqslant \frac{\phi(1) - \phi(0)}{1}.$$

由 $\phi(1) = f(\boldsymbol{x}) \leqslant A$, $\phi(0) = f(\mathbf{0}) \leqslant A$ 得

$$f(a\boldsymbol{x}) - f(\mathbf{0}) \leqslant 2Aa. \tag{4.3.3-2}$$

在引理 4.3.3 中再取 $a_0 = -1, a_1 = 0, a_2 = a$, 得

$$\frac{\phi(0) - \phi(-1)}{0 - (-1)} \leqslant \frac{\phi(a) - \phi(0)}{a}.$$

故

$$-2Aa \leqslant f(a\boldsymbol{x}) - f(\mathbf{0}). \tag{4.3.3-3}$$

由 (4.3.3-2) 和 (4.3.3-3) 式得

$$|f(a\boldsymbol{x}) - f(\mathbf{0})| \leqslant 2Aa. \tag{4.3.3-4}$$

取 $a = \varepsilon/(2A) < 1$, $U(\boldsymbol{x}_0, a\varepsilon) = aU(\boldsymbol{x}_0, \varepsilon)$. 对 $\boldsymbol{y} \in aU(\boldsymbol{x}_0, \varepsilon)$, 则 $\exists \boldsymbol{x} \in U(\boldsymbol{x}_0, \varepsilon)$ 使得 $\boldsymbol{y} = a\boldsymbol{x}$, 由 (4.3.3-4) 式得

$$|f(\boldsymbol{y}) - f(\mathbf{0})| = |f(a\boldsymbol{x}) - f(\mathbf{0})| \leqslant 2Aa = \varepsilon.$$

上式说明 f 在 $\boldsymbol{x}_0 = \mathbf{0}$ 连续, 上述过程对 $\boldsymbol{x}_0 \neq \mathbf{0}$ 也成立. □

例 4.3.5 单变量函数 $f(x) = \begin{cases} 0, & |x| \leqslant 1, \\ +\infty, & |x| > 1 \end{cases}$ 是正常凸函数, $f(x)$ 在 $x = 1$ 处邻域上无界, $f(x)$ 在 $x = 1$ 处不连续, 但 f 在 $\mathrm{ri}(\mathrm{dom}f)$ 上连续.

因此, 由前面两个定理的结论可证明一个正常凸函数 f 在 $\mathrm{ri}(\mathrm{dom}f)$ 上连续.

定理 4.3.11 设 f 在 $\mathbf{R}^n$ 上是正常凸函数, 则 f 在 $\mathrm{ri}(\mathrm{dom}f)$ 上连续.

证明 设 $\forall \boldsymbol{x}_0 \in \mathrm{ri}(\mathrm{dom}f)$, 则 $\boldsymbol{x}_0$ 是单纯形 $\mathrm{Sim}(\boldsymbol{y}_0, \cdots, \boldsymbol{y}_k)$ 的内点, 其中 $\boldsymbol{y}_0, \cdots, \boldsymbol{y}_k \in \mathrm{dom}f$, $k = \dim(\mathrm{dom}f)$. 由定理 4.3.9 知, f 在 $\boldsymbol{x}_0$ 的一个邻域内有上界. 再由定理 4.3.10 知 $f(\boldsymbol{x})$ 在 $\boldsymbol{x}_0$ 连续, 故 $f(\boldsymbol{x})$ 在 $\mathrm{ri}(\mathrm{dom}f)$ 连续. □

由此, 下面两个推论显然.

推论 4.3.5 设 f 是 $\mathbf{R}^n$ 上处处有限的凸函数, 则 f 在 $\mathbf{R}^n$ 上处处连续.

证明 根据性质 4.3.3(1) 可得. □

推论 4.3.6 设 $f(\boldsymbol{x}, t)$ 是 $\mathbf{R}^n \times T$ 上的实值函数, 其中 $T \subset \mathbf{R}$ 是任意非空子集. 如果对每一个固定的 $t \in T$, $f(\boldsymbol{x}, t)$ 是关于 $\boldsymbol{x}$ 的凸函数且有界, 则 $h(\boldsymbol{x}) = \sup\{f(\boldsymbol{x}, t) | t \in T\}$ 是有限连续凸函数.

本小节得到关于连续凸函数的几个结论:

1) $\mathbf{R}^n$ 上的凸函数在有效定义域相对内部中的任何紧子集上是有界的;

2) 如果一个正常凸函数在有效定义域内一个点的邻域内具有上界, 则这个正常凸函数在该点处连续;

3) 一个正常凸函数 f 在有效定义域相对内部上连续;

4) 在 $\mathbf{R}^n$ 上处处有限的凸函数是处处连续的;

5) 一个有界凸函数族的集合逐点上确界复合函数是有限连续凸函数.

最后, 通过本小节的结论知, 如果一个有界函数在 $\mathbf{R}^n$ 上满足一点凸性, 那么它在 $\mathbf{R}^n$ 上几乎处处连续.

4.3.4 Lipschitz 连续凸函数

本小节讨论一种更强的连续凸函数, 称为 Lipschitz 连续凸函数.

定义 4.3.5 设 f 在 $X \subset \mathbf{R}^n$ 上是广义函数. 如果 $\exists \alpha > 0$, 满足

$$|f(\boldsymbol{y}) - f(\boldsymbol{x})| \leqslant \alpha \|\boldsymbol{y} - \boldsymbol{x}\|, \quad \forall \boldsymbol{x}, \boldsymbol{y} \in X, \tag{4.3.4-1}$$

则称 f 在 X 上满足 Lipschitz 条件, 也称 Lipschitz 连续或在 X 上一致连续. 如果对于 $\boldsymbol{x} \in X$, 存在 $\boldsymbol{x}$ 的一个邻域 $B(\boldsymbol{x}, \varepsilon)$ 使得 f 在 $B(\boldsymbol{x}, \varepsilon)$ 满足 Lipschitz 条件, 则称 f 在 $\boldsymbol{x}$ 附近满足 Lipschitz 条件. 如果 f 在凸集 $C \subset \mathbf{R}^n$ 上是广义凸函数, 且满足 Lipschitz 条件, 则称 f 在 C 上是 Lipschitz 连续凸函数.

下面证明一个正常凸函数在有效定义域的相对内部的任何凸紧子集上是满足 Lipschitz 条件的.

定理 4.3.12 设 f 在 $\mathbf{R}^n$ 上是正常凸函数, $C \subset \mathbf{R}^n$ 是凸紧子集且 $C \subset \operatorname{int}(\operatorname{dom} f)$, 则 f 在 X 上满足 Lipschitz 条件.

证明 设 C 的一个邻域为 $B(C, \varepsilon)$, 其中 B 是凸集 C 的单位球, $\varepsilon > 0$, 由 B 的闭性及 $C \subset \operatorname{int}(\operatorname{dom} f)$, 有

$$\bigcap_{\varepsilon > 0} (B(C, \varepsilon) \cap (\mathbf{R}^n \backslash \operatorname{int}(\operatorname{dom} f))) = \varnothing. \tag{4.3.4-2}$$

上式说明 $\exists \varepsilon > 0$ 使得

$$B(C, \varepsilon) \cap (\mathbf{R}^n \backslash \operatorname{int}(\operatorname{dom} f)) = \varnothing,$$

即 $B(C, \varepsilon) \subset \operatorname{int}(\operatorname{dom} f)$. 则由定理 4.3.11 知 $f(\boldsymbol{x})$ 在 $C + \varepsilon B$ 上连续, 故 $f(\boldsymbol{x})$ 在 $B(C, \varepsilon)$ 上可达到最大值和最小值, 分别记为 $\bar{\alpha}$ 和 $\underline{\alpha}$. 设 $\forall \boldsymbol{x}, \boldsymbol{y} \in X, \boldsymbol{x} \neq \boldsymbol{y}$, 令

$$\boldsymbol{z} = \boldsymbol{y} + \frac{\varepsilon(\boldsymbol{y} - \boldsymbol{x})}{\|\boldsymbol{y} - \boldsymbol{x}\|}.$$

则

$$\|\boldsymbol{z}\| \leqslant \|\boldsymbol{y}\| + \varepsilon,$$

故 $\boldsymbol{z} \in B(C, \varepsilon)$. 设 $\lambda = \dfrac{\|\boldsymbol{y}-\boldsymbol{x}\|}{\|\boldsymbol{y}-\boldsymbol{x}\|+\varepsilon}$, 则 $0 < \lambda < 1$, $\boldsymbol{y} = (1-\lambda)\boldsymbol{x} + \lambda\boldsymbol{z}$. 因为 $f(\boldsymbol{x})$ 是凸函数, 于是

$$f(\boldsymbol{y}) \leqslant (1-\lambda) f(\boldsymbol{x}) + \lambda f(\boldsymbol{z}) = f(\boldsymbol{x}) + \lambda (f(\boldsymbol{z}) - f(\boldsymbol{x})).$$

故

$$f(\boldsymbol{y}) - f(\boldsymbol{x}) \leqslant \lambda(\bar{\alpha} - \underline{\alpha}) = \frac{\|\boldsymbol{y}-\boldsymbol{x}\|}{\|\boldsymbol{y}-\boldsymbol{x}\|+\varepsilon}(\bar{\alpha} - \underline{\alpha}) < \alpha' \|\boldsymbol{y}-\boldsymbol{x}\|,$$

其中 $\alpha' = \dfrac{\bar{\alpha} - \underline{\alpha}}{\varepsilon}$. 由 $\boldsymbol{x}, \boldsymbol{y}$ 在 X 中的任意性, 得到 f 在 X 上满足 Lipschitz 条件. □

根据文献 [12] 知, 如果对于实值凸函数 f 在开集 X 上某点邻域上有界, 那么函数 $f(\boldsymbol{x})$ 在 X 中任何一点的附近满足 Lipschitz 条件. 因此, 我们有下面的推论.

推论 4.3.7 设 f 在 $\mathbf{R}^n$ 上是正常凸函数, f 在开凸集 $C \subset \mathbf{R}^n$ 某点邻域上有界, 则 f 在 C 中任何一点的附近满足 Lipschitz 条件 (称 f 在 C 上是满足局部 Lipschitz 的).

当定理 4.3.12 证明过程中的 C 退化为一点时, 即为推论 4.3.7 的证明. 推论 4.3.7 说明对于任何实值凸函数总存在次梯度 [12].

例 4.3.6 单变量函数 $f(x) = \begin{cases} 0, & |x| \leqslant 1, \\ +\infty, & |x| > 1 \end{cases}$ 是正常凸函数, $f(x)$ 在 $x = 1$ 处不满足 Lipschitz 条件, 但 f 在 $\operatorname{int}(\operatorname{dom} f)$ 上满足 Lipschitz 条件.

本小节得到一个重要结论: 一个正常凸函数在有效定义域的相对内部的任何凸紧子集上是满足 Lipschitz 条件的, 即它是 Lipschitz 连续函数. 进一步, 我们有更弱的结论: 如果一个正常凸函数在某点邻域上有界, 则它在该点附近都满足 Lipschitz 条件. 也即说明了有界正常凸函数是 Lipschitz 连续的.

4.4 共轭凸函数

凸函数存在一种对偶函数, 称之为共轭函数. 下面证明实值凸函数的共轭函数总是存在的.

定义 4.4.1 设 f: $\mathbf{R}^n \to [-\infty, +\infty]$ 是凸函数, 则称

$$f^*(\boldsymbol{x}^*) = \sup\{\langle \boldsymbol{x}, \boldsymbol{x}^* \rangle - f(\boldsymbol{x}) \mid \boldsymbol{x} \in \mathbf{R}^n\} \tag{4.4.1-1}$$

为 f 的共轭凸函数 (简称共轭函数).

注 $f^{**}(\boldsymbol{x})$ 是函数 $f^*(\boldsymbol{x}^*)$ 的共轭函数.

例 4.4.1 示性函数 $\theta_C(\boldsymbol{x})=\begin{cases}0, & \boldsymbol{x}\in C,\\ +\infty, & \boldsymbol{x}\notin C,\end{cases}$ 其中 C 是凸集. 显然它的共轭函数:

$$f^*(\boldsymbol{x}^*)=\sup\{\langle\boldsymbol{x},\boldsymbol{x}^*\rangle-\theta_C(\boldsymbol{x})\,|\boldsymbol{x}\in\mathbf{R}^n\}=\sup\{\langle\boldsymbol{x},\boldsymbol{x}^*\rangle\,|\boldsymbol{x}\in C\}=\delta_C(\boldsymbol{x}^*),$$

即示性函数的共轭函数是 C 的支撑函数. 我们容易求得

$$f^{**}(\boldsymbol{x})=\sup\{\langle\boldsymbol{x},\boldsymbol{x}^*\rangle-\delta_C(\boldsymbol{x}^*)\,|\boldsymbol{x}^*\in\mathbf{R}^n\}=\theta_C(\boldsymbol{x}).$$

下面是共轭函数的一个重要结果.

性质 4.4.1 (Fenchel 不等式) 设 $f:\mathbf{R}^n\to[-\infty,+\infty]$ 是凸函数, 则 f^* 是闭凸函数, 并且

$$f(\boldsymbol{x})+f^*(\boldsymbol{x}^*)\geqslant\langle\boldsymbol{x},\boldsymbol{x}^*\rangle,\quad\forall\boldsymbol{x},\boldsymbol{x}^*\in\mathbf{R}^n \tag{4.4.1-2}$$

成立.

证明 设固定变量 $\boldsymbol{x}$, 那么 $\phi_{\boldsymbol{x}}(\boldsymbol{x}^*)=\langle\boldsymbol{x},\boldsymbol{x}^*\rangle-f(\boldsymbol{x})$ 是关于 $\boldsymbol{x}^*$ 的仿射函数, 所以 $\phi_{\boldsymbol{x}}$ 是闭凸函数. 由 (4.4.1-1) 式和性质 4.2.12 知, $f^*(\boldsymbol{x}^*)$ 是函数族 $\phi_{\boldsymbol{x}}$ 关于 $\boldsymbol{x}$ 的逐点上确界函数, 有

$$\mathrm{epi}f^*=\bigcap_{\boldsymbol{x}\in\mathbf{R}^n}\mathrm{epi}\phi_{\boldsymbol{x}}.$$

因为 $\mathrm{epi}\phi_{\boldsymbol{x}}$ 是闭凸集, 故 $\mathrm{epi}f^*$ 是闭凸集. 所以 $f^*(\boldsymbol{x}^*)$ 是闭凸函数.

由 (4.4.1-1) 式可得

$$f(\boldsymbol{x})+f^*(\boldsymbol{x}^*)\geqslant\langle\boldsymbol{x},\boldsymbol{x}^*\rangle,\quad\forall\boldsymbol{x},\boldsymbol{x}^*\in\mathbf{R}^n.$$

设集合 $C=\mathrm{epi}f\subset\mathbf{R}^n\times\mathbf{R}$ 的支撑函数是 $\delta_C(\boldsymbol{x}^*)=\sup\{\langle\boldsymbol{x}^*,\boldsymbol{y}\rangle\,|\boldsymbol{y}\in C\}$, 那么有

$$\begin{aligned}f^*(\boldsymbol{x}^*)&=\delta_{\mathrm{epi}f}(\boldsymbol{x}^*,-1)=\sup\{\langle(\boldsymbol{x}^*,-1),(\boldsymbol{x},t)\rangle\,|\boldsymbol{x}\in\mathbf{R}^n,t\geqslant f(\boldsymbol{x})\}\\&=\sup\{\langle\boldsymbol{x}^*,\boldsymbol{x}\rangle-t\,|\boldsymbol{x}\in\mathbf{R}^n,t\geqslant f(\boldsymbol{x})\}\\&=\sup\{\langle\boldsymbol{x}^*,\boldsymbol{x}\rangle-f(\boldsymbol{x})\,|\boldsymbol{x}\in\mathbf{R}^n\}.\end{aligned}$$

□

进一步, 我们知道闭正常凸函数的共轭函数还是闭正常凸函数, 共轭函数的共轭函数则是原函数.

性质 4.4.2 设 f 是 $\mathbf{R}^n$ 上的闭正常凸函数, 则 $f^*(\boldsymbol{x}^*)=\delta_{\mathrm{epi}f}(\boldsymbol{x}^*,-1)$ 是闭正常凸函数, 且

$$f(\boldsymbol{x})=f^{**}(\boldsymbol{x}). \tag{4.4.1-3}$$

证明 根据性质 4.4.1, 知 $f^*(\boldsymbol{x}^*)$ 是闭凸函数, 因为 $f(\boldsymbol{x})$ 是正常的, 故 $\exists \boldsymbol{x}_1\in \mathrm{dom}f$, $f(\boldsymbol{x}_1)$ 有限. 由 (4.4.1-2) 式有

$$f^*(\boldsymbol{x}^*)\geqslant \langle \boldsymbol{x}^*,\boldsymbol{x}_1\rangle - f(\boldsymbol{x}_1),$$

所以 $f^*(\boldsymbol{x}^*)>-\infty$. 假设 $f^*(\boldsymbol{x}^*)\equiv+\infty$, 因为 $f(\boldsymbol{x})$ 是正常的, 对任意的 $\boldsymbol{x}\in\mathbf{R}^n$, $\exists a\in\mathbf{R}$ 使得 $f(\boldsymbol{x})>a$, 有

$$\langle \boldsymbol{x}^*,\boldsymbol{x}\rangle - f(\boldsymbol{x}) < \langle \boldsymbol{x}^*,\boldsymbol{x}\rangle - a.$$

则有 $+\infty<\sup\{\langle \boldsymbol{x},\boldsymbol{x}^*\rangle - a\,|\boldsymbol{x}\in\mathbf{R}^n\}$, 得出矛盾. 因此 $f^*(\boldsymbol{x}^*)\not\equiv+\infty$, 故 $f^*(\boldsymbol{x}^*)$ 是正常的.

由共轭函数定义有

$$f(\boldsymbol{x})+f^*(\boldsymbol{x}^*)\geqslant\langle \boldsymbol{x},\boldsymbol{x}^*\rangle,\quad \forall \boldsymbol{x},\boldsymbol{x}^*\in\mathbf{R}^n.$$

则

$$f(\boldsymbol{x})\geqslant\sup\{\langle \boldsymbol{x},\boldsymbol{x}^*\rangle - f^*(\boldsymbol{x}^*)\,|\boldsymbol{x}^*\in\mathbf{R}^n\}=f^{**}(\boldsymbol{x}),$$

且 $\mathrm{dom}f\subset\mathrm{dom}f^{**}$. 下面只需要证明反向不等式, 分两种情况进行讨论.

(1) 设 $\boldsymbol{x}_0\in\mathrm{dom}f$, 集合

$$B_\varepsilon(\boldsymbol{x}_0,\alpha)=\{(\boldsymbol{x},t)\,|\boldsymbol{x}\in\boldsymbol{x}_0+\varepsilon B,\alpha\geqslant t\}, \tag{4.4.1-4}$$

其中 $f(\boldsymbol{x}_0)>\alpha$, $\varepsilon>0$, B 是单位球, 则 $\exists\varepsilon>0$ 使得 $B_\varepsilon(\boldsymbol{x}_0,\alpha)\cap\mathrm{epi}f=\varnothing$. 否则存在序列 $\varepsilon_i\to0$, 总是有 $(\boldsymbol{x}_i,t)\in B_{\varepsilon_i}(\boldsymbol{x}_0,\alpha)\cap\mathrm{epi}f$, 即有

$$f(\boldsymbol{x}_0)>\alpha\geqslant t\geqslant f(\boldsymbol{x}_i)\to f(\boldsymbol{x}_0),$$

得到矛盾. 根据凸集分离定理, 在 $\mathbf{R}^{n+1}$ 中存在非零向量 $(\boldsymbol{x}^*,t^*)$ 使得

$$\langle \boldsymbol{x},\boldsymbol{x}^*\rangle+\langle t,t^*\rangle\geqslant\langle \boldsymbol{x}',\boldsymbol{x}^*\rangle+\langle t',t^*\rangle,\quad\forall(\boldsymbol{x},t)\in B_\varepsilon(\boldsymbol{x}_0,\alpha),\quad\forall(\boldsymbol{x}',t')\in\mathrm{epi}f. \tag{4.4.1-5}$$

上式可取 $\boldsymbol{x}=\boldsymbol{x}'=\boldsymbol{x}_0$ 和 $t'=f(\boldsymbol{x}_0)>t$, 得 $t^*\leqslant0$. 设 $t^*=0$, 则由 (4.4.1-5) 式得

$$\langle \boldsymbol{x}_0+\varepsilon\boldsymbol{b},\boldsymbol{x}^*\rangle\geqslant\langle \boldsymbol{x}_0,\boldsymbol{x}^*\rangle,\quad\forall\boldsymbol{b}\in B. \tag{4.4.1-6}$$

在上式中取非零 $\boldsymbol{b}\in B$, $-\boldsymbol{b}\in B$ 有 $\langle \boldsymbol{b},\boldsymbol{x}^*\rangle=0$, 因为 $\boldsymbol{x}^*$, $\boldsymbol{b}\in B$ 均非零, 显然得到矛盾. 所以 $t^*<0$. 不妨设 $t^*=-1$, 有

$$\langle \boldsymbol{x}_0+\varepsilon\boldsymbol{b},\boldsymbol{x}^*\rangle-t\geqslant\langle \boldsymbol{x}',\boldsymbol{x}^*\rangle-t',\quad \forall\boldsymbol{b}\in B,\quad t'\geqslant f(\boldsymbol{x}'). \tag{4.4.1-7}$$

在上式中取 $\boldsymbol{b}=\boldsymbol{0},t'=f(\boldsymbol{x}'),t=f(\boldsymbol{x}_0)-\epsilon(f(\boldsymbol{x}_0)-\alpha)$ 得

$$\langle \boldsymbol{x}_0,\boldsymbol{x}^*\rangle-f(\boldsymbol{x}_0)+\epsilon(f(\boldsymbol{x}_0)-\alpha)\geqslant\sup_{\boldsymbol{x}'\in\mathbf{R}^n}\{\langle \boldsymbol{x}',\boldsymbol{x}^*\rangle-f(\boldsymbol{x}')\}=f^*(\boldsymbol{x}^*),$$

其中 $\epsilon>0$ 是任意给定的. 上式即为

$$(1-\epsilon)f(\boldsymbol{x}_0)+\epsilon\alpha\leqslant\langle \boldsymbol{x}_0,\boldsymbol{x}^*\rangle-f^*(\boldsymbol{x}^*),$$

有

$$(1-\epsilon)f(\boldsymbol{x}_0)+\epsilon\alpha\leqslant\langle \boldsymbol{x}_0,\boldsymbol{x}^*\rangle-f^*(\boldsymbol{x}^*)\leqslant\sup_{\boldsymbol{x}^*\in\mathbf{R}^n}\{\langle \boldsymbol{x}_0,\boldsymbol{x}^*\rangle-f^*(\boldsymbol{x}^*)\}=f^{**}(\boldsymbol{x}_0).$$

令 $\epsilon\to0$, 得到

$$f(\boldsymbol{x}_0)\leqslant f^{**}(\boldsymbol{x}_0),$$

表明当 $\boldsymbol{x}\in\mathrm{dom}f$ 时, $f(\boldsymbol{x})=f^{**}(\boldsymbol{x})$.

(2) 设 $\boldsymbol{x}_0\notin\mathrm{dom}f$, 则 $f(\boldsymbol{x}_0)=+\infty$, 类似 (4.4.1-4) 式定义 $B_\varepsilon(\boldsymbol{x}_0,\alpha)$, 其中 $f(\boldsymbol{x}_0)>\alpha$(有限值) , $\varepsilon>0$ 和 B 是单位球, 那么 $\exists\varepsilon>0$ 使得

$$B_\varepsilon(\boldsymbol{x}_0,\alpha)\cap\mathrm{epi}f=\varnothing.$$

同样由凸集分离定理得到 (4.4.1-5) 式成立, 由 $f(\boldsymbol{x}_0)=+\infty$ 得 $t^*<0$. 事实上, 在 (1) 中已经证明了若 $t^*=0$ 有 (4.4.1-6) 式成立, 导致矛盾. 不妨设 $t^*=-1$, 有 (4.4.1-7) 式成立. 因为 $f(\boldsymbol{x})$ 是正常凸函数, 故 $\exists\boldsymbol{x}_1$, $f(\boldsymbol{x}_1)$ 是有限值, 取 $\alpha<f(\boldsymbol{x}_1)$. 在 (4.4.1-7) 式取

$$\boldsymbol{b}=\boldsymbol{0},\quad t'=f(\boldsymbol{x}'),\ t=f(\boldsymbol{x}_1)-\epsilon(f(\boldsymbol{x}_1)-\alpha),$$

得

$$\langle \boldsymbol{x}_0,\boldsymbol{x}^*\rangle-f(\boldsymbol{x}_1)+\epsilon(f(\boldsymbol{x}_1)-\alpha)\geqslant\sup_{\boldsymbol{x}'\in\mathbf{R}^n}\{\langle \boldsymbol{x}',\boldsymbol{x}^*\rangle-f(\boldsymbol{x}')\}=f^*(\boldsymbol{x}^*),$$

其中 $\epsilon>0$ 是任意给定的. 上式即为

$$(1-\epsilon)f(\boldsymbol{x}_1)+\epsilon\alpha\leqslant\langle \boldsymbol{x}_0,\boldsymbol{x}^*\rangle-f^*(\boldsymbol{x}^*),$$

即

$$(1-\epsilon) f(\boldsymbol{x}_1)+\epsilon\alpha \leqslant \langle \boldsymbol{x}_0, \boldsymbol{x}^*\rangle - f^*(\boldsymbol{x}^*) \leqslant f^{**}(\boldsymbol{x}_0).$$

当 $\alpha \to +\infty$ 时, 上式左边趋于 $+\infty$, 故 $f^{**}(\boldsymbol{x}_0)=+\infty$.

所以当 $\boldsymbol{x} \neq \mathrm{dom} f$ 时, $f(\boldsymbol{x})=f^{**}(\boldsymbol{x})$. □

注 若 f 是 $\mathbf{R}^n$ 上的非正常凸函数, 由于 $(\mathrm{cl} f)^*(\boldsymbol{x}^*)=f^*(\boldsymbol{x}^*)$, 所以当 $f(\boldsymbol{x})$ 是非闭凸函数时, 则有 $f^{**}=(\mathrm{cl} f)^{**}=\mathrm{cl} f$.

所以对于一般凸函数, 下面的公式也成立:

$$f^*(\boldsymbol{x}^*)=\sup\{\langle \boldsymbol{x}, \boldsymbol{x}^*\rangle - f(\boldsymbol{x}) \,|\, \boldsymbol{x} \in \mathrm{ri}(\mathrm{dom} f)\}.$$

由定理 4.3.5 和性质 4.4.2 知下面推论成立.

推论 4.4.1 设 f 是 $\mathbf{R}^n$ 中的正常凸函数. 如果对于 $\boldsymbol{x}_0 \in \mathrm{dom} f$ 有 f 在 $\boldsymbol{x}_0$ 处是下半连续的, 则

$$f(\boldsymbol{x}_0)=f^{**}(\boldsymbol{x}_0).$$

例 4.4.2 已知函数 $f(x)=\begin{cases}-\dfrac{1}{2}-\ln x, & x>0,\\ +\infty, & x \leqslant 0,\end{cases}$ 由定义 4.4.1 可得它的共轭函数为

$$f^*(x^*)=\begin{cases}-\dfrac{1}{2}-\ln(-x^*), & x^*<0,\\ +\infty, & x^* \geqslant 0.\end{cases}$$

例 4.4.3 设 L 是 $\mathbf{R}^n$ 的子空间, 求子空间 L 的示性函数 $\theta_L(\boldsymbol{x})$ 的共轭函数 ${\theta_L}^*(\boldsymbol{x}^*)$. 由定义 4.4.1 可得

$$\begin{aligned}{\theta_L}^*(\boldsymbol{x}^*) &= \sup_{\boldsymbol{x}\in\mathbf{R}^n}\{\langle \boldsymbol{x}, \boldsymbol{x}^*\rangle - \theta_L(\boldsymbol{x})\} = \sup_{\boldsymbol{x}\in L}\{\langle \boldsymbol{x}, \boldsymbol{x}^*\rangle\}\\ &= \begin{cases}0, & \boldsymbol{x}^* \in L^{\perp},\\ +\infty, & \boldsymbol{x}^* \notin L^{\perp}.\end{cases}\end{aligned}$$

显然, 上面的结果与凸集示性函数的共轭函数一致.

设函数 g 是 $\mathbf{R}^n$ 上的实值凹函数, $f=-g$ 是 $\mathbf{R}^n$ 上的实值凸函数, 它的共轭函数为

$$f^*(\boldsymbol{x}^*)=\sup\{\langle \boldsymbol{x}, \boldsymbol{x}^*\rangle + g(\boldsymbol{x}) \,|\, \boldsymbol{x} \in \mathbf{R}^n\} = -\inf\{\langle \boldsymbol{x}, -\boldsymbol{x}^*\rangle - g(\boldsymbol{x}) \,|\, \boldsymbol{x} \in \mathbf{R}^n\}.$$

如果设 $g^*(\boldsymbol{x}^*)=\inf\{\langle \boldsymbol{x}, \boldsymbol{x}^*\rangle - g(\boldsymbol{x}) \,|\, \boldsymbol{x} \in \mathbf{R}^n\}$, 则 $f^*(\boldsymbol{x}^*)=-g^*(-\boldsymbol{x}^*)$.

下面是凸函数简单运算后的共轭函数关系.

性质 4.4.3 设 f 是 $\mathbf{R}^n$ 上的正常凸函数, 有下面的结论成立:

(1) 如果 $g(\boldsymbol{x}) = f(\boldsymbol{x}) + \alpha$, 其中 $\alpha \in \mathbf{R}$, 则 $g^*(\boldsymbol{x}^*) = f^*(\boldsymbol{x}^*) - \alpha$;

(2) 如果 $g(\boldsymbol{x}) = f(\boldsymbol{x}+\boldsymbol{y})$, 其中 $\boldsymbol{y} \in \mathbf{R}^n$, 则 $g^*(\boldsymbol{x}^*) = f^*(\boldsymbol{x}^*) - \langle \boldsymbol{x}^*, \boldsymbol{y}\rangle$;

(3) 如果 $g(\boldsymbol{x}) = f(\alpha\boldsymbol{x})$, 其中 $\alpha \neq 0$, 则 $g^*(\boldsymbol{x}^*) = f^*(\boldsymbol{x}^*/\alpha)$;

(4) 如果 $g(\boldsymbol{x}) = \alpha f(\boldsymbol{x})$, 其中 $\alpha > 0$, 则 $g^*(\boldsymbol{x}^*) = \alpha f^*(\boldsymbol{x}^*/\alpha)$.

证明 (1) 根据共轭函数定义得

$$g^*(\boldsymbol{x}^*) = \sup\{\langle \boldsymbol{x}, \boldsymbol{x}^*\rangle - f(\boldsymbol{x}) - \alpha \,|\boldsymbol{x} \in \mathbf{R}^n\} = f^*(\boldsymbol{x}^*) - \alpha.$$

(2) 固定 $\boldsymbol{y} \in \mathbf{R}^n$, 根据共轭函数定义得

$$\begin{aligned} g^*(\boldsymbol{x}^*) &= \sup\{\langle \boldsymbol{x}, \boldsymbol{x}^*\rangle - f(\boldsymbol{x}+\boldsymbol{y}) \,|\boldsymbol{x} \in \mathbf{R}^n\} \\ &= \sup\{\langle \boldsymbol{x}+\boldsymbol{y}, \boldsymbol{x}^*\rangle - f(\boldsymbol{x}+\boldsymbol{y}) - \langle \boldsymbol{y}, \boldsymbol{x}^*\rangle \,|\boldsymbol{x}+\boldsymbol{y} \in \mathbf{R}^n\} \\ &= f^*(\boldsymbol{x}^*) - \langle \boldsymbol{y}, \boldsymbol{x}^*\rangle. \end{aligned}$$

(3) 根据共轭函数定义得

$$g^*(\boldsymbol{x}^*) = \sup\{\langle \alpha\boldsymbol{x}, \boldsymbol{x}^*/\alpha\rangle - f(\alpha\boldsymbol{x}) \,|\boldsymbol{x} \in \mathbf{R}^n\} = f^*(\boldsymbol{x}^*/\alpha).$$

(4) 根据共轭函数定义得

$$g^*(\boldsymbol{x}^*) = \sup\{\langle \boldsymbol{x}, \boldsymbol{x}^*\rangle - \alpha f(\boldsymbol{x}) \,|\boldsymbol{x} \in \mathbf{R}^n\} = \alpha f^*(\boldsymbol{x}^*/\alpha). \qquad \square$$

下面的定理揭示了闭正齐次凸函数的共轭函数是共轭函数有效定义域的示性函数.

定理 4.4.1 设 f 是 $\mathbf{R}^n$ 中的闭正齐次凸函数, 则 f 的共轭函数是示性函数:

$$f^*(\boldsymbol{x}^*) = \theta_{\mathrm{dom} f^*}(\boldsymbol{x}^*).$$

证明 根据定理假设, 对于任意的 $\boldsymbol{x}, \boldsymbol{y} \in \mathbf{R}^n$, 有

$$f(\boldsymbol{x}+\boldsymbol{y}) \leqslant f(\boldsymbol{x}) + f(\boldsymbol{y}).$$

令 $\boldsymbol{x} \in \mathrm{ri}(\mathrm{dom} f), \boldsymbol{y} = \boldsymbol{0}$, 由上式显然有 $f(\boldsymbol{0}) \geqslant 0$. 再因为 $f(\boldsymbol{x})$ 是闭凸函数, 且由推论 4.3.3 得

$$f(\boldsymbol{0}) = \mathrm{cl} f(\boldsymbol{0}) = \lim_{\lambda \uparrow 1} f[(1-\lambda)\boldsymbol{x}] = \lim_{\lambda \uparrow 1} (1-\lambda) f(\boldsymbol{x}) = 0.$$

故 $f(\mathbf{0})=0$. 根据共轭函数定义得

$$f^*(\boldsymbol{x}^*)=\sup\{\langle \boldsymbol{x},\boldsymbol{x}^*\rangle-f(\boldsymbol{x})\,|\boldsymbol{x}\in\mathbf{R}^n\}\geqslant -f(\mathbf{0})=0.$$

假设 $\boldsymbol{x}_1^*\in\mathbf{R}^n$ 使得 $f^*(\boldsymbol{x}_1^*)>0$, 则存在 $\boldsymbol{x}_1\in\mathbf{R}^n$, 使

$$\langle \boldsymbol{x}_1,\boldsymbol{x}_1^*\rangle-f(\boldsymbol{x}_1)>0.$$

从而当 $t>0$ 时, 有

$$\begin{aligned}f^*(\boldsymbol{x}_1^*)&\geqslant \sup_{t>0}\ \{\langle t\boldsymbol{x}_1,\boldsymbol{x}_1^*\rangle-f(t\boldsymbol{x}_1)\}\\&=\sup_{t>0}\ t\{\langle \boldsymbol{x}_1,\boldsymbol{x}_1^*\rangle-f(\boldsymbol{x}_1)\}=+\infty.\end{aligned}$$

所以 $f^*(\boldsymbol{x}^*)$ 只取两个值: 0 和 $+\infty$, 故

$$f^*(\boldsymbol{x}^*)=\theta_{\mathrm{dom}f^*}(\boldsymbol{x}^*)=\begin{cases}0, & x^*\in\mathrm{dom}f^*,\\+\infty, & x^*\notin\mathrm{dom}f^*.\end{cases}$$ □

例 4.4.4 支撑函数的 $\delta_C(\boldsymbol{x}^*)$ 是正齐次闭凸函数, 容易求得它的共轭函数是示性函数 $\theta_C(\boldsymbol{x})$.

定义 4.4.2 如果映射 $A:\mathbf{R}^n\to\mathbf{R}^m$, 对于任意 $\boldsymbol{x}\in\mathbf{R}^n$, 任意 $\boldsymbol{y}\in\mathbf{R}^m$, 存在一个映射 $A^*:\mathbf{R}^m\to\mathbf{R}^n$, 且满足 $\langle A\boldsymbol{x},\boldsymbol{y}\rangle=\langle \boldsymbol{x},A^*\boldsymbol{y}\rangle$, 则称 A^* 是 A 的伴随映射.

下面得到凸函数在线性变换后的复合函数的共轭函数关系.

定理 4.4.2 设线性变换 $A:\mathbf{R}^n\to\mathbf{R}^m$, f 和 g 分别是 $\mathbf{R}^n$ 和 $\mathbf{R}^m$ 上的凸函数, 则有

$$(Af)^*=f^*A^*,\quad ((\mathrm{cl}g)A^*)=\mathrm{cl}(A^*g^*),$$

其中 A^* 是 A 的伴随映射, $Af(\boldsymbol{y})=\inf\limits_{A\boldsymbol{x}=\boldsymbol{y}}f(\boldsymbol{x})$, $A^*g^*(\boldsymbol{x}^*)=\inf\limits_{\boldsymbol{x}^*=A^*\boldsymbol{y}^*}g^*(\boldsymbol{y}^*)$.

如果 $\exists\boldsymbol{x}$ 满足 $A\boldsymbol{x}\in\mathrm{ri}(\mathrm{dom}g)$, $(gA)(\boldsymbol{x})=g(\boldsymbol{y})$, 则有

$$(gA)^*=\mathrm{cl}(A^*g^*).$$

证明 利用共轭函数的定义, 有

$$\begin{aligned}(Af)^*(\boldsymbol{y}^*)&=\sup_{\boldsymbol{y}\in\mathbf{R}^m}\left\{\langle\boldsymbol{y},\boldsymbol{y}^*\rangle-\inf_{A\boldsymbol{x}=\boldsymbol{y}}f(\boldsymbol{x})\right\}\\&=\sup_{\boldsymbol{y}}\sup_{A\boldsymbol{x}=\boldsymbol{y}}\ \{\langle\boldsymbol{y},\boldsymbol{y}^*\rangle-f(\boldsymbol{x})\}\end{aligned}$$

$$
\begin{aligned}
&= \sup_{\boldsymbol{x}} \{\langle A\boldsymbol{x}, \boldsymbol{y}^* \rangle - f(\boldsymbol{x})\} \\
&= \sup_{\boldsymbol{x}} \{\langle \boldsymbol{x}, A^*\boldsymbol{y}^* \rangle - f(\boldsymbol{x})\} \\
&= (f^* A^*)(\boldsymbol{y}^*).
\end{aligned}
$$

然后将上式应用于 A^*, g^* 有

$$(A^* g^*)^* = g^{**} A^* = (\mathrm{cl} g) A.$$

两边再取一次共轭, 有

$$((\mathrm{cl} g) A)^* = (A^* g^*)^{**} = \mathrm{cl}(A^* g^*).$$

设 $\boldsymbol{y}_0 \in \mathrm{ri}(\mathrm{dom} g)$, 如果 $g(\boldsymbol{y}_0) = -\infty$ 成立, 则由性质 4.3.2(1) 可知, 对任意的 $\boldsymbol{y} \in \mathrm{ri}(\mathrm{dom} g)$ 都有 $g(\boldsymbol{y}_0) = -\infty$, 得 $g^* = (gA)^{**} = +\infty$, 定理结论显然成立. 因此, 不妨设 $g(\boldsymbol{y}) > -\infty\,(\forall \boldsymbol{y} \in \mathbf{R}^m)$, 且由假设 $A\boldsymbol{x} \in \mathrm{ri}(\mathrm{dom} g)$ 及定理 4.3.8(2) 知 $(\mathrm{cl} g) A = \mathrm{cl}(gA)$, 所以

$$((\mathrm{cl} g) A)^* = (gA)^*.$$

□

进一步, 我们得到下卷积函数的共轭函数性质 (定义见性质 4.2.7) , m 个正常凸函数下卷积函数的共轭函数等于它们的共轭函数之和, m 个闭正常凸函数之和的共轭函数等于各个正常凸函数的共轭函数的下卷积函数.

定理 4.4.3 设 $f_1, \cdots, f_m$ 是 $\mathbf{R}^n$ 上的正常凸函数, 下卷积函数及其共轭函数分别为

$$
\begin{aligned}
\phi(\boldsymbol{x}) &= \inf\{f_1(\boldsymbol{x}_1) + \cdots + f_m(\boldsymbol{x}_m) \,|\, \boldsymbol{x}_i \in \mathbf{R}^n, \boldsymbol{x}_1 + \cdots + \boldsymbol{x}_m = \boldsymbol{x}\}, \\
\varphi^*(\boldsymbol{x}) &= \inf\{f_1^*(\boldsymbol{x}_1) + \cdots + f_m^*(\boldsymbol{x}_m) \,|\, \boldsymbol{x}_i \in \mathbf{R}^n, \boldsymbol{x}_1 + \cdots + \boldsymbol{x}_m = \boldsymbol{x}\},
\end{aligned}
$$

则

$$
\begin{aligned}
\phi^* &= f_1^* + \cdots + f_m^*, \\
(\mathrm{cl} f_1 + \cdots + \mathrm{cl} f_m)^* &= \mathrm{cl}\varphi^*,
\end{aligned}
$$

进一步, 如果 $\bigcap_{i=1}^m \mathrm{ri}(\mathrm{dom} f_i) \neq \varnothing$, 则 $(f_1 + \cdots + f_m)^* = \mathrm{cl}\varphi^*$.

证明 利用共轭函数的定义, 有

$$\phi(\boldsymbol{x}^*) = \sup_{\boldsymbol{x}} \left\{ \langle \boldsymbol{x}, \boldsymbol{x}^* \rangle - \inf_{\boldsymbol{x} = \boldsymbol{x}_1 + \cdots + \boldsymbol{x}_m} [f_1(\boldsymbol{x}_1) + \cdots + f_m(\boldsymbol{x}_m)] \right\}$$

$$= \sup_{\boldsymbol{x}} \sup_{\boldsymbol{x}=\boldsymbol{x}_1+\cdots+\boldsymbol{x}_m} \{\langle \boldsymbol{x}, \boldsymbol{x}^* \rangle - f_1(\boldsymbol{x}_1) - \cdots - f_m(\boldsymbol{x}_m)\}$$

$$= \sup_{\boldsymbol{x}_1+\cdots+\boldsymbol{x}_m} \{\langle \boldsymbol{x}_1, \boldsymbol{x}^* \rangle + \cdots + \langle \boldsymbol{x}_m, \boldsymbol{x}^* \rangle - f_1(\boldsymbol{x}_1) - \cdots - f_m(\boldsymbol{x}_m)\}$$

$$= f_1^*(\boldsymbol{x}^*) + \cdots + f_m^*(\boldsymbol{x}^*).$$

对 $f_1^*, \cdots, f_m^*$ 应用上式和性质 4.4.2 可得

$$\varphi^* = f_1^{**} + \cdots + f_m^{**} = \mathrm{cl} f_1 + \cdots + \mathrm{cl} f_m,$$

即为

$$(\mathrm{cl} f_1 + \cdots + \mathrm{cl} f_m)^* = (\varphi^*)^{**} = \mathrm{cl} \varphi^*.$$

如果 $\bigcap_{i=1}^m \mathrm{ri}(\mathrm{dom} f_i) \neq \varnothing$, 则由定理 4.3.6(2) 得

$$\mathrm{cl} f_1 + \cdots + \mathrm{cl} f_m = \mathrm{cl}(f_1 + \cdots + f_m),$$

上式两边取共轭有

$$(\mathrm{cl} f_1 + \cdots + \mathrm{cl} f_m)^* = [\mathrm{cl}(f_1 + \cdots + f_m)]^* = (f_1 + \cdots + f_m)^*.$$ □

本小节得到了共轭凸函数的几个重要性质:

1) 凸函数在 $\boldsymbol{x}$ 处的值加上它的共轭函数在 $\boldsymbol{x}^*$ 处的值不小于它们自变量之间的内积 $\langle \boldsymbol{x}, \boldsymbol{x}^* \rangle$;

2) 闭正常凸函数的共轭函数等于它的上方图支撑函数 $\delta_{\mathrm{epi}f}(\boldsymbol{x}^*, -1)$, 且闭正常凸函数的共轭函数的共轭函数是原函数;

3) 闭正齐次凸函数的共轭函数等于共轭函数有效定义域的示性函数;

4) 凸函数线性变换的复合函数的共轭函数等于伴随变换与凸函数共轭函数的复合函数;

5) m 个正常凸函数的下卷积函数的共轭函数等于它们的共轭函数之和, m 个闭正常凸函数之和的共轭函数等于各个正常凸函数的共轭函数的下卷积函数.

4.5 支撑凸函数

凸集的支撑凸函数在凸分析中是一类非常重要的特殊函数. 本质上, 支撑凸函数等价于一个支撑超平面, 具有一些重要性质.

定义 4.5.1 设非空凸集 $C \subset \mathbf{R}^n$, 称

$$\delta_C(\boldsymbol{x}^*) = \sup\{\langle \boldsymbol{x}, \boldsymbol{x}^* \rangle \mid \boldsymbol{x} \in C\} \tag{4.5.1-1}$$

是集合 C 的支撑凸函数, 或简称支撑函数.

根据 (4.5.1-1) 式有下面关系:

$$\inf\{\langle \boldsymbol{x}, \boldsymbol{x}^*\rangle \,|\boldsymbol{x}\in C\} = -\delta_C(-\boldsymbol{x}^*).$$

所以, 支撑函数是在凸集 C 上达到的上确界值或下确界值. 由上节知道 $\delta_C(\boldsymbol{x}^*)$ 的共轭函数是凸集 C 上的示性函数.

下面是凸集支撑函数的一些基本性质, 揭示了凸集与支撑函数之间的关系.

性质 4.5.1 设非空凸集 $C\subset \mathbf{R}^n$, $\forall \boldsymbol{x}^*\in\mathbf{R}^n$, $t\in\mathbf{R}$, 则下面结论成立:

(1) $\boldsymbol{x}\in \mathrm{cl}C$ 等价于 $\langle \boldsymbol{x}, \boldsymbol{x}^*\rangle \leqslant \delta_C(\boldsymbol{x}^*)$;

(2) $C\subset\{\boldsymbol{x}\,|\langle \boldsymbol{x}, \boldsymbol{x}^*\rangle\leqslant t\}$ 等价于 $\delta_C(\boldsymbol{x}^*)\leqslant t$.

证明 (1) 设 $\boldsymbol{x}\in \mathrm{cl}C$, $\exists \boldsymbol{x}_i\to\boldsymbol{x}, \boldsymbol{x}_i\in C$, 根据 (4.5.1-1) 式得

$$\langle \boldsymbol{x}_i, \boldsymbol{x}^*\rangle \leqslant \delta_C(\boldsymbol{x}^*),\quad i=1,2,\cdots.$$

令 $i\to+\infty$, 有 $\langle \boldsymbol{x}, \boldsymbol{x}^*\rangle \leqslant \delta_C(\boldsymbol{x}^*)$.

反之, 设 $\boldsymbol{x}_0$ 满足 $\langle \boldsymbol{x}_0, \boldsymbol{x}^*\rangle \leqslant \delta_C(\boldsymbol{x}^*)\,(\forall \boldsymbol{x}^*\in\mathbf{R}^n)$. 但 $\boldsymbol{x}_0\notin \mathrm{cl}C$, 则 $\{\boldsymbol{x}_0\}$ 和 C 可以严格分离, 由凸集分离定理知, 存在 $\boldsymbol{x}_0^*$ 和 $\varepsilon>0$, $\forall \boldsymbol{x}\in C$, 满足

$$\langle \boldsymbol{x}, \boldsymbol{x}_0^*\rangle \leqslant \langle \boldsymbol{x}_0, \boldsymbol{x}_0^*\rangle - \varepsilon.$$

在上式左边对 $\boldsymbol{x}\in C$ 取上确界得

$$\delta_C(\boldsymbol{x}^*) \leqslant \langle \boldsymbol{x}_0, \boldsymbol{x}_0^*\rangle - \varepsilon < \langle \boldsymbol{x}_0, \boldsymbol{x}_0^*\rangle,$$

即得矛盾.

(2) 设 $C\subset\{\boldsymbol{x}\,|\langle \boldsymbol{x}, \boldsymbol{x}^*\rangle\leqslant t\}$, 则当 $\boldsymbol{x}\in C$ 时, 由 $\langle \boldsymbol{x}, \boldsymbol{x}^*\rangle\leqslant t$ 得

$$\delta^*(\boldsymbol{x}^*\,|C) = \sup\{\langle \boldsymbol{x}, \boldsymbol{x}^*\rangle\,|\boldsymbol{x}\in C\}\leqslant t.$$

反之, 当 $\boldsymbol{x}\in C$ 时, 有 $\langle \boldsymbol{x}, \boldsymbol{x}^*\rangle\leqslant\delta_C(\boldsymbol{x}^*)\leqslant t$, 得 $C\subset\{\boldsymbol{x}\,|\langle \boldsymbol{x}, \boldsymbol{x}^*\rangle\leqslant t\}$. □

凸集的支撑函数与支撑超平面之间有密切的关系, 如凸紧集的支撑函数是某个线性函数构成的超平面在该凸集上的支撑超平面, 并且凸紧集的支撑函数等于该凸紧集相对内部和闭包分别构成的支撑函数.

性质 4.5.2 设凸紧集 $C\subset\mathbf{R}^n$, $\boldsymbol{x}^*\in\mathbf{R}^n, \boldsymbol{x}^*\neq\mathbf{0}$, 则有下面结论成立:

(1) 在 C 中 $\exists\bar{\boldsymbol{x}}$, 满足 $\delta_C(\boldsymbol{x}^*) = \langle\bar{\boldsymbol{x}}, \boldsymbol{x}^*\rangle$;

(2) 超平面 $H(\boldsymbol{x}^*, \delta_C(\boldsymbol{x}^*)) = \{\boldsymbol{x}\,|\langle \boldsymbol{x}, \boldsymbol{x}^*\rangle = \delta_C(\boldsymbol{x}^*)\}$ 在 $\bar{\boldsymbol{x}}$ 处支撑 C;

(3) 从超平面 $H(\boldsymbol{x}^*, \delta_C(\boldsymbol{x}^*))$ 到原点 $\mathbf{0}$ 的距离等于 $\delta_C\left(\dfrac{\boldsymbol{x}^*}{\|\boldsymbol{x}^*\|}\right)$;

(4) $\delta_C(\boldsymbol{x}^*) = \delta_{\mathrm{ri}C}(\boldsymbol{x}^*) = \delta_{\mathrm{cl}C}(\boldsymbol{x}^*)$.

证明 (1) 因为 $\delta_C(\boldsymbol{x}^*)$ 是线性函数 $\langle \boldsymbol{x}, \boldsymbol{x}^* \rangle$ 在 C 上的上确界, 且 C 是紧集, 所以根据定理 1.2.2 可知, 上确界一定在 C 上的某一点 $\bar{\boldsymbol{x}}$ 达到, 即

$$\delta_C(\boldsymbol{x}^*) = \langle \bar{\boldsymbol{x}}, \boldsymbol{x}^* \rangle.$$

(2) 因为超平面 $H(\boldsymbol{x}^*, \delta_C(\boldsymbol{x}^*))$ 是半空间 $H^{\leqslant}(\boldsymbol{x}^*, \delta_C(\boldsymbol{x}^*)) = \{\boldsymbol{x} | \langle \boldsymbol{x}, \boldsymbol{x}^* \rangle \leqslant \delta_C(\boldsymbol{x}^*)\}$ 的边界, 由性质 4.5.1(2) 知

$$C \subset \{\boldsymbol{x} \,|\langle \boldsymbol{x}, \boldsymbol{x}^* \rangle \leqslant \delta_C(\boldsymbol{x}^*)\}.$$

根据 (1) 有 $\delta_C(\boldsymbol{x}^*) = \langle \bar{\boldsymbol{x}}, \boldsymbol{x}^* \rangle$, 所以 $H(\boldsymbol{x}^*, \delta_C(\boldsymbol{x}^*))$ 是 C 的支撑超平面.

(3) 因为 $\boldsymbol{x}^*$ 是超平面 $H(\boldsymbol{x}^*, \delta_C(\boldsymbol{x}^*)) = \{\boldsymbol{x} \,|\langle \boldsymbol{x}, \boldsymbol{x}^* \rangle = \delta_C(\boldsymbol{x}^*)\}$ 的法向量, 那么 $\exists t > 0$ 使得 $t\boldsymbol{x}^* \in H(\boldsymbol{x}^*, \delta_C(\boldsymbol{x}^*))$. 所以 $\|t\boldsymbol{x}^*\|$ 是超平面 $H(\boldsymbol{x}^*, \delta_C(\boldsymbol{x}^*))$ 到原点 $\mathbf{0}$ 的距离, 且有 $\langle t\boldsymbol{x}^*, \boldsymbol{x}^* \rangle = \delta_C(\boldsymbol{x}^*)$, 得

$$\delta_C\left(\frac{\boldsymbol{x}^*}{\|\boldsymbol{x}^*\|}\right) = \frac{1}{\|\boldsymbol{x}^*\|}\delta_C(\boldsymbol{x}^*) = \frac{1}{\|\boldsymbol{x}^*\|}\langle t\boldsymbol{x}^*, \boldsymbol{x}^* \rangle = \|t\boldsymbol{x}^*\|.$$

(4) 这个结论显然成立. □

性质 4.5.2 说明了 $\delta_C(\boldsymbol{x}^*)$ 称为支撑凸函数的原因.

例 4.5.1 $\delta_B(\boldsymbol{x}^*)$ 是单位球 $B \subset \mathbf{R}^n$ 的支撑函数. 由性质 4.5.2(1) 知, $\exists \bar{\boldsymbol{x}} \in B$ 使得

$$\delta_B(\boldsymbol{x}^*) = \langle \boldsymbol{x}^*, \bar{\boldsymbol{x}} \rangle \geqslant \langle \boldsymbol{x}^*, \boldsymbol{x} \rangle, \quad \forall \boldsymbol{x} \in B.$$

有 $\|\boldsymbol{x}^*\|\|\bar{\boldsymbol{x}}\| \geqslant \langle \boldsymbol{x}^*, \bar{\boldsymbol{x}} \rangle \geqslant \left\langle \boldsymbol{x}^*, \dfrac{\boldsymbol{x}^*}{\|\boldsymbol{x}^*\|} \right\rangle = \|\boldsymbol{x}^*\|$, 得 $\|\bar{\boldsymbol{x}}\| = 1$, 因此有

$$\delta_B(\boldsymbol{x}^*) = \sup\{\langle \boldsymbol{x}, \boldsymbol{x}^* \rangle \,| \boldsymbol{x} \in B\} = \left\langle \frac{\boldsymbol{x}^*}{\|\boldsymbol{x}^*\|}, \boldsymbol{x}^* \right\rangle = \|\boldsymbol{x}^*\|.$$

单位球的支撑函数等于其对偶变量 $\boldsymbol{x}^*$ 的范数.

例 4.5.2 设集合 $C = \left\{ (x_1, x_2, \cdots, x_n)^{\mathrm{T}} \,| x_1, x_2, \cdots, x_n \geqslant 0, x_1 + x_2 + \cdots + x_n = 1 \right\}$, 求 $\delta_C(\boldsymbol{x}^*)$. 设 $\boldsymbol{x}^* = (x_1^*, x_2^*, \cdots, x_n^*)^{\mathrm{T}} \in \mathbf{R}^n$ 和 $\alpha = \max\{x_1^*, x_2^*, \cdots, x_n^*\}$. 由性质 4.5.2(1) 知, $\exists \bar{\boldsymbol{x}} \in C$ 使得

$$\delta_C(\boldsymbol{x}^*) = \langle \bar{\boldsymbol{x}}, \boldsymbol{x}^* \rangle \leqslant \sum_{i=1}^{n} \bar{x}_i x_i^* \leqslant \alpha \sum_{i=1}^{n} \bar{x}_i = \alpha.$$

因为存在某个 x_j^* 满足 $\alpha = x_j^*$, 令

$$x_i = 0\ (i = 1, 2, \cdots, n, i \neq j), \quad x_j = 1,$$

有 $\delta_C(\boldsymbol{x}^*) = \langle \bar{\boldsymbol{x}}, \boldsymbol{x}^* \rangle \geqslant \alpha$, 得 $\delta_C(\boldsymbol{x}^*) = \alpha$. 因此, 此例说明 1 的组合比例构成的集合 C 的支撑函数等于其对偶变量中的最大值.

定理 4.5.1 设凸集 $C_1, C_2 \subset \mathbf{R}^n$, 则 $\mathrm{cl}C_1 \subset \mathrm{cl}C_2$ 当且仅当

$$\delta_{C_1}(\boldsymbol{x}^*) \leqslant \delta_{C_2}(\boldsymbol{x}^*).$$

证明 设 $\mathrm{cl}C_1 \subset \mathrm{cl}C_2$, 则对任意 $\boldsymbol{x}^* \in \mathbf{R}^n$, 有

$$\delta_{C_1}(\boldsymbol{x}^*) = \sup\{\langle \boldsymbol{x}, \boldsymbol{x}^* \rangle \,|\, \boldsymbol{x} \in \mathrm{cl}C_1\} \leqslant \sup\{\langle \boldsymbol{x}, \boldsymbol{x}^* \rangle \,|\, \boldsymbol{x} \in \mathrm{cl}C_2\} = \delta_{C_2}(\boldsymbol{x}^*).$$

反之, 设 $\delta_{C_1}(\boldsymbol{x}^*) \leqslant \delta_{C_2}(\boldsymbol{x}^*)$. 若 $\boldsymbol{x} \in \mathrm{cl}C_1$, 则有

$$\langle \boldsymbol{x}, \boldsymbol{x}^* \rangle \leqslant \delta_{C_1}(\boldsymbol{x}^*) \leqslant \delta_{C_2}(\boldsymbol{x}^*),$$

由性质 4.5.1(1) 知 $\boldsymbol{x} \in \mathrm{cl}C_2$. □

定理 4.5.1 说明闭集的包含关系等价于对应支撑函数值的关系. 下面的定理 4.5.2 则说明闭凸集的支撑函数与示性函数彼此共轭.

定理 4.5.2 设闭凸集 $C \subset \mathbf{R}^n$, 则 $\theta_C(\boldsymbol{x})$ 和 $\delta_C(\boldsymbol{x}^*)$ 是彼此的共轭凸函数.

证明 根据示性函数与支撑函数的定义知

$$\theta_C^*(\boldsymbol{x}^*) = \sup\{\langle \boldsymbol{x}, \boldsymbol{x}^* \rangle - \theta_C(\boldsymbol{x}) \,|\, \boldsymbol{x} \in \mathbf{R}^n\} = \sup_{\boldsymbol{x} \in C} \langle \boldsymbol{x}, \boldsymbol{x}^* \rangle = \delta_C(\boldsymbol{x}^*).$$

由上式和性质 4.4.2 得

$$\delta_C^*(\boldsymbol{x}) = \theta_C^{**}(\boldsymbol{x}) = \mathrm{cl}\theta_C(\boldsymbol{x}) = \theta_{\mathrm{cl}C}(\boldsymbol{x}) = \theta_C(\boldsymbol{x}).$$ □

定理 4.5.3 设非空凸集 $C \subset \mathbf{R}^n$, 则 $\delta_C(\boldsymbol{x}^*)$ 是闭正齐次正常凸函数, 且 $\mathrm{dom}\delta_C$ 是包含原点的凸锥, $\delta_{tC}(\boldsymbol{x}^*) = t\delta_C(\boldsymbol{x}^*)$.

证明 由性质 4.1.5 知 $\delta_C(\boldsymbol{x}^*)$ 是闭正常凸函数, 根据支撑函数的定义, 当 $0 < t < +\infty$ 时, 有

$$\delta_C(t\boldsymbol{x}^*) = \sup\{\langle \boldsymbol{x}, t\boldsymbol{x}^* \rangle \,|\, \boldsymbol{x} \in C\} = t\sup\{\langle \boldsymbol{x}, \boldsymbol{x}^* \rangle \,|\, \boldsymbol{x} \in C\} = t\delta_C(\boldsymbol{x}^*),$$

得 $\delta_C(\boldsymbol{x}^*)$ 是正齐次函数, 由凸锥定义显然得 $\mathrm{dom}\delta_C$ 是凸锥. 再由性质 4.5.1 得 $\delta_C(\mathbf{0}) = 0$. □

因为支撑函数是正齐次函数, 所以有下面 3 个推论成立.

推论 4.5.1 设非空凸集 $C \subset \mathbf{R}^n$, 则对任意 $\boldsymbol{x}_1^*, \boldsymbol{x}_2^* \in \mathbf{R}$, 有

$$\delta_C(\boldsymbol{x}_1^* + \boldsymbol{x}_2^*) \leqslant \delta_C(\boldsymbol{x}_1^*) + \delta_C(\boldsymbol{x}_2^*).$$

推论 4.5.2 设非空有界闭凸集 $C \subset \mathbf{R}^n$, 则 $\delta_C(\boldsymbol{x}^*)$ 是有限正齐次凸函数.

证明 由定理 4.5.3 知, 对任意 $\boldsymbol{x}^* \in \mathbf{R}^n$, 线性函数 $\delta_C(\boldsymbol{x}^*) < +\infty$. □

推论 4.5.3 设 f 是 $\mathbf{R}^n$ 上的正齐次凸函数, $f(\boldsymbol{x}) \not\equiv +\infty\,(\boldsymbol{x} \in \mathbf{R}^n)$, 则 $\mathrm{cl}f(\boldsymbol{x}^*) = \delta_C(\boldsymbol{x}^*)$, 且 $C = \{\boldsymbol{x}^* \,|\langle \boldsymbol{x}, \boldsymbol{x}^*\rangle \leqslant f(\boldsymbol{x}), \forall \boldsymbol{x} \in \mathbf{R}^n\}$.

证明 因为 $f(\boldsymbol{x}) \not\equiv +\infty$, 故 $\mathrm{cl}f(\boldsymbol{x})$ 是闭正齐次正常凸函数或常量函数 $-\infty$. 如果 $f(\boldsymbol{x}) \equiv -\infty$, 则 $C = \varnothing$, 结论明显成立. 如果 f 是正常的, 设 $\mathrm{cl}f(\boldsymbol{x}^*) = \delta_C(\boldsymbol{x}^*)$, 由定理 4.5.2 和推论 4.5.2 有 $f^*(\boldsymbol{x}^*) = (\mathrm{cl}f)^*(\boldsymbol{x}^*) = \theta_C(\boldsymbol{x}^*)$, 还有

$$C = \{\boldsymbol{x}^* \,| f^*(\boldsymbol{x}^*) \leqslant 0\}$$

等价于对任意 $\boldsymbol{x} \in \mathbf{R}^n, \langle \boldsymbol{x}, \boldsymbol{x}^*\rangle - f(\boldsymbol{x}) \leqslant 0$, 所以

$$C = \{\boldsymbol{x}^* | \langle \boldsymbol{x}, \boldsymbol{x}^*\rangle \leqslant f(\boldsymbol{x}), \forall \boldsymbol{x} \in \mathbf{R}^n\}.$$ □

推论 4.5.4 设 f 是 $\mathbf{R}^n$ 上的闭正齐次凸函数, 则

$$f(\boldsymbol{x}) = \delta_{\mathrm{dom} f^*}(\boldsymbol{x}).$$

下面设 C^- 是 C 的极锥, 则由示性函数定义

$$\begin{aligned} C^- &= \{\boldsymbol{x}^* \,|\langle \boldsymbol{x}, \boldsymbol{x}^*\rangle \leqslant \theta_C(\boldsymbol{x}), \forall \boldsymbol{x} \in \mathbf{R}^n\} \\ &= \{\boldsymbol{x}^* \,|\langle \boldsymbol{x}, \boldsymbol{x}^*\rangle \leqslant 0, \forall \boldsymbol{x} \in C\} = -C^+. \end{aligned}$$

由定理 4.5.2 得下面结论, 如果非空闭凸锥 $C \subset \mathbf{R}^n$, 它的极锥 C^- 也是非空闭凸锥, 且 C^- 和 C 的示性函数彼此共轭.

定理 4.5.4 设非空闭凸锥 $C \subset \mathbf{R}^n$, 则 C^- 也是非空闭凸锥, 且 C^- 和 C 的示性函数彼此共轭, $C^{--} = (C^-)^- = C$.

证明 由性质 3.5.1 得 C^- 也是非空闭凸锥, 且 $C^{--} = (C^-)^- = C$, 再由定理 4.5.2 知 C^- 和 C 的示性函数彼此共轭. □

本小节得到了非空凸集 C 的支撑函数若干性质:

1) $\boldsymbol{x} \in \mathrm{cl}C$ 等价于在 $\forall \boldsymbol{x}^* \in \mathbf{R}^n$ 上的线性函数 $\langle \boldsymbol{x}, \boldsymbol{x}^*\rangle$ 不大于其支撑函数值 $\delta_C(\boldsymbol{x}^*)$, C 属于在 $\forall \boldsymbol{x}^* \in \mathbf{R}^n$ 上的线性函数 $\langle \boldsymbol{x}, \boldsymbol{x}^*\rangle$ 水平集等价于支撑函数 $\delta_C(\boldsymbol{x}^*)$ 不大于水平值 t;

2) C 的支撑函数是线性函数 $\langle \boldsymbol{x}, \boldsymbol{x}^*\rangle$ 构成的超平面在 C 上的支撑超平面, 凸紧集 C 的支撑函数等于凸集相对内部和闭包分别构成的支撑函数;

3) $\mathrm{cl}C_1 \subset \mathrm{cl}C_2$ 等价于对应支撑函数值 $\delta_{C_1}(\boldsymbol{x}^*) \leqslant \delta_{C_2}(\boldsymbol{x}^*)$;

4) $\delta_C(\boldsymbol{x}^*)$ 是闭正齐次正常凸函数, 且 $\mathrm{dom}\delta_C$ 是包含原点的凸锥;

5) 非空闭凸锥的极锥 C^- 也是非空闭凸锥, 且 C^- 和 C 的示性函数彼此共轭.

4.6 规范凸函数

本节讨论凸集上的另一个特殊凸函数——规范凸函数的一些性质.

定义 4.6.1 设非空凸集 $C \subset \mathbf{R}^n$, 称

$$v_C(\boldsymbol{x}) = \inf\{\lambda > 0 \,|\, \boldsymbol{x} \in \lambda C\} \tag{4.6.1-1}$$

是 C 的规范凸函数, 简称规范函数.

定理 4.6.1 设凸紧集 $C \subset \mathbf{R}^n$, $\mathbf{0} \in \text{int}C$, C° 是 C 的配极, 则有

$$\delta_C(\boldsymbol{x}) = v_{C^\circ}(\boldsymbol{x}), \quad \delta_{C^\circ}(\boldsymbol{x}) = v_C(\boldsymbol{x}),$$

且 $\delta_C(\boldsymbol{x})$ 与 $v_{C^\circ}(\boldsymbol{x})$ 是彼此共轭凸函数, $\delta_{C^\circ}(\boldsymbol{x})$ 与 $v_C(\boldsymbol{x})$ 是彼此共轭凸函数.

证明 根据推论 2.5.4 知 C° 是凸紧集, 且 $\mathbf{0} \in \text{int}C^\circ$. 假设 $\boldsymbol{x} \neq \mathbf{0}$, 作从 $\mathbf{0}$ 出发通过 $\boldsymbol{x}$ 的射线, 它与 C° 的边界交点是 $\boldsymbol{x}_0$. 因 $\boldsymbol{x}_0 \in C^\circ$, 根据配极的定义知, 对任意的 $\boldsymbol{x} \in C$ 有

$$\langle \boldsymbol{x}, \boldsymbol{x}_0 \rangle \leqslant 1.$$

从而

$$\delta_C(\boldsymbol{x}_0) = \sup\{\langle \boldsymbol{x}, \boldsymbol{x}_0 \rangle \,|\, \boldsymbol{x} \in C\} \leqslant 1.$$

因为 $\boldsymbol{x} = \dfrac{\|\boldsymbol{x}\|}{\|\boldsymbol{x}_0\|}\boldsymbol{x}_0$, 故有

$$\delta_C(\boldsymbol{x}) = \delta_C\left(\frac{\|\boldsymbol{x}\|}{\|\boldsymbol{x}_0\|}\boldsymbol{x}_0\right) = \frac{\|\boldsymbol{x}\|}{\|\boldsymbol{x}_0\|}\delta_C(\boldsymbol{x}_0) \leqslant \frac{\|\boldsymbol{x}\|}{\|\boldsymbol{x}_0\|} = v_{C^\circ}(\boldsymbol{x}). \tag{4.6.1-2}$$

因为 C 也是 C° 的配极, 同理可得

$$\delta_{C^\circ}(\boldsymbol{x}) \leqslant v_C(\boldsymbol{x}). \tag{4.6.1-3}$$

另一方面, 由 $\boldsymbol{x} \neq \mathbf{0}$ 得 $v_C(\boldsymbol{x}) > 0$, 选取 t 使得 $0 < t < v_C(\boldsymbol{x})$, 有 $\boldsymbol{x} \notin tC$. 由性质 2.5.7 得

$$tC = (tC^{\circ\circ}) = \left[\frac{1}{t}C^\circ\right]^\circ.$$

这说明在 $\dfrac{1}{t}C^\circ$ 中至少存在一点 $\dfrac{1}{t}\boldsymbol{x}^\circ$ 使得 $\left\langle \dfrac{1}{t}\boldsymbol{x}^\circ, \boldsymbol{x} \right\rangle > 1$, 即

$$\langle \boldsymbol{x}^\circ, \boldsymbol{x} \rangle > t, \quad \boldsymbol{x}^\circ \in C^\circ.$$

所以, 有

$$\delta_{C^\circ}(\boldsymbol{x}) = \sup\{\langle \boldsymbol{x}^\circ, \boldsymbol{x}\rangle \,|\boldsymbol{x}^\circ \in C^\circ\} > t. \tag{4.6.1-4}$$

因为 $v_C(\boldsymbol{x}) = \sup\{t|t < v_C(\boldsymbol{x})\}$, 由 (4.6.1-4) 式得

$$\delta_{C^\circ}(\boldsymbol{x}) \geqslant v_C(\boldsymbol{x}). \tag{4.6.1-5}$$

由于 C 和 C° 的对称性, 由 (4.5.2-5) 式又得

$$\delta_C(\boldsymbol{x}) \geqslant v_{C^\circ}(\boldsymbol{x}).$$

由于 $\delta_C(\mathbf{0}) = \delta_{C^\circ}(\mathbf{0}) = v_C(\mathbf{0}) = v_{C^\circ}(\mathbf{0}) = 0$, 所以由 (4.6.1-2), (4.6.1-3), (4.6.2-4) 和 (4.6.1-5) 式得定理结论成立. □

定理 4.6.1 说明 C 的支撑函数 $\delta_C(\boldsymbol{x})$ 与配极 C° 的规范函数 $v_{C^\circ}(\boldsymbol{x})$ 是彼此共轭凸函数, 对应的 $\delta_{C^\circ}(\boldsymbol{x})$ 与 $v_C(\boldsymbol{x})$ 也是彼此共轭凸函数.

4.7 三种严格凸函数

本节讨论有关严格凸函数、半严格凸函数和显凸函数的性质 [90-93].

4.7.1 严格凸函数

根据定义 4.1.1 知, 如果 f 在凸集 $C \subset \mathbf{R}^n$ 上是一个实函数, 对于 C 中任意两个不相同的点有 (4.1.1-2) 式成立, 那么 f 是严格凸函数. 严格凸函数具有全局唯一最小点的性质, 对于寻找严格凸函数的最优解具有重要的意义. 如果 f 是凸集 $C \subset \mathbf{R}^n$ 上的一个凸函数, 我们有下面的一个判定定理.

定理 4.7.1 设 f 在凸集 $C \subset \mathbf{R}^n$ 上是凸函数, 如果 f 在 C 上关于 t 满足一点严格凸性, 则 f 在 C 上是严格凸函数.

证明 根据已知, 存在一点 $t \in (0,1)$ 使得对任何不同两点 $\boldsymbol{x}, \boldsymbol{y} \in C$ 有

$$f[(1-t)\boldsymbol{x} + t\boldsymbol{y}] < (1-t)f(\boldsymbol{x}) + tf(\boldsymbol{y}). \tag{4.7.1-1}$$

采用反证法. 假设 f 在 C 上不是严格凸函数, 则存在不同的 $\boldsymbol{x}, \boldsymbol{y} \in C$ 和一个 $\lambda \in (0,1)$ 有

$$f[(1-\lambda)\boldsymbol{x} + \lambda\boldsymbol{y}] \geqslant (1-\lambda)f(\boldsymbol{x}) + \lambda f(\boldsymbol{y}).$$

设 $\boldsymbol{z} = (1-\lambda)\boldsymbol{x} + \lambda\boldsymbol{y}$, 有 $f(\boldsymbol{z}) \geqslant (1-\lambda)f(\boldsymbol{x}) + \lambda f(\boldsymbol{y})$, 但 f 在凸集 $C \subset \mathbf{R}^n$ 上是凸函数, 有

$$f(\boldsymbol{z}) \leqslant (1-\lambda)f(\boldsymbol{x}) + \lambda f(\boldsymbol{y}).$$

因此, 有

$$f(\boldsymbol{z})=(1-\lambda)f(\boldsymbol{x})+\lambda f(\boldsymbol{y}). \tag{4.7.1-2}$$

取 λ_1 和 λ_2 满足 $0<\lambda_1<\lambda<\lambda_2<1$ 和

$$\lambda=(1-t)\lambda_1+t\lambda_2. \tag{4.7.1-3}$$

令 $\boldsymbol{z}_1=(1-\lambda_1)\boldsymbol{x}+\lambda_1\boldsymbol{y}$, $\boldsymbol{z}_2=(1-\lambda_2)\boldsymbol{x}+\lambda_2\boldsymbol{y}$, 有

$$(1-t)\boldsymbol{z}_1+t\boldsymbol{z}_2=(1-\lambda)\boldsymbol{x}+\lambda\boldsymbol{y}=\boldsymbol{z}. \tag{4.7.1-4}$$

再由函数 f 在凸集 C 上是凸函数, 有

$$f(\boldsymbol{z}_1)\leqslant(1-\lambda_1)f(\boldsymbol{x})+\lambda_1 f(\boldsymbol{y}),\quad f(\boldsymbol{z}_2)\leqslant(1-\lambda_2)f(\boldsymbol{x})+\lambda_2 f(\boldsymbol{y}). \tag{4.7.1-5}$$

再由定理的假设、(4.7.1-4) 及 (4.3.1-5) 式得

$$\begin{aligned}f(\boldsymbol{z})&<(1-t)f(\boldsymbol{z}_1)+tf(\boldsymbol{z}_2)\\&\leqslant(1-t)(\lambda_1 f(\boldsymbol{x})+(1-\lambda_1)f(\boldsymbol{y}))+t(\lambda_2 f(\boldsymbol{x})+(1-\lambda_2)f(\boldsymbol{y}))\\&=(1-\lambda)f(\boldsymbol{x})+\lambda f(\boldsymbol{y}),\end{aligned}$$

与 (4.7.1-2) 式矛盾. □

例 4.7.1 线性函数 $f(\boldsymbol{x})=\boldsymbol{a}^{\mathrm{T}}\boldsymbol{x}+b$ 是凸的, 它不满足一点严格凸性.

如果一个 f 在凸集 C 上是严格凸函数, 根据一点严格凸性和严格凸函数的定义, 不可能存在不同两点 $\boldsymbol{x},\boldsymbol{y}\in C$ 使得

$$f[(1-t)\boldsymbol{x}+t\boldsymbol{y}]=(1-t)f(\boldsymbol{x})+tf(\boldsymbol{y}).$$

因此, 我们有下面结论.

定理 4.7.2 设 f 在凸集 $C\subset\mathbf{R}^n$ 上是凸函数, 则 f 在凸集 C 上是严格凸函数当且仅当 $\operatorname{epi}f$ 是凸集且当它的边界不含任何直线段.

证明 若 f 在凸集 C 上是严格凸函数, 显然 $\operatorname{epi}f$ 是凸集. 假设 $\operatorname{epi}f$ 是它的边界且至少含一条直线段, 即存在连接两点 $(\boldsymbol{x},f(\boldsymbol{x})),(\boldsymbol{y},f(\boldsymbol{y}))\in\operatorname{epi}f$ 的线段属于边界

$$\begin{aligned}&(t\boldsymbol{x}+(1-t)\boldsymbol{y},f(t\boldsymbol{x}+(1-t)\boldsymbol{y}))\\=&t(\boldsymbol{x},f(\boldsymbol{x}))+(1-t)(\boldsymbol{y},f(\boldsymbol{y}))\in\operatorname{epi}f,\quad\forall t\in(0,1).\end{aligned}$$

即有 $f(t\boldsymbol{x}+(1-t)\boldsymbol{y})=tf(\boldsymbol{x})+(1-t)f(\boldsymbol{y})$, 这与 f 在凸集 C 上是严格凸函数的定义矛盾.

反之, 从上面证明可知结论成立. □

定义 4.7.1 设闭区间 $[a,b]\subset\mathbf{R}$. 如果对 $[a,b]$ 内的子区间 I, 均有 $S\cap I\neq\varnothing$, 则称 S 在 $[a,b]$ 内是稠密的.

对于单变量可微严格凸函数, 具有下面的一个重要性质 [94], f 在开区间上的严格凸函数等价于其导函数 f' 在开区间上是严格单增的.

引理 4.7.1 设 f 在 $[a,b]\subset\mathbf{R}$ 上是可微函数, 则 f 在 (a,b) 上是严格凸函数的充要条件是导函数 f' 在 (a,b) 上是严格单增的.

证明 若 f 在 (a,b) 上是严格凸函数. 设 $x_1,x_2\in(a,b)$, 且 $x_1>x_2$, 由性质 4.1.2(2) 得

$$f\left(x_2\right)-f\left(x_1\right)>f'\left(x_1\right)\left(x_2-x_1\right),\quad f\left(x_1\right)-f\left(x_2\right)>f'\left(x_2\right)\left(x_1-x_2\right).$$

将上面两个不等式相加得 $f'\left(x_1\right)>f'\left(x_2\right)$.

反过来, 设 $x_1,x_2\in(a,b)$, 当 $x_1>x_2$ 时有 $f'\left(x_1\right)>f'\left(x_2\right)$, 得

$$f\left(x_1\right)-f\left(x_2\right)-f'\left(x_2\right)\left(x_1-x_2\right)=\int_{x_2}^{x_1}\left[f'(s)-f'\left(x_2\right)\right]\mathrm{d}s>0.$$

由性质 4.1.2(1) 和 (2) 知 f 在 (a,b) 上是严格凸函数. □

下面证明, 在 (a,b) 上单变量严格二阶可微凸函数的充要条件是其二阶导数在 (a,b) 上几乎处处大于 0.

定理 4.7.3 设 f 在 $[a,b]\subset\mathbf{R}$ 上是连续函数, 在 (a,b) 上二阶可微, 且在 (a,b) 上 $f''(x)\geqslant 0$, $S=\{x\in(a,b)\,|f''(x)>0\}$, 则 f 在 $[a,b]$ 上是严格凸函数的充要条件是 S 在 $[a,b]$ 内是稠密的.

证明 设 f 在 $[a,b]$ 上是严格凸函数, 由引理 4.7.1 知导函数 f' 在 (a,b) 上是严格单增的. 如果 S 在 $[a,b]$ 内不是稠密的, 那么存在一个区间 $I\subset(a,b)$ 使得 $I\cap S=\varnothing$, 有 $f''(x)=0,\forall x\in I$ 成立, 因此得 f' 在区间上 $I\subset(a,b)$ 是常数, 这与 f' 在 (a,b) 上是严格单增的矛盾.

反之, 设 S 在 $[a,b]$ 内是稠密的, 但 f 在 $[a,b]$ 上不是严格凸函数. 于是, f' 在 (a,b) 上不是严格单增的. 因此, $\exists x_1,x_2\in(a,b)$ 和 $x_1>x_2$ 使得

$$f'\left(x_1\right)=f'\left(x_2\right)=f'(x),\quad\forall x\in\left(x_2,x_1\right).$$

即得 $f(x)=0,\forall x\in\left(x_1,x_2\right)$, 这与 S 在 $[a,b]$ 内是稠密的矛盾. □

例 4.7.2 设单变量函数 $f(x)=x^3$, 有 $f'(x)=3x^2$, $f''(x)=6x$. 当 $0<a<b$ 时, f 在 $[a,b]\subset\mathbf{R}$ 上是严格凸函数. 当 $a<0$ 时, $f(x)=x^3$ 在 $[a,b]$ 上不是凸函数.

例 4.7.3 设单变量函数 $f(x)=-x^{\frac{1}{3}}$, 有 $f'(x)=-\dfrac{1}{3}x^{-\frac{2}{3}}$, $f''(x)=\dfrac{2}{9}x^{-\frac{4}{3}}$. 当 $0<a<b$ 时, f 在 $[a,b]\subset\mathbf{R}$ 上是严格凸函数. 当 $a<0$ 时, $f(x)=-x^{\frac{1}{3}}$ 在 $[a,b]$ 上不是凸函数.

上述例子说明, 在某个区域上的函数不是凸函数, 但是在另一个区域上却可能是严格凸函数.

本小节得到严格凸函数的几个性质:

1) 满足一点严格凸性的凸函数是严格凸函数;

2) 严格凸函数等价于其上方图是凸集且不含直线段;

3) 在区间 (a,b) 上单变量严格二阶可微凸函数等价于其二阶导数在 (a,b) 上几乎处处大于 0.

4.7.2 半严格凸函数

本小节讨论半严格凸函数性质, 它不同于严格凸函数, 因此需要专门进行研究 [90-93].

定义 4.7.2 设非空开凸集 $C\subset\mathbf{R}^n$, 函数 $f:C\to\mathbf{R}$, 如果 $\forall\boldsymbol{x},\boldsymbol{y}\in C$, $f(\boldsymbol{x})\neq f(\boldsymbol{y})$ 和 $\forall\lambda\in(0,1)$ 有

$$f[(1-\lambda)\boldsymbol{x}+\lambda\boldsymbol{y}]<(1-\lambda)f(\boldsymbol{x})+\lambda f(\boldsymbol{y}), \tag{4.7.2-1}$$

则称 f 在 C 上是半严格凸函数. 如果 $-f$ 是半严格凸函数, 称 f 在 C 上是半严格凹函数.

如果 $\forall\boldsymbol{x},\boldsymbol{y}\in C$, $f(\boldsymbol{x})\neq f(\boldsymbol{y})$, $\exists\lambda\in(0,1)$ 满足 (4.7.2-1) 式, 则称 f 满足关于 λ 的一点半严格凸性.

值得注意的是, 半严格凸函数已经不是凸函数了, 容易举例一个半严格凸函数不一定是凸函数, 反过来, 凸函数也不一定是半严格凸函数了.

例 4.7.4 $f(x)=\begin{cases}1, & x=0,\\ 0, & x\neq 0\end{cases}$ 是半严格凸函数, 但不是凸函数. $f(x)=\begin{cases}x, & x\geqslant 0,\\ 0, & x<0\end{cases}$ 是凸函数, 但不是半严格凸函数.

定理 4.7.4 设非空开凸集 $C\subset\mathbf{R}^n$, 函数 $f:C\to\mathbf{R}$, 则 f 在 C 上是半严格凸函数的充要条件是对任意的 $\boldsymbol{x}_i\in C$, $f(\boldsymbol{x}_1)<f(\boldsymbol{x}_2)\leqslant\cdots\leqslant f(\boldsymbol{x}_m)\,(m\geqslant 2)$ 和 $\lambda_i>0\,(i=1,\cdots,m)$ 有

$$f(\lambda_1\boldsymbol{x}_1+\cdots+\lambda_m\boldsymbol{x}_m)<\lambda_1 f(\boldsymbol{x}_1)+\cdots+\lambda_m f(\boldsymbol{x}_m) \tag{4.7.2-2}$$

成立, 其中 $\sum_{i=1}^{m}\lambda_i = 1$.

证明 假设 f 在 C 上是半严格凸函数, 对任意 $\boldsymbol{x}_i \in C$, $f(\boldsymbol{x}_1) < f(\boldsymbol{x}_2) \leqslant \cdots \leqslant f(\boldsymbol{x}_m)\,(m \geqslant 2)$ 和 $\lambda_i > 0\,(i = 1, \cdots, m)$, 且 $\sum_{i=1}^{m}\lambda_i = 1$, 设

$$
\begin{aligned}
&\boldsymbol{y}_1 = \boldsymbol{x}_1,\\
&\boldsymbol{y}_2 = \frac{\lambda_1\boldsymbol{y}_1 + \lambda_2\boldsymbol{x}_2}{\lambda_1+\lambda_2} = \frac{\lambda_1\boldsymbol{x}_1 + \lambda_2\boldsymbol{x}_2}{\lambda_1+\lambda_2},\\
&\boldsymbol{y}_2 = \frac{(\lambda_1+\lambda_2)\,\boldsymbol{y}_2 + \lambda_3\boldsymbol{x}_3}{\lambda_1+\lambda_2+\lambda_3} = \frac{\lambda_1\boldsymbol{x}_1 + \lambda_2\boldsymbol{x}_2 + \lambda_3\boldsymbol{x}_3}{\lambda_1+\lambda_2+\lambda_3},\\
&\qquad\cdots\cdots\\
&\boldsymbol{y}_m = \frac{(\lambda_1+\lambda_2+\cdots+\lambda_{m-1})\,\boldsymbol{y}_{m-1} + \lambda_m\boldsymbol{x}_m}{\lambda_1+\lambda_2+\cdots+\lambda_m} = \frac{\lambda_1\boldsymbol{x}_1 + \lambda_2\boldsymbol{x}_2 + \cdots + \lambda_m\boldsymbol{x}_m}{\lambda_1+\lambda_2+\cdots+\lambda_m}.
\end{aligned}
$$

则有

$$
\begin{aligned}
f(\boldsymbol{y}_2) &< \frac{\lambda_1}{\lambda_1+\lambda_2} f(\boldsymbol{x}_1) + \frac{\lambda_2}{\lambda_1+\lambda_2} f(\boldsymbol{x}_2) < f(\boldsymbol{x}_2) \leqslant f(\boldsymbol{x}_3),\\
f(\boldsymbol{y}_3) &< \frac{\lambda_1+\lambda_2}{\lambda_1+\lambda_2+\lambda_3} f(\boldsymbol{y}_2) + \frac{\lambda_3}{\lambda_1+\lambda_2+\lambda_3} f(\boldsymbol{x}_3)\\
&< \frac{\lambda_1}{\lambda_1+\lambda_2+\lambda_3} f(\boldsymbol{x}_1) + \frac{\lambda_2}{\lambda_1+\lambda_2} f(\boldsymbol{x}_2) + \frac{\lambda_3}{\lambda_1+\lambda_2+\lambda_3} f(\boldsymbol{x}_3)\\
&< f(\boldsymbol{x}_3) \leqslant f(\boldsymbol{x}_4),\\
&\qquad\cdots\cdots\\
f(\boldsymbol{y}_m) &< \frac{1}{\lambda_1+\lambda_2+\cdots+\lambda_{m-1}} \sum_{i=1}^{m-1} \lambda_i f(\boldsymbol{x}_i) < f(\boldsymbol{x}_{m-1}) \leqslant f(\boldsymbol{x}_m).
\end{aligned}
$$

综上所得:

$$
\begin{aligned}
& f(\lambda_1\boldsymbol{x}_1 + \lambda_2\boldsymbol{x}_2 + \cdots + \lambda_m\boldsymbol{x}_m)\\
<& \frac{\lambda_1+\lambda_2+\cdots+\lambda_{m-1}}{\lambda_1+\lambda_2+\cdots+\lambda_m} f(\boldsymbol{y}_{m-1}) + \frac{\lambda_m}{\lambda_1+\lambda_2+\cdots+\lambda_m} f(\boldsymbol{x}_m)\\
<& \lambda_1 f(\boldsymbol{x}_1) + \cdots + \lambda_m f(\boldsymbol{x}_m).
\end{aligned}
$$

反过来, 在 (4.7.2-2) 式取 $m = 2$, 结论成立. □

例 4.7.5 $f(x)=\begin{cases}1, & x=0,\\ 0, & x\neq 0\end{cases}$ 是半严格凸函数, 取值 $x_1=-0.5, x_2=1$, $x_3=0$, $\lambda_1=\lambda_3=0.5$, $\lambda_2=0.25$, 因此有

$$f(x_1)=f(x_2)<f(x_3),$$

$$f(\lambda_1x_1+\lambda_2x_2+\lambda_3x_3)=f(0)=1>\lambda_1f(x_1)+\lambda_2f(x_2)+\lambda_3f(x_3)=0.5.$$

因此, 定理的必要条件中的 $f(\boldsymbol{x}_1)<f(\boldsymbol{x}_2)\leqslant\cdots\leqslant f(\boldsymbol{x}_m)\,(m\geqslant 2)$ 是必需的.

定理 4.7.5 设非空开凸集 $C\subset\mathbf{R}^n$, $f:C\to\mathbf{R}$ 是在 C 上的半严格凸函数. 如果 f 满足关于 λ 的一点凸性, 则 f 在 C 上是凸函数.

证明 根据假设 $\exists\lambda\in(0,1)$, 对任意 $\boldsymbol{x},\boldsymbol{y}\in C$ 有

$$f[(1-\lambda)\boldsymbol{x}+\lambda\boldsymbol{y}]\leqslant(1-\lambda)f(\boldsymbol{x})+\lambda f(\boldsymbol{y}). \tag{4.7.2-3}$$

假设 f 在 C 上不是凸函数, 则 $\exists\boldsymbol{x},\boldsymbol{y}\in C$, $t\in(0,1)$ 使得

$$f[(1-t)\boldsymbol{x}+t\boldsymbol{y}]>(1-t)f(\boldsymbol{x})+tf(\boldsymbol{y}). \tag{4.7.2-4}$$

不妨设 $f(\boldsymbol{x})\leqslant f(\boldsymbol{y})$. 记 $(1-t)\boldsymbol{x}+t\boldsymbol{y}=\boldsymbol{z}$, 如果 $f(\boldsymbol{x})<f(\boldsymbol{y})$, 由函数 $f:C\to\mathbf{R}$ 在 C 上的半严格凸函数定义得

$$f[(1-t)\boldsymbol{x}+t\boldsymbol{y}]<(1-t)f(\boldsymbol{x})+tf(\boldsymbol{y}),$$

上式与 (4.7.2-4) 式矛盾. 那么有 $f(\boldsymbol{x})=f(\boldsymbol{y})$, 则有 $f(\boldsymbol{z})>f(\boldsymbol{x})=f(\boldsymbol{y})$. 分两种情形讨论.

(1) 若 $0<t<\lambda<1$, 设 $\boldsymbol{z}_1=\left(1-\dfrac{t}{\lambda}\right)\boldsymbol{x}+\dfrac{t}{\lambda}\boldsymbol{y}$, 则有

$$\boldsymbol{z}=(1-t)\boldsymbol{x}+t\boldsymbol{y}=\lambda\left[\left(1-\frac{t}{\lambda}\right)\boldsymbol{x}+\frac{t}{\lambda}\boldsymbol{y}\right]+(1-\lambda)\boldsymbol{x}=\lambda\boldsymbol{z}_1+(1-\lambda)\boldsymbol{x}.$$

由上式和 (4.7.2-3) 式有 $f(\boldsymbol{z})\leqslant(1-\lambda)f(\boldsymbol{z}_1)+\lambda f(\boldsymbol{x})$. 再由 (4.7.2-4) 式得

$$f(\boldsymbol{z})\leqslant(1-\lambda)f(\boldsymbol{z}_1)+\lambda f(\boldsymbol{x})<(1-\lambda)f(\boldsymbol{z}_1)+\lambda f(\boldsymbol{z}),$$

即得 $f(\boldsymbol{z})<f(\boldsymbol{z}_1)$. 再设 $\beta=\dfrac{t(1-\lambda)}{\lambda(1-t)}$, 由 $0<t<\lambda<1$ 得 $0<\beta<1$, 则

$$\boldsymbol{z}_1=\left(1-\frac{t}{\lambda}\right)\boldsymbol{x}+\frac{t}{\lambda}\boldsymbol{y}=\left(1-\frac{t}{\lambda}\right)\left(\frac{\boldsymbol{z}}{1-t}-\frac{t\boldsymbol{y}}{1-t}\right)+\frac{t}{\lambda}\boldsymbol{y}=(1-\beta)\boldsymbol{z}+\beta\boldsymbol{y}.$$

再由 $f(\boldsymbol{z}) > f(\boldsymbol{x}) = f(\boldsymbol{y})$ 和半严格凸函数的定义得

$$f(\boldsymbol{z}_1) \leqslant (1-\beta) f(\boldsymbol{z}) + \beta f(\boldsymbol{y}) < (1-\beta) f(\boldsymbol{z}) + \beta f(\boldsymbol{z}) = f(\boldsymbol{z}).$$

因此, 这与 $f(\boldsymbol{z}) < f(\boldsymbol{z}_1)$ 矛盾.

(2) 若 $0 < \lambda < t < 1$, 设 $\boldsymbol{z}_2 = \left(1 - \dfrac{1-t}{1-\lambda}\right)\boldsymbol{y} + \dfrac{1-t}{1-\lambda}\boldsymbol{x}$, 则有

$$\boldsymbol{z} = (1-t)\boldsymbol{x} + t\boldsymbol{y} = (1-\lambda)\boldsymbol{z}_2 + \lambda\boldsymbol{y}.$$

由上式和 (4.7.2-3) 式有 $f(\boldsymbol{z}) \leqslant (1-\lambda) f(\boldsymbol{z}_2) + \lambda f(\boldsymbol{y})$, 再由 (4.7.2-4) 式得

$$f(\boldsymbol{z}) \leqslant (1-\lambda) f(\boldsymbol{z}_2) + \lambda f(\boldsymbol{y}) < (1-\lambda) f(\boldsymbol{z}_2) + \lambda f(\boldsymbol{z}),$$

即得 $f(\boldsymbol{z}) < f(\boldsymbol{z}_2)$. 再设 $\gamma = \dfrac{t-\lambda}{t(1-\lambda)}$, 由 $0 < \lambda < t < 1$ 得 $0 < \gamma < 1$, 则

$$\boldsymbol{z}_2 = \frac{1}{1-\lambda}\boldsymbol{z} - \frac{\lambda}{1-\lambda}\boldsymbol{y} = \frac{1}{1-\lambda}\boldsymbol{z} - \frac{\lambda}{1-\lambda}\left(\frac{\boldsymbol{z}}{t} - \frac{(1-t)\boldsymbol{x}}{t}\right) = \gamma\boldsymbol{z} + (1-\gamma)\boldsymbol{x}.$$

再由 $f(\boldsymbol{z}) > f(\boldsymbol{x}) = f(\boldsymbol{y})$ 和半严格凸函数的定义得

$$f(\boldsymbol{z}_2) \leqslant (1-\gamma) f(\boldsymbol{x}) + \gamma f(\boldsymbol{z}) < f(\boldsymbol{z}).$$

因此, 这与 $f(\boldsymbol{z}) < f(\boldsymbol{z}_2)$ 矛盾. □

定理 4.7.5 说明若半严格凸函数满足一点凸性, 则它是凸函数. 反过来, 下面证明如果凸函数满足一点半严格凸性, 则它是半严格凸函数.

定理 4.7.6 设非空开凸集 $C \subset \mathbf{R}^n$, $f: C \to \mathbf{R}$ 在 C 上是凸函数. 如果 f 满足关于 λ 的一点半严格凸性, 则 f 在 C 上是半严格凸函数.

证明 根据假设 $\exists\lambda \in (0,1)$, 对任意 $\boldsymbol{x}, \boldsymbol{y} \in C$ 和 $f(\boldsymbol{x}) \neq f(\boldsymbol{y})$, 有

$$f[(1-\lambda)\boldsymbol{x} + \lambda\boldsymbol{y}] < (1-\lambda) f(\boldsymbol{x}) + \lambda f(\boldsymbol{y}). \tag{4.7.2-5}$$

假设 f 在 C 上不是半严格凸函数, 即 $\exists t \in (0,1)$, $\boldsymbol{x}, \boldsymbol{y} \in C$ 和 $f(\boldsymbol{x}) \neq f(\boldsymbol{y})$ 有

$$f[(1-t)\boldsymbol{x} + t\boldsymbol{y}] \geqslant (1-t) f(\boldsymbol{x}) + t f(\boldsymbol{y}). \tag{4.7.2-6}$$

不妨设 $f(\boldsymbol{x}) > f(\boldsymbol{y})$, 记 $\boldsymbol{z} = (1-t)\boldsymbol{x} + t\boldsymbol{y}$, 上式即为 $f(\boldsymbol{z}) \geqslant (1-t) f(\boldsymbol{x}) + t f(\boldsymbol{y}) > f(\boldsymbol{y})$. 由 f 是 C 上的凸函数和 $f(\boldsymbol{z}) \leqslant (1-t) f(\boldsymbol{x}) + t f(\boldsymbol{y})$ 可得 $f(\boldsymbol{z}) = (1-t) f(\boldsymbol{x}) + t f(\boldsymbol{y})$. 再由定理假设, 得

$$f[(1-\lambda)\boldsymbol{z} + \lambda\boldsymbol{y}] < (1-\lambda) f(\boldsymbol{z}) + \lambda f(\boldsymbol{y}) < f(\boldsymbol{z}),$$

$$f\left[\left(1-\lambda^{2}\right)\boldsymbol{z}+\lambda^{2}\boldsymbol{y}\right]=f\left[\lambda\left(\left(1-\lambda\right)\boldsymbol{z}+\lambda\boldsymbol{y}\right)+\left(1-\lambda\right)\boldsymbol{z}\right]$$
$$<\lambda f\left(\left(1-\lambda\right)\boldsymbol{z}+\lambda\boldsymbol{y}\right)+\left(1-\lambda\right)f\left(\boldsymbol{z}\right)$$
$$<\lambda^{2}f\left(\boldsymbol{y}\right)+\left(1-\lambda^{2}\right)f\left(\boldsymbol{z}\right)<f\left(\boldsymbol{z}\right),$$

......

$$f\left[\left(1-\lambda^{k}\right)\boldsymbol{z}+\lambda^{k}\boldsymbol{y}\right]<\lambda^{k}f\left(\boldsymbol{y}\right)+\left(1-\lambda^{k}\right)f\left(\boldsymbol{z}\right)<f\left(\boldsymbol{z}\right),\quad k=1,2,\cdots. \tag{4.7.2-7}$$

再由 $\boldsymbol{z}=(1-t)\boldsymbol{x}+t\boldsymbol{y}$, 得

$$\left(1-\lambda^{k}\right)\boldsymbol{z}+\lambda^{k}\boldsymbol{y}=\left(1-\lambda^{k}\right)\left[(1-t)\boldsymbol{x}+t\boldsymbol{y}\right]+\lambda^{k}\boldsymbol{y}$$
$$=\left(1-t-\lambda^{k}+\lambda^{k}t\right)\boldsymbol{x}+\left(t-\lambda^{k}t+\lambda^{k}\right)\boldsymbol{y}.$$

选取充分大得 k_1 使得 $\dfrac{\lambda^{k_1}}{1-\lambda}<\dfrac{t}{1-t}$, 记

$$\beta_1=t-t\lambda^{k_1}+\lambda^{k_1},\quad \beta_2=t-\frac{\lambda^{k_1+1}}{1-\lambda}+\frac{t\lambda^{k_1+1}}{1-\lambda},$$
$$\bar{\boldsymbol{x}}=(1-\beta_1)\boldsymbol{x}+\beta_1\boldsymbol{y},\quad \bar{\boldsymbol{y}}=(1-\beta_2)\boldsymbol{x}+\beta_2\boldsymbol{y}.$$

则有

$$\left(1-\lambda^{k_1}\right)\boldsymbol{z}+\lambda^{k_1}\boldsymbol{y}=(1-\beta_1)\boldsymbol{x}+\beta_1\boldsymbol{y}=\bar{\boldsymbol{x}}.$$

由上式和 (4.7.2-7) 式可得

$$f\left(\bar{\boldsymbol{x}}\right)=f\left(\left(1-\lambda^{k_1}\right)\boldsymbol{z}+\lambda^{k_1}\boldsymbol{y}\right)<f\left(\boldsymbol{z}\right). \tag{4.7.2-8}$$

下面再分两种情形讨论.

(i) 若 $f(\bar{\boldsymbol{x}})\geqslant f(\bar{\boldsymbol{y}})$, 则由 $\boldsymbol{z}=(1-t)\boldsymbol{x}+t\boldsymbol{y}=(1-\lambda)\bar{\boldsymbol{y}}+\lambda\bar{\boldsymbol{x}}$ 和 f 是 C 上凸函数有

$$f\left(\boldsymbol{z}\right)\leqslant(1-\lambda)f\left(\bar{\boldsymbol{y}}\right)+\lambda f\left(\bar{\boldsymbol{x}}\right)\leqslant f\left(\bar{\boldsymbol{x}}\right),$$

与 (4.7.2-8) 式矛盾.

(ii) 若 $f(\bar{\boldsymbol{x}})<f(\bar{\boldsymbol{y}})$, 则由 $\boldsymbol{z}=(1-t)\boldsymbol{x}+t\boldsymbol{y}=(1-\lambda)\bar{\boldsymbol{y}}+\lambda\bar{\boldsymbol{x}}$ 和条件 (4.7.2-5) 式得

$$f\left(\boldsymbol{z}\right)<(1-\lambda)f\left(\bar{\boldsymbol{y}}\right)+\lambda f\left(\bar{\boldsymbol{x}}\right). \tag{4.7.2-9}$$

再由 f 是 C 上的凸函数有

$$f\left(\bar{\boldsymbol{x}}\right)\leqslant(1-\beta_1)f\left(\boldsymbol{x}\right)+\beta_1 f\left(\boldsymbol{y}\right),\quad f\left(\bar{\boldsymbol{y}}\right)\leqslant(1-\beta_2)f\left(\boldsymbol{x}\right)+\beta_2 f\left(\boldsymbol{y}\right). \tag{4.7.2-10}$$

于是, 由 (4.7.2-9) 和 (4.7.2-10) 式得

$$
\begin{aligned}
f(\boldsymbol{z}) &< (1-\lambda) f(\bar{\boldsymbol{y}}) + \lambda f(\bar{\boldsymbol{x}}) \\
&\leqslant [(1-\lambda)(1-\beta_2) + \lambda(1-\beta_1)] f(\boldsymbol{x}) + [(1-\lambda)\beta_2 + \lambda\beta_1] f(\boldsymbol{y}) \\
&= (1-t) f(\boldsymbol{x}) + t f(\boldsymbol{y}).
\end{aligned}
$$

与 (4.7.2-6) 式中 $f(\boldsymbol{z}) \geqslant (1-t) f(\boldsymbol{x}) + t f(\boldsymbol{y})$ 矛盾. □

推论 4.7.1　设非空闭凸集 $C \subset \mathbf{R}^n$, $f: C \to \mathbf{R}$ 在 C 上是下半连续函数. 如果 f 满足关于 λ 的一点半严格凸性, 则 f 在 C 上是半严格凸函数.

证明　由定理 4.3.3 得 f 在 C 上是凸函数, 再由定理 4.7.6 得 f 在 C 上是半严格凸函数. □

推论 4.7.2　设非空闭凸集 $C \subset \mathbf{R}^n$, $f: C \to \mathbf{R}$ 在 C 上是下半连续函数. 如果 f 满足关于 λ 的一点严格凸性, 则 f 是在 C 上的半严格凸函数.

证明　由推论 4.3.1 知 f 在 C 上是凸函数, 再由定理 4.7.6 得 f 在 C 上是半严格凸函数. □

进一步证明, 如果下半连续半严格凸函数满足一点严格凸性, 则它是严格凸函数.

定理 4.7.7　设非空开凸集 $C \subset \mathbf{R}^n$, $f: C \to \mathbf{R}$ 是在 C 上的半严格凸函数. 如果 f 满足关于 λ 的一点严格凸性, 则 f 在 C 上是严格凸函数.

证明　根据假设 $\exists\lambda \in (0,1)$, 对任意的 $\boldsymbol{x}, \boldsymbol{y} \in C$ 和 $\boldsymbol{x} \neq \boldsymbol{y}$ 有

$$
f[(1-\lambda)\boldsymbol{x} + \lambda\boldsymbol{y}] < (1-\lambda) f(\boldsymbol{x}) + \lambda f(\boldsymbol{y}). \tag{4.7.2-11}
$$

因函数 f 在 C 上是半严格凸函数, 只需证明当 $f(\boldsymbol{x}) = f(\boldsymbol{y})$ 和 $\boldsymbol{x} \neq \boldsymbol{y}$ 时有

$$
f[(1-t)\boldsymbol{x} + t\boldsymbol{y}] < (1-t) f(\boldsymbol{x}) + t f(\boldsymbol{y}), \quad \forall t \in (0,1).
$$

任意的 $\boldsymbol{x}, \boldsymbol{y} \in C$ 满足 $\boldsymbol{x} \neq \boldsymbol{y}$ 和 $f(\boldsymbol{x}) = f(\boldsymbol{y})$, 由条件 (4.7.2-11) 式有

$$
f[(1-\lambda)\boldsymbol{x} + \lambda\boldsymbol{y}] < (1-\lambda) f(\boldsymbol{x}) + \lambda f(\boldsymbol{y}) = f(\boldsymbol{x}) = f(\boldsymbol{y}).
$$

记 $(1-\lambda)\boldsymbol{x} + \lambda\boldsymbol{y} = \bar{\boldsymbol{x}}$, 则 $f(\bar{\boldsymbol{x}}) < f(\boldsymbol{x}) = f(\boldsymbol{y})$.

设 $\forall t \in (0,1)$, 如果 $t < \lambda$, 则 $\exists\mu \in (0,1)$, 满足 $(1-t)\boldsymbol{x} + t\boldsymbol{y} = \mu\boldsymbol{y} + (1-\mu)\bar{\boldsymbol{x}}$. 再由 f 在 C 上是半严格凸函数, 有

$$
f((1-t)\boldsymbol{x} + t\boldsymbol{y}) = f(\mu\boldsymbol{y} + (1-\mu)\bar{\boldsymbol{x}}) < \mu f(\boldsymbol{y}) + (1-\mu) f(\bar{\boldsymbol{x}}) < f(\boldsymbol{y}).
$$

设 $\forall t \in (0,1)$, 如果 $t > \lambda$, 则 $\exists\nu \in (0,1)$, 满足 $(1-t)\boldsymbol{x} + t\boldsymbol{y} = (1-\nu)\boldsymbol{x} + \nu\bar{\boldsymbol{x}}$, 再由 f 在 C 上是半严格凸函数, 有

$$f\left((1-t)\boldsymbol{x}+t\boldsymbol{y}\right)=f\left((1-\nu)\boldsymbol{x}+\nu\bar{\boldsymbol{x}}\right)<(1-\nu)f(\boldsymbol{x})+\nu f(\bar{\boldsymbol{x}})<f(\boldsymbol{x}). \quad \square$$

本小节得到了有关半严格凸函数的判定性质:

1) 半严格凸函数等价于条件 (4.7.2-2) 式成立;

2) 满足一点半严格凸性的下半连续函数是半严格凸函数;

3) 满足一点严格凸性的下半连续函数是半严格凸函数.

4.7.3 显凸函数

本小节定义了显凸函数, 同时具有了凸函数和半严格凸函数的性质.

定义 4.7.3 设非空开凸集 $C\subset\mathbf{R}^n$, 如果 $f:C\to\mathbf{R}$ 在 C 上是半严格凸函数且也是凸函数, 则称 f 在 C 上是显凸函数. 如果 $-f$ 在 C 上是显凸函数, 则称 f 在 C 上是显凹函数.

通过定义我们知道凸函数、严格凸函数、半严格凸函数和显凸函数具有以下关系 (见图 4.7-1):

1) 严格凸函数是凸函数, 显凸函数是半严格凸函数和凸函数, 但反过来都不一定成立.

2) 凸函数满足一点严格凸性是严格凸函数, 严格凸函数满足一点半严格凸性是半严格凸函数, 半严格凸函数满足一点凸性是凸函数.

3) 凸函数满足一点半严格凸性是半严格凸函数, 半严格凸函数满足一点严格凸性是严格凸函数.

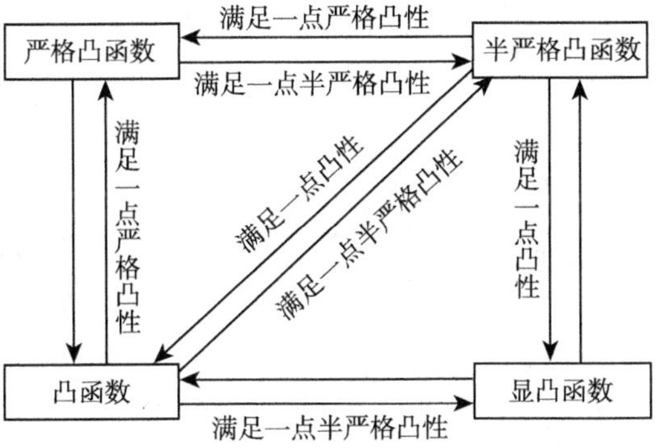

图 4.7-1 4 种凸函数关系

同时, 上图还表明, 满足一点凸性的半严格凸函数是显凸函数, 满足一点半严格凸性的凸函数是显凸函数.

显凸函数是凸函数和半严格凸函数, 因此, 它具有比定理 4.7.1 更强的条件, 见定理 4.7.8.

定理 4.7.8 设非空开凸集 $C\subset \mathbf{R}^n$, 函数 $f:C\to \mathbf{R}$, 则 f 在 C 上是显凸函数的充要条件是对任意 $\boldsymbol{x}_i,\boldsymbol{y}_i\in C$, 使得

$f(\boldsymbol{x}_1)\leqslant\cdots\leqslant f(\boldsymbol{x}_k)<f(\boldsymbol{x}_{k+1})\leqslant\cdots\leqslant f(\boldsymbol{x}_m)\,(m\geqslant 2)$ 和 $\lambda_i>0\,(i=1,\cdots,m)$,

有

$$f(\lambda_1\boldsymbol{x}_1+\cdots+\lambda_m\boldsymbol{x}_m)<\lambda_1 f(\boldsymbol{x}_1)+\cdots+\lambda_m f(\boldsymbol{x}_m), \tag{4.7.3-1}$$

$$f(\lambda_1\boldsymbol{y}_1+\cdots+\lambda_m\boldsymbol{y}_m)\leqslant\lambda_1 f(\boldsymbol{y}_1)+\cdots+\lambda_m f(\boldsymbol{y}_m) \tag{4.7.3-2}$$

成立, 其中 $\sum\limits_{i=1}^{m}\lambda_i=1$.

证明 充分性是显然的, 因为当 $m=2$ 时, 根据 (4.7.3-1) 和 (4.7.3-2) 式知 f 在 C 上是显凸函数.

下证必要性. 反过来, 设 f 在 C 上是显凸函数, 则 f 在 C 上是凸函数, 由定理 4.1.1 知 (4.7.3-2) 式成立. 若

$$f(\boldsymbol{x}_1)\leqslant\cdots\leqslant f(\boldsymbol{x}_k)<f(\boldsymbol{x}_{k+1})\leqslant\cdots\leqslant f(\boldsymbol{x}_m)\quad(m\geqslant 2),$$

当 $k=1$ 时由定理 4.7.4 知 (4.7.3-1) 式成立.

下面设 $k\geqslant 2$, 令

$$t=\sum_{i=1}^{k-1}\lambda_i,\quad \beta_i=\frac{\lambda_i}{t}\,(1\leqslant i\leqslant k-1);\quad \beta_i=\frac{\lambda_i}{1-t}\,(k\leqslant i\leqslant m),$$

有

$$f(\lambda_1\boldsymbol{x}_1+\cdots+\lambda_m\boldsymbol{x}_m)=f\left(t\sum_{i=1}^{k-1}\beta_i\boldsymbol{x}_i+(1-t)\sum_{i=k}^{m}\beta_i\boldsymbol{x}_i\right). \tag{4.7.3-3}$$

如果 $f\left(\sum\limits_{i=1}^{k-1}\beta_i\boldsymbol{x}_i\right)<f\left(\sum\limits_{i=k}^{m}\beta_i\boldsymbol{x}_i\right)$, 则由 (4.7.3-3) 式和 f 在 C 上是半严格凸函数, 有

$$\begin{aligned}f(\lambda_1\boldsymbol{x}_1+\cdots+\lambda_m\boldsymbol{x}_m)&<tf\left(\sum_{i=1}^{k-1}\beta_i\boldsymbol{x}_i\right)+(1-t)f\left(\sum_{i=k}^{m}\beta_i\boldsymbol{x}_i\right)\\&\leqslant t\sum_{i=1}^{k-1}\beta_i f(\boldsymbol{x}_i)+(1-t)\sum_{i=k}^{m}\beta_i f(\boldsymbol{x}_i)=\sum_{i=1}^{m}\lambda_i f(\boldsymbol{x}_i);\end{aligned}$$

如果 $f\left(\sum\limits_{i=1}^{k-1}\beta_i\boldsymbol{x}_i\right)\geqslant f\left(\sum\limits_{i=k}^{m}\beta_i\boldsymbol{x}_i\right)$, 则由 (4.7.3-3) 式和定理 4.7.4 得

$$\begin{aligned}f(\lambda_1\boldsymbol{x}_1+\cdots+\lambda_m\boldsymbol{x}_m)&\leqslant tf\left(\sum_{i=1}^{k-1}\beta_i\boldsymbol{x}_i\right)+(1-t)f\left(\sum_{i=k}^{m}\beta_i\boldsymbol{x}_i\right)\\&<t\sum_{i=1}^{k-1}\beta_i f(\boldsymbol{x}_i)+(1-t)\sum_{i=k}^{m}\beta_i f(\boldsymbol{x}_i)=\sum_{i=1}^{m}\lambda_i f(\boldsymbol{x}_i).\end{aligned}$$

上述过程表明 (4.7.3-1) 式成立. □

定理 4.7.9 设非空开凸集 $C\subset\mathbf{R}^n$, 如果 $f_i:C\to\mathbf{R}\,(i\in I)$ 在 C 上是显凸函数族, 其中 I 是指标集, 则 $f(\boldsymbol{x})=\max\{f_i(\boldsymbol{x})\,|i\in I\}$ 在 C 上是显凸函数.

证明 因为所有的 $f_i\,(i\in I)$ 在 C 上是凸函数, 那么由性质 4.2.12 知 f 在 C 上是凸函数. 下面只需要证明 f 在 C 上是半严格凸函数即可. 否则, $\exists\boldsymbol{x},\boldsymbol{y}\in C$, $f(\boldsymbol{x})\neq f(\boldsymbol{y})$ 和 $\lambda\in(0,1)$ 有

$$f[(1-\lambda)\boldsymbol{x}+\lambda\boldsymbol{y}]\geqslant(1-\lambda)f(\boldsymbol{x})+\lambda f(\boldsymbol{y}).$$

由于 f 是 C 上的凸函数, 有 $f[(1-\lambda)\boldsymbol{x}+\lambda\boldsymbol{y}]\leqslant(1-\lambda)f(\boldsymbol{x})+\lambda f(\boldsymbol{y})$, 得

$$f[(1-\lambda)\boldsymbol{x}+\lambda\boldsymbol{y}]=(1-\lambda)f(\boldsymbol{x})+\lambda f(\boldsymbol{y}).\tag{4.7.3-4}$$

令 $\bar{\boldsymbol{x}}=(1-\lambda)\boldsymbol{x}+\lambda\boldsymbol{y}$, 由 $f(\boldsymbol{x})=\max\{f_i(\boldsymbol{x})\,|i\in I\}$ 知, $\exists i,j,k\in I$ 使得

$$f(\bar{\boldsymbol{x}})=f_i(\bar{\boldsymbol{x}}),\quad f(\boldsymbol{x})=f_j(\boldsymbol{x}),\quad f(\boldsymbol{y})=f_k(\boldsymbol{y}).$$

由 (4.7.3-4) 式得

$$f_i(\bar{\boldsymbol{x}})=(1-\lambda)f_j(\boldsymbol{x})+\lambda f_k(\boldsymbol{y}).\tag{4.7.3-5}$$

如果 $f_i(\boldsymbol{x})\neq f_i(\boldsymbol{y})$, 由 f_i 是半严格凸函数, 有

$$f_i(\bar{\boldsymbol{x}})<(1-\lambda)f_i(\boldsymbol{x})+\lambda f_i(\boldsymbol{y}).$$

因为 $f_i(\boldsymbol{x})\leqslant f(\boldsymbol{x})=f_j(\boldsymbol{x})$, $f_i(\boldsymbol{y})\leqslant f(\boldsymbol{y})=f_k(\boldsymbol{y})$, 所以根据上式得

$$f_i(\bar{\boldsymbol{x}})<(1-\lambda)f_j(\boldsymbol{x})+\lambda f_k(\boldsymbol{y}),$$

与 (4.7.3-5) 式矛盾.

如果 $f_i(\boldsymbol{x})=f_i(\boldsymbol{y})$, 由 f_i 是凸函数, 有

$$f_i(\bar{\boldsymbol{x}})\leqslant(1-\lambda)f_i(\boldsymbol{x})+\lambda f_i(\boldsymbol{y})=f_i(\boldsymbol{x})=f_i(\boldsymbol{y}).$$

因为 $f(\boldsymbol{x}) \neq f(\boldsymbol{y})$, $f_i(\boldsymbol{x}) \leqslant f(\boldsymbol{x}) = f_j(\boldsymbol{x})$ 和 $f_i(\boldsymbol{y}) \leqslant f(\boldsymbol{y}) = f_k(\boldsymbol{y})$ 中至少有一个是严格不等式，所以根据上式得

$$f_i(\bar{\boldsymbol{x}}) < (1-\lambda) f(\boldsymbol{x}) + \lambda f(\boldsymbol{y}),$$

与 (4.7.3-5) 式矛盾. □

例 4.7.6 考虑下面两个半严格凸函数:

$$f_1(x) = \begin{cases} 1, & x = 0, \\ 0, & x \neq 0, \end{cases} \quad f_2(x) = \begin{cases} 1, & x = 1, \\ 0, & x \neq 1. \end{cases}$$

但是, 它们的上确界函数 $\sup\{f_1(x), f_2(x)\} = \begin{cases} 1, & x = 0, 1, \\ 0, & x \neq 0, 1 \end{cases}$ 不是半严格凸函数. 这个例子说明, 显凸函数族的上确界函数是显凸函数, 但是对于半严格凸函数不成立.

本小节得到了显凸函数的两个性质:

1) 显凸函数的充要条件是 (4.7.1-1) 和 (4.7.2-2) 式成立;

2) 显凸函数族的上确界函数是显凸函数.

4.8 强凸函数

强凸函数是比凸函数具有更强性质的一种函数 [48]. 当强凸函数具有二阶可微性时, 对应的 Hessian 矩阵是正定的.

定义 4.8.1 设非空开凸集 $C \subset \mathbf{R}^n$, 函数 $f: C \to \mathbf{R}$. 如果 $\exists t > 0$, 对任意 $\boldsymbol{x}, \boldsymbol{y} \in C$ 和任意 $\lambda \in (0,1)$ 有

$$f[(1-\lambda)\boldsymbol{x} + \lambda \boldsymbol{y}] \leqslant (1-\lambda) f(\boldsymbol{x}) + \lambda f(\boldsymbol{y}) - \frac{1}{2}\lambda(1-\lambda) t\|\boldsymbol{x} - \boldsymbol{y}\|^2, \quad (4.8.1\text{-}1)$$

则称 f 在 C 上是关于 t 的强凸函数. 如果 $-f$ 在 C 上是关于 t 的强凸函数, 则称 f 在 C 上是关于 t 的强凹函数.

容易举例一个凸函数不是强凸函数, 例如线性函数. 显然, 强凸函数是严格凸函数.

引理 4.8.1 设非空凸集 $C \subset \mathbf{R}^n$, 函数 $f: C \to \mathbf{R}$, 则 f 是关于 t 的强凸函数当且仅当 $g(\boldsymbol{x}) := f(\boldsymbol{x}) - \frac{1}{2}t\|\boldsymbol{x}\|^2$ 在 C 上是凸函数.

证明 若 f 在 C 上是关于 t 的强凸函数, 对任意 $\boldsymbol{x},\boldsymbol{y}\in C$, 任意 $\lambda\in(0,1)$, 有 (4.8.1-1) 式成立. 因此有下面式子成立:

$$\begin{aligned}
&g\left[(1-\lambda)\boldsymbol{x}+\lambda\boldsymbol{y}\right]\\
&=f\left[(1-\lambda)\boldsymbol{x}+\lambda\boldsymbol{y}\right]-\frac{1}{2}t\|(1-\lambda)\boldsymbol{x}+\lambda\boldsymbol{y}\|^2\\
&\leqslant(1-\lambda)f(\boldsymbol{x})+\lambda f(\boldsymbol{y})-\frac{1}{2}\lambda(1-\lambda)t\|\boldsymbol{x}-\boldsymbol{y}\|^2-\frac{1}{2}t\|(1-\lambda)\boldsymbol{x}+\lambda\boldsymbol{y}\|^2\\
&=(1-\lambda)g(\boldsymbol{x})+\lambda g(\boldsymbol{y})+\frac{1}{2}t(1-\lambda)\|\boldsymbol{x}\|^2+\frac{1}{2}t\lambda\|\boldsymbol{y}\|^2\\
&\quad-\frac{1}{2}\lambda(1-\lambda)t\left(\|\boldsymbol{x}\|^2-2\boldsymbol{x}^{\mathrm{T}}\boldsymbol{y}+\|\boldsymbol{y}\|^2\right)\\
&\quad-\frac{1}{2}t((1-\lambda)^2\|\boldsymbol{x}\|^2+2(1-\lambda)\lambda\boldsymbol{x}^{\mathrm{T}}\boldsymbol{y}+\lambda^2\|\boldsymbol{y}\|^2)\|\\
&=(1-\lambda)g(\boldsymbol{x})+\lambda g(\boldsymbol{y}),
\end{aligned}$$

上式说明 g 在 C 上是凸函数. 反之, 若 g 在 C 上是凸函数, 且有下面式子成立:

$$\begin{aligned}
f\left[(1-\lambda)\boldsymbol{x}+\lambda\boldsymbol{y}\right]&=g\left[(1-\lambda)\boldsymbol{x}+\lambda\boldsymbol{y}\right]+\frac{1}{2}t\|(1-\lambda)\boldsymbol{x}+\lambda\boldsymbol{y}\|^2\\
&\leqslant(1-\lambda)g(\boldsymbol{x})+\lambda g(\boldsymbol{y})+\frac{1}{2}t\|(1-\lambda)\boldsymbol{x}+\lambda\boldsymbol{y}\|^2\\
&=(1-\lambda)f(\boldsymbol{x})+\lambda f(\boldsymbol{y})-\frac{1}{2}t(1-\lambda)\|\boldsymbol{x}\|^2-\frac{1}{2}t\lambda\|\boldsymbol{y}\|^2\\
&\quad+\frac{1}{2}t\left((1-\lambda)^2\|\boldsymbol{x}\|^2+2(1-\lambda)\lambda\boldsymbol{x}^{\mathrm{T}}\boldsymbol{y}+\lambda^2\|\boldsymbol{y}\|^2\right)\\
&=(1-\lambda)f(\boldsymbol{x})+\lambda f(\boldsymbol{y})-\frac{1}{2}t(1-\lambda)\lambda\|\boldsymbol{x}-\boldsymbol{y}\|^2.
\end{aligned}$$

则 f 在 C 上是关于 t 的强凸函数. □

引理 4.8.1 说明了任何一个凸函数加上 $\frac{1}{2}t\|\boldsymbol{x}\|^2$ 构成一个强凸函数.

例 4.8.1 设线性函数 $f(\boldsymbol{x})=\boldsymbol{a}^{\mathrm{T}}\boldsymbol{x}+b$, 其中给定 $\boldsymbol{a}\in\mathbf{R}^n$ 和 $b\in\mathbf{R}$. 则函数

$$f(\boldsymbol{x})=\frac{1}{2}\|\boldsymbol{x}\|^2+\boldsymbol{a}^{\mathrm{T}}\boldsymbol{x}+b$$

是强凸函数.

定义 4.8.2 设非空凸集 $C\subset\mathbf{R}^n$, 函数 $f:C\to\mathbf{R}$. 如果存在函数 $h:C\to\mathbf{R}$ 满足条件:

$$h(\boldsymbol{x}_0)=f(\boldsymbol{x}_0),\boldsymbol{x}_0\in C;\quad h(\boldsymbol{x})\leqslant f(\boldsymbol{x}),\forall\boldsymbol{x}\in C,$$

则称 h 是 f 在 C 上的支持函数.

任何一个函数都存在无数个支持函数. 我们有下面的结论, 任何一个强凸函数都存在一个点上的二次强凸函数是该函数的支持函数, 利用这个结论可构造 SQP 算法.

定理 4.8.1 设非空开凸集 $C \subset \mathbf{R}^n$, 函数 $f : C \to \mathbf{R}$, 则 f 是关于 $t > 0$ 的强凸函数当且仅当函数对于每一个 $\boldsymbol{x}_0 \in C$, $\exists \boldsymbol{a} \in \mathbf{R}^n$ 使得 f 在 C 上的支持函数为

$$h(\boldsymbol{x}) = f(\boldsymbol{x}_0) + \frac{1}{2}t\left(\|\boldsymbol{x}\|^2 - \|\boldsymbol{x}_0\|^2\right) + \langle \boldsymbol{a}, \boldsymbol{x} - \boldsymbol{x}_0 \rangle, \quad \forall \boldsymbol{x} \in C. \tag{4.8.1-2}$$

证明 设 f 在 C 上是关于 t 的强凸函数, 那么由引理 4.8.1 知函数

$$g(\boldsymbol{x}) := f(\boldsymbol{x}) - \frac{1}{2}t\|\boldsymbol{x}\|^2$$

在 C 上是凸函数. 对于给定的 $\boldsymbol{x}_0 \in C$, 根据性质 4.4.1 知 g 的共轭函数 g^* 是闭凸函数, 并且

$$\langle \boldsymbol{x}, \boldsymbol{x}^* \rangle - g^*(\boldsymbol{x}^*) \leqslant g(\boldsymbol{x}), \quad \forall \boldsymbol{x}, \boldsymbol{x}^* \in \mathbf{R}^n.$$

设 $h_1(\boldsymbol{x}) = \langle \boldsymbol{x}, \boldsymbol{x}^* \rangle - g^*(\boldsymbol{x}^*)$, $g^*(\boldsymbol{x}^*) = \langle \boldsymbol{x}_0, \boldsymbol{x}^* \rangle - g(\boldsymbol{x}_0)$, 取对应的 $\boldsymbol{a} = \boldsymbol{x}^*$, 那么有

$$h_1(\boldsymbol{x}) = g(\boldsymbol{x}_0) + \langle \boldsymbol{a}, \boldsymbol{x} - \boldsymbol{x}_0 \rangle, \quad \forall \boldsymbol{x} \in C,$$

显然有 $h_1(\boldsymbol{x}_0) = g(\boldsymbol{x}_0)$, $h_1(\boldsymbol{x}) \leqslant g(\boldsymbol{x}), \forall \boldsymbol{x} \in C$, 说明 h_1 是 g 在 C 上的支持函数. 设

$$h(\boldsymbol{x}) = f(\boldsymbol{x}_0) + \frac{1}{2}t\left(\|\boldsymbol{x}\|^2 - \|\boldsymbol{x}_0\|^2\right) + \langle \boldsymbol{a}, \boldsymbol{x} - \boldsymbol{x}_0 \rangle, \quad \forall \boldsymbol{x} \in C.$$

有 $h(\boldsymbol{x}_0) = f(\boldsymbol{x}_0)$, 且

$$h(\boldsymbol{x}) = h_1(\boldsymbol{x}) + \frac{1}{2}t\|\boldsymbol{x}\|^2 \leqslant g(\boldsymbol{x}) + \frac{1}{2}t\|\boldsymbol{x}\|^2 = f(\boldsymbol{x}),$$

则 h 是函数 f 在 C 上的支持函数.

反过来, 对于每一个 $\boldsymbol{x}_0 \in C$, 都存在 $\boldsymbol{a} \in \mathbf{R}^n$ 使得 f 在 C 上的支持函数为 (4.8.1-2) 式. 设对于任意 $\boldsymbol{x}, \boldsymbol{y} \in C$, 任意 $\lambda \in (0,1)$, 不妨取 $\boldsymbol{x}_0 = (1-\lambda)\boldsymbol{x} + \lambda \boldsymbol{y}$ 使得

$$h(\boldsymbol{z}) = f(\boldsymbol{x}_0) + \frac{1}{2}t\left(\|\boldsymbol{z}\|^2 - \|\boldsymbol{x}_0\|^2\right) + \langle \boldsymbol{a}, \boldsymbol{z} - \boldsymbol{x}_0 \rangle, \quad \forall \boldsymbol{z} \in C.$$

因此, 上式中取 $\boldsymbol{z}=\boldsymbol{x}$ 或 $\boldsymbol{z}=\boldsymbol{y}$ 得

$$f(\boldsymbol{x}_0)+\frac{1}{2}t\left(\|\boldsymbol{x}\|^2-\|\boldsymbol{x}_0\|^2\right)+\langle\boldsymbol{a},\boldsymbol{x}-\boldsymbol{x}_0\rangle\leqslant f(\boldsymbol{x}),$$

$$f(\boldsymbol{x}_0)+\frac{1}{2}t\left(\|\boldsymbol{y}\|^2-\|\boldsymbol{x}_0\|^2\right)+\langle\boldsymbol{a},\boldsymbol{y}-\boldsymbol{x}_0\rangle\leqslant f(\boldsymbol{y}).$$

对上述两个不等式两端分别乘 $1-\lambda$ 和 λ, 然后相加得

$$f(\boldsymbol{x}_0)+\frac{1}{2}(1-\lambda)t\left(\|\boldsymbol{x}\|^2-\|\boldsymbol{x}_0\|^2\right)+\frac{1}{2}\lambda t\left(\|\boldsymbol{y}\|^2-\|\boldsymbol{x}_0\|^2\right)\leqslant(1-\lambda)f(\boldsymbol{x})+\lambda f(\boldsymbol{y}).$$

由于 $\frac{1}{2}(1-\lambda)t(\|\boldsymbol{x}\|^2-\|\boldsymbol{x}_0\|^2)+\frac{1}{2}\lambda t(\|\boldsymbol{y}\|^2-\|\boldsymbol{x}_0\|^2)-\frac{1}{2}(1-\lambda)\lambda t\|\boldsymbol{x}-\boldsymbol{y}\|^2=0$, 得

$$f[(1-\lambda)\boldsymbol{x}+\lambda\boldsymbol{y}]\leqslant(1-\lambda)f(\boldsymbol{x})+\lambda f(\boldsymbol{y})-\frac{1}{2}t(1-\lambda)\lambda\|\boldsymbol{x}-\boldsymbol{y}\|^2.$$ □

注 (4.8.1-2) 式也可以写为

$$h(\boldsymbol{x})=f(\boldsymbol{x}_0)+\frac{1}{2}t\|\boldsymbol{x}-\boldsymbol{x}_0\|^2+\langle\boldsymbol{a}+t\boldsymbol{x}_0,\boldsymbol{x}-\boldsymbol{x}_0\rangle-\langle t\boldsymbol{x}_0,\boldsymbol{x}_0\rangle,\quad\forall\boldsymbol{x}\in C.$$

上式表明函数 $f:C\to\mathbf{R}$ 是关于 $t>0$ 的强凸函数, 那么对于 $\boldsymbol{x}_0\in C$, 存在一个二次函数 h 是 f 在 C 上的支持函数, 有 $\min\limits_{\boldsymbol{x}\in C}h(\boldsymbol{x})\leqslant\min\limits_{\boldsymbol{x}\in C}f(\boldsymbol{x})$. 这是设计邻近点算法具有二阶收敛性的理论依据之一.

凸函数的若干变量的凸组合不大于对应函数值的凸组合. 进一步, 强凸函数也具有这个类似的性质.

定理 4.8.2 设非空开凸集 $C\subset\mathbf{R}^n$, $f:C\to\mathbf{R}$ 是关于 t 的强凸函数, 则

$$f\left(\sum_{i=1}^{n}\lambda_i\boldsymbol{x}_i\right)\leqslant\sum_{i=1}^{n}\lambda_i f(\boldsymbol{x}_i)-t\sum_{i=1}^{n}\lambda_i\|\boldsymbol{x}_i-\bar{\boldsymbol{x}}\|^2,\tag{4.8.1-3}$$

其中 $\boldsymbol{x}_i\in C$, $\lambda_i\in(0,1)$, $i=1,2,\cdots,n$, $\sum\limits_{i=1}^{n}\lambda_i=1$, $\bar{\boldsymbol{x}}=\sum\limits_{i=1}^{n}\lambda_i\boldsymbol{x}_i$.

证明 设 $\boldsymbol{x}_i\in C$, $\lambda_i\in(0,1)$, $i=1,2,\cdots,n$, $\sum\limits_{i=1}^{n}\lambda_i=1$ 和 $\bar{\boldsymbol{x}}=\sum\limits_{i=1}^{n}\lambda_i\boldsymbol{x}_i$. 存在 $\boldsymbol{a}\in\mathbf{R}^n$ 使得 f 在 C 上的支持函数为

$$h(\boldsymbol{z})=f(\bar{\boldsymbol{x}})+\frac{1}{2}t\left(\|\boldsymbol{z}\|^2-\|\bar{\boldsymbol{x}}\|^2\right)+\langle\boldsymbol{a},\boldsymbol{z}-\bar{\boldsymbol{x}}\rangle,\quad\forall\boldsymbol{z}\in C.$$

因此, 上式取 $\boldsymbol{z}=\boldsymbol{x}_i$, 可得

$$f(\bar{\boldsymbol{x}})+\frac{1}{2}t\left(\|\boldsymbol{x}_i\|^2-\|\bar{\boldsymbol{x}}\|^2\right)+\langle\boldsymbol{a},\boldsymbol{x}_i-\bar{\boldsymbol{x}}\rangle\leqslant f(\boldsymbol{x}_i).$$

对上式两端分别乘 λ_i, 再相加可得

$$f(\bar{\boldsymbol{x}})+\frac{1}{2}t\sum_{i=1}^{n}\lambda_i\left(\|\boldsymbol{x}_i\|^2-\|\bar{\boldsymbol{x}}\|^2\right)\leqslant\sum_{i=1}^{n}\lambda_i f(\boldsymbol{x}_i),$$

由上式得 (4.8.1-3) 式. □

当 $f:C\to\mathbf{R}$ 是可微强凸函数时, 则有下面关于强凸函数与梯度的结论.

定理 4.8.3 设非空开凸集 $C\subset\mathbf{R}^n$, $f:C\to\mathbf{R}$ 是可微函数, 则下面结论等价:

(1) f 是关于 t 的在 C 上的强凸函数;

(2) 对于任意 $\boldsymbol{x},\boldsymbol{y}\in C$, $\exists t>0$ 有

$$f(\boldsymbol{y})-f(\boldsymbol{x})\geqslant\nabla f(\boldsymbol{x})^{\mathrm{T}}(\boldsymbol{y}-\boldsymbol{x})+\frac{1}{2}t\|\boldsymbol{y}-\boldsymbol{x}\|^2; \tag{4.8.1-4}$$

(3) 对于任意 $\boldsymbol{x},\boldsymbol{y}\in C$, $\exists t>0$ 有

$$(\nabla f(\boldsymbol{y})-\nabla f(\boldsymbol{x}))^{\mathrm{T}}(\boldsymbol{y}-\boldsymbol{x})\geqslant t\|\boldsymbol{y}-\boldsymbol{x}\|^2. \tag{4.8.1-5}$$

证明 (1) ⇒ (2): 如果 $f(\boldsymbol{x})$ 在 C 上是强凸函数, 则对任意 $\boldsymbol{x},\boldsymbol{y}\in C$ 和任意 $\lambda\in(0,1)$ 有

$$(1-\lambda)f(\boldsymbol{x})+\lambda f(\boldsymbol{y})\geqslant f[(1-\lambda)\boldsymbol{x}+\lambda\boldsymbol{y}]+\frac{1}{2}\lambda(1-\lambda)t\|\boldsymbol{y}-\boldsymbol{x}\|^2.$$

由上式得到

$$f(\boldsymbol{y})-f(\boldsymbol{x})\geqslant\frac{f[\boldsymbol{x}+\lambda(\boldsymbol{y}-\boldsymbol{x})]-f(\boldsymbol{x})}{\lambda}+\frac{1}{2}(1-\lambda)t\|\boldsymbol{y}-\boldsymbol{x}\|^2.$$

在上式中令 $\lambda\to0$ 取极限, 得

$$f(\boldsymbol{y})-f(\boldsymbol{x})\geqslant\nabla f(\boldsymbol{x})^{\mathrm{T}}(\boldsymbol{y}-\boldsymbol{x})+\frac{1}{2}t\|\boldsymbol{y}-\boldsymbol{x}\|^2. \tag{4.8.1-6}$$

(2) ⇒ (1): 若 (4.8.1-4) 式成立, 设任意 $\boldsymbol{x},\boldsymbol{y}\in C$, $\boldsymbol{z}=(1-\lambda)\boldsymbol{x}+\lambda\boldsymbol{y}\,(0<\lambda<1)$, 则有

$$f(\boldsymbol{x})-f(\boldsymbol{z})\geqslant\nabla f(\boldsymbol{z})^{\mathrm{T}}(\boldsymbol{x}-\boldsymbol{z})+\frac{1}{2}t\lambda^2\|\boldsymbol{y}-\boldsymbol{x}\|^2,$$

$$f(\boldsymbol{y})-f(\boldsymbol{z})\geqslant\nabla f(\boldsymbol{z})^{\mathrm{T}}(\boldsymbol{y}-\boldsymbol{z})+\frac{1}{2}(1-\lambda)^2t\|\boldsymbol{y}-\boldsymbol{x}\|^2.$$

将上两式的两边分别乘 $1-\lambda$ 和 λ, 然后相加得

$$(1-\lambda)f(\boldsymbol{x})+\lambda f(\boldsymbol{y})\geqslant f[(1-\lambda)\boldsymbol{x}+\boldsymbol{y}]+\frac{1}{2}(1-\lambda)\lambda t\|\boldsymbol{y}-\boldsymbol{x}\|^2.$$

即 (4.8.1-1) 式成立.

(2) $\Rightarrow$ (3): 将 (4.8.1-6) 式中 $\boldsymbol{x}$ 和 $\boldsymbol{y}$ 互换, 得

$$f(\boldsymbol{x})-f(\boldsymbol{y})\geqslant\nabla f(\boldsymbol{y})^{\mathrm{T}}(\boldsymbol{x}-\boldsymbol{y})+\frac{1}{2}t\|\boldsymbol{y}-\boldsymbol{x}\|^2,\tag{4.8.1-7}$$

将 (4.8.1-6) 和 (4.8.1-7) 式相加得 (4.8.1-5) 式.

(3) $\Rightarrow$ (2): 对任意 $\lambda\in(0,1)$, 设函数 $\varphi(\lambda)=f(\boldsymbol{x}+\lambda(\boldsymbol{y}-\boldsymbol{x}))$, 则有

$$\varphi'(\lambda)=f'(\boldsymbol{x}+\lambda(\boldsymbol{y}-\boldsymbol{x}))=\nabla(\boldsymbol{x}+\lambda(\boldsymbol{y}-\boldsymbol{x}))^{\mathrm{T}}(\boldsymbol{y}-\boldsymbol{x}).\tag{4.8.1-8}$$

由 (4.8.1-5) 式得

$$(\nabla f(\boldsymbol{x}+\lambda(\boldsymbol{y}-\boldsymbol{x}))-\nabla f(\boldsymbol{x}))^{\mathrm{T}}(\boldsymbol{y}-\boldsymbol{x})\geqslant t\lambda\|\boldsymbol{y}-\boldsymbol{x}\|^2.$$

由上式和 (4.8.1-8) 式得

$$\begin{aligned}f(\boldsymbol{y})-f(\boldsymbol{x})&=\varphi(1)-\varphi(0)=\int_0^1\varphi'(\lambda)\,\mathrm{d}\lambda\\&=\int_0^1\nabla f(\boldsymbol{x}+\lambda(\boldsymbol{y}-\boldsymbol{x}))^{\mathrm{T}}(\boldsymbol{y}-\boldsymbol{x})\,\mathrm{d}\lambda\\&\geqslant\int_0^1\left(\nabla f(\boldsymbol{x})^{\mathrm{T}}(\boldsymbol{y}-\boldsymbol{x})+t\lambda\|\boldsymbol{y}-\boldsymbol{x}\|^2\right)\mathrm{d}\lambda\\&=\nabla f(\boldsymbol{x})^{\mathrm{T}}(\boldsymbol{y}-\boldsymbol{x})+\frac{1}{2}t\|\boldsymbol{y}-\boldsymbol{x}\|^2,\end{aligned}$$

即 (4.8.1-4) 式成立.

因此, 上述各结论互相等价. □

下面的结论说明, 如果函数 $f:C\to\mathbf{R}$ 是二阶可微强凸函数, 那么 f 在 C 上每一点的 Hessian 矩阵都是正定的.

定理 4.8.4 设非空开凸集 $C\subset\mathbf{R}^n$, 二阶可微函数 $f:C\to\mathbf{R}$, 则 f 关于 $t>0$ 在 C 上是强凸函数的充要条件是 $\exists t>0$, 对 $\forall\boldsymbol{x},\boldsymbol{y}\in C$ 有

$$(\boldsymbol{y}-\boldsymbol{x})^{\mathrm{T}}\nabla^2 f(\boldsymbol{x})(\boldsymbol{y}-\boldsymbol{x})\geqslant t\|\boldsymbol{y}-\boldsymbol{x}\|^2.\tag{4.8.1-9}$$

证明　设 f 是在 C 上的强凸函数, 由定理 4.8.3(3) 得, 对于任意 $\boldsymbol{x},\boldsymbol{y}\in C$, 任意 $\lambda\in(0,1)$, $\exists t>0$ 有

$$(\nabla f(\boldsymbol{x}+\lambda(\boldsymbol{y}-\boldsymbol{x}))-\nabla f(\boldsymbol{x}))^{\mathrm{T}}(\boldsymbol{x}+\lambda(\boldsymbol{y}-\boldsymbol{x})-\boldsymbol{x})\geqslant t\|\boldsymbol{x}+\lambda(\boldsymbol{y}-\boldsymbol{x})-\boldsymbol{x}\|^2,$$

即有

$$\left(\frac{\nabla f(\boldsymbol{x}+\lambda(\boldsymbol{y}-\boldsymbol{x}))-\nabla f(\boldsymbol{x})}{\lambda}\right)^{\mathrm{T}}(\boldsymbol{y}-\boldsymbol{x})\geqslant t\|\boldsymbol{y}-\boldsymbol{x}\|^2.$$

令 $\lambda\to0$ 取极限, 得 (4.8.1-9) 式成立.

反过来, 设 $\exists t>0$, 对任意 $\boldsymbol{x},\boldsymbol{y}\in C$ 有 (4.8.1-9) 式成立. 对于任意 $\lambda\in(0,1)$, 设函数

$$\phi(\lambda)=\nabla f(\boldsymbol{x}+\lambda(\boldsymbol{y}-\boldsymbol{x}))^{\mathrm{T}}(\boldsymbol{y}-\boldsymbol{x}),$$

那么有

$$\phi'(\lambda)=(\boldsymbol{y}-\boldsymbol{x})^{\mathrm{T}}\nabla^2(\boldsymbol{x}+\lambda(\boldsymbol{y}-\boldsymbol{x}))(\boldsymbol{y}-\boldsymbol{x}).\tag{4.8.1-10}$$

由上式和 (4.8.1-9) 式得

$$\begin{aligned}(\nabla f(\boldsymbol{y})-\nabla f(\boldsymbol{x}))^{\mathrm{T}}(\boldsymbol{y}-\boldsymbol{x})&=\phi(1)-\phi(0)=\int_0^1\phi'(\lambda)\,\mathrm{d}\lambda\\&=\int_0^1(\boldsymbol{y}-\boldsymbol{x})^{\mathrm{T}}\nabla^2(\boldsymbol{x}+\lambda(\boldsymbol{y}-\boldsymbol{x}))(\boldsymbol{y}-\boldsymbol{x})\,\mathrm{d}\lambda,\\&\geqslant\int_0^1\left(t\|\boldsymbol{y}-\boldsymbol{x}\|^2\right)\mathrm{d}\lambda=t\|\boldsymbol{y}-\boldsymbol{x}\|^2.\end{aligned}$$

即 (4.8.1-2) 式成立, 则 f 是在 C 上的强凸函数. □

本小节得到了强凸函数的一些重要性质:

1) 强凸函数等价于一个凸函数加一个二次函数;

2) 任何一个强凸函数都存在一个点上的二次强凸函数是该函数的支持函数;

3) 强凸函数的若干变量的凸组合的函数值不大于对应函数值的凸组合减去二次函数项;

4) 可微强凸函数等价于条件 (4.8.1-1) 或 (4.8.1-2) 式成立;

5) 二次可微强凸函数等价于它的 Hessian 矩阵正定性.

4.9 习　　题

4.9-1　根据凸函数定义和推论 4.3.2 判定下面函数是否是 (严格) 凸函数:

(1) $f(x)=\mathrm{e}^{ax},a\in\mathbf{R}$;　　　　(2) $f(x)=\mathrm{e}^x+x^2$;

(3) $f(x)=\begin{cases}x^a, & x\geqslant 0,\\ +\infty, & x<0,\end{cases}\quad a>1;$

(4) $f(x)=\begin{cases}-x^a, & x\geqslant 0,\\ +\infty, & x<0,\end{cases}\quad 0\leqslant a\leqslant 1.$

4.9-2 设下面函数 $f:\mathbf{R}^n\to\mathbf{R}$, 根据一点凸性判定下面函数是 (严格) 凸函数:

(1) $f(\boldsymbol{x})=\mathrm{e}^{\|\boldsymbol{x}\|}$; (2) $f(\boldsymbol{x})=\|\boldsymbol{x}\|_0$; (3) $f(\boldsymbol{x})=\sum\limits_{i=1}^{n-1}(x_i^2+x_{i+1})$;

(4) $f(\boldsymbol{x})=\begin{cases}\|\boldsymbol{x}\|^a, & \|\boldsymbol{x}\|\leqslant a,\\ +\infty, & \|\boldsymbol{x}\|>a,\end{cases}\quad a\geqslant 1;$

(5) $f(\boldsymbol{x})=\begin{cases}-(x_1x_2\cdots x_n)^{\frac{1}{n}}, & x_i\geqslant 0,\quad i=1,2,\cdots,n,\\ +\infty, & \text{其他};\end{cases}$

(6) $f(\boldsymbol{x})=\begin{cases}0, & 0\in C,\\ +\infty, & 0\notin C,\end{cases}$ 其中 $C\subset\mathbf{R}^n$ 是凸集.

4.9-3 证明一个凸集上的支撑函数、规范函数、距离函数和正齐次函数是凸函数 (性质 4.1.5).

4.9-4 解释正常凸函数与非正常凸函数的区别.

4.9-5 设函数 $f:\mathbf{R}^n\to\mathbf{R}$ 是正常凸函数, 证明 e^f 也是正常凸函数.

4.9-6 求下列凸函数的闭包, 判定是否为正常闭凸函数.

(1) $f(x)=\begin{cases}x^2-1, & x<1,\\ 2, & x=1,\\ +\infty, & x>1;\end{cases}$ (2) $f(\boldsymbol{x})=\begin{cases}0, & x_1^2+x_2^2<1,\\ 2, & x_1^2+x_2^2=1,\\ +\infty, & x_1^2+x_2^2>1;\end{cases}$

(3) $f(\boldsymbol{x})=\begin{cases}-(1-\|\boldsymbol{x}\|^2)^{\frac{1}{2}}, & \|\boldsymbol{x}\|\leqslant 1,\\ +\infty, & \|\boldsymbol{x}\|>1.\end{cases}$

4.9-7 设 $f:\mathbf{R}^n\times T\to\mathbf{R}$ 是有限函数, 其中集合 T 是 $\mathbf{R}$ 的任意集合. 对于任意 $t\in T$, $f(\boldsymbol{x},t)$ 关于 $\boldsymbol{x}$ 是凸函数. 对于任意 $\boldsymbol{x}$, $f(\boldsymbol{x},t)$ 是关于 t 的有界函数. 证明 $\phi(\boldsymbol{x})=\sup\limits_{t\in T}f(\boldsymbol{x},t)$ 关于 $\boldsymbol{x}$ 是有限的连续凸函数.

4.9-8 已知函数 $f(x)=\mathrm{e}^x$, 证明它的共轭函数:

$$f^*(x^*)=\begin{cases}x^*\lg x^*-x^*, & x^*>0,\\ 0, & x^*=0,\\ +\infty, & x^*<0.\end{cases}$$

4.9-9 已知 $\boldsymbol{x},\boldsymbol{a}\in\mathbf{R}^n, b\in\mathbf{R}$, 求下列函数的共轭函数:

(1) $f(\boldsymbol{x})=\boldsymbol{a}^{\mathrm{T}}\boldsymbol{x}+b$; (2) $f(\boldsymbol{x})=\dfrac{1}{2}\|\boldsymbol{x}\|^2$; (3) $f(x)=\begin{cases}-\lg x, & x>0,\\ +\infty, & x\leqslant 0.\end{cases}$

4.9-10 设 $f:\mathbf{R}^n\to\mathbf{R}$ 是凸函数, $A:\mathbf{R}^n\to\mathbf{R}^n$ 是一对一线性变换, 设 $\boldsymbol{a},\boldsymbol{b}\in\mathbf{R}^n$ 和 $\alpha\in\mathbf{R}$, 求函数: $f(A(\boldsymbol{x}-\boldsymbol{a}))+\langle\boldsymbol{x},\boldsymbol{b}\rangle+\alpha$ 的共轭函数.

4.9-11 设集合 $C\subset\mathbf{R}^2$ 是由 4 个顶点 $(0,1)^{\mathrm{T}},(0,-1)^{\mathrm{T}},(1,0)^{\mathrm{T}},(-1,0)^{\mathrm{T}}$ 构成的集合, 求支撑函数 $\delta_C(\boldsymbol{x}^*)$.

4.9-12 设集合 $C=\left\{(x_1,x_2)^{\mathrm{T}}|0\leqslant x_2\leqslant x_1-1\right\}$, 求支撑函数 $\delta_C(\boldsymbol{x}^*)$.

4.9-13 设集合 $C\subset\mathbf{R}^2$ 是由 4 个顶点 $(1,1)^{\mathrm{T}},(1,-1)^{\mathrm{T}},(-1,1)^{\mathrm{T}},(-1,-1)^{\mathrm{T}}$ 构成的集合, 求支撑函数 $\delta_C(\boldsymbol{x}^*)$.

4.9-14 设 $f(x_1,x_2)$ 是凸函数, 令函数 $g(x_1)=\inf\{f(x_1,x_2)|x_2\in\mathbf{R}\}$, 求共轭函数 $g^*(x_1^*)$.

4.9-15 设 $g_1,g_2,\cdots,g_m:\mathbf{R}\to\mathbf{R}$ 是正常闭凸函数, 设 $\boldsymbol{a}_1,\boldsymbol{a}_2,\cdots,\boldsymbol{a}_m\in\mathbf{R}^n$, 求函数 $\phi(\boldsymbol{x})=\sum\limits_{i=1}^{m}g(\langle\boldsymbol{x},\boldsymbol{a}_i\rangle)$ 的共轭函数 $\phi^*(\boldsymbol{x}^*)$.

4.9-16 证明性质 4.1.3 和定理 4.1.9.

4.9-17 证明性质 4.2.8 和性质 4.2.9.

4.9-18 如果函数 $f:\mathbf{R}^n\to\mathbf{R}$ 在凸集 C 上满足上半连续和一点半严格凸性, 证明 f 在 C 上是半严格凸函数.

4.9-19 如果函数 $f:\mathbf{R}^n\to\mathbf{R}$ 在凸集 C 上满足上半连续、一点半严格凸性和一点凸性, 证明 f 在 C 上是显凸函数.

4.9-20 如果函数 $f_1,f_2:\mathbf{R}^n\to\mathbf{R}$ 在凸集 C 上是半严格凸函数, 那么 f_1+f_2 是否在凸集 C 上是半严格凸函数? 如果不是, 请举出反例.

4.9-21 如果函数 $f_1,f_2:\mathbf{R}^n\to\mathbf{R}$ 在凸集 C 上是严格凸函数, 那么 f_1+f_2 是否在凸集 C 上是严格凸函数? 如果不是, 请举出反例.

4.9-22 如果函数 $f_1,f_2:\mathbf{R}^n\to\mathbf{R}$ 在凸集 C 上是显凸函数, 那么 f_1+f_2 是否在凸集 C 上是显凸函数? 如果不是, 请举出反例.

4.9-23　如果函数 $f:\mathbf{R}^n\to\mathbf{R}$ 在凸集 C 上是半严格凸函数, $g:\mathbf{R}\to\mathbf{R}$ 是单调增加函数, 那么复合函数 gf 是否在凸集 C 上是半严格凸函数? 如果不是, 请举出反例.

4.9-24　如果函数 $f:\mathbf{R}^n\to\mathbf{R}$ 是半严格凸函数, 那么 f 的共轭函数 f^* 是否为半严格凸函数? 如果不是, 请举出反例.

4.9-25　如果函数 $f:\mathbf{R}^n\to\mathbf{R}$ 是半严格闭凸函数, 那么 $\mathrm{cl}f$ 是否为半严格凸函数? 如果不是, 请举出反例.

4.9-26　如果函数 $f:\mathbf{R}^n\to\mathbf{R}$ 是强凸函数, 那么 f 是否为半严格凸函数和显凸函数? 如果不是, 请举出反例.

第 5 章　广义凸函数

广义凸函数是凸分析和凸优化中最重要的组成部分, 也是凸分析理论成果中内容最丰富的部分. 文献 [66-95] 得到了许多广义凸函数成果, 如拟凸、伪凸、不变凸、预不变凸、半严格不变凸等广义凸函数. 凸函数通常都是广义凸函数, 反之, 则不一定. 因此, 凸函数的一些性质对广义凸函数不一定成立, 如凸函数的稳定点一定是全局最优解, 而拟凸函数的稳定点不一定是全局最优解. 故广义凸函数比凸函数具有更弱的性质. 另一方面, 虽然许多广义凸函数不是凸函数, 但是凸函数的一些重要性质对广义凸函数仍然成立. 例如, 连续凹凸函数的鞍点定理对连续拟凹凸函数的情形仍然成立. 反过来, 拟凸函数的性质对于凸函数不一定成立. 例如, 拟凸函数等价于其水平集是凸集, 但函数的水平集是凸集却不一定导出它是凸函数, 这使得广义凸函数比凸函数应用得更广泛.

本章介绍了拟凸函数、半严格拟凸函数、伪凸函数和广义单调性等基本概念和性质, 重点讨论了这些广义凸函数之间的关系. 图 5.0-1 给出了凸函数和广义凸函数之间的蕴含关系, 除此之外, 本章还将讨论这些蕴含关系反过来成立的条件: 下半连续性、上半连续性和一点 (伪、拟) 凸性等, 其中一些性质的成立还与广义凸函数的稳定点、局部极小点和全局极小点有关. 这些性质都深刻地反映了与凸函数不同的广义凸函数的特征.

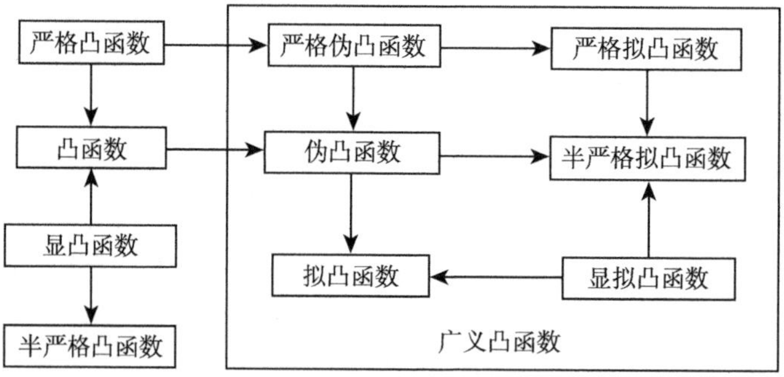

图 5.0-1

5.1 拟凸函数

拟凸函数是广义凸函数中最重要的一类. 拟凸函数的几何特性等价于其水平集是凸集, 并且定义域内任意两点值的凸组合函数值小于或等于两端点最大值. 拟凸函数不一定是凸函数, 例如对数函数: $f(x) = \ln x\,(x > 0)$ 是拟凸函数, 但不是凸函数. De Finietti 于 1949 年首次研究了具有凸水平集的函数 [21], 1951 年 Fenchel 推广和校正了 De Finietti 的结果, 并将这一概念命名为拟凸函数 [25], 后来拟凸函数成为广义凸函数中研究与应用最多的一种类型 [2,10,11,17-33,37,41,51-59,69].

5.1.1 基本性质

下面给出拟凸函数和一点拟凸性的定义.

定义 5.1.1 设 $C \subset \mathbf{R}^n$ 是非空凸集和函数 $f: C \to \mathbf{R}$. 如果对于任意 $\boldsymbol{x}, \boldsymbol{y} \in C$ 和任意 $\lambda \in (0,1)$ 有

$$f[(1-\lambda)\boldsymbol{x} + \lambda\boldsymbol{y}] \leqslant \max\{f(\boldsymbol{x}), f(\boldsymbol{y})\}, \tag{5.1.1-1}$$

则称 f 在 C 上是拟凸函数. 如果对于任意不同的两点 $\boldsymbol{x}, \boldsymbol{y} \in C$ 和 $\lambda \in (0,1)$ 有

$$f[(1-\lambda)\boldsymbol{x} + \lambda\boldsymbol{y}] < \max\{f(\boldsymbol{x}), f(\boldsymbol{y})\}, \tag{5.1.1-2}$$

则称 f 在 C 上是严格拟凸函数. 如果 $-f$ 在 C 上是拟凸函数, 则称 f 在 C 上是拟凹函数, 如果 $-f$ 在 C 上是严格凸函数, 则称 f 在 C 上是严格凹函数.

如果对于任意的 $\boldsymbol{x}, \boldsymbol{y} \in C$, 存在一点 $\lambda \in (0,1)$ 有 (5.1.1-1) 式成立, 则称 f 在 C 上满足一点拟凸性. 如果对于任意的 $\boldsymbol{x}, \boldsymbol{y} \in C$, 存在一点 $\lambda \in (0,1)$ 有 (5.1.1-2) 式成立, 则称 f 在 C 上满足一点严格拟凸性.

通过上述定义知, 拟凸函数的定义域中任意两点连线上的值不大于两个端点最大的值, 存在许多不连续函数是拟凸函数, 见下例.

例 5.1.1 设一元函数:

$$f(x) = \begin{cases} |x|/x, & \text{当 } x \neq 0, \\ 0, & \text{当 } x = 0, \end{cases}$$

则 f 是拟凸函数, 但不是严格拟凸函数.

一元函数 $f(x) = \begin{cases} 1, & x > 0, \\ 0, & x \leqslant 0 \end{cases}$ 是拟凸函数.

注 单变量单调函数都是拟凸函数, 单变量严格单调函数都是严格拟凸函数. 凸函数是拟凸函数, 严格凸函数是严格拟凸函数, 严格拟凸函数是拟凸函数. 但

是, 以上反过来都不一定成立, 拟凸函数不一定是凸函数, 甚至拟凸函数不一定是连续的. 见下面性质.

性质 5.1.1　设 $C \subset \mathbf{R}^n$ 是非空凸集和函数 $f: C \to \mathbf{R}$, 有下面结论成立:

(1) 如果 f 在 C 上是凸函数, 则 f 在 C 上是拟凸函数;

(2) 如果 f 在 C 上是严格凸函数, 则 f 在 C 上是严格拟凸函数;

(3) 如果 f 在 C 上是严格拟凸函数, 则 f 在 C 上是拟凸函数;

(4) 如果 f 在 C 上是拟凸函数, 且它的线段上的限制函数不是常数, 则 f 在 C 上是严格拟凸函数.

证明　(1) 对于任意 $\boldsymbol{x}, \boldsymbol{y} \in C$ 和任意 $\lambda \in (0,1)$ 有

$$f[(1-\lambda)\boldsymbol{x} + \lambda \boldsymbol{y}] \leqslant (1-\lambda) f(\boldsymbol{x}) + \lambda f(\boldsymbol{y}) \leqslant \max\{f(\boldsymbol{x}), f(\boldsymbol{y})\},$$

由定义 5.1.1 得结论.

(2) 对于任意两个不同的 $\boldsymbol{x}, \boldsymbol{y} \in C$ 和 $\lambda \in (0,1)$ 有

$$f[(1-\lambda)\boldsymbol{x} + \lambda \boldsymbol{y}] < (1-\lambda) f(\boldsymbol{x}) + \lambda f(\boldsymbol{y}) \leqslant \max\{f(\boldsymbol{x}), f(\boldsymbol{y})\},$$

由定义 5.1.1 得结论.

(3) 结论显然.

(4) 假设存在不同的两点 $\boldsymbol{x}, \boldsymbol{y} \in C$ 和 $\lambda \in (0,1)$, 有

$$f[(1-\lambda)\boldsymbol{x} + \lambda \boldsymbol{y}] \geqslant \max\{f(\boldsymbol{x}), f(\boldsymbol{y})\},$$

由 f 是在 C 上的拟凸函数, 得 $f[(1-\lambda)\boldsymbol{x} + \lambda \boldsymbol{y}] = \max\{f(\boldsymbol{x}), f(\boldsymbol{y})\}$. 不妨设 $f(\boldsymbol{x}) \geqslant f(\boldsymbol{y})$, 以及 $\boldsymbol{x}_\lambda = (1-\lambda)\boldsymbol{x} + \lambda \boldsymbol{y}$, 有 $f(\boldsymbol{x}_\lambda) = f(\boldsymbol{x})$. 根据假设 f 是在 C 上的线段上的限制函数且不是常数, 分别存在 $t \in (0, \lambda)$ 和 $\eta \in (\lambda, 1)$ 满足

$$\begin{aligned} & f(\boldsymbol{x}_t) < f(\boldsymbol{x}_\lambda) = f(\boldsymbol{x}), \quad \text{其中 } \boldsymbol{x}_t = (1-t)\boldsymbol{x} + t\boldsymbol{y}, \quad \boldsymbol{x}_t \in [\boldsymbol{x}, \boldsymbol{x}_\lambda], \\ & f(\boldsymbol{x}_\eta) < f(\boldsymbol{x}_\lambda) = f(\boldsymbol{x}), \quad \text{其中 } \boldsymbol{x}_\eta = (1-\eta)\boldsymbol{x} + \eta\boldsymbol{y}, \quad \boldsymbol{x}_\eta \in [\boldsymbol{x}_\lambda, \boldsymbol{y}]. \end{aligned}$$

由于 $\lambda \in (t, \eta)$, 有 $\boldsymbol{x}_\lambda \in [\boldsymbol{x}_t, \boldsymbol{x}_\eta]$, 再由 f 是在 C 上的拟凸函数, 有

$$f(\boldsymbol{x}_\lambda) \leqslant \max\{f(\boldsymbol{x}_t), f(\boldsymbol{x}_\eta)\} < f(\boldsymbol{x}),$$

上式与前面两个不等式矛盾. □

下面的定理是拟凸函数定义的一种推广, 表明了拟凸函数定义域中有限多点的凸组合值不大于这些点值中最大的值.

定理 5.1.1　设 $C \subset \mathbf{R}^n$ 是非空凸集, 函数 $f: C \to \mathbf{R}$, 则 f 在 C 上是拟凸函数的充要条件是对于任意 $\boldsymbol{x}_i \in C$ 和 $\lambda_i > 0\,(i = 1, \cdots, m)$ 有

$$f(\lambda_1 \boldsymbol{x}_1 + \cdots + \lambda_m \boldsymbol{x}_m) \leqslant \max\{f(\boldsymbol{x}_1), \cdots, f(\boldsymbol{x}_m)\} \tag{5.1.1-3}$$

成立, 其中 $\sum\limits_{i=1}^{m} \lambda_i = 1$.

证明 先证充分性. 当 $m = 2$ 时, 结论显然成立.

再证必要性. 设 f 是在 C 上的拟凸函数, 对于任意 $\boldsymbol{x}_i \in C$ 和任意 $\lambda_i > 0\,(i = 1, \cdots, m)$, 当 $m = 2$ 时, 由定义 5.1.1 知 (4.1.1-3) 式成立. 采用归纳法, 假设 $m-1$ 时 (4.1.1-3) 式成立, 设

$$\alpha \boldsymbol{y} = \lambda_1 \boldsymbol{x}_1 + \cdots + \lambda_{m-1} \boldsymbol{x}_{m-1}, \quad \alpha = \lambda_1 + \cdots + \lambda_{m-1}.$$

那么有 $\alpha + \lambda_m = 1$, $\sum\limits_{i=1}^{m-1} \lambda_i / \alpha = 1$ 和

$$f\left(\lambda_1 \boldsymbol{x}_1 + \cdots + \lambda_m \boldsymbol{x}_m\right) = f\left(\alpha \boldsymbol{y} + \lambda_m \boldsymbol{x}_m\right) \leqslant \max\left\{f\left(\boldsymbol{y}\right), f\left(\boldsymbol{x}_m\right)\right\}.$$

由归纳法假设得

$$\begin{aligned} f\left(\lambda_1 \boldsymbol{x}_1 + \cdots + \lambda_m \boldsymbol{x}_m\right) &\leqslant \max\left\{f\left(\frac{\lambda_1}{\alpha} \boldsymbol{x}_1 + \cdots + \frac{\lambda_{m-1}}{\alpha} \boldsymbol{x}_{m-1}\right), f\left(\boldsymbol{x}_m\right)\right\} \\ &\leqslant \max\left\{f\left(\boldsymbol{x}_1\right), \cdots, f\left(\boldsymbol{x}_{m-1}\right), f\left(\boldsymbol{x}_m\right)\right\}. \end{aligned}$$ □

定理 5.1.2 设 $C \subset \mathbf{R}^n$ 是非空凸集, 函数 $f:\ C \to \mathbf{R}$, 则 $f(\boldsymbol{x})$ 在 C 上是严格拟凸函数的充要条件是对任意 $\boldsymbol{x}_i \in C$ 和任意 $\lambda_i > 0\,(i = 1, \cdots, m)$ 有

$$f\left(\lambda_1 \boldsymbol{x}_1 + \cdots + \lambda_m \boldsymbol{x}_m\right) < \max\left\{f\left(\boldsymbol{x}_1\right), \cdots, f\left(\boldsymbol{x}_m\right)\right\} \tag{5.1.1-3}$$

成立, 其中 $\sum\limits_{i=1}^{m} \lambda_i = 1$.

证明 由定义 5.1.1 易知, 证明过程完全与定理 5.1.1 的证明一样. □

下面证明拟凸函数一个非常重要的性质: 拟凸函数等价于其水平集是凸集. 这个结论对于凸函数不成立. 例如, 函数 $f(x) = x^3$ 的水平集是凸集, 但它在 $\mathbf{R}$ 上不是凸函数. 这表明了广义凸函数具有的性质对于凸函数不一定成立.

定理 5.1.3 设 $C \subset \mathbf{R}^n$ 是非空凸集和函数 $f:\ C \to \mathbf{R}$, 则 f 在 C 上是拟凸函数的充要条件是 f 的水平集 $C_{f \leqslant \alpha} = \{\boldsymbol{x} \in C \,|\, f(\boldsymbol{x}) \leqslant \alpha\}$ 是凸集, 其中 $\forall \alpha \in \mathbf{R}$.

证明 假设 f 在 C 上是拟凸函数. 对于任意 $\alpha \in \mathbf{R}$, 任意 $\boldsymbol{x}, \boldsymbol{y} \in C_{f \leqslant \alpha}$ 和任意 $\lambda \in (0,1)$ 有 $f(\boldsymbol{x}) \leqslant \alpha$ 和 $f(\boldsymbol{y}) \leqslant \alpha$, 那么由定义 5.1.1 得

$$f\left[(1-\lambda)\,\boldsymbol{x} + \lambda \boldsymbol{y}\right] \leqslant \max\left\{f\left(\boldsymbol{x}\right), f\left(\boldsymbol{y}\right)\right\} \leqslant \alpha.$$

则 $(1-\lambda)\,\boldsymbol{x} + \lambda \boldsymbol{y} \in C_{f \leqslant \alpha}$, 所以水平集 $C_{f \leqslant \alpha}$ 是凸集.

反过来, 设水平集 $C_{f\leqslant\alpha}$ 是凸集, 其中任意 $\alpha\in\mathbf{R}$. 对任意 $\boldsymbol{x}_1,\boldsymbol{x}_2\in C$ 和任意 $\lambda\in(0,1)$, 令 $\alpha=\max\{f(\boldsymbol{x}_1),f(\boldsymbol{x}_2)\}$, 取水平集 $C_{f\leqslant\alpha}$, 显然 $\boldsymbol{x}_1,\boldsymbol{x}_2\in C_{f\leqslant\alpha}$, 且 $(1-\lambda)\boldsymbol{x}_1+\lambda\boldsymbol{x}_2\in\{\boldsymbol{x}\in C\,|\,f(\boldsymbol{x})\leqslant\alpha\}$, 即有

$$f[(1-\lambda)\boldsymbol{x}_1+\lambda\boldsymbol{x}_2]\leqslant\alpha=\max\{f(\boldsymbol{x}_1),f(\boldsymbol{x}_2)\}.$$

由于 $\boldsymbol{x}_1,\boldsymbol{x}_2\in C$ 的任意性, 上式说明 f 在 C 上是拟凸函数. □

下面的定理揭示了拟凸函数等价于函数任意子集上的函数值的上确界等于该函数子集凸包上的上确界值.

定理 5.1.4[10] 设 $C\subset\mathbf{R}^n$ 是非空凸集和函数 $f:\ C\to\mathbf{R}$, 则 f 在 C 上是拟凸函数的充要条件是 $\sup\{f(\boldsymbol{x})\,|\,\boldsymbol{x}\in D\}=\sup\{f(\boldsymbol{x})\,|\,\boldsymbol{x}\in \mathrm{co}D\}$, 其中 D 是 C 的任意子集.

证明 设 f 在 C 上是拟凸函数, D 是 C 的任意子集. 显然有

$$\sup\{f(\boldsymbol{x})\,|\,\boldsymbol{x}\in D\}\leqslant\sup\{f(\boldsymbol{x})\,|\,\boldsymbol{x}\in \mathrm{co}D\}.$$

设 $\alpha=\sup\{f(\boldsymbol{x})\,|\,\boldsymbol{x}\in D\}$, 定义水平集 $C_{f\leqslant\alpha}=\{\boldsymbol{x}\in C\,|\,f(\boldsymbol{x})\leqslant\alpha\}$. 由 $D\subset\mathrm{co}D\subset C$, 得 $D\subset C_{f\leqslant\alpha}$. 由定理 5.1.3 知 $C_{f\leqslant\alpha}$ 是凸集, 故 $\mathrm{co}D\subset C_{f\leqslant\alpha}$, 那么有

$$\sup\{f(\boldsymbol{x})\,|\,\boldsymbol{x}\in \mathrm{co}D\}\leqslant\alpha.$$

因此, $\sup\{f(\boldsymbol{x})\,|\,\boldsymbol{x}\in D\}=\sup\{f(\boldsymbol{x})\,|\,\boldsymbol{x}\in \mathrm{co}D\}$.

反过来, 若 $\sup\{f(\boldsymbol{x})\,|\,\boldsymbol{x}\in D\}=\sup\{f(\boldsymbol{x})\,|\,\boldsymbol{x}\in \mathrm{co}D\}$, 其中 D 是 C 的任意子集. 设 $\forall\alpha\in\mathbf{R}$, 定义水平集 $C_{f\leqslant\alpha}=\{\boldsymbol{x}\in C\,|\,f(\boldsymbol{x})\leqslant\alpha\}$. 任取 $\boldsymbol{x}_1,\boldsymbol{x}_2\in C_{f\leqslant\alpha}$ 和 $\lambda\in(0,1)$, 有

$$\sup\{f(\boldsymbol{x})\,|\,\boldsymbol{x}\in C_{f\leqslant\alpha}\}=\sup\{f(\boldsymbol{x})\,|\,\boldsymbol{x}\in \mathrm{co}C_{f\leqslant\alpha}\}\leqslant\alpha.$$

由上式得 $(1-\lambda)\boldsymbol{x}_1+\lambda\boldsymbol{x}_2\in\mathrm{co}C_{f\leqslant\alpha}$, 则 $f((1-\lambda)\boldsymbol{x}_1+\lambda\boldsymbol{x}_2)\leqslant\alpha$, 表明 $C_{f\leqslant\alpha}$ 是凸集. 再由定理 5.1.3 知 f 是 C 上的拟凸函数. □

由定理 5.1.4 直接得下面结论成立.

定理 5.1.5[10] 设 $C\subset\mathbf{R}^n$ 是非空凸集, 函数 $f:\ C\to\mathbf{R}$, 则 f 在 C 上是拟凹函数的充要条件是 $\inf\{f(\boldsymbol{x})\,|\,\boldsymbol{x}\in C\}=\inf\{f(\boldsymbol{x})\,|\,\boldsymbol{x}\in \mathrm{co}D\}$, 其中 D 是 C 的任意子集.

下面性质表明了拟凸函数复合后仍然保持拟凸性的结果, 对于判断拟凸函数具有重要作用, 例如 $f:\ C\to\mathbf{R}$ 是在 C 上的拟凸函数, 则 $\mathrm{e}^{f(\boldsymbol{x})}$ 是在 C 上的拟凸函数.

性质 5.1.2 设 $C\subset\mathbf{R}^n$ 是非空凸集, $f:\ C\to\mathbf{R}$ 在 C 上是拟凸函数, 非空集合 $S\subset\mathbf{R}$ 和 $\phi:\ S\to\mathbf{R}$ 是单调递增函数, 且 $f(C)\subset S$, 则有下面结论成立:

(1) $tf(\boldsymbol{x})$ 是在 C 上的拟凸函数, 其中 $t>0$;

(2) $tf(\boldsymbol{x})$ 是在 C 上的严格拟凸函数, 其中 $t>0$ 和 f 在 C 上是严格拟凸函数;

(3) $\phi(f(\boldsymbol{x}))$ 是在 C 上的拟凸函数;

(4) $\phi(f(\boldsymbol{x}))$ 是在 C 上的严格拟凸函数, 其中 f 在 C 上是严格拟凸函数.

证明 (1) 和 (2) 根据定义 5.1.1 是显然的.

(3) 对于任意 $\boldsymbol{x},\boldsymbol{y}\in C$ 和任意 $\lambda\in(0,1)$, 不妨设 $f(\boldsymbol{x})\leqslant f(\boldsymbol{y})$, 有 $\phi(f(\boldsymbol{x}))\leqslant\phi(f(\boldsymbol{y}))$, 则

$$\phi(f[(1-\lambda)\boldsymbol{x}+\lambda\boldsymbol{y}])\leqslant\phi(f(\boldsymbol{y})),$$

即说明 $\phi(f(\boldsymbol{x}))$ 是在 C 上的拟凸函数.

(4) 与 (3) 的证明完全类似. □

下面是线性函数与拟凸函数复合后的函数仍是拟凸函数的性质.

性质 5.1.3 设 $g(\boldsymbol{x})=\boldsymbol{A}\boldsymbol{x}+\boldsymbol{b}$, 其中 $\boldsymbol{x}\in\mathbf{R}^n$, $\boldsymbol{A}$ 是 $m\times n$ 的矩阵, $\boldsymbol{b}\in\mathbf{R}^m$ 为给定的. 若 f 是在凸集 $C\subset g(\mathbf{R}^n)$ 上的拟凸函数, 则 $f(g(\boldsymbol{x}))=f(\boldsymbol{A}\boldsymbol{x}+\boldsymbol{b})$ 是在 C 上的拟凸函数.

证明 对于任意 $\boldsymbol{x},\boldsymbol{y}\in C$ 和任意 $\lambda\in(0,1)$, 有

$$\begin{aligned}f[g((1-\lambda)\boldsymbol{x}+\lambda\boldsymbol{y})]&=f[\boldsymbol{A}((1-\lambda)\boldsymbol{x}+\lambda\boldsymbol{y})+\boldsymbol{b}]\\&\leqslant\max\{f(\boldsymbol{A}\boldsymbol{x}+\boldsymbol{b}),f(\boldsymbol{A}\boldsymbol{y}+\boldsymbol{b})\},\end{aligned}$$

说明 $f(g(\boldsymbol{x}))=f(\boldsymbol{A}\boldsymbol{x}+\boldsymbol{b})$ 是在 C 上的拟凸函数. □

性质 5.1.4 设 $f:\ \mathbf{R}^n\to\mathbf{R}$ 在 $\mathbf{R}^n$ 上是严格拟凸函数, $\boldsymbol{A}$ 是 $n\times m$ 的矩阵以及 $\boldsymbol{b}\in\mathbf{R}^n$, 则 $f(\boldsymbol{A}\boldsymbol{y}+\boldsymbol{b})$ 在 $\mathbf{R}^m$ 上是严格拟凸函数.

下面的结果是拟凸函数的限制函数仍然是拟凸函数的等价性结果.

定理 5.1.6 设 $C\subset\mathbf{R}^n$ 是非空凸集和函数 $f:\ C\to\mathbf{R}$, 则 f 在 C 上是 (严格) 拟凸函数当且仅当对任意 $\boldsymbol{x}\in C$ 和任意 $\boldsymbol{d}\in\mathbf{R}^n$, $\varphi_{\boldsymbol{x},\boldsymbol{d}}(t)$ 是 $t\in T(\boldsymbol{x},\boldsymbol{d})$ 上的 (严格) 拟凸函数, 其中

$$\varphi_{\boldsymbol{x},\boldsymbol{d}}(t)=f(\boldsymbol{x}+t\boldsymbol{d})\quad 和\quad T(\boldsymbol{x},\boldsymbol{d})=\{t\in\mathbf{R}|\boldsymbol{x}+t\boldsymbol{d}\in C\}.$$

证明 设 f 在 C 上是 (严格) 拟凸函数, 对于任意 $t_1,t_2\in T(\boldsymbol{x},\boldsymbol{d})$ 和任意 $\lambda\in(0,1)$, 有

$$\begin{aligned}\varphi_{\boldsymbol{x},\boldsymbol{d}}((1-\lambda)t_1+\lambda t_2)&=f((1-\lambda)(\boldsymbol{x}+t_1\boldsymbol{d})+\lambda(\boldsymbol{x}+t_2\boldsymbol{d}))\\&\leqslant(<)\max\{f(\boldsymbol{x}+t_1\boldsymbol{d}),f(\boldsymbol{x}+t_2\boldsymbol{d})\}\\&=\max\{\varphi_{\boldsymbol{x},\boldsymbol{d}}(t_1),\varphi_{\boldsymbol{x},\boldsymbol{d}}(t_2)\},\end{aligned}$$

由定义 5.1.1 知 $\varphi_{\boldsymbol{x},\boldsymbol{d}}(t)$ 是 $t\in T(\boldsymbol{x},\boldsymbol{d})$ 的 (严格) 拟凸函数.

反过来, 对任意 $\boldsymbol{x}\in C$ 和任意 $\boldsymbol{d}\in\mathbf{R}^n$, $\varphi_{\boldsymbol{x},\boldsymbol{d}}(t)$ 是 $t\in T(\boldsymbol{x},\boldsymbol{d})$ 上的 (严格) 拟凸函数. 对于任意 $\boldsymbol{x}_1,\boldsymbol{x}_2\in C$ 和任意 $\lambda\in(0,1)$, 设 $T(\boldsymbol{x}_1,\boldsymbol{x}_2-\boldsymbol{x}_1)=\{t\in\mathbf{R}|\boldsymbol{x}_1+t(\boldsymbol{x}_2-\boldsymbol{x}_1)\in C\}$, 有

$$\begin{aligned}f[(1-\lambda)\boldsymbol{x}_1+\lambda\boldsymbol{x}_2]&=f(\boldsymbol{x}_1+[(1-\lambda)\cdot1+\lambda\cdot0](\boldsymbol{x}_2-\boldsymbol{x}_1))\\&=\varphi_{\boldsymbol{x}_1,\boldsymbol{x}_2-\boldsymbol{x}_1}((1-\lambda)\cdot1+\lambda\cdot0)\\&\leqslant(<)\max\{\varphi_{\boldsymbol{x}_1,\boldsymbol{x}_2-\boldsymbol{x}_1}(1),\varphi_{\boldsymbol{x}_1,\boldsymbol{x}_2-\boldsymbol{x}_1}(0)\}\\&=\max\{f(\boldsymbol{x}_1),f(\boldsymbol{x}_2)\}.\end{aligned}$$

故 f 是在 C 上是拟凸函数. □

推论 5.1.1 设 $C\subset\mathbf{R}^n$ 是非空凸集和函数 $f:C\to\mathbf{R}$, 则 f 在 C 上是 (严格) 拟凸函数当且仅当对任意 $\boldsymbol{x},\boldsymbol{y}\in C$, $\varphi_{\boldsymbol{x},\boldsymbol{y}}(t)=f(\boldsymbol{x}+t(\boldsymbol{y}-\boldsymbol{x}))$ 是 $t\in[0,1]$ 上的 (严格) 拟凸函数.

下面结果表明: 拟凸函数在其定义域中如果满足一点凸性, 则等价于凸函数. 例如, 拟凸函数 $f(x)=\begin{cases}1, & x>0,\\ 0, & x\leqslant0\end{cases}$ 不满足一点凸性.

定理 5.1.7[95] 设 $C\subset\mathbf{R}^n$ 是非空凸集和函数 $f:C\to\mathbf{R}$, 则 f 在 C 上是凸函数当且仅当 f 在 C 上是拟凸函数且关于 $\alpha\in(0,1)$ 满足一点凸性.

证明 必要性显然成立.

下面证明充分性. 设 f 在 C 上是拟凸函数, 对于任意 $\boldsymbol{x},\boldsymbol{y}\in C$, 存在一个 $\alpha\in(0,1)$ 有

$$f[\alpha\boldsymbol{x}+(1-\alpha)\boldsymbol{y}]\leqslant\alpha f(\boldsymbol{x})+(1-\alpha)f(\boldsymbol{y}).\tag{5.1.1-4}$$

设任意 $\boldsymbol{x},\boldsymbol{y}\in C$ 和任意 $\lambda\in[0,1]$, 记 $\boldsymbol{z}_\lambda=\lambda\boldsymbol{x}+(1-\lambda)\boldsymbol{y}$. 下面采用反证法, 假设 f 在 C 上不是凸函数.

(1) 设 $f(\boldsymbol{x})=f(\boldsymbol{y})$, 假设存在 $\beta\in(0,1)$ 满足

$$f(\boldsymbol{z}_\beta)=f[\beta\boldsymbol{x}+(1-\beta)\boldsymbol{y}]>\beta f(\boldsymbol{x})+(1-\beta)f(\boldsymbol{y})=f(\boldsymbol{y}).\tag{5.1.1-5}$$

如果 $0<\beta<\alpha\leqslant1$, 设 $\gamma=\dfrac{\beta-\alpha}{1-\alpha}$, 有 $\boldsymbol{z}_\beta=\alpha\boldsymbol{x}+(1-\alpha)\boldsymbol{z}_\gamma$, 由 (5.1.1-4) 式和 (5.1.1-5) 式得到

$$f(\boldsymbol{z}_\beta)\leqslant\alpha f(\boldsymbol{x})+(1-\alpha)f(\boldsymbol{z}_\gamma)<f(\boldsymbol{z}_\gamma).\tag{5.1.1-6}$$

再设 $t=\dfrac{\beta-\gamma}{\beta}$, 有 $\boldsymbol{z}_\gamma=t\boldsymbol{y}+(1-t)\boldsymbol{z}_\beta$. 据此, 由 $f(\boldsymbol{y})<f(\boldsymbol{z}_\beta)$ 和 f 在 C 上是拟凸函数得

$$f(\boldsymbol{z}_\gamma)=f(t\boldsymbol{y}+(1-t)\boldsymbol{z}_\beta)\leqslant f(\boldsymbol{z}_\beta).$$

上式与 (5.1.1-6) 式矛盾.

如果 $0<\alpha<\beta\leqslant 1$, 设 $\gamma=\dfrac{\beta}{\alpha}$, 有 $\boldsymbol{z}_\beta=\alpha\boldsymbol{x}+(1-\alpha)\boldsymbol{z}_\gamma$, 由 (5.1.1-4) 式和 (5.1.1-5) 式得到

$$f(\boldsymbol{z}_\beta)\leqslant\alpha f(\boldsymbol{x})+(1-\alpha)f(\boldsymbol{z}_\gamma)<f(\boldsymbol{z}_\gamma). \tag{5.1.1-7}$$

再设 $t=\dfrac{\gamma-\beta}{1-\beta}$, 有 $\boldsymbol{z}_\gamma=t\boldsymbol{x}+(1-t)\boldsymbol{z}_\beta$. 据此, 由 $f(\boldsymbol{x})<f(\boldsymbol{z}_\beta)$ 和 f 在 C 上是拟凸函数得

$$f(\boldsymbol{z}_\gamma)=f(t\boldsymbol{x}+(1-t)\boldsymbol{z}_\beta)\leqslant f(\boldsymbol{z}_\beta).$$

上式与 (5.1.1-7) 式矛盾.

(2) 设 $f(\boldsymbol{x})\neq f(\boldsymbol{y})$, 假设存在 $\beta\in(0,1)$ 满足

$$f(\boldsymbol{z}_\beta)=f[\beta\boldsymbol{x}+(1-\beta)\boldsymbol{y}]>\beta f(\boldsymbol{x})+(1-\beta)f(\boldsymbol{y}). \tag{5.1.1-8}$$

由引理 4.3.2 得

$$f[\lambda\boldsymbol{x}+(1-\lambda)\boldsymbol{y}]\leqslant\lambda f(\boldsymbol{x})+(1-\lambda)f(\boldsymbol{y}),\quad \lambda\in A,$$

其中 A 由引理 4.3.2 定义. 下面进一步分两种情形证明.

(i) 设 $f(\boldsymbol{x})<f(\boldsymbol{y})$. 由 (5.1.1-8) 式和 A 在 $[0,1]$ 中的稠密性知, 存在 $\gamma\in A$ 和 $\gamma<\beta$ 使得

$$f(\boldsymbol{z}_\beta)=f[\beta\boldsymbol{x}+(1-\beta)\boldsymbol{y}]>\gamma f(\boldsymbol{x})+(1-\gamma)f(\boldsymbol{y}).$$

据此, 由 $\gamma\in A$ 可得

$$\begin{aligned}f(\boldsymbol{z}_\gamma)&=f[\gamma\boldsymbol{x}+(1-\gamma)\boldsymbol{y}]\leqslant\gamma f(\boldsymbol{x})+(1-\gamma)f(\boldsymbol{y})\\&<f(\boldsymbol{z}_\beta)=f[\beta\boldsymbol{x}+(1-\beta)\boldsymbol{y}].\end{aligned} \tag{5.1.1-9}$$

记 $t=\dfrac{\beta-\gamma}{1-\gamma}$, 有 $0<t<1$, 得 $\boldsymbol{z}_\beta=t\boldsymbol{x}+(1-t)\boldsymbol{z}_\gamma$.

如果 $f(\boldsymbol{x})\leqslant f(\boldsymbol{z}_\gamma)$, 由 f 在 C 上是拟凸函数得 $f(\boldsymbol{z}_\beta)=f(t\boldsymbol{x}+(1-t)\boldsymbol{z}_\gamma)\leqslant f(\boldsymbol{z}_\gamma)$, 这与 (5.1.1-9) 式矛盾.

如果 $f(\boldsymbol{x}) > f(\boldsymbol{z}_\gamma)$, 再由 f 在 C 上是拟凸函数和 $f(\boldsymbol{x}) < f(\boldsymbol{y})$ 得

$$f(\boldsymbol{z}_\beta) = f(t\boldsymbol{x} + (1-t)\boldsymbol{z}_\gamma) \leqslant f(\boldsymbol{x}) < \beta f(\boldsymbol{x}) + (1-\beta) f(\boldsymbol{y}) < f(\boldsymbol{z}_\beta),$$

上式矛盾.

(ii) 设 $f(\boldsymbol{x}) > f(\boldsymbol{y})$. 由 (5.1.1-8) 式和 A 在 $[0,1]$ 中的稠密性知, 存在 $\gamma \in A$ 和 $\gamma > \beta$ 使得

$$f(\boldsymbol{z}_\beta) = f[\beta\boldsymbol{x} + (1-\beta)\boldsymbol{y}] > \gamma f(\boldsymbol{x}) + (1-\gamma) f(\boldsymbol{y}).$$

据此, 由 $\gamma \in A$ 可得

$$\begin{aligned} f(\boldsymbol{z}_\gamma) = f[\gamma\boldsymbol{x} + (1-\gamma)\boldsymbol{y}] &\leqslant \gamma f(\boldsymbol{x}) + (1-\gamma) f(\boldsymbol{y}) \\ &< f(\boldsymbol{z}_\beta) = f[\beta\boldsymbol{x} + (1-\beta)\boldsymbol{y}]. \end{aligned} \tag{5.1.1-10}$$

记 $t = \dfrac{\gamma-\beta}{\gamma}$, 有 $0 < t < 1$, 得 $\boldsymbol{z}_\beta = t\boldsymbol{y} + (1-t)\boldsymbol{z}_\gamma$.

如果 $f(\boldsymbol{y}) \leqslant f(\boldsymbol{z}_\gamma)$, 由 f 在 C 上是拟凸函数得 $f(\boldsymbol{z}_\beta) = f(t\boldsymbol{x} + (1-t)\boldsymbol{z}_\gamma) \leqslant f(\boldsymbol{z}_\gamma)$, 这与 (5.1.1-9) 式矛盾.

如果 $f(\boldsymbol{y}) > f(\boldsymbol{z}_\gamma)$, 再由 f 在 C 上是拟凸函数和 $f(\boldsymbol{x}) > f(\boldsymbol{y})$ 得

$$f(\boldsymbol{z}_\beta) = f(t\boldsymbol{x} + (1-t)\boldsymbol{z}_\gamma) \leqslant f(\boldsymbol{y}) < \beta f(\boldsymbol{x}) + (1-\beta) f(\boldsymbol{y}) < f(\boldsymbol{z}_\beta),$$

上式矛盾.

最后, 由 (1) 和 (2) 的证明得到

$$f[\lambda\boldsymbol{x} + (1-\lambda)\boldsymbol{y}] \leqslant \lambda f(\boldsymbol{x}) + (1-\lambda) f(\boldsymbol{y}), \quad \forall \lambda \in (0,1). \qquad \square$$

在定理 5.1.7 中取 $\alpha = \dfrac{1}{2}$, 得下面推论.

推论 5.1.2 设 $C \subset \mathbf{R}^n$ 是非空开凸集和函数 $f: C \to \mathbf{R}$, 则 f 在 C 上是凸函数当且仅当 f 在 C 上是拟凸函数, 且对于任意 $\boldsymbol{x}, \boldsymbol{y} \in C$ 有

$$f\left[\frac{1}{2}\boldsymbol{x} + \frac{1}{2}\boldsymbol{y}\right] \leqslant \frac{1}{2} f(\boldsymbol{x}) + \frac{1}{2} f(\boldsymbol{y}).$$

本小节得到了凸函数、拟凸函数和严格拟凸之间的蕴含关系, 其中有两个重要结论:

1) 拟凸函数等价于它的限制函数的拟凸性;

2) 凸函数等价于它是拟凸的且满足一点凸性.

5.1.2 连续拟凸函数

本小节讨论具有下半连续、上半连续或连续的拟凸函数性质, 这些结果深刻地揭示了拟凸函数的特性 (如拟凸函数的充分性条件和必要性条件)[68-72,94].

定理 5.1.8 设 $C \subset \mathbf{R}^n$ 是非空凸集, f: $\mathrm{cl}C \to \mathbf{R}$ 是连续函数, 且在 $\mathrm{int}C$ 上是拟凸函数, 则 f 在 $\mathrm{cl}C$ 上是拟凸函数.

证明 设 $\boldsymbol{x}, \boldsymbol{y} \in \mathrm{cl}C$ 和任意 $\lambda \in (0,1)$, 若 $\boldsymbol{x}, \boldsymbol{y} \in \mathrm{int}C$, 则由定理假设, 有

$$f\left[(1-\lambda)\boldsymbol{x} + \lambda\boldsymbol{y}\right] \leqslant \max\left\{f(\boldsymbol{x}), f(\boldsymbol{y})\right\}.$$

否则, 在 $\mathrm{int}C$ 中存在序列 $\{\boldsymbol{x}_i\}$ 和 $\{\boldsymbol{y}_i\}$ 分别收敛到 $\boldsymbol{x}$ 和 $\boldsymbol{y}$, 有

$$f\left[(1-\lambda)\boldsymbol{x}_i + \lambda\boldsymbol{y}_i\right] \leqslant \max\left\{f(\boldsymbol{x}_i), f(\boldsymbol{y}_i)\right\}.$$

由于 f 是连续函数, 令 $i \to +\infty$ 得到

$$f\left[(1-\lambda)\boldsymbol{x} + \lambda\boldsymbol{y}\right] \leqslant \max\left\{f(\boldsymbol{x}), f(\boldsymbol{y})\right\}.$$

上式说明 f 在 $\mathrm{cl}C$ 上是拟凸函数. □

下面的引理 5.1.1 说明, 如果 f 在 C 上满足一点拟凸性, 则 f 关于区间 $(0,1)$ 中任意一点在 C 上几乎处处满足一点拟凸性. 换言之, f 在 C 上的拟凸条件 (5.1.1-1) 式在区间 $(0,1)$ 上是几乎处处成立的.

引理 5.1.1 设 $C \subset \mathbf{R}^n$ 是非空凸集, 函数 f: $C \to \mathbf{R}$, 且 f 满足一点拟凸性, 则集合

$$A = \left\{\lambda \in [0,1] \,|\, f\left[\lambda\boldsymbol{x} + (1-\lambda)\boldsymbol{y}\right] \leqslant \max\left\{f(\boldsymbol{x}), f(\boldsymbol{y})\right\}, \forall \boldsymbol{x}, \boldsymbol{y} \in C\right\} \quad (5.1.2\text{-}1)$$

在 $[0,1]$ 中稠密.

证明 设存在一点 $\alpha \in (0,1)$ 和任意 $\boldsymbol{x}, \boldsymbol{y} \in C$ 满足

$$f\left[\alpha\boldsymbol{x} + (1-\alpha)\boldsymbol{y}\right] \leqslant \min\left\{f(\boldsymbol{x}), f(\boldsymbol{y})\right\}. \quad (5.1.2\text{-}2)$$

显然 $0, 1 \in A$, 即集合 A 不是单点集. 下面假设 A 在 $[0,1]$ 中不稠密, 则存在 $\lambda_0 \in (0,1)$ 和它的一个邻域 $U(\lambda_0)$ 满足 $U(\lambda_0) \cap A = \varnothing$. 据此, 令

$$\lambda_1 = \inf\left\{\lambda \in A | \lambda \geqslant \lambda_0\right\}, \quad \lambda_2 = \sup\left\{\lambda \in A | \lambda \leqslant \lambda_0\right\}, \quad (5.1.2\text{-}3)$$

则 $0 \leqslant \lambda_2 < \lambda_1 \leqslant 1$. 因 $\alpha \in (0,1)$, 我们可选 $t_1, t_2 \in A$ 满足 $t_1 \geqslant \lambda_1$ 和 $t_2 \leqslant \lambda_2$, 且

$$\max\left\{\alpha, (1-\alpha)\right\}(t_1 - t_2) < \lambda_1 - \lambda_2. \quad (5.1.2\text{-}4)$$

记 $\bar{\lambda} = \alpha t_1 + (1-\alpha) t_2$, 因 $\forall \boldsymbol{x}, \boldsymbol{y} \in C$, 有

$$\bar{\lambda}\boldsymbol{x} + \left(1-\bar{\lambda}\right)\boldsymbol{y} = \alpha\left(t_1\boldsymbol{x} + (1-t_1)\boldsymbol{y}\right) + (1-\alpha)\left(t_2\boldsymbol{x} + (1-t_2)\boldsymbol{y}\right).$$

由 A 的定义得

$$\begin{aligned} f\left(\bar{\lambda}\boldsymbol{x} + \left(1-\bar{\lambda}\right)\boldsymbol{y}\right) &= f\left(\alpha\left(t_1\boldsymbol{x} + (1-t_1)\boldsymbol{y}\right) + (1-\alpha)\left(t_2\boldsymbol{x} + (1-t_2)\boldsymbol{y}\right)\right) \\ &\leqslant \max\left\{f\left(t_1\boldsymbol{x} + (1-t_1)\boldsymbol{y}\right), f\left(t_2\boldsymbol{x} + (1-t_2)\boldsymbol{y}\right)\right\} \\ &\leqslant \max\left\{f(\boldsymbol{x}), f(\boldsymbol{y})\right\}, \end{aligned}$$

即得 $\bar{\lambda} \in A$. 如果 $\bar{\lambda} \geqslant \lambda_0$, 则由 (5.1.2-3) 式有 $\lambda_1 \leqslant \bar{\lambda}$. 另外, 由 (5.1.2-4) 式有

$$\bar{\lambda} - t_2 = \alpha(t_1 - t_2) < \lambda_1 - \lambda_2,$$

于是得 $\lambda_1 > \bar{\lambda} - t_2 + \lambda_2 \geqslant \bar{\lambda} - \lambda_2 + \lambda_2 = \bar{\lambda}$, 因此, 得矛盾. 如果 $\bar{\lambda} \leqslant \lambda_0$, 类似地, 由 (5.1.2-3) 和 (5.1.2-5) 式导出同样的矛盾. 因此, 引理成立. □

下面可根据引理 5.1.1 证明任何一个上半连续或下半连续函数, 函数的拟凸性等价于该函数满足一点拟凸性, 见下面的结果.

定理 5.1.9 *设 $C \subset \mathbf{R}^n$ 是非空凸集, $f: C \to \mathbf{R}$ 是上半连续函数, 则 f 在 C 上是拟凸函数当且仅当 f 满足一点拟凸性.*

证明 因为必要性显然, 这里仅证明充分性.

设对于任意 $\boldsymbol{x}, \boldsymbol{y} \in C$, 存在一个 $\alpha \in (0,1)$ 有

$$f[\alpha\boldsymbol{x} + (1-\alpha)\boldsymbol{y}] \leqslant \max\{f(\boldsymbol{x}), f(\boldsymbol{y})\}. \tag{5.1.2-5}$$

假设 f 在 C 上不是拟凸函数, 则存在 $\boldsymbol{x}, \boldsymbol{y} \in C$ 和 $\bar{\lambda} \in (0,1)$ 使得

$$f\left[\bar{\lambda}\boldsymbol{x} + \left(1-\bar{\lambda}\right)\boldsymbol{y}\right] > \max\{f(\boldsymbol{x}), f(\boldsymbol{y})\}. \tag{5.1.2-6}$$

记 $\boldsymbol{z} = \bar{\lambda}\boldsymbol{x} + \left(1-\bar{\lambda}\right)\boldsymbol{y}$, 设 A 由 (5.1.2-1) 式定义, 因此存在一个收敛于的 $\bar{\lambda}$ 序列 $\{\lambda_i\} \subset A$ 满足 $\lambda_i \leqslant \bar{\lambda}$. 记

$$\boldsymbol{x}_i = \frac{\bar{\lambda} - \lambda_i}{1-\lambda_i}\boldsymbol{x} + \left(1 - \frac{\bar{\lambda} - \lambda_i}{1-\lambda_i}\right)\boldsymbol{y},$$

则有 $\boldsymbol{x}_i \to \boldsymbol{y}\,(i \to +\infty)$. 于是得到下式

$$\boldsymbol{z} = \lambda_i\boldsymbol{x} + (1-\lambda_i)\boldsymbol{x}_i, \tag{5.1.2-7}$$

由 $0<\dfrac{\bar{\lambda}-\lambda_i}{1-\lambda_i}<1$ 有 $\boldsymbol{x}_i\in C$. 由 f 是上半连续函数, 对于任意给定的 $\varepsilon>0$, 存在一个正数 N 使得对于任意 $i>N$ 有 $f(\boldsymbol{x}_i)\leqslant f(\boldsymbol{y})+\varepsilon$. 于是由 (5.2.1-7) 式和 A 的定义, 对于 $i>N$, 有

$$f(\boldsymbol{z})=f(\lambda_i\boldsymbol{x}+(1-\lambda_i)\boldsymbol{x}_i)\leqslant\max\{f(\boldsymbol{x}),f(\boldsymbol{x}_i)\}\leqslant\max\{f(\boldsymbol{x}),f(\boldsymbol{y})+\varepsilon\}.$$

在上式中, 令 $\varepsilon\to0$, 则可得到 $f(\boldsymbol{z})\leqslant\max\{f(\boldsymbol{x}),f(\boldsymbol{y})\}$, 与 (5.1.2-6) 式矛盾. 因此, 对于任意 $\boldsymbol{x},\boldsymbol{y}\in C$ 和任意 $\lambda\in(0,1)$ 有 $f[(1-\lambda)\boldsymbol{x}+\lambda\boldsymbol{y}]\leqslant\max\{f(\boldsymbol{x}),f(\boldsymbol{y})\}$ 成立. □

对于下半连续函数, 我们有类似定理 5.1.9 的结论.

定理 5.1.10 设 $C\subset\mathbf{R}^n$ 是非空凸集, $f: C\to\mathbf{R}$ 是下半连续函数, 有下面结论成立:

(1) 如果 f 满足一点拟凸性, 则 f 在 C 上是拟凸函数;

(2) 如果 f 满足一点半严格拟凸性, 则 f 在 C 上是拟凸函数.

证明 (1) 设对于任意 $\boldsymbol{x},\boldsymbol{y}\in C$, 存在一个 $\lambda\in(0,1)$ 有

$$f[(1-\lambda)\boldsymbol{x}+\lambda\boldsymbol{y}]\leqslant\max\{f(\boldsymbol{x}),f(\boldsymbol{y})\}. \tag{5.1.2-8}$$

根据定理 5.1.3, 只要证明水平集 $C_{f\leqslant\alpha}=\{\boldsymbol{x}\in C\,|\,f(\boldsymbol{x})\leqslant\alpha\}$ 是凸集即可. 由于 $f: C\to\mathbf{R}$ 是下半连续函数, 则 $C_{f\leqslant\alpha}$ 是闭集. 根据定理假设, 对于任意 $\boldsymbol{x},\boldsymbol{y}\in C_{f\leqslant\alpha}$, 存在一个 $\lambda\in(0,1)$ 使得

$$f[(1-\lambda)\boldsymbol{x}+\lambda\boldsymbol{y}]\leqslant\max\{f(\boldsymbol{x}),f(\boldsymbol{y})\}\leqslant\alpha,$$

即有 $(1-\lambda)\boldsymbol{x}+\lambda\boldsymbol{y}\in C_{f\leqslant\alpha}$, 根据引理 4.3.1 得 $C_{f\leqslant\alpha}$ 是凸集.

(2) 设对于任意 $\boldsymbol{x},\boldsymbol{y}\in C$ 且 $f(\boldsymbol{x})\neq f(\boldsymbol{y})$, 存在一个 $\lambda\in(0,1)$ 有

$$f[(1-\lambda)\boldsymbol{x}+\lambda\boldsymbol{y}]<\max\{f(\boldsymbol{x}),f(\boldsymbol{y})\}. \tag{5.1.2-9}$$

根据 (1) 的结论, 只需证明对于任意 $\boldsymbol{x},\boldsymbol{y}\in C$, 存在一个 $\lambda\in(0,1)$ 使 (5.1.2-8) 式成立. 采用反证法, 假设存在 $\boldsymbol{x},\boldsymbol{y}\in C$ 和任意 $\alpha\in(0,1)$ 使得

$$f[(1-\alpha)\boldsymbol{x}+\alpha\boldsymbol{y}]>\max\{f(\boldsymbol{x}),f(\boldsymbol{y})\}. \tag{5.1.2-10}$$

若 $f(\boldsymbol{x})\neq f(\boldsymbol{y})$, 则在 (5.1.2-10) 式中取 $\alpha=\lambda$ 时, (5.1.2-10) 与 (5.1.2-9) 式矛盾.

若 $f(\boldsymbol{x})=f(\boldsymbol{y})$, 则 (5.1.2-10) 式为

$$f[(1-\alpha)\boldsymbol{x}+\alpha\boldsymbol{y}]>f(\boldsymbol{x})=f(\boldsymbol{y}),\quad\forall\alpha\in(0,1). \tag{5.1.2-11}$$

注意 $\lambda(1-\alpha), 1-\lambda(1-\alpha) \in (0,1)$，根据 (5.1.2-9), (5.1.2-10) 和 (5.1.2-11) 式得, 对任意的 $\alpha \in (0,1)$, 有

$$\begin{aligned} f(\boldsymbol{y}) &< f[(1-\lambda(1-\alpha))\boldsymbol{x}+\lambda(1-\alpha)\boldsymbol{y}] \\ &= f[\lambda(\alpha\boldsymbol{x}+(1-\alpha)\boldsymbol{y})+(1-\lambda)\boldsymbol{x}] \\ &< \max\{f(\alpha\boldsymbol{x}+(1-\alpha)\boldsymbol{y}), f(\boldsymbol{x})\} \\ &< f(\alpha\boldsymbol{x}+(1-\alpha)\boldsymbol{y}). \end{aligned}$$

同样再结合上式, 还可以得到对任意的 $\alpha \in (0,1)$ 有

$$\begin{aligned} & f[\lambda((1-\lambda(1-\alpha))\boldsymbol{x}+\lambda(1-\alpha)\boldsymbol{y})+(1-\lambda)\boldsymbol{y}] \\ <& \max\{f((1-\lambda(1-\alpha))\boldsymbol{x}+\lambda(1-\alpha)\boldsymbol{y}), f(\boldsymbol{y})\} \\ =& f((1-\lambda(1-\alpha))\boldsymbol{x}+\lambda(1-\alpha)\boldsymbol{y}) \\ <& f(\alpha\boldsymbol{x}+(1-\alpha)\boldsymbol{y}). \end{aligned}$$

在上式中取 $\alpha=\dfrac{\lambda}{1+\lambda}$ 得到下面一个矛盾的不等式:

$$f\left(\frac{\lambda}{1+\lambda}\boldsymbol{x}+\frac{1}{1+\lambda}\boldsymbol{y}\right)<f\left(\frac{\lambda}{1+\lambda}\boldsymbol{x}+\frac{1}{1+\lambda}\boldsymbol{y}\right).$$

上面已经证明 (1) 的假设成立, 所以 f 在 C 上是拟凸函数. □

定理 5.1.11　设 $C \subset \mathbf{R}^n$ 是非空凸集, $f: C \to \mathbf{R}$ 在 C 上是拟凸函数, 且 f 满足一点严格拟凸性, 则 f 在 C 上是严格拟凸函数.

证明　首先, 设对于任意 $\boldsymbol{x}, \boldsymbol{y} \in C$ 且 $\boldsymbol{x} \neq \boldsymbol{y}$, 存在一个 $\lambda \in (0,1)$ 有

$$f[(1-\lambda)\boldsymbol{x}+\lambda\boldsymbol{y}]<\max\{f(\boldsymbol{x}), f(\boldsymbol{y})\}.$$

采用反证法. f 在 C 上不是严格拟凸函数, 那么存在 $\boldsymbol{x}, \boldsymbol{y} \in C$ 且 $\boldsymbol{x} \neq \boldsymbol{y}$, 存在一个 $\alpha \in (0,1)$, 有

$$f[(1-\alpha)\boldsymbol{x}+\alpha\boldsymbol{y}] \geqslant \max\{f(\boldsymbol{x}), f(\boldsymbol{y})\}.$$

记 $\boldsymbol{z}=(1-\alpha)\boldsymbol{x}+\alpha\boldsymbol{y}$, 因为 f 在 C 上是拟凸函数, 则有 $f(\boldsymbol{z}) \leqslant \max\{f(\boldsymbol{x}), f(\boldsymbol{y})\}$, 得

$$f(\boldsymbol{z})=\max\{f(\boldsymbol{x}), f(\boldsymbol{y})\}. \tag{5.1.2-12}$$

选取 λ_1, λ_2 满足 $0<\lambda_1<\alpha<\lambda_2<1$ 和

$$\alpha=(1-\lambda)\lambda_1+\lambda\lambda_2.$$

令 $\boldsymbol{z}_1=(1-\lambda_1)\boldsymbol{x}+\lambda_1\boldsymbol{y}$, $\boldsymbol{z}_2=(1-\lambda_2)\boldsymbol{x}+\lambda_2\boldsymbol{y}$, 有

$$(1-\lambda)\boldsymbol{z}_1+\lambda\boldsymbol{z}_2=(1-\alpha)\boldsymbol{x}+\alpha\boldsymbol{y}=\boldsymbol{z}. \tag{5.1.2-13}$$

再由函数 f 在凸集 $C\subset\mathbf{R}^n$ 上是拟凸函数得

$$f(\boldsymbol{z}_1)\leqslant\max\{f(\boldsymbol{x}),f(\boldsymbol{y})\},\quad f(\boldsymbol{z}_2)\leqslant\max\{f(\boldsymbol{x}),f(\boldsymbol{y})\}. \tag{5.1.2-14}$$

由定理的假设、(5.1.2-13) 和 (5.1.2-14) 式得

$$f(\boldsymbol{z})<\max\{f(\boldsymbol{z}_1),f(\boldsymbol{z}_2)\}\leqslant\max\{f(\boldsymbol{x}),f(\boldsymbol{y})\},$$

则上式与 (5.1.2-12) 式矛盾. □

下面是有关拟凸函数优化的全局最优性性质, 即判定拟凸函数优化的局部最优性是其全局最优性的结果.

性质 5.1.5 设 $C\subset\mathbf{R}^n$ 是非空凸集, $f: C\to\mathbf{R}$ 在 C 上是拟凸函数. 如果 $\boldsymbol{x}^*\in C$ 是 $\min\limits_{\boldsymbol{x}\in C}f(\boldsymbol{x})$ 的严格局部最优解, 即存在一个 $\boldsymbol{x}^*$ 的邻域 $O(\boldsymbol{x}^*)$ 有

$$f(\boldsymbol{x}^*)<f(\boldsymbol{x}),\quad\forall\boldsymbol{x}\in O(\boldsymbol{x}^*)\cap C,$$

则 $\boldsymbol{x}^*$ 是 $\min\limits_{\boldsymbol{x}\in C}f(\boldsymbol{x})$ 的严格全局最优解, 其所有严格全局最优解构成的集合是一个凸集.

证明 采用反证法. 假设 $\boldsymbol{x}^*$ 不是 $\min\limits_{\boldsymbol{x}\in C}f(\boldsymbol{x})$ 的严格全局最优解, 则存在一个 $\boldsymbol{x}'\in C$ 和 $\boldsymbol{x}^*\neq\boldsymbol{x}'$ 使得 $f(\boldsymbol{x}^*)\geqslant f(\boldsymbol{x}')$. 由 f 在 C 上是拟凸函数得

$$f((1-\lambda)\boldsymbol{x}^*+\lambda\boldsymbol{x}')\leqslant f(\boldsymbol{x}^*). \tag{5.1.2-15}$$

对于充分小的 $\lambda\in(0,1)$, 使得 $(1-\lambda)\boldsymbol{x}^*+\lambda\boldsymbol{x}'\in O(\boldsymbol{x}^*)\cap C$. 则说明 (5.1.2-15) 式与 $\min\limits_{\boldsymbol{x}\in C}f(\boldsymbol{x})$ 的严格局部最优解矛盾.

设 $\alpha=\min\limits_{\boldsymbol{x}\in C}f(\boldsymbol{x})$, 定义水平集合 $\{\boldsymbol{x}\in C|\alpha=f(\boldsymbol{x})\}=\{\boldsymbol{x}\in C|\alpha\geqslant f(\boldsymbol{x})\}$ 是由全部严格全局最优解构成的集合, 据此, 由定理 5.1.3 知定理结论成立. □

下面是近似拟凸函数的定义.

定义 5.1.2 设 $C\subset\mathbf{R}^n$ 是一个非空集合. 如果 $f: C\to\mathbf{R}$ 的水平集 $C_{f\leqslant\alpha}=\{\boldsymbol{x}\in C\,|\,f(\boldsymbol{x})\leqslant\alpha\}$ 的闭包是凸集, 则称 f 在 C 上是近似拟凸函数.

通过定义 4.3.2 知, 集合 C 满足一点凸性, 则称 C 是一个近似凸集. 容易证明近似凸集的闭包也是近似凸集, 并且是凸集 (见文献 [93]). 下面的定理表明一个函数满足一点拟凸性, 其水平集也满足一点拟凸性; 如果函数是下半连续的, 它也是拟凸函数.

定理 5.1.12　设 $C \subset \mathbf{R}^n$ 是一个非空凸集, 函数 $f: C \to \mathbf{R}$ 满足一点拟凸性, 则 f 的水平集 $C_{f \leqslant \alpha} = \{\boldsymbol{x} \in C \,|\, f(\boldsymbol{x}) \leqslant \alpha\}$ 是近似凸集, 且 $C_{f \leqslant \alpha}$ 的闭包是凸集. 进一步, 如果 $f: C \to \mathbf{R}$ 是下半连续函数, 则 f 是拟凸函数.

证明　设任意 $\boldsymbol{x}, \boldsymbol{y} \in C_{f \leqslant \alpha}$, 有 $f(\boldsymbol{x}) \leqslant \alpha$, $f(\boldsymbol{y}) \leqslant \alpha$, 由定理所设可得

$$f[(1-\lambda)\boldsymbol{x} + \lambda\boldsymbol{y}] \leqslant \max\{f(\boldsymbol{x}), f(\boldsymbol{y})\} \leqslant \alpha,$$

有 $(1-\lambda)\boldsymbol{x} + \lambda\boldsymbol{y} \in C_{f \leqslant \alpha}$, 由定义 5.1.2 知 $C_{f \leqslant \alpha}$ 是近似凸集. 如果 $f: C \to \mathbf{R}$ 是下半连续函数, 由定理 5.1.10 知 f 是拟凸函数. □

由定理 5.1.12 知下面结论成立, 且表明拟凸函数与凸函数有类似的特性 (见定理 4.3.2).

性质 5.1.6　下半连续的近似拟凸函数是拟凸函数.

定理 5.1.13[94]　设 $C \subset \mathbf{R}^n$ 是一个非空开凸集, $f: C \to \mathbf{R}$ 是上半连续函数. 如果 f 在 C 上满足一点严格拟凸性, 则 f 在 C 上是严格拟凸函数.

证明　由定理 5.1.9 和定理 5.1.11 得结论成立. □

本小节, 得到下面拟凸函数的几个重要性质:

1) 拟凸函数满足上半连续性, 等价于该函数满足一点拟凸性;

2) 当函数满足下半连续性且满足一点拟凸性或一点半严格拟凸性时, 则它是拟凸函数;

3) 满足一点严格拟凸性的拟凸函数是严格拟凸的;

4) 拟凸函数的严格局部极小点也是全局严格极小点;

5) 函数满足一点拟凸性, 则它的水平集是近似凸集 (一点凸集);

6) 下半连续的近似拟凸函数是拟凸函数;

7) 满足一点严格拟凸性的上半连续函数是严格拟凸函数.

5.1.3　可微拟凸函数

本小节讨论可微拟凸函数的性质, 其中包含可微拟凸函数判定的必要性和充分性条件.

定义 5.1.3　设 $C \subset \mathbf{R}^n$ 是非空凸集. 如果拟凸函数 $f: C \to \mathbf{R}$ 是可微函数, 则称 f 在 C 上是可微拟凸函数. 如果拟凹函数 $f: C \to \mathbf{R}$ 是可微函数, 则称 f 在 C 上是可微拟凹函数.

引理 5.1.2 设 $I\subset \mathbf{R}$ 是一个非空区间, $\varphi: I\to \mathbf{R}$ 是可微函数, 则 φ 在 I 上是拟凸函数的充要条件是对于任意 $x,y\in I$ 有

$$\varphi(x)\geqslant \varphi(y)\Rightarrow \varphi'(x)(y-x)\leqslant 0. \tag{5.1.3-1}$$

证明 设 φ 在 I 上是拟凸函数, 对于任意的 $x,y\in I$, 且 $x<y$, 若 $\varphi(x)\geqslant \varphi(y)$, 有

$$\varphi(x)\geqslant \varphi(t),\quad \forall t\in[x,y].$$

于是 φ 在 x 处是局部单调递减函数, 所以有 $\varphi'(x)(y-x)\leqslant 0$.

反过来, 若 (5.1.3-1) 式成立. 假设 φ 在 I 上不是拟凸函数, 则存在 $x,y\in I$, 且 $x<y$, 有

$$M=\max\{\varphi(t)\,|t\in[x,y]\}>\max\{\varphi(x),\varphi(y)\}.$$

设 $\tau=\inf\{t\in[x,y]\ |\varphi(t)=M\}$, 由 φ 的连续性, 存在 $\varepsilon>0$, 对于任意的 $t\in(\tau-\varepsilon,\tau)$ 满足

$$\varphi(y)<\varphi(t)<M.$$

在区间 $[\tau-\varepsilon,\tau]$ 上使用微分中值定理, 存在 $t^*\in(\tau-\varepsilon,\tau)$ 使得

$$\varepsilon\varphi'(t^*)=\varphi'(t^*)(\tau-(\tau-\varepsilon))=\varphi(\tau)-\varphi(\tau-\varepsilon)=M-\varphi(\tau-\varepsilon)>0,$$

由上式得 $\varphi'(t^*)>0$. 同时, 由 $\varphi(y)<\varphi(t^*)$ 和 (5.1.3-1) 式得 $\varphi'(t^*)\leqslant 0$, 导致矛盾. □

引理 5.1.2 说明了单变量单调函数都是拟凸函数.

定理 5.1.14 设 $C\subset \mathbf{R}^n$ 是非空凸集, $f:\ C\to \mathbf{R}$ 是可微函数, 则 f 在 C 上是拟凸函数的充要条件是对于任意 $\boldsymbol{x},\boldsymbol{y}\in C$ 有

$$f(\boldsymbol{x})\geqslant f(\boldsymbol{y})\Rightarrow \nabla f(\boldsymbol{x})^{\mathrm{T}}(\boldsymbol{y}-\boldsymbol{x})\leqslant 0. \tag{5.1.3-2}$$

证明 假设 f 在 C 上是拟凸函数, 再设 $\boldsymbol{x},\boldsymbol{y}\in C$, $f(\boldsymbol{x})\geqslant f(\boldsymbol{y})$ 以及一个单变量限制函数

$$\varphi(t)=f(\boldsymbol{x}+t(\boldsymbol{y}-\boldsymbol{x})),\quad \forall t\in[0,1].$$

有 $\varphi(0)=f(\boldsymbol{x})\geqslant f(\boldsymbol{y})=\varphi(1)$. 由引理 5.1.2 得 $\varphi'(0)(1-0)\leqslant 0$, 即

$$\nabla f(\boldsymbol{x})^{\mathrm{T}}(\boldsymbol{y}-\boldsymbol{x})\leqslant 0.$$

反过来, 假设 (5.1.3-2) 式成立. 如果 f 在 C 上不是拟凸函数, 设 $\boldsymbol{x},\boldsymbol{y}\in C$, $f(\boldsymbol{x})\geqslant f(\boldsymbol{y})$, 再设单变量限制函数

$$\varphi(t)=f(\boldsymbol{x}+t(\boldsymbol{y}-\boldsymbol{x})),\quad \forall t\in[0,1].$$

由推论 5.1.1 知 $\varphi(t)$ 不是拟凸函数, 则存在 $t_1, t_2 \in [0,1], \varphi(t_1) \geqslant \varphi(t_2)$, 且

$$\varphi'(t_1)(t_2 - t_1) > 0.$$

令 $\boldsymbol{x}_1 = \boldsymbol{x} + t_1(\boldsymbol{y} - \boldsymbol{x}), \boldsymbol{x}_2 = \boldsymbol{x} + t_2(\boldsymbol{y} - \boldsymbol{x})$, 则有 $f(\boldsymbol{x}_1) \geqslant f(\boldsymbol{x}_2)$ 和

$$\varphi'(t_1) = \nabla f(\boldsymbol{x}_1)^{\mathrm{T}}(\boldsymbol{y} - \boldsymbol{x}) = \frac{1}{t_2 - t_1}\nabla f(\boldsymbol{x}_1)^{\mathrm{T}}(\boldsymbol{x}_2 - \boldsymbol{x}_1).$$

即有 $\nabla f(\boldsymbol{x}_1)^{\mathrm{T}}(\boldsymbol{x}_2 - \boldsymbol{x}_1) = \varphi'(t_1)(t_2 - t_1) > 0$, 与 (5.1.3-2) 式矛盾. □

定理 5.1.14 说明拟凸函数当一点函数值 $f(\boldsymbol{x})$ 不小于另一点的函数值 $f(\boldsymbol{y})$ 时, 可以确定该点的梯度 $\nabla f(\boldsymbol{x})$ 与两点方向 $\boldsymbol{y} - \boldsymbol{x}$ 的内积不大于 0.

定义 5.1.4　设 $C \subset \mathbf{R}^n$ 是非空凸集, $f: C \to \mathbf{R}$ 是可微函数, $\boldsymbol{x}, \boldsymbol{d} \in C$, 有

(1) 如果点 $\boldsymbol{x}$ 满足 $\nabla f(\boldsymbol{x}) = 0$, 则称点 $\boldsymbol{x}$ 是 f 的稳定点或平稳点;

(2) 如果存在 $\eta > 0$ 满足

$$f(\boldsymbol{x} + t\boldsymbol{d}) < f(\boldsymbol{x}), \quad \forall t \in (0, \eta),$$

则 $\boldsymbol{d}$ 是 f 在点 $\boldsymbol{x}$ 处的下降方向.

可微拟凸函数关于稳定点和下降方向具有下面性质: 若 $\boldsymbol{x}$ 不是 f 的稳定点, 且 $f(\boldsymbol{x}) > f(\boldsymbol{y})$, 则 $f(\boldsymbol{x})$ 沿着方向 $\nabla f(\boldsymbol{x})$ 和 $\boldsymbol{y} - \boldsymbol{x}$ 下降, 具体见下面两个定理.

定理 5.1.15　设 $C \subset \mathbf{R}^n$ 是一个非空开凸集, $f: C \to \mathbf{R}$ 是可微拟凸函数, 则对于任意 $\boldsymbol{x}, \boldsymbol{y} \in C$ 和 $f(\boldsymbol{x}) > f(\boldsymbol{y})$ 有

$$\nabla f(\boldsymbol{x}) \neq 0 \Rightarrow \nabla f(\boldsymbol{x})^{\mathrm{T}}(\boldsymbol{y} - \boldsymbol{x}) < 0. \tag{5.1.3-3}$$

证明　采用反证法. 假设存在 $\boldsymbol{x}, \boldsymbol{y} \in C$ 和 $f(\boldsymbol{x}) > f(\boldsymbol{y})$ 使 (5.1.3-3) 式不成立, 即有

$$\nabla f(\boldsymbol{x}) \neq 0 \Rightarrow \nabla f(\boldsymbol{x})^{\mathrm{T}}(\boldsymbol{y} - \boldsymbol{x}) \geqslant 0.$$

根据 $f: C \to \mathbf{R}$ 是可微拟凸函数和定理 5.1.14 得 $\nabla f(\boldsymbol{x})^{\mathrm{T}}(\boldsymbol{y} - \boldsymbol{x}) = 0$. 由 f 的连续性, 存在 $\varepsilon > 0$ 使得 $\boldsymbol{z} = \varepsilon\nabla f(\boldsymbol{x}) + \boldsymbol{y} \in C$, 且 $f(\boldsymbol{x}) > f(\boldsymbol{z})$. 再由 f 的拟凸性得

$$\nabla f(\boldsymbol{x})^{\mathrm{T}}(\boldsymbol{z} - \boldsymbol{x}) \leqslant 0$$

及

$$\nabla f(\boldsymbol{x})^{\mathrm{T}}(\boldsymbol{z} - \boldsymbol{x}) = \nabla f(\boldsymbol{x})^{\mathrm{T}}(\boldsymbol{z} - \boldsymbol{y}) + \nabla f(\boldsymbol{x})^{\mathrm{T}}(\boldsymbol{y} - \boldsymbol{x})$$

$$= \varepsilon \nabla f(\boldsymbol{x})^{\mathrm{T}} \nabla f(\boldsymbol{x}) > 0,$$

显然, 上述两个不等式矛盾. □

进一步, 对于严格拟凸函数, 若 f 不存在稳定点, 则不稳定点与其附近点之差是下降方向 (见 (5.1.3-4) 式).

定理 5.1.16 设 $C \subset \mathbf{R}^n$ 是非空开凸集和 $f: C \to \mathbf{R}$ 是可微函数, 且 $\nabla f(\boldsymbol{x}) \neq 0\,(\forall \boldsymbol{x} \in C)$, 则 f 在 C 上是严格拟凸函数的充要条件是对于任意 $\boldsymbol{x}, \boldsymbol{y} \in C$ 有

$$f(\boldsymbol{x}) \geqslant f(\boldsymbol{y}) \Rightarrow \nabla f(\boldsymbol{x})^{\mathrm{T}} (\boldsymbol{y} - \boldsymbol{x}) < 0. \tag{5.1.3-4}$$

证明 假设 f 在 C 上是严格拟凸函数, 若 $f(\boldsymbol{x}) > f(\boldsymbol{y})$, 则由定理 5.1.15 得 (5.1.3-4) 式成立. 若 $f(\boldsymbol{x}) = f(\boldsymbol{y})$, 由 f 在 C 上是严格拟凸函数, 取 $\boldsymbol{x}_t = \boldsymbol{x} + t(\boldsymbol{y} - \boldsymbol{x}), \forall t \in (0,1)$, 满足

$$f(\boldsymbol{x}) > f(\boldsymbol{x}_t).$$

那么再由定理 5.1.15 得

$$\nabla f(\boldsymbol{x})^{\mathrm{T}} (\boldsymbol{x}_t - \boldsymbol{x}) < 0 \Rightarrow t \nabla f(\boldsymbol{x})^{\mathrm{T}} (\boldsymbol{y} - \boldsymbol{x}) < 0,$$

上式即说明 (5.1.3-4) 式成立.

反过来, 假设 (5.1.3-4) 式成立. 由定理 5.1.15 知 f 在 C 上是拟凸函数, 采用反证法. 假设 f 在 C 上不是严格拟凸函数, 则存在 $\boldsymbol{x}, \boldsymbol{y} \in C$ 和 $t \in (0,1)$ 使得

$$f(\boldsymbol{x} + t(\boldsymbol{y} - \boldsymbol{x})) \geqslant \max\{f(\boldsymbol{x}), f(\boldsymbol{y})\}.$$

设限制函数

$$\varphi(t) = f(\boldsymbol{x} + t(\boldsymbol{y} - \boldsymbol{x})), \quad \forall t \in [0,1].$$

令 $\boldsymbol{x}_t = \boldsymbol{x} + t(\boldsymbol{y} - \boldsymbol{x}), \forall t \in (0,1)$, 有 $f(\boldsymbol{x}_t) \geqslant \max\{f(\boldsymbol{x}), f(\boldsymbol{y})\}$, 那么由假设可得

$$\varphi'(t) = \nabla f(\boldsymbol{x}_t)^{\mathrm{T}} (\boldsymbol{y} - \boldsymbol{x}_t) < 0.$$

因此, $\varphi(t)$ 在 t 的附近是严格单调递减函数, 于是存在 $\varphi(t)$ 在 $\tau \in (0,t)$ 使得 $\varphi(t) < \varphi(\tau)$. 由于 $\varphi(t)$ 在 $[0,t]$ 上是拟凸函数, $\varphi(0) = f(\boldsymbol{x}) \leqslant \varphi(t) = f(\boldsymbol{x}_t)$, 则根据拟凸性得

$$\varphi(\tau) \leqslant \max\{\varphi(0), \varphi(t)\} = \varphi(t),$$

这显然与 $\varphi(t) < \varphi(\tau)$ 矛盾. □

本小节得到可微拟凸函数的两个重要结论:

1) 可微拟凸函数成立的充要条件是 (5.1.3-2) 式成立;

2) 不存在稳定点的严格可微拟凸函数的充要条件是 (5.1.3-3) 式成立.

5.2 半严格拟凸函数

半严格拟凸函数是一类特殊的广义凸函数 [90-92], 因为拟凸函数不一定是半严格拟凸函数, 半严格拟凸函数不一定是拟凸函数. 本节给出它们之间的蕴涵关系, 以及半严格拟凸函数与严格拟凸函数之间的关系.

5.2.1 基本性质

下面的半严格拟凸函数定义不同于严格拟凸函数定义, 类似于半严格凸函数定义.

定义 5.2.1 设 $C \subset \mathbf{R}^n$ 是非空凸集和函数 $f:\ C \to \mathbf{R}$. 如果对于任意的 $\boldsymbol{x}, \boldsymbol{y} \in C$, $f(\boldsymbol{x}) \neq f(\boldsymbol{y})$ 和对任意的 $\lambda \in (0,1)$, 有

$$f[(1-\lambda)\boldsymbol{x} + \lambda \boldsymbol{y}] < \max\{f(\boldsymbol{x}), f(\boldsymbol{y})\}, \tag{5.2.1-1}$$

或等价地

$$\forall \boldsymbol{x}, \boldsymbol{y} \in C, \quad \forall \lambda \in (0,1), \quad f(\boldsymbol{x}) > f(\boldsymbol{y}) \Rightarrow f[(1-\lambda)\boldsymbol{x} + \lambda \boldsymbol{y}] < f(\boldsymbol{x}),$$

则称 f 在 C 上是半严格拟凸函数. 如果 $-f$ 在 C 上是半严格拟凸函数, 则称 f 在 C 上是半严格拟凹函数.

如果对于任意的 $\boldsymbol{x}, \boldsymbol{y} \in C$ 和 $f(\boldsymbol{x}) \neq f(\boldsymbol{y})$, 存在一点 $\lambda \in (0,1)$ 满足 (5.2.1-1) 式, 则称 f 在 C 上满足一点半严格拟凸性.

凸函数、严格拟凸函数都与半严格拟凸函数有蕴含关系, 见下面定理.

定理 5.2.1 设 $C \subset \mathbf{R}^n$ 是非空凸集, 函数 $f:\ C \to \mathbf{R}$, 有下面结论成立:

(1) 如果 f 是在 C 上的严格拟凸函数, 则 f 是在 C 上的半严格拟凸函数;

(2) 如果 f 是在 C 上的凸函数, 则 f 是在 C 上的半严格拟凸函数.

证明 由定义 5.1.1 和定义 5.2.1 直接可得. □

例 5.2.1 设 $C = [-1,1]$ 和函数

$$f(x) = \begin{cases} 1, & -1 \leqslant x \leqslant 1, x \neq 0, \\ 2, & x = 0. \end{cases}$$

因为 $f(0) = 2 > 1 = \max\{f(1), f(-1)\}$, 所以 f 不是拟凸函数; 但它是半严格拟凸函数.

例 5.2.2 设 $C=[0,2]$ 和函数

$$f(x)=\begin{cases}x, & 0\leqslant x\leqslant 1,\\ 1, & 1<x\leqslant 2.\end{cases}$$

f 是拟凸函数, 但 f 不是半严格拟凸函数.

例 5.2.1 和例 5.2.2 表明拟凸函数与半严格拟凸函数之间不具有蕴含关系. 后面的定理 5.2.9 证明了如果函数是下半连续的半严格拟凸函数, 则函数也是拟凸函数. 接下来, 证明半严格拟凸函数的限制函数上的半严格拟凸性的等价性.

定理 5.2.2 设 $C\subset\mathbf{R}^n$ 是非空凸集, 函数 $f: C\to\mathbf{R}$, 则 f 在 C 上是半严格拟凸函数当且仅当对任意的 $\boldsymbol{x}\in C$ 和 $\boldsymbol{d}\in\mathbf{R}^n$, $\varphi_{\boldsymbol{x},\boldsymbol{d}}(t)$ 是 $t\in T(\boldsymbol{x},\boldsymbol{d})$ 的半严格拟凸函数, 其中

$$\varphi_{\boldsymbol{x},\boldsymbol{d}}(t)=f(\boldsymbol{x}+t\boldsymbol{d}),\quad T(\boldsymbol{x},\boldsymbol{d})=\{t\in\mathbf{R}|\boldsymbol{x}+t\boldsymbol{d}\in C\}.$$

证明 设 f 在 C 上是半严格拟凸函数, 对于任意的 $t_1,t_2\in T(\boldsymbol{x},\boldsymbol{d})$, $\varphi_{\boldsymbol{x},\boldsymbol{d}}(t_1)\neq\varphi_{\boldsymbol{x},\boldsymbol{d}}(t_2)$ 和任意的 $\lambda\in(0,1)$, 有 $f(\boldsymbol{x}+t_1\boldsymbol{d})\neq f(\boldsymbol{x}+t_2\boldsymbol{d})$ 和

$$\begin{aligned}\varphi_{\boldsymbol{x},\boldsymbol{d}}((1-\lambda)t_1+\lambda t_2)&=f((1-\lambda)(\boldsymbol{x}+t_1\boldsymbol{d})+\lambda(\boldsymbol{x}+t_2\boldsymbol{d}))\\&<\max\{f(\boldsymbol{x}+t_1\boldsymbol{d}),f(\boldsymbol{x}+t_2\boldsymbol{d})\}\\&=\max\{\varphi_{\boldsymbol{x},\boldsymbol{d}}(t_1),\varphi_{\boldsymbol{x},\boldsymbol{d}}(t_2)\},\end{aligned}$$

由定义 5.1.1 知 $\varphi_{\boldsymbol{x},\boldsymbol{d}}(t)$ 是 $t\in T(\boldsymbol{x},\boldsymbol{d})$ 的半严格拟凸函数.

反过来, 对任意的 $\boldsymbol{x}\in C$ 和 $\boldsymbol{d}\in\mathbf{R}^n$, $\varphi_{\boldsymbol{x},\boldsymbol{d}}(t)$ 是 $t\in T(\boldsymbol{x},\boldsymbol{d})$ 的半严格拟凸函数. 对于任意的 $\boldsymbol{x}_1,\boldsymbol{x}_2\in C$, $f(\boldsymbol{x}_1)\neq f(\boldsymbol{x}_2)$ 和 $\lambda\in(0,1)$, 设

$$T(\boldsymbol{x}_1,\boldsymbol{x}_2-\boldsymbol{x}_1)=\{t\in\mathbf{R}|\boldsymbol{x}_1+t(\boldsymbol{x}_2-\boldsymbol{x}_1)\in C\},$$

有 $\varphi_{\boldsymbol{x}_1,\boldsymbol{x}_2-\boldsymbol{x}_1}(1)\neq\varphi_{\boldsymbol{x}_1,\boldsymbol{x}_2-\boldsymbol{x}_1}(0)$ 和

$$\begin{aligned}f[(1-\lambda)\boldsymbol{x}_1+\lambda\boldsymbol{x}_2]&=f(\boldsymbol{x}_1+[(1-\lambda)\cdot 1+\lambda\cdot 0](\boldsymbol{x}_2-\boldsymbol{x}_1))\\&=\varphi_{\boldsymbol{x}_1,\boldsymbol{x}_2-\boldsymbol{x}_1}((1-\lambda)\cdot 1+\lambda\cdot 0)\\&<\max\{\varphi_{\boldsymbol{x}_1,\boldsymbol{x}_2-\boldsymbol{x}_1}(1),\varphi_{\boldsymbol{x}_1,\boldsymbol{x}_2-\boldsymbol{x}_1}(0)\}\\&=\max\{f(\boldsymbol{x}_1),f(\boldsymbol{x}_2)\}.\end{aligned}$$

故 f 是在 C 上是半严格拟凸函数. □

半严格拟凸函数等价于该函数线段上的限制函数也是半严格拟凸性.

推论 5.2.1 设 $C \subset \mathbf{R}^n$ 是非空凸集, 函数 $f: C \to \mathbf{R}$, 则 f 在 C 上是半严格拟凸函数当且仅当对任意的 $\boldsymbol{x}, \boldsymbol{y} \in C$, $\varphi_{\boldsymbol{x},\boldsymbol{y}}(t) = f(\boldsymbol{x} + t(\boldsymbol{y} - \boldsymbol{x}))$ 是 $t \in [0,1]$ 的半严格拟凸函数.

下面是半严格拟凸函数的复合函数性质, 半严格拟凸函数在单变量递增函数复合下仍然保持半严格拟凸性.

定理 5.2.3 设 $C \subset \mathbf{R}^n$ 是非空凸集, $f: C \to \mathbf{R}$ 是半严格拟凸函数, 非空集合 $S \subset \mathbf{R}$ 和 $\phi: S \to \mathbf{R}$ 是单调递增函数, 且 $f(C) \subset S$, 则有下面结论成立:

(1) $tf(\boldsymbol{x})$ 是在 C 上的半严格拟凸函数, 其中 $t > 0$;

(2) $\phi(f(\boldsymbol{x}))$ 是在 C 上的半严格拟凸函数.

证明 由定义 5.2.1 容易证明. □

定理 5.2.4 设 $C \subset \mathbf{R}^n$ 是一个非空凸集, 函数 $g, f: C \to \mathbf{R}$, 则有下面结论成立:

(1) 如果 f 在 C 上是非负凸函数且 g 在 C 上是正凹函数, 则 $\dfrac{f}{g}$ 在 C 上是半严格拟凸函数;

(2) 如果 f 在 C 上是非正凸函数且 g 在 C 上是正凸函数, 则 $\dfrac{f}{g}$ 在 C 上是半严格拟凸函数;

(3) 如果 f 在 C 上是凸函数且 g 在 C 上是正仿射函数, 则 $\dfrac{f}{g}$ 在 C 上是半严格拟凸函数.

证明 (1) 设 $\forall \boldsymbol{x}, \boldsymbol{y} \in C$, $\forall \lambda \in (0,1)$ 和 $\dfrac{f(\boldsymbol{x})}{g(\boldsymbol{x})} > \dfrac{f(\boldsymbol{y})}{g(\boldsymbol{y})}$, 由假设有

$$
\begin{aligned}
f[(1-\lambda)\boldsymbol{x} + \lambda \boldsymbol{y}] &\leqslant (1-\lambda) f(\boldsymbol{x}) + \lambda f(\boldsymbol{y}) \\
&< (1-\lambda) f(\boldsymbol{x}) + \lambda \frac{f(\boldsymbol{x})}{g(\boldsymbol{x})} g(\boldsymbol{y}) \\
&= \frac{f(\boldsymbol{x})}{g(\boldsymbol{x})} ((1-\lambda) g(\boldsymbol{x}) + \lambda g(\boldsymbol{y})) \\
&\leqslant \frac{f(\boldsymbol{x})}{g(\boldsymbol{x})} g[(1-\lambda)\boldsymbol{x} + \lambda \boldsymbol{y}],
\end{aligned}
$$

上式即为 $\dfrac{f[(1-\lambda)\boldsymbol{x} + \lambda \boldsymbol{y}]}{g[(1-\lambda)\boldsymbol{x} + \lambda \boldsymbol{y}]} < \dfrac{f(\boldsymbol{x})}{g(\boldsymbol{x})}$, 根据定义 5.2.1 知结论成立.

(2) 和 (3) 的证明过程与 (1) 的证明类似. □

下面证明满足一点拟凸性的半严格拟凸函数是拟凸函数.

定理 5.2.5 设 $C \subset \mathbf{R}^n$ 是非空凸集, $f: C \to \mathbf{R}$ 是半严格拟凸函数. 如果 f 在 C 上满足一点拟凸性, 则 f 在 C 上是拟凸函数.

证明 由定理所设知, 对于任意 $\boldsymbol{x}, \boldsymbol{y} \in C$, 存在一个 $\lambda \in (0,1)$ 使

$$f[(1-\lambda)\boldsymbol{x} + \lambda\boldsymbol{y}] \leqslant \max\{f(\boldsymbol{x}), f(\boldsymbol{y})\}. \tag{5.2.1-2}$$

采用反证法. 假设 f 在 C 上不是拟凸函数, 则存在 $\boldsymbol{x}, \boldsymbol{y} \in C$ 和 $t \in (0,1)$ 使得

$$f[(1-t)\boldsymbol{x} + t\boldsymbol{y}] > \max\{f(\boldsymbol{x}), f(\boldsymbol{y})\}. \tag{5.2.1-3}$$

不妨设 $f(\boldsymbol{x}) \leqslant f(\boldsymbol{y})$, 记 $(1-t)\boldsymbol{x} + t\boldsymbol{y} = \boldsymbol{z}$, 如果 $f(\boldsymbol{x}) < f(\boldsymbol{y})$, 由 (5.2.1-3) 式得 $f(\boldsymbol{z}) > f(\boldsymbol{y})$. 由函数 f 是 C 上的半严格拟凸函数定义知 $f(\boldsymbol{z}) < f(\boldsymbol{y})$, 得到矛盾. 那么有 $f(\boldsymbol{x}) = f(\boldsymbol{y})$, 则有

$$f(\boldsymbol{z}) > f(\boldsymbol{x}) = f(\boldsymbol{y}).$$

下面分两种情形讨论.

(1) 若 $0 < t < \lambda < 1$, 设 $\boldsymbol{z}_1 = \left(1 - \dfrac{t}{\lambda}\right)\boldsymbol{x} + \dfrac{t}{\lambda}\boldsymbol{y}$, 则有

$$\boldsymbol{z} = (1-t)\boldsymbol{x} + t\boldsymbol{y} = \lambda\left[\left(1 - \frac{t}{\lambda}\right)\boldsymbol{x} + \frac{t}{\lambda}\boldsymbol{y}\right] + (1-\lambda)\boldsymbol{x} = \lambda\boldsymbol{z}_1 + (1-\lambda)\boldsymbol{x}.$$

由上式和 (5.2.1-3) 式有

$$f(\boldsymbol{z}) \leqslant \max\{f(\boldsymbol{z}_1), f(\boldsymbol{x})\} = f(\boldsymbol{z}_1).$$

再设 $\beta = \dfrac{t(1-\lambda)}{\lambda(1-t)}$, 由 $0 < t < \lambda < 1$ 得 $0 < \beta < 1$, 则

$$\boldsymbol{z}_1 = \left(1 - \frac{t}{\lambda}\right)\boldsymbol{x} + \frac{t}{\lambda}\boldsymbol{y} = \left(1 - \frac{t}{\lambda}\right)\left(\frac{\boldsymbol{z}}{1-t} - \frac{t\boldsymbol{y}}{1-t}\right) + \frac{t}{\lambda}\boldsymbol{y} = (1-\beta)\boldsymbol{z} + \beta\boldsymbol{y}.$$

再由 $f(\boldsymbol{z}) > f(\boldsymbol{x}) = f(\boldsymbol{y})$ 和半严格拟凸函数的定义有

$$f(\boldsymbol{z}_1) < \max\{f(\boldsymbol{z}), f(\boldsymbol{y})\} = f(\boldsymbol{z}).$$

因此, 这与 $f(\boldsymbol{z}) \leqslant f(\boldsymbol{z}_1)$ 矛盾.

(2) 若 $0<\lambda<t<1$, 设 $\boldsymbol{z}_2=\left(1-\frac{1-t}{1-\lambda}\right)\boldsymbol{y}+\frac{1-t}{1-\lambda}\boldsymbol{x}$, 则有

$$\boldsymbol{z}=(1-t)\,\boldsymbol{x}+t\boldsymbol{y}=(1-\lambda)\,\boldsymbol{z}_2+\lambda\boldsymbol{y}.$$

由上式有 $f(\boldsymbol{z})\leqslant\max\{f(\boldsymbol{z}_2),f(\boldsymbol{y})\}=f(\boldsymbol{z}_2)$. 再设 $\gamma=\frac{t-\lambda}{t(1-\lambda)}$, 由 $0<\lambda<t<1$ 得 $0<\gamma<1$, 则

$$\boldsymbol{z}_2=\frac{1}{1-\lambda}\boldsymbol{z}-\frac{\lambda}{1-\lambda}\boldsymbol{y}=\frac{1}{1-\lambda}\boldsymbol{z}-\frac{\lambda}{1-\lambda}\left(\frac{\boldsymbol{z}}{t}-\frac{(1-t)\,\boldsymbol{x}}{t}\right)=\gamma\boldsymbol{z}+(1-\gamma)\,\boldsymbol{x}.$$

再由 $f(\boldsymbol{z})>f(\boldsymbol{x})=f(\boldsymbol{y})$ 和半严格拟凸函数的定义有

$$f(\boldsymbol{z}_2)<\max\{f(\boldsymbol{x}),f(\boldsymbol{z})\}=f(\boldsymbol{z}).$$

因此, 这与 $f(\boldsymbol{z})\leqslant f(\boldsymbol{z}_2)$ 矛盾. □

下面结果表明, 满足一点半严格拟凸性的拟凸函数是半严格拟凸函数. 而前面例 5.2.2 中的函数不满足一点半严格拟凸性.

定理 5.2.6 设 $C\subset\mathbf{R}^n$ 是非空凸集, $f: C\to\mathbf{R}$ 是拟凸函数. 如果 f 在 C 上满足一点半严格拟凸性, 则 f 在 C 上是半严格拟凸函数.

证明 由定理所设, 对于任何 $\boldsymbol{x},\boldsymbol{y}\in C$ 和 $f(\boldsymbol{x})\neq f(\boldsymbol{y})$, 存在一个 $\lambda\in(0,1)$ 使

$$f[(1-\lambda)\,\boldsymbol{x}+\lambda\boldsymbol{y}]<\max\{f(\boldsymbol{x}),f(\boldsymbol{y})\}.$$

假设函数 f 在 C 上不是半严格拟凸函数, 即存在 $t\in(0,1)$, $\boldsymbol{x},\boldsymbol{y}\in C$ 和 $f(\boldsymbol{x})\neq f(\boldsymbol{y})$ 使

$$f[(1-t)\,\boldsymbol{x}+t\boldsymbol{y}]\geqslant\max\{f(\boldsymbol{x}),f(\boldsymbol{y})\}. \tag{5.2.1-4}$$

不妨设 $f(\boldsymbol{x})>f(\boldsymbol{y})$, 记 $\boldsymbol{z}=(1-t)\,\boldsymbol{x}+t\boldsymbol{y}$, 上式即为 $f(\boldsymbol{z})\geqslant f(\boldsymbol{x})>f(\boldsymbol{y})$. 由于 f 是在 C 上的拟凸函数, 得 $f(\boldsymbol{z})\leqslant\max\{f(\boldsymbol{x}),f(\boldsymbol{y})\}=f(\boldsymbol{x})f(\boldsymbol{z})=f(\boldsymbol{x})>f(\boldsymbol{y})$. 再由定理假设有

$$f[(1-\lambda)\,\boldsymbol{z}+\lambda\boldsymbol{y}]<\max\{f(\boldsymbol{z}),f(\boldsymbol{y})\}=f(\boldsymbol{z}),$$

$$\begin{aligned}f\left[\left(1-\lambda^2\right)\boldsymbol{z}+\lambda^2\boldsymbol{y}\right]&=f[\lambda((1-\lambda)\,\boldsymbol{z}+\lambda\boldsymbol{y})+(1-\lambda)\,\boldsymbol{z}]\\&<\max\{f((1-\lambda)\,\boldsymbol{z}+\lambda\boldsymbol{y}),f(\boldsymbol{z})\}=f(\boldsymbol{z}),\\&\cdots\cdots\end{aligned}$$

由归纳法得

$$f\left[\left(1-\lambda^k\right)\boldsymbol{z}+\lambda^k\boldsymbol{y}\right]<f(\boldsymbol{z}),\quad k=1,2,\cdots. \tag{5.2.1-5}$$

由 $\boldsymbol{z}=(1-t)\boldsymbol{x}+t\boldsymbol{y}$, 得

$$\begin{aligned}\left(1-\lambda^k\right)\boldsymbol{z}+\lambda^k\boldsymbol{y}&=\left(1-\lambda^k\right)[(1-t)\boldsymbol{x}+t\boldsymbol{y}]+\lambda^k\boldsymbol{y}\\&=\left(1-t-\lambda^k+\lambda^k t\right)\boldsymbol{x}+\left(t-\lambda^k t+\lambda^k\right)\boldsymbol{y}.\end{aligned}$$

选取充分大得 k_1 使得 $\dfrac{\lambda^{k_1}}{1-\lambda}<\dfrac{t}{1-t}$, 设

$$\beta_1=t-t\lambda^{k_1}+\lambda^{k_1},\quad \beta_2=t-\frac{\lambda^{k_1+1}}{1-\lambda}+\frac{t\lambda^{k_1+1}}{1-\lambda},$$

$$\bar{\boldsymbol{x}}=(1-\beta_1)\boldsymbol{x}+\beta_1\boldsymbol{y},\quad \bar{\boldsymbol{y}}=(1-\beta_2)\boldsymbol{x}+\beta_2\boldsymbol{y}.$$

则有

$$\left(1-\lambda^{k_1}\right)\boldsymbol{z}+\lambda^{k_1}\boldsymbol{y}=(1-\beta_1)\boldsymbol{x}+\beta_1\boldsymbol{y}=\bar{\boldsymbol{x}}.$$

由上式和 (5.2.1-5) 式得到

$$f(\bar{\boldsymbol{x}})=f\left(\left(1-\lambda^{k_1}\right)\boldsymbol{z}+\lambda^{k_1}\boldsymbol{y}\right)<f(\boldsymbol{z}). \tag{5.2.1-6}$$

下面分两种情形讨论.

(1) 若 $f(\bar{\boldsymbol{x}})\geqslant f(\bar{\boldsymbol{y}})$, 则由 $\boldsymbol{z}=(1-t)\boldsymbol{x}+t\boldsymbol{y}=(1-\lambda)\bar{\boldsymbol{y}}+\lambda\bar{\boldsymbol{x}}$ 和 f 在 C 上是拟凸函数有

$$f(\boldsymbol{z})\leqslant\max\{f(\bar{\boldsymbol{y}}),f(\bar{\boldsymbol{x}})\}=f(\bar{\boldsymbol{x}}),$$

上式与 (5.2.1-6) 式矛盾.

(2) 若 $f(\bar{\boldsymbol{x}})<f(\bar{\boldsymbol{y}})$, 则由 $\boldsymbol{z}=(1-t)\boldsymbol{x}+t\boldsymbol{y}=(1-\lambda)\bar{\boldsymbol{y}}+\lambda\bar{\boldsymbol{x}}$ 和定理假设得

$$f(\boldsymbol{z})<\max\{f(\bar{\boldsymbol{y}}),f(\bar{\boldsymbol{x}})\}=f(\bar{\boldsymbol{y}}). \tag{5.2.1-7}$$

再由 f 在 C 上是拟凸函数有

$$f(\bar{\boldsymbol{x}})\leqslant f(\boldsymbol{x}),\quad f(\bar{\boldsymbol{y}})\leqslant f(\boldsymbol{x}). \tag{5.2.1-8}$$

这样由 (5.2.1-7) 和 (5.2.1-8) 式得

$$f(\boldsymbol{z})<f(\bar{\boldsymbol{y}})\leqslant f(\boldsymbol{x}),$$

上式与 $f(\boldsymbol{z})=f(\boldsymbol{x})$ 矛盾. □

进一步, 下面的结果表明满足一点严格拟凸性的半严格拟凸函数是严格拟凸函数.

定理 5.2.7　设 $C \subset \mathbf{R}^n$ 是非空凸集, $f: C \to \mathbf{R}$ 在 C 上是半严格拟凸函数. 如果满足一点严格拟凸性, 则 f 在 C 上是严格拟凸函数.

证明　由定理所设, 对于任意 $\boldsymbol{x}, \boldsymbol{y} \in C$ 和 $\boldsymbol{x} \neq \boldsymbol{y}$, 存在一个 $\lambda \in (0,1)$ 使

$$f[(1-\lambda)\boldsymbol{x} + \lambda\boldsymbol{y}] < \max\{f(\boldsymbol{x}), f(\boldsymbol{y})\}. \tag{5.2.1-9}$$

因 f 在 C 上是半严格拟凸函数, 对于任意的 $\boldsymbol{x}, \boldsymbol{y} \in C$, $f(\boldsymbol{x}) \neq f(\boldsymbol{y})$ 和任意的 $t \in (0,1)$, 有

$$f[(1-t)\boldsymbol{x} + t\boldsymbol{y}] < \max\{f(\boldsymbol{x}), f(\boldsymbol{y})\}. \tag{5.2.1-10}$$

下面只需要证明, 对于任意不同的两点 $\boldsymbol{x}, \boldsymbol{y} \in C$ 和 $f(\boldsymbol{x}) = f(\boldsymbol{y})$ 也有 (5.2.1-10) 式成立. 由 (5.2.1-9) 式得

$$f[(1-\lambda)\boldsymbol{x} + \lambda\boldsymbol{y}] < \max\{f(\boldsymbol{x}), f(\boldsymbol{y})\} = f(\boldsymbol{x}).$$

对于任意的 $t \in (0,1)$, 由上式和 (5.2.1-10) 式得

$$\begin{aligned} f[(1-t)\boldsymbol{x} + t((1-\lambda)\boldsymbol{x} + \lambda\boldsymbol{y})] &< \max\{f(\boldsymbol{x}), f((1-\lambda)\boldsymbol{x} + \lambda\boldsymbol{y})\} \\ &< \max\{f(\boldsymbol{x}), f(\boldsymbol{y})\}. \end{aligned}$$

由于 $(1-t)\boldsymbol{x} + t((1-\lambda)\boldsymbol{x} + \lambda\boldsymbol{y}) = (1-t\lambda)\boldsymbol{x} + t\lambda\boldsymbol{y}$, 根据 $t\lambda \in (0,1)$, 当 $f(\boldsymbol{x}) = f(\boldsymbol{y})$ 时也有 (5.2.1-10) 式成立. □

定理 5.2.8　设 $C \subset \mathbf{R}^n$ 是非空凸集, 函数 $f: C \to \mathbf{R}$, 则 f 在 C 上是凸函数的充要条件是 f 在 C 上是半严格拟凸函数, 且满足一点凸性.

证明　如果 f 在 C 上是凸函数, 由定理 5.2.1(2) 知 f 在 C 上是半严格拟凸函数. 反过来, 由定理 5.2.5 知 f 在 C 上是拟凸函数, 再由定理 5.1.7 即可知 f 在 C 上是凸函数. □

本小节得到了凸函数、严格凸函数、拟凸函数、严格拟凸函数与半严格拟凸函数之间的关系:

1) 满足一点凸性的半严格凸函数等价于它是凸函数;

2) 满足一点严格拟凸性的半严格拟凸函数等价于严格拟凸函数;

3) 满足一点拟凸性的半严格拟凸函数等价于拟凸函数;

4) 满足一点半严格拟凸性的拟凸函数等价于半严格拟凸函数.

具体见下面图 5.2-1.

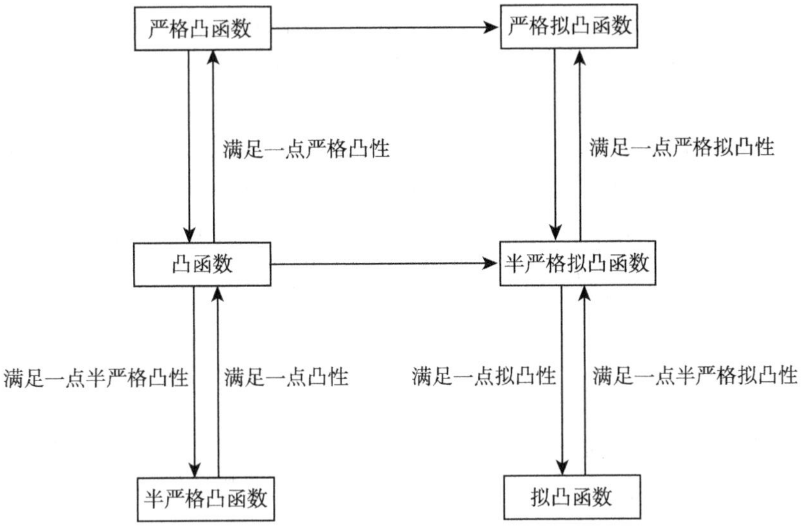

图 5.2-1 凸性与拟凸性之间关系

5.2.2 连续半严格拟凸函数

本小节讨论满足下半连续、上半连续或连续的半严格拟凸函数性质. 例如, 下面的定理说明下半连续半严格拟凸函数是拟凸函数; 例 5.2.1 说明了半严格凸函数不是拟凸函数, 因为它不是下半连续函数.

定理 5.2.9 设 $C \subset \mathbf{R}^n$ 是非空凸集, $f: C \to \mathbf{R}$ 是下半连续函数. 如果 f 在 C 上是半严格拟凸函数, 则 f 在 C 上是拟凸函数.

证明 由 f 在 C 上是半严格拟凸函数, 对于任意的 $\boldsymbol{x}, \boldsymbol{y} \in C$, $f(\boldsymbol{x}) \neq f(\boldsymbol{y})$ 和任意的 $\lambda \in (0,1)$, 有

$$f[(1-\lambda)\boldsymbol{x} + \lambda\boldsymbol{y}] < \max\{f(\boldsymbol{x}), f(\boldsymbol{y})\}.$$

再由定理 5.1.10(2) 知 f 是在 C 上的拟凸函数. □

设水平集合 $C_{f\leqslant\alpha} = \{\boldsymbol{x} \in C \,|\, f(\boldsymbol{x}) \leqslant \alpha\}$ 和 $C_{f=\alpha} = \{\boldsymbol{x} \in C \,|\, f(\boldsymbol{x}) = \alpha\}$, 其中 $\forall \alpha \in \mathbf{R}$. 则易知前面例 5.2.2 中条件 $C_{f=\alpha} = C_{f\leqslant\alpha}$ 与 $C_{f=\alpha} \subset \partial C_{f\leqslant\alpha}$ 不成立.

定理 5.2.10 设 C 凸集和 $f: C \to \mathbf{R}$ 是连续函数, $\forall \alpha \in \mathbf{R}$, 则 f 是半严格拟凸函数的充要条件是 f 的水平集 $C_{f\leqslant\alpha}$ 是凸集, 且 $C_{f=\alpha} = C_{f\leqslant\alpha}$ 或 $C_{f=\alpha} \subset \partial C_{f\leqslant\alpha}$ 成立.

证明 如果 f 是半严格拟凸函数, 则由定理 5.2.9 知 f 是拟凸函数, 再由定理 5.1.3 知 $C_{f\leqslant\alpha}$ 是凸集. 显然有 $C_{f=\alpha} \subset \partial C_{f\leqslant\alpha}$, 但如果 $C_{f=\alpha} \not\subset \partial C_{f\leqslant\alpha}$ 成立, 则存在一个 $\alpha_0 \in \mathbf{R}$ 和 $\boldsymbol{x}_0 \in C_{f=\alpha_0}$ 有 $\boldsymbol{x}_0 \notin \partial C_{f\leqslant\alpha_0}$. 那么有

$$\boldsymbol{x}_0 \in C_{f\leqslant\alpha_0} \backslash \partial C_{f\leqslant\alpha_0} = \mathrm{int} C_{f\leqslant\alpha_0},$$

即 $\boldsymbol{x}_0$ 是 $C_{f\leqslant\alpha_0}$ 的内点. 由定理 2.3.2 和定理 2.3.3 知 $\boldsymbol{x}_0$ 可以表示成 $C_{f\leqslant\alpha_0}$ 中的两个不同点的凸组合, 即有对任意 $\boldsymbol{x}_1\in C_{f\leqslant\alpha_0}$, 存在 $\boldsymbol{x}_2\in C_{f\leqslant\alpha_0}$ 满足 $[\boldsymbol{x}_1,\boldsymbol{x}_0]\subset[\boldsymbol{x}_1,\boldsymbol{x}_2]\subset C_{f\leqslant\alpha_0}$. 若 $f(\boldsymbol{x}_1)\neq f(\boldsymbol{x}_2)$, 不妨设 $f(\boldsymbol{x}_1)>f(\boldsymbol{x}_2)$, 由 f 是半严格拟凸函数得 $\alpha_0\geqslant f(\boldsymbol{x}_1)>f(\boldsymbol{x}_0)$, 这与 $\boldsymbol{x}_0\in C_{f=\alpha_0}$ 矛盾. 因此, $f(\boldsymbol{x}_1)=f(\boldsymbol{x}_2)$, 由 f 是拟凸函数得 $\alpha_0\geqslant f(\boldsymbol{x}_1)\geqslant f(\boldsymbol{x}_0)=\alpha_0$, 有 $f(\boldsymbol{x}_1)=f(\boldsymbol{x}_0)=\alpha_0$. 但是 $\boldsymbol{x}_1\neq\boldsymbol{x}_0$ 是 $C_{f\leqslant\alpha_0}$ 中的任意点, 所以 f 是等于 α_0 的常数函数, 即有 $C_{f=\alpha}=C_{f\leqslant\alpha}$.

反过来, 若水平集 $C_{f\leqslant\alpha}$ 是凸集, 则知 f 是拟凸函数, 由拟凸函数定义知, 对于任意的 $\boldsymbol{x},\boldsymbol{y}\in C$ 和任意的 $\lambda\in(0,1)$ 有

$$f(\boldsymbol{x})>f(\boldsymbol{y})\Rightarrow f[\lambda\boldsymbol{x}+(1-\lambda)\boldsymbol{y}]\leqslant f(\boldsymbol{x}).\tag{5.2.2-1}$$

再由半严格拟凸函数定义知, 只要证明 (5.2.2-1) 式是严格不等式. 记 $f(\boldsymbol{x})=\alpha_1$, 再由 (5.2.2-1) 式得 $\boldsymbol{y}\in C_{f\leqslant\alpha_1}$ 且 f 不是等于 α_1 的常数, 由定理所设知

$$C_{f=\alpha_1}=\partial C_{f\leqslant\alpha_1}$$

不成立, 那么有 $C_{f=\alpha_1}\subset\partial C_{f\leqslant\alpha_1}$ 成立. 由于 $f(\boldsymbol{y})<\alpha_1$, 得 $\boldsymbol{y}\in\mathrm{int}C_{f\leqslant\alpha_1}$. 再由 f 是连续函数, 则存在 $\boldsymbol{y}$ 的开邻域 $B(\boldsymbol{y},\varepsilon)\subset\mathrm{int}C_{f\leqslant\alpha_1}$, 满足

$$f(\boldsymbol{u})<\alpha_1,\quad\forall\boldsymbol{u}\in B(\boldsymbol{y},\varepsilon)\subset\mathrm{int}C_{f\leqslant\alpha_1},\quad\varepsilon>0.$$

设 $\boldsymbol{x}_\lambda=\lambda\boldsymbol{x}+(1-\lambda)\boldsymbol{y},\forall\lambda\in(0,1)$. 对于任意点 $\boldsymbol{u}\in B(\boldsymbol{x}_\lambda,(1-\lambda)\varepsilon)$, 取点 $\boldsymbol{z}=\dfrac{\boldsymbol{u}-\lambda\boldsymbol{x}}{1-\lambda}$, 则有 $\boldsymbol{u}=\lambda\boldsymbol{x}+(1-\lambda)\boldsymbol{z}$, 即 $\boldsymbol{u}$ 是线段 $(\boldsymbol{x},\boldsymbol{z})$ 上的点. 但有

$$||\boldsymbol{z}-\boldsymbol{y}||=\frac{||\boldsymbol{u}-\boldsymbol{x}_\lambda||}{1-\lambda}<\varepsilon,$$

即有 $\boldsymbol{z}\in B(\boldsymbol{y},\varepsilon)$, 可得到 $\boldsymbol{u}\in\mathrm{co}(B(\boldsymbol{y},\varepsilon)\cup\{\boldsymbol{x}\})$, 即有 $B(\boldsymbol{x}_\lambda,(1-\lambda)\varepsilon)\subset\mathrm{co}(B(\boldsymbol{y},\varepsilon)\cup\{\boldsymbol{x}\})$. 因 $C_{f\leqslant\alpha_1}$ 是凸集, 所以 $\boldsymbol{u}\in\mathrm{co}(B(\boldsymbol{y},\varepsilon)\cup\{\boldsymbol{x}\})\subset C_{f\leqslant\alpha_1}$, 从而 $B(\boldsymbol{x}_\lambda,(1-\lambda)\varepsilon)\subset C_{f\leqslant\alpha_1}$, 表明 $\boldsymbol{x}_\lambda$ 是 $C_{f\leqslant\alpha_1}$ 的内点. 因此, 有

$$f[\lambda\boldsymbol{x}+(1-\lambda)\boldsymbol{y}]<\alpha_1.$$

□

下面结论表明: 满足一点半严格拟凸性的下半连续函数是半严格拟凸函数.

定理 5.2.11　设 $C\subset\mathbf{R}^n$ 是非空凸集, $f:C\to\mathbf{R}$ 是下半连续函数. 如果满足一点半严格拟凸性, 则 f 在 C 上是半严格拟凸函数.

证明　由定理 5.1.10(2) 得 f 在 C 上是拟凸函数, 再由定理 5.2.6 知 f 是半严格拟凸函数. □

定义 5.2.2 设 $C \subset \mathbf{R}^n$ 是非空凸集. 如果对于任意 $\boldsymbol{x}, \boldsymbol{y} \in \partial C$ 和任意 $\lambda \in (0,1)$ 有

$$(1-\lambda)\boldsymbol{x} + \lambda \boldsymbol{y} \in \mathrm{int} C,$$

则称 C 是一个严格凸集.

严格凸集的几何意义是凸集的任何边界点两点相连的线段上的点都是内点, 严格凸集内部非空. 因此, 开集不是严格凸集, 因为开集没有边界.

定理 5.2.12 设函数 $f: \mathbf{R}^n \to \mathbf{R}$ 是连续严格拟凸函数, 则水平集 $C_{f \leqslant \alpha}$ 是严格凸集, 且 $C_{f=\alpha} \subset \partial C_{f \leqslant \alpha}$, 其中 $\forall \alpha \in \mathbf{R}$.

证明 由定理 5.2.1(1) 知 f 也是半严格拟凸函数, 则由定理 5.2.10 知 $C_{f \leqslant \alpha}$ 是凸集, 且 $C_{f=\alpha} \subset \partial C_{f \leqslant \alpha}$ 或 $C_{f=\alpha} = C_{f \leqslant \alpha}$ 成立. 下面首先证明 $C_{f \leqslant \alpha}$ 是严格凸集. 采用反证法. 假设存在某个 $\alpha_0 \in \mathbf{R}$ 使得 $C_{f \leqslant \alpha_0}$ 不是严格凸集, 存在两点 $\boldsymbol{x}_1, \boldsymbol{x}_2 \in \partial C_{f \leqslant \alpha_0}$ 和一个 $t \in (0,1)$ 使得

$$\boldsymbol{x}_t = t\boldsymbol{x}_1 + (1-t)\boldsymbol{x}_2 \in \partial C_{f \leqslant \alpha_0}. \tag{5.2.2-2}$$

由 f 是严格拟凸函数知

$$f(\boldsymbol{x}_t) < \max\{f(\boldsymbol{x}_1), f(\boldsymbol{x}_2)\} \leqslant \alpha_0.$$

因 f 是连续函数, 所以有 $\boldsymbol{x}_t \in \mathrm{int} C_{f \leqslant \alpha_0}$, 显然与 (5.2.2-2) 式矛盾.

另一方面, 假设有 $C_{f=\alpha} = C_{f \leqslant \alpha}$ 成立. 因 $C_{f \leqslant \alpha}$ 是严格凸集, 设对任意两个不同的 $\boldsymbol{x}_1, \boldsymbol{x}_2 \in \partial C_{f \leqslant \alpha}$ 和任意 $t \in (0,1)$ 有 $\boldsymbol{x}_t = t\boldsymbol{x}_1 + (1-t)\boldsymbol{x}_2 \in \mathrm{int} C_{f \leqslant \alpha}$, 由 f 是严格拟凸函数知

$$f(\boldsymbol{x}_t) < \max\{f(\boldsymbol{x}_1), f(\boldsymbol{x}_2)\} = \alpha.$$

由于 $C_{f=\alpha} = C_{f \leqslant \alpha}$, 则 $f(\boldsymbol{x}_t) = \alpha$, 得矛盾. □

下面定理说明了下半连续半严格拟凸函数具有全局最优性.

定理 5.2.13 设 $C \subset \mathbf{R}^n$ 是非空凸集和 $f: C \to \mathbf{R}$ 是下半连续函数, 则 f 在 C 上是半严格拟凸函数的充要条件 f 在 C 上是拟凸函数, 且 f 的任一限制函数的局部极小点也是全局极小点.

证明 先证必要性. 若 f 在 C 上是半严格拟凸函数, 由定理 5.2.9 知 f 在 C 上是拟凸函数. 由推论 5.2.1 得对任意的 $\boldsymbol{x} \in C$ 和 $\boldsymbol{d} \in \mathbf{R}^n$, $\varphi_{\boldsymbol{x},\boldsymbol{d}}(t)$ 是 $t \in T_1(\boldsymbol{x},\boldsymbol{d})$ 的半严格拟凸函数, 其中

$$\varphi_{\boldsymbol{x},\boldsymbol{d}}(t) = f(\boldsymbol{x} + t\boldsymbol{d}), \quad T_1(\boldsymbol{x},\boldsymbol{d}) = \{t \in [0,1] \,|\, \boldsymbol{x} + t\boldsymbol{d} \in C\}.$$

设 $t^* \in T_1(\boldsymbol{x},\boldsymbol{d})$ 是 $\varphi_{\boldsymbol{x},\boldsymbol{d}}(t)$ 在 $T_1(\boldsymbol{x},\boldsymbol{d})$ 上的局部极小点, 但 $t^* \in T_1(\boldsymbol{x},\boldsymbol{d})$ 不是 $\varphi_{\boldsymbol{x},\boldsymbol{d}}(t)$ 在 $T_1(\boldsymbol{x},\boldsymbol{d})$ 上的全局极小点, 则存在一个 $t_0 \in T_1(\boldsymbol{x},\boldsymbol{d})$ 和 $t^* \neq t_0$ 使得

$\varphi_{\boldsymbol{x},\boldsymbol{d}}(t_0) < \varphi_{\boldsymbol{x},\boldsymbol{d}}(t^*)$. 由 $\varphi_{\boldsymbol{x},\boldsymbol{d}}(t)$ 是 $t \in T_1(\boldsymbol{x},\boldsymbol{d})$ 的半严格拟凸函数, 对任意的 $s \in (0,1)$ 有

$$\varphi_{\boldsymbol{x},\boldsymbol{d}}(t^* + s(t^* - t_0)) < \varphi_{\boldsymbol{x},\boldsymbol{d}}(t^*). \tag{5.2.2-3}$$

对于充分小的 $s \in (0,1)$, 由 $t^* \in T_1(\boldsymbol{x},\boldsymbol{d})$ 是 $\varphi_{\boldsymbol{x},\boldsymbol{d}}(t)$ 在 $T_1(\boldsymbol{x},\boldsymbol{d})$ 上局部极小点得

$$\varphi_{\boldsymbol{x},\boldsymbol{d}}(t^* + s(t^* - t_0)) \geqslant \varphi_{\boldsymbol{x},\boldsymbol{d}}(t^*),$$

上式与 (5.2.2-3) 式矛盾.

再证充分性. 假设 f 在 C 上不是半严格拟凸函数, 则存在 $\boldsymbol{x},\boldsymbol{y} \in C$ 和一个 $s \in (0,1)$ 使得 $f(\boldsymbol{x}) > f(\boldsymbol{y})$, 且

$$f(\boldsymbol{x}_s) = f[(1-s)\boldsymbol{x} + s\boldsymbol{y}] \geqslant \max\{f(\boldsymbol{x}), f(\boldsymbol{y})\} = f(\boldsymbol{x}).$$

但因为 f 是在 C 上的拟凸函数, 有 $f(\boldsymbol{x}_s) = f[(1-s)\boldsymbol{x} + s\boldsymbol{y}] \leqslant \max\{f(\boldsymbol{x}), f(\boldsymbol{y})\} = f(\boldsymbol{x})$, 得 $f(\boldsymbol{x}_s) = f(\boldsymbol{x})$. 设 f 的限制函数

$$\varphi_{\boldsymbol{x},\boldsymbol{y}-\boldsymbol{x}}(t) = f(\boldsymbol{x} + t(\boldsymbol{y} - \boldsymbol{x})), \quad \forall t \in [0,1].$$

假设 $t^* \in [0,1]$ 是 $\varphi_{\boldsymbol{x},\boldsymbol{y}-\boldsymbol{x}}(t)$ 在 $[0,1]$ 上全局极小点, 因为 $f(\boldsymbol{x}) > f(\boldsymbol{y})$, 那么一定有

$$\varphi_{\boldsymbol{x},\boldsymbol{y}-\boldsymbol{x}}(s) = \varphi_{\boldsymbol{x},\boldsymbol{y}-\boldsymbol{x}}(0) > \varphi_{\boldsymbol{x},\boldsymbol{y}-\boldsymbol{x}}(1) \geqslant \varphi_{\boldsymbol{x},\boldsymbol{y}-\boldsymbol{x}}(t^*).$$

若 $\varphi_{\boldsymbol{x},\boldsymbol{y}-\boldsymbol{x}}(t)$ 在 $[0,s]$ 上的局部极小点不是全局的, 则有 $s < t^*$, 那么一定存在 $s' \in (0,s)$ 满足

$$\varphi_{\boldsymbol{x},\boldsymbol{y}-\boldsymbol{x}}(s') < \varphi_{\boldsymbol{x},\boldsymbol{y}-\boldsymbol{x}}(s).$$

由 $\varphi_{\boldsymbol{x},\boldsymbol{y}-\boldsymbol{x}}(t)$ 在 $[s',t^*]$ 上是拟凸函数, 有 $s \in (s',t^*)$ 和

$$\varphi_{\boldsymbol{x},\boldsymbol{y}-\boldsymbol{x}}(s') < \varphi_{\boldsymbol{x},\boldsymbol{y}-\boldsymbol{x}}(s) \leqslant \max\{\varphi_{\boldsymbol{x},\boldsymbol{y}-\boldsymbol{x}}(s'), \varphi_{\boldsymbol{x},\boldsymbol{y}-\boldsymbol{x}}(t^*)\} = \varphi_{\boldsymbol{x},\boldsymbol{y}-\boldsymbol{x}}(s'),$$

得出矛盾. 因此, 有 $t^* < s$, 且 $\varphi_{\boldsymbol{x},\boldsymbol{y}-\boldsymbol{x}}(t)$ 在 $[s,1]$ 上的局部极小点不是全局的, 那么一定存在 $s' \in (s,1)$ 满足

$$\varphi_{\boldsymbol{x},\boldsymbol{y}-\boldsymbol{x}}(s') < \varphi_{\boldsymbol{x},\boldsymbol{y}-\boldsymbol{x}}(s).$$

再由 $\varphi_{\boldsymbol{x},\boldsymbol{y}-\boldsymbol{x}}(t)$ 在 $[t^*,s']$ 上是拟凸函数, 有 $s \in (t^*,s')$ 和

$$\varphi_{\boldsymbol{x},\boldsymbol{y}-\boldsymbol{x}}(s') < \varphi_{\boldsymbol{x},\boldsymbol{y}-\boldsymbol{x}}(s) \leqslant \max\{\varphi_{\boldsymbol{x},\boldsymbol{y}-\boldsymbol{x}}(s'), \varphi_{\boldsymbol{x},\boldsymbol{y}-\boldsymbol{x}}(t^*)\} = \varphi_{\boldsymbol{x},\boldsymbol{y}-\boldsymbol{x}}(s'),$$

同样得矛盾. □

因为定理 5.2.13 的第 2 个条件在例 5.2.2 中不成立, 所以该例中的 f 是拟凸函数, 但不是半严格拟凸函数. 进一步, 下面的定理说明连续的半严格拟凸函数也具有全局最优性. 同样, 定理 5.2.14 中的必要性条件对例 5.2.2 也不成立.

定理 5.2.14 设 $C \subset \mathbf{R}^n$ 是非空凸集, $f: C \to \mathbf{R}$ 是连续拟凸函数, 则 f 在 C 上是半严格拟凸函数的充要条件是 f 在 C 上的任何局部极小点也是 $\min\limits_{\boldsymbol{x}\in C} f(\boldsymbol{x})$ 的全局极小点.

证明 若 f 是在 C 上的半严格拟凸函数. 设 $\boldsymbol{x}^* \in C$ 是 $\min\limits_{\boldsymbol{x}\in C} f(\boldsymbol{x})$ 的局部极小点, 假设 $\boldsymbol{x}^* \in C$ 不是 $\min\limits_{\boldsymbol{x}\in C} f(\boldsymbol{x})$ 的全局极小点, 则存在一个 $\boldsymbol{x}' \in C$ 和 $\boldsymbol{x}^* \neq \boldsymbol{x}'$ 使得 $f(\boldsymbol{x}^*) > f(\boldsymbol{x}')$. 由 f 是 C 上的半严格拟凸函数, 对于充分小的 $\lambda \in (0,1)$, 使得

$$f((1-\lambda)\boldsymbol{x}^* + \lambda \boldsymbol{x}') < f(\boldsymbol{x}^*).$$

上式与 $\min\limits_{\boldsymbol{x}\in C} f(\boldsymbol{x})$ 的严格局部极小点矛盾.

反过来, 设 f 在 C 上的任何局部极小点也是全局极小点. 由定理 5.2.10 来证明水平集 $C_{f\leqslant\alpha}$ 是凸集, 且 $C_{f=\alpha} = C_{f\leqslant\alpha}$ 或 $C_{f=\alpha} \subset \partial C_{f\leqslant\alpha}$ 成立, 其中 $\alpha \in \mathbf{R}$. 由于 f 是 C 上的拟凸函数, 知 $C_{f\leqslant\alpha}$ 是凸集, 所以 $\partial C_{f\leqslant\alpha}$ 是凸集, 剩下只需证明 $C_{f=\alpha} = C_{f\leqslant\alpha}$ 或 $C_{f=\alpha} \subset \partial C_{f\leqslant\alpha}$ 成立. 假设 f 在 C 上不是常数函数, 否则结论直接成立. 不妨设, 存在 $\alpha \in \mathbf{R}$ 使得

$$C_{f<\alpha} = \{\boldsymbol{x} \in C \,|\, f(\boldsymbol{x}) < \alpha\} \neq \varnothing,$$

且假设 $C_{f=\alpha} \subset \partial C_{f\leqslant\alpha}$ 不成立, 则需证明 $C_{f=\alpha} = C_{f\leqslant\alpha}$. 因此, 存在一个 $\boldsymbol{x}_0 \in C_{f=\alpha}$, 但 $\boldsymbol{x}_0 \notin \partial C_{f\leqslant\alpha}$, 故 $\boldsymbol{x}_0 \in \mathrm{int} C_{f\leqslant\alpha}$. 再由 f 在 C 上是连续拟凸函数, 则 $C_{f<\alpha}$ 是开凸集. 因为 $\boldsymbol{x}_0 \in C_{f=\alpha}$, 所以 $\boldsymbol{x}_0 \notin C_{f<\alpha}$. 又因 $\boldsymbol{x}_0 \in \mathrm{int} C_{f\leqslant\alpha}$, 取一个严格递减序列 $\varepsilon_i > 0\,(i=1,2,\cdots)$, 当 $i \to +\infty$ 时有 $\varepsilon_i \to 0$, 对应有 $B(\boldsymbol{x}_0, \varepsilon_i)$ 是 $\boldsymbol{x}_0$ 的一个序列闭邻域, 存在一个 I 使得对于 $i \geqslant I$ 满足 $B(\boldsymbol{x}_0, \varepsilon_i) \subset \mathrm{int} C_{f\leqslant\alpha}$. 如果对于 $i \geqslant I$, 存在某个 i 使得

$$f(\boldsymbol{x}) = \alpha, \quad \forall \boldsymbol{x} \in B(\boldsymbol{x}_0, \varepsilon_i) \subset \mathrm{int} C_{f\leqslant\alpha}$$

成立, 那么 $\boldsymbol{x}_0$ 是 f 在 C 上的一个局部极小点. 由定理假设得 f 在 C 上是常数函数, 这与前面所设矛盾. 因此, 由 f 在 C 上是连续函数, 对于 $i \geqslant I$, 优化问题 $\max\{f(\boldsymbol{x}) \,|\forall \boldsymbol{x} \in B(\boldsymbol{x}_0, \varepsilon_i)\}$ 存在极小点 $\boldsymbol{x}_i \in C_{f<\alpha}\,(i > I)$, 且 $f(\boldsymbol{x}_i) < \alpha$. 因为 f 在 C 上的任何局部极小点也是全局极小点, 所以有 $f(\boldsymbol{x}_i) = f(\boldsymbol{x}_{i+1}) = \cdots < \alpha\,(i > I)$, 显然 $i \to +\infty$, 有 $\boldsymbol{x}_i \to \boldsymbol{x}_0$, 再由 f 在 C 上的连续性, 得 $f(\boldsymbol{x}_0) < \alpha$, 因此得 $\boldsymbol{x}_0 \notin C_{f=\alpha}$, 得到矛盾. □

下面定理说明了在下半连续拟凸条件下, 严格拟凸函数与半严格拟凸函数之间的蕴含关系.

定理 5.2.15 设 $C \subset \mathbf{R}^n$ 是非空凸集, $f: C \to \mathbf{R}$ 是下半连续拟凸函数, 则 f 在 C 上是严格拟凸函数的充要条件是 f 是半严格拟凸函数, 且在 C 内的任何线段上的限制函数的极小点不超过一个.

证明 若 f 在 C 上是严格拟凸函数, 显然 f 是半严格拟凸函数, 且在 C 内的任何线段上的极小点不超过一个. 反过来, 假设 f 不是 C 上的严格拟凸函数, 则存在两个不同 $\boldsymbol{x}, \boldsymbol{y} \in C$ 和一个 $s \in (0,1)$ 使得

$$f(\boldsymbol{x}_s) = f[(1-s)\boldsymbol{x} + s\boldsymbol{y}] \geqslant \max\{f(\boldsymbol{x}), f(\boldsymbol{y})\}.$$

因为 f 在 C 上是拟凸函数, 则 $f(\boldsymbol{x}_s) = f[(1-s)\boldsymbol{x} + s\boldsymbol{y}] = \max\{f(\boldsymbol{x}), f(\boldsymbol{y})\}$. 因为 f 还是半严格拟凸函数, 若 $f(\boldsymbol{x}) > f(\boldsymbol{y})$, 则有

$$f(\boldsymbol{x}_s) = f[(1-s)\boldsymbol{x} + s\boldsymbol{y}] < \max\{f(\boldsymbol{x}), f(\boldsymbol{y})\},$$

得到矛盾. 因此, 有 $f(\boldsymbol{x}_s) = f(\boldsymbol{x}) = f(\boldsymbol{y})$, 设限制函数

$$\varphi(t) = f(\boldsymbol{x} + t(\boldsymbol{y} - \boldsymbol{x})), \quad \forall t \in [0,1], \quad \boldsymbol{x} + t(\boldsymbol{y} - \boldsymbol{x}) \in C.$$

那么 $\varphi(t)$ 不是常数函数, 否则与 C 内任何线段上的限制函数的极小点不超过一个相矛盾. 那么存在一个 $t' \in (0,1)$ 是在 C 内的任何线段上的限制函数 $\varphi(t)$ 的极小点, 则有

$$\varphi(t') < \varphi(0) = \varphi(1) = \varphi(s). \tag{5.2.2-4}$$

若 $s < t'$, 则由 f 是半严格拟凸函数, $\varphi(t)$ 是在 $[0,t']$ 上的半严格拟凸函数, 有 $\varphi(s) < \varphi(0)$, 与 (5.2.2-4) 式矛盾. 若 $s > t'$, 则由 f 是半严格拟凸函数, $\varphi(t)$ 是 $[t',1]$ 上的半严格拟凸函数, 有 $\varphi(s) < \varphi(1)$, 也与 (5.2.2-4) 式矛盾. 因此, f 是 C 上的严格拟凸函数. □

本小节得到半连续情形下的半严格拟凸函数的几个重要性质:

1) 下半连续半严格拟凸函数是拟凸函数;

2) 连续半严格拟凸函数 f 等价于该函数的水平集 $C_{f \leqslant \alpha}$ 是凸集, 且 $C_{f=\alpha} = C_{f \leqslant \alpha}$ 或 $C_{f=\alpha} \subset \partial C_{f \leqslant \alpha}$ 成立;

3) 满足一点半严格拟凸性的下半连续函数是半严格拟凸函数;

4) 下半连续半严格拟凸函数 f 等价于 f 是拟凸函数且它的任一限制函数的局部极小点也是全局极小点;

5) 连续拟凸的半严格拟凸函数 f 等价于 f 在 C 上的任何子集 $C_1 \subset C$ 上的局部极小点也是 $\min\limits_{\boldsymbol{x} \in C} f(\boldsymbol{x})$ 的全局极小点;

6) 下半连续拟凸函数 f 是严格拟凸函数等价于 f 是半严格拟凸函数, 且在定义域内的任何线段上的限制函数的极小点不超过一个.

5.2.3 显拟凸函数

由于拟凸函数与半严格拟凸函数没有直接的蕴含关系, 下面把同时具有拟凸性和半严格拟凸性的函数称为显拟凸函数, 定义如下.

定义 5.2.3 设 $C \subset \mathbf{R}^n$ 是非空凸集. 如果 $f: C \to \mathbf{R}$ 在 C 上是拟凸和半严格拟凸函数, 则称 f 在 C 上是显拟凸函数. 如果 $-f$ 在 C 上是显拟凸函数, 则称 f 在 C 上是显拟凹函数.

下面是显拟凸函数的一个必要条件.

定理 5.2.16 设 $C \subset \mathbf{R}^n$ 是非空凸集, $f: C \to \mathbf{R}$ 是显拟凸函数. 如果对任意的 $\boldsymbol{x}_i \in C(i=1,\cdots,m)$ 满足 $f(\boldsymbol{x}_j) < \max\limits_{i \neq j}\{f(\boldsymbol{x}_i)\}$, 则当 $\lambda_i > 0\,(i=1,\cdots,m)$ 时有

$$f(\lambda_1 \boldsymbol{x}_1 + \cdots + \lambda_m \boldsymbol{x}_m) < \max\{f(\boldsymbol{x}_1), \cdots, f(\boldsymbol{x}_m)\} \tag{5.2.3-1}$$

成立, 其中 $\sum\limits_{i=1}^{m} \lambda_i = 1$.

证明 设对于任意固定的 $j \in \{1,\cdots,m\}$, 记

$$\boldsymbol{y}_m = \sum_{i=1}^{m} \lambda_i \boldsymbol{x}_i, \quad \boldsymbol{y}_{m-1} = \frac{1}{1-\lambda_j} \sum_{i=1, i\neq j}^{m} \lambda_i \boldsymbol{x}_i. \tag{5.2.3-2}$$

若 $\boldsymbol{y}_{m-1} = \boldsymbol{x}_j$, 由 (5.2.3-2) 式有

$$\boldsymbol{y}_m = \sum_{i=1}^{m} \lambda_i \boldsymbol{x}_i = \sum_{i=1, i\neq j}^{m} \lambda_i \boldsymbol{x}_i + \frac{\lambda_j}{1-\lambda_j} \sum_{i=1, i\neq j}^{m} \lambda_i \boldsymbol{x}_i = \boldsymbol{y}_{m-1} = \boldsymbol{x}_j.$$

由定理假设得

$$f(\boldsymbol{y}_m) = f(\boldsymbol{x}_j) < \max_{i\neq j}\{f(\boldsymbol{x}_i)\}.$$

若 $\boldsymbol{y}_{m-1} \neq \boldsymbol{x}_j$, 由 (5.2.3-2) 式有

$$\boldsymbol{y}_m = \sum_{i=1}^{m} \lambda_i \boldsymbol{x}_i = \frac{1-\lambda_j}{1-\lambda_j} \sum_{i=1, i\neq j}^{m} \lambda_i \boldsymbol{x}_i + \lambda_j \boldsymbol{x}_j = (1-\lambda_j)\, \boldsymbol{y}_{m-1} + \lambda_j \boldsymbol{x}_j. \tag{5.2.3-3}$$

对上式分两种情形:

(1) 当 $f(\boldsymbol{x}_j) < f(\boldsymbol{y}_{m-1})$ 时, 因 f 在 C 上是拟凸和半严格拟凸函数, 则有

$$f(\boldsymbol{y}_m) < \max\{f(\boldsymbol{y}_{m-1}), f(\boldsymbol{x}_j)\} = f(\boldsymbol{y}_{m-1}).$$

由上式和定理 5.1.1 可得

$$f\left(\boldsymbol{y}_{m-1}\right)=f\left(\frac{1}{1-\lambda_j}\sum_{i=1,i\neq j}^{m}\lambda_i\boldsymbol{x}_i\right)<\max_{i\neq j}\left\{f\left(\boldsymbol{x}_i\right)\right\}=\max_{i=1,2,\cdots,m}\left\{f\left(\boldsymbol{x}_i\right)\right\}.$$

(2) 当 $f\left(\boldsymbol{x}_j\right)\geqslant f\left(\boldsymbol{y}_{m-1}\right)$ 时, 由 (5.2.3-3) 式和 f 在 C 上是拟凸和半严格拟凸函数得

$$\begin{aligned}f\left(\boldsymbol{y}_m\right)&<\max\left\{f\left(\boldsymbol{y}_{m-1}\right),f\left(\boldsymbol{x}_j\right)\right\}=f\left(\boldsymbol{x}_j\right)\\&<\max_{i\neq j}\left\{f\left(\boldsymbol{x}_i\right)\right\}=\max_{i=1,2,\cdots,m}\left\{f\left(\boldsymbol{x}_i\right)\right\}.\end{aligned}$$

□

例 5.2.3[98]　设函数

$$f(x)=\begin{cases}-x^2, & x\leqslant 0,\\0, & 0<x<2,\\x-2, & x\geqslant 2.\end{cases}$$

容易验证上面的函数是拟凸函数, 但不是半严格拟凸函数. 令

$$x_1=-\frac{1}{2},\quad x_2=-\frac{1}{4},\quad x_3=1,\quad \lambda_1=\lambda_2=\frac{1}{4},\quad \lambda_3=\frac{1}{2},$$

满足 $f\left(x_1\right)<\max\left\{f\left(x_2\right),f\left(x_3\right)\right\}$, 但是

$$f\left(\sum_{i=1}^{3}\lambda_i x_i\right)=f\left(\frac{5}{16}\right)=0=\max\left\{f\left(x_1\right),f\left(x_2\right),f\left(x_3\right)\right\}.$$

上面的例子表明, 如果 f 在 C 上不是显拟凸函数, 则定理 5.2.16 的结论不成立.

例 5.2.4[95]　设函数

$$f(x)=\begin{cases}-1, & x\neq 0,\\0, & x=0.\end{cases}$$

容易验证上面的函数是半严格拟凸函数, 但不是拟凸函数. 令

$$x_1=-\frac{1}{2},\quad x_2=0,\quad x_3=\frac{1}{2},\quad \lambda_1=\lambda_2=\frac{1}{4},\quad \lambda_3=\frac{1}{2},$$

满足 $f\left(x_1\right)<\max\left\{f\left(x_2\right),f\left(x_3\right)\right\}$, 但是

$$f\left(\sum_{i=1}^{3}\lambda_i x_i\right)=f(0)=0=\max\left\{f\left(x_1\right),f\left(x_2\right),f\left(x_3\right)\right\}.$$

上面的例子表明, 如果 f 在 C 上不是显拟凸函数, 则定理 5.2.16 的结论不成立.

在定理 5.2.16 中, 令 f 为 $-f$, 则对于显拟凹函数有类似的结论.

定理 5.2.17 设 $C\subset \mathbf{R}^n$ 是非空凸集和函数 $f: C\to \mathbf{R}$ 是显拟凹函数. 如果对任意的 $\boldsymbol{x}_i\in C(i=1,\cdots,m)$ 满足 $f(\boldsymbol{x}_j)>\min\limits_{i\neq j}\{f(\boldsymbol{x}_i)\}$, 则当 $\lambda_i>0$ $(i=1,\cdots,m)$ 时有

$$f(\lambda_1\boldsymbol{x}_1+\cdots+\lambda_m\boldsymbol{x}_m)>\min\{f(\boldsymbol{x}_1),\cdots,f(\boldsymbol{x}_m)\} \tag{5.2.3-4}$$

成立, 其中 $\sum\limits_{i=1}^{m}\lambda_i=1$.

本小节得到了显拟凸函数的一个必要条件 (5.2.3-1) 式, 以及显拟凹函数的一个必要条件 (5.2.3-4) 式.

5.3 伪凸函数

伪凸函数是研究得比较多的另一种广义凸函数[6,7,10,11,14,15,18,19,26,27,30,39,46,54-61]. 伪凸函数的性质比较接近可微凸函数的性质. 例如, 可微凸函数的稳定点是全局最优点, 伪凸函数也具有这个良好的性质. 但是, 这个性质对于上节的 (严格、半严格) 拟凸函数却不一定成立.

5.3.1 基本性质

一般伪凸函数均假设是可微函数. 因此, 伪凸函数是介于可微凸函数与可微拟凸函数之间的广义凸函数, 伪凸函数不一定是凸函数.

定义 5.3.1 设 $C\subset \mathbf{R}^n$ 是非空开凸集, $f: C\to \mathbf{R}$ 是可微函数. 如果对于任意 $\boldsymbol{x},\boldsymbol{y}\in C$ 有

$$f(\boldsymbol{x})>f(\boldsymbol{y})\Rightarrow \nabla f(\boldsymbol{x})^{\mathrm{T}}(\boldsymbol{y}-\boldsymbol{x})<0, \tag{5.3.1-1}$$

则称 f 在 C 上是伪凸函数. 如果对于任意 $\boldsymbol{x},\boldsymbol{y}\in C$ 和 $\boldsymbol{x}\neq\boldsymbol{y}$ 有

$$f(\boldsymbol{x})\geqslant f(\boldsymbol{y})\Rightarrow \nabla f(\boldsymbol{x})^{\mathrm{T}}(\boldsymbol{y}-\boldsymbol{x})<0, \tag{5.3.1-2}$$

则称 f 在 C 上是严格伪凸函数. 如果 $-f$ 在 C 上是伪凸函数, 则称 f 在 C 上是伪凹函数. 如果 $-f$ 在 C 上是严格伪凸函数, 则称 f 在 C 上是严格伪凹函数.

注 当 (5.3.1-1) 和 (5.3.1-2) 式成立时, 点 $\boldsymbol{y}-\boldsymbol{x}$ 是 f 在点 $\boldsymbol{x}$ 处的下降方向, 即存在 $t_0>0$, 有

$$f(\boldsymbol{x}+t(\boldsymbol{y}-\boldsymbol{x}))-f(\boldsymbol{x})=t\nabla f(\boldsymbol{x})^{\mathrm{T}}(\boldsymbol{y}-\boldsymbol{x})+o(t)<0,\quad \forall t\in(0,t_0).$$

上式表明当 f 在 C 上是伪凸函数时, 如果 $\boldsymbol{x}$ 不是局部极小点, 总是存在下降方向.

伪凸函数定义 (5.3.1-1) 式可以等价于下面条件:

$$\nabla f(\boldsymbol{x})^{\mathrm{T}}(\boldsymbol{y}-\boldsymbol{x}) \geqslant 0 \Rightarrow f(\boldsymbol{x}) \leqslant f(\boldsymbol{y});$$

严格伪凸函数定义 (5.3.1-2) 式也可以等价于下面条件:

$$\nabla f(\boldsymbol{x})^{\mathrm{T}}(\boldsymbol{y}-\boldsymbol{x}) \geqslant 0 \Rightarrow f(\boldsymbol{x}) < f(\boldsymbol{y}).$$

例 5.3.1　多项式函数: $f(x_1,x_2)=x_1^{-2}x_2^{-0.5}$ 是在集合 $C=\left\{(x_1,x_2)^{\mathrm{T}}\,|x_1,x_2>0\right\}$ 上的拟凸函数, 也是伪凸函数.

伪凸函数不一定用导数形式刻画, 见下面的性质 5.3.1 和性质 5.3.2.

性质 5.3.1　设 $C\subset\mathbf{R}^n$ 是非空开凸集, $f: C\to\mathbf{R}$ 是可微函数, 则 f 在 C 上是伪凸函数的充要条件是对于任意 $\boldsymbol{x},\boldsymbol{y}\in C$, 存在 $b=b(\boldsymbol{x},\boldsymbol{y})>0$, 有

$$f(\boldsymbol{x})>f(\boldsymbol{y}) \Rightarrow f((1-\lambda)\boldsymbol{x}+\lambda\boldsymbol{y}) \leqslant f(\boldsymbol{x})-\lambda(1-\lambda)b, \quad \forall\lambda\in(0,1). \tag{5.3.1-3}$$

证明　假设 (5.3.1-3) 式成立, 则对任意的 $\boldsymbol{x},\boldsymbol{y}\in C$, $f(\boldsymbol{x})>f(\boldsymbol{y})$ 和任意的 $\lambda\in(0,1)$, 存在 $b=b(\boldsymbol{x},\boldsymbol{y})>0$, 有

$$f[(1-\lambda)\boldsymbol{x}+\lambda\boldsymbol{y}]-f(\boldsymbol{x}) \leqslant -\lambda(1-\lambda)b.$$

于是得

$$\frac{f[\boldsymbol{x}+\lambda(\boldsymbol{y}-\boldsymbol{x})]-f(\boldsymbol{x})}{\lambda} \leqslant -(1-\lambda)b.$$

在上式中令 $\lambda\to 0$ 取极限, 得

$$\nabla f(\boldsymbol{x})^{\mathrm{T}}(\boldsymbol{y}-\boldsymbol{x}) < -b.$$

上式说明 f 是在 C 上的伪凸函数.

反过来, 设 f 在 C 上是伪凸函数. 对于任意 $\boldsymbol{x},\boldsymbol{y}\in C$ 有

$$f(\boldsymbol{x})>f(\boldsymbol{y}) \Rightarrow \nabla f(\boldsymbol{x})^{\mathrm{T}}(\boldsymbol{y}-\boldsymbol{x})<0.$$

设 f 的限制函数 $\varphi(\lambda)=f((1-\lambda)\boldsymbol{x}+\lambda\boldsymbol{y})\,(\forall\lambda\in[0,1])$, 有 $\varphi(0)=f(\boldsymbol{x})>\varphi(1)=f(\boldsymbol{y})$, 由拉格朗日中值定理 (定理 1.3.1) 有

$$\varphi(\lambda)-\varphi(0)=\varphi'(\xi)\lambda, \quad \exists\xi\in(0,\lambda), \tag{5.3.1-4}$$

$$\varphi(\lambda)-\varphi(1)=\varphi'(\zeta)(\lambda-1), \quad \exists\zeta\in(\lambda,1), \tag{5.3.1-5}$$

其中, $\xi=\xi(\boldsymbol{x},\boldsymbol{y}),\zeta=\zeta(\boldsymbol{x},\boldsymbol{y})$. 由定理 5.3.2 的第一个结论知 f 在 C 上是拟凸函数, 据 (5.3.1-4) 式有

$$\varphi(\lambda)\leqslant\max\{\varphi(0),\varphi(1)\}=\varphi(0),$$

可得 $\varphi'(\xi)\leqslant 0$.

再由 (5.3.1-4) 和 (5.3.1-5) 式有

$$\begin{aligned}\varphi'(\xi)-\varphi'(\zeta)&=\frac{\varphi(\lambda)-\varphi(0)}{\lambda}+\frac{\varphi(\lambda)-\varphi(1)}{1-\lambda}\\&=\frac{\varphi(\lambda)-(1-\lambda)\varphi(0)-\lambda\varphi(1)}{\lambda(1-\lambda)}\\&\leqslant\frac{\varphi(0)-(1-\lambda)\varphi(0)-\lambda\varphi(1)}{\lambda(1-\lambda)}=\frac{\varphi(0)-\varphi(1)}{(1-\lambda)}<0.\end{aligned}$$

令 $(1-\lambda)\times(5.3.1\text{-}4)+\lambda\times(5.3.1\text{-}5)$, 有

$$\begin{aligned}\varphi(\lambda)&=(1-\lambda)\varphi(0)+\lambda\varphi(1)+\lambda(1-\lambda)\varphi'(\xi)-\lambda(1-\lambda)\varphi'(\zeta)\\&<\varphi(0)+\lambda(1-\lambda)(\varphi'(\xi)-\varphi'(\zeta)),\end{aligned}$$

取 $b=b(\boldsymbol{x},\boldsymbol{y})=-(\varphi'(\xi)-\varphi'(\zeta))>0$, 即得 (5.3.1-3) 式成立. □

类似地可以证明下面结论成立.

性质 5.3.2 设 $C\subset\mathbf{R}^n$ 是一个非空开凸集, $f:\ C\to\mathbf{R}$ 是可微函数, 则 $f(\boldsymbol{x})$ 是在 C 上的严格伪凸函数的充要条件是对于任意 $\boldsymbol{x},\boldsymbol{y}\in C$ 和 $\boldsymbol{x}\neq\boldsymbol{y}$, 存在 $b=b(\boldsymbol{x},\boldsymbol{y})>0$ 有

$$f(\boldsymbol{x})\geqslant f(\boldsymbol{y})\Rightarrow f((1-\lambda)\boldsymbol{x}+\lambda\boldsymbol{y})<f(\boldsymbol{x})-\lambda(1-\lambda)b,\quad\forall\lambda\in(0,1).\quad(5.3.1\text{-}6)$$

性质 5.3.1 和性质 5.3.2 说明伪凸函数可以用不等式 (5.3.1-3) 和 (5.3.1-6) 来定义. 设一元函数: $f(x)=\begin{cases}x^3, & x>0,\\ x^3-1, & x\leqslant 0,\end{cases}$ 容易验证该函数满足条件 (5.3.1-3), 却不是伪凸函数. 因为它不是可微函数, 只有当可微函数满足条件 (5.3.1-3) 时才是伪凸函数. 以下讨论的伪凸函数均是可微函数.

性质 5.3.3 设 $C\subset\mathbf{R}^n$ 是空开凸集, $f:\ C\to\mathbf{R}$ 是伪凸函数. 如果存在点 $\boldsymbol{x}^*\in C$ 满足 $\nabla f(\boldsymbol{x}^*)=0$, 则 $\boldsymbol{x}^*$ 是 $\min\limits_{\boldsymbol{x}\in C}f(\boldsymbol{x})$ 的全局最优解. 若进一步, 假设 f 在 C 上是严格伪凸函数, 那么 $\boldsymbol{x}^*$ 是 $\min\limits_{\boldsymbol{x}\in C}f(\boldsymbol{x})$ 的唯一全局最优解.

证明　假设 $\boldsymbol{x}^* \in C$ 不是 $\min\limits_{\boldsymbol{x}\in C} f(\boldsymbol{x})$ 的全局最优解, 则存在一个 $\boldsymbol{x}' \in C$ 使得 $f(\boldsymbol{x}^*) > f(\boldsymbol{x}')$. 由 f 是可微伪凸函数, 则 $0 = \nabla f(\boldsymbol{x}^*)^{\mathrm{T}}(\boldsymbol{x}' - \boldsymbol{x}^*) < 0$, 得到矛盾. 如果 $\min\limits_{\boldsymbol{x}\in C} f(\boldsymbol{x})$ 存在另一个全局最优解 $\bar{\boldsymbol{x}} \in C$, 且 $\bar{\boldsymbol{x}} \neq \boldsymbol{x}^*$ 和 $f(\bar{\boldsymbol{x}}) = f(\boldsymbol{x}^*)$, 再由 f 是严格伪凸函数, 则得下面的矛盾式子:

$$0 = \nabla f(\boldsymbol{x}^*)^{\mathrm{T}}(\bar{\boldsymbol{x}} - \boldsymbol{x}^*) < 0. \qquad \square$$

性质 5.3.3 说明伪凸函数的所有稳定点都是全局最优解.

定理 5.3.1　设 $C \subset \mathbf{R}^n$ 是非空凸集, $f: C \to \mathbf{R}$ 是可微函数, 则 f 在 C 上是 (严格) 伪凸函数当且仅当对任意的 $\boldsymbol{x} \in C$ 和 $\boldsymbol{d} \in \mathbf{R}^n$, $\varphi_{\boldsymbol{x},\boldsymbol{d}}(t)$ 是 $t \in T(\boldsymbol{x},\boldsymbol{d})$ 的 (严格) 伪凸函数, 其中 $\varphi_{\boldsymbol{x},\boldsymbol{d}}(t) = f(\boldsymbol{x} + t\boldsymbol{d})$ 和 $T(\boldsymbol{x},\boldsymbol{d}) = \{t \in \mathbf{R} | \boldsymbol{x} + t\boldsymbol{d} \in C\}$.

证明　设 f 在 C 上是伪凸函数, 对于任意的 $t_1, t_2 \in T(\boldsymbol{x},\boldsymbol{d})$ 和

$$f(\boldsymbol{x} + t_1\boldsymbol{d}) = \varphi_{\boldsymbol{x},\boldsymbol{d}}(t_1) > \varphi_{\boldsymbol{x},\boldsymbol{d}}(t_2) = f(\boldsymbol{x} + t_2\boldsymbol{d}),$$

有

$$\varphi'_{\boldsymbol{x},\boldsymbol{d}}(t_1)(t_2 - t_1) = \nabla f(\boldsymbol{x} + t_1\boldsymbol{d})^{\mathrm{T}}(\boldsymbol{x} + t_2\boldsymbol{d} - (\boldsymbol{x} + t_1\boldsymbol{d})) < 0,$$

由定义 5.3.1 知 $\varphi_{\boldsymbol{x},\boldsymbol{d}}(t)$ 是 $t \in T(\boldsymbol{x},\boldsymbol{d})$ 的伪凸函数.

反过来, 对任意的 $\boldsymbol{x} \in C$ 和 $\boldsymbol{d} \in \mathbf{R}^n$, $\varphi_{\boldsymbol{x},\boldsymbol{d}}(t)$ 是 $t \in T(\boldsymbol{x},\boldsymbol{d})$ 的伪凸函数. 对于任意的 $\boldsymbol{x}_1, \boldsymbol{x}_2 \in C$, 设 $T(\boldsymbol{x}_1, \boldsymbol{x}_2 - \boldsymbol{x}_1) = \{t \in \mathbf{R} | \boldsymbol{x}_1 + t(\boldsymbol{x}_2 - \boldsymbol{x}_1) \in C\}$, 若 $f(\boldsymbol{x}_1) = \varphi_{\boldsymbol{x}_1,\boldsymbol{x}_2-\boldsymbol{x}_1}(0) > f(\boldsymbol{x}_2) = \varphi_{\boldsymbol{x}_1,\boldsymbol{x}_2-\boldsymbol{x}_1}(1)$, 则

$$\varphi'_{\boldsymbol{x}_1,\boldsymbol{x}_2-\boldsymbol{x}_1}(0)(1-0) = \nabla f(\boldsymbol{x}_1)^{\mathrm{T}}(\boldsymbol{x}_2 - \boldsymbol{x}_1) < 0.$$

故 f 在 C 上是伪凸函数.

严格伪凸的情形类似可证. $\square$

与凸函数等价于在线段上限制函数是凸函数类似, 伪凸函数也等价于在线段上限制函数是伪凸函数.

推论 5.3.1　设 $C \subset \mathbf{R}^n$ 是非空凸集, $f: C \to \mathbf{R}$ 是可微函数, 则 f 在 C 上是 (严格) 伪凸函数当且仅当对任意的 $\boldsymbol{x}, \boldsymbol{y} \in C$, $\varphi_{\boldsymbol{x},\boldsymbol{y}}(t) = f(\boldsymbol{x} + t(\boldsymbol{y} - \boldsymbol{x}))$ 是 $t \in [0,1]$ 的 (严格) 伪凸函数.

本小节得到伪凸函数的重要结论:

1) 伪凸函数的任何稳定点均是全局极小点;

2) 伪凸函数等价于它的限制函数也是伪凸的.

5.3.2 与其他广义凸函数的关系

一般情形下, 伪凸函数是拟凸函数和半严格拟凸函数. 但反过来, 拟凸函数和半严格拟凸函数不一定是伪凸函数.

定理 5.3.2 设 $C \subset \mathbf{R}^n$ 是非空开凸集, $f: C \to \mathbf{R}$ 是伪凸函数, 则 f 在 C 上是拟凸函数和半严格拟凸函数. 进一步, 若 f 在 C 上是严格伪凸函数, 则 f 在 C 上是严格拟凸函数.

证明 假设 f 在 C 上不是拟凸函数, 则根据定理 5.1.14 知, 存在 $\boldsymbol{x}, \boldsymbol{y} \in C$ 和 $f(\boldsymbol{x}) \geqslant f(\boldsymbol{y})$ 有

$$\nabla f(\boldsymbol{x})^{\mathrm{T}}(\boldsymbol{y}-\boldsymbol{x}) > 0.$$

设一个限制函数 $\varphi(t) = f(\boldsymbol{x} + t(\boldsymbol{y}-\boldsymbol{x})), \forall t \in [0,1]$, 则有

$$\varphi'(0) = \nabla f(\boldsymbol{x})^{\mathrm{T}}(\boldsymbol{y}-\boldsymbol{x}) > 0.$$

因此, 存在 $t \in (0,1)$ 使得 $\varphi(t) > \varphi(0)$, 那么 $\max\{\varphi(t) | \forall t \in [0,1]\}$ 存在一个最大点 $t^* \in (0,1)$ 使得 $\varphi'(t^*) = 0$. 记 $\boldsymbol{x}^* = \boldsymbol{x} + t^*(\boldsymbol{y}-\boldsymbol{x})$, 则有

$$f(\boldsymbol{x}^*) > f(\boldsymbol{x}) \geqslant f(\boldsymbol{y}).$$

再由 f 在 C 上是伪凸函数, 那么有 $\nabla f(\boldsymbol{x}^*)^{\mathrm{T}}(\boldsymbol{y}-\boldsymbol{x}^*) < 0$. 因为 $\boldsymbol{y}-\boldsymbol{x}^* = (1-t^*)(\boldsymbol{y}-\boldsymbol{x})$, 得

$$\varphi'(t^*) = \nabla f(\boldsymbol{x}^*)^{\mathrm{T}}(\boldsymbol{y}-\boldsymbol{x}) = \frac{1}{1-t^*}\nabla f(\boldsymbol{x}^*)^{\mathrm{T}}(\boldsymbol{y}-\boldsymbol{x}^*) < 0.$$

显然, 推出矛盾. 因此, f 在 C 上是拟凸函数.

下面证明第 2 个结论. 由 f 在 C 上是伪凸函数, 根据性质 5.3.1, 对于任意 $\boldsymbol{x}, \boldsymbol{y} \in C$, 存在 $b = b(\boldsymbol{x}, \boldsymbol{y}) > 0$, 有

$$\begin{aligned} f(\boldsymbol{x}) > f(\boldsymbol{y}) \Rightarrow f((1-\lambda)\boldsymbol{x} + \lambda\boldsymbol{y}) &\leqslant f(\boldsymbol{x}) - \lambda(1-\lambda)b < f(\boldsymbol{x}) \\ &= \max\{f(\boldsymbol{x}), f(\boldsymbol{y})\}, \forall \lambda \in (0,1), \end{aligned}$$

上式说明 f 在 C 上是半严格拟凸函数.

若 f 在 C 上是严格伪凸函数, 根据性质 5.3.2 得, 对于任意 $\boldsymbol{x}, \boldsymbol{y} \in C$ 和 $\boldsymbol{x} \neq \boldsymbol{y}$, 存在 $b = b(\boldsymbol{x}, \boldsymbol{y}) > 0$, 有

$$\begin{aligned} f(\boldsymbol{x}) \geqslant f(\boldsymbol{y}) \Rightarrow f((1-\lambda)\boldsymbol{x} + \lambda\boldsymbol{y}) &\leqslant f(\boldsymbol{x}) - \lambda(1-\lambda)b \\ &< \max\{f(\boldsymbol{x}), f(\boldsymbol{y})\}, \quad \forall \lambda \in (0,1). \end{aligned}$$

上式说明 f 在 C 上是严格拟凸函数. □

例 5.3.2　下面几个例子表明了凸函数、拟凸函数、半严格拟凸函数和伪凸函数之间的关系:

(1) 二元函数: $f(x_1,x_2)=x_1+x_2$ 是凸函数、拟凸函数和伪凸函数, 但不是严格凸函数.

(2) 二元函数: $f(x_1,x_2)=x_1+x_2+(x_1+x_2)^3$ 是伪凸函数, 不是凸函数.

(3) 二元函数: $f(x_1,x_2)=(x_1+x_2)^3$ 是半严格拟凸函数, 但不是伪凸函数和严格拟凸函数.

(4) 二元函数: $f(x_1,x_2)=\begin{cases}-(x_1+x_2)^2, & x_1+x_2<0,\\ 0, & x_1+x_2\geqslant 0\end{cases}$ 是拟凸函数, 但不是半严格拟凸函数.

上述例子说明, 伪凸函数、拟凸函数、严格拟凸函数和半严格拟凸函数之间不一定是蕴含关系. 但在一些条件下, 它们之间会有等价关系. 例如, 一个函数在开凸集上是连续可微的, 那么严格伪凸函数与严格拟凸函数是等价的.

定理 5.3.3　设 $C\subset\mathbf{R}^n$ 是非空开凸集, $f:\ C\to\mathbf{R}$ 是连续可微函数, 则 f 在 C 上是严格伪凸函数的充要条件是 f 在 C 上是严格拟凸函数和伪凸函数.

证明　根据定理 5.3.2 知, 只需证明 f 在 C 上是严格拟凸函数和伪凸函数, 则 f 在 C 上就是严格伪凸函数. 假设 f 在 C 上不是严格伪凸函数, 那么存在 $\boldsymbol{x},\boldsymbol{y}\in C$ 和 $\boldsymbol{x}\neq\boldsymbol{y}$ 有

$$f(\boldsymbol{x})\geqslant f(\boldsymbol{y})\Rightarrow\nabla f(\boldsymbol{x})^{\mathrm{T}}(\boldsymbol{y}-\boldsymbol{x})\geqslant 0. \tag{5.3.2-1}$$

若 $\nabla f(\boldsymbol{x})=0$, 则 $\boldsymbol{x}\in C$ 是 $\min\limits_{\boldsymbol{x}\in C}f(\boldsymbol{x})$ 的全局最优解, 有 $f(\boldsymbol{x})\leqslant f(\boldsymbol{y})$, 因此可得 $f(\boldsymbol{x})=f(\boldsymbol{y})$. 根据 $\boldsymbol{x}\neq\boldsymbol{y}$ 和 f 在 C 上是严格拟凸函数, 有

$$f((1-\lambda)\boldsymbol{x}+\lambda\boldsymbol{y})<\max\{f(\boldsymbol{x}),f(\boldsymbol{y})\}=f(\boldsymbol{x}),\quad \forall\lambda(0,1),$$

上式与 $\boldsymbol{x}\in C$ 是 $\min\limits_{\boldsymbol{x}\in C}f(\boldsymbol{x})$ 的全局最优解矛盾. 因此, $\nabla f(\boldsymbol{x})\neq 0$, 根据定理 5.1.16, 对于任意 $\boldsymbol{x},\boldsymbol{y}\in C$, 有

$$f(\boldsymbol{x})\geqslant f(\boldsymbol{y})\Rightarrow\nabla f(\boldsymbol{x})^{\mathrm{T}}(\boldsymbol{y}-\boldsymbol{x})<0.$$

上式与 (5.3.2-1) 式矛盾. □

下面定理说明, 如果可微函数在开凸集上不存在稳定点, 则伪凸函数与拟凸函数等价. 例如 $f(x)=x^3$ 在 $\mathbf{R}$ 上是拟凸函数, 不是伪凸函数. 但 $f(x)=x^3$ 在 $(0,+\infty)$ 与 $(-\infty,0)$ 上是伪凸函数.

定理 5.3.4　设 $C\subset\mathbf{R}^n$ 是非空开凸集, $f:\ C\to\mathbf{R}$ 是可微函数. 如果不存在点 $\boldsymbol{x}^*\in C$ 满足 $\nabla f(\boldsymbol{x}^*)=0$, 则 f 在 C 上是伪凸函数的充要条件是 f 在 C 上是拟凸函数.

证明 根据定理 5.3.2 只要证明充分性即可. 设 f 在 C 上是拟凸函数, 假设 f 在 C 上不是伪凸函数, 则存在 $\boldsymbol{x},\boldsymbol{y}\in C$ 有

$$f(\boldsymbol{x})>f(\boldsymbol{y})\Rightarrow\nabla f(\boldsymbol{x})^{\mathrm{T}}(\boldsymbol{y}-\boldsymbol{x})\geqslant 0,$$

根据定理 5.1.14 得 $\nabla f(\boldsymbol{x})^{\mathrm{T}}(\boldsymbol{y}-\boldsymbol{x})\leqslant 0$, 有 $\nabla f(\boldsymbol{x})^{\mathrm{T}}(\boldsymbol{y}-\boldsymbol{x})=0$, 由定理 5.1.15 得矛盾. □

若 (严格) 拟凸函数的稳定点都是严格局部极小点, 则它也是 (严格) 伪凸函数.

定理 5.3.5 设 $C\subset\mathbf{R}^n$ 是非空开凸集, $f\colon C\to\mathbf{R}$ 是连续可微函数, 有下面结论成立:

(1) 若 f 在 C 上是拟凸函数, 且它的全部稳定点都是严格局部极小点, 则 f 在 C 上是伪凸函数;

(2) 若 f 在 C 上是严格拟凸函数, 且它的全部稳定点都是局部极小点, 则 f 在 C 上是严格伪凸函数.

证明 (1) 假设 f 在 C 上不是伪凸函数, 存在 $\boldsymbol{x},\boldsymbol{y}\in C$ 有

$$f(\boldsymbol{x})>f(\boldsymbol{y})\Rightarrow\nabla f(\boldsymbol{x})^{\mathrm{T}}(\boldsymbol{y}-\boldsymbol{x})\geqslant 0. \tag{5.3.2-2}$$

分两种情形讨论:

(i) 当 $\boldsymbol{x}$ 是稳定点时, 由定理所设知 $\boldsymbol{x}$ 是 f 在 C 上的严格局部极小点, 存在充分小的 $t>0$, 满足

$$f(\boldsymbol{x})<f(\boldsymbol{x}+t(\boldsymbol{y}-\boldsymbol{x}))=f((1-t)\boldsymbol{x}+t\boldsymbol{y}).$$

由 f 在 C 上是拟凸函数, 有

$$f(\boldsymbol{x})<f((1-t)\boldsymbol{x}+t\boldsymbol{y})\leqslant f(\boldsymbol{x}),$$

得矛盾.

(ii) 当 $\boldsymbol{x}$ 不是稳定点时, 有 $\nabla f(\boldsymbol{x})\neq 0$, 由定理 5.1.15 得, 对任意的 $\boldsymbol{x},\boldsymbol{y}\in C$, 有

$$f(\boldsymbol{x})>f(\boldsymbol{y})\Rightarrow\nabla f(\boldsymbol{x})^{\mathrm{T}}(\boldsymbol{y}-\boldsymbol{x})<0,$$

与 (5.3.2-2) 式矛盾. 因此, f 在 C 上是伪凸函数.

(2) 的证明过程与 (1) 的证明类似, 我们得到 f 在 C 上的局部极小点都是全局极小点, 由定理 5.3.3 得 f 在 C 上是严格伪凸函数. □

注 由性质 5.3.3 知 f 在 C 上是严格伪凸函数时, f 在 C 上的全局极小点是唯一的.

下面我们把定理 5.3.5 的条件降低到更弱的情形, 对于单变量函数有下面更弱的结果, 即通过稳定点的性质就可以判断其伪凸性.

引理 5.3.1 设 $I \subset \mathbf{R}$ 是非空开区间, $\varphi : I \to \mathbf{R}$ 是可微函数, 则 φ 在 I 上是 (严格) 伪凸函数的充要条件是 φ 在 I 上的任何稳定点都是 φ 在 I 上的 (严格) 局部极小点.

证明 若 φ 在 I 上是伪凸函数. 设 $t^* \in I$ 是 φ 的一个稳定点, 如果 $t^* \in I$ 不是 φ 的局部极小点, 则存在一个 $t' \in I$ 使得 $\varphi(t') < \varphi(t^*)$, 由伪凸性有

$$\varphi(t^*)(t' - t^*) < 0,$$

得 $\varphi(t^*) \neq 0$, 这与 $t^* \in I$ 是 φ 的一个稳定点矛盾.

反过来, 若 φ 在 I 上的任何稳定点都是 φ 在 I 上的局部极小点. 假设 φ 在 I 上不是伪凸函数, 则存在 $t_1, t_2 \in I$, $\varphi(t_1) > \varphi(t_2)$, 且 $\varphi'(t_1)(t_2 - t_1) \geqslant 0$. 不妨设 $t_1 < t_2$, 有 $\varphi'(t_1) \geqslant 0$. 如果 $\varphi'(t_1) = 0$, 即 t_1 是 φ 在 I 上的局部极小点, 那么存在一个 $\varepsilon > 0$, 使得对于任意的 $t \in (t_1, t_1 + \varepsilon)$ 有

$$\varphi(t_2) < \varphi(t_1) \leqslant \varphi(t).$$

如果 $\varphi'(t_1) > 0$, 也存在一个 $t \in (t_1, t_2)$ 使得 $\varphi(t_2) < \varphi(t_1) < \varphi(t)$. 因此, φ 在 (t_1, t_2) 上达到一个最大点 $\bar{t} \in (t_1, t_2)$, 显然有 $\varphi'(\bar{t}) = 0$, 但由于 φ 在 I 上的任何稳定点都是 φ 在 I 上的局部极小点, 因此, 存在一个充分小的 $\varepsilon > 0$, 满足

$$\varphi(\bar{t}) = \varphi(t), \quad \forall t \in [\bar{t} - \varepsilon, \bar{t} + \varepsilon];$$

$$\varphi(t) < \varphi(\bar{t}), \quad \forall t \in (t_1, \bar{t} - \varepsilon), \quad \forall t \in (\bar{t} + \varepsilon, t_2).$$

那么有 $\varphi'(t) = 0, \forall t \in [\bar{t} - \varepsilon, \bar{t} + \varepsilon]$. 于是得到边界点 $\bar{t} - \varepsilon$ 或 $\bar{t} + \varepsilon$ 也是 φ 在 I 上的局部极小点, 显然与前面的不等式矛盾.

同理, 对于严格伪凸函数的证明类似可证. □

下面的结论说明, 如果一个拟凸函数的任何稳定点都是 (严格) 局部极小点, 则它的限制函数的任何稳定点都是 (严格) 局部极小点.

引理 5.3.2 设 $C \subset \mathbf{R}^n$ 是非空开凸集, $f : C \to \mathbf{R}$ 是可微拟凸函数. 如果 f 在 C 上的任何稳定点都是 f 在 C 上的 (严格) 局部极小点, 则对任意的 $\boldsymbol{x} \in C$ 和 $\boldsymbol{d} \in \mathbf{R}^n$, $\varphi_{\boldsymbol{x},\boldsymbol{d}}(t)$ 在 $T(\boldsymbol{x},\boldsymbol{d})$ 上的任何稳定点都是 f 在 $T(\boldsymbol{x},\boldsymbol{d})$ 上的 (严格) 局部极小点, 其中

$$\varphi_{\boldsymbol{x},\boldsymbol{d}}(t) = f(\boldsymbol{x} + t\boldsymbol{d}), \quad T(\boldsymbol{x},\boldsymbol{d}) = \{t \in \mathbf{R} | \boldsymbol{x} + t\boldsymbol{d} \in C\}.$$

证明 设 t_0 是 $\varphi_{\boldsymbol{x},\boldsymbol{d}}(t)$ 的一个稳定点, 则 $\varphi'_{\boldsymbol{x},\boldsymbol{d}}(t_0) = \nabla f(\boldsymbol{x} + t_0\boldsymbol{d})^{\mathrm{T}} \boldsymbol{d} = 0$. 记 $\boldsymbol{x}_0 = \boldsymbol{x} + t_0\boldsymbol{d}$, 假设 t_0 不是 $\varphi_{\boldsymbol{x},\boldsymbol{d}}(t)$ 的一个局部极小点, 对于任意给定的充分

小的 $\varepsilon > 0$, 存在一个 $t \in (t_0 - \varepsilon, t_0 + \varepsilon) \subset T(\boldsymbol{x}, \boldsymbol{d})$ 使得 $\varphi_{\boldsymbol{x},\boldsymbol{d}}(t) < \varphi_{\boldsymbol{x},\boldsymbol{d}}(t_0)$. 假设 $\boldsymbol{x}_0 = \boldsymbol{x} + t_0\boldsymbol{d}$ 不是 f 在 C 上的稳定点, 有 $\nabla f(\boldsymbol{x}_0) \neq 0$, 由定理 5.1.15 得

$$\nabla f(\boldsymbol{x}_0)^{\mathrm{T}} (t - t_0) \boldsymbol{d} = \nabla f(\boldsymbol{x}_0)^{\mathrm{T}} (\boldsymbol{x} + t\boldsymbol{d} - \boldsymbol{x}_0) < 0,$$

显然上式与 $\nabla f(\boldsymbol{x} + t_0\boldsymbol{d})^{\mathrm{T}} \boldsymbol{d} = 0$ 矛盾. 因此, $\boldsymbol{x}_0$ 是 f 在 C 上的稳定点, 再由定理假设, 得 $\boldsymbol{x}_0$ 是 f 在 C 上的局部极小点. 由于 $\varepsilon > 0$ 可以任意小, 那么有 $f(\boldsymbol{x}_0) \leqslant f(\boldsymbol{x})$, 即 $\varphi_{\boldsymbol{x},\boldsymbol{d}}(t_0) \leqslant \varphi_{\boldsymbol{x},\boldsymbol{d}}(t)$, 又得到一个矛盾. 因此, t_0 是 $\varphi_{\boldsymbol{x},\boldsymbol{d}}(t)$ 的一个局部极小点.

同理, 对于严格的情形也类似可证. □

下面有更好的结果, 伪凸函数等价于拟凸函数且它的任何稳定点都是局部极小点. 换句话说, 拟凸函数具有全局最优性等价于该拟凸函数是伪凸函数.

定理 5.3.6 设 $C \subset \mathbf{R}^n$ 是非空开凸集, $f: C \to \mathbf{R}$ 是连续可微函数, 则 f 在 C 上是 (严格) 伪凸函数的充要条件是 f 在 C 上是 (严格) 拟凸函数, 且 f 在 C 上的任何稳定点都是 f 在 C 上的 (严格) 局部极小点.

证明 由定理 5.3.2 和性质 5.3.3 知必要性条件成立.

下面证明充分性. 若 f 在 C 上是 (严格) 拟凸函数, 且 f 在 C 上的任何稳定点都是 f 在 C 上的 (严格) 局部极小点. 由引理 5.3.2 有, 对任意的 $\boldsymbol{x} \in C$ 和 $\boldsymbol{d} \in \mathbf{R}^n$, $\varphi_{\boldsymbol{x},\boldsymbol{d}}(t)$ 在 $T(\boldsymbol{x}, \boldsymbol{d})$ 上的任何稳定点都是 f 在 $T(\boldsymbol{x}, \boldsymbol{d})$ 上的 (严格) 局部极小点, 其中 $\varphi_{\boldsymbol{x},\boldsymbol{d}}(t) = f(\boldsymbol{x} + t\boldsymbol{d})$ 和 $T(\boldsymbol{x}, \boldsymbol{d}) = \{t \in \mathbf{R} | \boldsymbol{x} + t\boldsymbol{d} \in C\}$. 再由引理 5.3.1 有 $\varphi_{\boldsymbol{x},\boldsymbol{d}}(t)$ 在 $T(\boldsymbol{x}, \boldsymbol{d})$ 上是伪凸函数, 根据定理 5.3.1 得 f 在 C 上是 (严格) 伪凸函数. □

例 5.3.3 易知 $f(x_1, x_2) = -x_1x_2$ 在集合 $C = \left\{(x_1, x_2)^{\mathrm{T}} | x_1, x_2 > 0\right\}$ 上是拟凸函数也是伪凸函数. 另一方面, 它在集合 $C = \left\{(x_1, x_2)^{\mathrm{T}} | x_1, x_2 \geqslant 0\right\}$ 上是拟凸函数, 但不是伪凸函数. 因为伪凸函数的稳定点是局部极小点, 但拟凸函数不一定成立, 说明了定理 5.3.2~ 定理 5.3.6 中的函数定义域为开凸集是必要的. 另一方面, 也说明了稳定点是全局最优解是决定伪凸函数的必要条件.

下面是两个函数相除后的伪凸性结论.

定理 5.3.7 设 $C \subset \mathbf{R}^n$ 是非空凸集和 $g, f: C \to \mathbf{R}$ 在 C 上是可微函数, 有下面结论成立:

(1) 如果 f 在 C 上是非负凸函数且 g 在 C 上是正凹函数, 则 $\dfrac{f}{g}$ 在 C 上是伪凸函数;

(2) 如果 f 在 C 上是凸函数且 g 在 C 上是正仿射函数, 则 $\dfrac{f}{g}$ 在 C 上是伪凸函数;

(3) 如果 f 在 C 上是正严格凸函数且 g 在 C 上是正凹函数, 则 $\dfrac{f}{g}$ 在 C 上是严格伪凸函数;

(4) 如果 f 在 C 上是非负凸函数且 g 在 C 上是正严格凹函数, 则 $\dfrac{f}{g}$ 在 C 上是严格伪凸函数.

证明 设 $h=\dfrac{f}{g}$, 根据定理 5.2.4 知 h 在 C 上是半严格拟凸函数, 因为 h 在 C 上是连续函数, 再由定理 5.2.9 知 h 在 C 上是拟凸函数. 若 h 在 C 上不存在稳定点, 由定理 5.3.4 知 h 在 C 上是伪凸函数. 下面证明 h 在 C 上的任意稳定点都是 h 在 C 上的 (严格) 局部极小点. 设 $\boldsymbol{x}_0\in C$ 是 h 的稳定点, 即有 $\nabla h(\boldsymbol{x}_0)=0$, 有

$$\nabla h(\boldsymbol{x}_0)=\frac{g(\boldsymbol{x}_0)\nabla f(\boldsymbol{x}_0)-f(\boldsymbol{x}_0)\nabla g(\boldsymbol{x}_0)}{g(\boldsymbol{x}_0)^2}=0\Leftrightarrow\nabla f(\boldsymbol{x}_0)=h(\boldsymbol{x}_0)\nabla g(\boldsymbol{x}_0),$$

即有

$$\nabla f(\boldsymbol{x}_0)^{\mathrm{T}}(\boldsymbol{x}-\boldsymbol{x}_0)=h(\boldsymbol{x}_0)\nabla g(\boldsymbol{x}_0)^{\mathrm{T}}(\boldsymbol{x}-\boldsymbol{x}_0),\quad\forall\boldsymbol{x}\in C.\tag{5.3.2-3}$$

再由定理假设得

$$h(\boldsymbol{x}_0)\nabla g(\boldsymbol{x}_0)^{\mathrm{T}}(\boldsymbol{x}-\boldsymbol{x}_0)\geqslant h(\boldsymbol{x}_0)(g(\boldsymbol{x})-g(\boldsymbol{x}_0)),\quad\forall\boldsymbol{x}\in C.\tag{5.3.2-4}$$

若 g 在 C 上是正严格凹函数, 上式为严格不等式.

下面由 f 在 C 上是 (严格) 凸函数, 有

$$f(\boldsymbol{x})\geqslant(>)f(\boldsymbol{x}_0)+\nabla f(\boldsymbol{x}_0)^{\mathrm{T}}(\boldsymbol{x}-\boldsymbol{x}_0),\quad\forall\boldsymbol{x}\in C.$$

由上式、(5.3.2-3) 和 (5.3.2-4) 式得

$$f(\boldsymbol{x})\geqslant(>)f(\boldsymbol{x}_0)+h(\boldsymbol{x}_0)(g(\boldsymbol{x})-g(\boldsymbol{x}_0))=h(\boldsymbol{x}_0)g(\boldsymbol{x}),\quad\forall\boldsymbol{x}\in C.$$

上式说明 $\boldsymbol{x}_0\in C$ 是 $h(\boldsymbol{x})$ 的 (严格) 全局极小点.

根据定理 5.3.6 知, 上面过程证明了定理的 4 种情形. □

下面的定理说明, 严格单调递增函数的复合伪凸函数保持伪凸性.

定理 5.3.8 设 $C\subset\mathbf{R}^n$ 是非空凸集和 $f:C\to\mathbf{R}$ 是 (严格) 伪凸函数, 非空集合 $S\subset\mathbf{R}$ 和 $\phi:S\to\mathbf{R}$ 是可微函数, 且满足 $\phi'(t)>0$, 则 $\phi(f(\boldsymbol{x}))$ 在 C 上是 (严格) 伪凸函数.

证明 由 $\phi'(t)>0$ 知 ϕ: $S\to\mathbf{R}$ 是严格单调递增函数, 有 $\nabla\phi(f(\boldsymbol{x}))=\phi'(f(\boldsymbol{x}))\nabla f(\boldsymbol{x})$. 对于任意 $\boldsymbol{x},\boldsymbol{y}\in C$ 和 $\phi(f(\boldsymbol{x}))>(\geqslant)\phi(f(\boldsymbol{y}))$, 有 $f(\boldsymbol{x})>(\geqslant)f(\boldsymbol{y})$, 根据 f 是 (严格) 伪凸函数有

$$\nabla f(\boldsymbol{x})^{\mathrm{T}}(\boldsymbol{y}-\boldsymbol{x})<0.$$

即有 $\nabla\phi(f(\boldsymbol{x}))^{\mathrm{T}}(\boldsymbol{y}-\boldsymbol{x})=\phi'(f(\boldsymbol{x}))\nabla f(\boldsymbol{x})^{\mathrm{T}}(\boldsymbol{y}-\boldsymbol{x})<0$, 即得定理结论. □

本小节得到可微函数情形下在开凸集上的伪凸函数与 (严格) 拟凸函数之间的一些重要关系:

1) 可微伪凸函数是拟凸函数和半严格拟凸函数, 严格伪凸函数是严格拟凸函数;

2) 严格伪凸函数等价于它同时是严格拟凸函数和伪凸函数;

3) 若 f 是 (严格) 拟凸函数, 且它的全部稳定点都是严格局部极小点等价于 f 是 (严格) 伪凸函数;

4) 非负凸函数 f 与正凹函数 g 的商 $\dfrac{f}{g}$ 是伪凸函数;

5) 凸函数 f 与正仿射函数 g 的商 $\dfrac{f}{g}$ 是伪凸函数;

6) 正严格凸函数 f 与正凹函数 g 的商 $\dfrac{f}{g}$ 是严格伪凸函数;

7) 非负凸函数 f 与正严格凹函数 g 的商 $\dfrac{f}{g}$ 是严格伪凸函数;

8) 伪凸函数的可微严格单调函数的复合还是伪凸函数.

5.3.3 拟线性函数

线性 (仿射) 函数是凸函数也是凹函数, 本小节介绍拟线性函数的一些性质, 既是拟凸函数也是拟凹函数, 但是拟线性函数不一定是线性函数.

定义 5.3.2 设 $C\subset\mathbf{R}^n$ 是非空凸集. 若 f: $C\to\mathbf{R}$ 在 C 上既是 (半严格) 拟凸函数, 又是 (半严格) 拟凹函数, 则 f 在 C 上是 (半严格) 拟线性函数.

例 5.3.4 单变量函数 $f(x)=x|x|-x^2=\begin{cases}0, & x\geqslant 0,\\ -2x^2, & x\leqslant 0\end{cases}$ 是拟线性函数, 但不是半严格拟线性函数. 本例也说明拟线性函数并不一定是线性函数.

性质 5.3.4 设 $C\subset\mathbf{R}^n$ 是非空凸集, 函数 f: $C\to\mathbf{R}$, 则 f 在 C 上是拟线性函数当且仅当下列条件之一成立:

(1) 对于任意 $\boldsymbol{x},\boldsymbol{y}\in C$ 和 $\lambda\in(0,1)$, 有

$$\min\{f(\boldsymbol{x}),f(\boldsymbol{y})\}\leqslant f[(1-\lambda)\boldsymbol{x}+\lambda\boldsymbol{y}]\leqslant\max\{f(\boldsymbol{x}),f(\boldsymbol{y})\};\tag{5.3.3-1}$$

(2) 水平集 $\{x \in C \,|\, f(x) \leqslant \alpha\}$, $\{x \in C \,|\, f(x) \geqslant \alpha\}$ 和 $\{x \in C \,|\, f(x) = \alpha\}$ 都是凸集;

(3) f 在 C 上的任何线段限制函数都是单调递增或单调递减函数.

证明 根据定义 5.1.1、定义 5.3.2, 以及定理 5.1.3 和定理 5.1.6 知上述结论成立. □

性质 5.3.4 说明, 拟线性函数等价于任何两端点的凸组合值不大于两个端点的最大值且不小于两个端点的最小值, 或者 3 个水平集 $C_{f=\alpha}$, $C_{f\leqslant\alpha}$ 和 $C_{f\geqslant\alpha}$ 都是凸集, 或者任何线段限制函数都是单调递增或单调递减函数. 下面的性质 5.3.5 说明, 半严格拟线性函数等价于任何两端点的凸组合值小于两个端点的最大值并大于两个端点的最小值, 或者任何线段限制函数都是单调递增或单调递减函数.

性质 5.3.5 设 $C \subset \mathbf{R}^n$ 是非空凸集, 函数 $f: C \to \mathbf{R}$, 则 f 在 C 上是半严格拟线性函数当且仅当下列条件之一成立:

(1) 对于任意 $x, y \in C$, $f(x) \neq f(y)$ 和 $\lambda \in (0,1)$, 有

$$\min\{f(x), f(y)\} < f[(1-\lambda)x + \lambda y] < \max\{f(x), f(y)\}; \tag{5.3.3-2}$$

(2) f 在 C 上的任何线段限制函数都是严格单调递增或严格单调递减函数, 或者是常数函数.

证明 根据定义 5.1.1、定义 5.3.2, 以及定理 5.1.6 可知上面结论成立. □

定理 5.3.9 连续半严格拟线性函数在非空凸集上是拟线性函数.

证明 由定理 5.2.9 可知结论成立. □

对于非常值连续半严格拟线性函数, 可以得到下面的性质.

定理 5.3.10 设 $C \subset \mathbf{R}^n$ 是非空凸集, 函数 $f: C \to \mathbf{R}$, 则 f 在 C 上是非常值连续半严格拟线性函数, 且有下面结论成立:

(1) 对于任何 $\alpha \in \mathbf{R}$ 有 $\mathrm{int}C_{f=\alpha} = \varnothing$;

(2) f 的水平集 $C_{f=\alpha}$ 是集合 $C_{f\leqslant\alpha}$ 和 $C_{f\geqslant\alpha}$ 的边界点.

证明 (1) 假设 $\mathrm{int}C_{f=\alpha} \neq \varnothing$, 则存在点 $x_0 \in \mathrm{int}C_{f=\alpha}$, 那么对任意一点 $x \in C$, 存在一点 $\bar{x} \in C$ 使得

$$f((1-\lambda)x_0 + \lambda\bar{x}) = \alpha, \quad \forall\lambda \in [0,1], \quad [x_0, \bar{x}] \subset [x_0, x].$$

但根据性质 5.3.5(2) 得

$$f((1-\lambda)x_0 + \lambda x) = \alpha, \quad \forall\lambda \in [0,1].$$

由于点 $x \in C$ 是任意的, 所以 f 在 C 上是常值函数. 这与已知矛盾.

(2) 设 $\boldsymbol{x}_0 \in C_{f=\alpha}$, 由 (1) 知对于 $\boldsymbol{x}_0$ 的任何开邻域 $U(\boldsymbol{x}_0, \varepsilon)$, 存在一点 $\boldsymbol{x}_1 \in U(\boldsymbol{x}_0, \varepsilon)$ 使得 $f(\boldsymbol{x}_0) \neq f(\boldsymbol{x}_1)$. 设一个限制函数:

$$\varphi(\lambda) = f((1-\lambda)\boldsymbol{x}_0 + \lambda \boldsymbol{x}_1) = \alpha, \quad \forall \lambda \in [0,1].$$

那么由性质 5.3.5(2) 知, f 在线段 $(\boldsymbol{x}_0 - \varepsilon(\boldsymbol{x}_1 - \boldsymbol{x}_0), \boldsymbol{x}_0 + \varepsilon(\boldsymbol{x}_1 - \boldsymbol{x}_0))$ 上是严格单调递增函数或严格单调递减函数. 因此, $\boldsymbol{x}_0$ 是集合 $C_{f\leqslant\alpha}$ 和 $C_{f\geqslant\alpha}$ 的边界点. □

下面的定理说明, 连续半严格拟线性函数等价于其等式水平集是在函数定义域上的一个超平面的部分.

定理 5.3.11 设 $C \subset \mathbf{R}^n$ 是非空开凸集. 如果 $f: C \to \mathbf{R}$ 在 C 上是非常值连续函数, 则 f 在 C 上是半严格拟线性函数当且仅当水平集 $C_{f=\alpha} = C \cap H_\alpha$, 其中 H_α 是分离 $C_{f\leqslant\alpha}$ 和 $C_{f\geqslant\alpha}$ 的一个超平面.

证明 设 f 在 C 上是半严格拟线性函数, 由定理 5.3.10 得

$$\mathrm{int}C_{f\geqslant\alpha} \cap \mathrm{int}C_{f\leqslant\alpha} = \mathrm{int}C_{f=\alpha} = \varnothing.$$

由于 f 在 C 上是拟线性函数, 知集合 $C_{f\leqslant\alpha}$ 和 $C_{f\geqslant\alpha}$ 是凸集. 根据凸集分离定理, 存在分离集合 $C_{f\leqslant\alpha}$ 和 $C_{f\geqslant\alpha}$ 的超平面 H_α, 即存在一个非零点 $\boldsymbol{a} \in \mathbf{R}^n$, $\boldsymbol{x}_0 \in C_{f=\alpha}$ 使得超平面

$$H_\alpha = \{\boldsymbol{x} \in \mathbf{R}^n \,|\, \langle \boldsymbol{a}, \boldsymbol{x} - \boldsymbol{x}_0 \rangle = 0\}$$

满足

$$\langle \boldsymbol{a}, \boldsymbol{x} - \boldsymbol{x}_0 \rangle \geqslant 0, \quad \forall \boldsymbol{x} \in C_{f\geqslant\alpha},$$

$$\langle \boldsymbol{a}, \boldsymbol{x} - \boldsymbol{x}_0 \rangle \leqslant 0, \quad \forall \boldsymbol{x} \in C_{f\leqslant\alpha}.$$

由上式和 $C_{f\leqslant\alpha} \cap C_{f\geqslant\alpha} = C_{f=\alpha}$ 得

$$\boldsymbol{x} \in C_{f=\alpha} \Rightarrow \boldsymbol{x} \in H_\alpha,$$

即有 $C_{f=\alpha} \subset C \cap H_\alpha$. 另外, 上述超平面也严格分离 $\mathrm{int}C_{f\geqslant\alpha}$ 和 $\mathrm{int}C_{f\leqslant\alpha}$:

$$\langle \boldsymbol{a}, \boldsymbol{x} - \boldsymbol{x}_0 \rangle > 0, \quad \forall \boldsymbol{x} \in \mathrm{int}C_{f\geqslant\alpha},$$

$$\langle \boldsymbol{a}, \boldsymbol{x} - \boldsymbol{x}_0 \rangle < 0, \quad \forall \boldsymbol{x} \in \mathrm{int}C_{f\leqslant\alpha}.$$

因此, 若 $\boldsymbol{x} \in C \cap H_\alpha$, 则 $\boldsymbol{x}$ 不属于 $\mathrm{int}C_{f\geqslant\alpha}$ 和 $\mathrm{int}C_{f\leqslant\alpha}$, 即有 $C_{f=\alpha} \supset C \cap H_\alpha$.

反过来, 设 f 在 C 上不是半严格拟线性函数. 则由性质 5.3.5(1) 得, 存在 $\boldsymbol{x}, \boldsymbol{y} \in C$, $f(\boldsymbol{x}) \neq f(\boldsymbol{y})$ 和 $\lambda \in (0,1)$ 有

$$\min\{f(\boldsymbol{x}), f(\boldsymbol{y})\} \geqslant f[(1-\lambda)\boldsymbol{x} + \lambda \boldsymbol{y}],$$

或

$$f\left[(1-\lambda)\boldsymbol{x}+\lambda\boldsymbol{y}\right]\geqslant\max\left\{f(\boldsymbol{x}),f(\boldsymbol{y})\right\},$$

或者上述两个不等式都成立. 设 f 在 $[\boldsymbol{x},\boldsymbol{y}]\subset C$ 上的限制函数:

$$\varphi(t)=f\left((1-t)\boldsymbol{x}+t\boldsymbol{y}\right),\quad \forall t\in[0,1],$$

则有 $\min\limits_{t\in[0,1]}\varphi(t)<\varphi(\lambda)$ 或 $\max\limits_{t\in[0,1]}\varphi(t)>\varphi(\lambda)$. 否则有

$$\min_{t\in[0,1]}\varphi(t)=\varphi(\lambda)=\max_{t\in[0,1]}\varphi(t),$$

导致 $\varphi(t)$ 在 $[0,1]$ 上取常值, 这样有 $\varphi(0)=f(\boldsymbol{x})=\varphi(1)=f(\boldsymbol{y})$, 与 $f(\boldsymbol{x})\neq f(\boldsymbol{y})$ 矛盾. 因此得到

$$\min\left\{f(\boldsymbol{x}),f(\boldsymbol{y})\right\}\geqslant\varphi(\lambda)>\min_{t\in[0,1]}\varphi(t)$$

或

$$\max_{t\in[0,1]}\varphi(t)>\varphi(\lambda)\geqslant\max\left\{f(\boldsymbol{x}),f(\boldsymbol{y})\right\}$$

或者两者都成立. 考虑第一种情况, 由 f 的连续性得 $\varphi(t)$ 在 $[0,1]$ 上也是连续的, 设 $t^*\in[0,1]$ 满足

$$\min\left\{f(\boldsymbol{x}),f(\boldsymbol{y})\right\}\geqslant\varphi(\lambda)>\min_{t\in[0,1]}\varphi(t)=\varphi(t^*).$$

记 $\boldsymbol{x}^*=(1-t^*)\boldsymbol{x}+t^*\boldsymbol{y}$, 则 $\boldsymbol{x}^*\in[\boldsymbol{x},\boldsymbol{y}]\subset C$. 设一个非空水平集

$$C_{f=f(\boldsymbol{x}^*)}=\left\{\boldsymbol{x}\in C\,|\,f(\boldsymbol{x})=f(\boldsymbol{x}^*)\right\}.$$

由此, 可设 H 是满足 $C_{f=f(\boldsymbol{x}^*)}=C\cap H$ 的超平面, 由于

$$\min\left\{f(\boldsymbol{x}),f(\boldsymbol{y})\right\}>f(\boldsymbol{x}^*)=\varphi(t^*),$$

得 $\boldsymbol{x},\boldsymbol{y}\in C_{f>f(\boldsymbol{x}^*)}=\{\boldsymbol{x}\in C\,|\,f(\boldsymbol{x})>f(\boldsymbol{x}^*)\}$, 有 $\boldsymbol{x},\boldsymbol{y}\notin H$, 且它们位于同一个半空间. 此外显然有 $\boldsymbol{x}^*\in H$, 这与 $\boldsymbol{x},\boldsymbol{y}$ 位于相对两个半空间矛盾. □

如果函数 f 的定义域 $C=\mathbf{R}^n$, 则连续半严格拟线性函数等价于其等式水平集是一个超平面.

定理 5.3.12 如果 $f:\mathbf{R}^n\to\mathbf{R}$ 是非常值连续函数, 则 f 是半严格拟线性函数当且仅当水平集 $C_{f=\alpha}=\{\boldsymbol{x}\in\mathbf{R}^n\,|\,f(\boldsymbol{x})=\alpha\}$ 是一个超平面.

证明 从定理 5.3.11 证明可得. □

定理 5.3.13 设 $C \subset \mathbf{R}^n$ 是非空开凸集, $f: C \to \mathbf{R}$ 是可微函数, 则 f 在 C 上是拟线性函数当且仅对任意 $\boldsymbol{x}, \boldsymbol{y} \in C$ 有

$$f(\boldsymbol{x}) = f(\boldsymbol{y}) \Rightarrow \nabla f(\boldsymbol{x})^{\mathrm{T}} (\boldsymbol{y} - \boldsymbol{x}) = 0. \tag{5.3.3-3}$$

证明 若 f 在 C 上是拟线性函数, 根据性质 5.3.4(2) 得, f 在 C 上的任何水平集都是凸集. 若 $f(\boldsymbol{x}) = f(\boldsymbol{y})$, 于是有 $f((1-\lambda)\boldsymbol{x} + \lambda\boldsymbol{y}) = f(\boldsymbol{y}), \forall\lambda \in [0,1]$. 说明 f 在 C 上的限制函数

$$\varphi(\lambda) = f((1-\lambda)\boldsymbol{x} + \lambda\boldsymbol{y}) = f(\boldsymbol{y}), \quad \forall\lambda \in [0,1]$$

是常值函数, 即 (5.3.3-3) 式成立.

反过来, 设 (5.3.3-3) 式成立, 并假设 f 在 C 上不是拟线性函数. 则由定理 5.1.14, 存在两点 $\boldsymbol{x}, \boldsymbol{y} \in C$ 使得

$$f(\boldsymbol{x}) \geqslant f(\boldsymbol{y}) \Rightarrow \nabla f(\boldsymbol{x})^{\mathrm{T}} (\boldsymbol{y} - \boldsymbol{x}) > 0. \tag{5.3.3-4}$$

设 f 在 C 上的限制函数为

$$\varphi(\lambda) = f((1-\lambda)\boldsymbol{x} + \lambda\boldsymbol{y}) = f(\boldsymbol{y}), \quad \forall\lambda \in [0,1],$$

则有 $\varphi'(0) = \nabla f(\boldsymbol{x})^{\mathrm{T}} (\boldsymbol{y} - \boldsymbol{x}) > 0$. 因此, $\varphi(\lambda)$ 在 $(0,1)$ 上存在一个最大点 λ^* 满足

$$\varphi(\lambda^*) > \varphi(0) \geqslant \varphi(1).$$

因此, 再由 $\varphi(\lambda)$ 在 $(0,1)$ 上的连续性知, 存在一点 λ' 满足 $\varphi(\lambda') = \varphi(0)$ 和

$$\varphi(\lambda') = f((1-\lambda')\boldsymbol{x} + \lambda'\boldsymbol{y}) = f(\boldsymbol{x}).$$

由上式和 (5.3.3-3) 式得

$$\nabla f(\boldsymbol{x})^{\mathrm{T}} (((1-\lambda')\boldsymbol{x} + \lambda'\boldsymbol{y}) - \boldsymbol{x}) = 0 \Rightarrow \nabla f(\boldsymbol{x})^{\mathrm{T}} (\boldsymbol{y} - \boldsymbol{x}) = 0,$$

这与 (5.3.3-4) 式矛盾. □

本小节得到在开凸集上的拟线性函数的几个重要结果:

1) 拟线性函数等价于任何两端点的凸组合值不大于两个端点的最大值, 不小于两个端点的最小值;

2) 半严格拟线性函数等价于任何两端点的凸组合值小于两个端点的最大值, 大于两个端点的最小值;

3) 连续半严格拟线性函数等价于其等式水平集是有效定义域中一个超平面的一部分;

4) 可微拟线性函数等价于 (5.3.3-4) 式.

5.3.4 伪线性函数

伪线性函数是拟线性函数的推广, 拟凸函数不一定是伪凸函数. 因此, 拟线性函数不一定是伪线性函数.

定义 5.3.3 设 $C \subset \mathbf{R}^n$ 是非空开凸集. 若 $f: C \to \mathbf{R}$ 同时在 C 上是伪凸函数和伪凹函数, 则称 f 是在 C 上的伪线性函数.

拟线性函数不一定是线性函数, 同样伪线性函数也不一定是线性函数, 见下例.

例 5.3.5 函数

$$f(\boldsymbol{x}) = \ln(x_1 + x_2 + x_3 + 1)$$

在开凸集 $C = \left\{(x_1, x_2, x_3)^{\mathrm{T}} \,|x_1 + x_2 + x_3 + 1 > 0\right\}$ 上是伪线性函数.

定理 5.3.14 设 $C \subset \mathbf{R}^n$ 是非空开凸集. 如果 $f: C \to \mathbf{R}$ 是在 C 上的伪线性函数, 则下面结论成立:

(1) 在 C 上不存在点 $\boldsymbol{x} \in C$ 使得 $\nabla f(\boldsymbol{x}) = 0$, 或 f 是常数函数;

(2) f 是在 C 上的半严格拟线性函数;

(3) 当且仅当 f 是在 C 上线段的限制函数导数保号.

证明 (1) 由于 f 是在 C 上的伪线性函数, 则 f 是 C 上的伪凸函数和伪凹函数. 由性质 5.3.3 知, 如果存在点 $\boldsymbol{x}_0 \in C$ 使得 $\nabla f(\boldsymbol{x}_0) = 0$, 那么点 $\boldsymbol{x}_0$ 是 f 在 C 上的极小点和极大点. 因此, 有

$$f(\boldsymbol{x}_0) \leqslant f(\boldsymbol{x}) \leqslant f(\boldsymbol{x}_0), \quad \forall \boldsymbol{x} \in C,$$

表明 f 是常数函数.

(2) 由定理 5.3.2 知 f 是在 C 上的半严格拟线性函数.

(3) 如果 f 是在 C 上的伪线性函数, 由引理 5.3.2 和定理 5.3.6 知 f 在 C 上的限制函数也是伪线性函数. 再由 (2) 知 f 是在 C 上的半严格拟线性函数, 根据性质 5.3.5(2) 得非常值函数 f 在 C 上任何线段的限制函数是严格单调递增或严格单调递减函数.

下面只需要证明函数单变量情形即可. 对于任意的 $x \in C$, 有 $f'(x) \leqslant 0$ 或 $f'(x) \geqslant 0$. 由 (1) 得 f 在 C 上不存在稳定点, 所以对于任意的 $x \in C$, 有 $f'(x) < 0$ 或 $f'(x) > 0$.

反过来, 不妨设对任意的 $x \in C$, 有 $f'(x) < 0$. 则 f 在 C 上不存在稳定点, 且是严格单调递减函数, 由性质 5.3.5(2) 知 f 是在 C 上的半严格拟线性函数, 由定理 5.3.4 知 f 是在 C 上的伪线性函数. □

例 5.3.6 单变量函数 $f(x) = x|x| - x^2 = \begin{cases} 0, & x > 0, \\ -2x^2, & x \leqslant 0 \end{cases}$ 是拟线性函数, 不是伪线性函数. 定理 5.3.14 的结论对这个单变量函数均不成立.

相反, 定理 5.3.14 的结论对例 5.3.5 中的函数均成立.

定理 5.3.15 设 $C\subset\mathbf{R}^n$ 是非空开凸集. 如果 $f: C\to\mathbf{R}$ 在 C 上是非常值连续函数, 则 f 在 C 上是伪线性函数当且仅当水平集 $\{\boldsymbol{x}\in C\,|\,f(\boldsymbol{x})=\alpha\}=C\cap H_\alpha$, 其中 H_α 是分离 $\{\boldsymbol{x}\in C\,|\,f(\boldsymbol{x})\leqslant\alpha\}$ 和 $\{\boldsymbol{x}\in C\,|\,f(\boldsymbol{x})\geqslant\alpha\}$ 的一个超平面, 并且不存在稳定点.

证明 由定理 5.3.14 和定理 5.3.11 可知结论成立. □

定理 5.3.16 设 $C\subset\mathbf{R}^n$ 是非空开凸集, $f: C\to\mathbf{R}$ 是可微函数, 则 f 在 C 上是伪线性函数当且仅对于任意 $\boldsymbol{x},\boldsymbol{y}\in C$ 有

$$f(\boldsymbol{x})=f(\boldsymbol{y})\Longleftrightarrow\nabla f(\boldsymbol{x})^{\mathrm{T}}(\boldsymbol{y}-\boldsymbol{x})=0. \tag{5.3.4-1}$$

证明 若 f 在 C 上是伪线性函数, 则由定理 5.3.2 知 f 在 C 上是拟线性函数, 再由定理 5.3.13 得 (5.3.3-3) 式成立, 只要再证明 $\nabla f(\boldsymbol{x})^{\mathrm{T}}(\boldsymbol{y}-\boldsymbol{x})=0\Rightarrow f(\boldsymbol{x})=f(\boldsymbol{y})$ 即可, 设限制函数:

$$\varphi(\lambda)=f((1-\lambda)\boldsymbol{x}+\lambda\boldsymbol{y}),\quad\forall\lambda\in[0,1],$$

有 $\varphi'(0)=\nabla f(\boldsymbol{x})^{\mathrm{T}}(\boldsymbol{y}-\boldsymbol{x})=0$. 因 $\varphi(\lambda)$ 也是伪线性函数, 再由定理 5.1.14 知 $\varphi(\lambda)$ 是常值函数, 那么有 $f(\boldsymbol{x})=f(\boldsymbol{y})$.

反过来, 假设对于任意 $\boldsymbol{x},\boldsymbol{y}\in C$ 有 (5.3.4-1) 式成立. 由定理 5.3.13 知 f 在 C 上是拟线性函数, 若 f 在 C 上不存在稳定点, 则由定理 5.3.4 知 f 在 C 上是伪线性函数. 若 f 在 C 上存在稳定点, 再由 (5.3.4-1) 式成立知, f 在 C 上是常数函数, 也是伪线性函数. □

定理 5.3.16 表明了伪线性函数的两端点等值时等价于其中限制函数的端点导数等于 0, 对于拟线性函数 (5.3.4-1) 式逆方向不成立 (见定理 5.3.13).

定理 5.3.17 设 $C\subset\mathbf{R}^n$ 是非空开凸集, $f: C\to\mathbf{R}$ 是可微函数. 若不存在点 $\boldsymbol{x}\in C$ 使得 $\nabla f(\boldsymbol{x})=0$, 则 f 在 C 上是伪线性函数当且仅当它的正规化梯度映射 $\boldsymbol{x}\to\dfrac{\nabla f(\boldsymbol{x})}{\|\nabla f(\boldsymbol{x})\|}$ 在每个水平集是常值映射.

证明 设 f 在 C 上是伪线性函数, 对任意 $\boldsymbol{x},\boldsymbol{y}\in C$ 和 $f(\boldsymbol{x})=f(\boldsymbol{y})$, 定义集合:

$$S_1=\left\{\boldsymbol{d}\in\mathbf{R}^n|\nabla f(\boldsymbol{x})^{\mathrm{T}}\boldsymbol{d}=0\right\},\quad S_2=\left\{\boldsymbol{d}\in\mathbf{R}^n|\nabla f(\boldsymbol{y})^{\mathrm{T}}\boldsymbol{d}=0\right\}.$$

设 $\boldsymbol{d}\in S_1$, 则对任意 $t>0$ 和 $\boldsymbol{x}+t\boldsymbol{d}\in C$, 有 $\nabla f(\boldsymbol{x})^{\mathrm{T}}(\boldsymbol{x}+t\boldsymbol{d}-\boldsymbol{x})=0$. 由定理 5.3.16 的蕴含关系 "$\Leftarrow$" 知 $f(\boldsymbol{x}+t\boldsymbol{d})=f(\boldsymbol{x})=f(\boldsymbol{y})$. 再由定理 5.3.16 的蕴含关系 "$\Rightarrow$" 得

$$\nabla f(\boldsymbol{y})^{\mathrm{T}}(\boldsymbol{x}+t\boldsymbol{d}-\boldsymbol{y})=0\quad 和\quad\nabla f(\boldsymbol{y})^{\mathrm{T}}(\boldsymbol{x}-\boldsymbol{y})=0,$$

得 $\nabla f(\boldsymbol{y})^{\mathrm{T}}(\boldsymbol{d})=0$, 有 $\boldsymbol{d}\in S_2$, 即 $S_1\subset S_2$. 同样可证明 $S_1\supset S_2$, 得 $S_1=S_2$. 则由 S_1 和 S_2 的定义得 $\dfrac{\nabla f(\boldsymbol{x})}{\|\nabla f(\boldsymbol{x})\|}=\pm\dfrac{\nabla f(\boldsymbol{y})}{\|\nabla f(\boldsymbol{y})\|}$. 如果 $\boldsymbol{a}=\dfrac{\nabla f(\boldsymbol{y})}{\|\nabla f(\boldsymbol{y})\|}$, 那么 $\dfrac{\nabla f(\boldsymbol{x})}{\|\nabla f(\boldsymbol{x})\|}=-\boldsymbol{a}$, 得到当 $\boldsymbol{x},\boldsymbol{y}\in C$ 时, $\nabla f(\boldsymbol{x})$ 与 $\nabla f(\boldsymbol{y})$ 符号相反. 设 $\forall\varepsilon>0, t\in(0,\varepsilon)$, $\boldsymbol{z}_1=\boldsymbol{x}+t\boldsymbol{a}$ 和 $\boldsymbol{z}_2=\boldsymbol{y}+t\boldsymbol{a}$ 满足

$$f(\boldsymbol{z}_1)<f(\boldsymbol{x})=f(\boldsymbol{y})<f(\boldsymbol{z}_2).$$

由上式和 f 的连续性知, 存在 $\alpha\in(0,1)$ 使得

$$f(\boldsymbol{z})=f(\boldsymbol{x})=f(\boldsymbol{y}),\quad 其中\boldsymbol{z}=\alpha\boldsymbol{z}_1+(1-\alpha)\boldsymbol{z}_2.$$

根据定理 5.3.16 的蕴含关系“$\Rightarrow$”有 $(\boldsymbol{z}-\boldsymbol{y})^{\mathrm{T}}\boldsymbol{a}=0$ 和 $(\boldsymbol{x}-\boldsymbol{y})^{\mathrm{T}}\boldsymbol{a}=0$. 因此,

$$(\boldsymbol{z}-\boldsymbol{y})^{\mathrm{T}}\boldsymbol{a}=(\alpha\boldsymbol{z}_1+(1-\alpha)\boldsymbol{z}_2-\boldsymbol{y})^{\mathrm{T}}\boldsymbol{a}=(\alpha(\boldsymbol{x}-\boldsymbol{y})+t\boldsymbol{a})^{\mathrm{T}}\boldsymbol{a}=t\|\boldsymbol{a}\|^2>0,$$

得到矛盾, 于是有 $\dfrac{\nabla f(\boldsymbol{x})}{\|\nabla f(\boldsymbol{x})\|}=\dfrac{\nabla f(\boldsymbol{y})}{\|\nabla f(\boldsymbol{y})\|}$.

反过来, 假设对任意 $\boldsymbol{x},\boldsymbol{y}\in C$ 和 $f(\boldsymbol{x})=f(\boldsymbol{y})$ 有 $\dfrac{\nabla f(\boldsymbol{x})}{\|\nabla f(\boldsymbol{x})\|}=\dfrac{\nabla f(\boldsymbol{y})}{\|\nabla f(\boldsymbol{y})\|}$ 成立. 设限制函数

$$\varphi(\lambda)=f((1-\lambda)\boldsymbol{x}+\lambda\boldsymbol{y}),\quad\forall\lambda\in[0,1]$$

在 $[0,1]$ 上不是单调函数, 则存在 $\lambda_1,\lambda_2\in(0,1)$, 使得

$$\varphi(\lambda_1)=\varphi(\lambda_2)\quad 且\quad\varphi'(\lambda_1)\varphi'(\lambda_2)<0.$$

记

$$\boldsymbol{z}_1=\boldsymbol{x}+\lambda_1(\boldsymbol{y}-\boldsymbol{x})\quad 和\quad\boldsymbol{z}_2=\boldsymbol{x}+\lambda_2(\boldsymbol{y}-\boldsymbol{x}),$$

有 $f(\boldsymbol{z}_1)=f(\boldsymbol{z}_2)$, 根据假设又有 $\dfrac{\nabla f(\boldsymbol{z}_1)}{\|\nabla f(\boldsymbol{z}_1)\|}=\dfrac{\nabla f(\boldsymbol{z}_2)}{\|\nabla f(\boldsymbol{z}_2)\|}$ 和

$$\varphi'(\lambda_2)=\nabla f(\boldsymbol{z}_2)^{\mathrm{T}}(\boldsymbol{y}-\boldsymbol{x})=\frac{\|\nabla f(\boldsymbol{z}_2)\|\nabla f(\boldsymbol{z}_1)^{\mathrm{T}}(\boldsymbol{y}-\boldsymbol{x})}{\|\nabla f(\boldsymbol{z}_1)\|}=\varphi'(\lambda_1)\frac{\|\nabla f(\boldsymbol{z}_2)\|}{\|\nabla f(\boldsymbol{z}_1)\|},$$

由上式得 $\varphi'(\lambda_1)\varphi'(\lambda_2)>0$, 得到矛盾. 故 $\varphi(\lambda)$ 在 $[0,1]$ 上是单调函数. 因此, 由性质 5.3.4(3) 得 $\varphi(\lambda)$ 是拟线性的, 从而 f 在 C 上是拟线性函数. 由于不存在点 $\boldsymbol{x}\in C$ 使得 $\nabla f(\boldsymbol{x})=0$, 由定理 5.3.4 得 f 在 C 上是伪线性函数. □

例 5.3.7 设 $f(\boldsymbol{x}) = \ln(x_1+x_2+x_3+1)$ 和开凸集 $C = \left\{(x_1,x_2,x_3)^{\mathrm{T}} \mid x_1+x_2+x_3+1>0\right\}$, 则 f 在 C 上是伪线性函数, f 的梯度为

$$\nabla f(\boldsymbol{x})^{\mathrm{T}} = \left(\frac{1}{x_1+x_2+x_3+1}, \frac{1}{x_1+x_2+x_3+1}, \frac{1}{x_1+x_2+x_3+1}\right),$$

有 $\|\nabla f(\boldsymbol{x})\| = \dfrac{\sqrt{3}}{x_1+x_2+x_3+1}$, 得 $\dfrac{\nabla f(\boldsymbol{x})^{\mathrm{T}}}{\|\nabla f(\boldsymbol{x})\|} = (\sqrt{3},\sqrt{3},\sqrt{3})$ 是一个常值.

定理 5.3.18 设 $f\colon \mathbf{R}^n \to \mathbf{R}$ 是非常值可微函数, 则 f 在 $\mathbf{R}^n$ 上是伪线性函数当且仅当它的正规化梯度映射在 $\mathbf{R}^n$ 上是常值映射.

证明 设 f 在 $\mathbf{R}^n$ 上是伪线性函数, 假设它的正规化梯度映射在 $\mathbf{R}^n$ 上不是常值映射. 即存在 $\boldsymbol{x},\boldsymbol{y} \in \mathbf{R}^n$ 有 $\dfrac{\nabla f(\boldsymbol{x})}{\|\nabla f(\boldsymbol{x})\|} \neq \dfrac{\nabla f(\boldsymbol{y})}{\|\nabla f(\boldsymbol{y})\|}$, 则由定理 5.3.17 得 $f(\boldsymbol{x}) \neq f(\boldsymbol{y})$, 定义集合:

$$S_1 = \left\{\boldsymbol{d} \in \mathbf{R}^n \mid \nabla f(\boldsymbol{x})^{\mathrm{T}} \boldsymbol{d} = 0\right\}, \quad S_2 = \left\{\boldsymbol{d} \in \mathbf{R}^n \mid \nabla f(\boldsymbol{y})^{\mathrm{T}} \boldsymbol{d} = 0\right\}.$$

则有 $(\boldsymbol{x}+S_1)\cap(\boldsymbol{y}+S_2)=\varnothing$, 否则, 存在一点 $\boldsymbol{z}\in(\boldsymbol{x}+S_1)\cap(\boldsymbol{y}+S_2)$ 满足

$$\nabla f(\boldsymbol{x})^{\mathrm{T}}(\boldsymbol{z}-\boldsymbol{x}) = 0 \quad 和 \quad \nabla f(\boldsymbol{y})^{\mathrm{T}}(\boldsymbol{z}-\boldsymbol{y}) = 0,$$

由定理 5.3.16 得 $f(\boldsymbol{z}) = f(\boldsymbol{x}) = f(\boldsymbol{y})$, 这与 $f(\boldsymbol{x}) \neq f(\boldsymbol{y})$ 矛盾.

另一方面, S_1 和 S_2 是正交子空集, 由于 $\dfrac{\nabla f(\boldsymbol{x})}{\|\nabla f(\boldsymbol{x})\|} \neq \dfrac{\nabla f(\boldsymbol{y})}{\|\nabla f(\boldsymbol{y})\|}$, 有 $(\boldsymbol{x}+S_1)\cap(\boldsymbol{y}+S_2)\neq\varnothing$, 得到矛盾.

反过来, 直接由定理 5.3.17 得. □

定理 5.3.19 设 $C\subset\mathbf{R}^n$ 是非空开凸集, $f\colon C\to\mathbf{R}$ 在 C 上是伪线性函数. 如果对于任意 $t\in\mathbf{R}$, 函数 $\phi\colon \mathbf{R}\to\mathbf{R}$ 满足 $\phi'(t)>0$ 或 $\phi'(t)<0$, 则 $\phi(f(\boldsymbol{x}))$ 在 C 上是伪线性函数.

本小节得到伪线性函数的几个重要结论:

1) 伪线性函数的必要性条件是在 C 上不存在稳定点或 f 是常数函数; f 是半严格拟线性函数当且仅当 f 是在 C 上线段的限制函数导数保号 (即定理 5.3.14).

2) 伪线性函数等价于其有效定义域属于超平面, 且不存在稳定点 (即定理 5.3.15).

3) 伪线性函数等价于它的正规化梯度映射在每个水平集上是常值映射 (即定理 5.3.17、定理 5.3.18).

5.4　二次可微广义凸函数

本节讨论二阶可微函数与广义凸性之间的关系. 下面的定理证明了二阶连续可微拟凸函数的二阶 Hessian 矩阵的特征值最多只有一个负值, 且在限制函数的零点方向上是半正定的.

定理 5.4.1　设 $f: C \to \mathbf{R}$ 在凸集 $C \subset \mathbf{R}^n$ 上是二阶连续可微拟凸函数, 则下面结论成立:

(1) 对任意的 $\boldsymbol{x} \in C$, Hessian 矩阵 $\nabla^2 f(\boldsymbol{x})$ 至多有一个负特征值;

(2) 对任意的 $\boldsymbol{x} \in C$ 和任意的 $\boldsymbol{d} \in \mathbf{R}^n$, 有 $\boldsymbol{d}^{\mathrm{T}} \nabla f(\boldsymbol{x}) = 0 \Rightarrow \boldsymbol{d}^{\mathrm{T}} \nabla^2 f(\boldsymbol{x}) \boldsymbol{d} \geqslant 0$.

证明　(1) 假设存在一个 $\boldsymbol{x}_0 \in C$ 的二阶矩阵 $\nabla^2 f(\boldsymbol{x}_0)$, 其有多于一个的负特征值, 则存在 $\nabla^2 f(\boldsymbol{x}_0)$ 的 n 个正交特征向量: $\boldsymbol{d}_1, \boldsymbol{d}_2, \cdots, \boldsymbol{d}_n \in \mathbf{R}^n$. 不妨设 $\boldsymbol{d}_1$ 和 $\boldsymbol{d}_2$ 相关联对应的特征值是负值, 又设 S 是由 $\boldsymbol{d}_1$ 和 $\boldsymbol{d}_2$ 生成的子空间, 那么存在向量 $\boldsymbol{y} \in S \backslash \{\boldsymbol{0}\}$ 使得 $\boldsymbol{y}^{\mathrm{T}} \nabla^2 f(\boldsymbol{x}_0) \boldsymbol{y} < 0$. 设限制函数:

$$\varphi(t) = f(\boldsymbol{x}_0 + t\boldsymbol{y}), \quad \forall t \in T(\boldsymbol{x}_0, \boldsymbol{y}) = \{t \in \mathbf{R} | \boldsymbol{x}_0 + t\boldsymbol{y} \in C\}.$$

由定理 5.1.6 得 $\varphi(t)$ 在 $T(\boldsymbol{x}_0, \boldsymbol{y})$ 上是拟凸函数. 下面分两种情形讨论.

(i) 若 $\nabla f(\boldsymbol{x}_0) = 0$, 则有

$$\varphi'(0) = \nabla f(\boldsymbol{x}_0)^{\mathrm{T}} \boldsymbol{y} = 0, \quad \varphi''(0) = \boldsymbol{y}^{\mathrm{T}} \nabla^2 f(\boldsymbol{x}_0) \boldsymbol{y} < 0.$$

上式表明, $t = 0$ 是 $\varphi(t)$ 在 $T(\boldsymbol{x}_0, \boldsymbol{y})$ 上的严格局部极大点, 即存在 0 的一个邻域 $O(0) \subset T(\boldsymbol{x}_0, \boldsymbol{y})$ 满足 $\varphi(t) < \varphi(0), \forall t \in O(0)$. 取 $t_1, t_2 \in O(0)$ 满足 $t_1 < 0 < t_2$, 由 $\varphi(t)$ 的拟凸性得

$$\varphi((1-\lambda) t_1 + \lambda t_2) < \max\{\varphi(t_1), \varphi(t_2)\} < \varphi(0), \quad \forall \lambda \in (0,1).$$

上式表明存在 λ 满足 $(1-\lambda) t_1 + \lambda t_2 = 0$, 有 $\varphi(0) < \varphi(0)$, 得出矛盾.

(ii) 若 $\nabla f(\boldsymbol{x}_0) \neq 0$, 那么 S 与 $\nabla f(\boldsymbol{x}_0)$ 的正交子空间非空, 且维数等于 1 或 2, 所以存在向量 $\boldsymbol{y} \in S \backslash \{\boldsymbol{0}\}$ 使得 $\nabla f(\boldsymbol{x}_0)^{\mathrm{T}} \boldsymbol{y} = 0$ 和 $\boldsymbol{y}^{\mathrm{T}} \nabla^2 f(\boldsymbol{x}_0) \boldsymbol{y} < 0$ 成立. 根据前面的证明知 $t = 0$ 是 $\varphi(t)$ 在 $T(\boldsymbol{x}_0, \boldsymbol{y})$ 上的严格局部极大点, 由 $\varphi(t)$ 的拟凸性类似有 $\varphi(0) < \varphi(0)$, 得出矛盾.

(2) 假设存在 $\boldsymbol{x}_0 \in C$ 和 $\boldsymbol{d} \in \mathbf{R}^n$, 满足 $\boldsymbol{d}^{\mathrm{T}} \nabla f(\boldsymbol{x}_0) = 0 \Rightarrow \boldsymbol{d}^{\mathrm{T}} \nabla^2 f(\boldsymbol{x}_0) \boldsymbol{d} < 0$. 由上面的证明得到一个矛盾. □

例 5.4.1　设函数 $f(\boldsymbol{x}) = x_1^2 - x_2^2$, 其中 $C = \left\{(x_1, x_2)^{\mathrm{T}} \in \mathbf{R}^2 | x_1, x_2 > 0\right\}$, 它不是拟凸函数. 容易计算对任意的 $\boldsymbol{x} \in C$ 的 Hessian 矩阵: $\nabla^2 f(\boldsymbol{x}) = \begin{pmatrix} 2 & 0 \\ 0 & -2 \end{pmatrix}$,

它的两个特征值为 2 和 -2. 定理 5.4.1 中第一个结论对此例成立, 第二个结论却不成立.

定理 5.4.1 的条件是必要的, 但如果不存在稳定点的话, 条件也是充分的. 即若一个二阶连续可微的函数不含稳定点且在限制函数的零点方向上是半正定的, 则它是拟凸函数.

定理 5.4.2 设 $f: C \to \mathbf{R}$ 在开凸集 $C \subset \mathbf{R}^n$ 上是二阶连续可微函数. 如果对任意的 $\boldsymbol{x} \in C$ 和任意的 $\boldsymbol{d} \in \mathbf{R}^n$, 有 $\nabla f(\boldsymbol{x}) \neq 0$ 和

$$\boldsymbol{d}^{\mathrm{T}} \nabla f(\boldsymbol{x}) = 0 \Rightarrow \boldsymbol{d}^{\mathrm{T}} \nabla^2 f(\boldsymbol{x}) \boldsymbol{d} \geqslant 0,$$

则 f 在凸集 C 上是拟凸函数.

证明 假设 f 在 C 上不是拟凸函数, 则存在两点 $\boldsymbol{x}_1, \boldsymbol{x}_2 \in C$ 和一个 $\lambda \in (0,1)$ 使得

$$f(\boldsymbol{x}_2) \leqslant f(\boldsymbol{x}_1), \quad f(\boldsymbol{x}_1) < f(\lambda \boldsymbol{x}_2 + (1-\lambda)\boldsymbol{x}_1).$$

记 $\boldsymbol{x}_t = t\boldsymbol{x}_2 + (1-t)\boldsymbol{x}_1$, 设限制函数 $\varphi(t) = f(t\boldsymbol{x}_2 + (1-t)\boldsymbol{x}_1), \forall t \in [0,1]$. 由此得

$$\varphi(1) = f(\boldsymbol{x}_2) \leqslant \varphi(0) = f(\boldsymbol{x}_1) < \varphi(\lambda) = f(\lambda \boldsymbol{x}_2 + (1-\lambda)\boldsymbol{x}_1).$$

由于 $\varphi(t)$ 在 $[0,1]$ 上连续, 则 $\varphi(t)$ 在 $(0,1)$ 上一定取最大值, 即存在一个 $t_0 = \arg\max\limits_{t \in (0,1)} \varphi(t)$, 使得

$$\varphi'(t_0) = \nabla f(\boldsymbol{x}_{t_0})^{\mathrm{T}} (\boldsymbol{x}_2 - \boldsymbol{x}_1) = 0,$$

$$f(t\boldsymbol{x}_2 + (1-t)\boldsymbol{x}_1) < f(t_0\boldsymbol{x}_2 + (1-t_0)\boldsymbol{x}_1), \quad \forall t > t_0. \tag{5.4.1-1}$$

给定一个充分小的 $\varepsilon > 0$, 设严格单调减小的序列 $\varepsilon_i \downarrow 0\,(i \to +\infty)$, 则存在 N, 对于 $i > N$ 满足 $\varepsilon_i < \varepsilon$, 运用二阶泰勒中值定理, 有对应的 $t_i \in (t_0, t_0 + \varepsilon_i)$, 对于任意的 $\tau \in [0, \varepsilon_i]$, 有

$$f(\boldsymbol{x}_{t_0} + \tau(\boldsymbol{x}_2 - \boldsymbol{x}_1)) = f(\boldsymbol{x}_{t_0}) + \frac{1}{2}\tau^2 (\boldsymbol{x}_2 - \boldsymbol{x}_1)^{\mathrm{T}} \nabla^2 f(\boldsymbol{x}_{t_i}) (\boldsymbol{x}_2 - \boldsymbol{x}_1) < f(\boldsymbol{x}_{t_0}).$$

由上式得

$$(\boldsymbol{x}_2 - \boldsymbol{x}_1)^{\mathrm{T}} \nabla^2 f(\boldsymbol{x}_{t_i}) (\boldsymbol{x}_2 - \boldsymbol{x}_1) < 0, \quad i > N, \tag{5.4.1-2}$$

显然 $\boldsymbol{x}_{t_i} \xrightarrow{t_i \downarrow t_0} \boldsymbol{x}_{t_0}$, 根据定理假设知 $\nabla f(\boldsymbol{x}_{t_i})^{\mathrm{T}} (\boldsymbol{x}_2 - \boldsymbol{x}_1) \neq 0$ 成立. 设一个二元函数 ϕ: $\mathbf{R}^2 \to \mathbf{R}$,

$$\phi(\tau, s) = f(s\nabla f(\boldsymbol{x}_{t_0}) + \tau(\boldsymbol{x}_2 - \boldsymbol{x}_1) + \boldsymbol{x}_{t_0}) - f(\boldsymbol{x}_{t_0}),$$

$$\forall \tau \in [t_0-\varepsilon, t_0+\varepsilon], \quad \forall s \in [-\varepsilon, \varepsilon],$$

其中 $\varepsilon > \varepsilon_i\,(i > N)$. 显然有 $\phi(0,0)=0$, $\phi'_s(0,0) = ||\nabla f(\boldsymbol{x}_{t_0})||^2 \neq 0$, 因为 ε 是充分小的, ϕ 在 $[t_0-t_1, t_1-t_0] \times [-\varepsilon, \varepsilon]$ 上具有连续偏导, 则由隐函数存在定理, 存在一个在 0 点的邻域 U 上的连续可微函数 $s(\tau)$ 满足

$$s(0)=0, \quad \phi(\tau, s(\tau))=0, \quad \forall \tau \in U \subset [t_0-\varepsilon, t_0+\varepsilon].$$

由于 $\boldsymbol{x}_{t_i} \xrightarrow{t_i \downarrow t_0} \boldsymbol{x}_{t_0}$, 存在 N' 使得当 $i > N'$ 时有 $t_i - t_0 \in U$. 设 $\tau_i = t_i - t_0\,(i > N')$, 有 $\tau_i \to 0\,(i \to +\infty)$ 和 $s(\tau_i) \to 0\,(i \to +\infty)$. 设 $\boldsymbol{d}(\tau) = s(\tau)\nabla f(\boldsymbol{x}_{t_0}) + \tau(\boldsymbol{x}_2 - \boldsymbol{x}_1) + \boldsymbol{x}_{t_0}$, 由于 C 是开集, 则存在充分小的 τ 使得 $\boldsymbol{d}(\tau) \in C$. 不妨设, 对 $\forall \tau \in U$ 有 $\boldsymbol{d}(\tau) \in C$, 则有 $\boldsymbol{d}(\tau_i) = s(\tau_i)\nabla f(\boldsymbol{x}_{t_0}) + \boldsymbol{x}_{t_i}\,(i > N')$. 由 f 是在开凸集 C 上的二阶连续可微函数和 (5.4.1-2) 式知, 存在 $\boldsymbol{x}_{t_i}$ 的邻域 $O_r(\boldsymbol{x}_{t_i})$ 使得 $\boldsymbol{d}(\tau_i) \in O_r(\boldsymbol{x}_{t_i})$ 满足

$$(\boldsymbol{x}_2 - \boldsymbol{x}_1)^{\mathrm{T}} \nabla^2 f(\boldsymbol{d}(\tau_i))(\boldsymbol{x}_2 - \boldsymbol{x}_1) < 0. \tag{5.4.1-3}$$

对函数 $\phi(\tau, s(\tau))$ 关于 τ 求导得

$$\phi'(\tau, s(\tau)) = \nabla f(\boldsymbol{d}(\tau))^{\mathrm{T}} (s'(\tau)\nabla f(\boldsymbol{x}_{t_0}) + (\boldsymbol{x}_2 - \boldsymbol{x}_1)) = 0, \tag{5.4.1-4}$$

因为 $\boldsymbol{y}(0) = \boldsymbol{x}_{t_0}$ 和 $\nabla f(\boldsymbol{x}_{t_0}) = \nabla f(\boldsymbol{d}(0))$, 于是在 (5.4.1-4) 式中令 $\tau = 0$ 得 $s'(0) = 0$, 再由 $s'(\tau)$ 的连续性知 $s'(\tau) \to 0\,(\tau \to 0)$. 因此, 由 (5.4.1-3) 式知, 存在充分小的 τ_i 使得

$$(s'(\tau_i)\nabla f(\boldsymbol{x}_{t_0}) + \boldsymbol{x}_2 - \boldsymbol{x}_1)^{\mathrm{T}} \nabla^2 f(\boldsymbol{d}(\tau_i))(s'(\tau_i)\nabla f(\boldsymbol{x}_{t_0}) + \boldsymbol{x}_2 - \boldsymbol{x}_1) < 0.$$

再根据 (5.4.1-4) 式和定理假设知

$$(s'(\tau_i)\nabla f(\boldsymbol{x}_{t_0}) + \boldsymbol{x}_2 - \boldsymbol{x}_1)^{\mathrm{T}} \nabla^2 f(\boldsymbol{d}(\tau_i))(s'(\tau_i)\nabla f(\boldsymbol{x}_{t_0}) + \boldsymbol{x}_2 - \boldsymbol{x}_1) \geqslant 0,$$

不难发现, 上面两个不等式是矛盾的. □

类似再证明下面一个结论: 一个函数为二阶连续可微伪凸函数当且仅当其在限制函数的零点方向上是半正定的, 且其每一个稳定点是局部最优解.

定理 5.4.3　设 $f\colon C \to \mathbf{R}$ 在开凸集 $C \subset \mathbf{R}^n$ 上是二阶连续可微函数, 则 f 是在 C 上的 (严格) 伪凸函数当且仅当下面条件成立:

(1) 若 $\forall \boldsymbol{x} \in C$ 和 $\forall \boldsymbol{d} \in \mathbf{R}^n$, 有 $\boldsymbol{d}^{\mathrm{T}}\nabla f(\boldsymbol{x}) = 0 \Rightarrow \boldsymbol{d}^{\mathrm{T}}\nabla^2 f(\boldsymbol{x})\boldsymbol{d} \geqslant 0$;

(2) 若 $\boldsymbol{x} \in C$ 满足 $\nabla f(\boldsymbol{x}) = 0$, 有 $\boldsymbol{x}$ 是 $\min\limits_{\boldsymbol{x} \in C} f(\boldsymbol{x})$ 在 C 上的 (严格) 局部最优解.

证明 若 f 是在 C 上的 (严格) 伪凸函数, 由定理 5.3.2 知 f 是在 C 上的 (严格) 拟凸函数. 再由定理 5.4.2 和性质 5.3.3 知条件 (1) 和 (2) 成立.

反过来, 若条件 (1) 和 (2) 成立, 对任意给定的 $\boldsymbol{x} \in C$ 和 $\boldsymbol{d} \in \mathbf{R}^n$, 设限制函数 $\varphi_{\boldsymbol{x},\boldsymbol{d}}(t) = f(\boldsymbol{x}+t\boldsymbol{d})$ 和 $T(\boldsymbol{x},\boldsymbol{d}) = \{t \in \mathbf{R} | \boldsymbol{x}+t\boldsymbol{d} \in C\}$. 由于 C 是开凸集, 那么 $T(\boldsymbol{x},\boldsymbol{d})$ 是一个开区间. 对于任意的 $t \in T(\boldsymbol{x},\boldsymbol{d})$, 有

$$\varphi'_{\boldsymbol{x},\boldsymbol{d}}(t) = \nabla f(\boldsymbol{x}+t\boldsymbol{d})^{\mathrm{T}} \boldsymbol{d} \quad 和 \quad \varphi''_{\boldsymbol{x},\boldsymbol{d}}(t) = \boldsymbol{d}^{\mathrm{T}} \nabla^2 f(\boldsymbol{x}+t\boldsymbol{d})\boldsymbol{d}.$$

若对于任意的 $t \in T(\boldsymbol{x},\boldsymbol{d})$ 有 $\varphi'_{\boldsymbol{x},\boldsymbol{d}}(t) = \nabla f(\boldsymbol{x}+t\boldsymbol{d})^{\mathrm{T}} \boldsymbol{d} \neq 0$, 即 $\varphi_{\boldsymbol{x},\boldsymbol{d}}(t)$ 不存在稳定点, 有 $\varphi'_{\boldsymbol{x},\boldsymbol{d}}(t) > 0$ 或 $\varphi'_{\boldsymbol{x},\boldsymbol{d}}(t) < 0$. 根据伪凸函数定义, $\varphi_{\boldsymbol{x},\boldsymbol{d}}(t)$ 在 $T(\boldsymbol{x},\boldsymbol{d})$ 上是伪凸的. 如果存在 $t_0 \in T(\boldsymbol{x},\boldsymbol{d})$ 使得

$$\varphi'_{\boldsymbol{x},\boldsymbol{d}}(t_0) = \nabla f(\boldsymbol{x}+t_0\boldsymbol{d})^{\mathrm{T}} \boldsymbol{d} = 0,$$

但对于任意的 $t \in T(\boldsymbol{x},\boldsymbol{d})$ 有 $\nabla f(\boldsymbol{x}+t\boldsymbol{d}) \neq 0$. 记集合:

$$C_0 = \{\boldsymbol{x}+t\boldsymbol{d} \in C | \nabla f(\boldsymbol{x}+t\boldsymbol{d}) \neq 0, \quad \forall t \in T(\boldsymbol{x},\boldsymbol{d})\}.$$

根据定理 5.4.2 知函数 $f(\boldsymbol{x}+t\boldsymbol{d})$ 在 C_0 上是拟凸的, 再由定理 5.3.4 知 $f(\boldsymbol{x}+t\boldsymbol{d})$ 在 C_0 上是伪凸的, 又由定理 5.3.1 知 $\varphi_{\boldsymbol{x},\boldsymbol{d}}(t)$ 在 $T(\boldsymbol{x},\boldsymbol{d})$ 上是伪凸的. 若存在 $t_0 \in T(\boldsymbol{x},\boldsymbol{d})$, 则有

$$\varphi'_{\boldsymbol{x},\boldsymbol{d}}(t_0) = \nabla f(\boldsymbol{x}+t_0\boldsymbol{d})^{\mathrm{T}} \boldsymbol{d} = 0, \quad 且 \quad \nabla f(\boldsymbol{x}+t_0\boldsymbol{d}) = 0.$$

由条件 (2) 知 t_0 是 $\min\limits_{t \in T(\boldsymbol{x},\boldsymbol{d})} \varphi_{\boldsymbol{x},\boldsymbol{d}}(t)$ 在 $T(\boldsymbol{x},\boldsymbol{d})$ 上的 (严格) 局部最优解, 根据引理 5.3.1 知 $\varphi_{\boldsymbol{x},\boldsymbol{d}}(t)$ 在 $T(\boldsymbol{x},\boldsymbol{d})$ 上是伪凸的. 因此, 我们证明了对任意的 $\boldsymbol{x} \in C$ 和 $\boldsymbol{d} \in \mathbf{R}^n$, $\varphi_{\boldsymbol{x},\boldsymbol{d}}(t)$ 在 $T(\boldsymbol{x},\boldsymbol{d})$ 上是伪凸的. 根据定理 5.3.1 知 f 是在 C 上的 (严格) 伪凸函数. □

本小节得到三个重要结论:

1) 二阶连续可微拟凸函数的二阶 Hessian 矩阵的特征值最多只有一个负值, 且在限制函数的零点方向上是半正定的;

2) 如果一个二阶连续可微的函数不含稳定点且在限制函数的零点方向上是半正定的, 则它是拟凸函数;

3) 一个函数为二阶连续可微伪凸函数当且仅当在限制函数的零点方向上是半正定的, 且其每一个稳定点是局部最优解.

5.5 广义单调性

本节介绍与广义凸性密切相关的广义单调性这一性质. 广义单调性是针对向量函数而定义的, 如果广义凸函数是可微的, 那么它的梯度与广义单调性存在等

价性.

5.5.1 基本性质

下面给出向量函数 6 种不同的广义单调性定义.

定义 5.5.1 设 $C \subset \mathbf{R}^n$ 是非空集合, $F: C \to \mathbf{R}^n$ 是向量映射或值函数, 则

(1) 如果对于任意 $\boldsymbol{x}, \boldsymbol{y} \in C$ 有

$$(\boldsymbol{y}-\boldsymbol{x})^{\mathrm{T}}(F(\boldsymbol{y})-F(\boldsymbol{x})) \geqslant 0, \tag{5.5.1-1}$$

则称 F 在 C 上是单调映射;

(2) 如果对于任意 $\boldsymbol{x}, \boldsymbol{y} \in C$ 有

$$(\boldsymbol{y}-\boldsymbol{x})^{\mathrm{T}}(F(\boldsymbol{y})-F(\boldsymbol{x})) > 0, \tag{5.5.1-2}$$

则称 F 在 C 上是严格单调映射;

(3) 如果对于任意 $\boldsymbol{x}, \boldsymbol{y} \in C$ 有

$$(\boldsymbol{y}-\boldsymbol{x})^{\mathrm{T}} F(\boldsymbol{x}) \geqslant 0 \Rightarrow (\boldsymbol{y}-\boldsymbol{x})^{\mathrm{T}} F(\boldsymbol{y}) \geqslant 0, \tag{5.5.1-3}$$

或等价于

$$(\boldsymbol{y}-\boldsymbol{x})^{\mathrm{T}} F(\boldsymbol{x}) > 0 \Rightarrow (\boldsymbol{y}-\boldsymbol{x})^{\mathrm{T}} F(\boldsymbol{y}) > 0, \tag{5.5.1-4}$$

则称 F 在 C 上是伪单调映射;

(4) 如果对于任意 $\boldsymbol{x}, \boldsymbol{y} \in C$ 有

$$(\boldsymbol{y}-\boldsymbol{x})^{\mathrm{T}} F(\boldsymbol{x}) > 0 \Rightarrow (\boldsymbol{y}-\boldsymbol{x})^{\mathrm{T}} F(\boldsymbol{y}) \geqslant 0, \tag{5.5.1-5}$$

则称 F 在 C 上是拟单调映射;

(5) 若 F 在 C 上是拟单调映射, 且对于任意 $\boldsymbol{x}, \boldsymbol{y} \in C$ 存在 $\boldsymbol{z} \in \operatorname{ri}[\boldsymbol{x}, \boldsymbol{y}]$ 有

$$(\boldsymbol{y}-\boldsymbol{x})^{\mathrm{T}} F(\boldsymbol{z}) \neq 0, \tag{5.5.1-6}$$

则称 F 在 C 上是严格拟单调映射;

(6) 若 F 在 C 上是拟单调映射, 且对于任意 $\boldsymbol{x}, \boldsymbol{y} \in C$ 和 $\boldsymbol{x} \neq \boldsymbol{y}$ 有

$$(\boldsymbol{y}-\boldsymbol{x})^{\mathrm{T}} F(\boldsymbol{x}) > 0 \Rightarrow \boldsymbol{z} \in \operatorname{ri}\left[\frac{\boldsymbol{x}+\boldsymbol{y}}{2}, \boldsymbol{y}\right], \quad (\boldsymbol{y}-\boldsymbol{x})^{\mathrm{T}} F(\boldsymbol{z}) > 0, \tag{5.5.1-7}$$

则称 F 在 C 上是半严格拟单调映射.

下面的性质 5.5.1 将给出单调映射、拟单调映射、伪单调映射、半严格拟单调映射和严格单调映射之间的蕴含关系.

性质 5.5.1 设 $C \subset \mathbf{R}^n$ 是非空凸集合, $F: C \to \mathbf{R}^n$ 是向量映射, 有下面结论:

(1) 若 F 在 C 上是单调映射, 则 F 在 C 上是伪单调映射;

(2) 若 F 在 C 上是伪单调映射, 则 F 在 C 上是半严格拟单调映射;

(3) 若 F 在 C 上是半严格拟单调映射, 则 F 在 C 上是拟单调映射;

(4) 若 F 在 C 上是严格单调映射, 则 F 在 C 上是严格拟单调映射;

(5) 若 F 在 C 上是严格单调映射, 则 F 在 C 上是半严格拟单调映射;

(6) 若 F 在 C 上是伪单调映射, 则 F 在 C 上是拟单调映射.

证明 (1) 若 F 在 C 上是单调映射, 由 (5.5.1-1) 式得, 对于任意 $\boldsymbol{x}, \boldsymbol{y} \in C$ 有

$$(\boldsymbol{y}-\boldsymbol{x})^{\mathrm{T}} F(\boldsymbol{y}) \geqslant (\boldsymbol{y}-\boldsymbol{x})^{\mathrm{T}} F(\boldsymbol{x}),$$

因此有 (5.5.1-3) 和 (5.5.1-4) 式成立.

(2) 若 F 在 C 上是伪单调映射, 那么根据定义 5.5.1 知, F 在 C 上是拟单调映射. 对于任意 $\boldsymbol{x}, \boldsymbol{y} \in C$ 和 $\boldsymbol{x} \neq \boldsymbol{y}$ 有 $(\boldsymbol{y}-\boldsymbol{x})^{\mathrm{T}} F(\boldsymbol{x}) > 0$. 取 $\boldsymbol{z} = \boldsymbol{x} + \dfrac{3}{4}(\boldsymbol{y}-\boldsymbol{x})$, 有

$$\boldsymbol{z} \in \operatorname{ri}\left[\frac{\boldsymbol{x}+\boldsymbol{y}}{2}, \boldsymbol{y}\right], \quad (\boldsymbol{z}-\boldsymbol{x})^{\mathrm{T}} F(\boldsymbol{x}) = \frac{3}{4}(\boldsymbol{y}-\boldsymbol{x})^{\mathrm{T}} F(\boldsymbol{x}) > 0.$$

再由伪单调性: $(\boldsymbol{z}-\boldsymbol{x})^{\mathrm{T}} F(\boldsymbol{z}) = \dfrac{3}{4}(\boldsymbol{y}-\boldsymbol{x})^{\mathrm{T}} F(\boldsymbol{z}) > 0$, 即有 (5.5.1-7) 式成立.

(3) 若 F 在 C 上是半严格拟单调映射, 则根据定义 5.5.1(6) 知 F 在 C 上是拟单调映射.

(4) 若 F 在 C 上是严格单调映射, 则对于任意 $\boldsymbol{x}, \boldsymbol{y} \in C$ 有

$$(\boldsymbol{y}-\boldsymbol{x})^{\mathrm{T}} (F(\boldsymbol{y}) - F(\boldsymbol{x})) > 0,$$

上式说明 F 在 C 上是拟单调映射. 如果 F 在 C 上不是严格拟单调映射, 那么存在 $\boldsymbol{x}, \boldsymbol{y} \in C$ 和 $\forall \boldsymbol{z} \in \operatorname{ri}[\boldsymbol{x}, \boldsymbol{y}]$ 有

$$(\boldsymbol{y}-\boldsymbol{x})^{\mathrm{T}} F(\boldsymbol{z}) = 0.$$

设 $0 < \lambda_1 < \lambda_2 < 1$, $\boldsymbol{x}_1 = \boldsymbol{x} + \lambda_1(\boldsymbol{y}-\boldsymbol{x})$, $\boldsymbol{x}_2 = \boldsymbol{x} + \lambda_2(\boldsymbol{y}-\boldsymbol{x})$, 有 $\boldsymbol{x}_1, \boldsymbol{x}_2 \in \operatorname{ri}[\boldsymbol{x}, \boldsymbol{y}]$, 那么再由 F 的严格单调性得

$$(\boldsymbol{x}_2-\boldsymbol{x}_1)^{\mathrm{T}} (F(\boldsymbol{x}_2) - F(\boldsymbol{x}_1)) > 0.$$

即有 $(\boldsymbol{y}-\boldsymbol{x})^{\mathrm{T}} (F(\boldsymbol{x}_2) - F(\boldsymbol{x}_1)) > 0$, 得 $0 = (\boldsymbol{y}-\boldsymbol{x})^{\mathrm{T}} F(\boldsymbol{x}_2) > (\boldsymbol{y}-\boldsymbol{x})^{\mathrm{T}} F(\boldsymbol{x}_1) = 0$, 得到矛盾.

(5) 若 F 在 C 上是严格单调映射, 由 (4) 知 F 在 C 上是严格拟单调映射, 且对于任意 $\boldsymbol{x}, \boldsymbol{y} \in C$ 有

$$(\boldsymbol{y}-\boldsymbol{x})^{\mathrm{T}} F(\boldsymbol{x})>0 \Rightarrow(\boldsymbol{y}-\boldsymbol{x})^{\mathrm{T}} F(\boldsymbol{y}) \geqslant 0.$$

对于 $\boldsymbol{x}_1 \in \operatorname{ri}[\boldsymbol{x}, \boldsymbol{y}]$, 从上面不等式可得 $(\boldsymbol{y}-\boldsymbol{x})^{\mathrm{T}} F(\boldsymbol{x}_1) \geqslant 0$, 因此, 对于任意的 $\boldsymbol{z} \in \operatorname{ri}\left[\dfrac{\boldsymbol{x}+\boldsymbol{y}}{2}, \boldsymbol{y}\right] \subset \operatorname{ri}[\boldsymbol{x}, \boldsymbol{y}]$ 有

$$(\boldsymbol{y}-\boldsymbol{x})^{\mathrm{T}} F(\boldsymbol{z}) \geqslant 0.$$

再由 F 在 C 上是严格单调映射知, 上式存在 $\boldsymbol{z} \in \operatorname{ri}\left[\dfrac{\boldsymbol{x}+\boldsymbol{y}}{2}, \boldsymbol{y}\right]$ 使得

$$(\boldsymbol{y}-\boldsymbol{x})^{\mathrm{T}} F(\boldsymbol{z})>0.$$

(6) 由定义 5.5.1 显然. □

下面的性质说明非零点的连续伪单调映射等价于拟单调映射.

性质 5.5.2 设 $C \subset \mathbf{R}^n$ 是非空开凸集合和 $F: C \rightarrow \mathbf{R}^n$ 是连续向量映射, 且 $F(\boldsymbol{x}) \neq 0\,(\forall \boldsymbol{x} \in C)$, 则 F 在 C 上是伪单调映射当且仅当 F 在 C 上是拟单调映射.

证明 只要证明充分性即可. 假设 F 在 C 上是拟单调映射, 但不是伪单调映射, 那么存在 $\boldsymbol{x}, \boldsymbol{y} \in C$ 有

$$(\boldsymbol{y}-\boldsymbol{x})^{\mathrm{T}} F(\boldsymbol{x}) \geqslant 0, \quad (\boldsymbol{y}-\boldsymbol{x})^{\mathrm{T}} F(\boldsymbol{y})<0.$$

因 $F(\boldsymbol{x}) \neq 0\,(\forall \boldsymbol{x} \in C)$, 存在某个 $\boldsymbol{u} \in \mathbf{R}^n$ 满足 $\boldsymbol{u}^{\mathrm{T}} F(\boldsymbol{x})>0$. 由 F 是连续向量映射, 则存在 $\varepsilon>0$ 使得 $(\boldsymbol{y}+\varepsilon \boldsymbol{u}-\boldsymbol{x})^{\mathrm{T}} F(\boldsymbol{y}+\varepsilon \boldsymbol{u})<0$, 再根据 F 在 C 上是拟单调映射, 有

$$(\boldsymbol{y}+\varepsilon \boldsymbol{u}-\boldsymbol{x})^{\mathrm{T}} F(\boldsymbol{x}) \leqslant 0.$$

由上式得 $\boldsymbol{u}^{\mathrm{T}} F(\boldsymbol{x}) \leqslant(\boldsymbol{x}-\boldsymbol{y})^{\mathrm{T}} F(\boldsymbol{x}) \leqslant 0$, 这与 $\boldsymbol{u}^{\mathrm{T}} F(\boldsymbol{x})>0$ 矛盾. □

性质 5.5.3 设 $C \subset \mathbf{R}^n$ 是内部非空凸集合和 $F: \mathrm{cl} C \rightarrow \mathbf{R}^n$ 是连续向量映射. 若 F 在 $\operatorname{int} C$ 上是拟单调映射, 则 F 在 $\mathrm{cl} C$ 上是拟单调映射.

证明 对于任意 $\boldsymbol{x}, \boldsymbol{y} \in \mathrm{cl} C$, 设 $(\boldsymbol{y}-\boldsymbol{x})^{\mathrm{T}} F(\boldsymbol{x})>0$, 则存在收敛于 $\boldsymbol{x}, \boldsymbol{y} \in \mathrm{cl} C$ 的序列 $\{\boldsymbol{x}_i\},\{\boldsymbol{y}_i\} \subset \operatorname{int} C$. 由于 $F: \mathrm{cl} C \rightarrow \mathbf{R}^n$ 是连续向量映射, 因此 $(\boldsymbol{y}-\boldsymbol{x})^{\mathrm{T}} F(\boldsymbol{x})$ 也是连续函数. 存在一个 I, 对于 $i>I$ 有 $(\boldsymbol{y}_i-\boldsymbol{x}_i)^{\mathrm{T}} F(\boldsymbol{x}_i)>0$, 再由 F 在 $\operatorname{int} C$ 上是拟单调映射, 得

$$(\boldsymbol{y}_i-\boldsymbol{x}_i)^{\mathrm{T}} F(\boldsymbol{y}_i) \geqslant 0,$$

在上式中令 $i \to +\infty$ 有 $(\boldsymbol{y}-\boldsymbol{x})^{\mathrm{T}} F(\boldsymbol{y}) \geqslant 0$, 则 F 在 cl$C$ 上是拟单调映射. □

性质 5.5.3 说明, 如果连续向量映射在有效定义域内部 int$C$ 是拟单调映射, 则它在 cl$C$ 上是拟单调映射. 进一步, 如果可微向量映射在有效定义域上是拟单调映射, 则它在 int$C$ 上限制函数方向取值为 0 的 Jacobi 矩阵是半正定的, 见下面性质的结果.

性质 5.5.4 设 $C \subset \mathbf{R}^n$ 是非空凸集合和 $F:$ cl$C \to \mathbf{R}^n$ 是可微向量映射. 若 F 在 C 上是拟单调映射, 则对 $\boldsymbol{x} \in$ int$C$ 和 $\boldsymbol{d} \in \mathbf{R}^n$ 有

$$\boldsymbol{d}^{\mathrm{T}} F(\boldsymbol{x})=0 \Rightarrow \boldsymbol{d}^{\mathrm{T}} \boldsymbol{J} F(\boldsymbol{x}) \boldsymbol{d} \geqslant 0,$$

其中 $\boldsymbol{J} F(\boldsymbol{x})$ 表示 F 在 $\boldsymbol{x}$ 处的 Jacobi 矩阵.

证明 采用反证法. 存在 $\boldsymbol{x} \in$ int$C$ 和 $\boldsymbol{d} \in \mathbf{R}^n$ 使得

$$\boldsymbol{d}^{\mathrm{T}} F(\boldsymbol{x})=0 \Rightarrow \boldsymbol{d}^{\mathrm{T}} \boldsymbol{J} F(\boldsymbol{x}) \boldsymbol{d}<0.$$

设存在充分小的 $t>0$, 使得 $\boldsymbol{x}+t \boldsymbol{d} \in C$. 由 $F:$ cl$C \to \mathbf{R}^n$ 是可微向量映射得

$$F(\boldsymbol{x}+t \boldsymbol{d})=F(\boldsymbol{x})+t \boldsymbol{J} F(\boldsymbol{x}) \boldsymbol{d}+o(t).$$

即对于充分小的 t, 有

$$\boldsymbol{d}^{\mathrm{T}} F(\boldsymbol{x}+t \boldsymbol{d})=\boldsymbol{d}^{\mathrm{T}} F(\boldsymbol{x})+t \boldsymbol{d}^{\mathrm{T}} \boldsymbol{J} F(\boldsymbol{x}) \boldsymbol{d}+o(t)<0,$$

由此得到 $\boldsymbol{d}^{\mathrm{T}} F(\boldsymbol{x}+t \boldsymbol{d})<0$. 类似地有 $\boldsymbol{d}^{\mathrm{T}} F(\boldsymbol{x}-t \boldsymbol{d})>0$. 因此, 设 $\boldsymbol{x}_1=\boldsymbol{x}+t \boldsymbol{d}$ 和 $\boldsymbol{x}_2=\boldsymbol{x}-t \boldsymbol{d}$, 再根据 F 在 C 上是拟单调映射得

$$\begin{aligned} &(-2 t \boldsymbol{d})^{\mathrm{T}} F(\boldsymbol{x}_1)=(\boldsymbol{x}_2-\boldsymbol{x}_1)^{\mathrm{T}} F(\boldsymbol{x}_1)>0 \\ &\qquad \Rightarrow(-2 t \boldsymbol{d})^{\mathrm{T}} F(\boldsymbol{x}_2)=(\boldsymbol{x}_2-\boldsymbol{x}_1)^{\mathrm{T}} F(\boldsymbol{x}_2) \geqslant 0, \end{aligned}$$

即有 $\boldsymbol{d}^{\mathrm{T}} F(\boldsymbol{x}-t \boldsymbol{d}) \leqslant 0$, 因此得到矛盾. □

本小节得到广义单调性的几个基本性质:

1) F 在 C 上有下面关系：单调映射 $\Rightarrow$ 伪单调映射 $\Rightarrow$ 半严格拟单调映射 $\Rightarrow$ 拟单调映射，严格单调映射 $\Rightarrow$ 严格拟单调映射，严格单调映射 $\Rightarrow$ 半严格拟单调映射，伪单调映射 $\Rightarrow$ 拟单调映射 (见性质 5.5.1);

2) 如果 F 在 C 上具有非零值的连续向量映射, 则 F 是伪单调映射等于其是拟单调映射 (见性质 5.5.2);

3) 若 $F:$ cl$C \to \mathbf{R}^n$ 是连续向量映射, 且 F 在 int$C$ 上是拟单调映射, 则 F 在 cl$C$ 上是拟单调映射 (见性质 5.5.3);

4) 如果可微向量映射在有效定义域上是拟单调映射, 则它在 int$C$ 上限制函数方向取值为 0 的 Jacobi 矩阵是半正定的 (见性质 5.5.4).

5.5.2　与广义凸函数的关系

事实上, 当凸函数或广义凸函数具有可微性时, 其梯度的单调性与凸性之间存在着等价关系, 见下面的性质.

性质 5.5.5　设 $C \subset \mathbf{R}^n$ 是空凸集合和 $f: C \to \mathbf{R}^n$ 是可微函数, 则下面结论成立:

(1) f 在 C 上是凸函数当且仅当 ∇f 在 C 上是单调映射;

(2) f 在 C 上是严格凸函数当且仅当 ∇f 在 C 上是严格单调映射;

(3) f 在 C 上是伪凸函数当且仅当 ∇f 在 C 上是伪单调映射;

(4) f 在 C 上是拟凸函数当且仅当 ∇f 在 C 上是拟单调映射;

(5) f 在 C 上是严格拟凸函数当且仅当 ∇f 在 C 上是严格拟单调映射;

(6) f 在 C 上是半严格拟凸函数当且仅当 ∇f 在 C 上是半严格拟单调映射.

证明　(1) 若 f 在 C 上是凸函数, 则由性质 4.1.1(1)(2) 得

$$f(\boldsymbol{y}) - f(\boldsymbol{x}) \geqslant \nabla f(\boldsymbol{x})^{\mathrm{T}}(\boldsymbol{y} - \boldsymbol{x}), \quad \forall \boldsymbol{x}, \boldsymbol{y} \in C,$$

$$f(\boldsymbol{x}) - f(\boldsymbol{y}) \geqslant \nabla f(\boldsymbol{y})^{\mathrm{T}}(\boldsymbol{x} - \boldsymbol{y}), \quad \forall \boldsymbol{x}, \boldsymbol{y} \in C.$$

将两式相加得到 $(\nabla f(\boldsymbol{y}) - \nabla f(\boldsymbol{x}))^{\mathrm{T}}(\boldsymbol{y} - \boldsymbol{x}) \geqslant 0, \forall \boldsymbol{x}, \boldsymbol{y} \in C$, 由定义 5.5.1 知 ∇f 在 C 上是单调映射.

反过来, 若 f 在 C 上不是凸函数, 则存在 $\boldsymbol{x}, \boldsymbol{y} \in C$, 使得

$$f(\boldsymbol{y}) - f(\boldsymbol{x}) < \nabla f(\boldsymbol{x})^{\mathrm{T}}(\boldsymbol{y} - \boldsymbol{x}).$$

设 f 在 C 上的一个限制函数为

$$\varphi(t) = f(\boldsymbol{x} + t(\boldsymbol{y} - \boldsymbol{x})), \quad t \in [0, 1].$$

由拉格朗日中值定理, 存在一个 $\bar{t} \in (0, 1)$ 满足

$$\varphi(1) - \varphi(0) = \varphi'(\bar{t}) = \nabla f(\boldsymbol{x} + \bar{t}(\boldsymbol{y} - \boldsymbol{x}))^{\mathrm{T}}(\boldsymbol{y} - \boldsymbol{x}).$$

取 $\boldsymbol{x} + \bar{t}(\boldsymbol{y} - \boldsymbol{x}) = \bar{\boldsymbol{x}}$, 有

$$\nabla f(\bar{\boldsymbol{x}})^{\mathrm{T}}(\boldsymbol{y} - \boldsymbol{x}) = f(\boldsymbol{y}) - f(\boldsymbol{x}) < \nabla f(\boldsymbol{x})^{\mathrm{T}}(\boldsymbol{y} - \boldsymbol{x}),$$

由上式可得 $(\nabla f(\bar{\boldsymbol{x}}) - \nabla f(\boldsymbol{x}))^{\mathrm{T}}(\bar{\boldsymbol{x}} - \boldsymbol{x}) < 0$, 这与 ∇f 在 C 上是单调映射矛盾.

(2) 证明过程与 (1) 完全类似.

(3) 若 f 在 C 上是伪凸函数, 则对任意 $\boldsymbol{x}, \boldsymbol{y} \in C$, 有

$$\nabla f(\boldsymbol{x})^{\mathrm{T}}(\boldsymbol{y} - \boldsymbol{x}) \geqslant 0 \Rightarrow f(\boldsymbol{x}) \leqslant f(\boldsymbol{y}).$$

由定理 5.3.2 知 f 在 C 上是拟凸函数, 再由定理 5.1.14 得 $\nabla f(\boldsymbol{y})^{\mathrm{T}}(\boldsymbol{y}-\boldsymbol{x}) \geqslant 0$, 最后由伪单调映射定义知 ∇f 在 C 上是伪单调映射.

反之, 当 ∇f 在 C 上是伪单调映射时, 假设 f 在 C 上不是伪凸函数, 则存在 $\boldsymbol{x}, \boldsymbol{y} \in C$, 使得

$$f(\boldsymbol{x}) > f(\boldsymbol{y}) \Rightarrow \nabla f(\boldsymbol{x})^{\mathrm{T}}(\boldsymbol{y}-\boldsymbol{x}) \geqslant 0.$$

根据 ∇f 的伪单调性, 有 $\nabla f(\boldsymbol{y})^{\mathrm{T}}(\boldsymbol{y}-\boldsymbol{x}) \geqslant 0$. 定义 f 在 C 上的限制函数:

$$\varphi(t)=f(\boldsymbol{x}+t(\boldsymbol{y}-\boldsymbol{x})), \quad t \in[0,1].$$

记 $\boldsymbol{x}_t=\boldsymbol{x}+t(\boldsymbol{y}-\boldsymbol{x})$, 有

$$\nabla f(\boldsymbol{x})^{\mathrm{T}}(\boldsymbol{x}_t-\boldsymbol{x})=t \nabla f(\boldsymbol{x})^{\mathrm{T}}(\boldsymbol{y}-\boldsymbol{x}) \geqslant 0,$$

于是得 $\nabla f(\boldsymbol{x}_t)^{\mathrm{T}}(\boldsymbol{x}_t-\boldsymbol{x}) \geqslant 0$, 即 $\varphi'(t) \geqslant 0$. 说明 $\varphi(t)$ 在 $[0,1]$ 上是单调增加的, 即有 $\varphi(1) \geqslant \varphi(0)$, 因此可得 $f(\boldsymbol{y}) \geqslant f(\boldsymbol{x})$, 这与 $f(\boldsymbol{x})>f(\boldsymbol{y})$ 矛盾.

(4) 若 f 在 C 上是拟凸函数, 设对任意 $\boldsymbol{x}, \boldsymbol{y} \in C$ 使得 $\nabla f(\boldsymbol{x})^{\mathrm{T}}(\boldsymbol{y}-\boldsymbol{x})>0$, 由定理 5.1.14 得 $f(\boldsymbol{x})<f(\boldsymbol{y})$. 再由定理 5.1.14 得 $\nabla f(\boldsymbol{y})^{\mathrm{T}}(\boldsymbol{y}-\boldsymbol{x}) \geqslant 0$, 因此, ∇f 在 C 上是拟单调映射.

反过来, 当 ∇f 在 C 上是拟单调映射时, 若 f 在 C 上不是拟凸函数, 则存在 $\boldsymbol{x}, \boldsymbol{y} \in C$ 使得

$$f(\boldsymbol{x}) \geqslant f(\boldsymbol{y}) \Rightarrow \nabla f(\boldsymbol{x})^{\mathrm{T}}(\boldsymbol{y}-\boldsymbol{x})>0.$$

定义 f 在 C 上的限制函数:

$$\varphi(t)=f(\boldsymbol{x}+t(\boldsymbol{y}-\boldsymbol{x})), \quad t \in[0,1].$$

记 $\boldsymbol{x}_t=\boldsymbol{x}+t(\boldsymbol{y}-\boldsymbol{x})$, 有

$$\nabla f(\boldsymbol{x})^{\mathrm{T}}(\boldsymbol{x}_t-\boldsymbol{x})=t \nabla f(\boldsymbol{x})^{\mathrm{T}}(\boldsymbol{y}-\boldsymbol{x})>0,$$

得 $\nabla f(\boldsymbol{x}_t)^{\mathrm{T}}(\boldsymbol{x}_t-\boldsymbol{x})>0$, 即 $\varphi'(t)>0$. 说明 $\varphi(t)$ 在 $[0,1]$ 上是严格单调增加的, 即有 $\varphi(1)>\varphi(0)$, 因此可得 $f(\boldsymbol{y})>f(\boldsymbol{x})$, 这与 $f(\boldsymbol{x}) \geqslant f(\boldsymbol{y})$ 矛盾.

(5) 若 f 在 C 上是严格拟凸函数, 由 (4) 知 ∇f 在 C 上是拟单调映射. 假设 ∇f 在 C 上不是严格拟单调映射, 即存在 $\boldsymbol{x}, \boldsymbol{y} \in C$ 使得对于任意的 $\bar{\boldsymbol{x}} \in[\boldsymbol{x}, \boldsymbol{y}]$ 有

$$\nabla f(\bar{\boldsymbol{x}})^{\mathrm{T}}(\boldsymbol{y}-\boldsymbol{x})=0.$$

上式表明 f 在 C 上的限制函数

$$\varphi(t)=f(\boldsymbol{x}+t(\boldsymbol{y}-\boldsymbol{x})), \quad t \in[0,1]$$

在 $[0,1]$ 上是一个常数函数, 根据性质 5.1.1(4) 知 f 在 C 上不是拟凸函数, 得到矛盾.

反过来, 若 ∇f 在 C 上是严格拟单调映射, 则它也是拟单调映射, 由 (4) 知 f 在 C 上是拟凸函数. 根据 ∇f 在 C 上是严格拟单调映射的定义, f 在 C 上任何线段的限制函数不是常数, 再根据性质 5.1.1(4) 知 f 在 C 上是严格拟凸函数.

(6) 若 f 在 C 上是半严格拟凸函数, 由定理 5.2.9 知 f 在 C 上是拟凸函数, 再由 (4) 知 ∇f 在 C 上是拟单调映射. 假设 ∇f 在 C 上不是半严格拟单调映射, 那么存在 $\boldsymbol{x},\boldsymbol{y}\in C$ 和 $\boldsymbol{x}\neq\boldsymbol{y}$ 有

$$(\boldsymbol{y}-\boldsymbol{x})^{\mathrm{T}}\nabla f(\boldsymbol{x})>0\Rightarrow(\boldsymbol{y}-\boldsymbol{x})^{\mathrm{T}}\nabla f(\boldsymbol{z})\leqslant 0,\quad \forall \boldsymbol{z}\in \operatorname{ri}\left[\frac{\boldsymbol{x}+\boldsymbol{y}}{2},\boldsymbol{y}\right].$$

但从拟单调映射定义得 $(\boldsymbol{y}-\boldsymbol{x})^{\mathrm{T}}\nabla f(\boldsymbol{z})\geqslant 0,\forall \boldsymbol{z}\in \operatorname{ri}\left[\frac{\boldsymbol{x}+\boldsymbol{y}}{2},\boldsymbol{y}\right]$. 因此, 有

$$(\boldsymbol{y}-\boldsymbol{x})^{\mathrm{T}}\nabla f(\boldsymbol{z})=0,\quad \forall \boldsymbol{z}\in \operatorname{ri}\left[\frac{\boldsymbol{x}+\boldsymbol{y}}{2},\boldsymbol{y}\right].$$

由定理 5.1.14 得 $f(\boldsymbol{x})<f(\boldsymbol{y})$, 再由 f 在 C 上是半严格拟凸函数定义知

$$f(\boldsymbol{x}+\lambda(\boldsymbol{y}-\boldsymbol{x}))<f(\boldsymbol{y}),\quad \lambda\in(0,1).$$

在上式中取 $\lambda=\frac{1}{2}$ 得 $f\left(\frac{\boldsymbol{x}+\boldsymbol{y}}{2}\right)<f(\boldsymbol{y})$, 定义线段 $\left[\frac{\boldsymbol{x}+\boldsymbol{y}}{2},\boldsymbol{y}\right]$ 上的限制函数:

$$\varphi(\lambda)=f\left(\frac{\boldsymbol{x}+\boldsymbol{y}}{2}+\lambda\left(\frac{\boldsymbol{y}-\boldsymbol{x}}{2}\right)\right),\quad \forall\lambda\in[0,1].$$

由上式得 $\varphi'(\lambda)=\frac{1}{2}(\boldsymbol{y}-\boldsymbol{x})^{\mathrm{T}}\nabla f(\boldsymbol{z})=0,\forall \boldsymbol{z}\in \operatorname{ri}\left[\frac{\boldsymbol{x}+\boldsymbol{y}}{2},\boldsymbol{y}\right],\forall\lambda\in[0,1]$, 知 $\varphi(\lambda)$ 在线段 $\left[\frac{\boldsymbol{x}+\boldsymbol{y}}{2},\boldsymbol{y}\right]$ 上是常值函数, 这与 $f\left(\frac{\boldsymbol{x}+\boldsymbol{y}}{2}\right)<f(\boldsymbol{y})$ 矛盾.

反过来, 设 ∇f 在 C 上是半严格拟单调映射. 由性质 5.5.1 知 ∇f 在 C 上是拟单调映射, 由 (4) 知 f 在 C 上是拟凸函数. 假设 f 在 C 上不是半严格拟凸函数, 根据半严格拟凸函数定义, 一定存在 $\boldsymbol{x},\boldsymbol{y}\in C$ 和 $\delta\in(0,1)$, 使得 $f(\boldsymbol{x})>f(\boldsymbol{y})$ 且 $f[\delta\boldsymbol{x}+(1-\delta)\boldsymbol{y}]\geqslant f(\boldsymbol{x})$. 由 f 在 C 上是拟凸函数和连续函数知 $f[\delta\boldsymbol{x}+(1-\delta)\boldsymbol{y}]=f(\boldsymbol{x})$, 一定存在一个 $\lambda\leqslant\delta$ 满足下面的式子:

$$f(\boldsymbol{y}+t(\boldsymbol{x}-\boldsymbol{y}))=f(\boldsymbol{y}+\lambda(\boldsymbol{x}-\boldsymbol{y}))=f(\boldsymbol{x}),\quad \forall t\in[\lambda,1],\tag{5.5.2-1}$$

$$f(\boldsymbol{y}+t(\boldsymbol{x}-\boldsymbol{y}))<f(\boldsymbol{y}+\lambda(\boldsymbol{x}-\boldsymbol{y}))=f(\boldsymbol{x}),\quad \forall t\in[0,\lambda). \tag{5.5.2-2}$$

设 $\lambda=\min\{t|f(\boldsymbol{y}+t(\boldsymbol{x}-\boldsymbol{y}))=f(\boldsymbol{x}),t\in[0,1]\}$. 因为 $f(\boldsymbol{x})>f(\boldsymbol{y})$, 显然有 $0<\lambda\leqslant\delta$. 如果 (5.5.2-1) 式不成立, 则存在某个 $t\in(\lambda,1)$ 满足 $f(\boldsymbol{y}+t(\boldsymbol{x}-\boldsymbol{y}))<f(\boldsymbol{x})$. 根据 f 在 $[\boldsymbol{y},\boldsymbol{y}+t(\boldsymbol{x}-\boldsymbol{y})]$ 上是拟凸的, 且 $\boldsymbol{y}+\lambda(\boldsymbol{x}-\boldsymbol{y})\in[\boldsymbol{y},\boldsymbol{y}+t(\boldsymbol{x}-\boldsymbol{y})]$, 有

$$f(\boldsymbol{x})=f(\boldsymbol{y}+\lambda(\boldsymbol{x}-\boldsymbol{y}))\leqslant\max\{f(\boldsymbol{y}),f(\boldsymbol{y}+t(\boldsymbol{x}-\boldsymbol{y}))\}<f(\boldsymbol{x}),$$

得到矛盾, 因此, (5.5.2-1) 式成立. 如果 (5.5.2-2) 式不成立, 则存在某个 $t\in[0,\lambda)$ 满足

$$f(\boldsymbol{y}+t(\boldsymbol{x}-\boldsymbol{y}))=f(\boldsymbol{x}),$$

显然上式与 λ 所设矛盾. 如果 $1>\lambda\geqslant\dfrac{1}{2}$, 取充分小的 $\varepsilon>0$ 和 $t=1-\lambda-\varepsilon$, 有 $t\in\left(0,\dfrac{1}{2}\right)$, 这时用 $\boldsymbol{y}+t(\boldsymbol{x}-\boldsymbol{y})$ 替换 $\boldsymbol{y}$. 因此, 不妨设 $\lambda<\dfrac{1}{2}$. 再设限制函数:

$$\varphi(t)=f(\boldsymbol{y}+t(\boldsymbol{x}-\boldsymbol{y})),\quad \forall t\in[0,1]. \tag{5.5.2-3}$$

由于 f 在 C 上是可微拟凸函数, 由定理 5.1.14 得

$$\nabla f(\boldsymbol{y}+\lambda(\boldsymbol{x}-\boldsymbol{y}))^{\mathrm{T}}(\lambda-t)(\boldsymbol{x}-\boldsymbol{y})\leqslant 0. \tag{5.5.2-4}$$

求导 $\varphi'(t)=\nabla f(\boldsymbol{y}+t(\boldsymbol{x}-\boldsymbol{y}))^{\mathrm{T}}(\boldsymbol{x}-\boldsymbol{y}),\forall t\in[0,1]$, 由 (5.5.2-1) 式得

$$\varphi'(t)=\nabla f(\boldsymbol{y}+t(\boldsymbol{x}-\boldsymbol{y}))^{\mathrm{T}}(\boldsymbol{x}-\boldsymbol{y})=0,\quad \forall t\in(\lambda,1). \tag{5.5.2-5}$$

取充分小的 $\varepsilon>0$, 对区间 $[\lambda-\varepsilon,\lambda]$ 使用拉格朗日中值定理, 即由 (5.5.2-2) 式存在 $\tau\in(\lambda-\varepsilon,\lambda)$ 使得

$$\varphi(\lambda)-\varphi(\lambda-\varepsilon)=\varepsilon\nabla f(\boldsymbol{y}+\tau(\boldsymbol{x}-\boldsymbol{y}))^{\mathrm{T}}(\boldsymbol{x}-\boldsymbol{y})>0.$$

设 $\boldsymbol{x}_\tau=\boldsymbol{y}+\tau(\boldsymbol{x}-\boldsymbol{y})$, 上式即为

$$\nabla f(\boldsymbol{x}_\tau)^{\mathrm{T}}(\boldsymbol{x}-\boldsymbol{x}_\tau)=(1-\tau)\nabla f(\boldsymbol{x}_\tau)^{\mathrm{T}}(\boldsymbol{x}-\boldsymbol{y})>0.$$

再由 ∇f 在 C 上是半严格拟单调映射知, 存在

$$\begin{aligned}\boldsymbol{z}\in\operatorname{ri}\left[\frac{\boldsymbol{y}+\tau(\boldsymbol{x}-\boldsymbol{y})+\boldsymbol{x}}{2},\boldsymbol{x}\right]&=\operatorname{ri}\left[\boldsymbol{y}+\frac{(1+\tau)}{2}(\boldsymbol{x}-\boldsymbol{y}),\boldsymbol{x}\right]\\&\subset\operatorname{ri}[\boldsymbol{y}+\lambda(\boldsymbol{x}-\boldsymbol{y}),\boldsymbol{x}],\end{aligned}$$

使得 $\nabla f(\boldsymbol{z})^{\mathrm{T}}(\boldsymbol{x}-\boldsymbol{x}_\tau)=(1-\xi)\nabla f(\boldsymbol{z})^{\mathrm{T}}(\boldsymbol{x}-\boldsymbol{y})>0$, 其中 $\boldsymbol{z}=\boldsymbol{y}+\xi(\boldsymbol{x}-\boldsymbol{y})$ 和 $\xi\in(\lambda,1)$, 根据 (5.5.2-5) 式有 $\nabla f(\boldsymbol{z})^{\mathrm{T}}(\boldsymbol{x}-\boldsymbol{y})=0$, 得到矛盾. □

因此, 本小节得到了单调性与凸性的之间的关系 (见性质 5.5.5):

1) f 是凸函数等价于 ∇f 是单调映射;

2) f 是严格凸函数等价于 ∇f 是严格单调映射;

3) f 是伪凸函数等价于 ∇f 是伪单调映射;

4) f 是拟凸函数等价于 ∇f 是拟单调映射;

5) f 是严格拟凸函数等价于 ∇f 在 C 上是严格拟单调映射;

6) f 是半严格拟凸函数等价于 ∇f 是半严格拟单调映射;

因此, 可微函数的凸性可以通过其微分单调性来判定.

5.5.3 仿射映射

最后讨论一类特殊映射——仿射映射的广义单调性性质.

定义 5.5.2 已知 $\boldsymbol{A}$ 是 $n\times n$ 的矩阵和点 $\boldsymbol{b}\in\mathbf{R}^n$, 称 $F(\boldsymbol{x})=\boldsymbol{A}\boldsymbol{x}+\boldsymbol{b}$ 是仿射映射.

事实上, 有与性质 5.5.4 类似的结论如下, 即仿射映射的伪单调性与拟单调性的等价关系.

性质 5.5.6 设 $C\subset\mathbf{R}^n$ 是非空开凸集合和 $F(\boldsymbol{x})=\boldsymbol{A}\boldsymbol{x}+\boldsymbol{b}$ 是仿射映射, 则 F 在 C 上是伪单调映射当且仅当 F 在 C 上是拟单调映射或者对任意的 $\boldsymbol{x}\in C$ 和任意的 $\boldsymbol{d}\in\mathbf{R}^n$ 有

$$\boldsymbol{d}^{\mathrm{T}}(\boldsymbol{A}\boldsymbol{x}+\boldsymbol{b})=0\Rightarrow\boldsymbol{d}^{\mathrm{T}}\boldsymbol{A}\boldsymbol{d}\geqslant 0. \tag{5.5.3-1}$$

证明 设 F 在 C 上是伪单调的, 则 F 在 C 上是拟单调的. 下面证明 (5.5.3-1) 式成立. 设

$$\boldsymbol{d}^{\mathrm{T}}(\boldsymbol{A}\boldsymbol{x}+\boldsymbol{b})=0,\quad \forall\boldsymbol{x}\in C,\quad \forall\boldsymbol{d}\in\mathbf{R}^n.$$

令 $\boldsymbol{y}=\boldsymbol{x}+t\boldsymbol{d}\in C$, 其中 $t\in\mathbf{R}\backslash\{0\}$, 有

$$(\boldsymbol{y}-\boldsymbol{x})^{\mathrm{T}}(\boldsymbol{A}\boldsymbol{x}+\boldsymbol{b})=t\boldsymbol{d}^{\mathrm{T}}(\boldsymbol{A}\boldsymbol{x}+\boldsymbol{b})=0.$$

由 F 在 C 上是拟单调映射, 有 $(\boldsymbol{y}-\boldsymbol{x})^{\mathrm{T}}(\boldsymbol{A}\boldsymbol{y}+\boldsymbol{b})\geqslant 0$. 因此, 得

$$(\boldsymbol{y}-\boldsymbol{x})^{\mathrm{T}}(\boldsymbol{A}\boldsymbol{y}+\boldsymbol{b})=t\boldsymbol{d}^{\mathrm{T}}(\boldsymbol{A}\boldsymbol{x}+\boldsymbol{b}+t\boldsymbol{A}\boldsymbol{d})=t^2\boldsymbol{d}^{\mathrm{T}}\boldsymbol{A}\boldsymbol{d}\geqslant 0.$$

反过来, 先假设 F 在 C 上是拟单调映射. 如果对于任意的 $\boldsymbol{x}\in C$ 有 $F(\boldsymbol{x})\neq\mathbf{0}$, 由性质 5.5.2 知 F 在 C 上是伪单调映射. 如果存在点 $\boldsymbol{x}_0\in C$ 使得 $F(\boldsymbol{x}_0)=\mathbf{0}$, 则对于任意的 $\boldsymbol{x}\in C$ 有

$$(\boldsymbol{x}-\boldsymbol{x}_0)^{\mathrm{T}}(\boldsymbol{A}\boldsymbol{x}_0+\boldsymbol{b})=0.$$

下面用反证法. 若 F 在 C 上不是伪单调映射, 那么存在 $\boldsymbol{x}_1 \in C$ 使得

$$(\boldsymbol{x}_1-\boldsymbol{x}_0)^{\mathrm{T}}(\boldsymbol{A}\boldsymbol{x}_1+\boldsymbol{b})<0. \tag{5.5.3-2}$$

因为 $C \subset \mathbf{R}^n$ 是一个开凸集, 设 $\boldsymbol{x}_2=\boldsymbol{x}_0+t(\boldsymbol{x}_1-\boldsymbol{x}_0)$, 取 $t<0$ 使得 $\boldsymbol{x}_2 \in C$. 那么有

$$(1-t)(\boldsymbol{x}_1-\boldsymbol{x}_0)=\boldsymbol{x}_1-\boldsymbol{x}_2,$$

得 $(\boldsymbol{x}_1-\boldsymbol{x}_2)^{\mathrm{T}}(\boldsymbol{A}\boldsymbol{x}_1+\boldsymbol{b})<0$. 则再由 F 在 C 上是拟单调映射得

$$(\boldsymbol{x}_1-\boldsymbol{x}_2)^{\mathrm{T}}(\boldsymbol{A}\boldsymbol{x}_2+\boldsymbol{b}) \leqslant 0.$$

那么, 通过上式和 $\boldsymbol{A}\boldsymbol{x}_0+\boldsymbol{b}=\mathbf{0}$ 得

$$\begin{aligned}
0 &\geqslant (\boldsymbol{x}_1-\boldsymbol{x}_2)^{\mathrm{T}}(\boldsymbol{A}\boldsymbol{x}_2+\boldsymbol{b}) \\
&=(1-t)(\boldsymbol{x}_1-\boldsymbol{x}_0)^{\mathrm{T}}(\boldsymbol{A}(\boldsymbol{x}_0+t(\boldsymbol{x}_1-\boldsymbol{x}_0))+\boldsymbol{b}) \\
&=(1-t)t(\boldsymbol{x}_1-\boldsymbol{x}_0)^{\mathrm{T}}\boldsymbol{A}(\boldsymbol{x}_1-\boldsymbol{x}_0) \\
&=(1-t)t(\boldsymbol{x}_1-\boldsymbol{x}_0)^{\mathrm{T}}(\boldsymbol{A}\boldsymbol{x}_1+\boldsymbol{b}).
\end{aligned}$$

因此, 得 $(\boldsymbol{x}_1-\boldsymbol{x}_0)^{\mathrm{T}}(\boldsymbol{A}\boldsymbol{x}_1+\boldsymbol{b}) \geqslant 0$, 这与 (5.5.3-2) 式矛盾.

最后, 设 $\forall \boldsymbol{x} \in C, \forall \boldsymbol{d} \in \mathbf{R}^n$ 有 (5.5.3-1) 式成立, 再假设 F 在 C 上不是伪单调映射, 则存在 $\boldsymbol{x}, \boldsymbol{y} \in C$ 有

$$(\boldsymbol{y}-\boldsymbol{x})^{\mathrm{T}}(\boldsymbol{A}\boldsymbol{x}+\boldsymbol{b}) \geqslant 0, \quad 且 \quad (\boldsymbol{y}-\boldsymbol{x})^{\mathrm{T}}(\boldsymbol{A}\boldsymbol{y}+\boldsymbol{b})<0.$$

因仿射映射是连续的, 存在一个 $\boldsymbol{x}_t=\boldsymbol{x}+t(\boldsymbol{y}-\boldsymbol{x})$, $t \in [0,1)$ 使得

$$(\boldsymbol{y}-\boldsymbol{x})^{\mathrm{T}}(\boldsymbol{A}\boldsymbol{x}_t+\boldsymbol{b})=0.$$

因此有

$$\begin{aligned}
(\boldsymbol{y}-\boldsymbol{x})^{\mathrm{T}}\boldsymbol{A}(\boldsymbol{y}-\boldsymbol{x}) &=(\boldsymbol{y}-\boldsymbol{x})^{\mathrm{T}}\boldsymbol{A}\frac{\boldsymbol{y}-\boldsymbol{x}_t}{1-t} \\
&=\frac{1}{1-t}(\boldsymbol{y}-\boldsymbol{x})^{\mathrm{T}}(\boldsymbol{A}\boldsymbol{y}+\boldsymbol{b}-\boldsymbol{A}\boldsymbol{x}_t-\boldsymbol{b}) \\
&=\frac{1}{1-t}(\boldsymbol{y}-\boldsymbol{x})^{\mathrm{T}}(\boldsymbol{A}\boldsymbol{y}+\boldsymbol{b})<0,
\end{aligned}$$

令 $\boldsymbol{d}=\boldsymbol{y}-\boldsymbol{x}$, 可得上式与 (5.5.3-1) 式矛盾. □

进一步, 若仿射映射是拟单调的, 则有零点边界点判定条件, 若无零边界点的话, 则该仿射映射是伪单调的.

性质 5.5.7 设 $C \subset \mathbf{R}^n$ 是内部非空凸集合和 $F(\boldsymbol{x}) = \boldsymbol{A}\boldsymbol{x} + \boldsymbol{b}$ 在 C 上是拟单调映射, 则

(1) 如果存在 $\boldsymbol{x} \in C, \boldsymbol{d} \in \mathbf{R}^n$ 使得 $\boldsymbol{d}^{\mathrm{T}}(\boldsymbol{A}\boldsymbol{x}+\boldsymbol{b}) = 0 \Rightarrow \boldsymbol{d}^{\mathrm{T}}\boldsymbol{A}\boldsymbol{d} < 0$, 则 $\boldsymbol{x}$ 是边界点, 且 $F(\boldsymbol{x}) = \mathbf{0}$;

(2) 如果 F 在 C 的边界上均是非零点, 则 F 在 C 上是伪单调映射.

证明 (1) 假设 $\boldsymbol{x} \in \mathrm{int}C$, 根据性质 5.5.6 有 $\boldsymbol{d}^{\mathrm{T}}(\boldsymbol{A}\boldsymbol{x}+\boldsymbol{b}) = 0 \Rightarrow \boldsymbol{d}^{\mathrm{T}}\boldsymbol{A}\boldsymbol{d} \geqslant 0$. 因此, 若 $\boldsymbol{x} \in C, \boldsymbol{d} \in \mathbf{R}^n$ 使得 $\boldsymbol{d}^{\mathrm{T}}(\boldsymbol{A}\boldsymbol{x}+\boldsymbol{b}) = 0 \Rightarrow \boldsymbol{d}^{\mathrm{T}}\boldsymbol{A}\boldsymbol{d} < 0$, 得到矛盾. 所以 $\boldsymbol{x}$ 是边界点.

假设 $F(\boldsymbol{x}) = \boldsymbol{A}\boldsymbol{x}+\boldsymbol{b} \neq \mathbf{0}$, 得 $\boldsymbol{d} \neq \mathbf{0}$. 不妨任取 $\boldsymbol{y} \in \mathrm{int}C$, 如果 $\boldsymbol{d}^{\mathrm{T}}(\boldsymbol{A}\boldsymbol{y}+\boldsymbol{b}) = 0$, 根据性质 5.5.6 有 $\boldsymbol{d}^{\mathrm{T}}(\boldsymbol{A}\boldsymbol{y}+\boldsymbol{b}) = 0 \Rightarrow \boldsymbol{d}^{\mathrm{T}}\boldsymbol{A}\boldsymbol{d} \geqslant 0$, 这与假设矛盾. 因此, 根据性质 5.5.6 有 $\boldsymbol{d}^{\mathrm{T}}(\boldsymbol{A}\boldsymbol{y}+\boldsymbol{b}) \neq 0$. 下面分两种情形.

(i) 若 $\boldsymbol{d}^{\mathrm{T}}(\boldsymbol{A}\boldsymbol{y}+\boldsymbol{b}) > 0$, 令 $\boldsymbol{x}_t = \boldsymbol{y} + t(\boldsymbol{x}-\boldsymbol{y})$, 其中 $\forall t \in [0,1]$. 对于任意的 $t \in (0,1)$ 有

$$\begin{aligned}\boldsymbol{d}^{\mathrm{T}}(\boldsymbol{A}\boldsymbol{x}_t+\boldsymbol{b}) &= \boldsymbol{d}^{\mathrm{T}}(\boldsymbol{A}(\boldsymbol{y}+t(\boldsymbol{x}-\boldsymbol{y}))+\boldsymbol{b}) \\ &= (1-t)\,\boldsymbol{d}^{\mathrm{T}}(\boldsymbol{A}\boldsymbol{y}+\boldsymbol{b}) > 0. \end{aligned} \tag{5.5.3-3}$$

设 $\boldsymbol{y}_t = \boldsymbol{x}_t + t\boldsymbol{d}\,(\forall t \in [0,1])$, 得

$$\boldsymbol{d}^{\mathrm{T}}(\boldsymbol{A}\boldsymbol{y}_0+\boldsymbol{b}) > 0, \quad \boldsymbol{d}^{\mathrm{T}}(\boldsymbol{A}\boldsymbol{y}_1+\boldsymbol{b}) = \boldsymbol{d}^{\mathrm{T}}(\boldsymbol{A}\boldsymbol{x}+\boldsymbol{b}+\boldsymbol{A}\boldsymbol{d}) < 0,$$

由上式知存在 $t_1 \in (0,1)$ 使得 $\boldsymbol{d}^{\mathrm{T}}(\boldsymbol{A}\boldsymbol{y}_{t_1}+\boldsymbol{b}) = 0$. 由 F 在 C 上是拟单调映射知

$$(\boldsymbol{y}_t-\boldsymbol{x}_t)^{\mathrm{T}}(\boldsymbol{A}\boldsymbol{x}_t+\boldsymbol{b}) > 0 \Rightarrow (\boldsymbol{y}_t-\boldsymbol{x}_t)^{\mathrm{T}}(\boldsymbol{A}\boldsymbol{y}_t+\boldsymbol{b}) \geqslant 0, \quad \forall t \in (0,1),$$

即 $\boldsymbol{d}^{\mathrm{T}}(\boldsymbol{A}\boldsymbol{y}_t+\boldsymbol{b}) \geqslant 0, \forall t \in (0,1)$. 显然函数 $\boldsymbol{d}^{\mathrm{T}}(\boldsymbol{A}\boldsymbol{y}_t+\boldsymbol{b})$ 关于变量 t 是连续的, 令 $t \to 1$ 得

$$\boldsymbol{d}^{\mathrm{T}}(\boldsymbol{A}\boldsymbol{y}_1+\boldsymbol{b}) \geqslant 0,$$

这与 $\boldsymbol{d}^{\mathrm{T}}(\boldsymbol{A}\boldsymbol{y}_1+\boldsymbol{b}) < 0$ 矛盾.

(ii) 若 $\boldsymbol{d}^{\mathrm{T}}(\boldsymbol{A}\boldsymbol{y}+\boldsymbol{b}) < 0$, 由 (5.5.3-3) 式有 $\boldsymbol{d}^{\mathrm{T}}(\boldsymbol{A}\boldsymbol{x}_t+\boldsymbol{b}) < 0\,(\forall t \in (0,1))$. 设

$$\boldsymbol{z}_t = \boldsymbol{x}_t - t\boldsymbol{d} \quad (\forall t \in [0,1]),$$

得

$$\boldsymbol{d}^{\mathrm{T}}(\boldsymbol{A}\boldsymbol{z}_0+\boldsymbol{b}) < 0, \quad \boldsymbol{d}^{\mathrm{T}}(\boldsymbol{A}\boldsymbol{z}_1+\boldsymbol{b}) = \boldsymbol{d}^{\mathrm{T}}(\boldsymbol{A}\boldsymbol{x}+\boldsymbol{b}-\boldsymbol{A}\boldsymbol{d}) > 0.$$

存在 $t_2 \in (0,1)$ 使得 $\boldsymbol{d}^{\mathrm{T}}(\boldsymbol{A}\boldsymbol{z}_{t_2}+\boldsymbol{b})=0$. 由 $\boldsymbol{d}^{\mathrm{T}}(\boldsymbol{A}\boldsymbol{x}_t+\boldsymbol{b})<0\,(\forall t\in(0,1))$ 得

$$(\boldsymbol{x}_t-\boldsymbol{z}_t)^{\mathrm{T}}(\boldsymbol{A}\boldsymbol{z}_t+\boldsymbol{b})>0, \quad \forall t\in(0,1),$$

这与 $\boldsymbol{d}^{\mathrm{T}}(\boldsymbol{A}\boldsymbol{z}_{t_2}+\boldsymbol{b})=0$ 矛盾.

(2) 设 $\boldsymbol{x},\boldsymbol{y}\in C$ 和 $(\boldsymbol{y}-\boldsymbol{x})^{\mathrm{T}}F(\boldsymbol{x})\geqslant 0$. 由于已知 $F(\boldsymbol{x})=\boldsymbol{A}\boldsymbol{x}+\boldsymbol{b}$ 在 C 上是拟单调映射, 若 $(\boldsymbol{y}-\boldsymbol{x})^{\mathrm{T}}F(\boldsymbol{x})>0$, 则有 $(\boldsymbol{y}-\boldsymbol{x})^{\mathrm{T}}F(\boldsymbol{y})>0$, 得 F 在 C 上是伪单调映射. 因此, 根据伪单调定义, 只要证明: $(\boldsymbol{y}-\boldsymbol{x})^{\mathrm{T}}F(\boldsymbol{x})\geqslant 0\Rightarrow(\boldsymbol{y}-\boldsymbol{x})^{\mathrm{T}}F(\boldsymbol{y})\geqslant 0$ 即可. 采用反证法, 假设

$$(\boldsymbol{y}-\boldsymbol{x})^{\mathrm{T}}F(\boldsymbol{x})\geqslant 0\Rightarrow(\boldsymbol{y}-\boldsymbol{x})^{\mathrm{T}}F(\boldsymbol{y})<0$$

成立. 因为已知 $F(\boldsymbol{x})=\boldsymbol{A}\boldsymbol{x}+\boldsymbol{b}$ 在 C 上是拟单调映射, 若 $(\boldsymbol{y}-\boldsymbol{x})^{\mathrm{T}}F(\boldsymbol{x})>0$, 则有 $(\boldsymbol{y}-\boldsymbol{x})^{\mathrm{T}}F(\boldsymbol{y})\geqslant 0$, 这与 $(\boldsymbol{y}-\boldsymbol{x})^{\mathrm{T}}F(\boldsymbol{y})<0$ 矛盾. 因此, 我们仅需要证明

$$(\boldsymbol{y}-\boldsymbol{x})^{\mathrm{T}}F(\boldsymbol{x})=0\Rightarrow(\boldsymbol{y}-\boldsymbol{x})^{\mathrm{T}}F(\boldsymbol{y})<0 \tag{5.5.3-4}$$

成立, 并在此情形下推出矛盾.

由性质 5.5.5 知 F 在 $\mathrm{int}C$ 上是伪单调映射, 如果 $\boldsymbol{x},\boldsymbol{y}\in\mathrm{int}C$, 则有

$$(\boldsymbol{y}-\boldsymbol{x})^{\mathrm{T}}F(\boldsymbol{x})=0\Rightarrow(\boldsymbol{y}-\boldsymbol{x})^{\mathrm{T}}F(\boldsymbol{y})\geqslant 0$$

成立, 这与 (5.5.3-4) 式矛盾. 只要再证明下面两种情况与 (5.5.3-4) 式矛盾即可.

(i) 若 $\boldsymbol{x},\boldsymbol{y}\in C$ 都是边界点, 且 $F(\boldsymbol{x})\neq\boldsymbol{0}$ 和 $F(\boldsymbol{y})\neq\boldsymbol{0}$, (5.5.3-4) 式成立, 得

$$(\boldsymbol{y}-\boldsymbol{x})^{\mathrm{T}}F(\boldsymbol{y})=(\boldsymbol{y}-\boldsymbol{x})^{\mathrm{T}}(\boldsymbol{A}\boldsymbol{y}+\boldsymbol{A}\boldsymbol{x}-\boldsymbol{A}\boldsymbol{x}+\boldsymbol{b})=(\boldsymbol{y}-\boldsymbol{x})^{\mathrm{T}}\boldsymbol{A}(\boldsymbol{y}-\boldsymbol{x})<0,$$

则由 (1) 得 $F(\boldsymbol{x})=\boldsymbol{0}$, 矛盾.

(ii) 若在 $\boldsymbol{x},\boldsymbol{y}\in C$ 中有一点是边界点和一个内点. 设 $\boldsymbol{x}\in\mathrm{bd}C$, $F(\boldsymbol{x})\neq\boldsymbol{0}$ 和 $\boldsymbol{y}\in\mathrm{int}C$, 再令 $\boldsymbol{x}_\lambda=\boldsymbol{x}+\lambda(\boldsymbol{y}-\boldsymbol{x})$, 则存在 $\lambda\in(0,1)$ 使得 $\boldsymbol{x}_\lambda\in\mathrm{int}C$, 有

$$(\boldsymbol{y}-\boldsymbol{x}_\lambda)^{\mathrm{T}}F(\boldsymbol{x}_\lambda)\geqslant(>)\,0\Rightarrow(\boldsymbol{y}-\boldsymbol{x}_\lambda)^{\mathrm{T}}F(\boldsymbol{y})\geqslant(>)\,0,$$

即

$$\begin{aligned}(\boldsymbol{y}-\boldsymbol{x}_\lambda)^{\mathrm{T}}F(\boldsymbol{x}_\lambda)&=(1-\lambda)(\boldsymbol{y}-\boldsymbol{x})^{\mathrm{T}}(\boldsymbol{A}\boldsymbol{x}+\lambda\boldsymbol{A}(\boldsymbol{y}-\boldsymbol{x})+b)\\&=(1-\lambda)(\boldsymbol{y}-\boldsymbol{x})^{\mathrm{T}}(\boldsymbol{A}\boldsymbol{x}+b)+(1-\lambda)\lambda(\boldsymbol{y}-\boldsymbol{x})^{\mathrm{T}}\boldsymbol{A}(\boldsymbol{y}-\boldsymbol{x}).\end{aligned} \tag{5.5.3-5}$$

因为 $(\boldsymbol{y}-\boldsymbol{x})^{\mathrm{T}}F(\boldsymbol{x})=0$, 如果 $(\boldsymbol{y}-\boldsymbol{x}_\lambda)^{\mathrm{T}}F(\boldsymbol{x}_\lambda)<0$, 则由 (5.5.3-5) 式有

$$(\boldsymbol{y}-\boldsymbol{x})^{\mathrm{T}}\boldsymbol{A}(\boldsymbol{y}-\boldsymbol{x})<0,$$

根据 (1) 知 $F(\boldsymbol{x})=\mathbf{0}$, 得矛盾. 因此有 $(\boldsymbol{y}-\boldsymbol{x}_\lambda)^{\mathrm{T}}F(\boldsymbol{x}_\lambda)\geqslant 0$, 得 $(\boldsymbol{y}-\boldsymbol{x}_\lambda)^{\mathrm{T}}F(\boldsymbol{y})\geqslant 0$, 所以有

$$(\boldsymbol{y}-\boldsymbol{x})^{\mathrm{T}}F(\boldsymbol{y})\geqslant 0,$$

与 (5.5.3-4) 式矛盾. □

最后, 仿射映射的复合映射具有下面的性质.

性质 5.5.8 设 $D\subset\mathbf{R}^m$ 是非空凸集合, 映射 $G: D\to\mathbf{R}^m$, $C=\{\boldsymbol{x}\in\mathbf{R}^n|\boldsymbol{A}\boldsymbol{x}+\boldsymbol{b}\in D\}$ 和 $F(\boldsymbol{x})=\boldsymbol{A}^{\mathrm{T}}G(\boldsymbol{A}\boldsymbol{x}+\boldsymbol{b})$, 有下面结论成立:

(1) 若 G 在 D 上是拟单调映射, 则 F 在 C 上是拟单调映射;

(2) 若 G 在 D 上是伪单调映射, 则 F 在 C 上是伪单调映射;

(3) 若 G 在 D 上是严格拟单调映射, 则 F 在 C 上是严格拟单调映射;

(4) 若 G 在 D 上是半严格拟单调映射, 则 F 在 C 上是半严格拟单调映射.

证明 设 $\boldsymbol{x}_1,\boldsymbol{x}_2\in C$, $\boldsymbol{y}_1=\boldsymbol{A}\boldsymbol{x}_1+\boldsymbol{b}$ 和 $\boldsymbol{y}_2=\boldsymbol{A}\boldsymbol{x}_2+\boldsymbol{b}$, 得

$$\begin{aligned}(\boldsymbol{x}_2-\boldsymbol{x}_1)^{\mathrm{T}}F(\boldsymbol{x}_1)&=(\boldsymbol{x}_2-\boldsymbol{x}_1)^{\mathrm{T}}\boldsymbol{A}^{\mathrm{T}}G(\boldsymbol{A}\boldsymbol{x}_1+\boldsymbol{b})\\&=(\boldsymbol{A}\boldsymbol{x}_2-\boldsymbol{A}\boldsymbol{x}_1)^{\mathrm{T}}G(\boldsymbol{A}\boldsymbol{x}_1+\boldsymbol{b})\\&=(\boldsymbol{y}_2-\boldsymbol{y}_1)^{\mathrm{T}}G(\boldsymbol{y}_1),\end{aligned}$$

同上得

$$(\boldsymbol{x}_2-\boldsymbol{x}_1)^{\mathrm{T}}F(\boldsymbol{x}_2)=(\boldsymbol{y}_2-\boldsymbol{y}_1)^{\mathrm{T}}G(\boldsymbol{y}_2),$$

根据上面两个等式, 由定义 5.5.1 知所有结论成立. □

本小节得到了有关仿射映射的几个单调性性质:

1) 仿射映射是伪单调映射等价于它是拟单调映射或者 (5.5.3-1) 式成立 (性质 5.5.6);

2) 若仿射映射是拟单调映射, 则含有零点的边界点, 否则 F 是伪单调映射 (性质 5.5.7);

3) 含仿射映射的复合映射与单调性等价 (性质 5.5.8).

5.6 习　　题

5.6-1 判定下面单变量函数是拟凸函数、半严格拟凸函数还是严格拟凸函数. 同时验证下面函数是否满足一点拟凸性、一点半严格拟凸性和一点严格拟凸性.

(1) $f(x)=x|x|$;

(2) $f(x)=x|x|+x^2$;

(3) $f(x)=x|x|-x^2$;

(4) $f(x) = x|x| + x^3$;

(5) $f(x) = \begin{cases} \dfrac{1}{x-1}, & 0 \leqslant x < 1, \\ 0, & x = 1, \\ \dfrac{1}{x+1}, & x > 1; \end{cases}$

(6) $f(x) = \begin{cases} -x^2, & x < 0, \\ 0, & x = 0, \\ x + x^3, & x > 0. \end{cases}$

5.6-2　设函数 f: $[a, b] \to \mathbf{R}$ 和点 $x_1 \in [a, b]$, 且 $f(x_1) = a$, $\lim\limits_{x \uparrow x_1} f(x) = -\infty$, $\lim\limits_{x \downarrow x_1} f(x) = -\infty$, 证明该函数不是拟凸函数.

5.6-3　判定下面多变量函数是拟凸函数、半严格拟凸函数还是严格拟凸函数. 同时验证下面函数是否满足一点拟凸性、一点半严格拟凸性和一点严格拟凸性.

(1) $f(\boldsymbol{x}) = \ln \sum\limits_{i=1}^{n} x_i^2$;

(2) $f(\boldsymbol{x}) = \sum\limits_{i=1}^{n} (x_i + x_i^3)$;

(3) $f(\boldsymbol{x}) = \ln \sum\limits_{i=1}^{n} (1 + x_i^2)$;

(4) $f(\boldsymbol{x}) = \sum\limits_{i=1}^{n} x_i |x_i|$;

(5) $f(\boldsymbol{x}) = \ln \sum\limits_{i=1}^{n} x_i$, $x_i > 0, i = 1, 2, \cdots, n$;

5.6-4　设凸集 $C \subset \mathbf{R}^n$ 和函数 f: $C \to \mathbf{R}$, 证明下面几个结论:

(1) 如果 f 在 C 上是正值凸函数, 则 $-1/f$ 在 C 上是拟凸函数;

(2) 如果 f 在 C 上是负值拟凸函数, 则 $1/f$ 在 C 上是拟凹函数;

(3) 如果 f 在 C 上是负值严格拟凸函数, 则 $1/f$ 在 C 上是严格拟凹函数.

5.6-5　证明分式函数: $f(\boldsymbol{x}) = \dfrac{\boldsymbol{a}^{\mathrm{T}}\boldsymbol{x} + \alpha}{\boldsymbol{b}^{\mathrm{T}}\boldsymbol{x} + \beta}$ 在 $C = \left\{\boldsymbol{x} \in \mathbf{R}^n | \boldsymbol{b}^{\mathrm{T}}\boldsymbol{x} + \beta > 0\right\}$ 上是拟凸和拟凹的.

5.6-6　证明分式函数: $f(\boldsymbol{x}) = \dfrac{\boldsymbol{a}^{\mathrm{T}}\boldsymbol{x} + \alpha}{\boldsymbol{b}^{\mathrm{T}}\boldsymbol{x} + \beta}$ 在 $C = \left\{\boldsymbol{x} \in \mathbf{R}^n | \boldsymbol{b}^{\mathrm{T}}\boldsymbol{x} + \beta < 0\right\}$ 上是拟凸和拟凹的.

5.6-7　通过一点拟凸性、一点半严格拟凸性和一点严格拟凸性判定函数:

$f(\boldsymbol{x}) = \ln\dfrac{x_1^4+1}{x_2+1}$ 在 $C = \{\boldsymbol{x}\in\mathbf{R}^2|x_2>-1\}$ 上是拟凸、半严格拟凸还是严格拟凸. 运用限制函数定理证明该函数的拟凸性. 设任意 $\boldsymbol{x},\boldsymbol{y}\in C$, 是否有 $f(\boldsymbol{x})\geqslant f(\boldsymbol{y})\Rightarrow\nabla f(\boldsymbol{x})^{\mathrm{T}}(\boldsymbol{y}-\boldsymbol{x})\leqslant(<)0$ 成立?

5.6-8　设 $C=[-1,1]$ 和函数

$$f(x)=\begin{cases}1, & -1\leqslant x\leqslant 1, x\neq 0,\\ 2, & x=0,\end{cases}$$

证明水平集 $C_{f\leqslant\alpha}$ 是凸集, 且 $C_{f=\alpha}=C_{f\leqslant\alpha}$ 或 $C_{f=\alpha}\subset\partial C_{f\leqslant\alpha}$ 成立.

5.6-9　判定习题 5.6-1 中 6 个函数是否为伪凸函数.

5.6-10　设 $f:(a,b)\to\mathbf{R}$ 是可微函数, 证明 f 是在 (a,b) 上的拟凸函数当且仅当 f 的任何一个稳定点都不是 f 的严格局部极大点.

5.6-11　在 5.6-10 中开区间 (a,b) 改为闭区间, 那么其结论还成立吗?

5.6-12　验证下面函数是否为伪凸函数:

(1) $f(\boldsymbol{x})=\ln\sum\limits_{i=1}^{n}x_i$, $x_i>0, i=1,2,\cdots,n$;

(2) $f(\boldsymbol{x})=\ln\sum\limits_{i=1}^{n}x_i^3$, $x_i>0, i=1,2,\cdots,n$;

(3) $f(\boldsymbol{x})=\ln(x_1^4+x_2^2+x_3^6)-\ln(x_2-1)$, $x_2>1$.

5.6-13　验证下面函数是否为伪线性函数.

(1) $f(\boldsymbol{x})=\dfrac{x_1+2-x_2\sqrt{(x_1+2)^2+x_2^2-1}}{(x_1+2)^2+x_2^2}$;

(2) $f(\boldsymbol{x})=2x_1+x_2-x_3-\dfrac{1}{2x_1+x_2-x_3+4}$, 其中 $C=\left\{(x_1,x_2,x_3)^{\mathrm{T}}|2x_1+x_2-x_3+4>0\right\}$;

(3) $f(\boldsymbol{x})=\mathrm{e}^{x_1+\cdots+x_n}$, $x_i>0, i=1,2,\cdots,n$.

5.6-14　使用定理 5.3.4~定理 5.3.6 和定理 5.4.2、定理 5.4.3 验证函数 $f(\boldsymbol{x})=-x_1^2-x_1x_2$ 在 $\mathrm{int}\mathbf{R}_+^2$ 上是伪凸, 而在 $\mathbf{R}_+^2$ 上是拟凸. 解释伪凸函数为什么定义在开凸集上.

5.6-15　使用性质 5.5.5 验证习题 5.6-3、习题 5.6-11 的拟凸性、半严格拟凸性、严格拟凸性和伪凸性.

5.6-16　证明定理 5.2.1、定理 5.3.19.

第 6 章　凸函数微分

函数微分学在最优化理论和算法研究中起到了关键作用, 其中凸函数微分所具有的独特性质, 更是有效推动了凸优化理论的研究和发展. 例如, 凸函数微分用于描述和刻画伪凸函数、拟凸函数和强凸函数等性质; 有界凸函数总是存在次微分, 几乎是处处可微的, 而非凸函数不具备这种性质. 凸函数微分理论也逐渐发展成为次可微优化与算法研究的一个独立分支 [52,95]. 凸函数微分既包含了通常意义下的可微理论, 也包含了一般情形实值凸函数的上 (下) 方向导数和对应的次微分性质, 以及次微分连续性等理论. 另外, 广义凸函数微分也是凸函数微分的一个重要方向. 例如, 拟凸函数对应的 Dini 导数可用于刻画拟凸函数的性质, 有关拟凸函数微分研究构成了广义凸优化研究的一个重要分支 [46,57,62,81].

本章分别讨论关于凸函数的可微性、方向导数、次梯度与次微分等定义和性质, 以及凸函数次微分连续性和运算性质等结果. 最后, 介绍一阶 Dini 方向导数和二阶 Dini 方向导数关于 (拟) 凸函数的性质.

6.1　可　微　性

凸函数微分比非凸函数微分具有更多的独特性质, 并在揭示凸函数的性质时发挥了特有的作用. 例如, 凸函数梯度等于 0 的稳定点是凸函数的全局最优点这一性质, 表明找到了凸函数的稳定点就找到了凸函数的全局最优点, 这在设计凸优化算法中非常重要. 下面给出一般广义实值函数的梯度或微分定义.

定义 6.1.1　设广义实值函数 $f\colon \mathbf{R}^n \to [-\infty, +\infty]$. 对于 $f(\boldsymbol{x}) < +\infty$, 如果存在 $\boldsymbol{x}$ 处一个邻域中的任意向量 $\boldsymbol{z}$ 满足

$$\lim_{\boldsymbol{z}\to\boldsymbol{x}} \frac{o(\|\boldsymbol{z}-\boldsymbol{x}\|)}{\|\boldsymbol{z}-\boldsymbol{x}\|} = 0,$$

$$f(\boldsymbol{z}) = f(\boldsymbol{x}) + \langle \nabla f(\boldsymbol{x}), \boldsymbol{z}-\boldsymbol{x}\rangle + o(\|\boldsymbol{z}-\boldsymbol{x}\|), \tag{6.1.1-1}$$

则称 f 在 $\boldsymbol{x}$ 处是可微的, 称 $\nabla f(\boldsymbol{x})$ 为 f 在 $\boldsymbol{x}$ 处的梯度 (或微分). 如果有

$$\lim_{\boldsymbol{y}\to\boldsymbol{x}} \nabla f(\boldsymbol{y}) = \nabla f(\boldsymbol{x}),$$

则称 f 在 $\boldsymbol{x}$ 处是连续可微的.

注　如果 f 在 $\boldsymbol{x}$ 处是连续可微的, 那么存在 $\boldsymbol{x}$ 的一个邻域, f 在这个邻域内的所有点都是可微的.

如果 (6.1.1-1) 式成立, 则 f 在点 $\boldsymbol{x}$ 处的梯度是唯一的. 假设 $\exists \boldsymbol{y}_1, \boldsymbol{y}_2 \in \mathbf{R}^n$ 使得

$$f(\boldsymbol{z}) = f(\boldsymbol{x}) + \langle \boldsymbol{y}_1, \boldsymbol{z}-\boldsymbol{x}\rangle + o_1(\|\boldsymbol{z}-\boldsymbol{x}\|),$$

$$f(\boldsymbol{z}) = f(\boldsymbol{x}) + \langle \boldsymbol{y}_2, \boldsymbol{z}-\boldsymbol{x}\rangle + o_2(\|\boldsymbol{z}-\boldsymbol{x}\|),$$

其中 $\lim\limits_{\boldsymbol{z}\to\boldsymbol{x}} \dfrac{o_1(\|\boldsymbol{z}-\boldsymbol{x}\|)}{\|\boldsymbol{z}-\boldsymbol{x}\|} = 0$, $\lim\limits_{\boldsymbol{z}\to\boldsymbol{x}} \dfrac{o_2(\|\boldsymbol{z}-\boldsymbol{x}\|)}{\|\boldsymbol{z}-\boldsymbol{x}\|} = 0$, 则有 $\boldsymbol{y}_1 = \boldsymbol{y}_2$. 否则, 若 $\boldsymbol{y}_1 \neq \boldsymbol{y}_2$, 令 $\boldsymbol{z} = \boldsymbol{x} + t(\boldsymbol{y}_1 - \boldsymbol{y}_2)$, 其中 $t > 0$ 充分小, 则上两式相减得

$$\frac{t\langle \boldsymbol{y}_1-\boldsymbol{y}_2, \boldsymbol{y}_1-\boldsymbol{y}_2\rangle}{t\|\boldsymbol{y}_1-\boldsymbol{y}_2\|} = \frac{o_1(\|\boldsymbol{z}-\boldsymbol{x}\|)}{\|\boldsymbol{z}-\boldsymbol{x}\|} - \frac{o_2(\|\boldsymbol{z}-\boldsymbol{x}\|)}{\|\boldsymbol{z}-\boldsymbol{x}\|},$$

令 $\boldsymbol{z} \to \boldsymbol{x}$, 由上式得 $\|\boldsymbol{y}_1-\boldsymbol{y}_2\| = 0$, 即有 $\boldsymbol{y}_1 = \boldsymbol{y}_2$, 得到矛盾.

例 6.1.1　设凸函数 $f(\boldsymbol{x}) = (x_1+x_2)^2$, 得 $\nabla f(\boldsymbol{x}) = (2(x_1+x_2), 2(x_1+x_2))^{\mathrm{T}}$, 有

$$f(\boldsymbol{z}) - f(\boldsymbol{x}) - \langle \nabla f(\boldsymbol{x}), \boldsymbol{z}-\boldsymbol{x}\rangle = (z_1+z_2-x_1-x_2)^2.$$

由上式得

$$\lim_{\boldsymbol{z}\to\boldsymbol{x}} \frac{f(\boldsymbol{z}) - f(\boldsymbol{x}) - \langle \nabla f(\boldsymbol{x}), \boldsymbol{z}-\boldsymbol{x}\rangle}{\|\boldsymbol{z}-\boldsymbol{x}\|} = \lim_{\boldsymbol{z}\to\boldsymbol{x}} \frac{(z_1+z_2-x_1-x_2)^2}{\sqrt{(z_1-x_1)^2+(z_2-x_2)^2}}.$$

设 $s_1 = z_1 - x_1$, $s_2 = z_2 - x_2$, 令 $t = s_1^2 + s_2^2$, 可得 $(s_1+s_2)^2 = t + 2s_1s_2$, 则有

$$\lim_{\boldsymbol{z}\to\boldsymbol{x}} \frac{f(\boldsymbol{z}) - f(\boldsymbol{x}) - \langle \nabla f(\boldsymbol{x}), \boldsymbol{z}-\boldsymbol{x}\rangle}{\|\boldsymbol{z}-\boldsymbol{x}\|} = \lim_{s_1,s_2\to 0} \frac{(s_1+s_2)^2}{\sqrt{s_1^2+s_2^2}} = 0.$$

下面是有关实值函数微分或梯度的几个等价的结论.

定理 6.1.1　设 f : $\mathbf{R}^n \to [-\infty, +\infty]$ 是广义函数, 对于满足 $-\infty < f(\boldsymbol{x}) < +\infty$ 的任意 $\boldsymbol{d}, \boldsymbol{x} \in \mathbf{R}^n$, 以下结论等价:

(1) $f(\boldsymbol{x})$ 在点 $\boldsymbol{x}$ 处是可微的;

(2) $\lim\limits_{\boldsymbol{z}\to\boldsymbol{x}} \dfrac{f(\boldsymbol{z}) - f(\boldsymbol{x}) - \langle \nabla f(\boldsymbol{x}), \boldsymbol{z}-\boldsymbol{x}\rangle}{\|\boldsymbol{z}-\boldsymbol{x}\|} = 0$;

(3) $f'(\boldsymbol{x}, \boldsymbol{d}) = \langle \nabla f(\boldsymbol{x}), \boldsymbol{d}\rangle$;

(4) (Gateaux 可微) $\lim\limits_{t\to 0} \dfrac{f(\boldsymbol{x}+t\boldsymbol{d}) - f(\boldsymbol{x}) - \langle \nabla f(\boldsymbol{x}), t\boldsymbol{d}\rangle}{t} = 0$;

(5) (Frechet 可微) $\lim\limits_{\boldsymbol{d}\to\boldsymbol{0}}\dfrac{f(\boldsymbol{x}+\boldsymbol{d})-f(\boldsymbol{x})-\langle\nabla f(\boldsymbol{x}),\boldsymbol{d}\rangle}{||\boldsymbol{d}||}=0.$

证明 (1)⇒(2): 根据 (6.1.1-1) 式, (2) 显然成立. 反之, (2)⇒(1) 也是显然的.

(2)⇒(3): 令 $\boldsymbol{z}=\boldsymbol{x}+t\boldsymbol{d}$, 根据 1.3.2 节的方向导数定义和 (6.1.1-1) 式, 有

$$\begin{aligned}f'(\boldsymbol{x},\boldsymbol{d})&=\lim_{t\to0}\frac{f(\boldsymbol{x}+t\boldsymbol{d})-f(\boldsymbol{x})}{t}\\&=\lim_{t\to0}\frac{\langle\nabla f(\boldsymbol{x}),t\boldsymbol{d}\rangle+o(||t\boldsymbol{d}||)}{t}=\langle\nabla f(\boldsymbol{x}),\boldsymbol{d}\rangle.\end{aligned}$$

(3)⇒(4): 根据 (3) 有

$$f'(\boldsymbol{x},\boldsymbol{d})=\lim_{t\to0}\frac{f(\boldsymbol{x}+t\boldsymbol{d})-f(\boldsymbol{x})-t\langle\nabla f(\boldsymbol{x}),\boldsymbol{d}\rangle}{t},$$

因此, 结论 (4) 成立.

(4)⇒(5): 令 $t\boldsymbol{d}=\boldsymbol{d}_1$, 有 $t||\boldsymbol{d}||=||\boldsymbol{d}_1||$, 其中 $\boldsymbol{d}\neq\boldsymbol{0}$, 根据 (4) 有

$$\begin{aligned}0&=\lim_{t\to0}\frac{f(\boldsymbol{x}+t\boldsymbol{d})-f(\boldsymbol{x})-\langle\nabla f(\boldsymbol{x}),t\boldsymbol{d}\rangle}{t}\\&=||\boldsymbol{d}||\lim_{\boldsymbol{d}_1\to\boldsymbol{0}}\frac{f(\boldsymbol{x}+\boldsymbol{d}_1)-f(\boldsymbol{x})-\langle\nabla f(\boldsymbol{x}),\boldsymbol{d}_1\rangle}{||\boldsymbol{d}_1||}.\end{aligned}$$

因此, 结论 (5) 成立.

(5)⇒(2): 取 $\boldsymbol{x}+\boldsymbol{d}=\boldsymbol{z}$, 当 $\boldsymbol{d}\to\boldsymbol{0}$ 时, 有 $\boldsymbol{z}\to\boldsymbol{x}$, 因此得

$$\begin{aligned}&\lim_{\boldsymbol{d}\to\boldsymbol{0}}\frac{f(\boldsymbol{x}+\boldsymbol{d})-f(\boldsymbol{x})-\langle\nabla f(\boldsymbol{x}),\boldsymbol{d}\rangle}{||\boldsymbol{d}||}\\=&\lim_{\boldsymbol{z}\to\boldsymbol{x}}\frac{f(\boldsymbol{z})-f(\boldsymbol{x})-\langle\nabla f(\boldsymbol{x}),\boldsymbol{z}-\boldsymbol{x}\rangle}{||\boldsymbol{z}-\boldsymbol{x}||}.\end{aligned}$$

□

下面的结论告诉我们, 如果 f 在点 $\boldsymbol{x}$ 处连续可微, 则 f 在 $\boldsymbol{x}$ 处严格可微.

定理 6.1.2[12] 设 $f:\ \mathbf{R}^n\to[-\infty,+\infty]$ 是广义函数. 对于满足 $-\infty<f(\boldsymbol{x})<+\infty$ 的任意 $\boldsymbol{d},\boldsymbol{x}\in\mathbf{R}^n$, 如果 f 在点 $\boldsymbol{x}$ 处是连续可微的, 则 f 在 $\boldsymbol{x}$ 处严格可微:

$$\lim_{t\to0^+,\boldsymbol{z}\to\boldsymbol{x}}\frac{f(\boldsymbol{z}+t\boldsymbol{d})-f(\boldsymbol{z})}{t}=\langle\nabla f(\boldsymbol{x}),\boldsymbol{d}\rangle.\tag{6.1.1-2}$$

证明 若 f 在点 $\boldsymbol{x}$ 处是连续可微的, 即对任意给定的 $\varepsilon>0$, 存在 $\boldsymbol{x}$ 的一个邻域 $U(\boldsymbol{x},\delta)\,(\delta>0)$, 满足

$$||\nabla f(\boldsymbol{z})-\nabla f(\boldsymbol{x})||<\varepsilon,\quad\forall\boldsymbol{z}\in U(\boldsymbol{x},\delta).\tag{6.1.1-3}$$

对任意的 $\boldsymbol{d} \in \mathbf{R}^n$ 有

$$|\langle \nabla f(\boldsymbol{z}), \boldsymbol{d}\rangle - \langle \nabla f(\boldsymbol{x}), \boldsymbol{d}\rangle| \leqslant \varepsilon||\boldsymbol{d}||, \tag{6.1.1-4}$$

表明对任意的 $\boldsymbol{z} \in U(\boldsymbol{x}, \delta)$, $\nabla f(\boldsymbol{z})$ 和 $\nabla f(\boldsymbol{x})$ 都存在. 所以, 根据定理 6.1.1, 对于充分小的 t 有

$$\left|\frac{f(\boldsymbol{z}+t\boldsymbol{d}) - f(\boldsymbol{z})}{t} - \langle \nabla f(\boldsymbol{z}), \boldsymbol{d}\rangle\right| < \varepsilon. \tag{6.1.1-5}$$

由 (6.1.1-3) 式和上式得

$$\begin{aligned}
&\left|\frac{f(\boldsymbol{z}+t\boldsymbol{d}) - f(\boldsymbol{z})}{t} - \langle \nabla f(\boldsymbol{x}), \boldsymbol{d}\rangle\right| \\
=&\left|\frac{f(\boldsymbol{z}+t\boldsymbol{d}) - f(\boldsymbol{z})}{t} - \langle \nabla f(\boldsymbol{z}), \boldsymbol{d}\rangle + \langle \nabla f(\boldsymbol{z}), \boldsymbol{d}\rangle - \langle \nabla f(\boldsymbol{x}), \boldsymbol{d}\rangle\right| \\
<&\varepsilon + \varepsilon||\boldsymbol{d}||,
\end{aligned}$$

令 $\boldsymbol{z} \to \boldsymbol{x}$, $\varepsilon \to 0$, 根据上式可得 (6.1.1-2) 式成立. □

根据 (6.1.1-2) 式和定理 6.1.1(2) 知, 严格可微函数一定是可微函数.

例 6.1.2 设函数 $f: \mathbf{R} \to \mathbf{R}$,

$$f(x) = \begin{cases} x^2 \sin \dfrac{1}{x}, & x \neq 0, \\ 0, & x = 0, \end{cases}$$

我们有

$$f'(x) = \begin{cases} 2x \sin \dfrac{1}{x} - \cos \dfrac{1}{x}, & x \neq 0, \\ 0, & x = 0. \end{cases}$$

在 $x = 0$ 处, 函数 f 的导数存在且导函数不连续. 但有

$$\lim_{z \to 0} \frac{f(z) - f(0) - \langle \nabla f(0), z - 0\rangle}{\|z - 0\|} = \lim_{z \to 0} \frac{z^2 \sin \frac{1}{z}}{\|z\|} = 0.$$

设 $d \neq 0$, 求得 f 在 $x = 0$ 处的方向导数为

$$f'(0, d) = \lim_{t \to 0} \frac{f(0 + td) - f(0)}{t} = \lim_{t \to 0} td^2 \sin \frac{1}{td} = 0,$$

因此, 函数 f 在 $x = 0$ 处不是连续可微的.

上述例子表明连续 Gateaux 可微函数才是严格可微函数. 在有限维空间中, Gateaux 可微与方向导数等价. 连续可微函数还满足局部 Lipschitz 条件 [12].

定理 6.1.3 设 $f:\ \mathbf{R}^n \to [-\infty, +\infty]$ 是 $\mathbf{R}^n$ 上的凸函数. 如果 $f(\boldsymbol{x}_0)$ 在 $\boldsymbol{x}_0$ 处有限且可微, 则 f 是正常凸函数且 $\boldsymbol{x}_0 \in \operatorname{int}(\operatorname{dom} f)$.

证明 根据性质 4.1.1 得

$$f(\boldsymbol{x}_2) - f(\boldsymbol{x}_0) \geqslant \langle \boldsymbol{x}_2 - \boldsymbol{x}_0, \nabla f(\boldsymbol{x}_0) \rangle, \quad \forall \boldsymbol{x}_0, \boldsymbol{x}_2 \in \mathbf{R}^n.$$

如果存在 $f(\boldsymbol{x}_2) = -\infty$, 则由上式有 $f(\boldsymbol{x}_0) = -\infty$, 与假设矛盾. 因此, f 是正常的凸函数. 由假设知 $\boldsymbol{x}_0 \in \operatorname{dom} f$, 又由 $f(\boldsymbol{x}_0)$ 在 $\boldsymbol{x}_0$ 处是连续函数, 显然有 $\boldsymbol{x}_0 \in \operatorname{int}(\operatorname{dom} f)$. □

本小节得到了可微函数的几个结论:

1) 可微函数等价于方向导数、Gateaux 导数均存在;

2) 连续 Gateaux 可微函数是严格可微函数;

3) 如果广义凸函数在某个点上是有界可微的, 则该函数是正常凸函数.

6.2 次 微 分

凸函数不一定是可微函数, 例如 $f(x) = |x|\ (x \in \mathbf{R})$. 但是当凸函数不可微时, 它存在单边方向导数和次微分. 反过来, 当凸函数可微时, 凸函数的梯度就是它的次微分.

6.2.1 方向导数

用 $t \downarrow 0$ 表示变量 t 是从 $(0, +\infty)$(正向) 趋向于 0, 也可表示为 $t \to 0^+$; 用 $t \uparrow 0$ 表示变量 t 是从 $(-\infty, 0)$(负向) 趋向于 0, 也可表示为 $t \to 0^-$.

定义 6.2.1 设 $f:\ \mathbf{R}^n \to [-\infty, +\infty]$ 是广义函数, $f(\boldsymbol{x}) < +\infty$ 和 $\forall \boldsymbol{d}, \boldsymbol{x} \in \mathbf{R}^n$, 有

(1) 如果

$$f'_+(\boldsymbol{x}, \boldsymbol{d}) = \lim_{t \downarrow 0} \frac{f(\boldsymbol{x} + t\boldsymbol{d}) - f(\boldsymbol{x})}{t} \tag{6.2.1-1}$$

存在 (可以取 $+\infty$ 或 $-\infty$), 则称 $f'_+(\boldsymbol{x}, \boldsymbol{d})$ 是 f 在 $\boldsymbol{x}$ 沿方向 $\boldsymbol{d}$ 的正 (右) 方向导数;

(2) 如果

$$-f'_-(\boldsymbol{x}, -\boldsymbol{d}) = \lim_{t \uparrow 0} \frac{f(\boldsymbol{x} + t\boldsymbol{d}) - f(\boldsymbol{x})}{t} \tag{6.2.1-2}$$

存在 (可以取 $+\infty$ 或 $-\infty$), 则称 $f'_-(\boldsymbol{x}, \boldsymbol{d})$ 是 f 在 $\boldsymbol{x}$ 沿方向 $\boldsymbol{d}$ 的负 (左) 方向导数;

(3) 如果

$$f'(\boldsymbol{x},\boldsymbol{d})=\lim_{t\to 0}\frac{f(\boldsymbol{x}+t\boldsymbol{d})-f(\boldsymbol{x})}{t} \tag{6.2.1-3}$$

存在 (可以取 $+\infty$ 或 $-\infty$), 则称 $f'(\boldsymbol{x},\boldsymbol{d})$ 是 f 在 $\boldsymbol{x}$ 沿方向 $\boldsymbol{d}$ 的方向导数;

(4) 特别地, 当 $n=1$ 时, 设 $\boldsymbol{d}=1$, $x\in\mathbf{R}$, 若

$$f'_+(x)=f'_+(x,1)=\lim_{t\downarrow 0}\frac{f(x+t)-f(x)}{t}$$

或

$$f'_-(x)=-f'_-(x,-1)=\lim_{t\uparrow 0}\frac{f(x+t)-f(x)}{t},$$

则 $f'_+(x)$, $f'_-(x)$ 分别称为 f 在 x 的正 (右) 导数、负 (左) 导数.

注　如果 (6.2.1-1) 和 (6.2.1-2) 式成立, 有

$$\begin{aligned}f'_+(\boldsymbol{x},-\boldsymbol{d})&=\lim_{t\downarrow 0}\frac{f(\boldsymbol{x}-t\boldsymbol{d})-f(\boldsymbol{x})}{t}\\&=-\lim_{-t\uparrow 0}\frac{f(\boldsymbol{x}+(-t)\boldsymbol{d})-f(\boldsymbol{x})}{-t}=f'_-(\boldsymbol{x},-\boldsymbol{d})\end{aligned}$$

和

$$\begin{aligned}-f'_-(\boldsymbol{x},-\boldsymbol{d})&=\lim_{t\uparrow 0}\frac{f(\boldsymbol{x}-t\boldsymbol{d})-f(\boldsymbol{x})}{t}\\&=-\lim_{-t\downarrow 0}\frac{f(\boldsymbol{x}+(-t)\boldsymbol{d})-f(\boldsymbol{x})}{-t}=-f'_+(\boldsymbol{x},-\boldsymbol{d}).\end{aligned}$$

因此, 一维情形下, 有 $f'_+(x,-1)=f'_-(x,-1)$.

下面的例子表明左方向导数与右方向导数不一定相等.

例 6.2.1　设凸函数 $f(x)=|x|$, $x,d\in\mathbf{R}$, 若 $x>0$, 对于充分小的 t 有 $x+td>0$, 得

$$f'_+(x,d)=\lim_{t\downarrow 0}\frac{f(x+td)-f(x)}{t}=\lim_{t\downarrow 0}\frac{x+td-x}{t}=d,$$

$$-f'_-(x,-d)=\lim_{t\uparrow 0}\frac{f(x+td)-f(x)}{t}=\lim_{t\uparrow 0}\frac{x+td-x}{t}=d,$$

即 $f'_+(x,d)=-f'_-(x,-d)=d$.

若 $x<0$, 同理也可得 $f'_+(x,d)=-f'_-(x,-d)=d$.

若 $x=0$, 且 $d \geqslant 0$, 则有

$$f'_+(0,d)=\lim_{t\downarrow 0}\frac{f(0+td)-f(0)}{t}=\lim_{t\downarrow 0}\frac{td}{t}=d,$$

$$-f'_-(0,-d)=\lim_{t\uparrow 0}\frac{f(0+td)-f(0)}{t}=\lim_{t\uparrow 0}\frac{-td}{t}=-d.$$

若 $d<0$, 有

$$f'_+(0,d)=\lim_{t\downarrow 0}\frac{f(0+td)-f(0)}{t}=\lim_{t\downarrow 0}\frac{-td}{t}=-d,$$

$$-f'_-(0,-d)=\lim_{t\uparrow 0}\frac{f(0+td)-f(0)}{t}=\lim_{t\uparrow 0}\frac{td}{t}=d.$$

可得 $f'_+(0,d)>-f'_-(0,-d)$, $f'_+(x,-d)=f'_-(x,-d)\,(x\neq 0)$.

下面证明方向导数存在的充要条件是左、右方向导数同时存在且相等.

定理 6.2.1 设 $f:\mathbf{R}^n\to[-\infty,+\infty]$ 是广义函数. 对于满足 $-\infty<f(\boldsymbol{x})<+\infty$ 的任意 $\boldsymbol{d},\boldsymbol{x}\in\mathbf{R}^n$, 则 $f'(\boldsymbol{x},\boldsymbol{d})$ 存在等价于 $f'_+(\boldsymbol{x},\boldsymbol{d})$ 和 $f'_-(\boldsymbol{x},\boldsymbol{d})$ 均存在, 且 $f'_+(\boldsymbol{x},\boldsymbol{d})=-f'_-(\boldsymbol{x},-\boldsymbol{d})$.

证明 如果 $f'(\boldsymbol{x},\boldsymbol{d})$ 存在, 显然有 $f'_+(\boldsymbol{x},\boldsymbol{d})$, $f'_-(\boldsymbol{x},\boldsymbol{d})$ 存在, 且 $f'_+(\boldsymbol{x},\boldsymbol{d})=-f'_-(\boldsymbol{x},-\boldsymbol{d})$.

反过来, 设任何序列 $\{t_i\}\subset\mathbf{R}$, 且 $t_i\to 0\,(i\to+\infty)$, 在序列中至少存在一个单调下降的子序列 $\{t_{i_k}\}\subset\{t_i\}$, 其中的元素全部大于零或全部小于零, 即 $t_{i_k}\downarrow 0\,(i_k\to+\infty)$, 那么有

$$\begin{aligned}f'_+(\boldsymbol{x},\boldsymbol{d})&=\lim_{t_{i_k}\downarrow 0}\frac{f(\boldsymbol{x}+t_{i_k}\boldsymbol{d})-f(\boldsymbol{x})}{t_{i_k}}\\&=-\lim_{-t_{i_k}\uparrow 0}\frac{f(\boldsymbol{x}+(-t_{i_k})(-\boldsymbol{d}))-f(\boldsymbol{x})}{-t_{i_k}}=-f'_-(\boldsymbol{x},-\boldsymbol{d}),\end{aligned}$$

即有

$$\begin{aligned}f'_+(\boldsymbol{x},\boldsymbol{d})&=\lim_{t_i\to 0}\frac{f(\boldsymbol{x}+t_i\boldsymbol{d})-f(\boldsymbol{x})}{t_i}=-f'_-(\boldsymbol{x},-\boldsymbol{d})\\&=-\lim_{-t_i\to 0}\frac{f(\boldsymbol{x}-t_i(-\boldsymbol{d}))-f(\boldsymbol{x})}{-t_i}.\end{aligned}$$

上式说明 $f'_+(\boldsymbol{x},\boldsymbol{d})=-f'_-(\boldsymbol{x},-\boldsymbol{d})=f'(\boldsymbol{x},\boldsymbol{d})$ 成立. □

对于正常凸函数, 它在有效定义域内的点的正方向导数总是存在的.

定理 6.2.2 设 f 在 $\mathbf{R}^n$ 上是正常凸函数, 对任意 $\boldsymbol{x}\in\mathrm{dom}f$, 任意 $\boldsymbol{d}\in\mathbf{R}^n$, 有下面结论成立:

(1) $f'_+(\boldsymbol{x},\boldsymbol{d})$ 存在;

(2) 若对于 $t\in[-\varepsilon,\varepsilon]$ 和 $\varepsilon>0$ 有 $\boldsymbol{x}+t\boldsymbol{d}\in\mathrm{dom}f$, 则 $f'_+(\boldsymbol{x},\boldsymbol{d})$ 有限;

(3) $f'_+(\boldsymbol{x},\boldsymbol{d})$ 是 $\boldsymbol{d}$ 的正齐次凸函数, 且 $f'(\boldsymbol{x},\boldsymbol{0})=0$, $f'_+(\boldsymbol{x},\boldsymbol{d})\geqslant -f'_-(\boldsymbol{x},-\boldsymbol{d})$.

证明 (1) 对任意 $t>0$, $\boldsymbol{x}+t\boldsymbol{d}\notin\mathrm{dom}f$, 有 $f(\boldsymbol{x}+t\boldsymbol{d})=+\infty$, $f'_+(\boldsymbol{x},\boldsymbol{d})=+\infty$. 如果 t 充分小时, $\boldsymbol{x}+t\boldsymbol{d}\in\mathrm{dom}f$, 则由引理 4.3.3 知, $\phi(t)=\dfrac{f(\boldsymbol{x}+t\boldsymbol{d})-f(\boldsymbol{x})}{t}$ 是 t 的不减单调函数. 所以极限

$$f'_+(\boldsymbol{x},\boldsymbol{d})=\lim_{t\downarrow 0}\frac{f(\boldsymbol{x}+t\boldsymbol{d})-f(\boldsymbol{x})}{t}=\lim_{t>0}\frac{f(\boldsymbol{x}+t\boldsymbol{d})-f(\boldsymbol{x})}{t}$$

存在.

(2) 由 (1) 的证明过程可知结论成立.

(3) 由定义 6.2.1 知, 若 $\alpha>0$, 有

$$\begin{aligned}f'_+(\boldsymbol{x},\alpha\boldsymbol{d})&=\lim_{t\downarrow 0}\frac{f(\boldsymbol{x}+\alpha\boldsymbol{d})-f(\boldsymbol{x})}{t}\\&=\alpha\lim_{t\alpha\downarrow 0}\frac{f[\boldsymbol{x}+(t\alpha)\boldsymbol{d}]-f(\boldsymbol{x})}{t\alpha}\\&=\alpha f'_+(\boldsymbol{x},\boldsymbol{d}),\end{aligned}$$

当 $0\leqslant\lambda\leqslant 1$ 时, 有

$$\begin{aligned}&\frac{f[\boldsymbol{x}+t(\lambda\boldsymbol{d}_1+(1-\lambda)\boldsymbol{d}_2)]-f(\boldsymbol{x})}{t}\\=&\frac{f[\lambda(\boldsymbol{x}+t\boldsymbol{d}_1)+(1-\lambda)(\boldsymbol{x}+t\boldsymbol{d}_2)]-\lambda f(\boldsymbol{x})-(1-\lambda)f(\boldsymbol{x})}{t}\\\leqslant&\lambda\frac{f(\boldsymbol{x}+t\boldsymbol{d}_1)-f(\boldsymbol{x})}{t}+(1-\lambda)\frac{f(\boldsymbol{x}+t\boldsymbol{d}_2)-f(\boldsymbol{x})}{t},\end{aligned}$$

对上式按 $\lambda\downarrow 0$ 取极限, 得

$$f'_+(\boldsymbol{x},\lambda\boldsymbol{d}_1+(1-\lambda)\boldsymbol{d}_2)\leqslant\lambda f'_+(\boldsymbol{x},\boldsymbol{d}_1)+(1-\lambda)f'_+(\boldsymbol{x},\boldsymbol{d}_2).$$

所以, $f'_+(\boldsymbol{x},\boldsymbol{d})$ 是关于 $\boldsymbol{d}$ 的正齐次凸函数.

显然, $f'_+(\boldsymbol{x},\boldsymbol{0})=0$. 任意给定 $t_1>f'_+(\boldsymbol{x},-\boldsymbol{d})$, $t_2>f'_+(\boldsymbol{x},\boldsymbol{d})$, 因为 $f'_+(\boldsymbol{x},\boldsymbol{d})$ 是 $\boldsymbol{d}$ 的凸函数, 所以 $f'_+(\boldsymbol{x},\boldsymbol{d})$ 是 $\boldsymbol{d}$ 的拟凸函数, 根据定理 5.1.3 的必要性有

$$\frac{1}{2}t_1+\frac{1}{2}t_2>f'_+\left[\boldsymbol{x},\frac{1}{2}(-\boldsymbol{d})+\frac{1}{2}\boldsymbol{d}\right]=f'_+(\boldsymbol{x},\boldsymbol{0})=0.$$

故对于任意 $\boldsymbol{d}\in\mathbf{R}^n$, 有 $\dfrac{1}{2}f'_+(\boldsymbol{x},-\boldsymbol{d})+\dfrac{1}{2}f'_+(\boldsymbol{x},\boldsymbol{d})\geqslant 0$. 即对 $\boldsymbol{d}$ 有

$$f'_+(\boldsymbol{x},\boldsymbol{d})\geqslant -f'_+(\boldsymbol{x},-\boldsymbol{d})=-f'_-(\boldsymbol{x},-\boldsymbol{d}).\qquad\square$$

由定理 6.2.2(2) 和 (3) 可得下面结论.

推论 6.2.1 设 f 是 $\mathbf{R}^n$ 上的正常凸函数, 则对 $\boldsymbol{x}\in\mathrm{dom}f$, 关于 $\boldsymbol{d}$ 的函数 $f'_+(\boldsymbol{x},\boldsymbol{d})$ 有

$$\mathrm{dom}f'_+(\boldsymbol{x},\boldsymbol{d})=\mathrm{cone}\,(\mathrm{dom}f-\boldsymbol{x}),\quad \forall\boldsymbol{d}\in\mathbf{R}^n.$$

例 6.2.2 在 $\mathbf{R}$ 中, 设单变量函数

$$f(x)=\begin{cases}+\infty, & x<-1,\\ 2, & x=-1,\\ x^2, & -1<x\leqslant 0,\\ x, & 0<x\leqslant 1,\\ +\infty, & x>1\end{cases}$$

是正常凸函数, $\mathrm{dom}f=[-1,1]$, 可得它的方向导数为

$$f'_+(x)=\begin{cases}-\infty, & x<-1,\\ -\infty, & x=-1,\\ 2x, & -1<x\leqslant 0,\\ 1, & 0<x<1,\\ +\infty, & x=1,\\ +\infty, & x>1,\end{cases}\qquad f'_-(x)=\begin{cases}-\infty, & x<-1,\\ -\infty, & x=-1,\\ 2x, & -1<x\leqslant 0,\\ 1, & 0<x\leqslant 1,\\ +\infty, & x>1.\end{cases}$$

我们有

$$f'_+(-1)=f'_-(-1)=-\infty,\quad f'_+(1)=+\infty>f'_-(1)=1.$$

当 $x\in\mathrm{int}\,(\mathrm{dom}f)=(-1,1)$ 时有

$$f'_+(x)=f'_-(x).$$

$f(x)$ 仅在 $\mathrm{int}\,(\mathrm{dom}f)$ 处是可微的, 在边界上左右方向导数存在, 但不一定可微.

本小节得到的函数方向导数的主要结论是: 从定理 6.1.2、定理 6.2.1 和定理 6.2.2 知正常凸函数仅在有效定义域内可微, 且在有限定义域内方向导数一定存在.

6.2.2 次梯度

由于正常凸函数的方向导数存在, 结合下面的性质可知, 正常凸函数的次微分也存在. 在存在的情况下讨论性质才有意义.

定义 6.2.2　设 $f:\mathbf{R}^n\to[-\infty,+\infty]$ 是正常凸函数. 如果存在向量 $\boldsymbol{x}^*\in\mathbf{R}^n$ 使得对任意 $\boldsymbol{z}\in\mathbf{R}^n$ 有

$$f(\boldsymbol{z})\geqslant f(\boldsymbol{x})+\langle\boldsymbol{x}^*,\boldsymbol{z}-\boldsymbol{x}\rangle, \tag{6.2.2-1}$$

则称 $\boldsymbol{x}^*$ 是 f 在 $\boldsymbol{x}$ 处的次梯度. f 在 $\boldsymbol{x}$ 处的所有次梯度构成的集合称为 f 在 $\boldsymbol{x}$ 处的次微分, 用 $\partial f(\boldsymbol{x})$ 表示. 如果 $\partial f(\boldsymbol{x})\neq\varnothing$, 则称 f 在 $\boldsymbol{x}$ 是次可微的.

由定义 6.2.2 可知, 当 $f(\boldsymbol{x})=-\infty$ 时, 有 $\partial f(\boldsymbol{x})=\mathbf{R}^n$; 当 $f(x)=+\infty$, $\mathrm{dom}f\neq\varnothing$ 时, $\partial f(\boldsymbol{x})=\varnothing$. 这说明凸函数的有限次微分仅在有效定义域内存在, 为避免上述情形出现, 下面的讨论仅考虑正常凸函数的情况.

例 6.2.3　设 $f(\boldsymbol{x})=\|\boldsymbol{x}\|$, $\boldsymbol{x}\in\mathbf{R}^2$, 仅求 $\boldsymbol{x}=\mathbf{0}$ 时的次梯度, 由 (6.2.2-1) 式可知, $\partial f(\mathbf{0})$ 中的 $\boldsymbol{x}^*$ 满足

$$\|\boldsymbol{z}\|\geqslant\langle\boldsymbol{x}^*,\boldsymbol{z}\rangle,\quad\forall\boldsymbol{z}\in\mathbf{R}^n.$$

因为 $|\langle\boldsymbol{x}^*,\boldsymbol{z}\rangle|\leqslant\|\boldsymbol{z}\|\,\|\boldsymbol{x}^*\|$, 所以 $\partial f(\mathbf{0})=B$, 其中 B 是单位球.

当 $\boldsymbol{x}\neq\mathbf{0}$ 时, 不难验证 $\partial f(\boldsymbol{x})=||\boldsymbol{x}||^{-1}\boldsymbol{x}$.

例 6.2.4　设正常凸函数:

$$f(\boldsymbol{x})=\begin{cases}\|\boldsymbol{x}\|, & \|\boldsymbol{x}\|\leqslant 1,\\ +\infty, & \|\boldsymbol{x}\|>1.\end{cases}$$

可以验证, 当 $\|\boldsymbol{x}\|\leqslant 1$ 时, $f(\boldsymbol{x})$ 是次可微; 而当 $\|\boldsymbol{x}\|>1$ 时, $\partial f(\boldsymbol{x})=\varnothing$.

例 6.2.5　设 $C\subset\mathbf{R}^n$ 是非空凸集, 则示性函数 $f(\boldsymbol{x})=\theta_C(\boldsymbol{x})$ 的次梯度 $\boldsymbol{x}^*\in\partial f(\boldsymbol{x})$ 的等价条件是

$$\theta_C(\boldsymbol{z})\geqslant\theta_C(\boldsymbol{x})+\langle\boldsymbol{x}^*,\boldsymbol{z}-\boldsymbol{x}\rangle,\quad\forall\boldsymbol{z}\in\mathbf{R}^n.$$

对于 $\boldsymbol{x}\in C$, 任意 $\boldsymbol{z}\in\mathbf{R}^n$, 有 $0\geqslant\langle\boldsymbol{x}^*,\boldsymbol{z}-\boldsymbol{x}\rangle$. 由定义 3.6.3 知 $\boldsymbol{x}^*$ 是 C 在 $\boldsymbol{x}$ 处的法线, 所以 $\partial\theta_C(\boldsymbol{x})$ 是 C 在 $\boldsymbol{x}$ 处的法线锥.

性质 6.2.1　设 $f:\mathbf{R}^n\to[-\infty,+\infty]$ 是正常凸函数, 有下面结论成立:

(1) 如果 f 在 $\boldsymbol{x}$ 处可微, 则 $\nabla f(\boldsymbol{x})$ 是 f 在 $\boldsymbol{x}$ 处的唯一次梯度, 且

$$f(\boldsymbol{z})\geqslant f(\boldsymbol{x})+\langle\nabla f(\boldsymbol{x}),\boldsymbol{z}-\boldsymbol{x}\rangle,\quad\forall\boldsymbol{z}\in\mathbf{R}^n; \tag{6.2.2-2}$$

(2) 如果 f 在 $\boldsymbol{x}$ 处有唯一次梯度, 则 f 在 $\boldsymbol{x}$ 处可微;

(3) $\partial f(\boldsymbol{x})$ 是闭凸集;

(4) 设 $\boldsymbol{x}\in\mathrm{ri}(\mathrm{dom}f)$, 则 $\mathrm{cl}f'_+(\boldsymbol{x},\boldsymbol{d})=\delta_{\partial f(\boldsymbol{x})}(\boldsymbol{d})$, $\forall\boldsymbol{d}\in\mathbf{R}^n$.

证明　(1) 由性质 4.1.1 和定义 6.2.2 知结论成立.

(2) 设 $\boldsymbol{x}^*$ 是 f 在 $\boldsymbol{x}$ 处的唯一次梯度, 任意给定非零 $\boldsymbol{d}\in\mathbf{R}^n$, 令

$$\phi(t)=f(\boldsymbol{x}+t\boldsymbol{d})-f(\boldsymbol{x})-\langle\boldsymbol{x}^*,t\boldsymbol{d}\rangle,\quad\forall t\in\mathbf{R}. \tag{6.2.2-3}$$

根据定理 6.1.1, 只要证明 $\lim\limits_{t\to 0}\dfrac{\phi(t)}{t}=0$, 即可得 f 在 $\boldsymbol{x}$ 处可微. 根据定义 6.2.2 得 $\phi(t)\geqslant 0$, 设 $t>0$, 则有 $\dfrac{\phi(t)}{t}\geqslant 0$, 由引理 4.3.3 得 $\dfrac{\phi(t)}{t}$ 是单调递增函数. 令 $t\downarrow 0$ 有 $\dfrac{\phi(t)}{t}\to 0$, 即得

$$\phi'_+(t,1)=0.$$

再设 $t>0$, 令

$$\varphi(t)=\phi(-t)=f(\boldsymbol{x}-t\boldsymbol{d})-f(\boldsymbol{x})+\langle \boldsymbol{x}^*,t\boldsymbol{d}\rangle$$

是关于 t 的凸函数, 有 $\varphi(t)\geqslant 0$, 则 $\dfrac{\varphi(t)}{t}\geqslant 0$. 令 $t\downarrow 0$, 有 $\dfrac{\varphi(t)}{t}\to 0$, 即

$$0=\lim_{t\downarrow 0}\frac{\varphi(t)}{t}=\lim_{t\downarrow 0}\frac{\phi(-t)}{t}=\phi'_+(0,-1)=\phi'_-(0,-1).$$

因此, $\phi'_+(t,1)=-\phi'_+(t,-1)$, 根据定理 6.2.1 得 $\lim\limits_{t\to 0}\dfrac{\phi(t)}{t}=0$.

(3) 结论显然.

(4) 由性质 4.3.4 有

$$\mathrm{cl}f'_+(\boldsymbol{x},\boldsymbol{d})=\lim_{t\downarrow 0}f'(\boldsymbol{x},\boldsymbol{d}+t(\boldsymbol{x}-\boldsymbol{d})).$$

由定义 6.2.2 得 $f(\boldsymbol{x}+t\boldsymbol{d})-f(\boldsymbol{x})\geqslant t\langle \boldsymbol{x}^*,\boldsymbol{d}\rangle$. 设 $t>0$, 再由定理 6.2.2(3), 并令 $t\downarrow 0$ 得

$$f'_+(\boldsymbol{x},\boldsymbol{d})\geqslant\langle \boldsymbol{x}^*,\boldsymbol{d}\rangle,\quad \forall \boldsymbol{d}\in\mathbf{R}^n,$$

即 $f'_+(\boldsymbol{x},\boldsymbol{d}+t(\boldsymbol{x}-\boldsymbol{d}))\geqslant\langle \boldsymbol{x}^*,\boldsymbol{d}+t(\boldsymbol{x}-\boldsymbol{d})\rangle$. 因 $t\downarrow 0$, 所以有 $\mathrm{cl}f'_+(\boldsymbol{x},\boldsymbol{d})\geqslant \delta_{\partial f(\boldsymbol{x})}(\boldsymbol{d})$, 再由推论 4.5.3 知结论成立. □

性质 6.2.1 告诉我们, 正常可微凸函数等价于它的次梯度唯一且等于它的梯度; 正常凸函数的次梯度集由一个闭凸集构成; 若 $\boldsymbol{x}\in\mathrm{ri}(\mathrm{dom}f)$, 则 f 正方向导数的上方图闭集构成的闭凸函数等于 f 次梯度集合的支撑函数, 即 $\mathrm{cl}f'_+(\boldsymbol{x},\boldsymbol{d})=\delta_{\partial f(\boldsymbol{x})}(\boldsymbol{d})$. 事实上, 还可以得到 $\partial f(\boldsymbol{x})$ 与 $f'_+(\boldsymbol{x},\boldsymbol{d})$ 的等价关系 (定理 6.2.3).

定理 6.2.3 设 $f:\ \mathbf{R}^n\to[-\infty,+\infty]$ 是正常凸函数, 则 $\boldsymbol{x}^*\in\partial f(\boldsymbol{x})$ 的充要条件是对任意 $\boldsymbol{d}\in\mathbf{R}^n$ 有

$$f'_+(\boldsymbol{x},\boldsymbol{d})\geqslant\langle \boldsymbol{x}^*,\boldsymbol{d}\rangle. \tag{6.2.2-4}$$

证明 若 $\boldsymbol{x}^*\in\partial f(\boldsymbol{x})$, 设 $\boldsymbol{z}=\boldsymbol{x}+t\boldsymbol{d}$, 由定义 6.2.2 有

$$\frac{f(\boldsymbol{x}+t\boldsymbol{d})-f(\boldsymbol{x})}{t}\geqslant\langle \boldsymbol{x}^*,\boldsymbol{d}\rangle,\quad t>0,\quad \forall \boldsymbol{d}\in\mathbf{R}^n.$$

令 $t \downarrow 0$ 取极限, 则有

$$f'_+(\boldsymbol{x}, \boldsymbol{d}) \geqslant \langle \boldsymbol{x}^*, \boldsymbol{d} \rangle, \quad \forall \boldsymbol{d} \in \mathbf{R}^n.$$

反之, 如果对任意 $\boldsymbol{d} \in \mathbf{R}^n$, 上式成立. 考虑到差商 $\dfrac{f(\boldsymbol{x}+t\boldsymbol{d})-f(\boldsymbol{x})}{t}$ 在 $t>0$ 时是 t 的不减函数, 故

$$\frac{f(\boldsymbol{x}+t\boldsymbol{d})-f(\boldsymbol{x})}{t} \geqslant f'_+(\boldsymbol{x}, \boldsymbol{d}) \geqslant \langle \boldsymbol{x}^*, \boldsymbol{d} \rangle, \quad t>0, \quad \forall \boldsymbol{d} \in \mathbf{R}^n.$$

即 (6.2.2-1) 式成立, 得 $\boldsymbol{x}^* \in \partial f(\boldsymbol{x})$. □

次梯度不等式 (6.2.2-4) 的几何意义是: 仿射函数 $h(\boldsymbol{z}) = f(\boldsymbol{x}) + \langle \boldsymbol{x}^*, \boldsymbol{z}-\boldsymbol{x} \rangle$ 的图形是 epi$f$ 在点 $(\boldsymbol{x}, f(\boldsymbol{x}))$ 的支撑超平面, 即正常凸函数的正方向导数的上方图的闭集构成的函数等于该凸函数的次微分的支撑函数. 对于一维的情形, 定理 6.2.3 说明次梯度是 $\mathbf{R}^2$ 中过 $(\boldsymbol{x}, f(\boldsymbol{x}))$ 而不与 ri(epi$f$) 相交的非垂直直线的斜率 $\boldsymbol{x}^*$, 它们构成了在 $f'_-(\boldsymbol{x})$ 和 $f'_+(\boldsymbol{x})$ 之间全体实数组成的闭区间 (见例 6.2.2 和后面的例 6.2.6).

性质 6.2.2 设 $f: \mathbf{R}^n \to [-\infty, +\infty]$ 是凸函数, $f(\boldsymbol{x}) < +\infty\ (\forall \boldsymbol{x} \in \mathbf{R}^n)$, 有下面结论成立:

(1) 如果 f 在 $\boldsymbol{x}$ 处次可微, 则 f 是正常凸的;

(2) 如果 f 在 $\boldsymbol{x}$ 处不是次可微的, 则 f 在 $\boldsymbol{x}$ 处一定存在无限的双边方向导数

$$f'_+(\boldsymbol{x}, \boldsymbol{d}) = -f'_-(\boldsymbol{x}, \boldsymbol{d}) = -\infty. \tag{6.2.2-5}$$

证明 (1) 如果 $f(\boldsymbol{x})$ 在 $\boldsymbol{x}$ 处次可微, 存在向量 $\boldsymbol{x}^* \in \mathbf{R}^n$ 使得对任意 $\boldsymbol{z} \in \mathbf{R}^n$ 有

$$+\infty > f(\boldsymbol{z}) \geqslant f(\boldsymbol{x}) + \langle \boldsymbol{x}^*, \boldsymbol{z}-\boldsymbol{x} \rangle > -\infty,$$

故 f 是正常的.

(2) 如果 f 在 $\boldsymbol{x}$ 处不是次可微的, 则有 $\partial f(\boldsymbol{x}) = \varnothing$. 但 $\partial f(\boldsymbol{x}) = \varnothing$ 的等价条件是

$$\mathrm{cl} f'_+(\boldsymbol{x}, \boldsymbol{d}) = \delta_{\partial f(\boldsymbol{x})}(\boldsymbol{d}) = -\infty.$$

因此, 存在某些 $\boldsymbol{d}$ 使得 $f'_+(\boldsymbol{x}, \boldsymbol{d}) = -\infty$. 由定理 6.2.2 得 $-f'(\boldsymbol{x}, -\boldsymbol{y}) = -\infty$. □

由性质 6.2.2 可知, 如果凸函数次可微存在, 则其函数值有限, 否则方向导数为负无穷大.

性质 6.2.3 设 $f(\boldsymbol{x})$ 是 $\mathbf{R}^n$ 上的正常凸函数, 有下面结论成立:

(1) 若 $\boldsymbol{x} \notin \mathrm{dom} f$, 则 $\partial f(\boldsymbol{x}) = \varnothing$;

(2) 若 $\boldsymbol{x}\in \mathrm{ri}\,(\mathrm{dom}f)$, 则 $\partial f(\boldsymbol{x})\neq\varnothing$, 且

$$f'_{+}(\boldsymbol{x},\boldsymbol{d})=\sup\{\langle\boldsymbol{x}^*,\boldsymbol{d}\rangle\,|\boldsymbol{x}^*\in\partial f(\boldsymbol{x})\}=\delta_{\partial f(\boldsymbol{x})}(\boldsymbol{d}) \tag{6.2.2-6}$$

是关于 $\boldsymbol{d}$ 的闭凸函数;

(3) $\partial f(\boldsymbol{x})$ 非空有界的充要条件是 $\boldsymbol{x}\in\mathrm{int}\,(\mathrm{dom}f)$, 这时 $f'_{+}(\boldsymbol{x},\boldsymbol{d})$ 处处有限, 在 (6.2.2-6) 式中的“sup”变为“max”;

(4) 若 $f(\boldsymbol{x})>\inf\{f(\boldsymbol{z})\,|\boldsymbol{z}\in\mathbf{R}^n\}$, 则 $C=\{\boldsymbol{z}\,|f(\boldsymbol{z})\leqslant f(\boldsymbol{x})\}$ 在 $\boldsymbol{x}$ 处的法线锥为

$$N_{\boldsymbol{x}}(C)=\mathrm{cl}\,(\mathrm{cone}\,[\partial f(\boldsymbol{x})]).$$

证明 (1) 若 $\boldsymbol{x}\notin\mathrm{dom}f$, 则有 $f(\boldsymbol{x})=+\infty$. 这时只要取 $\boldsymbol{z}\in\mathrm{dom}f$, 则 (6.2.2-1) 式对任何 $\boldsymbol{x}^*\in\partial f(\boldsymbol{x})$ 都不能成立, 所以有 $\partial f(\boldsymbol{x})=\varnothing$.

(2) 如果 $\boldsymbol{x}\in\mathrm{ri}\,(\mathrm{dom}f)$, 显然 $f'_{+}(\boldsymbol{x},\boldsymbol{0})=0$, 于是对 $\boldsymbol{d}\in\mathbf{R}^n$, $f'_{+}(\boldsymbol{x},\boldsymbol{d})$ 在 $\mathrm{dom}f'_{+}(\boldsymbol{x},\boldsymbol{d})$ 上不恒为 $+\infty$, 所以 $f'_{+}(\boldsymbol{x},\boldsymbol{d})$ 关于 $\boldsymbol{d}\in\mathbf{R}^n$ 是正常的. 由性质 6.2.1(4)、性质 6.2.2(1) 和定理 6.2.2(3) 得 $f'_{+}(\boldsymbol{x},\boldsymbol{d})$ 关于 $\boldsymbol{d}\in\mathbf{R}^n$ 是正常闭凸函数, 有 $\partial f(\boldsymbol{x})\neq\varnothing$. 再由性质 6.2.1(4) 知, (6.2.2-6) 式成立.

(3) 如果 $\boldsymbol{x}\in\mathrm{int}\,(\mathrm{dom}f)$, 有 $\mathrm{ri}\,(\mathrm{dom}f)=\mathrm{int}\,(\mathrm{dom}f)$. 由 (2) 知, $\mathrm{dom}f'_{+}(\boldsymbol{x},\boldsymbol{d})=\mathbf{R}^n$, 所以 $f'_{+}(\boldsymbol{x},\boldsymbol{d})$ 处处有限. 根据性质 4.5.1 和定理 6.2.1, 又因 $f'_{+}(\boldsymbol{x},\boldsymbol{d})=\delta_{\partial f(\boldsymbol{x})}(\boldsymbol{d})$, 于是 $\partial f(\boldsymbol{x})$ 非空有界.

反过来, 如果 $\partial f(\boldsymbol{x})$ 非空有界, 则由推论 4.5.2 和推论 4.5.4 得 $\delta_{\partial f(\boldsymbol{x})}(\boldsymbol{d})$ 处处有限, 且 $\mathrm{cl}f'_{+}(\boldsymbol{x},\boldsymbol{d})=\delta_{\partial f(\boldsymbol{x})}(\boldsymbol{d})$, 所以 $f'_{+}(\boldsymbol{x},\boldsymbol{d})$ 处处有限. 由方向导数定义知, 对任意的 $\boldsymbol{d}\in\mathbf{R}^n$, $\exists t>0$, 使得 $\boldsymbol{x}+t\boldsymbol{d}\in\mathrm{dom}f$, 所以 $\boldsymbol{x}\in\mathrm{int}\,(\mathrm{dom}f)$. 显然, 在有界闭集上正线性齐次函数达到上确界, 故 (6.2.2-6) 式中的“sup”由“max”代替.

(4) 由已知条件可得 $C\neq\varnothing$, 故由法锥的定义 (定义 3.6.3) 知, $\boldsymbol{x}^*$ 是 C 在 $\boldsymbol{x}$ 处的法线向量等价于

$$\begin{aligned}N_{\boldsymbol{x}}(C)&=\{\boldsymbol{x}^*\,|f(\boldsymbol{z})<f(\boldsymbol{x}),\langle\boldsymbol{z}-\boldsymbol{x},\boldsymbol{x}^*\rangle\leqslant 0\,,\ \boldsymbol{z}\in\mathbf{R}^n\}\\&=\{\boldsymbol{x}^*\,|f(\boldsymbol{z})<f(\boldsymbol{x}),\langle t(\boldsymbol{z}-\boldsymbol{x}),\boldsymbol{x}^*\rangle\leqslant 0\ ,t>0,\boldsymbol{z}\in\mathbf{R}^n\}.\end{aligned}$$

由定义 6.2.2 和定理 6.2.1, 有

$$\begin{aligned}K&:=\left\{\boldsymbol{d}\in\mathbf{R}^n\,\middle|f'_{+}(\boldsymbol{x},\boldsymbol{d})>0\right\}\\&=\{\boldsymbol{d}\,|\boldsymbol{d}=t(\boldsymbol{z}-\boldsymbol{x}),f(\boldsymbol{z})<f(\boldsymbol{x}),t>0,\forall\boldsymbol{z}\in\mathbf{R}^n\}.\end{aligned}$$

设 K 的极锥为

$$K^-=\{\boldsymbol{x}^*\,|\langle\boldsymbol{x}^*,\boldsymbol{d}\rangle\leqslant 0,\boldsymbol{d}\in K\}$$

$$= \{\boldsymbol{x}^* \,|\langle \boldsymbol{x}^*, t(\boldsymbol{z}-\boldsymbol{x})\rangle \leqslant 0, t(\boldsymbol{z}-\boldsymbol{x}) \in K, t>0, \boldsymbol{z} \in \mathbf{R}^n\}.$$

由性质 4.3.5 得, $\mathrm{cl}\{\boldsymbol{z}\,|f(\boldsymbol{z}) < f(\boldsymbol{x})\} = \{\boldsymbol{z}\,|\mathrm{cl}f(\boldsymbol{z}) < f(\boldsymbol{x})\}$, 且

$$\begin{aligned}\mathrm{cl}K &= \left\{\boldsymbol{d}\,\middle|\mathrm{cl}f'_+(\boldsymbol{x},\boldsymbol{d}) \leqslant 0\right\} = \left\{\boldsymbol{d}\,\middle|\delta_{\partial f(\boldsymbol{x})}(\boldsymbol{d}) \leqslant 0\right\} \\ &= \{\boldsymbol{d}\,|\langle \boldsymbol{d}, \boldsymbol{x}^*\rangle \leqslant 0, \forall \boldsymbol{x}^* \in \partial f(\boldsymbol{x})\} \\ &= \{\boldsymbol{d}\,|\langle \boldsymbol{d}, t\boldsymbol{x}^*\rangle \leqslant 0, \forall t\boldsymbol{x}^* \in t\partial f(\boldsymbol{x}), t>0\} \\ &= \mathrm{cl}(\mathrm{cone}\partial f(\boldsymbol{x}))^-,\end{aligned}$$

得

$$N_{\boldsymbol{x}}(C) = K^- = \mathrm{cl}K^- = \left(\mathrm{cl}(\mathrm{cone}\partial f(\boldsymbol{x}))^-\right)^- = \mathrm{cl}(\mathrm{cone}\partial f(\boldsymbol{x})). \qquad \square$$

例 6.2.6　在 $\mathbf{R}$ 中, 设单变量函数

$$f(x) = \begin{cases} +\infty, & x < -1, \\ 2, & x = -1, \\ x^2, & -1 < x \leqslant 0, \\ x, & 0 < x \leqslant 1, \\ +\infty, & x > 1 \end{cases}$$

是正常凸函数, $\mathrm{dom}f = [-1, 1]$, 它的次梯度为

$$\partial f(x) = \begin{cases} \varnothing, & x \leqslant -1, \\ 2x, & -1 < x < 0, \\ [0,1], & x = 0, \\ 1, & 0 < x < 1, \\ [1,+\infty], & x = 1, \\ \varnothing, & x > 1. \end{cases}$$

结合例 6.2.2 的方向导数知定理 6.2.1、性质 6.2.2 和性质 6.2.3 均成立.

本小节得到了正常凸函数次微分的几个重要结果:

1) 正常可微凸函数等价于它的唯一次梯度, 且正常凸函数的次梯度由一个闭凸集构成.

2) 凸函数次微分等价于它的次梯度与 $\mathbf{R}^n$ 中任何一个点的内积不大于在该点上的正方向导数.

3) 如果凸函数次可微存在, 则其函数值是有限的; 否则, 其方向导数为负无穷大.

4) 正常凸函数的非有效定义域点的次微分是一个空集; 正常凸函数的有效定义域相对内点的次微分非空, 且次微分的支撑函数等于其正方向导数.

6.3 次微分连续性

根据性质 6.2.2、性质 6.2.3 知, 凸函数次微分存在是仅当函数取有限值时, 次微分非空. 因此, 只有对次微分存在的情形讨论它的连续性才有意义.

定义 6.3.1 设 f 是 $\mathbf{R}^n$ 上的正常凸函数, 则称

$$\mathrm{dom}\partial f=\{\boldsymbol{x}\in\mathbf{R}^n\,|\partial f(\boldsymbol{x})\neq\varnothing\}$$

为次微分映射 ∂f 的有效定义域, 又称

$$\mathrm{rang}\partial f=\cup\{\partial f(\boldsymbol{x})\,|\forall\boldsymbol{x}\in\mathbf{R}^n\}$$

为 ∂f 的值域.

性质 6.3.1 设 ∂f 的值域是 ∂f^* 的有效定义域, 则

$$\mathrm{ri}\,(\mathrm{dom}f^*)\subset\mathrm{rang}\partial f\subset\mathrm{dom}f^*.$$

证明 由性质 6.2.3 得 $\mathrm{ri}\,(\mathrm{dom}f)\subset\partial f\subset\mathrm{dom}f$, 再由性质 4.4.1 的公式

$$f(\boldsymbol{x})+f^*(\boldsymbol{x}^*)\geqslant\langle\boldsymbol{x},\boldsymbol{x}^*\rangle,\quad\forall\boldsymbol{x},\boldsymbol{x}^*\in\mathbf{R}^n,$$

知结论成立. □

下面分别讨论单变量函数与多变量函数这两种情形的次微分连续性.

6.3.1 单变量情形

为方便起见, 将左、右导数 $f'_-(x)$ 和 $f'_+(x)$ 推广到 $\mathrm{dom}f$ 区间之外. 在 $\mathrm{dom}f$ 右边的点, 设 $f'_-=f'_+=+\infty$; 在 $\mathrm{dom}f$ 左边的点, 设 $f'_-=f'_+=-\infty$.

性质 6.3.2 设 $f(x)$ 是 $\mathbf{R}$ 上的闭正常凸函数, 则下面结论成立:

(1) f'_- 和 f'_+ 在 $\mathbf{R}$ 上是不减函数, 在 $\mathrm{int}\,(\mathrm{dom}f)$ 上有限;

(2) 当 $z_1<x<z_2$ 时, 有

$$f'_+(z_1)\leqslant f'_-(x)\leqslant f'_+(x)\leqslant f'_-(z_2);\tag{6.3.1-1}$$

(3) 对任意 $x\in\mathbf{R}$, 有

$$\lim_{z\downarrow x}f'_+(z)=f'_+(x),\quad\lim_{z\downarrow x}f'_-(z)=f'_+(x),\tag{6.3.1-2}$$

$$\lim_{z\uparrow x} f'_+(z) = f'_-(x), \quad \lim_{z\uparrow x} f'_-(z) = f'_-(x). \tag{6.3.1-3}$$

证明　(1) 根据单变量单边方向导数定义有

$$f'_+(x) = f'_+(x,1), \quad f'_-(x) = -f'_-(x,-1).$$

由定理 6.2.2, 有 $f'_-(x) \leqslant f'_+(x)$. 再由推广后的定义知, 当 $x \notin \mathrm{dom}f$ 时, 这个不等式仍成立.

对于 $t>0$, 差商 $\dfrac{f(x+t)-f(x)}{t}$ 是关于 t 单调增加的. 对于 $x \in \mathrm{int}(\mathrm{dom}f)$, 由定理 6.2.2 在单变量情形下有 $f'_+(x) < +\infty$. 因为 $f'_+(x,-1) = f'_-(x)$, 有 $f'_-(x) > -\infty$, 所以在 $\mathrm{int}(\mathrm{dom}f)$ 上 f'_- 和 f'_+ 是有限的.

当 $x,z \in \mathrm{dom}f$, 且 $x<z$ 时, 有

$$f'_+(x) \leqslant \frac{f(z)-f(x)}{z-x} = \frac{f(x)-f(z)}{x-z} \leqslant f'_-(z);$$

当 $x,z \notin \mathrm{dom}f$, 且 $x<z$ 时, 根据推广了的定义, 也有 $f'_+(x) \leqslant f'_-(z)$. 如果 x,z 中只有一个在 $\mathrm{dom}f$ 中, 显然有 $f'_+(x) \leqslant f'_-(z)$ 成立. 故 (1) 成立.

(2) 由 (1) 直接可得.

(3) 由 (2) 得

$$f'_+(x) \leqslant \lim_{z\downarrow x} f'_-(z) \leqslant \lim_{z\downarrow x} f'_+(z).$$

再证上式的反向不等式成立. 由推论 4.3.3, 得 $\lim\limits_{y\downarrow x} f(y) = f(x)$. 故 $\exists \varepsilon > 0$, 在 $x<z<x+\varepsilon$ 时, 根据引理 4.3.3 有

$$\frac{f(z)-f(x)}{z-x} = \lim_{y\downarrow x} \frac{f(z)-f(y)}{z-y} \geqslant \lim_{y\downarrow x} f'_+(y).$$

由上式得

$$\lim_{y\downarrow x} f'_+(y) \leqslant \lim_{z\downarrow x} \frac{f(z)-f(x)}{z-x} = f'_+(x),$$

所以 (6.3.1-2) 式成立, 类似可以证明 (6.3.1-3) 式.　□

性质 6.3.2 和定理 6.2.3 说明, 对任意 $x \in \mathrm{dom}f$ 有

$$\partial f(x) = \left\{x^* \in \mathbf{R} \,\middle|\, f'_-(x) \leqslant x^* \leqslant f'_+(x)\right\}. \tag{6.3.1-4}$$

下面利用 (6.3.1-4) 式计算单变量凸函数的次微分.

例 6.3.1 在 **R** 中, 设单变量函数为

$$f(x)=\begin{cases}+\infty, & x\leqslant -1,\\ -x, & -1<x<0,\\ 0, & x=0,\\ x, & 0<x<1,\\ +\infty, & x\geqslant 1.\end{cases}$$

下面求 $\partial f(x)$. $f(x)$ 是正常凸函数, 利用左、右导数定义得

$$f'_+(x)=\begin{cases}-\infty, & x\leqslant -1,\\ -1, & -1<x<0,\\ 1, & x=0,\\ 1, & 0<x<1,\\ +\infty, & x\geqslant 1,\end{cases}\qquad f'_-(x)=\begin{cases}-\infty, & x\leqslant -1,\\ -1, & -1<x<0,\\ -1, & x=0,\\ 1, & 0<x<1,\\ +\infty, & x\geqslant 1.\end{cases}$$

所以, 由 (6.3.1-4) 式有

$$\partial f(x)=\begin{cases}\varnothing, & x\leqslant -1,\\ -1, & -1<x<0,\\ [-1,1], & x=0,\\ 1, & 0<x<1,\\ \varnothing, & x\geqslant 1.\end{cases}$$

例 6.3.2 在 **R** 中, 设单变量函数为

$$f(x)=\begin{cases}0, & x>0,\\ -1, & x=0,\\ +\infty, & x<0,\end{cases}$$

求 $f'_+(x)$.

解 计算可得

$$f'_+(x)=\begin{cases}0, & x>0,\\ +\infty, & x=0,\\ -\infty, & x<0,\end{cases}\qquad f'_-(x)=\begin{cases}0, & x>0,\\ -\infty, & x=0,\\ -\infty, & x<0,\end{cases}$$

$$\partial f(x)=\begin{cases}0, & x>0,\\ [-\infty,+\infty], & x=0,\\ \varnothing, & x<0.\end{cases}$$

显然, $f'_+(x)$ 在 $x=0$ 处不是连续的, $f(x)$ 不是闭的.

下面的定理说明一个单调增加函数 φ 的积分函数是一个闭正常凸函数, 凸函数的方向导数构成了其被积分函数.

定理 6.3.1　设 φ: $\mathbf{R}\to[-\infty,+\infty]$ 是单调递增函数, 且 $\varphi(\alpha)<+\infty$, 其中 $\alpha\in\mathbf{R}$, 令

$$\varphi_+(x)=\lim_{z\downarrow x}\varphi(z),\quad \varphi_-(x)=\lim_{z\uparrow x}\varphi(z),\quad f(x)=\int_\alpha^x\varphi(t)\,\mathrm{d}t,$$

则 $f(x)$ 是闭正常凸函数, 且满足 $f'_-=\varphi_-\leqslant\varphi\leqslant\varphi_+=f'_+$. 如果 $g(x)$ 是另一个闭正常凸函数, 满足 $g'_-\leqslant\varphi\leqslant g'_+$, 则 $g=f+\beta$, 其中 β 是常数.

证明　设集合 $I=\{x|\varphi(x)<+\infty\}$, 则 $\alpha\in I$. 因为 $\varphi(x)$ 是单调递增函数, 当 $x\in\mathrm{cl}I$ 时, $f(x)$ 作为 Riemann 积分存在且有限. 当 $x\notin\mathrm{cl}I$ 时, 则 $f(x)=+\infty$.

下面证明 $f(x)$ 在 I 上是凸的, 设 $x,y\in I$, $0<\lambda<1$, $z=(1-\lambda)x+\lambda y\,(x<y)$, 则有

$$\lambda=\frac{z-x}{y-x},\quad 1-\lambda=\frac{y-z}{y-x},\quad x<z<y.$$

根据 $\varphi(x)$ 是单调递增函数, 有

$$f(z)-f(x)=\int_x^z\varphi(t)\,\mathrm{d}t\leqslant\int_x^z\varphi(z)\,\mathrm{d}t=(z-x)\varphi(z)\leqslant(z-x)\varphi(y),$$

$$f(y)-f(z)=\int_z^y\varphi(t)\,\mathrm{d}t\leqslant\int_z^y\varphi(y)\,\mathrm{d}t=(y-z)\varphi(y).$$

由上式得

$$(1-\lambda)[f(z)-f(x)]+\lambda[f(z)-f(y)]\leqslant[(1-\lambda)(z-x)-\lambda(y-z)]\varphi(y)=0,$$

从而得 $f(z)\leqslant(1-\lambda)f(x)+\lambda f(y)$, 所以 f 在 I 上是凸函数. 因为积分函数是连续的, 所以 f 是闭的.

设 $\forall x\in I$, 当 $\forall z>x$ 时, 有

$$\frac{f(z)-f(x)}{z-x}=\frac{1}{z-x}\int_x^z\varphi(t)\,\mathrm{d}t\geqslant\frac{1}{z-x}\varphi(x)(z-x)=\varphi(x).$$

令 $z\downarrow x$, 得 $f'_+(x)\geqslant\varphi(x)$. 类似可证 $\varphi(x)\geqslant f'_-(x)$.

当 $x\notin I$ 时, 这两个不等式显然也成立. 因此, 有 $f'_+=\varphi_+$ 和 $f'_-=\varphi_-$, 即 $f'_-=\varphi_-\leqslant\varphi\leqslant\varphi_+=f'_+$.

如果 $g(x)$ 是满足 $g'_-\leqslant\varphi\leqslant g'_+$ 的另一个闭正常凸函数, 那么同样有 $g'_+=\varphi_+$ 和 $g'_-=\varphi_-$. 所以得 $g'_+=f'_+$ 和 $g'_-=f'_-$. 根据左、右导数的定义得 $I\subset \mathrm{dom}f\subset \mathrm{cl}I$, 则

$$\mathrm{ri}\,(\mathrm{dom}g)=\mathrm{ri}\,(\mathrm{dom}f)=\mathrm{ri}I.$$

因为 $f(x)$ 和 $g(x)$ 都是闭的, 所以在 $\mathrm{ri}I$ 上 $f(x)$ 和 $g(x)$ 的值是存在的. 事实上, $\mathrm{ri}I=\mathrm{int}I\neq\varnothing$. 由性质 6.3.2, 在 $\mathrm{int}I$ 中 f 和 g 的左、右导数有限, 且

$$f'_+(x)-g'_+(x)=0,\quad f'_-(x)-g'_-(x)=0,\quad \forall x\in\mathrm{int}I.$$

上式表明, 在 $\mathrm{int}I$ 上, $f-g$ 的值为常数. □

由定理 6.3.1 知下面结论显然成立, 对于任意的 $x,y\in I$ 有

$$f(y)-f(x)=\int_x^y f'_+(t)\,\mathrm{d}t=\int_x^y f'_-(t)\,\mathrm{d}t.$$

本小节得到下面几个结论:

1) 对于单变量闭凸函数, 它的正、负方向导数是单调递增函数, 正方向导数是下半连续的, 负方向导数是上半连续的;

2) 通过一个单调递增函数定义的积分函数是一个闭正常凸函数.

6.3.2 多变量情形

下面证明当 f 是 $\mathbf{R}^n$ 上的闭正常凸函数时, $\partial f(\boldsymbol{x})$ 是一个闭映射 (上半连续映射), 即有下面结论.

定理 6.3.2 设 f 是 $\mathbf{R}^n$ 上的闭正常凸函数. 如果序列 $\{\boldsymbol{x}_i\}$ 和 $\{\boldsymbol{x}_i^*\}$ 满足 $\lim\limits_{i\to\infty}\boldsymbol{x}_i=\boldsymbol{x},\lim\limits_{i\to\infty}\boldsymbol{x}_i^*=\boldsymbol{x}^*$ 和 $\boldsymbol{x}_i^*\in\partial f(\boldsymbol{x}_i)$, 则 $\boldsymbol{x}^*\in\partial f(\boldsymbol{x})$.

证明 根据性质 4.4.1, 对任意的 $i=1,2,\cdots$, 有

$$\langle\boldsymbol{x}_i,\boldsymbol{x}_i^*\rangle\geqslant f(\boldsymbol{x}_i)+f^*(\boldsymbol{x}_i^*).\tag{6.3.2-1}$$

由推论 4.3.5 知 $f(\boldsymbol{x})$ 和 $f^*(\boldsymbol{x}^*)$ 是处处连续的. 当 $i\to\infty$ 时, 由定理假设可得

$$\langle\boldsymbol{x},\boldsymbol{x}^*\rangle\geqslant f(\boldsymbol{x})+f^*(\boldsymbol{x}^*),\tag{6.3.2-2}$$

由定理 6.4.1 得 $\boldsymbol{x}^*\in\partial f(\boldsymbol{x})$. □

下面结论说明在开凸集上的正常凸函数的次微分是一个上半连续映射.

定理 6.3.3　设 $C\subset\mathbf{R}^n$ 是开凸集, f 是 C 上的正常凸函数. 对于任意 $\boldsymbol{x}\in C$, 如果 $\{f_i(\boldsymbol{x})\}$ 是收敛到函数 $f(\boldsymbol{x})$ 的有限凸函数序列, 即 $\lim\limits_{i\to\infty}f_i(\boldsymbol{x})=f(\boldsymbol{x})$. 设序列 $\{\boldsymbol{x}_i\}\subset C$ 收敛到点 $\boldsymbol{x}\in C$, 序列 $\{\boldsymbol{d}_i\}\subset\mathbf{R}^n$ 收敛到点 $\boldsymbol{d}$, 则

$$\limsup_{i\to\infty}f_{i+}'(\boldsymbol{x}_i,\boldsymbol{d}_i)\leqslant f_+'(\boldsymbol{x},\boldsymbol{d}),\tag{6.3.2-3}$$

且对于任意给定的 $\varepsilon>0$, $\exists N>0$ 使得

$$\partial f_i(\boldsymbol{x}_i)\subset\partial f(\boldsymbol{x})+\varepsilon B,\quad\forall i>N,\tag{6.3.2-4}$$

其中 B 是单位球.

证明　设 $f_+'(\boldsymbol{x},\boldsymbol{d})<+\infty$, 由 $C\subset\mathbf{R}^n$ 是开凸集可知, $\exists\lambda>0$ 满足 $\boldsymbol{x}+\lambda\boldsymbol{d}\in C$. 由引理 4.3.3 和方向导数定义有

$$\frac{f(\boldsymbol{x}+\lambda\boldsymbol{d})-f(\boldsymbol{x})}{\lambda}<f_+'(\boldsymbol{x},\boldsymbol{d}).$$

由 $\lim\limits_{i\to\infty}f_i(\boldsymbol{x})=f(\boldsymbol{x})$ 和推论 4.3.5, 知

$$\lim_{i\to\infty}f_i(\boldsymbol{x}_i)=f(\boldsymbol{x}),$$

$$\lim_{i\to\infty}f_i(\boldsymbol{x}_i+\lambda\boldsymbol{d}_i)=f(\boldsymbol{x}+\lambda\boldsymbol{d}).$$

故当 i 充分大时, 根据前面两个式子有

$$\frac{f_i(\boldsymbol{x}_i+\lambda\boldsymbol{d}_i)-f_i(\boldsymbol{x}_i)}{\lambda}<f_+'(\boldsymbol{x},\boldsymbol{d}),$$

即得 $\limsup\limits_{i\to\infty}f_{i+}'(\boldsymbol{x}_i,\boldsymbol{d}_i)\leqslant f_+'(\boldsymbol{x},\boldsymbol{d})$. 因此, 取 $\forall\boldsymbol{d}_i=\boldsymbol{d}\in\mathbf{R}^n\ (i=1,2,\cdots)$ 有

$$\limsup_{i\to\infty}f_{i+}'(\boldsymbol{x}_i,\boldsymbol{d})\leqslant f_+'(\boldsymbol{x},\boldsymbol{d}).$$

因为 $C\subset\operatorname{int}(\operatorname{dom}f)$ 和 $C\subset\operatorname{int}(\operatorname{dom}f_i)$, 由性质 6.2.3 知, $\partial f(\boldsymbol{x})$ 和 $\partial f_i(\boldsymbol{x}_i)$ 是有界闭凸集, 得

$$f_{i+}'(\boldsymbol{x}_i,\boldsymbol{d})=\delta_{\partial f_i(\boldsymbol{x}_i)}(\boldsymbol{d}),\quad f_+'(\boldsymbol{x},\boldsymbol{d})=\delta_{\partial f(\boldsymbol{x})}(\boldsymbol{d}),$$

且在 $\mathbf{R}^n$ 中处处有限. 再由推论 4.3.5 可知, 任给 $\varepsilon>0,\exists i_0$, 在 $i\geqslant i_0$, $\forall\boldsymbol{d}\in B$ 时, 有

$$f_{i+}'(\boldsymbol{x}_i,\boldsymbol{d})\leqslant f_+'(\boldsymbol{x},\boldsymbol{d})+\varepsilon,$$

其中 B 是单位球. 由于方向导数关于 $\forall \boldsymbol{d} \in \mathbf{R}^n$ 是正齐次的, 此时若 $i \geqslant i_0$, 则有

$$f_{i+}'(\boldsymbol{x}_i, \boldsymbol{d}) \leqslant f_+'(\boldsymbol{x}, \boldsymbol{d}) + \varepsilon||\boldsymbol{d}||.$$

再由性质 6.2.3 知, 对于 $i \geqslant i_0$, $\forall \boldsymbol{d} \in \mathbf{R}^n$ 时, 有

$$\delta_{\partial f_i(\boldsymbol{x}_i)}(\boldsymbol{d}) \leqslant \delta_{\partial f(\boldsymbol{x})}(\boldsymbol{d}) + \varepsilon \delta_B(\boldsymbol{d}) = \delta_{\partial f(\boldsymbol{x})+\varepsilon B}(\boldsymbol{d}).$$

最后根据定理 4.5.1, 上面的不等式等价于

$$\partial f_i(\boldsymbol{x}_i) \subset \partial f(\boldsymbol{x}) + \varepsilon B. \qquad \square$$

下面得到正常凸函数连续性的另一条性质, 对于给定的一个非零方向 $\boldsymbol{d} \in \mathbf{R}^n$, 它的正方向导数 $f_+'(\boldsymbol{x}, \boldsymbol{d})$ 在负方向导数的集合上几乎处处连续.

定理 6.3.4 设 f 在 $\mathbf{R}^n$ 上是正常凸函数, $\boldsymbol{d} \in \mathbf{R}^n$ 是非零点. 令

$$D(\boldsymbol{d}) = \left\{\boldsymbol{x} \in \operatorname{int}(\operatorname{dom} f) \,\middle|\, f_+'(\boldsymbol{x}, \boldsymbol{d}) = -f_+'(\boldsymbol{x}, -\boldsymbol{d})\right\}, \tag{6.3.2-5}$$

则 $f_+'(\boldsymbol{x}, \boldsymbol{d})$ 在 $\boldsymbol{x} \in D(\boldsymbol{d})$ 处连续, 且 $D(\boldsymbol{d})$ 在 $\operatorname{int}(\operatorname{dom} f)$ 中处处稠密.

证明 设 $\boldsymbol{x} \in D$. 根据定理 6.3.3 知, $f_+'(\boldsymbol{x}, \boldsymbol{d})$ 在 $\operatorname{int}(\operatorname{dom} f)$ 中关于 $\boldsymbol{x}$ 是上半连续的. 下面证明 $f_+'(\boldsymbol{x}, \boldsymbol{d})$ 在 D 中是下半连续的, 即

$$\liminf_{\boldsymbol{z} \to \boldsymbol{x}} f_+'(\boldsymbol{z}, \boldsymbol{d}) = -f_+'(\boldsymbol{x}, -\boldsymbol{d}), \quad \forall \boldsymbol{x} \in \operatorname{int}(\operatorname{dom} f). \tag{6.3.2-6}$$

由定理 6.2.2(3), 有

$$f_+'(\boldsymbol{z}, \boldsymbol{d}) \geqslant -f_+'(\boldsymbol{z}, -\boldsymbol{d}).$$

而 $f_+'(\boldsymbol{z}, -\boldsymbol{d})$ 在 $\operatorname{int}(\operatorname{dom} f)$ 中关于 $\boldsymbol{z}$ 是上半连续的, 有

$$\begin{aligned}\liminf_{\boldsymbol{z} \to \boldsymbol{x}} f_+'(\boldsymbol{z}, \boldsymbol{d}) &\geqslant \liminf_{\boldsymbol{z} \to \boldsymbol{x}} \left(-f_+'(\boldsymbol{z}, -\boldsymbol{d})\right) \\ &\geqslant -\limsup_{\boldsymbol{z} \to \boldsymbol{x}} \left(f_+'(\boldsymbol{z}, -\boldsymbol{d})\right) \geqslant -f_+'(\boldsymbol{x}, -\boldsymbol{d}).\end{aligned}$$

反过来, 存在 $\lambda > 0$, $\boldsymbol{x} + \lambda(-\boldsymbol{d}) \in D(\boldsymbol{d})$, 由引理 4.3.3 和方向导数定义 (定义 6.2.1) 有

$$\frac{f(\boldsymbol{x} - \lambda \boldsymbol{d}) - f(\boldsymbol{x})}{\lambda} < f_+'(\boldsymbol{x}, -\boldsymbol{d}). \tag{6.3.2-7}$$

令 $t = -\lambda$, 因 $f_+'(\boldsymbol{x}, \boldsymbol{d}) = -f_+'(\boldsymbol{x}, -\boldsymbol{d})$, 令 $g(t) = f(\boldsymbol{x} + t\boldsymbol{d})$. 由 (6.3.2-7) 式, 有

$$\frac{g(t) - g(0)}{t} > -f_+'(\boldsymbol{x}, \boldsymbol{d}),$$

令 $t\uparrow 0$, 上式得 $g'_-(0)\geqslant -f'_+(\boldsymbol{x},\boldsymbol{d})$. 令 $\lambda\downarrow 0$, 再由 (6.3.2-7) 式有 $g'_+(0)\leqslant f'_+(\boldsymbol{x},-\boldsymbol{d})=-f'_+(\boldsymbol{x},\boldsymbol{d})$. 最后, 由定理 6.2.1 得 $g'_-(0)=g'_+(0)=-f'_+(\boldsymbol{x},\boldsymbol{d})$, 则有

$$\liminf_{\boldsymbol{z}\to\boldsymbol{x}} f'_+(\boldsymbol{x},\boldsymbol{d})\leqslant \lim_{t\uparrow 0} f'_+(\boldsymbol{x}+t\boldsymbol{d},\boldsymbol{d})=\lim_{t\uparrow 0} g'_-(t)=g'_-(0)=-f'_+(\boldsymbol{x},-\boldsymbol{d}).$$

因此, (6.3.2-6) 式成立.

现在, 假设 $D(\boldsymbol{d})$ 不在 $\operatorname{int}(\operatorname{dom}f)$ 中处处稠密, 则存在点 $\boldsymbol{x}_0\in\operatorname{int}(\operatorname{dom}f)$ 的一个邻域 $O_\varepsilon(\boldsymbol{x}_0)\subset\operatorname{int}(\operatorname{dom}f)$ 满足

$$f'_+(\boldsymbol{z},\boldsymbol{d})\neq -f'_+(\boldsymbol{z},-\boldsymbol{d}),\quad \forall \boldsymbol{z}\in O_\varepsilon(\boldsymbol{x}_0)\subset\operatorname{int}(\operatorname{dom}f).$$

设一个限制函数:

$$\phi_{\boldsymbol{z}}(t)=f(\boldsymbol{z}+t\boldsymbol{d}),\quad \forall \boldsymbol{z}\in O_\varepsilon(\boldsymbol{x}_0),\quad \forall t\in\mathbf{R}.$$

则 $\exists\varepsilon>0$ 使得 $\boldsymbol{z}_t=\boldsymbol{x}_0+t\boldsymbol{d}\in O_\varepsilon(\boldsymbol{x}_0),\forall t\in(-\varepsilon,\varepsilon)$, 且有 $f'_+(\boldsymbol{z}_t,\boldsymbol{d})>-f'_-(\boldsymbol{z}_t,-\boldsymbol{d})$. 设

$$\varphi(t)=\begin{cases}+\infty, & t\geqslant\varepsilon,\\ f'_+(\boldsymbol{z}_t,t\boldsymbol{d}), & \varepsilon>t>0,\\ f'_+(\boldsymbol{z}_t,-\boldsymbol{d}), & t=0,\\ -f'_-(\boldsymbol{z}_t,-t\boldsymbol{d}), & -\varepsilon<t<0,\\ -\infty, & t\leqslant-\varepsilon,\end{cases}$$

显然 φ 在 $[-\infty,+\infty]$ 上是单调递增函数, 则有

$$f'_+(\boldsymbol{z}_t,\boldsymbol{d})=\lim_{t\downarrow 0}\varphi(t)=\varphi_+(0)>\varphi(0)=-f'_+(\boldsymbol{z}_t,-\boldsymbol{d}),\tag{6.3.2-8}$$

令 $\alpha\in(-\varepsilon,0)$ 和 $\phi(x)=\displaystyle\int_\alpha^x\varphi(t)\,\mathrm{d}t$, 则由定理 6.3.1 知函数 ϕ 是闭正常凸函数, 对于任意 $x,y\in(-\varepsilon,\varepsilon)$ 有

$$\phi(y)-\phi(x)=\int_x^y f'_+(\boldsymbol{z}_t,\boldsymbol{d})\,\mathrm{d}t=-\int_x^y f'_-(\boldsymbol{z}_t,-\boldsymbol{d})\,\mathrm{d}t.$$

显然上式不成立, 与 (6.3.2-8) 式矛盾. □

下面的结论告诉我们, 正常凸函数在有效定义域内部几乎是处处可微的.

定理 6.3.5 设 $f(\boldsymbol{x})$ 在 $\mathbf{R}^n$ 上是正常凸函数和 D_0 是 f 所有可微的点构成的集合, 则 D_0 是 $\operatorname{int}(\operatorname{dom}f)$ 的稠密子集, 即 $\operatorname{int}(\operatorname{dom}f)\backslash D_0=\bar{D}$ 的测度是 0, 梯度 $\nabla f(\boldsymbol{x})$ 在 D_0 上是连续的.

证明 由定理 6.2.1 知 $D(\boldsymbol{d}) \subset D_0$, 其中 $\boldsymbol{d}$ 是 (6.3.2-5) 式中定义的向量, 再由定理 6.3.4 知 $D(\boldsymbol{d})$ 在 $\text{int}(\text{dom}f)$ 中处处稠密, 那么 D_0 是 $\text{int}(\text{dom}f)$ 的稠密子集, 且梯度 $\nabla f(\boldsymbol{x})$ 在 D_0 上是连续的. □

本小节得到正常凸函数次微分的几个重要结论:

1) 闭正常凸函数的次微分是一个上半连续映射;

2) 如果 $\{f_i(\boldsymbol{x})\}$ 是收敛到函数 $f(\boldsymbol{x})$ 的有限凸函数序列, 则对应序列正方向导数和次微分都是上半连续的;

3) 正常凸函数对于任何非零方向的正方向导数几乎处处连续, 即它在有效定义域内部几乎是处处可微的.

本小节内容告诉我们一个非常重要的结论, 正常凸函数在有效定义域内部几乎是处处可微的, 在有效定义域内部任何一点的邻域都存在可微点.

6.4 共轭函数次微分

如果正常凸函数的次微分总是存在, 那么它对应的共轭函数的次微分也存在. 正常凸函数与其共轭函数之间存在对偶关系, 并且与它们的次微分之间也存在对偶关系.

定理 6.4.1 设 $f(\boldsymbol{x})$ 在 $\mathbf{R}^n$ 上是正常凸函数, 则下面结论等价:

(1) $\boldsymbol{x}^* \in \partial f(\boldsymbol{x})$;

(2) $\langle \boldsymbol{x}^*, \boldsymbol{x}\rangle - f(\boldsymbol{x}) = \sup\{\langle \boldsymbol{x}^*, \boldsymbol{z}\rangle - f(\boldsymbol{z}) \,|\boldsymbol{z} \in \mathbf{R}^n\}$;

(3) $f(\boldsymbol{x}) + f^*(\boldsymbol{x}^*) = \langle \boldsymbol{x}, \boldsymbol{x}^*\rangle$;

(4) $f(\boldsymbol{x}) + f^*(\boldsymbol{x}^*) \leqslant \langle \boldsymbol{x}, \boldsymbol{x}^*\rangle$.

证明 (1)⇒(2): 若 $\boldsymbol{x}^* \in \partial f(\boldsymbol{x})$, 根据定义 6.2.2, 对于任意的 $\boldsymbol{z} \in \mathbf{R}^n$ 有

$$\langle \boldsymbol{x}^*, \boldsymbol{x}\rangle - f(\boldsymbol{x}) \geqslant \langle \boldsymbol{x}^*, \boldsymbol{z}\rangle - f(\boldsymbol{z}),$$

上式表明 (2) 成立.

(2)⇒(3): 若 (2) 成立, 由共轭函数定义 (定义 4.4.1) 知 (3) 成立.

(3)⇒(4): 若 (3) 成立, (4) 显然成立.

(4)⇒(1): 若 (4) 成立, 由共轭函数定义 (定义 4.4.1) 知

$$\sup\{\langle \boldsymbol{x}^*, \boldsymbol{z}\rangle - f(\boldsymbol{z}) \,|\boldsymbol{z} \in \mathbf{R}^n\} = f^*(\boldsymbol{x}^*) \leqslant \langle \boldsymbol{x}, \boldsymbol{x}^*\rangle - f(\boldsymbol{x}).$$

上式为

$$\langle \boldsymbol{x}^*, \boldsymbol{x}\rangle - f(\boldsymbol{x}) \geqslant \langle \boldsymbol{x}^*, \boldsymbol{z}\rangle - f(\boldsymbol{z}), \quad \forall \boldsymbol{z} \in \mathbf{R}^n,$$

表明 $\boldsymbol{x}^* \in \partial f(\boldsymbol{x})$. □

下面的定理说明正常凸函数的次微分支撑函数等价于它的共轭函数. 如果正常凸函数是闭凸的, 其共轭函数的次微分是原支撑凸函数的支撑.

定理 6.4.2　设 f 在 $\mathbf{R}^n$ 上是正常凸函数. 如果 $f(\boldsymbol{x}) = \mathrm{cl}f(\boldsymbol{x})$, 则下面结论等价:

(1) $\boldsymbol{x} \in \partial f^*(\boldsymbol{x}^*)$;

(2) $\langle \boldsymbol{x}^*, \boldsymbol{x} \rangle - f^*(\boldsymbol{x}^*) = \sup\{\langle \boldsymbol{z}^*, \boldsymbol{x} \rangle - f(\boldsymbol{z}^*) \,|\boldsymbol{z}^* \in \mathbf{R}^n\}$;

(3) $\boldsymbol{x}^* \in \partial(\mathrm{cl}f(\boldsymbol{x}))$.

证明　利用性质 4.4.1、性质 4.4.2 和定理 6.4.1, 有

$$f^{**}(\boldsymbol{x}) + f^*(\boldsymbol{x}^*) = \langle \boldsymbol{x}, \boldsymbol{x}^* \rangle,$$

当 $f(\boldsymbol{x}) = \mathrm{cl}f(\boldsymbol{x}) = f^{**}(\boldsymbol{x})$ 时, 根据定理 6.4.1 知结论均成立. □

下面的推论说明正常闭凸函数与其共轭函数的次微分是等价关系.

推论 6.4.1　设 $f(\boldsymbol{x})$ 是 $\mathbf{R}^n$ 上的闭正常凸函数, 则 $\boldsymbol{x} \in \partial f^*(\boldsymbol{x}^*)$ 的等价条件是 $\boldsymbol{x}^* \in \partial f(\boldsymbol{x})$.

证明　由定理 6.4.1 和定理 6.4.2 可知结论成立. □

定理 6.4.3　设 f 在 $\mathbf{R}^n$ 上是正常次可微凸函数, 则

$$\mathrm{cl}f(\boldsymbol{x}) = f(\boldsymbol{x}), \partial(\mathrm{cl}f(\boldsymbol{x})) = \partial f(\boldsymbol{x}).$$

证明　根据定义 4.4.3 有 $f(\boldsymbol{x}) \geqslant \mathrm{cl}f(\boldsymbol{x})$, 再由性质 4.4.1 和性质 4.4.2 的注, 得

$$f(\boldsymbol{x}) \geqslant \mathrm{cl}f(\boldsymbol{x}) = f^{**}(\boldsymbol{x}) \geqslant \langle \boldsymbol{x}, \boldsymbol{x}^* \rangle - f^*(\boldsymbol{x}^*).$$

如果 $f(\boldsymbol{x})$ 在 $\boldsymbol{x}$ 处次可微, 则至少存在一个 $\boldsymbol{x}^*$ 使得 $\langle \boldsymbol{x}, \boldsymbol{x}^* \rangle - f^*(\boldsymbol{x}^*) = f(\boldsymbol{x})$, 故 $\mathrm{cl}f(\boldsymbol{x}) = f(\boldsymbol{x})$. 再由定理 6.4.1 和定理 6.4.2 知 $\partial(\mathrm{cl}f(\boldsymbol{x})) = \partial f(\boldsymbol{x})$. □

例 6.4.1　设 $f(\boldsymbol{x}) = \dfrac{1}{2}||\boldsymbol{x}||^2\ (\boldsymbol{x} \in \mathbf{R}^n)$, 求 $\partial f(\boldsymbol{x})$. 易知 f 的共轭函数是

$$f^*(\boldsymbol{x}^*) = \frac{1}{2}||\boldsymbol{x}^*||^2, \quad \forall \boldsymbol{x}^* \in \mathbf{R}^n.$$

由定理 6.4.1 知 $\boldsymbol{x}^* \in \partial f(\boldsymbol{x})$ 与 $\boldsymbol{x} \in \partial f^*(x^*)$ 的等价条件是

$$\langle \boldsymbol{x}, \boldsymbol{x}^* \rangle = \frac{1}{2}||\boldsymbol{x}||^2 + \frac{1}{2}||\boldsymbol{x}^*||^2,$$

当且仅当 $\boldsymbol{x} = \boldsymbol{x}^*$ 时, 有 $\partial f(\boldsymbol{x}) = \boldsymbol{x}$. 事实上, f 在 $\mathbf{R}^n$ 上是可微的, 可得 $\partial f(\boldsymbol{x}) = \nabla f(\boldsymbol{x})$.

例 6.4.2 设 $f(\boldsymbol{x}) = ||\boldsymbol{x}||, \boldsymbol{x} \in \mathbf{R}^n$, 求 $\partial f(\boldsymbol{x})$. 因为 $f(\boldsymbol{x}) = \delta_B(\boldsymbol{x})$, B 是单位球, 则

$$f^*(\boldsymbol{x}^*) = \delta_B^*(\boldsymbol{x}^*) = \begin{cases} 0, & \boldsymbol{x}^* \in B, \\ +\infty, & \boldsymbol{x}^* \notin B. \end{cases}$$

由定理 6.4.1 知 $\boldsymbol{x}^* \in \partial f(\boldsymbol{x})$ 与 $\boldsymbol{x} \in \partial f^*(\boldsymbol{x}^*)$ 的等价条件是

$$||\boldsymbol{x}^*|| \leqslant 1, \quad \langle \boldsymbol{x}, \boldsymbol{x}^* \rangle = ||\boldsymbol{x}||,$$

即有 $\partial f(\boldsymbol{x}) = \boldsymbol{x}/||\boldsymbol{x}||$. 当 $\boldsymbol{x} = \mathbf{0}$ 时, $\partial f(\mathbf{0}) = B$, 其中 B 是单位球.

性质 6.4.1 设 $C \subset \mathbf{R}^n$ 是闭凸集, $f(\boldsymbol{x}) = \delta_C(\boldsymbol{x})$, 则

$$\partial f(\boldsymbol{x}) = \{\boldsymbol{x}^* \,|\boldsymbol{x}^* \in C, \langle \boldsymbol{x}, \boldsymbol{x}^* \rangle = f(\boldsymbol{x})\},$$

特别地, $\partial f(\mathbf{0}) = C$.

证明 设 $\boldsymbol{x}^* \in \{\boldsymbol{x}^* \,|\boldsymbol{x}^* \in C, \langle \boldsymbol{x}, \boldsymbol{x}^* \rangle = f(\boldsymbol{x})\}$, 对任意的 $\boldsymbol{z} \in \mathbf{R}^n$, 有

$$f(\boldsymbol{z}) - f(\boldsymbol{x}) \geqslant \langle \boldsymbol{z}, \boldsymbol{x}^* \rangle - \langle \boldsymbol{x}, \boldsymbol{x}^* \rangle = \langle \boldsymbol{z} - \boldsymbol{x}, \boldsymbol{x}^* \rangle,$$

由次微分定义得 $\boldsymbol{x}^* \in \partial f(\boldsymbol{x})$, 有 $\partial f(\boldsymbol{x}) \supset \{\boldsymbol{x}^* \,|\boldsymbol{x}^* \in C, \langle \boldsymbol{x}, \boldsymbol{x}^* \rangle = f(\boldsymbol{x})\}$.

反之, 设 $\boldsymbol{x}^* \in \partial f(\boldsymbol{x})$, 则由次微分定义, 对任意的 $\boldsymbol{z} \in \mathbf{R}^n$, 有

$$f(\boldsymbol{z}) - f(\boldsymbol{x}) \geqslant \langle \boldsymbol{z} - \boldsymbol{x}, \boldsymbol{x}^* \rangle. \tag{6.4.1-1}$$

由 $f(\boldsymbol{x}) = \delta_C(\boldsymbol{x})$ 得,

$$\begin{aligned} &\sup\{\langle \boldsymbol{z} - \boldsymbol{x}, \boldsymbol{x}' \rangle \,|\boldsymbol{x}' \in C\} + \sup\{\langle \boldsymbol{x}, \boldsymbol{x}' \rangle \,|\boldsymbol{x}' \in C\} \\ \geqslant &\langle \boldsymbol{z} - \boldsymbol{x}, \boldsymbol{x}' \rangle + \langle \boldsymbol{x}, \boldsymbol{x}' \rangle, \quad \forall \boldsymbol{x}' \in C. \end{aligned}$$

由上式即有 $f(\boldsymbol{z} - \boldsymbol{x}) + f(\boldsymbol{x}) \geqslant f(\boldsymbol{z})$, 结合 (6.4.4-1) 式得

$$\langle \boldsymbol{z} - \boldsymbol{x}, \boldsymbol{x}' \rangle \geqslant \langle \boldsymbol{z} - \boldsymbol{x}, \boldsymbol{x}^* \rangle, \quad \forall \boldsymbol{x}' \in C. \tag{6.4.1-2}$$

假设 $\boldsymbol{x}^* \notin C$, 那么 $\boldsymbol{x}^*$ 与 C 可以强分离, 由强分离定理得, 存在一个非零向量 $\boldsymbol{x}_0$, 满足

$$\langle \boldsymbol{x}_0, \boldsymbol{x} \rangle < \langle \boldsymbol{x}_0, \boldsymbol{x}^* \rangle, \quad \forall \boldsymbol{x} \in C, \tag{6.4.1-3}$$

取 $\boldsymbol{z} = \boldsymbol{x} + \boldsymbol{x}_0$, 得 (6.4.1-2) 与 (6.4.1-4) 式矛盾, 所以 $\boldsymbol{x}^* \in C$.

再证 $\langle \boldsymbol{x}, \boldsymbol{x}^* \rangle = f(\boldsymbol{x})$. 在 (6.4.1-1) 式中, 令 $\boldsymbol{z} = \mathbf{0}$, 得 $\langle \boldsymbol{x}, \boldsymbol{x}^* \rangle \geqslant f(\boldsymbol{x})$. 而 $\boldsymbol{x}^* \in C$, 又有 $\langle \boldsymbol{x}, \boldsymbol{x}^* \rangle \leqslant f(\boldsymbol{x})$, 所以 $\langle \boldsymbol{x}, \boldsymbol{x}^* \rangle = f(\boldsymbol{x})$, 则 $\partial f(\boldsymbol{x}) = \{\boldsymbol{x}^* | \boldsymbol{x}^* \in C, \langle \boldsymbol{x}, \boldsymbol{x}^* \rangle = f(\boldsymbol{x})\}$. 特别地, 令 $\boldsymbol{x} = \mathbf{0}$, 则对于任意 $\boldsymbol{x}^* \in C$, 有 $f(\mathbf{0}) = \langle \mathbf{0}, \boldsymbol{x}^* \rangle = 0$, 所以 $\partial f(\mathbf{0}) = C$. □

性质 6.4.2　设 f 是 $\mathbf{R}^n$ 上的正齐次闭凸函数, 则

$$\partial f(\boldsymbol{x}) = \{\boldsymbol{x}^* \in \mathrm{dom} f^* \,|\langle \boldsymbol{x}, \boldsymbol{x}^* \rangle = f(\boldsymbol{x})\}.$$

证明　由推论 4.5.4 得, $f(\boldsymbol{x}) = \delta_{\mathrm{dom} f^*}(\boldsymbol{x})$, 再由性质 6.4.1 直接得结论成立. □

本节得到了正常凸函数次微分与其共轭函数次微分之间的等价性, 即由定理 6.4.1 可知通过共轭函数去获得其次微分, 又即闭正常凸函数的次微分等价于它的共轭函数的次微分.

6.5　凸函数中值定理

中值定理是数学分析中的一个非常重要的结论, 本书也应用该定理对广义凸函数的性质进行了证明, 接下来将证明凸函数的一个中值定理. 在此之前, 先证明下面的性质.

性质 6.5.1　设 f 在 $\mathbf{R}^n$ 上是正常凸函数, 则下面 3 个结论等价:

(1) f 在 $\mathbf{R}^n$ 上有全局极小点 $\boldsymbol{x}^*$, 即 $f(\boldsymbol{x}) \geqslant f(\boldsymbol{x}^*), \forall \boldsymbol{x} \in \mathbf{R}^n$;

(2) $\mathbf{0} \in \partial f(\boldsymbol{x}^*)$;

(3) $f'_+(\boldsymbol{x}^*, \boldsymbol{d}) \geqslant 0, \forall \boldsymbol{x} \in \mathbf{R}^n$.

证明　由定义 6.2.2 知 (1) 和 (2) 是等价的, 再由定理 6.2.3 知 (2) 和 (3) 等价. □

性质 6.5.2　设 f 在 $\mathbf{R}^n$ 上是正常闭凸函数, $\boldsymbol{x}, \boldsymbol{y} \in \mathbf{R}^n$, $\forall t \in [0,1]$ 的限制函数:

$$\varphi(t) = f(\boldsymbol{x} + t(\boldsymbol{y} - \boldsymbol{x})),$$

则有 $\partial\varphi(t) = \{\langle \boldsymbol{u}, \boldsymbol{y} - \boldsymbol{x} \rangle \,|\forall \boldsymbol{u} \in \partial f(\boldsymbol{x}_t)\}$, 其中 $\boldsymbol{x}_t = \boldsymbol{x} + t(\boldsymbol{y} - \boldsymbol{x})$.

证明　分别由 $\varphi(t)$ 的正负导数定义得

$$\varphi'_+(t) = \lim_{\lambda \downarrow 0} \frac{f(\boldsymbol{x}_t + \lambda(\boldsymbol{y} - \boldsymbol{x})) - f(\boldsymbol{x}_t)}{\lambda} = f'_+(\boldsymbol{x}_t, \boldsymbol{y} - \boldsymbol{x}),$$

$$\varphi'_-(t) = \lim_{\lambda \uparrow 0} \frac{f(\boldsymbol{x}_t + \lambda(\boldsymbol{y} - \boldsymbol{x})) - f(\boldsymbol{x}_t)}{\lambda} = f'_-(\boldsymbol{x}_t, \boldsymbol{y} - \boldsymbol{x}).$$

再由性质 6.2.3、性质 6.3.2 得

$$f'_+(\boldsymbol{x}_t, \boldsymbol{y} - \boldsymbol{x}) = \max\{\langle \boldsymbol{u}, \boldsymbol{y} - \boldsymbol{x} \rangle \,|\forall \boldsymbol{u} \in \partial f(\boldsymbol{x}_t)\},$$

$$-f'_+(\boldsymbol{x}_t, \boldsymbol{y} - \boldsymbol{x}) = \min\{\langle \boldsymbol{u}, \boldsymbol{y} - \boldsymbol{x} \rangle \,|\forall \boldsymbol{u} \in \partial f(\boldsymbol{x}_t)\}.$$

由上面两式得 $\partial\varphi(t)=\left[\varphi'_-(t),\varphi'_+(t)\right]$. □

下面是凸函数的一个中值定理.

定理 6.5.1(中值定理) 设 f 在 $\mathbf{R}^n$ 上是正常闭凸函数, 两个不同的点 $\boldsymbol{x},\boldsymbol{y}\in\mathbf{R}^n$, 则存在一点 $\tau\in(0,1)$ 和 $\boldsymbol{u}\in\partial f(\boldsymbol{x}_\tau)$ 满足

$$f(\boldsymbol{y})-f(\boldsymbol{x})=\langle\boldsymbol{u},\boldsymbol{y}-\boldsymbol{x}\rangle,$$

其中 $\boldsymbol{x}_\tau=\boldsymbol{x}+\tau(\boldsymbol{y}-\boldsymbol{x})$.

证明 设 $\boldsymbol{x},\boldsymbol{y}\in\mathbf{R}^n$, $\forall t\in[0,1]$ 的限制函数 $\varphi(t)=f(\boldsymbol{x}+t(\boldsymbol{y}-\boldsymbol{x}))$, 再设函数:

$$\phi(t)=\varphi(t)-\varphi(0)-t(\varphi(1)-\varphi(0)).$$

则 $\phi(t)$ 在 $[0,1]$ 上是连续凸函数, 由于 $\phi(0)=\phi(1)=0$, 那么 $\phi(t)$ 在 $(0,1)$ 上达到一个最小点 τ, 由性质 6.5.1 得 $0\in\partial\phi(\tau)$. 再由性质 6.5.2, 存在 $\boldsymbol{u}\in\partial f(\boldsymbol{x}_\tau)$ 使得 $\partial\varphi(t)=\varphi(1)-\varphi(0)$, 即有

$$f(\boldsymbol{y})-f(\boldsymbol{x})=\langle\boldsymbol{u},\boldsymbol{y}-\boldsymbol{x}\rangle. \quad\square$$

显然, 定理 6.5.1 中的 f 如果是可微函数, 则结论即为可微函数的中值定理. 我们有下面推论.

推论 6.5.1 设 f 在 $\mathbf{R}^n$ 上是正常闭凸函数和两个不同的点 $\boldsymbol{x},\boldsymbol{y}\in\mathbf{R}^n$, 则

$$f(\boldsymbol{y})-f(\boldsymbol{x})=\int_0^1\langle\boldsymbol{u}_\tau,\boldsymbol{y}-\boldsymbol{x}\rangle\,\mathrm{d}\tau,$$

其中 $\tau\in(0,1),\boldsymbol{u}_\tau\in\partial f(\boldsymbol{x}_\tau)$, $\boldsymbol{x}_\tau=\boldsymbol{x}+\tau(\boldsymbol{y}-\boldsymbol{x})$.

本小节得到正常闭凸函数的一个中值定理, 显然它是拉格朗日中值定理的推广.

6.6 若干运算性质

6.6.1 次可加性

正常凸函数的次微分总是存在的, 而且在有效定义域内部几乎处处可微. 因此, 正常凸函数的次微分在运算性质方面与可微函数有很大的相似性. 例如, 两个函数之和的导数等于它们分别求导后的和. 但是次微分的和可能只存在包含关系, 即两个正常凸函数相加的次微分包含它们各自次微分的和.

定理 6.6.1 设 f_1 和 f_2 在 $\mathbf{R}^n$ 上是正常凸函数, 下面结论成立:

(1) $\partial f_1(\boldsymbol{x})+\partial f_2(\boldsymbol{x})\subset\partial(f_1(\boldsymbol{x})+f_2(\boldsymbol{x}))$;

(2) 如果存在 $\boldsymbol{x}_1 \in \mathrm{dom}f_1 \cap \mathrm{dom}f_2$ 使得 f_1 在 $\boldsymbol{x}_1$ 处连续, 则对 $\boldsymbol{x} \in \mathrm{dom}f_1 \cap \mathrm{dom}f_2$ 有

$$\partial\left(f_1(\boldsymbol{x})+f_2(\boldsymbol{x})\right)=\partial f_1(\boldsymbol{x})+\partial f_2(\boldsymbol{x}). \tag{6.6.1-1}$$

证明 (1) 设 $f(\boldsymbol{z})=f_1(\boldsymbol{z})+f_2(\boldsymbol{z})\,(\forall \boldsymbol{z} \in \mathbf{R}^n)$, $\boldsymbol{x}_1^* \in \partial f_1(\boldsymbol{x})$ 和 $\boldsymbol{x}_2^* \in \partial f_2(\boldsymbol{x})$. 由次微分的定义, 对任意的 $\boldsymbol{z} \in \mathbf{R}^n$, 有

$$\begin{aligned} f(\boldsymbol{z}) &= f_1(\boldsymbol{z})+f_2(\boldsymbol{z}) \\ &\geqslant f_1(\boldsymbol{x})+\langle \boldsymbol{z}-\boldsymbol{x}, \boldsymbol{x}_1^*\rangle+f_2(\boldsymbol{x})+\langle \boldsymbol{z}-\boldsymbol{x}, \boldsymbol{x}_2^*\rangle \\ &= f(\boldsymbol{x})+\langle \boldsymbol{z}-\boldsymbol{x}, \boldsymbol{x}_1^*+\boldsymbol{x}_2^*\rangle . \end{aligned}$$

上式说明 $\boldsymbol{x}_1^*+\boldsymbol{x}_2^* \in \partial f(\boldsymbol{x})$, 即 $\partial f_1(\boldsymbol{x})+\partial f_2(\boldsymbol{x}) \subset \partial f(\boldsymbol{x})$.

(2) 只需证明 $\partial f(\boldsymbol{x}) \subset \partial f_1(\boldsymbol{x})+\partial f_2(\boldsymbol{x})$. 先证 $\boldsymbol{x}=\mathbf{0} \in \mathrm{dom}f_1 \cap \mathrm{dom}f_2$ 时成立.

设 $\boldsymbol{x}^* \in \partial f(\mathbf{0})$, 其中 $f(\mathbf{0})=f_1(\mathbf{0})+f_2(\mathbf{0})$, 由次微分的定义, 对任意的 $\boldsymbol{x} \in \mathbf{R}^n$, 有

$$f_1(\boldsymbol{x})+f_2(\boldsymbol{x}) \geqslant f_1(\mathbf{0})+f_2(\mathbf{0})+\langle \boldsymbol{x}, \boldsymbol{x}^*\rangle, \tag{6.6.1-2}$$

即

$$f_1(\boldsymbol{x})-\langle \boldsymbol{x}, \boldsymbol{x}^*\rangle \geqslant f(\mathbf{0})-f_2(\boldsymbol{x}). \tag{6.6.1-3}$$

在 $\mathbf{R}^{n+1}$ 中设集合

$$A_1=\{(\boldsymbol{x}, \alpha)\,|\,\alpha>f_1(\boldsymbol{x})-\langle \boldsymbol{x}, \boldsymbol{x}^*\rangle\}, \quad A_2=\{(\boldsymbol{y}, \beta)\,|\,\beta<f(\mathbf{0})-f_2(\boldsymbol{y})\}.$$

因为 f_1 和 f_2 是凸函数, 所以 A_1 和 A_2 是两个不相交的凸集. 根据凸集分离定理, 存在非零向量 $(\boldsymbol{x}_0^*, \alpha^*)$, 满足

$$\langle \boldsymbol{y}, \boldsymbol{x}_0^*\rangle+\beta\alpha^* \leqslant \langle \boldsymbol{x}, \boldsymbol{x}_0^*\rangle+\alpha\alpha^*, \quad (\boldsymbol{x}, \alpha) \in A_1, \quad (\boldsymbol{y}, \beta) \in A_2. \tag{6.6.1-4}$$

因为 α 可以无限增大, 由 (6.6.1-4) 式知 $\alpha^* \geqslant 0$.

假设 $\alpha^*=0$, 则 (6.6.1-4) 式变成

$$\langle \boldsymbol{y}, \boldsymbol{x}_0^*\rangle \leqslant \langle \boldsymbol{x}, \boldsymbol{x}_0^*\rangle, \quad \boldsymbol{x} \in \mathrm{dom}f_1, \quad \boldsymbol{y} \in \mathrm{dom}f_2. \tag{6.6.1-5}$$

设 $\boldsymbol{y}=\boldsymbol{x}_1$, 因为 $\boldsymbol{x}_1 \in \mathrm{dom}f_1$ 和 $f_1(\boldsymbol{x})$ 在 $\boldsymbol{x}_1$ 连续, 则当 $\varepsilon>0$ 充分小时, 有 $\boldsymbol{x}_1+\varepsilon\boldsymbol{z} \in \mathrm{dom}f_1$ 和 $\boldsymbol{z} \in B$, 其中 B 是单位球. 在 (6.6.1-5) 式中代入 $\boldsymbol{y}=\boldsymbol{x}_1$, $\boldsymbol{x}=\boldsymbol{x}_1+\varepsilon\boldsymbol{z}$, 得

$$0 \leqslant \varepsilon\langle \boldsymbol{z}, \boldsymbol{x}_0^*\rangle, \quad \boldsymbol{z} \in B,$$

由上式得 $\boldsymbol{x}_0^* = \mathbf{0}$, 这与 $(\boldsymbol{x}_0^*, \alpha^*)$ 是非零向量矛盾.

不妨设 $\alpha^* = 1$. 根据 A_1 和 A_2 的定义, 由 (6.6.1-4) 式得

$$\begin{aligned} & f(\mathbf{0}) - f_2(\boldsymbol{y}) + \langle \boldsymbol{y}, \boldsymbol{x}_0^* \rangle \\ \leqslant & f_1(\boldsymbol{x}) - \langle \boldsymbol{x}, \boldsymbol{x}^* \rangle + \langle \boldsymbol{x}, \boldsymbol{x}_0^* \rangle, \quad \boldsymbol{x} \in \mathrm{dom} f_1, \quad \boldsymbol{y} \in \mathrm{dom} f_2. \end{aligned} \tag{6.6.1-6}$$

因为在 $\mathrm{dom} f_1$ 和 $\mathrm{dom} f_2$ 之外, f_1, f_2 均是 $+\infty$, 故对于 $\boldsymbol{x}, \boldsymbol{y} \in \mathbf{R}^n$, (6.6.1-6) 式仍成立. 在 (6.6.1-6) 式中, 设 $\boldsymbol{y} = \mathbf{0}$, 有

$$\langle \boldsymbol{x}, \boldsymbol{x}^* - \boldsymbol{x}_0^* \rangle \leqslant f_1(\boldsymbol{x}) - f_1(\mathbf{0}), \quad \forall \boldsymbol{x} \in \mathbf{R}^n.$$

由上式知 $\boldsymbol{x}^* - \boldsymbol{x}_0^* \in \partial f_1(\mathbf{0})$, 设 $\boldsymbol{x} = \mathbf{0}$, 由 (6.6.1-6) 式, 有

$$\langle \boldsymbol{y}, \boldsymbol{x}_0^* \rangle \leqslant f_2(\boldsymbol{y}) - f_2(\mathbf{0}), \quad \forall \boldsymbol{y} \in \mathbf{R}^n.$$

上式表明 $\boldsymbol{x}_0^* \in \partial f_2(\mathbf{0})$, 则

$$\boldsymbol{x}^* = (\boldsymbol{x}^* - \boldsymbol{x}_0^*) + \boldsymbol{x}_0^* \in \partial f_1(\mathbf{0}) + \partial f_2(\mathbf{0}),$$

即有

$$\partial f(\mathbf{0}) \subset \partial f_1(\mathbf{0}) + \partial f_2(\mathbf{0}).$$

最后, 固定 $\boldsymbol{x}$, 设 $h_1(\boldsymbol{y}) = f_1(\boldsymbol{y} + \boldsymbol{x})$, $h_2(\boldsymbol{y}) = f_2(\boldsymbol{y} + \boldsymbol{x})$, 有 $\partial f_1(\boldsymbol{x}) = \partial h_1(\mathbf{0})$ 和 $\partial f_2(\boldsymbol{x}) = \partial h_2(\mathbf{0})$. 由 $\partial h(\mathbf{0}) = \partial h_1(\mathbf{0}) + \partial h_2(\mathbf{0})$ 得 (6.6.1-1) 式成立.□

例 6.6.1 设函数 $f_1(x) = \begin{cases} -x, & x < 0, \\ 0, & x \geqslant 0, \end{cases}$ $f_2(x) = \begin{cases} 0, & x < 0, \\ x, & x \geqslant 0. \end{cases}$ 计算得

$$f_{1+}'(0) = \lim_{t \downarrow 0} \frac{f(t) - f(0)}{t} = 0, \quad f_{1-}'(0) = \lim_{t \uparrow 0} \frac{f(t) - f(0)}{t} = -1,$$

$$f_{2+}'(0) = \lim_{t \downarrow 0} \frac{f(t) - f(0)}{t} = 1, \quad f_{2-}'(0) = \lim_{t \uparrow 0} \frac{f(t) - f(0)}{t} = 0.$$

根据上述式子可得 $\partial f_1(0) = [-1, 0]$, $\partial f_2(0) = [0, 1]$. 通过

$$f_1(x) + f_2(x) = \begin{cases} -x, & x < 0, \\ x, & x \geqslant 0 \end{cases}$$

有

$$\partial [f_1(0) + f_2(0)] = [-1, 1],$$

因此, $\partial(f_1(0)+f_2(0))=\partial f_1(0)+\partial f_2(0)$.

进一步, 有另一个结论: 当两个正常凸函数有效定义域的相对内部相交非空时, 它们和的次微分等于各自函数的次微分之和.

定理 6.6.2 设 f_1 和 f_2 是 $\mathbf{R}^n$ 上的正常凸函数, 且 $\mathrm{ri}(\mathrm{dom} f_1)\cap\mathrm{ri}(\mathrm{dom} f_2)\neq\varnothing$, 则

$$\partial(f_1(\boldsymbol{x})+f_2(\boldsymbol{x}))=\partial f_1(\boldsymbol{x})+\partial f_2(\boldsymbol{x}),\quad \forall \boldsymbol{x}\in\mathbf{R}^n.$$

证明 设 $f(\boldsymbol{x})=f_1(\boldsymbol{x})+f_2(\boldsymbol{x})$, 由定理 6.6.1(1) 知, 只要证明

$$\partial f(\boldsymbol{x})\subset\partial f_1(\boldsymbol{x})+\partial f_2(\boldsymbol{x}),\quad \forall \boldsymbol{x}\in\mathbf{R}^n$$

即可. 由定理 4.4.3 知 $f^*(\boldsymbol{x}^*)$ 可以表示为

$$f^*(\boldsymbol{x}^*)=\inf\{f_1^*(\boldsymbol{x}_1^*)+f_2^*(\boldsymbol{x}_2^*)\,|\,\boldsymbol{x}^*=\boldsymbol{x}_1^*+\boldsymbol{x}_2^*\}. \tag{6.6.1-7}$$

设 $\boldsymbol{x}^*\in\partial f(\boldsymbol{x})$, 由定理 6.4.1 知 $\boldsymbol{x}^*\in\partial f(\boldsymbol{x})$ 的充要条件是

$$\langle\boldsymbol{x},\boldsymbol{x}^*\rangle=f_1(\boldsymbol{x})+f_2(\boldsymbol{x})+f^*(\boldsymbol{x}^*). \tag{6.6.1-8}$$

在 (6.6.1-8) 式中, 对每一个 $\boldsymbol{x}^*$, 有某些 $\boldsymbol{x}_1^*$, $\boldsymbol{x}_2^*$ 达到下确界, 由 (6.6.1-7) 式得

$$\langle\boldsymbol{x},\boldsymbol{x}_1^*\rangle+\langle\boldsymbol{x},\boldsymbol{x}_2^*\rangle=f_1(\boldsymbol{x})+f_2(\boldsymbol{x})+f_1^*(\boldsymbol{x}_1^*)+f_2^*(\boldsymbol{x}_2^*). \tag{6.6.1-9}$$

由定理 6.4.1 和 Fenchel 不等式, 总有

$$\langle\boldsymbol{x},\boldsymbol{x}_1^*\rangle\leqslant f_1(\boldsymbol{x})+f_1^*(\boldsymbol{x}_1^*),\quad \langle\boldsymbol{x},\boldsymbol{x}_2^*\rangle\leqslant f_2(\boldsymbol{x})+f_2^*(\boldsymbol{x}_2^*). \tag{6.6.1-10}$$

再由定理 6.4.1 知 $\boldsymbol{x}_1^*\in\partial f_1(\boldsymbol{x})$ 和 $\boldsymbol{x}_2^*\in\partial f_2(\boldsymbol{x})$. 故由 (6.6.1-9) 和 (6.6.1-10) 式, 有

$$\partial f(\boldsymbol{x})\subset\partial f_1(\boldsymbol{x})+\partial f_2(\boldsymbol{x}).$$ □

推论 6.6.1 设 K_1, K_2 是 $\mathbf{R}^n$ 中包含原点在内的凸锥, $\mathrm{ri}K_1\cap\mathrm{ri}K_2\neq\varnothing$, 则

$$N_{\mathbf{0}}(K_1\cap K_2)=N_{\mathbf{0}}(K_1)+N_{\mathbf{0}}(K_2).$$

证明 因为 $\delta_{K_1\cap K_2}(\boldsymbol{x})=\delta_{K_1}(\boldsymbol{x})+\delta_{K_2}(\boldsymbol{x})$, $\mathrm{dom}\delta_{K_i}(\boldsymbol{x})=K_i, i=1,2$, 有

$$\mathrm{ri}(\mathrm{dom}\delta_{K_1}(\boldsymbol{x}))\cap\mathrm{ri}(\mathrm{dom}\delta_{K_2}(\boldsymbol{x}))\neq\varnothing.$$

由例 6.2.3 得 $\partial\delta_{K_i}(\mathbf{0})=-N_{\mathbf{0}}(K_i)\,(i=1,2)$, 由定理 6.6.2 得

$$-N_{\mathbf{0}}(K_1\cap K_2)=\partial\delta_{K_1\cap K_2}(\mathbf{0})=\partial\delta_{K_1}(\mathbf{0})+\partial\delta_{K_2}(\mathbf{0})=-N_{\mathbf{0}}(K_1)-N_{\mathbf{0}}(K_2).$$ □

推论 6.6.2 设 $C_1, C_2 \subset \mathbf{R}^n$ 是非空凸集, $\mathrm{ri}C_1 \cap \mathrm{ri}C_2 \neq \varnothing$, $N_{\boldsymbol{x}}(C_1)$, $N_{\boldsymbol{x}}(C_2)$ 分别为 C_1, C_2 在 $\boldsymbol{x}$ 处的法锥, 则 $N_{\boldsymbol{x}}(C_1) + N_{\boldsymbol{x}}(C_2)$ 是 $C_1 \cap C_2$ 在 $\boldsymbol{x}$ 处的法锥.

证明过程完全类似于推论 6.6.1.

本小节得到两个正常凸函数的次微分的次可加性结论:

1) 两个正常凸函数的次微分的和被和的次微分包含, 仅在它们的有效定义域内部相等;

2) 当两个正常凸函数有效定义域的相对内部相交是非空时, 它们和的次微分等于各自函数的次微分之和.

6.6.2 复合凸函数次微分

本小节研究凸函数与线性变换或仿射映射的复合凸函数次微分的运算性质.

定理 6.6.3 设 $g(\boldsymbol{y})$ 在 $\mathbf{R}^m$ 上是正常凸函数, 线性变换 $A: \mathbf{R}^n \longrightarrow \mathbf{R}^m$, $f(\boldsymbol{x}) = g(A\boldsymbol{x})$, 有下面结论成立:

(1) $A^*\partial g(\boldsymbol{y}) \subset \partial f(\boldsymbol{x})$;

(2) 如果 $\mathrm{rang}A \cap \mathrm{ri}(\mathrm{dom}g) \neq \varnothing$, 则有

$$\partial f(\boldsymbol{x}) = A^*\partial g(\boldsymbol{y}),$$

其中: A^* 是 A 的伴随映射 (见定义 4.4.2), $\mathrm{rang}A$ 是 A 的值域, $\boldsymbol{y} = A\boldsymbol{x}$.

证明 (1) 设 $\boldsymbol{x}^* \in A^*\partial(gA)(\boldsymbol{x})$, 则存在一个 $\boldsymbol{y}^* \in \partial(gA)(\boldsymbol{x})$ 使得 $\boldsymbol{x}^* = A^*\boldsymbol{y}^*$. 所以根据次微分定义, 对于任意的 $\boldsymbol{z} \in \mathbf{R}^n$ 有

$$f(\boldsymbol{z}) = g(A\boldsymbol{z}) \geqslant g(A\boldsymbol{x}) + \langle \boldsymbol{y}^*, A\boldsymbol{z} - A\boldsymbol{x}\rangle = f(\boldsymbol{x}) + \langle \boldsymbol{x}^*, \boldsymbol{z} - \boldsymbol{x}\rangle,$$

上式说明, $\boldsymbol{x}^* \in \partial f(\boldsymbol{x})$.

(2) 由 (1), 只要证明 $\partial f(\boldsymbol{x}) \subset A^*\partial(gA)(\boldsymbol{x})$ 即可. 因为 $\mathrm{rang}A \cap \mathrm{ri}(\mathrm{dom}g) \neq \varnothing$, 则 $f(\boldsymbol{x})$ 是正常凸函数. 由定理 4.4.2, 有

$$f^*(\boldsymbol{x}^*) = \inf\{g^*(\boldsymbol{y}^*) \,|\, A^*\boldsymbol{y}^* = \boldsymbol{x}^*\}. \tag{6.6.2-1}$$

当 $f^*(\boldsymbol{x}^*) \neq +\infty$ 时, 对于每一个 $\boldsymbol{x}^*$, 使得在某一个 $\boldsymbol{y}^*$ 达到上式的下确界. 由定理 6.4.1, 有

$$f(\boldsymbol{x}) + f^*(\boldsymbol{x}^*) = \langle \boldsymbol{x}, \boldsymbol{x}^*\rangle, \tag{6.6.2-2}$$

由 (6.6.2-1) 式, 存在 $\boldsymbol{y}^*$ 满足 $A^*\boldsymbol{y}^* = \boldsymbol{x}^*$ 和 $f(\boldsymbol{x}) + g^*(\boldsymbol{y}^*) = \langle \boldsymbol{x}, A^*\boldsymbol{y}^*\rangle$, 即

$$(gA)(\boldsymbol{x}) + g^*(\boldsymbol{y}^*) = \langle A\boldsymbol{x}, \boldsymbol{y}^*\rangle. \tag{6.6.2-3}$$

由上式, 再根据定理 6.4.1 得 $\boldsymbol{y}^* \in \partial g(A\boldsymbol{x})$, $\boldsymbol{x}^* \in A^*\partial g(A\boldsymbol{x})$, 即

$$\partial f(\boldsymbol{x}) \subset A^*\partial(gA)(\boldsymbol{x}).$$

□

对于仿射映射, 我们也有类似定理 6.6.3 的结果.

定理 6.6.4 设 $A:\ \mathbf{R}^n \longrightarrow \mathbf{R}^m$ 是仿射映射 ($\boldsymbol{y} = A\boldsymbol{x} = A_0\boldsymbol{x} + \boldsymbol{b}, \boldsymbol{b} \in \mathbf{R}^m$), $g(\boldsymbol{y})$ 是 $\mathbf{R}^m$ 的正常凸函数, $f(\boldsymbol{x}) = g(A\boldsymbol{x})$, 则

$$\partial f(\boldsymbol{x}) = \partial (g \circ A)(\boldsymbol{x}) = A_0^* \partial g(A\boldsymbol{x}) = A_0^* \partial g(\boldsymbol{y}), \quad \forall \boldsymbol{x} \in \mathbf{R}^n, \tag{6.6.2-4}$$

其中 A_0^* 是 A_0 的伴随映射.

例 6.6.2 设 $f(\boldsymbol{x}) = ||\boldsymbol{A}\boldsymbol{x}||$, 其中 $\boldsymbol{x} \in \mathbf{R}^n$, $\boldsymbol{A}$ 是 $n \times n$ 的可逆矩阵. 令 $\boldsymbol{y} = \boldsymbol{A}\boldsymbol{x}$ 和 $g(\boldsymbol{y}) = ||\boldsymbol{y}||$. 由例 6.4.2 得 $\partial g(\boldsymbol{y}) = \dfrac{\boldsymbol{y}}{||\boldsymbol{y}||}(\boldsymbol{y} \neq \mathbf{0})$, 以及得 $\partial f(\boldsymbol{x}) = \dfrac{\boldsymbol{A}^*\boldsymbol{A}\boldsymbol{x}}{||\boldsymbol{A}\boldsymbol{x}||}(\boldsymbol{A}\boldsymbol{x} \neq \mathbf{0})$.

本小节得到复合映射凸函数的两个结论:

1) 正常凸函数线性变换后的复合函数的次微分等于正常凸函数次微分的伴随线性变换;

2) 正常凸函数仿射变换后的复合函数的次微分等于正常凸函数次微分的伴随仿射变换.

6.6.3 凸函数族上确界函数次微分

下面讨论在紧集上的凸函数族逐点上确界函数的次微分性质.

以下假设 $Y \subset \mathbf{R}^m$ 是一个非空集合, 对于每一个 $\boldsymbol{y} \in Y$, 函数 $g(\boldsymbol{x}, \boldsymbol{y}):\ \mathbf{R}^n \times \mathbf{R}^m \to \mathbf{R}$ 是关于 $\boldsymbol{x}$ 的凸函数, 记 $f(\boldsymbol{x}) = \sup\limits_{\boldsymbol{y} \in Y} g(\boldsymbol{x}, \boldsymbol{y})$, 称 f 为 g 在集合 Y 上的凸函数族逐点上确界函数.

先证明下面引理.

引理 6.6.1 设 $Y \subset \mathbf{R}^m$ 是紧集, 对于 $\boldsymbol{x}_0 \in \mathbf{R}^n$ 和 $\boldsymbol{y}_0 \in Y$. 如果存在 $(\boldsymbol{x}_0, \boldsymbol{y}_0)$ 的一个邻域 $O(\boldsymbol{x}_0, \boldsymbol{y}_0)$ 使得函数 g 在 $O(\boldsymbol{x}_0, \boldsymbol{y}_0)$ 上是连续函数, 则 f 在 $\boldsymbol{x}_0$ 处连续.

证明 设 $Y(\boldsymbol{x}) = \{\boldsymbol{y} \in Y \,|\, g(\boldsymbol{x}, \boldsymbol{y}) = f(\boldsymbol{x})\}$, $(\boldsymbol{x}, \boldsymbol{y}) \in O(\boldsymbol{x}_0, \boldsymbol{y}_0)$, 则当 $\boldsymbol{y} \in Y(\boldsymbol{x})$, $\boldsymbol{y}_0 \in Y(\boldsymbol{x}_0)$ 时, 有

$$g(\boldsymbol{x}, \boldsymbol{y}) = f(\boldsymbol{x}) \geqslant g(\boldsymbol{x}, \boldsymbol{y}_0), \tag{6.6.3-1}$$

$$g(\boldsymbol{x}_0, \boldsymbol{y}) \leqslant f(\boldsymbol{x}_0) = g(\boldsymbol{x}_0, \boldsymbol{y}_0). \tag{6.6.3-2}$$

上面两式相减, 得

$$g(\boldsymbol{x}, \boldsymbol{y}) - g(\boldsymbol{x}_0, \boldsymbol{y}) \geqslant f(\boldsymbol{x}) - f(\boldsymbol{x}_0) \geqslant g(\boldsymbol{x}, \boldsymbol{y}_0) - g(\boldsymbol{x}_0, \boldsymbol{y}_0). \tag{6.6.3-3}$$

由 g 在 $O(\boldsymbol{x}_0, \boldsymbol{y}_0)$ 上是连续函数, 当 $\boldsymbol{x} \to \boldsymbol{x}_0$ 时, 对于固定 $\boldsymbol{y}_0$, 有 $g(\boldsymbol{x}, \boldsymbol{y}_0) - g(\boldsymbol{x}_0, \boldsymbol{y}_0) \to 0$.

另一方面, 若存在序列 $\{\boldsymbol{x}_k\}$ 使得 $\lim\limits_{k\to\infty}\boldsymbol{x}_k=\boldsymbol{x}_0$, 对应有 $\boldsymbol{y}_k\in Y(\boldsymbol{x}_k)\subset Y$. 因为 Y 是紧集, 不妨设

$$\boldsymbol{y}_k\to\boldsymbol{y}_0\in Y\quad(k\to+\infty),$$

使

$$g(\boldsymbol{x}_k,\boldsymbol{y}_k)-g(\boldsymbol{x}_0,\boldsymbol{y}_k)\to g(\boldsymbol{x}_0,\boldsymbol{y}_0)-g(\boldsymbol{x}_0,\boldsymbol{y}_0)=0\quad(k\to+\infty,\boldsymbol{x}\to\boldsymbol{x}_0).$$

综上, 当 $\boldsymbol{x}\to\boldsymbol{x}_0$ 时, (6.6.3-3) 式左、右两边均趋于 0, 所以 $f(\boldsymbol{x})\to f(\boldsymbol{x}_0)$ 在 $\boldsymbol{x}_0$ 连续. □

引理 6.6.2 假设引理 6.6.1 的条件成立, 记

$$Y(\boldsymbol{x})=\{\boldsymbol{y}\in Y\,|g(\boldsymbol{x},\boldsymbol{y})=f(\boldsymbol{x})\}.$$

设 C 是开集, 且 $Y(\boldsymbol{x}_0)\subset C\subset Y$, 则存在 $\boldsymbol{x}_0$ 的一个邻域 $O(\boldsymbol{x}_0)$ 使得

$$Y(\boldsymbol{x})\subset C,\quad\forall\boldsymbol{x}\in O(\boldsymbol{x}_0).$$

证明 假设存在序列 $\{\boldsymbol{x}_k\}$, $\lim\limits_{k\to\infty}\boldsymbol{x}_k=\boldsymbol{x}_0$, 并存在 $\boldsymbol{y}_k\in Y(\boldsymbol{x}_k)$, $\boldsymbol{y}_k\notin C$. 由定义有

$$f(\boldsymbol{x}_k)=f(\boldsymbol{x}_k,\boldsymbol{y}_k),$$

在 $\boldsymbol{x}_k\to\boldsymbol{x}_0$ 时取极限, 根据引理 6.6.1 得 $f(\boldsymbol{x}_0)=g(\boldsymbol{x}_0,\boldsymbol{y}_0)$, 即有 $\boldsymbol{y}_0\in Y(\boldsymbol{x}_0)\subset C$. 则当 k 充分大时, $\boldsymbol{y}_k$ 属于 $\boldsymbol{y}_0$ 的某个邻域. 因 C 是开集, 那么有 $\boldsymbol{y}_k\in C$, 与假设矛盾. □

为了得到 $f(\boldsymbol{x})=\sup\limits_{\boldsymbol{y}\in Y}g(\boldsymbol{x},\boldsymbol{y})$ 的次微分, 首先求出它的方向导数关系, 见下面结论.

定理 6.6.5 设 $Y\subset\mathbf{R}^m$ 是紧集, 对于每一个 $\boldsymbol{y}\in Y$, $g(\boldsymbol{x},\boldsymbol{y})$ 关于 $\boldsymbol{x}$ 在 $\mathbf{R}^n$ 上是凸函数. 对于给定的 $\boldsymbol{x}_0,\boldsymbol{d}\in\mathbf{R}^n$, 如果存在区间 $[-\delta,\delta]\,(\delta>0)$ 使得 $g(\boldsymbol{x}_0+t\boldsymbol{d},\boldsymbol{y})$ 关于 $t\in[-\delta,\delta]$, $\boldsymbol{y}\in Y$ 连续, 则 $f(\boldsymbol{x})$ 在 $\boldsymbol{x}_0$ 沿方向 $\boldsymbol{d}$ 的方向导数存在且

$$f'_+(\boldsymbol{x}_0,\boldsymbol{d})=\max_{\boldsymbol{y}}\left\{g'_+(\boldsymbol{x}_0,\boldsymbol{y},\boldsymbol{d})\,|\boldsymbol{y}\in Y(\boldsymbol{x}_0)\right\},\tag{6.6.3-4}$$

其中: $g'_+(\boldsymbol{x}_0,\boldsymbol{y},\boldsymbol{d})$ 是 $g(\boldsymbol{x},\boldsymbol{y})$ 在 $\boldsymbol{x}_0$ 沿方向 $\boldsymbol{d}$ 的方向导数.

证明 固定变量 $(\boldsymbol{x}_0,\boldsymbol{y},\boldsymbol{d})$, 对于任意的 $t\in[-\delta,\delta]\,(\delta>0)$, 设

$$\varphi(t)=f(\boldsymbol{x}_0+t\boldsymbol{d}),\quad\phi(t,\boldsymbol{y})=g(\boldsymbol{x}_0+t\boldsymbol{d},\boldsymbol{y}),\quad Y(t)=Y(\boldsymbol{x}_0+t\boldsymbol{d}).$$

显然有 $f'_+(\boldsymbol{x}_0,\boldsymbol{d})=\varphi'_+(0)\,,g'_+(\boldsymbol{x}_0,\boldsymbol{y},\boldsymbol{d})=\phi'_+(0,\boldsymbol{y})$, 其中

$$\varphi'_+(0)=\lim_{\lambda\downarrow0}\frac{\varphi(t)-\varphi(0)}{t},\quad\phi'_+(0,\boldsymbol{y})=\lim_{\lambda\downarrow0}\frac{\phi(t,\boldsymbol{y})-\phi(0,\boldsymbol{y})}{t}.$$

(6.6.3-4) 式等价于下式:

$$\varphi'_+(0)=\max_{\boldsymbol{y}}\left\{\phi'_+(0,\boldsymbol{y})\,|\boldsymbol{y}\in Y(0)\right\}. \tag{6.6.3-5}$$

因为 $\phi(t,\boldsymbol{y})$ 关于 $\boldsymbol{y}$ 连续, Y 是紧集, 故有

$$\varphi(t)=\max\{\phi(t,\boldsymbol{y})\,|\boldsymbol{y}\in Y\}$$

成立. 因此, 在 $t\in[-\delta,\delta]$ 上 $\varphi(t)$ 是有限凸的. 由定理 6.2.2 知 $\varphi'_+(0)$ 存在且有限, 运用引理 4.3.3, 对于 $t>0$, 得

$$\begin{aligned}&\frac{\phi(t,\boldsymbol{y})-\phi(0,\boldsymbol{y})}{t}\geqslant\frac{\varphi(t)-\varphi(0)}{t}\\&\geqslant\frac{\phi(t,\boldsymbol{y}_0)-\phi(0,\boldsymbol{y}_0)}{t},\quad \boldsymbol{y}\in Y(t),\quad \boldsymbol{y}_0\in Y(0).\end{aligned} \tag{6.6.3-6}$$

由 φ 的凸性在 $t\downarrow 0$ 时单调不增可得

$$\frac{\varphi(t)-\varphi(0)}{t}\geqslant\phi'_+(0,\boldsymbol{y}_0),\quad \boldsymbol{y}_0\in Y(0).$$

令 $t\downarrow 0$, 由上式得

$$\varphi'_+(0)\geqslant\sup\left\{\phi'_+(0,\boldsymbol{y})\,|\boldsymbol{y}\in Y(0)\right\}. \tag{6.6.3-7}$$

由引理 6.6.2, 对于满足 $Y(0)\subset Y(\boldsymbol{x}_0)\subset C$ 的开集 C, 存在 $\varepsilon>0$, 在 $0\leqslant t<\varepsilon$ 时, 有 $Y(t)\subset C$. 故在 $t<\varepsilon$ 时, 由 (6.6.3-6) 式有

$$\varphi'_+(0)\leqslant\frac{\varphi(t)-\varphi(0)}{t}\leqslant\sup_{\boldsymbol{y}}\left\{\frac{\phi(t,\boldsymbol{y})-\phi(0,\boldsymbol{y})}{t}\middle|\boldsymbol{y}\in C\right\}.$$

在全体 $C\supset Y(0)$ 上, 对上式右边取下确界, 因为 $\phi(t,\boldsymbol{y})$ 关于 $\boldsymbol{y}$ 连续, $Y(0)$ 是紧集的闭子集, 则得

$$\inf_{C\supset Y(0)}\sup_{\boldsymbol{y}}\left\{\frac{\phi(t,\boldsymbol{y})-\phi(0,\boldsymbol{y})}{t}\middle|\boldsymbol{y}\in C\right\}=\max_{\boldsymbol{y}}\left\{\frac{\phi(t,\boldsymbol{y})-\phi(0,\boldsymbol{y})}{t}\middle|\boldsymbol{y}\in Y(0)\right\}.$$

所以通过前面两个不等式得

$$\varphi'_+(0)\leqslant\max_{\boldsymbol{y}}\left\{\frac{\phi(t,\boldsymbol{y})-\phi(0,\boldsymbol{y})}{t}\middle|\boldsymbol{y}\in Y(0)\right\}, \tag{6.6.3-8}$$

取序列 $t_k\to 0$ 和 $\boldsymbol{y}_k\in Y(0)$, (6.6.3-8) 式满足

$$\frac{\phi(t_k,\boldsymbol{y}_k)-\phi(0,\boldsymbol{y}_k)}{t_k}\leqslant\max_{y}\left\{\frac{\phi(t_k,\boldsymbol{y})-\phi(0,\boldsymbol{y})}{t_k}\,|\boldsymbol{y}\in Y(0)\right\}. \tag{6.6.3-9}$$

不妨设 $\boldsymbol{y}_k \to \boldsymbol{y}_0 \in Y(0)$, 再由引理 4.3.3 得

$$\frac{\phi(0,\boldsymbol{y}_k)-\phi(-\delta,\boldsymbol{y}_k)}{\delta} \leqslant \frac{\phi(t_k,\boldsymbol{y}_k)-\phi(0,\boldsymbol{y}_k)}{t_k} \leqslant \frac{\phi(\delta,\boldsymbol{y}_k)-\phi(0,\boldsymbol{y}_k)}{\delta}.$$

由 $\phi(t,\boldsymbol{y})$ 的连续性及 Y 的紧致性, 差商 $\dfrac{\phi(t_k,\boldsymbol{y}_k)-\phi(0,\boldsymbol{y}_k)}{t_k}$ 有界, 再由连续性可设

$$\lim_{t_k\to 0}\frac{\phi(t_k,\boldsymbol{y}_k)-\phi(0,\boldsymbol{y}_k)}{t_k}=\eta.$$

设 $t>0$ 固定, 则 k 充分大时, $t>t_k$. 再由引理 4.3.3 有

$$\frac{\phi(t,\boldsymbol{y}_k)-\phi(0,\boldsymbol{y}_k)}{t} \geqslant \frac{\phi(t_k,\boldsymbol{y}_k)-\phi(0,\boldsymbol{y}_k)}{t_k},$$

对上式在 $k\to\infty$ 时取极限, 得

$$\frac{\phi(t,\boldsymbol{y}_0)-\phi(0,\boldsymbol{y}_0)}{t} \geqslant \eta.$$

再令 $t\downarrow 0$ 得 $\phi'_+(0,\boldsymbol{y}_0)\geqslant\eta$. 由 (6.6.3-8) 和 (6.6.3-9) 式得

$$\varphi'_+(0)\leqslant\eta\leqslant\phi'_+(0,\boldsymbol{y}_0),\quad \boldsymbol{y}_0\in Y(0).$$

最后根据 (6.6.3-7) 式和上式, 则有

$$\phi'_+(0,\boldsymbol{y}_0)\geqslant\varphi'_+(0)\geqslant\sup_{\boldsymbol{y}}\left\{\phi'_+(0,\boldsymbol{y})\,|\boldsymbol{y}\in Y(0)\right\},\quad \boldsymbol{y}_0\in Y(0),$$

即 (6.6.3-5) 式成立. □

通过定理 6.6.5 可知 $f(\boldsymbol{x})=\sup\limits_{\boldsymbol{y}\in Y} g(\boldsymbol{x},\boldsymbol{y})$ 在点 $\boldsymbol{x}$ 的方向导数等于 $\boldsymbol{y}\in Y(\boldsymbol{x})$ 的所有方向导数的最大值, 即 (6.6.3-4) 式. 据此, 我们可以求出 $f(\boldsymbol{x})=\sup\limits_{\boldsymbol{y}\in Y} g(\boldsymbol{x},\boldsymbol{y})$ 在 $\boldsymbol{x}$ 处的次微分.

定理 6.6.6 设 Y 是紧集, $g(\boldsymbol{x},\boldsymbol{y})$ 在 $\boldsymbol{y}\in Y$ 时是 $\boldsymbol{x}$ 的凸函数, 定理 6.6.5 假设均成立. 如果对于 $\boldsymbol{x}_0$ 的一个邻域 $O(\boldsymbol{x}_0)$ 及 $\boldsymbol{y}\in Y$ 有 $g(\boldsymbol{x},\boldsymbol{y})$ 关于 $\boldsymbol{x}\in O(\boldsymbol{x}_0)$, $\boldsymbol{y}\in Y$ 连续, 则有

$$\partial f(\boldsymbol{x}_0)=\mathrm{cl}\left(\mathrm{co}\bigcup_{\boldsymbol{y}\in Y(\boldsymbol{x}_0)}\partial g(\boldsymbol{x}_0,\boldsymbol{y})\right). \tag{6.6.3-10}$$

证明 由性质 6.2.3(3), 方向导数等于次微分的支撑函数

$$g'_+(\boldsymbol{x}_0,\boldsymbol{y},\boldsymbol{d})=\max\{\langle\boldsymbol{d},\boldsymbol{x}^*\rangle\,|\boldsymbol{x}^*\in\partial g(\boldsymbol{x}_0,\boldsymbol{y})\},\tag{6.6.3-11}$$

再由定理 6.6.5 有

$$\begin{aligned}f'_+(\boldsymbol{x}_0,\boldsymbol{d})&=\max_{\boldsymbol{y}\in Y(\boldsymbol{x}_0)}\ \max_{\boldsymbol{x}^*\in\partial f(\boldsymbol{x}_0,\boldsymbol{y})}\langle\boldsymbol{d},\boldsymbol{x}^*\rangle=\max_{\boldsymbol{x}^*}\left\{\langle\boldsymbol{d},\boldsymbol{x}^*\rangle\,\middle|\,\boldsymbol{x}^*\in\bigcup_{\boldsymbol{y}\in Y(\boldsymbol{x}_0)}\partial g(\boldsymbol{x}_0,\boldsymbol{y})\right\}\\&=\sup_{\boldsymbol{x}^*}\left\{\langle\boldsymbol{d},\boldsymbol{x}^*\rangle\,\middle|\,\boldsymbol{x}^*\in\mathrm{cl}\left(\mathrm{co}\bigcup_{\boldsymbol{y}\in Y(\boldsymbol{x}_0)}\partial g(\boldsymbol{x}_0,\boldsymbol{y})\right)\right\}.\end{aligned}\tag{6.6.3-12}$$

再由性质 6.2.3(2), 得

$$f'_+(\boldsymbol{x}_0,\boldsymbol{d})=\sup_{\boldsymbol{x}^*}\{\langle\boldsymbol{d},\boldsymbol{x}^*\rangle\,|\boldsymbol{x}^*\in\partial f(\boldsymbol{x}_0)\},\tag{6.6.3-13}$$

由 (6.6.3-11), (6.6.3-12) 和 (6.6.3-13) 式知 (6.6.3-10) 式两边集合的支撑函数相等, 所以 (6.6.3-10) 式成立. □

根据定理 6.6.6 显然有下面的结论.

推论 6.6.3 设 m 个函数 $f_1,f_2,\cdots,f_m:\mathbf{R}^n\to\mathbf{R}$ 是连续凸函数, 设函数

$$f(\boldsymbol{x})=\max\{f_i(\boldsymbol{x})\,|i=1,2,\cdots,m\},\quad\forall\boldsymbol{x}\in\mathbf{R}^n,$$

则有 $\partial f(\boldsymbol{x})=\mathrm{co}\left\{\bigcup_{i\in I(\boldsymbol{x})}\partial f_i(\boldsymbol{x})\right\}$, 其中 $I(\boldsymbol{x})=\{i|f_i(\boldsymbol{x})=f(\boldsymbol{x}),i=1,2,\cdots,m\}$.

本小节主要得到 $f(\boldsymbol{x})=\sup_{\boldsymbol{y}\in Y}g(\boldsymbol{x},\boldsymbol{y})$ 在点 $\boldsymbol{x}$ 处的次微分是由所有 $\boldsymbol{y}\in Y(\boldsymbol{x})$ 的次微分 $\partial g(\boldsymbol{x}_0,\boldsymbol{y})$ 组成的闭凸包构成.

6.7 Dini 方向导数

前几节, 我们讨论了凸函数次微分的连续性、对偶性、中值定理及一些运算性质等. 当函数是广义凸函数时, 它的正负方向导数不一定存在 [37], 对应的次微分也不一定存在. 例如 $f(x)=\begin{cases}1, & x>0,\\0, & x\leqslant 0\end{cases}$ 不是凸函数, 是拟凸函数, 它的正、负方向导数分别为

$$f'_+(0)=\lim_{t\downarrow 0}\frac{f(t)-f(0)}{t}=+\infty,\quad f'_-(0)=\lim_{t\uparrow 0}\frac{f(t)-f(0)}{t}=0.$$

由此可见, 因为 $f(x)$ 不是凸函数, 它在 $x=0$ 处的次微分若根据定义 6.2.2 则可知不等式:

$$0 \geqslant \langle x^*, z\rangle, \quad 1 \geqslant \langle x^*, z\rangle, \quad \forall z \in \mathbf{R}$$

仅在 $x^*=0$ 时成立, 不符合定理 6.2.1 的充要性条件. 若根据定理 6.2.1, 则 $f(x)$ 在 $x=0$ 处的次微分应该是整个空间 $\mathbf{R}$. 为此, 对于广义凸函数需要定义新的方向导数和微分.

本节将定义一阶 Dini 方向导数和二阶 Dini 方向导数, 讨论在 Dini 方向导数情形下凸函数与拟凸函数的一些性质 [37].

6.7.1 一阶 Dini 方向导数

Dini 方向导数是比正负方向导数更弱的一种情形 [37,45,62,81], 可以刻画一些广义凸函数的性质, 如拟凸性或伪凸性的充要条件.

定义 6.7.1[37,81] 设实值函数 $f:\ \mathbf{R}^n \to \mathbf{R}$ 和 $\forall \boldsymbol{d} \in \mathbf{R}^n$. 如果

$$f_+^D(\boldsymbol{x}, \boldsymbol{d}) = \limsup_{t \downarrow 0} \frac{f(\boldsymbol{x}+t\boldsymbol{d})-f(\boldsymbol{x})}{t} \tag{6.7.1-1}$$

成立, 则称 $f_+^D(\boldsymbol{x}, \boldsymbol{d})$ 是 f 在 $\boldsymbol{x}$ 处沿方向 $\boldsymbol{d}$ 的上 (upper-)Dini 方向导数. 如果

$$f_-^D(\boldsymbol{x}, \boldsymbol{d}) = \liminf_{t \downarrow 0} \frac{f(\boldsymbol{x}+t\boldsymbol{d})-f(\boldsymbol{x})}{t} \tag{6.7.1-2}$$

成立, 则称 $f_-^D(\boldsymbol{x}, \boldsymbol{d})$ 是 f 在 $\boldsymbol{x}$ 处沿方向 $\boldsymbol{d}$ 的下 (lower-)Dini 方向导数.

注 显然 $f_-^D(\boldsymbol{x}, \boldsymbol{d}) \leqslant f_+^D(\boldsymbol{x}, \boldsymbol{d})$ 成立. 对于任意的 $\lambda>0$, Dini 方向导数是满足正齐次的:

$$f_-^D(\boldsymbol{x}, \lambda\boldsymbol{d}) = \lambda f_-^D(\boldsymbol{x}, \boldsymbol{d}), \quad f_+^D(\boldsymbol{x}, \lambda\boldsymbol{d}) = \lambda f_+^D(\boldsymbol{x}, \boldsymbol{d}).$$

如果 $f_-^D(\boldsymbol{x}, \boldsymbol{d}) = f_+^D(\boldsymbol{x}, \boldsymbol{d})$, 则 $f'_+(\boldsymbol{x}, \boldsymbol{d}) = f_+^D(\boldsymbol{x}, \boldsymbol{d})$, 即正方向导数存在等价于上 Dini 方向导数等于下 Dini 方向导数.

例 6.7.1 设函数 $f:\ C \to \mathbf{R}$, 其中 $C=[-1,1]$,

$$f(x) = \begin{cases} 0, & x \leqslant 0, \\ x, & \dfrac{1}{k+1} < x < \dfrac{1}{k}, k=1,2,\cdots, \\ 2x, & x = \dfrac{1}{k}, k=1,2,\cdots, \\ 2, & x>1. \end{cases}$$

我们有 $f_-^D(0,1)=1<f_+^D(0,1)=2$. 当 $k>1$ 时, 有

$$f_+^D\left(\frac{1}{k},1\right)=\limsup_{t\downarrow 0}\frac{f\left(\frac{1}{k}+t\right)-f\left(\frac{1}{k}\right)}{t}=\limsup_{t\downarrow 0}\frac{\frac{1}{k-1}-\frac{2}{k}}{t}=-\infty,$$

$$f_-^D\left(\frac{1}{k},1\right)=\liminf_{t\downarrow 0}\frac{f\left(\frac{1}{k}+t\right)-f\left(\frac{1}{k}\right)}{t}=\liminf_{t\downarrow 0}\frac{\frac{1}{k-1}-\frac{2}{k}}{t}=-\infty.$$

当 $\frac{1}{k+1}<x<\frac{1}{k}\,(k>1)$ 时, 有

$$f_+^D(x,1)=\limsup_{t\downarrow 0}\frac{f(x+t)-f(x)}{t}=\limsup_{t\downarrow 0}\frac{x-x}{t}=0,$$

$$f_-^D(x,1)=\liminf_{t\downarrow 0}\frac{f(x)-f(x)}{t}=\liminf_{t\downarrow 0}\frac{x-x}{t}=0.$$

例 6.7.1 说明不可微函数的上、下 Dini 方向导数存在, 但不一定相等.

类似于可微函数的中值定理, 下面我们有 Dini 方向导数中值定理.

引理 6.7.1(中值定理)[81]　设实值函数 $f:\ \mathbf{R}^n\to\mathbf{R}$, 线段 $[\boldsymbol{x}_1,\boldsymbol{x}_2]$ 上限制函数

$$\varphi(t)=f(\boldsymbol{x}_1+t(\boldsymbol{x}_2-\boldsymbol{x}_1))$$

在区间 $[0,1]$ 上是下半连续的. 若 f 在 $\boldsymbol{x}\in\mathbf{R}^n$ 沿方向 $\boldsymbol{d}\in\mathbf{R}^n$ 均存在下 Dini 方向导数, 则 $\exists\tau\in[0,1]$ 使得

$$f_-^D(\boldsymbol{x}_\tau,\boldsymbol{x}_2-\boldsymbol{x}_1)\geqslant f(\boldsymbol{x}_2)-f(\boldsymbol{x}_1),\tag{6.7.1-3}$$

其中 $\boldsymbol{x}_\tau=\boldsymbol{x}_1+\tau(\boldsymbol{x}_2-\boldsymbol{x}_1)$.

例 6.7.2　设函数 $f:\ C\to\mathbf{R}$, 其中 $C=[-1,1]$,

$$f(x)=\begin{cases}0, & 0<|x|\leqslant 1,\\ 1, & x=0,\end{cases}$$

令 $x_1=0,x_2=1$, 那么 $\varphi(t)=f(t)$ 在区间 $[0,1]$ 上不是下半连续的. 我们有

$$f_-^D(0,x_2-x_1)=\liminf_{t\downarrow 0}\frac{f(t)-f(0)}{t}=\liminf_{t\downarrow 0}\frac{0-1}{t}=-\infty<-1.$$

若 $\tau \in (0,1)$, 我们有

$$f_-^D(x_\tau, x_2 - x_1) = \liminf_{t\downarrow 0} \frac{f(\tau+t) - f(\tau)}{t}$$
$$= \liminf_{t\downarrow 0} \frac{0-0}{t} = 0 > f(x_2) - f(x_1) = -1.$$

这个例子说明引理 6.7.1 中的下半连续条件仅是一个充分条件.

引理 6.7.2 设 $f: C \to \mathbf{R}$ 是在凸集 $C \subset \mathbf{R}^n$ 上的实值函数, 任何线段 $[\boldsymbol{x}_1, \boldsymbol{x}_2] \subset C$ 上的函数 $\varphi(t) = f(\boldsymbol{x}_1 + t(\boldsymbol{x}_2 - \boldsymbol{x}_1))$ 在区间 $[0,1]$ 上都是下半连续的, 且 f 在任意 $\boldsymbol{x} \in C$ 处沿任意方向 $\boldsymbol{d} \in \mathbf{R}^n$ 的上、下 Dini 方向导数存在. 如果存在点 $\boldsymbol{x}_\tau = \boldsymbol{x}_1 + \tau(\boldsymbol{x}_2 - \boldsymbol{x}_1)\,(\tau \in (0,1))$, 使得

$$\varphi(\tau) = f(\boldsymbol{x}_\tau) > \max\{f(\boldsymbol{x}_1), f(\boldsymbol{x}_2)\},$$

则 $\exists t_1, t_2 \in [0,1]$ 使得 $f(\boldsymbol{x}_{t_2}) \leqslant f(\boldsymbol{x}_{t_1})$, $f_-^D(\boldsymbol{x}_{t_1}, \boldsymbol{x}_{t_2} - \boldsymbol{x}_{t_1}) > 0$. 进一步, 如果在 $C \subset \mathbf{R}^n$ 中不存在任何线段使函数 f 在线段上恒为常数, 则 $\exists t_1, t_2 \in [0,1]$ 使得

$$f(\boldsymbol{x}_{t_2}) < f(\boldsymbol{x}_{t_1}), \quad f_-^D(\boldsymbol{x}_{t_1}, \boldsymbol{x}_{t_2} - \boldsymbol{x}_{t_1}) > 0.$$

证明 设

$$a = \max\{t \in [0,\tau]\,|\,\varphi(t) \leqslant \max\{f(\boldsymbol{x}_1), f(\boldsymbol{x}_2)\}\},$$

$$b = \min\{t \in [\tau,1]\,|\,\varphi(t) \leqslant \max\{f(\boldsymbol{x}_1), f(\boldsymbol{x}_2)\}\}.$$

令 $\boldsymbol{x}_a = \boldsymbol{x}_1 + a(\boldsymbol{x}_2 - \boldsymbol{x}_1)$ 和 $\boldsymbol{x}_b = \boldsymbol{x}_1 + b(\boldsymbol{x}_2 - \boldsymbol{x}_1)$, 因为 $\varphi(t)$ 在区间 $[0,1]$ 是下半连续的, 有

$$a < \tau < b, \quad f(\boldsymbol{x}_a), f(\boldsymbol{x}_b) \leqslant \max\{f(\boldsymbol{x}_1), f(\boldsymbol{x}_2)\},$$

$$f(\boldsymbol{x}_t) > \max\{f(\boldsymbol{x}_1), f(\boldsymbol{x}_2)\}, \quad \forall t \in (a,b).$$

下面分两种情形讨论:

(1) 假设 f 在线段 $(\boldsymbol{x}_a, \boldsymbol{x}_b)$ 上恒为常数 K, 则 $K = f(\boldsymbol{x}_t) > \max\{f(\boldsymbol{x}_1), f(\boldsymbol{x}_2)\}, \forall t \in (a,b)$, 有

$$f_-^D(\boldsymbol{x}_a, \boldsymbol{x}_b - \boldsymbol{x}_a) = f_-^D(\boldsymbol{x}_b, \boldsymbol{x}_a - \boldsymbol{x}_b) = +\infty.$$

如果 $f(\boldsymbol{x}_a) \leqslant f(\boldsymbol{x}_b)$, 则令 $t_1 = b$, $t_2 = a$. 否则 $f(\boldsymbol{x}_a) \geqslant f(\boldsymbol{x}_b)$, 则令 $t_1 = a$, $t_2 = b$, 均有引理结论成立.

(2) 假设 f 在线段 $(\boldsymbol{x}_a, \boldsymbol{x}_b)$ 上不恒为常数, 那么在 $(\boldsymbol{x}_a, \boldsymbol{x}_b)$ 中存在两点 $\boldsymbol{x}_c$, $\boldsymbol{x}_d \in (\boldsymbol{x}_a, \boldsymbol{x}_b)$ 满足 $f(\boldsymbol{x}_c) < f(\boldsymbol{x}_d)$. 由引理 6.7.1 知, 存在 $\boldsymbol{x}_e \in [\boldsymbol{x}_c, \boldsymbol{x}_d)$ 使得

$$f_-^D(\boldsymbol{x}_e, \boldsymbol{x}_d - \boldsymbol{x}_c) \geqslant f(\boldsymbol{x}_d) - f(\boldsymbol{x}_c) > 0.$$

因为 $\boldsymbol{x}_e \in (\boldsymbol{x}_a, \boldsymbol{x}_b)$, 有 $f(\boldsymbol{x}_e) > \max\{f(\boldsymbol{x}_1), f(\boldsymbol{x}_2)\} \geqslant \max\{f(\boldsymbol{x}_a), f(\boldsymbol{x}_b)\}$. 如果 $\boldsymbol{x}_d \in (\boldsymbol{x}_c, \boldsymbol{x}_b)$, 则令 $t_1 = e$, $t_2 = b$; 如果 $\boldsymbol{x}_d \in (\boldsymbol{x}_a, \boldsymbol{x}_c)$, 则令 $t_1 = e$, $t_2 = c$. 得引理前半部分成立. 如果在 $C \subset \mathbf{R}^n$ 中不存在任何线段使得函数 f 在这个线段上恒为常数, 类似地可获得引理后半部分结论成立. □

下面得到一个 (严格) 凸函数与拟凸函数的等价性判定定理.

定理 6.7.1 设凸集 $C \subset \mathbf{R}^n$, $f\colon C \to \mathbf{R}$ 是实函数, 则 f 在 C 上是 (严格) 凸函数当且仅当对任意 $\boldsymbol{x}^* \in \mathbf{R}^n$ 函数

$$f^*(\boldsymbol{x}) = f(\boldsymbol{x}) + \langle \boldsymbol{x}^*, \boldsymbol{x} \rangle \tag{6.7.1-4}$$

在 C 上是 (严格) 拟凸的.

证明 如果 f 在 C 上是 (严格) 凸函数, 对于任意固定的 $\boldsymbol{x}^* \subset \mathbf{R}^n$, $\langle \boldsymbol{x}^*, \boldsymbol{x} \rangle$ 在 C 上是线性函数, 显然 f^* 在 C 上是 (严格) 拟凸的.

反过来, 假设对任意 $\boldsymbol{x}^* \in \mathbf{R}^n$, 有 (6.7.1-4) 式定义的函数 f^* 在 C 上是 (严格) 拟凸的. 因 $\langle \boldsymbol{x}^*, \boldsymbol{x} \rangle$ 在 C 上是线性函数, 则 f 在 C 上也是 (严格) 拟凸的. 如果 f 在 C 上不是 (严格) 凸函数, 那么 $\exists \boldsymbol{x}_1, \boldsymbol{x}_2 \in C$, $\boldsymbol{x}_1 \neq \boldsymbol{x}_2$, $\lambda \in (0,1)$ 使得

$$f((1-\lambda)\boldsymbol{x}_1 + \lambda \boldsymbol{x}_2) > (\geqslant)\, (1-\lambda) f(\boldsymbol{x}_1) + \lambda f(\boldsymbol{x}_2). \tag{6.7.1-5}$$

因为 $\langle \boldsymbol{x}^*, \boldsymbol{x} \rangle\, (\forall \boldsymbol{x}^* \in \mathbf{R}^n)$ 在 C 上是关于 $\boldsymbol{x}$ 的线性函数, 那么 $f^*(\boldsymbol{x}) = f(\boldsymbol{x}) + \langle \boldsymbol{x}^*, \boldsymbol{x} \rangle$ 在 C 上也不是 (严格) 凸函数. 由 (6.7.1-5) 式得

$$f^*((1-\lambda)\boldsymbol{x}_1 + \lambda \boldsymbol{x}_2) > (\geqslant)\, (1-\lambda) f^*(\boldsymbol{x}_1) + \lambda f^*(\boldsymbol{x}_2). \tag{6.7.1-6}$$

对于固定的 $\boldsymbol{x}_1, \boldsymbol{x}_2 \in C$, 一定存在 $\boldsymbol{x}^* \in \mathbf{R}^n$ 使得 $\langle \boldsymbol{x}^*, \boldsymbol{x}_1 - \boldsymbol{x}_2 \rangle = f(\boldsymbol{x}_2) - f(\boldsymbol{x}_1)$, 由 (6.7.1-6) 式得

$$f^*((1-\lambda)\boldsymbol{x}_1 + \lambda \boldsymbol{x}_2) > (\geqslant)\, f^*(\boldsymbol{x}_1) = f^*(\boldsymbol{x}_2). \tag{6.7.1-7}$$

但根据函数 f^* 在 C 上是 (严格) 拟凸的, 由 (6.7.1-7) 式得

$$f^*(\boldsymbol{x}_1) \geqslant (>)\, f^*((1-\lambda)\boldsymbol{x}_1 + \lambda \boldsymbol{x}_2) > (\geqslant)\, f^*(\boldsymbol{x}_1) = f^*(\boldsymbol{x}_2),$$

上式是矛盾的, 因此, f 在 C 上是 (严格) 凸函数. □

例 6.7.3 函数 $f(x)=x^3$ 不是凸的. 设 $x^*=-3$, 由 (6.7.1-4) 式得函数 $f(x)=x^3-3x$ 不是拟凸的. 该函数有稳定点 $x=\pm1$, 取 $x_1=4$, $x_2=0$, 我们有

$$f\left(\frac{x_1+x_2}{2}\right)=2>\max\{f(x_1),f(x_2)\}=0.$$

例 6.7.3 说明, 凸函数与拟凸函数关系密切. 进一步, 下面的结论是拟凸函数的一个必要条件.

定理 6.7.2 设 $f:\ C\to\mathbf{R}$ 在凸集 $C\subset\mathbf{R}^n$ 上是拟凸函数, 且 f 在任意 $\boldsymbol{x}\in C$ 沿任意方向 $\boldsymbol{d}\in\mathbf{R}^n$ 的上、下 Dini 方向导数存在, 则对任意 $\boldsymbol{x}_1,\boldsymbol{x}_2\in C$ 有下面式子成立:

$$f(\boldsymbol{x}_2)-f(\boldsymbol{x}_1)\leqslant 0\Rightarrow f_-^D(\boldsymbol{x}_1,\boldsymbol{x}_2-\boldsymbol{x}_1)\leqslant 0, \tag{6.7.1-8}$$

$$f(\boldsymbol{x}_2)-f(\boldsymbol{x}_1)\leqslant 0\Rightarrow f_+^D(\boldsymbol{x}_1,\boldsymbol{x}_2-\boldsymbol{x}_1)\leqslant 0. \tag{6.7.1-9}$$

证明 设 $f(\boldsymbol{x}_2)-f(\boldsymbol{x}_1)\leqslant 0$, $0<t<1$, 由 f 的拟凸性, 有

$$\frac{f(\boldsymbol{x}_1+t(\boldsymbol{x}_2-\boldsymbol{x}_1))-f(\boldsymbol{x}_1)}{t}\leqslant 0.$$

由上式可得 $f_+^D(\boldsymbol{x}_1,\boldsymbol{x}_2-\boldsymbol{x}_1)\leqslant 0$, 同时也有 $f_-^D(\boldsymbol{x}_1,\boldsymbol{x}_2-\boldsymbol{x}_1)\leqslant 0$. □

例 6.7.4 设函数 $f:\ C\to\mathbf{R}$, 其中 $C=[-1,1]$,

$$f(x)=\begin{cases}0, & 0<|x|\leqslant 1,\\ 1, & x=0,\end{cases}$$

易知 f 在凸集 C 上不是拟凸函数, 令 $x_1=0, 0<|x_2|\leqslant 1$, 那么 $\varphi(t)=f(tx_2)$ 在区间 $[0,1]$ 上不是下半连续的. 但是

$$f_+^D(0,x_2)=\limsup_{t\downarrow 0}\frac{f(t)-f(0)}{t}=\limsup_{t\downarrow 0}\frac{0-1}{t}=-\infty<0.$$

这说明定理 6.7.2 的结论是必要的.

在定理 5.1.14 中, 如果 $f:\ C\to\mathbf{R}$ 是在凸集 $C\subset\mathbf{R}^n$ 上的可微函数, 那么拟凸函数的充要条件是对任意的 $\boldsymbol{x},\boldsymbol{y}\in C$ 有 $f(\boldsymbol{x})\geqslant f(\boldsymbol{y})\Rightarrow\nabla f(\boldsymbol{x})^{\mathrm{T}}(\boldsymbol{y}-\boldsymbol{x})\leqslant 0$. 同样, 如果 $f:\ C\to\mathbf{R}$ 在凸集 $C\subset\mathbf{R}^n$ 上具有 Dini 方向导数, 那么也有类似结论.

定理 6.7.3 设 $f:\ C\to\mathbf{R}$ 是在凸集 $C\subset\mathbf{R}^n$ 上的实函数. 如果任何线段 $[\boldsymbol{x}_1,\boldsymbol{x}_2]\subset C$ 上的函数 $\varphi(t)=f(\boldsymbol{x}_1+t(\boldsymbol{x}_2-\boldsymbol{x}_1))$ 在区间 $[0,1]$ 是下半连续的,

且 f 在任意 $\boldsymbol{x} \in C$ 沿任意方向 $\boldsymbol{d} \in \mathbf{R}^n$ 的上、下 Dini 方向导数存在, 则 f 在 C 上是拟凸函数等价于对任意 $\boldsymbol{x}_1, \boldsymbol{x}_2 \in C$ 有

$$f(\boldsymbol{x}_2) - f(\boldsymbol{x}_1) \leqslant 0 \Rightarrow f_-^D(\boldsymbol{x}_1, \boldsymbol{x}_2 - \boldsymbol{x}_1) \leqslant 0$$

或

$$f(\boldsymbol{x}_2) - f(\boldsymbol{x}_1) \leqslant 0 \Rightarrow f_+^D(\boldsymbol{x}_1, \boldsymbol{x}_2 - \boldsymbol{x}_1) \leqslant 0.$$

证明　由定理 6.7.2 得必要性条件成立.

反过来, 假设 f 在 C 上不是拟凸函数, 则存在 $\boldsymbol{x}_1, \boldsymbol{x}_2 \in C$, $\boldsymbol{x}_\tau = \boldsymbol{x}_1 + \tau(\boldsymbol{x}_2 - \boldsymbol{x}_1)$ $(\tau \in (0,1))$, 使得

$$\varphi(\tau) = f(\boldsymbol{x}_\tau) > \max\{f(\boldsymbol{x}_1), f(\boldsymbol{x}_2)\}.$$

由引理 6.7.2 知, $\exists t_1, t_2 \in [0,1]$ 使得 $f(\boldsymbol{x}_{t_1}) \leqslant f(\boldsymbol{x}_{t_2})$ 和 $f_-^D(\boldsymbol{x}_{t_1}, \boldsymbol{x}_{t_2} - \boldsymbol{x}_{t_1}) > 0$, 与假设矛盾. □

进一步, 我们有下面更弱的结论.

定理 6.7.4　设 $f: C \to \mathbf{R}$ 是凸集 $C \subset \mathbf{R}^n$ 上的函数. 如果任何线段 $[\boldsymbol{x}_1, \boldsymbol{x}_2] \subset C$ 上的函数 $\varphi(t) = f(\boldsymbol{x}_1 + t(\boldsymbol{x}_2 - \boldsymbol{x}_1))$ 在区间 $[0,1]$ 都是下半连续的, 且

$$f(\boldsymbol{x}_2) - f(\boldsymbol{x}_1) < 0 \Rightarrow f_-^D(\boldsymbol{x}_1, \boldsymbol{x}_2 - \boldsymbol{x}_1) < 0$$

或

$$f(\boldsymbol{x}_2) - f(\boldsymbol{x}_1) < 0 \Rightarrow f_+^D(\boldsymbol{x}_1, \boldsymbol{x}_2 - \boldsymbol{x}_1) < 0$$

成立, 则 $f(\boldsymbol{x})$ 在 C 上是拟凸函数.

证明　类似于定理 6.7.3. □

例 6.7.5　设函数 $f: C \to \mathbf{R}$, 其中 $C = [-1,1]$,

$$f(x) = \begin{cases} 0, & 0 < |x| \leqslant 1, \\ -1, & x = 0, \end{cases}$$

则 f 在凸集 C 上是拟凸函数且下半连续的. 令 $x_2 = 0, 0 < |x_1| \leqslant 1$, 有

$$f_+^D(x_1, 0 - x_1) = \limsup_{t \downarrow 0} \frac{f(x_1 - tx_1) - f(x_1)}{t} = \limsup_{t \downarrow 0} \frac{0 - 0}{t} = 0,$$

说明定理 6.7.3 的必要条件成立, 但定理 6.7.4 的充分条件不一定成立. 如果 $\bar{x} = 0$ 和 $0 < |x| \leqslant 1$, 有

$$f_-^D(0, x) = \liminf_{t \downarrow 0} \frac{f(tx) - f(0)}{t} = \liminf_{t \downarrow 0} \frac{0 + 1}{t} = +\infty,$$

即 $\bar{x}=0$ 是 f 在 C 上的 inf-驻点.

定义 6.7.2 设 $f:\ C\to\mathbf{R}$ 是在 $C\subset\mathbf{R}^n$ 上的实值函数. 如果点 $\bar{\boldsymbol{x}}\in C$ 满足

$$f_{-}^{D}(\bar{\boldsymbol{x}},\boldsymbol{x}-\bar{\boldsymbol{x}})\geqslant 0\quad(\forall\boldsymbol{x}\in C),$$

则称 $\bar{\boldsymbol{x}}\in C$ 是 f 在 C 上的 inf-驻点 (平稳点或稳定点).

注 如果 f 在 C 上是可微凸的, $f_{-}^{D}(\bar{\boldsymbol{x}},\boldsymbol{x}-\bar{\boldsymbol{x}})\geqslant 0\,(\forall\boldsymbol{x}\in C)$ 为

$$\nabla f(\bar{\boldsymbol{x}})^{\mathrm{T}}(\boldsymbol{x}-\bar{\boldsymbol{x}})\geqslant 0\quad(\forall\boldsymbol{x}\in C)$$

成立, 则 $f(\boldsymbol{x})-f(\bar{\boldsymbol{x}})\geqslant\nabla f(\bar{\boldsymbol{x}})^{\mathrm{T}}(\boldsymbol{x}-\bar{\boldsymbol{x}})\geqslant 0\,(\forall\boldsymbol{x}\in C)$, 于是 $\bar{\boldsymbol{x}}\in C$ 是 $\min\limits_{\boldsymbol{x}\in C}f(\boldsymbol{x})$ 是 C 上的全局最优解.

下面给出全局最优解判定结论.

定理 6.7.5 设 $f:\ C\to\mathbf{R}$ 是在凸集 $C\subset\mathbf{R}^n$ 上的函数. 如果 $\bar{\boldsymbol{x}}\in C$ 是 $f(\boldsymbol{x})$ 在 C 上的 inf-驻点, 且

$$f(\boldsymbol{x}_2)-f(\boldsymbol{x}_1)<0\Rightarrow f_{-}^{D}(\boldsymbol{x}_1,\boldsymbol{x}_2-\boldsymbol{x}_1)<0,\quad\forall\boldsymbol{x}_1,\boldsymbol{x}_2\in C\tag{6.7.1-10}$$

成立, 则 $\bar{\boldsymbol{x}}\in C$ 是 $\min\limits_{\boldsymbol{x}\in C}f(\boldsymbol{x})$ 在 C 上的全局最优解.

证明 如果 $\bar{\boldsymbol{x}}\in C$ 不是 $\min\limits_{\boldsymbol{x}\in C}f(\boldsymbol{x})$ 在 C 上的全局最优解, 那么 $\exists\boldsymbol{x}\in C$ 使得 $f(\boldsymbol{x})<f(\bar{\boldsymbol{x}})$, 由定理所设知 $\bar{\boldsymbol{x}}\in C$ 不是 f 在 C 上的 inf-驻点, 得到矛盾. □

定理 6.7.6 设 $f:\ C\to\mathbf{R}$ 在开凸集 $C\subset\mathbf{R}^n$ 上是拟凸函数, 任何线段 $[\boldsymbol{x}_1,\boldsymbol{x}_2]\subset C$ 上的函数 $\varphi(t)=f(\boldsymbol{x}_1+t(\boldsymbol{x}_2-\boldsymbol{x}_1))$ 在区间 $[0,1]$ 上是上半连续的. 如果 $\bar{\boldsymbol{x}}\in C$ 是 $\min\limits_{\boldsymbol{x}\in C}f(\boldsymbol{x})$ 在 C 上的全局最优解, 且 $f_{-}^{D}(\bar{\boldsymbol{x}},\boldsymbol{x}-\bar{\boldsymbol{x}})\geqslant 0\,(\forall\boldsymbol{x}\in C)$, 则

$$f(\boldsymbol{x}_1)-f(\boldsymbol{x}_2)<0\Rightarrow f_{-}^{D}(\boldsymbol{x}_2,\boldsymbol{x}_1-\boldsymbol{x}_2)<0,\quad\forall\boldsymbol{x}_1,\boldsymbol{x}_2\in C.\tag{6.7.1-11}$$

证明 设 $\boldsymbol{x}_1,\boldsymbol{x}_2\in C$, $f(\boldsymbol{x}_1)<f(\boldsymbol{x}_2)$. 因 $\boldsymbol{x}_2$ 不是 $\min\limits_{\boldsymbol{x}\in C}f(\boldsymbol{x})$ 在 C 上的全局最优解, 所以 $\boldsymbol{x}_2$ 不是 f 在 C 上的 inf-驻点. 因此, $\exists\boldsymbol{x}'\in C$ 使得 $f_{-}^{D}(\boldsymbol{x}_2,\boldsymbol{x}'-\boldsymbol{x}_2)<0$. 令 $\boldsymbol{d}=\boldsymbol{x}'-\boldsymbol{x}_2$, 因 C 是开凸集, 有 $f(\boldsymbol{x}_1+t(\boldsymbol{x}_2-\boldsymbol{x}_1)-t\boldsymbol{d})$ 关于 t 在区间 $[0,1]$ 上是上半连续函数. 那么存在充分小的 $t>0$ 满足

$$f(\boldsymbol{x}_1+t(\boldsymbol{x}_2-\boldsymbol{x}_1)-t\boldsymbol{d})<f(\boldsymbol{x}_2),$$

且 $\boldsymbol{u}:=\boldsymbol{x}_1+t(\boldsymbol{x}_2-\boldsymbol{x}_1)-t\boldsymbol{d}\in C$. 对于 $0<\lambda<1-t<1$, 固定 t, 令

$$\boldsymbol{x}_\lambda:=\boldsymbol{x}_2+\lambda(\boldsymbol{x}_1-\boldsymbol{x}_2),\quad\boldsymbol{y}_\lambda:=\boldsymbol{x}_2+\frac{t\lambda}{1-t-\lambda}\boldsymbol{d}\in C.$$

不难验证 $\boldsymbol{x}_\lambda \in [\boldsymbol{u}, \boldsymbol{y}_\lambda]$, 根据 f 在 C 上是拟凸函数知

$$f(\boldsymbol{x}_\lambda) \leqslant \max\{f(\boldsymbol{u}), f(\boldsymbol{y}_\lambda)\}.$$

由上式得

$$\frac{f(\boldsymbol{x}_\lambda) - f(\boldsymbol{x}_2)}{\lambda} \leqslant \max\left\{\frac{f(\boldsymbol{u}) - f(\boldsymbol{x}_2)}{\lambda}, \frac{f(\boldsymbol{y}_\lambda) - f(\boldsymbol{x}_2)}{\lambda}\right\}.$$

由于 $f(\boldsymbol{u}) < f(\boldsymbol{x}_2)$, 令 $\lambda \downarrow 0$, 上式即 $\dfrac{f(\boldsymbol{u}) - f(\boldsymbol{x}_2)}{\lambda} \to -\infty$, 因此有

$$f_-^D(\boldsymbol{x}_2, \boldsymbol{x}_1 - \boldsymbol{x}_2) = \liminf_{\lambda\downarrow 0} \frac{f(\boldsymbol{x}_\lambda) - f(\boldsymbol{x}_2)}{\lambda} \leqslant \liminf_{\lambda\downarrow 0} \frac{f(\boldsymbol{y}_\lambda) - f(\boldsymbol{x}_2)}{\lambda}. \tag{6.7.1-12}$$

设 $\tau = \dfrac{t\lambda}{1-t-\lambda}$, 得 $\lambda = \dfrac{1-t}{\tau+t}\tau$. 令 $\lambda \downarrow 0$ 有 $\tau \downarrow 0$, 由 (6.7.1-12) 式得

$$\begin{aligned}\liminf_{\lambda\downarrow 0} \frac{f(\boldsymbol{y}_\lambda) - f(\boldsymbol{x}_2)}{\lambda} &= \liminf_{\tau\downarrow 0} \frac{f(\boldsymbol{x}_2 + \tau\boldsymbol{d}) - f(\boldsymbol{x}_2)}{\tau} \cdot \frac{1+\tau}{1-t} \\ &= \frac{1}{1-t} f_-^D(\boldsymbol{x}_2, \boldsymbol{d}) < 0,\end{aligned}$$

即有 $f_-^D(\boldsymbol{x}_2, \boldsymbol{x}_1 - \boldsymbol{x}_2) < 0$. □

定理 6.7.6 说明拟凸函数任意线段的限制函数都是上半连续的. 如果存在一点不是全局最优, 那么该点存在 Dini 导数方向小于 0 的方向. 这说明在定理 6.7.6 的假设下, 条件 (6.7.1-10) 是当 inf-驻点为全局极小点的充要条件.

本小节得到一阶 Dini 方向关于拟凸函数的几个重要性质:

1) f 是 (严格) 凸函数当且仅当对任意给定的点 $\boldsymbol{x}^*$ 使得 $f^*(\boldsymbol{x}) = f(\boldsymbol{x}) + \langle \boldsymbol{x}^*, \boldsymbol{x} \rangle$ 是拟凸函数;

2) f 是拟凸函数的必要条件是 (6.7.1-8) 式和 (6.8.1-9) 式成立;

3) 如果函数任意线段上的限制函数是下半连续的, f 为拟凸函数的充要条件是 (6.7.1-8) 式或 (6.8.1-9) 式成立;

4) 如果 f 满足条件 (6.7.1-10) 式, 那么它的 inf-驻点是全局极小点;

5) 如果一个拟凸函数任何线段上的限制函数是上半连续的, 且 f 全局极小点的任何方向的 Dini 导数不小于 0, 则条件 (6.7.1-10) 式成立.

6.7.2 二阶 Dini 方向导数

在 5.4 节中, 二阶可微函数与拟 (伪) 凸函数之间有着密切联系, 文献 [37,45, 62,81] 讨论了二阶 Dini 方向导数与 (拟) 凸函数之间的关系, 下面我们给出二阶 Dini 方向导数与 (拟) 凸函数的性质 [81].

定义 6.7.3 设实值函数 $f:\ \mathbf{R}^n \to \mathbf{R}, \forall \boldsymbol{d}, \boldsymbol{x} \in \mathbf{R}^n$. 如果极限

$$f_{+}^{DD}(\boldsymbol{x}, \boldsymbol{d}) = \limsup_{t \downarrow 0} \frac{\langle f_G'(\boldsymbol{x} - t\boldsymbol{d}), \boldsymbol{d}\rangle - \langle f_G'(\boldsymbol{x}), \boldsymbol{d}\rangle}{t} \tag{6.7.2-1}$$

存在, 其中 $f_G'(\boldsymbol{x})$ 是 Gateaux 导数, 则称 $f_{+}^{DD}(\boldsymbol{x}, \boldsymbol{d})$ 是 f 在 $\boldsymbol{x}$ 沿方向 $\boldsymbol{d}$ 的二阶上 Dini 方向导数.

由定理 6.1.1 知 $f_G'(\boldsymbol{x})$ 是 Gateaux 导数, 且对应的方向导数也存在. 根据定义 6.7.1 得

$$f_{+}^{DD}(\boldsymbol{x}, \boldsymbol{d}) = (\langle f_G'(\cdot), \boldsymbol{d}\rangle)_{+}^{D}(\boldsymbol{x}, \boldsymbol{d}).$$

由定理 6.1.1, 在有限维空间中, Gateaux 导数与方向导数是等价的. 如果实值函数 $f:\ \mathbf{R}^n \to \mathbf{R}$ 是二阶连续可微的, 则有

$$\begin{aligned}\langle f_G'(\boldsymbol{x} - t\boldsymbol{d}), \boldsymbol{d}\rangle - \langle f_G'(\boldsymbol{x}), \boldsymbol{d}\rangle &= \nabla f(\boldsymbol{x} - t\boldsymbol{d})^{\mathrm{T}} \boldsymbol{d} - \nabla f(\boldsymbol{x})^{\mathrm{T}} \boldsymbol{d} \\ &= -t\boldsymbol{d}^{\mathrm{T}} \nabla^2 f(\boldsymbol{x})^{\mathrm{T}} \boldsymbol{d} + o(t),\end{aligned}$$

此时有 $f_{+}^{DD}(\boldsymbol{x}, \boldsymbol{d}) = -\boldsymbol{d}^{\mathrm{T}} \nabla f(\boldsymbol{x})^{\mathrm{T}} \boldsymbol{d}$.

例 6.7.1 中的函数 Gateaux 导数不存在, 因此, 其二阶上 Dini 方向导数也不存在.

下面的定理 6.7.7 是一个二阶上 Dini 方向导数的中值定理.

定理 6.7.7 设实值函数 $f:\ \mathbf{R}^n \to \mathbf{R}$ 是连续 Gateaux 可导的. 如果 $f_{+}^{DD}(\boldsymbol{x}, \boldsymbol{d})$ 是 f 在 $\boldsymbol{x}$ 沿方向 $\boldsymbol{d} \in \mathbf{R}^n$ 的二阶上 Dini 方向导数, 则对任意的 $\boldsymbol{x}, \boldsymbol{y} \in \mathbf{R}^n$, $\exists \boldsymbol{z} \in [\boldsymbol{x}, \boldsymbol{y}]$ 使得

$$f(\boldsymbol{y}) \geqslant f(\boldsymbol{x}) + \langle f_G'(\boldsymbol{x}), \boldsymbol{y} - \boldsymbol{x}\rangle - \frac{1}{2} f_{+}^{DD}(\boldsymbol{z}, \boldsymbol{y} - \boldsymbol{x}). \tag{6.7.2-2}$$

证明 对任意 $\boldsymbol{x}, \boldsymbol{y} \in \mathbf{R}^n$, 任意 $t \in \mathbf{R}$, 设函数

$$\phi(t) = f(\boldsymbol{x} + t(\boldsymbol{y} - \boldsymbol{x})) - f(\boldsymbol{x}) - t\langle f_G'(\boldsymbol{x}), \boldsymbol{y} - \boldsymbol{x}\rangle.$$

再设函数

$$h(t) = f(\boldsymbol{y} + t(\boldsymbol{x} - \boldsymbol{y})) + t\langle f_G'(\boldsymbol{y} + t(\boldsymbol{x} - \boldsymbol{y})), \boldsymbol{y} - \boldsymbol{x}\rangle - f(\boldsymbol{y}) + \phi(1)t^2.$$

有 $h(0) = 0$ 和 $h(1) = 0$, 显然 h 是连续函数, 由引理 6.7.1 知 $\exists \tau \in [0, 1]$ 使得

$$h_{-}^{D}(\tau, 1) \geqslant h(1) - h(0) = 0.$$

由于 $h(t)=\phi(1-t)-\phi(1)+t\phi'(1-t)+\phi(1)t^2$, 因此得

$$h_-^D(\tau,1)=2\tau\phi(1)+\tau h_+^{DD}(1-\tau,1)\geqslant 0,$$

由上式直接可得 (6.7.2-2) 式成立. □

根据定理 6.7.7 知在有限维空间中, 二阶上 Dini 方向导数的中值定理成立, 进一步得到如下定理.

定理 6.7.8 设实值函数 $f:\ \mathbf{R}^n\to\mathbf{R}$ 是连续 Gateaux 可导的. 如果 $f_+^{DD}(\boldsymbol{x},\boldsymbol{d})$ 是 f 在 $\boldsymbol{x}$ 沿方向 $\boldsymbol{d}\in\mathbf{R}^n$ 的二阶上 Dini 方向导数, 则 f 在 C 上是凸函数当且仅当对任意 $\boldsymbol{x}\in C$ 和任意 $\boldsymbol{d}\in\mathbf{R}^n$ 有 $f_+^{DD}(\boldsymbol{x},\boldsymbol{d})\leqslant 0$.

证明 设 f 在 C 上是凸函数, 对任意 $t>0,\boldsymbol{x}\in C$ 和任意 $\boldsymbol{d}\in\mathbf{R}^n$, 由性质 6.2.1 得

$$f(\boldsymbol{x}-t\boldsymbol{d})\geqslant f(\boldsymbol{x})+\langle\nabla f(\boldsymbol{x}),-t\boldsymbol{d}\rangle,$$

$$f(\boldsymbol{x})\geqslant f(\boldsymbol{x}-t\boldsymbol{d})+\langle\nabla f(\boldsymbol{x}-t\boldsymbol{d}),t\boldsymbol{d}\rangle,$$

上面两式相加, 由定理 6.1.1 得

$$\langle f_G'(\boldsymbol{x}-t\boldsymbol{d}),\boldsymbol{d}\rangle-\langle f_G'(\boldsymbol{x}),\boldsymbol{d}\rangle\leqslant 0,\quad \forall t>0.$$

再由上式得

$$f_+^{DD}(\boldsymbol{x},\boldsymbol{d})=\limsup_{t\downarrow 0}\frac{\langle f_G'(\boldsymbol{x}-t\boldsymbol{d}),\boldsymbol{d}\rangle-\langle f_G'(\boldsymbol{x}),\boldsymbol{d}\rangle}{t}\leqslant 0.$$

反过来, $\forall\boldsymbol{x},\boldsymbol{y}\in C$ 有 $f_+^{DD}(\boldsymbol{x},\boldsymbol{y}-\boldsymbol{x})\leqslant 0$, 由定理 6.7.1 知 $\exists\boldsymbol{z}\in[\boldsymbol{x},\boldsymbol{y}]$ 使得

$$f(\boldsymbol{y})\geqslant f(\boldsymbol{x})+\langle f_G'(\boldsymbol{x}),\boldsymbol{y}-\boldsymbol{x}\rangle-\frac{1}{2}f_+^{DD}(\boldsymbol{z},\boldsymbol{y}-\boldsymbol{x}).$$

由上式和性质 4.1.1 知

$$f(\boldsymbol{y})\geqslant f(\boldsymbol{x})+\langle\nabla f(\boldsymbol{x}),\boldsymbol{y}-\boldsymbol{x}\rangle.$$

因此, f 是在 C 上的凸函数. □

例 6.7.6 设函数 $f(x)=x^{\frac{4}{3}}\ (\forall x\in\mathbf{R})$, 显然函数 f 是凸的. 导数 $f'(x)=\dfrac{4}{3}x^{\frac{1}{3}}\ (\forall x\in\mathbf{R})$ 是连续的, f 在 x 沿方向 d 的二阶上 Dini 方向导数为

$$f_+^{DD}(x,d)=\begin{cases}-4d^2/(9x^{2/3}), & x\neq 0,\\ -\infty, & x=0.\end{cases}$$

上式说明 $f_+^{DD}(x,d)\leqslant 0$.

例 6.7.7 设函数 $f(x)=x^3\,(\forall x\in\mathbf{R})$, 显然 f 不是凸的. 导数 $f'(x)=3x^2\,(\forall x\in\mathbf{R})$ 是连续的, f 在 x 沿方向 d 的二阶上 Dini 方向导数 $f_+^{DD}(x,d)=-6xd^2>0\,(x<0)$.

定理 6.7.9 设实值函数 $f:\ \mathbf{R}^n\to\mathbf{R}$ 是连续 Gateaux 可导的, f 在 $\boldsymbol{x}$ 沿方向 $\boldsymbol{d}\in\mathbf{R}^n$ 的二阶上 Dini 方向导数存在. 如果 $\exists\alpha>0$ 满足

$$\alpha\left(\left\langle f_G'(\boldsymbol{x}),\boldsymbol{y}\right\rangle\right)^2-\frac{1}{2}f_+^{DD}(\boldsymbol{x},\boldsymbol{y})\geqslant 0,\quad \forall\boldsymbol{x},\boldsymbol{y}\in\mathbf{R}^n, \tag{6.7.2-3}$$

则 $\mathrm{e}^{\alpha f(\boldsymbol{x})}$ 在 $\mathbf{R}^n$ 上是凸函数.

证明 设 $h(\boldsymbol{x})=\mathrm{e}^{\alpha f(\boldsymbol{x})}$, 有 $h_G'(\boldsymbol{x})=\alpha\mathrm{e}^{\alpha f(\boldsymbol{x})}f_G'(\boldsymbol{x})$. 根据定义 6.7.2 得

$$\begin{aligned}
h_+^{DD}(\boldsymbol{x},\boldsymbol{d})&=\limsup_{t\downarrow 0}\frac{\left\langle h_G'(\boldsymbol{x}-t\boldsymbol{d}),\boldsymbol{d}\right\rangle-\left\langle h_G'(\boldsymbol{x}),\boldsymbol{d}\right\rangle}{t}\\
&=\limsup_{t\downarrow 0}\frac{\alpha\mathrm{e}^{\alpha f(\boldsymbol{x}-t\boldsymbol{d})}\left\langle f_G'(\boldsymbol{x}-t\boldsymbol{d}),\boldsymbol{d}\right\rangle-\alpha\mathrm{e}^{\alpha f(\boldsymbol{x})}\left\langle f_G'(\boldsymbol{x}),\boldsymbol{d}\right\rangle}{t}\\
&\leqslant\limsup_{t\downarrow 0}\frac{\alpha\mathrm{e}^{\alpha f(\boldsymbol{x}-t\boldsymbol{d})}\left(\left\langle f_G'(\boldsymbol{x}-t\boldsymbol{d}),\boldsymbol{d}\right\rangle-\left\langle f_G'(\boldsymbol{x}),\boldsymbol{d}\right\rangle\right)}{t}\\
&\quad+\limsup_{t\downarrow 0}\frac{\left(\alpha\mathrm{e}^{\alpha f(\boldsymbol{x}-t\boldsymbol{d})}-\alpha\mathrm{e}^{\alpha f(\boldsymbol{x})}\right)\left\langle f_G'(\boldsymbol{x}),\boldsymbol{d}\right\rangle}{t}\\
&=\frac{1}{2}f_+^{DD}(\boldsymbol{x},\boldsymbol{y})-\alpha\left(\left\langle f_G'(\boldsymbol{x}),\boldsymbol{y}\right\rangle\right)^2\leqslant 0,
\end{aligned}$$

根据定理 6.7.8 得结论成立. □

根据 Gateaux 可导定义和定理 6.1.1 知, 在 $\mathbf{R}^n$ 中 Gateaux 可导与可微是等价的. 根据定理 6.7.9 得到伪凸函数和拟凸函数的判定定理.

定理 6.7.10 设实值函数 $f:\ \mathbf{R}^n\to\mathbf{R}$ 是连续 Gateaux 可导, f 在 $\boldsymbol{x}$ 沿方向 $\boldsymbol{d}\in\mathbf{R}^n$ 的二阶上 Dini 方向导数存在. 如果 $\exists\alpha>0$ 满足

$$\alpha\left(\left\langle f_G'(\boldsymbol{x}),\boldsymbol{y}\right\rangle\right)^2-\frac{1}{2}f_+^{DD}(\boldsymbol{x},\boldsymbol{y})\geqslant 0,\quad \forall\boldsymbol{x},\boldsymbol{y}\in\mathbf{R}^n,$$

则 f 在 $\mathbf{R}^n$ 上既是伪凸函数, 又是拟凸函数.

证明 由定理 6.7.9 知 $h(\boldsymbol{x})=\mathrm{e}^{\alpha f(\boldsymbol{x})}$ 在 $\mathbf{R}^n$ 上是凸函数, 那么对任意的 $\boldsymbol{x},\boldsymbol{y}\in\mathbf{R}^n$ 有

$$h(\boldsymbol{y})-h(\boldsymbol{x})\geqslant\left\langle h_G'(\boldsymbol{x}),\boldsymbol{y}-\boldsymbol{x}\right\rangle,$$

即有 $e^{\alpha f(\boldsymbol{y})}-e^{\alpha f(\boldsymbol{x})}\geqslant \alpha e^{\alpha f(\boldsymbol{x})}\langle f'_G(\boldsymbol{x}),\boldsymbol{y}-\boldsymbol{x}\rangle$. 如果 $\langle f'_G(\boldsymbol{x}),\boldsymbol{y}-\boldsymbol{x}\rangle\geqslant 0$, 显然有 $f(\boldsymbol{y})\geqslant f(\boldsymbol{x})$, 说明 f 在 $\mathbf{R}^n$ 上既是伪凸函数, 又是拟凸函数. □

本小节得到了一个重要结论: f 是凸函数等价于它的二阶上 Dini 方向导数不大于 0. 如果条件 (6.7.2-3) 式成立, f 在 $\mathbf{R}^n$ 上是伪凸函数和拟凸函数.

6.8 习 题

6.8-1 设函数 $f(x_1,x_2)=|x_1|+|x_2|$, 求该函数的正、负方向导数和次微分.

6.8-2 设 $f,g:\ \mathbf{R}^n\to\mathbf{R}$ 是凸函数, 如果对于 $\forall\boldsymbol{x}\in\mathbf{R}^n$ 有 $\partial f(\boldsymbol{x})\cap\partial g(\boldsymbol{x})\neq\varnothing$, 证明函数 $f-g$ 是常数.

6.8-3 设在 $\mathbf{R}$ 中, 单变量函数:

$$f(x)=\begin{cases}+\infty, & x<-1,\\ 1, & x=-1,\\ -x, & -1<x\leqslant 0,\\ x^2, & 0<x\leqslant 1,\\ +\infty, & x>1.\end{cases}$$

求该函数的正负方向导数和次微分.

6.8-4 设凸函数 $f:\ \mathbf{R}^n\to\mathbf{R}$, 对于任意给定的 $\boldsymbol{x}\in\mathbf{R}^n$, 设函数

$$g(\boldsymbol{x})=\min\left\{0,\sup_{\boldsymbol{x}\neq\boldsymbol{x}'}\frac{f(\boldsymbol{x}')-f(\boldsymbol{x})}{||\boldsymbol{x}'-\boldsymbol{x}||}\right\},$$

证明 $g(\boldsymbol{x})=\min\{||\boldsymbol{y}||:\boldsymbol{y}\in\partial f(\boldsymbol{x})\}$.

6.8-5 设一个单调递增函数 $\varphi:\ \mathbf{R}\to\mathbf{R}$ 和凸函数 $f(x)=\int_0^x\varphi(t)\,\mathrm{d}t$, 求函数 f 的正负方向导数和次微分.

6.8-6 设 $f,g:\ \mathbf{R}^n\to\mathbf{R}$ 是凸函数, 对于 $\forall\boldsymbol{x}\in\mathbf{R}^n$, 如果 $f+g$ 是可微的, 则 f,g 也是可微的.

6.8-7 设 $f,g:\ \mathbf{R}\to(0,+\infty)$ 是单调增加凸函数, 证明 fg 是单调增加凸函数, 且下式成立:

$$\partial[f(x)g(x)]=f(x)\partial g(x)+g(x)\partial f(x).$$

6.8-8 设 $f:\ \mathbf{R}^n\to\mathbf{R}$ 函数对于 $\forall\boldsymbol{x}\in\mathbf{R}^n$ 满足 $\partial f(x)\neq\varnothing$, 则证明 f 是凸函数.

6.8-9　证明性质 6.5.1.

6.8-10　设二元函数: $f(x,y)=x^2+2|xy|+y^2$, 求正负方向导数、Dini 方向导数和 $\partial f(x)$.

6.8-11　设二元函数: $f(x,y)=xy$, 求二阶上 Dini 方向导数, 根据定理 6.7.3 和定理 6.7.8 判定该函数是否为拟凸函数或凸函数.

6.8-12　设 $f: C\to \mathbf{R}$ 在开凸集 $C\subset \mathbf{R}^n$ 上存在二阶上 Dini 方向导数, 如果对任意 $\boldsymbol{x}\in C$ 和任意 $\boldsymbol{d}\in \mathbf{R}^n$, 有 $\nabla f(\boldsymbol{x})\neq 0$ 和 $\boldsymbol{d}^{\mathrm{T}}\nabla f(\boldsymbol{x})=0\Rightarrow f_{+}^{DD}(\boldsymbol{x},\boldsymbol{d})\leqslant 0$, 证明 f 是在凸集 C 上的拟凸函数是否成立. 如果不成立. 请举出反例.

参 考 文 献

[1] Kahtopobич Л B, Akилob Γ Π. 泛函分析: 上册 [M]. 刘证, 郑权, 张云生, 译. 北京: 高等教育出版社, 1982.

[2] Arrow K J, Enthoven A C. Quasi-concave programming[J]. Econometrica, 1961, 29(4): 779-800.

[3] Avriel M. R-convex functions[J]. Mathemntical Programming, 1972, 2: 309-323.

[4] Avriel M, Diewert W E, Schaible S, et al. Introduction to concave and generalized concave functions[C]//Schaible S, Ziemba W T, eds. Generalized Goncavity in Optimization and Economics. New York: Academic Press, 1981.

[5] Avriel M, Diewert W E, Schaible S, et al. Generalized Concavity[M]. New York: Plenum Press, 1988.

[6] Avriel M, Schaible S. Second-order conditions of peudoconvex C^2-functions[R]. Stanford University, Department of Operations Research, Technical Report No. TR-76-12, 1976.

[7] Avrnial M, Zang I. Generalized convex functions with applications on nonlinear programming[C]//Van Moeseke, ed. Mathematics Program for Activity Analysis. Amsterdam: North-Holland Publishing Company, 1974: 23-33.

[8] Bazaraa M S, Sherali H D, Shetty C M. Nonlinear Programming, Theory and Algorithms[M]. New York: John Wiley and Sons, 1993.

[9] Berge C. Espaces Topologiques, Fonctions Multivoques[M]. Paris: Dunod, 1959.

[10] Bereanu B. Quasi-convexity, strictly quasi-convexity and pseudo-convexity of composite objective functions[J]. Revue Française d' Automatique, Informatique, Recherche Opérationnelle, 1972, 6(R1): 15-26.

[11] Cambini A, Martein L. Generalized Convexity and Optimization: Theory and Applications[M]. Berlin: Springer-Verlag, 2009.

[12] Clarke F H. Optimization and Nonsmooth Analysis[M]. New York: Wiley-InterScience, 1983.

[13] Cottle R W. On the convexity of quadratic forms over convex sets[J]. Operations Research, 1967, 15(1): 170-172.

[14] Cottle R W, Ferland J A. On pseudo-convex functions of nonnegative variables[J]. Mathematical Programming, 1971, 1(1): 95-101.

[15] Cottle R W, Ferland J A. Matrix-theoretic criteria for the quasi-convexity and pseudo-convexity of quadratic functions[J]. Linear Algebra and Its Applications, 1972, 5(2): 123-136.

[16] Craven B D. Duality for generalized convex fractional programs[M]//Schaible S, Ziemba W T, eds. Generalized Concavity in Optimization and Economics. New York: Academic

Press, 1981.

[17] Crouzeix J P. On second order conditions for quasiconvexity[J]. Mathematical Programming, 1980, 18: 349-352.

[18] Crouzeix J P, Ferland J A. Criteria for quasi-convexity and pseudo-convexity: Relationships and comparisons[J]. Mathematical Programming, 1982, 23: 193-205.

[19] Crouzeix J P. Criteria for generalized convexity and generalized monotonicity in the differentiable case[C]// Hadjisavvas N, Komlosi S, Schaible S, eds. Handbook of Generalized Convexity and Generalized Monotonicity. New York: Springer, 2005: 89-119.

[20] La Torre D, Popovici N, Rocca M. Scalar characterizations of weakly cone-convex and weakly cone-quasiconvex functions[J]. Nonlinear Analysis: Theory, Methods & Applications, 2010, 72(3-4): 1909-1915.

[21] De Finetti B. Sulle stratificazioni convesse[J]. Annali di Matematica Pura ed Applicata, 1949, 30: 173-183.

[22] Diewert W E, Avriel M, Zang I. Nine kinds of quasiconcavity and concavity[J]. Journal of Economic Theory, 1981, 25(3): 397-420.

[23] Luc D T. Theory of Vector Optimization[M]. Berlin: Springer-Verlag, 1989.

[24] Eichberger J. Game Theory for Economists[M]. San Diego: Academic Press, 1993.

[25] Fenchel W. Convex Cones, Sets and Functions[M]. Mimeographed Lecture Notes. Princeton, New Jersey: Princeton University, 1951.

[26] Ferland J A. Quasi-convex and pseudo-convex functions on solid convex sets[D]. Palo Alto: Stanford University, 1971.

[27] Ferland J A. Maximal domains of quasi-convexity and pseudo-convexity for quadratic functions[J]. Mathematical Programming, 1972, 3(1): 178-192.

[28] Ferland J A. Mathematical programming problems with quasi-convex objective functions[J]. Mathematical Programming, 1972, 3(1): 296-301.

[29] Fudenberg D, Tirole J. Game Theory[M]. Cambridge: Massachussetts Institute of Technology Press, 1991.

[30] Hadjisavvas N, Schaible S. On strong pseudomonotonicity and (semi) strict quasimonotonicity[J]. Journal of Optimization Theory and Applications, 1995, 85(3): 741-742.

[31] Schaible S. Generalized monotone maps[C]//Karmann A, Mosler K, Schader M, et al, eds. Operations Research'92. Heidelberg: Physica, 1993.

[32] Hanson M A. On sufficiency of the Kuhn-Tucker conditions[J]. Journal of Mathematical Analysis and Applications, 1981, 80(2): 545-550.

[33] Hassouni A. Sous-différentiels des fonctions qunsi-convexes[R]. Toulouse: These de Troisieme Cicle, Universit'e Paul Sabatier, 1983.

[34] Jeyakumar V, Gwinner J. Inequality systems and optimization[J]. Journal of Mathematical Analysis and Applications, 1991, 159(1): 51-71.

[35] Karamardian S. Complementarity problems over cones with monotone and pseudomonotone maps[J]. Journal of Optimization Theory and Applications, 1976, 18(4): 445-454.

[36] Karamardian S, Schaible S. Seven kinds of monotone maps[J]. Journal of Optimization Theory and Applications, 1990, 66: 37-46.

[37] Komlósi S. Generalized convexity and generalized derivatives[C]//Hadjisavvas N, Komlósi, S, Schaible S, eds. Handbook of Generalized Convexity and Generalized Monotonicity: Nonconvex Optimization and Its Applications. New York: Springer, 2005: 421-463.

[38] Luc D T, Schaible S. Efficiency and generalized concavity[J]. Journal of Optimization Theory and Applications, 1997, 94(1): 147-153.

[39] Mangasarian O L. Pseudo-convex functions[J]. Journal of the Society for Industrial and Applied Mathematics Series A Control, 1965, 3(2): 281-290.

[40] Mangasarian O L. Nonlinear Programming[M]. New York: McGraw-Hill Book Company, 1969.

[41] Mangasarian O L. Convexity, pseudo-convexity and quasi-convexity of composite functions[J]. Stochastic Optimization Models in Finance, 1975, 5: 33-41.

[42] Martos B. Nonlinear Programming Theory and Methods[M]. Amsterdam: North-Holland Publishing Company, 1975.

[43] Martos B. Subdefinite matrices and quadratic forms[J]. SIAM Journal on Applied Mathematics, 1969, 17(6): 1215-1223.

[44] Martos B. Quadratic programming with a quasiconvex objective function[J]. Operations Research, 1971, 19(1): 87-97.

[45] Massam H, Zlobec S. Various definitions of the derivative in mathematical programming[J]. Mathematical Programming, 1974, 7: 144-161.

[46] Mereau P, Paquet J G. Second order conditions for pseudo-convex functions[J]. SIAM Journal on Applied Mathematics, 1974, 27(1): 131-137.

[47] Mishra S K, Wang S Y, Lai K K. Generalized Convexity and Vector Optimization[M]. Berlin: Springer-Verlag, 2009.

[48] Mordukhovich B S, Nam N M. Convex analysis and beyond[C]// Mikosch T V, Resnick S, Zwart B, et al, eds. Springer Series in Operations Research and Financial Engineering. New York: Springer, 2022.

[49] Nikodem K. On strongly convex functions and related classes of functions[C]//Rassias T M, ed. Handbook of Functional Equations. New York: Springer, 2014: 365-405.

[50] Ortega J M, Rheinboldt W C. Iterative Solution of Nonlinear Equations in Several Variables[M]. New York: Academic Press, 1970.

[51] Otani K. A characterization of quasi-concave functions[J]. Journal of Economic Theory, 1983, 31(1): 194-196.

[52] Rockafellar R T. Convex Analysis[M]. Princeton: Princeton University Press, 1970.

[53] Schaible S. Quasi-convex optimization in real linear spaces[J]. Zeitschrift für Operations Research, 1972, 16: 205-213.

[54] Schnible S. Quasi-concave, strictly quasi-concave and pseudo-concave functions[C]//Henn R, Künzi H P, Schubert H, eds. Operations Research Verfahren (Methods of Operations Research). Meisenheim: Verlag Anton Hain, 1973: 308-316.

[55] Schaible S. Generalized convexity of quadratic functions[C]//Schaible S, Ziemba W T, eds. Generalized Goncavity in Optimization and Economics. New York: Academic Press, 1981: 183-197.

[56] Schaible S. Quasiconvex, pseudoconvex, and strictly pseudoconvex quadratic functions[J]. Journal of Optimization Theory and Applications, 1981, 35(3): 303-338.

[57] Schaible S. Second-order characterizations of pseudoconvex quadratic functions[J]. Journal of Optimization Theory and Applications, 1977, 21(1): 15-26.

[58] Schaible S. Quasi-concavity and pseudo-concavity of cubic functions[J]. Mathematical Programming, 1973, 5(1): 243-247.

[59] Schaible S. Beiträge zur quasi-convexen programmierung[D]. Köln: Universität Köln, 1971.

[60] Schaible S, Cottle R W. On pseudoconvex quadratic forms[C]//Beckenbach E F, ed. General Inequalities II. Basel: Birkauser-Verlag, 1980.

[61] Schaible S, Ziemba W T. Generalized Concavity in Optimization and Economics[M]. New York: Academic Press, 1981.

[62] Shiraishi S. On connections between approximate second-order directional derivative and second-order Dini derivative for convex functions[J]. Mathematical Programming, 1993, 58: 257-262.

[63] Soltan V. Lectures on Convex Sets[M]. Singapore: World Scientific, 2015.

[64] Thompson W A, Jr, Parke D W. Some properties of generalized concave functions[J]. Operations Research, 1973, 21(1): 305-313.

[65] Weir T, Jeyakumar V. A class of nonconvex functions and mathematical programming[J]. Bulletin of the Australian Mathematical Society, 1988, 38(2): 177-189.

[66] Yang X M. Convexity of semi-continuous functions[J]. Opsearch, 1994, 31(4): 309-317.

[67] Yang X M. Semistrictly convex functions[J]. Opsearch, 1994, 31(1): 15-27.

[68] Yang X M. A note on convexity of upper semi-continuous functions[J]. Opsearch, 2001, 38(2): 235-237.

[69] Yang X M. A note on criteria of quasiconvex functions[J]. OR Transactions, 2001, 5(2): 55-56.

[70] Yang X M. Generalized convexity in optimization[D]. Hong Kong: The Hong Kong Polytechnie University, 2002.

[71] Yang X M, Teo K L, Yang X Q. A characterization of convex function[J]. Applied Mathematics Letters, 2000, 13(1): 27-30.

[72] Yang X M, Liu S Y. Three kinds of generalized convexity[J]. Journal of Optimization Theory and Applications, 1995, 86(2): 501-513.

[73] Yang X M, Yang X Q, Teo K L. On properties of semipreinvex functions[J]. Bulletin of the Australian Mathematical Society, 2003, 68(3): 449-459.

[74] Yang X M, Yang X Q, Teo K L. Some properties of prequasiinvex functions[J]. Indian Journal of Pure and Applied Mathematics, 2003, 34(12): 1689-1696.

[75] Yang X M, Yang X Q, Teo K L. Explicitly B-preinvex functions[J]. Journal of Computational and Applied Mathematics, 2002, 146(1): 25-36.

[76] Yang X M, Yang X Q, Teo K L. Two properties of semistrictly preinvex functions[J]. Indian Journal of Pure and Applied Mathematics, 2004, 35(11): 1285-1292.

[77] Yang X M, Yang X Q, Teo K L. Characterizations and applications of prequasi-invex functions[J]. Journal of Optimization Theory and Applications, 2001, 110(3): 645-668.

[78] Yang X M. A note on preinvexity[J]. Journal of Industrial and Management Optimization, 2014, 10(4): 1319-1321.

[79] Yang X M, Li D. On properties of preinvex functions[J]. Journal of Mathematical Analysis and Applications, 2001, 256(1): 229-241.

[80] Yang X M, Li D. Semistrictly preinvex functions[J]. Journal of Mathematical Analysis and Applications, 2001, 258(1): 287-308.

[81] Yang X Q. Generalized second-order characterizations of convex functions[J]. Journal of Optimization Theory and Applications, 1994, 82(1): 173-180.

[82] Yang X Q, Chen G Y. A class of nonconvex functions and pre-variational inequalities[J]. Journal of Mathematical Analysis and Applications, 1992, 169(2): 359-373.

[83] 李飞, 杨新民. 向量值映射的恰当锥拟凸性和锥半连续性的刻画 [J]. 应用数学学报, 2021, 44(5): 646-658.

[84] 李庆娜, 李萌萌, 于盼盼. 凸分析讲义 [M]. 北京: 科学出版社, 2019.

[85] 刘光中. 凸分析与极值问题 [M]. 北京: 高等教育出版社, 1991.

[86] 冯德兴. 凸分析基础 [M]. 北京: 科学出版社, 1995.

[87] 史树中. 凸分析 [M]. 上海: 上海科学技术出版社, 1990.

[88] 彭建文, 杨新民. 上 C-连续条件下集值映射的 C-拟凸性 [J]. 重庆师范学院学报 (自然科学版), 2003, 20(2): 86+92.

[89] 颜丽佳. 显拟凹函数的一个新性质 [J]. 重庆师范大学学报 (自然科学版), 2013, 30(5): 18-20.

[90] 杨新民. 三种凸性函数的判别准则 [J]. 重庆师范学院学报 (自然科学版), 1991, 8(2): 11-17.

[91] 杨新民. 显凸函数的一些性质 [J]. 运筹学学报, 1991, 10(1): 68-69.

[92] 杨新民. 显凸函数的两个新性质 [J]. 重庆师范学院学报 (自然科学版), 1992, 9(4): 94-96.

[93] 杨新民. 拟凸函数的某些性质 [J]. 工程数学学报, 1993, 10(1): 51-56.

[94] 杨新民. 上半连续函数的拟凸性 [J]. 运筹学学报, 1999, 3(1): 48-51.

[95] 杨新民. 凸函数的一个新特征性质 [J]. 重庆师范学院学报 (自然科学版), 2000, 17(1): 9-11+39.

[96] 杨新民, 邓群. 严格凸函数的一个特征性质 [J]. 重庆师范学院学报 (自然科学版), 1995, 12(3): 12.

[97] Helly E. Uber mengen konvexer korper mit gemeinschaftlichen punkte[J]. Jahresbericht der Deutschen Mathematiker-reinigung, 1923, 32: 175-176.

[98] 杨新民, 戎卫东. 广义凸性及其应用 [M]. 北京: 科学出版社, 2016.

[99] 旷华武. 弱近似凸集及其应用 [J]. 四川大学学报 (自然科学版), 2004, 41(2): 226-230.

[100] 旷华武. 凸分析若干结果探讨 [J]. 应用数学学报, 2012, 35(4): 595-607.

[101] Moffat S M, Moursi W M, Wang X F. Nearly convex sets: Fine properties and domains or ranges of subdifferentials of convex functions[J]. Mathematical Programming, 2016, 160:

193-223.

[102] 徐茂杨, 史小波. 近似凸集的某些性质 [J]. 数学的实践与认识, 2022, 52(1): 223-229.

[103] 刘歆, 刘亚峰. 凸分析 [M]. 北京: 科学出版社, 2024.

附录 1　常用数学符号列表

符号	说明	符号	说明
x^*	最优解	∇	函数梯度
$\oplus$	直和运算	$'$	单变量函数的一阶导数
$\varnothing$	空集	$''$	单变量函数的二阶导数
$\forall$	任意的	∞	无穷大
$\exists$	存在	$\lim$	极限
$\in$	属于	$\downarrow$	变量沿正数趋于 0
$\subset$	被包含于	$\uparrow$	变量沿负数趋于 0
$\cap$	集合相交	$\rightarrow$	函数或向量映射，或序列收敛
$\supset$	包含	$\int_a^b$	在区间 $[a,b]$ 的积分
$\cup$	集合并	$f(x)$	$f:\mathbf{R}\rightarrow\mathbf{R}$ 是单变量实数函数
$\perp$	集合或空间正交	$f(\boldsymbol{x})$	$f:\mathbf{R}^n\rightarrow\mathbf{R}$ 是多变量实数函数
∂	函数微分或集合边界	$f'(x)$	单变量函数 f 的导数
∂C	集合 C 所有边界点构成的集合	$\sup\limits_{\boldsymbol{x}\in C} f(\boldsymbol{x})$	$f(\boldsymbol{x})$ 在集合 C 上的上确界
$\approx$	约等于	$\inf\limits_{\boldsymbol{x}\in C} f(\boldsymbol{x})$	函数 $f(\boldsymbol{x})$ 在集合 C 上的下确界
$\notin$	不属于	$\operatorname{rng} f$	映射 $f:\mathbf{R}^n\rightarrow\mathbf{R}$ 的值域
$\not\subset$	不被包含于		

附录 2 本书定义符号列表

符号	说明	来源
A	$\mathbf{R}^n$ 中的仿射集	定义 2.1.4
aff$X$	包含集合 X 的仿射集	定义 2.1.4
ap$C$	锥 C 的所有顶点构成的集合	定义 3.1.4
B	原点半径为 1 的邻域或单位球	定义 1.2.1
$B(\boldsymbol{x},\varepsilon)$	$\boldsymbol{x}\in\mathbf{R}^n$ 的半径为 $\varepsilon>0$ 的闭邻域	定义 1.2.1
$B(C,\varepsilon)$	集合 $C\subset\mathbf{R}^n$ 的半径为 ε 的闭邻域	定义 1.2.1
bd$C$	凸集 C 的边界集	定义 2.3.2
C	$\mathbf{R}^n$ 中的凸集或凸锥	定义 2.2.1, 定义 3.1.1
C^+	集合 $C\subset\mathbf{R}^n$ 的正极锥或共轭锥	定义 3.4.1
C^-	集合 $C\subset\mathbf{R}^n$ 的极锥或负极锥	定义 3.5.1
S°	S° 是 S 的配极	定义 2.5.5
$C_{\boldsymbol{a}}(S)$	集合 S 的锥包	定义 3.2.1
$C_{f=\alpha}$	函数 f 的水平集：$\{\boldsymbol{x}\in C \mid f(\boldsymbol{x})=\alpha\}$	定义 4.1.6
$C_{f\leqslant\alpha}$	函数 f 的水平集：$\{\boldsymbol{x}\in C \mid f(\boldsymbol{x})\leqslant\alpha\}$	定义 4.1.6
$C_{f\geqslant\alpha}$	函数 f 的水平集：$\{\boldsymbol{x}\in C \mid f(\boldsymbol{x})\geqslant\alpha\}$	定义 4.1.6
clC	集合 C 的闭包	定义 2.3.2
cl $f(\boldsymbol{x})$	$f(\boldsymbol{x})$ 的上方图的闭凸函数	定义 4.3.4
co X	集合 X 的凸包	定义 2.2.5
cone S	集合 S 的凸锥包	定义 3.2.1
$\mathrm{cone}_{\boldsymbol{a}}\, S$	集合 S 的以 $\boldsymbol{a}$ 为顶点 $\boldsymbol{a}$ 的凸锥包	定义 3.2.1
co X	集合 X 的凸包	定义 2.2.5
$d(\boldsymbol{x},\boldsymbol{y})$	点 $\boldsymbol{x}$ 到点 $\boldsymbol{y}$ 的距离	1.1.2 节
$d(\boldsymbol{x},X)$	点 $\boldsymbol{x}$ 到集合 X 的距离	1.1.2 节
$D_{\boldsymbol{a}}(C)$	C 在点 $\boldsymbol{a}$ 的支撑锥	定义 3.6.3
$\dim(S)$	集合 S 的维数	定义 2.1.3
dir X	集合 X 的方向子空间	定义 2.1.8
dom f	函数 $f(\boldsymbol{x})$ 的有效定义域	定义 4.1.3
epi$f$	函数 f 的上方图	定义 4.1.3
exp C	集合 C 的暴露点集合	定义 2.5.4
ext C	集合 C 的极点集合	定义 2.5.2
$\nabla f(\boldsymbol{x})$	函数 f 在 $\boldsymbol{x}$ 处的梯度	定义 6.1.1
$\nabla^2 f(\boldsymbol{x})$	函数 f 在 $\boldsymbol{x}$ 处的二阶梯度，Hessian 矩阵	1.3.2 节
$\partial f(\boldsymbol{x})$	函数 f 在 $\boldsymbol{x}$ 处的次微分	定义 6.2.2
$f'(\boldsymbol{x},\boldsymbol{d})$	函数 f 在 $\boldsymbol{x}$ 处沿方向 $\boldsymbol{d}$ 的方向导数	定义 6.2.1
$f'_+(\boldsymbol{x},\boldsymbol{d})$	函数 f 在 $\boldsymbol{x}$ 处沿方向 $\boldsymbol{d}$ 的正（右）方向导数	定义 6.2.1
$f'_-(\boldsymbol{x},\boldsymbol{d})$	函数 f 在 $\boldsymbol{x}$ 处沿方向 $\boldsymbol{d}$ 的负（左）方向导数	定义 6.2.1
$f'_G(\boldsymbol{x})$	函数 f 在 $\boldsymbol{x}$ 处 Gateaux 可导	定理 6.1.1
$f^D_+(\boldsymbol{x},\boldsymbol{d})$	函数 f 在 $\boldsymbol{x}$ 处沿方向 $\boldsymbol{d}$ 的上 Dini 方向导数	定义 6.7.1

续表

符号	说明	来源
$f_{-}^{D}(\boldsymbol{x},\boldsymbol{d})$	函数 f 在 $\boldsymbol{x}$ 处沿方向 $\boldsymbol{d}$ 的下 Dini 方向导数	定义 6.7.1
$f_{+}^{DD}(\boldsymbol{x},\boldsymbol{d})$	函数 f 在 $\boldsymbol{x}$ 处沿方向 $\boldsymbol{d}$ 的二阶上 Dini 方向导数	定义 6.7.3
$f^{*}(\boldsymbol{x}^{*})$	函数 $f(\boldsymbol{x})$ 的共轭函数	定义 4.4.1
$f^{**}(\boldsymbol{x})$	函数 $f^{*}(\boldsymbol{x}^{*})$ 的共轭函数	定义 4.4.1
$h(\boldsymbol{x})$	函数 f 在 C 上的支持函数	定义 4.8.2
$H(\boldsymbol{a},\alpha)$	向量 $\boldsymbol{a}\in\mathbf{R}^n$ 和 $\alpha\in\mathbf{R}$ 的超平面	定理 2.1.4
$H^{\leqslant}(\boldsymbol{a},\alpha)$	向量 $\boldsymbol{a}\in\mathbf{R}^n$ 和 $\alpha\in\mathbf{R}$ 的超平面的闭半空间	2.4 节
$H^{\geqslant}(\boldsymbol{a},\alpha)$	向量 $\boldsymbol{a}\in\mathbf{R}^n$ 和 $\alpha\in\mathbf{R}$ 的超平面的闭半空间	2.4 节
$\operatorname{int} C$	集合 C 的内部，是所有内点构成的集合	定义 2.3.1
L	平面或线性变换	性质 2.1.5
$\operatorname{lin} X$	集合 X 的回收空间	定义 3.3.2
$l(\boldsymbol{x},\boldsymbol{y}\rangle$	以 $\boldsymbol{x}$ 为顶点过点 $\boldsymbol{y}$ 的开半直线，不含点 $\boldsymbol{x}$	2.2.1 节
$l\langle\boldsymbol{x},\boldsymbol{y})$	以 $\boldsymbol{y}$ 为顶点过点 $\boldsymbol{x}$ 的开半直线，不含点 $\boldsymbol{y}$	2.2.1 节
$l[\boldsymbol{x},\boldsymbol{y}\rangle$	以 $\boldsymbol{x}$ 为顶点过点 $\boldsymbol{y}$ 的闭半直线	2.2.1 节
$l\langle\boldsymbol{x},\boldsymbol{y}]$	以 $\boldsymbol{y}$ 为顶点过点 $\boldsymbol{x}$ 的闭半直线	2.2.1 节
$l\langle\boldsymbol{x},\boldsymbol{y}\rangle$	过点 $\boldsymbol{x}$ 和 $\boldsymbol{y}$ 的直线	2.2.1 节
$(\boldsymbol{x},\boldsymbol{y})$	不含两端点 $\boldsymbol{x}$ 和 $\boldsymbol{y}$ 的直线段	2.2.1 节
$[\boldsymbol{x},\boldsymbol{y})$	以点 $\boldsymbol{x}$ 和 $\boldsymbol{y}$ 为顶点的直线段，不含点 $\boldsymbol{y}$	2.2.1 节
$(\boldsymbol{x},\boldsymbol{y}]$	以点 $\boldsymbol{x}$ 和 $\boldsymbol{y}$ 为顶点的直线段，不含点 $\boldsymbol{x}$	2.2.1 节
$[\boldsymbol{x},\boldsymbol{y}]$	以点 $\boldsymbol{x}$ 和 $\boldsymbol{y}$ 为顶点的直线段	2.2.1 节
$N_{\boldsymbol{a}}(C)$	集合 C 在点 $\boldsymbol{a}$ 处的法锥	定义 3.6.3
$\operatorname{nor}\ C$	集合 C 的法锥	定义 3.6.4
$N_r(\boldsymbol{x})$	$\boldsymbol{x}\in\mathbf{R}^n$ 的半径为 $r>0$ 的闭邻域	定义 1.2.1
$\operatorname{ort} X$	$\operatorname{dir}X$ 的垂直子空间， X 的正交集合	定义 2.1.8
$O_r(\boldsymbol{x})$	$\boldsymbol{x}\in\mathbf{R}^n$ 的半径为 $r>0$ 的开邻域	定义 1.2.1
p_C	在集合 C 上的度量投影	定义 3.6.2
$Q_{\boldsymbol{a}}(C)$	集合 C 在点 $\boldsymbol{a}$ 处的法集	定义 3.6.3
$\operatorname{rb}C$	集合 C 的闭包去掉它的相对内部构成的集合	定义 2.3.3
$\operatorname{rec}X$	集合 X 的回收锥	定义 3.3.1
$\operatorname{ri}C$	集合 C 的相对内部	定义 2.3.3
$\operatorname{rng}L$	变换 $L:\ \mathbf{R}^n\to\mathbf{R}^m$ 的值域	定理 3.3.4
X	$\mathbf{R}^n$ 中的一般集合	1.1.2 节
$\\|\boldsymbol{x}\\|$	$\boldsymbol{x}$ 的范数	1.1.2 节
S	$\mathbf{R}^n$ 中的子空间	定义 2.1.1
S°	集合 S 的配极	定义 2.5.5
$S^{\circ\circ}$	$S^{\circ\circ}=(S^{\circ})^{\circ}$ 为 S 的双配极	定义 2.5.5
$\operatorname{Sim}(\boldsymbol{x}_0,\boldsymbol{x}_1,\cdots,\boldsymbol{x}_m)$	m 维单纯形	定义 2.2.6
$\operatorname{Simc}(\boldsymbol{a},\boldsymbol{x}_1,\cdots,\boldsymbol{x}_k)$	关于顶点 $\boldsymbol{a}$ 的 k-简单锥	定义 3.1.3
$\operatorname{span} S$	包含 S 的最小子空间	性质 2.1.3 的注
$\operatorname{Sup}C$	凸集 C 的支撑子空间	定义 2.3.3
$S(\boldsymbol{x},\varepsilon)$	$\boldsymbol{x}\in\mathbf{R}^n$ 的半径为 $\varepsilon>0$ 的邻域球面	定义 1.2.1
$T_{\boldsymbol{a}}(C)$	集合C 在点 $\boldsymbol{a}$ 的切锥	定义 3.6.3
$U(\boldsymbol{x},\varepsilon)$	$\boldsymbol{x}\in\mathbf{R}^n$ 的半径为 $\varepsilon>0$ 的开邻域	定义 1.2.1
$U(C,\varepsilon)$	C 的半径为 $\varepsilon>0$ 的开邻域	定义 1.2.1

续表

符号	说明	来源
$v_C(\boldsymbol{x})$	在集合 C 上的规范函数, 也可记为 $v(\boldsymbol{x} \mid C)$	定义 4.1.5
$\varphi_{\boldsymbol{x},\boldsymbol{d}}(t)$	f 在 $\boldsymbol{x}$ 处沿方向 $\boldsymbol{d}$ 的关于变量 $t \in \mathbf{R}$ 的限制函数	定理 4.1.3
$\theta_C(\boldsymbol{x})$	在集合 C 上的示性函数	定义 4.1.5
$\delta_C(\boldsymbol{x}^*)$	是集合C 的支撑函数, 也可记为 $\delta^*(\boldsymbol{x}^* \mid C)$	定义 3.9.2

“现代数学基础丛书”已出版书目

(按出版时间排序)

1 数理逻辑基础(上册) 1981.1 胡世华 陆钟万 著
2 紧黎曼曲面引论 1981.3 伍鸿熙 吕以辇 陈志华 著
3 组合论(上册) 1981.10 柯 召 魏万迪 著
4 数理统计引论 1981.11 陈希孺 著
5 多元统计分析引论 1982.6 张尧庭 方开泰 著
6 概率论基础 1982.8 严士健 王隽骧 刘秀芳 著
7 数理逻辑基础(下册) 1982.8 胡世华 陆钟万 著
8 有限群构造(上册) 1982.11 张远达 著
9 有限群构造(下册) 1982.12 张远达 著
10 环与代数 1983.3 刘绍学 著
11 测度论基础 1983.9 朱成熹 著
12 分析概率论 1984.4 胡迪鹤 著
13 巴拿赫空间引论 1984.8 定光桂 著
14 微分方程定性理论 1985.5 张芷芬 丁同仁 黄文灶 董镇喜 著
15 傅里叶积分算子理论及其应用 1985.9 仇庆久等 编
16 辛几何引论 1986.3 J. 柯歇尔 邹异明 著
17 概率论基础和随机过程 1986.6 王寿仁 著
18 算子代数 1986.6 李炳仁 著
19 线性偏微分算子引论(上册) 1986.8 齐民友 著
20 实用微分几何引论 1986.11 苏步青等 著
21 微分动力系统原理 1987.2 张筑生 著
22 线性代数群表示导论(上册) 1987.2 曹锡华等 著
23 模型论基础 1987.8 王世强 著
24 递归论 1987.11 莫绍揆 著
25 有限群导引(上册) 1987.12 徐明曜 著
26 组合论(下册) 1987.12 柯 召 魏万迪 著
27 拟共形映射及其在黎曼曲面论中的应用 1988.1 李 忠 著
28 代数体函数与常微分方程 1988.2 何育赞 著

29　同调代数　1988.2　周伯壎　著
30　近代调和分析方法及其应用　1988.6　韩永生　著
31　带有时滞的动力系统的稳定性　1989.10　秦元勋等　编著
32　代数拓扑与示性类　1989.11　马德森著　吴英青　段海豹译
33　非线性发展方程　1989.12　李大潜　陈韵梅　著
34　反应扩散方程引论　1990.2　叶其孝等　著
35　仿微分算子引论　1990.2　陈恕行等　编
36　公理集合论导引　1991.1　张锦文　著
37　解析数论基础　1991.2　潘承洞等　著
38　拓扑群引论　1991.3　黎景辉　冯绪宁　著
39　二阶椭圆型方程与椭圆型方程组　1991.4　陈亚浙　吴兰成　著
40　黎曼曲面　1991.4　吕以辇　张学莲　著
41　线性偏微分算子引论(下册)　1992.1　齐民友　徐超江　编著
42　复变函数逼近论　1992.3　沈燮昌　著
43　Banach 代数　1992.11　李炳仁　著
44　随机点过程及其应用　1992.12　邓永录等　著
45　丢番图逼近引论　1993.4　朱尧辰等　著
46　线性微分方程的非线性扰动　1994.2　徐登洲　马如云　著
47　广义哈密顿系统理论及其应用　1994.12　李继彬　赵晓华　刘正荣　著
48　线性整数规划的数学基础　1995.2　马仲蕃　著
49　单复变函数论中的几个论题　1995.8　庄圻泰　著
50　复解析动力系统　1995.10　吕以辇　著
51　组合矩阵论　1996.3　柳柏濂　著
52　Banach 空间中的非线性逼近理论　1997.5　徐士英　李　冲　杨文善　著
53　有限典型群子空间轨道生成的格　1997.6　万哲先　霍元极　著
54　实分析导论　1998.2　丁传松等　著
55　对称性分岔理论基础　1998.3　唐　云　著
56　Gel fond-Baker 方法在丢番图方程中的应用　1998.10　乐茂华　著
57　半群的 S-系理论　1999.2　刘仲奎　著
58　有限群导引(下册)　1999.5　徐明曜等　著
59　随机模型的密度演化方法　1999.6　史定华　著
60　非线性偏微分复方程　1999.6　闻国椿　著
61　复合算子理论　1999.8　徐宪民　著
62　离散鞅及其应用　1999.9　史及民　编著

63　调和分析及其在偏微分方程中的应用　1999. 10　苗长兴　著
64　惯性流形与近似惯性流形　2000. 1　戴正德　郭柏灵　著
65　数学规划导论　2000. 6　徐增堃　著
66　拓扑空间中的反例　2000. 6　汪　林　杨富春　编著
67　拓扑空间论　2000. 7　高国士　著
68　非经典数理逻辑与近似推理　2000. 9　王国俊　著
69　序半群引论　2001. 1　谢祥云　著
70　动力系统的定性与分支理论　2001. 2　罗定军　张　祥　董梅芳　编著
71　随机分析学基础(第二版)　2001. 3　黄志远　著
72　非线性动力系统分析引论　2001. 9　盛昭瀚　马军海　著
73　高斯过程的样本轨道性质　2001. 11　林正炎　陆传荣　张立新　著
74　数组合地图论　2001. 11　刘彦佩　著
75　光滑映射的奇点理论　2002. 1　李养成　著
76　动力系统的周期解与分支理论　2002. 4　韩茂安　著
77　神经动力学模型方法和应用　2002. 4　阮炯　顾凡及　蔡志杰　编著
78　同调论——代数拓扑之一　2002. 7　沈信耀　著
79　金兹堡-朗道方程　2002. 8　郭柏灵等　著
80　排队论基础　2002. 10　孙荣恒　李建平　著
81　算子代数上线性映射引论　2002. 12　侯晋川　崔建莲　著
82　微分方法中的变分方法　2003. 2　陆文端　著
83　周期小波及其应用　2003. 3　彭思龙　李登峰　谌秋辉　著
84　集值分析　2003. 8　李　雷　吴从炘　著
85　数理逻辑引论与归结原理　2003. 8　王国俊　著
86　强偏差定理与分析方法　2003. 8　刘　文　著
87　椭圆与抛物型方程引论　2003. 9　伍卓群　尹景学　王春朋　著
88　有限典型群子空间轨道生成的格(第二版)　2003. 10　万哲先　霍元极　著
89　调和分析及其在偏微分方程中的应用(第二版)　2004. 3　苗长兴　著
90　稳定性和单纯性理论　2004. 6　史念东　著
91　发展方程数值计算方法　2004. 6　黄明游　编著
92　传染病动力学的数学建模与研究　2004. 8　马知恩　周义仓　王稳地　靳　祯　著
93　模李超代数　2004. 9　张永正　刘文德　著
94　巴拿赫空间中算子广义逆理论及其应用　2005. 1　王玉文　著
95　巴拿赫空间结构和算子理想　2005. 3　钟怀杰　著
96　脉冲微分系统引论　2005. 3　傅希林　闫宝强　刘衍胜　著

97 代数学中的 Frobenius 结构　2005. 7　汪明义　著
98 生存数据统计分析　2005. 12　王启华　著
99 数理逻辑引论与归结原理(第二版)　2006. 3　王国俊　著
100 数据包络分析　2006. 3　魏权龄　著
101 代数群引论　2006. 9　黎景辉　陈志杰　赵春来　著
102 矩阵结合方案　2006. 9　王仰贤　霍元极　麻常利　著
103 椭圆曲线公钥密码导引　2006. 10　祝跃飞　张亚娟　著
104 椭圆与超椭圆曲线公钥密码的理论与实现　2006. 12　王学理　裴定一　著
105 散乱数据拟合的模型方法和理论　2007. 1　吴宗敏　著
106 非线性演化方程的稳定性与分歧　2007 .4　马　天　汪守宏　著
107 正规族理论及其应用　2007.4　顾永兴　庞学诚　方明亮　著
108 组合网络理论　2007. 5　徐俊明　著
109 矩阵的半张量积:理论与应用　2007. 5　程代展　齐洪胜　著
110 鞅与 Banach 空间几何学　2007. 5　刘培德　著
111 戴维-斯特瓦尔松方程　2007. 5　戴正德　蒋慕蓉　李栋龙　著
112 非线性常微分方程边值问题　2007. 6　葛渭高　著
113 广义哈密顿系统理论及其应用　2007. 7　李继彬　赵晓华　刘正荣　著
114 Adams 谱序列和球面稳定同伦群　2007. 7　林金坤　著
115 矩阵理论及其应用　2007. 8　陈公宁　著
116 集值随机过程引论　2007. 8　张文修　李寿梅　汪振鹏　高　勇　著
117 偏微分方程的调和分析方法　2008.1　苗长兴　张　波　著
118 拓扑动力系统概论　2008. 1　叶向东　黄　文　邵　松　著
119 线性微分方程的非线性扰动(第二版)　2008.3　徐登洲　马如云　著
120 数组合地图论(第二版)　2008. 3　刘彦佩　著
121 半群的 *S*-系理论(第二版)　2008. 3　刘仲奎　乔虎生　著
122 巴拿赫空间引论(第二版)　2008. 4　定光桂　著
123 拓扑空间论(第二版)　2008. 4　高国士　著
124 非经典数理逻辑与近似推理(第二版)　2008. 5　王国俊　著
125 非参数蒙特卡罗检验及其应用　2008. 8　朱力行　许王莉　著
126 Camassa-Holm 方程　2008. 8　郭柏灵　田立新　杨灵娥　殷朝阳　著
127 环与代数(第二版)　2009. 1　刘绍学　郭晋云　朱　彬　韩　阳　著
128 泛函微分方程的相空间理论及应用　2009. 4　王　克　范　猛　著
129 概率论基础(第二版)　2009. 8　严士健　王隽骧　刘秀芳　著
130 自相似集的结构　2010. 1　周作领　瞿成勤　朱智伟　著

131　现代统计研究基础　2010.3　王启华　史宁中　耿　直　主编
132　图的可嵌入性理论(第二版)　2010.3　刘彦佩　著
133　非线性波动方程的现代方法(第二版)　2010.4　苗长兴　著
134　算子代数与非交换 L^p空间引论　2010.5　许全华　吐尔德别克　陈泽乾　著
135　非线性椭圆型方程　2010.7　王明新　著
136　流形拓扑学　2010.8　马　天　著
137　局部域上的调和分析与分形分析及其应用　2011.6　苏维宜　著
138　Zakharov 方程及其孤立波解　2011.6　郭柏灵　甘在会　张景军　著
139　反应扩散方程引论(第二版)　2011.9　叶其孝　李正元　王明新　吴雅萍　著
140　代数模型论引论　2011.10　史念东　著
141　拓扑动力系统——从拓扑方法到遍历理论方法　2011.12　周作领　尹建东　许绍元　著
142　Littlewood-Paley 理论及其在流体动力学方程中的应用　2012.3　苗长兴　吴家宏　章志飞　著
143　有约束条件的统计推断及其应用　2012.3　王金德　著
144　混沌、Mel′nikov 方法及新发展　2012.6　李继彬　陈凤娟　著
145　现代统计模型　2012.6　薛留根　著
146　金融数学引论　2012.7　严加安　著
147　零过多数据的统计分析及其应用　2013.1　解锋昌　韦博成　林金官　编著
148　分形分析引论　2013.6　胡家信　著
149　索伯列夫空间导论　2013.8　陈国旺　编著
150　广义估计方程估计方法　2013.8　周　勇　著
151　统计质量控制图理论与方法　2013.8　王兆军　邹长亮　李忠华　著
152　有限群初步　2014.1　徐明曜　著
153　拓扑群引论(第二版)　2014.3　黎景辉　冯绪宁　著
154　现代非参数统计　2015.1　薛留根　著
155　三角范畴与导出范畴　2015.5　章　璞　著
156　线性算子的谱分析(第二版)　2015.6　孙　炯　王　忠　王万义　编著
157　双周期弹性断裂理论　2015.6　李　星　路见可　著
158　电磁流体动力学方程与奇异摄动理论　2015.8　王　术　冯跃红　著
159　算法数论(第二版)　2015.9　裴定一　祝跃飞　编著
160　有限集上的映射与动态过程——矩阵半张量积方法　2015.11　程代展　齐洪胜　贺风华　著
161　偏微分方程现代理论引论　2016.1　崔尚斌　著
162　现代测量误差模型　2016.3　李高荣　张　君　冯三营　著

163 偏微分方程引论 2016.3 韩丕功 刘朝霞 著
164 半导体偏微分方程引论 2016.4 张凯军 胡海丰 著
165 散乱数据拟合的模型、方法和理论(第二版) 2016.6 吴宗敏 著
166 交换代数与同调代数(第二版) 2016.12 李克正 著
167 Lipschitz 边界上的奇异积分与 Fourier 理论 2017.3 钱 涛 李澎涛 著
168 有限 p 群构造(上册) 2017.5 张勤海 安立坚 著
169 有限 p 群构造(下册) 2017.5 张勤海 安立坚 著
170 自然边界积分方法及其应用 2017.6 余德浩 著
171 非线性高阶发展方程 2017.6 陈国旺 陈翔英 著
172 数理逻辑导引 2017.9 冯 琦 编著
173 简明李群 2017.12 孟道骥 史毅茜 著
174 代数 K 理论 2018.6 黎景辉 著
175 线性代数导引 2018.9 冯 琦 编著
176 基于框架理论的图像融合 2019.6 杨小远 石 岩 王敬凯 著
177 均匀试验设计的理论和应用 2019.10 方开泰 刘民千 覃 红 周永道 著
178 集合论导引(第一卷：基本理论) 2019.12 冯 琦 著
179 集合论导引(第二卷：集论模型) 2019.12 冯 琦 著
180 集合论导引(第三卷：高阶无穷) 2019.12 冯 琦 著
181 半单李代数与 BGG 范畴 $\mathcal{O}$ 2020.2 胡 峻 周 凯 著
182 无穷维线性系统控制理论(第二版) 2020.5 郭宝珠 柴树根 著
183 模形式初步 2020.6 李文威 著
184 微分方程的李群方法 2021.3 蒋耀林 陈 诚 著
185 拓扑与变分方法及应用 2021.4 李树杰 张志涛 编著
186 完美数与斐波那契序列 2021.10 蔡天新 著
187 李群与李代数基础 2021.10 李克正 著
188 混沌、Melnikov 方法及新发展(第二版) 2021.10 李继彬 陈凤娟 著
189 一个大跳准则——重尾分布的理论和应用 2022.1 王岳宝 著
190 Cauchy-Riemann 方程的 L^2 理论 2022.3 陈伯勇 著
191 变分法与常微分方程边值问题 2022.4 葛渭高 王宏洲 庞慧慧 著
192 可积系统、正交多项式和随机矩阵——Riemann-Hilbert 方法 2022.5 范恩贵 著
193 三维流形组合拓扑基础 2022.6 雷逢春 李风玲 编著
194 随机过程教程 2022.9 任佳刚 著
195 现代保险风险理论 2022.10 郭军义 王过京 吴 荣 尹传存 著
196 代数数论及其通信应用 2023.3 冯克勤 刘凤梅 杨 晶 著

197　临界非线性色散方程　2023.3　苗长兴　徐桂香　郑继强　著
198　金融数学引论(第二版)　2023.3　严加安　著
199　丛代数理论导引　2023.3　李　方　黄　敏　著
200　$\mathcal{PT}$ 对称非线性波方程的理论与应用　2023.12　闫振亚　陈　勇　沈雨佳　温子超　李　昕　著
201　随机分析与控制简明教程　2024.3　熊　捷　张帅琪　著
202　动力系统中的小除数理论及应用　2024.3　司建国　司　文　著
203　凸分析　2024.3　刘　歆　刘亚锋　著
204　周期系统和随机系统的分支理论　2024.3　任景莉　唐点点　著
205　多变量基本超几何级数理论　2024.9　张之正　著
206　广义函数与函数空间导论　2025. 1　张　平　邵瑞杰　编著
207　凸分析基础　2025.3　杨新民　孟志青　编著